Advanced Starch Modification

This book comprehensively reviews modification techniques for starch and applications in food packaging and 3D/4D printing. The initial chapter provides an overview of the primary sources of starch, including cereals, tubers, and roots, and discusses global trends in starch production and consumption. The book examines the molecular architecture of starch granules and their inherent functional properties. It covers traditional chemical, enzymatic, and biotechnological modification methods, alongside innovative techniques such as nanotechnology, microwave, cold plasma, and ultrasound treatment. Additionally, the book presents the development of smart starches and starch nanoparticles, discussing their extraction, modification, characterization, and applications, particularly in food packaging and 3D/4D printing. The book also explores responsive and smart starches that change properties in response to environmental stimuli, discussing their development and applications. Towards the end, the book presents the functional and nutritional benefits of modified starches, as well as their regulatory and safety aspects. This book is an important resource for researchers, industry professionals, and students in food science, materials science, and biotechnology.

Key Features:

- Discusses traditional chemical, enzymatic, and biotechnological methods and innovative approaches for starch modification
- Provides an overview of primary starch sources, such as cereals, tubers, and roots, and discusses global production and consumption trends
- Analyzes the molecular architecture and inherent functional properties of starch granules, setting a foundation for modification techniques
- Presents the development of smart starches and starch nanoparticles, including their extraction, modification, characterization, and applications, particularly in food packaging and 3D/4D printing
- Explores the functional and nutritional benefits of modified starches, emphasizing their roles in health and wellness
- Provides a comprehensive overview of regulatory standards and safety assessments for the use of modified starches in food and other applications

Advanced Starch Modification

Emerging Technologies and Industrial Applications

Edited by
Prem Prakash Srivastav, Rakesh Kumar Gupta, and
Sangeetha Karunanithi

CRC Press
Taylor & Francis Group
Boca Raton London New York

CRC Press is an imprint of the
Taylor & Francis Group, an **informa** business

First edition published 2026
by CRC Press
2385 NW Executive Center Drive, Suite 320, Boca Raton FL 33431

and by CRC Press
4 Park Square, Milton Park, Abingdon, Oxon, OX14 4RN

CRC Press is an imprint of Taylor & Francis Group, LLC

ISBN: 978-1-032-91593-7 (hbk)
ISBN: 978-1-032-91631-6 (pbk)
ISBN: 978-1-003-56428-7 (ebk)

DOI: 10.1201/9781003564287

Typeset in Times
by KnowledgeWorks Global Ltd.

This book is dedicated

- *to the global scientific community committed to advancing knowledge at the crossroads of food science, material innovation, and sustainability*
- *to the trailblazers in starch research, whose pioneering contributions laid the groundwork for today's innovations*
- *to our mentors and colleagues, whose guidance, collaboration, and encouragement have enriched this academic journey*
- *to the next generation of researchers and students, whose curiosity and passion will shape the future of sustainable materials and intelligent food systems*
- *to our families, whose unwavering support, patience, and belief have been our greatest strength*

Contents

Foreword

It is with great pleasure that I write this foreword for the book ***Advanced Starch Modification: Emerging Technologies and Industrial Applications***. This timely and well-curated volume addresses a rapidly evolving field that bridges food science, materials engineering, and sustainable technology. The editors and contributors have succeeded in presenting a comprehensive and forward-looking exploration of starch science, from its molecular foundations to its cutting-edge applications. This book stands out for its holistic approach. It begins by tracing the origins of starch from various botanical sources—cereals, tubers, and roots—while contextualizing global trends in production and consumption. This provides a strong foundation for the in-depth discussion that follows on the molecular structure and inherent properties of starch granules, which are critical to understanding starch's behavior and functional versatility. The subsequent chapters offer a thorough review of both conventional and emerging starch modification techniques. Traditional chemical, enzymatic, and biotechnological approaches are presented alongside innovative, sustainable technologies such as microwave-assisted treatment, ultrasound, cold plasma, and nanotechnology. These sections are particularly relevant in today's context, where environmentally responsible processing and enhanced material functionality are of paramount importance. One of the most compelling aspects of this book is its focus on smart starch systems and starch nanoparticles. Their development and application in intelligent food packaging and 3D/4D printing are presented with scientific clarity and practical insight. These topics not only reflect the current state of research but also anticipate future directions in food design and functional biomaterials. Furthermore, the book addresses the nutritional and functional benefits of modified starches, as well as their safety and regulatory considerations. This balanced perspective ensures that the reader gains both a technical and ethical understanding of starch modification and its applications. As a scholar engaged in the field of food and biomaterials science, I find this book to be a significant and timely contribution. It will undoubtedly serve as a valuable reference for researchers, graduate students, and industry professionals alike. I commend the editors for their scholarly rigor and vision in compiling this important work, which I believe will inspire continued innovation and interdisciplinary collaboration in the years ahead.

Dr. Roberto Castro Muñoz

Professor
Gdansk University of Technology,
Gdansk, Poland

Acknowledgments

We extend our heartfelt gratitude to everyone who contributed their time, knowledge, and effort to make this edited volume a reality. The depth and quality of this book are a reflection of the collaborative spirit and dedication of many individuals. First and foremost, we thank the authors for their exceptional contributions. Their expertise and willingness to share cutting-edge research have been crucial in creating a comprehensive and insightful resource that advances understanding of starch modification and its innovative applications in food packaging and 3D/4D printing. We are also deeply appreciative of the anonymous reviewers whose careful evaluation and constructive feedback significantly improved the scholarly rigor and clarity of the chapters. Their valuable suggestions ensured that the content meets the highest standards of academic excellence. Our sincere thanks go to our colleagues and mentors, whose encouragement, critical discussions, and insightful guidance have profoundly influenced the scope and depth of this work. Their intellectual support has been both inspiring and motivating throughout this journey. We gratefully acknowledge the tireless efforts of the editorial and production teams. Their meticulous attention to detail and dedication to the editorial process have been essential in transforming the manuscripts into a polished and cohesive volume. We would also like to express our deepest appreciation to our families and friends for their constant encouragement, patience, and understanding during the course of this project. Their support provided the personal strength needed to persevere through the challenges encountered along the way. Finally, we are thankful to the Indian Institute of Technology Kharagpur for providing a supportive research environment and access to resources that made this project possible. The institution's commitment to fostering innovation and academic excellence has been invaluable. This volume is a testament to the power of collaboration, knowledge sharing, and the collective pursuit of scientific advancement. We are honored to have worked alongside such dedicated individuals and look forward to continued progress in this exciting and impactful field.

Thank you.

The Editors

Dr. Prem Prakash Srivastav is a distinguished Professor of Food Science and Technology at the Agricultural and Food Engineering Department, Indian Institute of Technology (IIT) Kharagpur, West Bengal, India. He holds a B.Sc. degree from Gorakhpur University, an M.Sc. degree in Food Technology with a minor in Process Engineering from GBPUA&T, Pantnagar, and a Ph.D. from IIT Kharagpur. In his academic role, he teaches various undergraduate-, postgraduate-, and Ph.D.-level courses. Dr. Srivastav has guided numerous research projects at the Ph.D. (16), master's (80), and undergraduate levels. His research interests include the development of convenience, functional, and therapeutic foods; nutraceutical extraction; and low-cost food processing machinery design. He has a substantial publication record, with over 200 research and review papers, numerous technical bulletins, and monographs. Dr. Srivastav has attended, chaired, and presented at numerous international and national conferences and delivered invited lectures at various schools. With 7 books, 22 book chapters, and 4 patents to his credit, his contributions are widely recognised. He has received several awards and is a life member of professional bodies, including ISTE, AFST (I), IDA, and AMI and, a member of ASABE and the Institute of Food Technologists (USA).

Dr. Rakesh Kumar Gupta earned his Ph.D. and M.Tech. degrees from the Agricultural and Food Engineering Department at the Indian Institute of Technology Kharagpur, India, and his B.Tech. in Agricultural Engineering from Assam University Silchar, India. His research focuses on the modification of starch using cold plasma and other emerging technologies for applications in sustainable food packaging systems. Dr. Gupta's contributions to the field have been recognised with several awards, including Best Poster, Best Oral Presentation Awards, AFSTI Springer Nature Travel Grant Award at national and international conferences. He has authored more than 31 peer-reviewed research and review articles, along with over 15 book chapters. His scholarly work has received wide recognition, accumulating more than 745 citations with an h-index of 14. Notably, two of his publications have been recognised as Top-Cited Papers in *Food Chemistry Advances* and *Measurement: Food* (Elsevier). He has presented his research at numerous national and international forums, including the prestigious International Union of Food Science and Technology (IUFoST) Congress, reflecting the global relevance of his work. His research interests broadly span starch modification, non-thermal technologies, essential oil extraction, and the development of eco-friendly active packaging systems aimed at addressing sustainability challenges in the food sector.

Dr. Sangeetha Karunanithi is an emerging leader and innovator in food process engineering, recognised for advancing sustainable, next-generation food technologies that bridge traditional practices with cutting-edge science. She is Assistant Professor in the Department of Food Technology at Vignan's Foundation for Science, Technology & Research (Deemed to be University), Vadlamudi, Andhra Pradesh, India, and earned her Ph.D. in Food Process Engineering from the prestigious Indian Institute of Technology Kharagpur, following B. Tech and M. Tech degrees from NIFTEM-T, India. Her pioneering research focuses on sustainability-driven innovations, including plasma- and microwave-assisted essential oil extraction, nanoemulsion-based delivery systems, lipid structuring for functional foods, and valorisation of food by-products. By promoting circular economy principles, minimising food waste, and enhancing nutritional retention, she develops eco-friendly, scalable solutions that strengthen global food security. Dr. Sangeetha has authored 10 peer-reviewed research papers, 18 book chapters, 7 review papers, 2 technical articles, and edited 2 books with leading international publishers including CRC Press and Springer Nature. With over 430 citations (h-index: 12), her work has gained international recognition and has been showcased at premier scientific forums such as the IUFoST World Congress. A multiple award-winning emerging researcher, she continues to shape the global food science community by driving next-generation food preservation technologies, lipid structuring innovations, strengthening food security, and championing sustainable solutions for the future of food systems worldwide.

1 Sources of Starch and Global Trends in Production and Consumption

Rakesh Kumar Gupta, Proshanta Guha, and Prem Prakash Srivastav
Agricultural and Food Engineering Department, Indian Institute of Technology Kharagpur, West Bengal, India

1.1 INTRODUCTION

Starch, a naturally prevalent polysaccharide, is essential in the worldwide food and non-food sectors owing to its varied functional attributes and extensive accessibility. Composed mostly of amylose and amylopectin, starch is the principal energy reserve in plants and serves as a biodegradable, renewable resource with several industrial uses (Gupta et al., 2022). Its relevance in food composition relies on its capacity to adjust texture, viscosity, and stability, while in non-food industries, starch has applications in bioplastics, medicines, adhesives, and textiles (Zarski et al., 2021). The rising need for sustainable and eco-friendly materials has heightened the significance of starch as a multifaceted biopolymer. The main sources of starch are grains (including maize, wheat, and rice), tubers (such as potato, cassava, and taro), and legumes (such peas and lentils) (Carvalho et al., 2024; Nath & Dutta, 2024). Corn is the most extensively farmed and used crop worldwide, followed by cassava, which is significantly important in tropical and subtropical areas. Global starch production is propelled by increasing populations, heightened demand for processed meals, and the broadening industrial use of starch-based products. Currently, the Asia-Pacific region and North America dominate starch production and consumption, with growing interest in neglected crops such as taro and yam due to their distinctive starch characteristics and nutritional advantages.

Native starch is often modified to enhance its physicochemical, thermal, and mechanical characteristics in order to satisfy the functional needs of many applications. Conventional alteration methods, including chemical and enzymatic treatments, although successful, raise issues about environmental damage and regulatory limitations (Gupta et al., 2024d; Shrivastava et al., 2024). Consequently, there is an increasing transition toward environmentally friendly and sustainable modification methods. Innovative green technologies, including cold plasma treatment, plasma-activated water, ultrasound, high-pressure processing, and ozone treatment, provide environmentally sustainable options that improve starch functioning without residual toxicity (Olawoye et al., 2024). These non-thermal methods not only conform to sustainable development objectives but also maintain the natural integrity of starch, making them appropriate for clean-label product formulation.

Therefore, the worldwide use of starch is experiencing a significant transformation due to the rising demand for functional meals, biodegradable materials, and clean-label additives. In food product development, starch remains a crucial element for texture enhancement, fat substitution, and prolonging shelf life. Furthermore, modified starches, particularly those sourced from sustainable technology, are becoming popular in health-oriented and ecologically sensitive industries. The starch sector is positioned for sustained expansion due to continuous research into new starch sources and improved modification techniques, and thus for enhancing food security, promoting environmental conservation, and fostering innovation in material science. This evolving environment highlights the need for multidisciplinary collaboration to fully exploit the potential of starch for future uses.

1.2 SOURCES OF STARCH

Starch is a predominant and essential carbohydrate in global human diets, functioning as a principal energy source and industrial raw material. Starch, a complex carbohydrate composed of several interconnected glucose molecules, serves as nature's effective energy storage mechanism in plants. This chapter examines the many sources of starch (Figure 1.1), its compositions, nutritional importance, and industrial uses. Starch is extensively present throughout the plant world, exhibiting variable concentrations in various plant tissues. Plants synthesize glucose via photosynthesis and save surplus glucose as starch in specialized structures known as starch granules or amyloplasts (Butler & Tetlow, 2024; Lal & Bhatla, 2023). Energy reserves are located in several plant organs, contingent upon the species. The major sources of starch are discussed in this section.

1.2.1 Cereal Grains

Cereal grains provide the primary source of starch worldwide in the diet and the starch industry. These crops accumulate starch mostly in their seeds. Corn is the primary commercial source, representing over 80% of the worldwide starch industry. The United States is the leading producer, with substantial industrial processing capabilities. Cornmeal has about 74% starch by weight, making it very

DOI: 10.1201/9781003564287-1

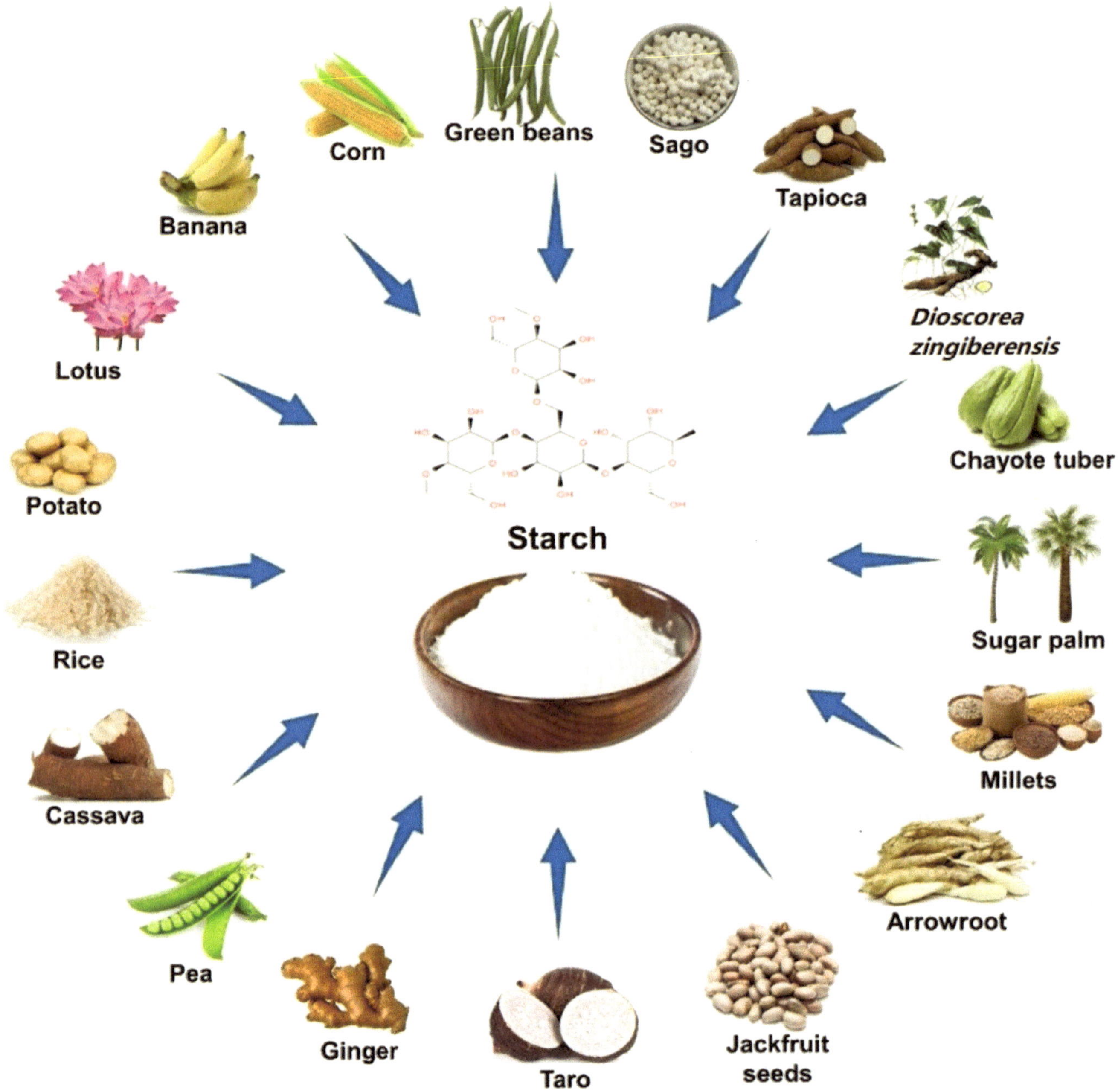

FIGURE 1.1 Sources of starch.

starch-rich (Raman, 2023). Another important source of starch, with unique qualities, is wheat. The two kinds of granules that distinguish wheat starch are A-type, which has granules bigger than 10 µm in diameter, and B-type, which has granules no larger than 10 µm (Guo et al., 2023). This unique distribution affects its functional properties in food applications. Despite this, among typical cereals, rice has the largest starch concentration (67%–70% total starch content). The smooth texture of rice is a result of its granules, which are the smallest of the main starches and measure less than 10 µm (Guo et al., 2023; Verma et al., 2018).

Other noteworthy sources of cereal starch include the following:

- Sorghum, also known as milo, is a crop that can be produced in more dry areas and has a composition similar to maize.
- Barley has been grown for thousands of years for starch production.
- Several millets have 57%–59% starch content, include finger millet, barnyard millet, and foxtail millet.
- Oats are a source of soluble fiber and carbohydrates.
- Substitute grain sources include buckwheat and rye.

1.2.2 Root and Tuber Crops

Root or tuber plants make up the second major category of starch sources. These plants have underground organs that store starch as energy reserves. The potato is one of the most widely consumed starchy vegetables worldwide. A medium-sized potato has about 31 grams of starch (WebMD, 2024). Potato starch granules have a notably large size of 2.7–70.7 μm and a distinctive lenticular shape. Because of this, potato starch has unique properties, such a high boiling viscosity (Guo et al., 2023). Furthermore, the tropical root crop cassava, also referred to as tapioca, is a significant source of starch and a staple diet in tropical areas. Sago pearls are made in India from a large portion of the extracted tapioca starch, particularly in the Salem area of Tamil Nadu (Phukan & Nongkhlaw, 2022). Another important source of tuber starch is sweet potatoes. Sweet potato starch has unique characteristics that influence its digestion and cooking behavior, although with certain similarities to potatoes. One study suggests that granules of sweet potato starch with a particle size of less than 2.27 μm are classified as tiny, those with a particle size between 2.27 and 17.51 μm as medium, and those with a particle size more than 17.51 μm as big (Lv et al., 2019).

Some significant sources of root and tuber starch are as follows:

- Historically, arrowroot has been a significant source of readily absorbed starch.
- Taro (*Colocasia*) is frequently grown in tropical areas and contains 70%–80% starch (Gupta et al., 2024b).
- There are many yam species, such as smaller yams and elephant foot yams. One report suggests that purple yam starch granules range in size from 24.11 to 45.37 μm (Shrivastava et al., 2024).
- Dasheen and other root crops with specific uses are also significant sources.

1.2.3 Legumes

Legumes, which include beans, peas, and lentils, are a significant but sometimes disregarded source of starch with special nutritional qualities. When consumed, legume starch suddenly increases blood glucose and insulin levels because it is digested more slowly than starch from cereals and tubers (Obadi & Xu, 2024). Blood sugar regulation may benefit from this slower pace of digestion. The unique botanical origin of legume-derived starches is responsible for the variation in granule size, which ranges from 0.4 to 103 μm (Singh, 2020).

Some important sources of legume starch are as follows:

- A variety of bean types, including kidney beans, are important sources.
- Traditionally, peas have been used to extract starch.
- Lentils and chickpeas are substantial sources of resistant starch.
- Fava beans and other indigenous legume types are also significant.

1.2.4 Palm Sources and Fruits

Southeast Asia and the Pacific Islands are home to the remarkable starch-producing sago palm (*Metroxylon sagu*). Moreover, an individual sago plant may produce between 200 and 250 kg of wet sago annually, or 25 tons of dry starch per hectare. There are 84.7 g of carbohydrates in 100 g of this starch (Susanto et al., 2024). Parts of Indonesia, Malaysia, and Papua New Guinea have historically relied heavily on sago starch as a staple food. Other palm sources include species of *Arenga* and *Mauritia*, which are collected in certain areas for their starch. Globally, Malaysia ranks third in sago production, after Indonesia and Papua New Guinea, which together generate about 94.6% of the world's sago starch (Phukan & Nongkhlaw, 2022).

Several fruits include significant levels of starch, even though they are not usually thought of as substantial sources. Especially when unripe, bananas have a comparatively high starch content (78.8%) (Cheng et al., 2024). As bananas mature, their starch turns into sugars.

Other fruits with quantifiable starch include the following:

- Among common fruits, the mamey sapote has the greatest starch content (6.33 g).
- Jackfruit is a significant source of starch (1.47 g).
- Plantains, even when ripe, retain a greater starch content.
- Breadfruit is an essential starchy staple in the Pacific Islands.
- The starch contents of mangoes, jackfruit, litchi, longans, and loquat kernels are 64%, 56%, 53%, 59%, and 71%, respectively (Phukan & Nongkhlaw, 2022).

1.3 GLOBAL PRODUCTION OF STARCH

It is difficult to evaluate the global output of starch because the data provided vary greatly across researchers and authors. Numerous private companies conduct in-depth research based on different raw materials, starch types, and geographic locations. Unfortunately, owing to their cost, these studies are not accessible for consultation. A recent study cited by Vilpoux and Junior (2022) reported that global starch production between 2011 and 2020 was estimated to be between 20 and 113.5 million tons, excluding the starch used to produce ethanol. According to a report by the province of Manitoba, Canada, in 2023, the global production of starch was 134.5 million tonnes in 2022 and during the next eight years, it increased at a compound annual growth rate (CAGR) of 5.1%, reaching 199.8 million tonnes

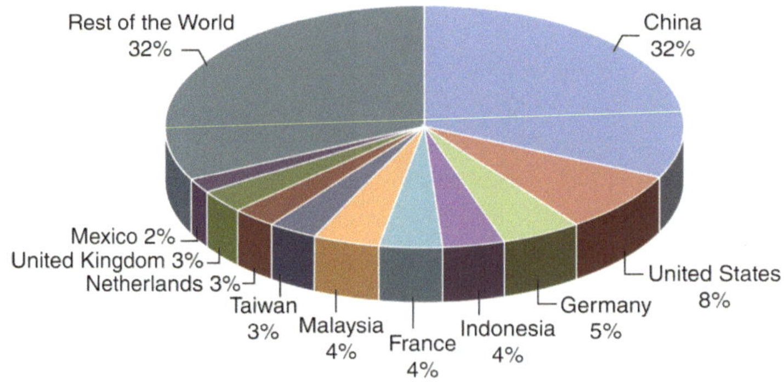

FIGURE 1.2 Top 10 countries importing native starch in 2022.

by 2030 (Manitoba, 2023). Figure 1.2 and Figure 1.3 demonstrate the top ten countries importing the native and modified starch in 2022. A 2024 statistic from the European Union indicates the production of starch alone grew from 8.7 million tonnes in 2004 to 9.2 million tonnes in 2023 (Starch Europe, 2024).

The majority of the world's starch is produced in the Asia-Pacific area, especially China. China is the world's greatest producer of starch, with an output of \$37.1 billion in 2021, or 59.4% of the Asia-Pacific region's total. By 2030, it is anticipated that China's starch market will have grown by 5.9% from 2022 to reach 74.4 million tonnes (Manitoba, 2023). India contributes less, about 3.5% of the Asia-Pacific region's output, valued at \$2.1 billion in 2021. Particularly for cassava starch, Thailand and Indonesia are two other major producers in the area. Despite this, the US market is projected to reach 28.2 million tonnes in 2020 and 30.6 million tonnes in 2022. The US output is expected to continue growing at a consistent rate of about 2.7% until 2030. In 2021, the starch sector in Canada grew by 26.1%, and exports now make up 50.1% of overall manufacturing output. The Canadian market is dominated by Ingredion Canada Inc., which accounts for 30.1% of total output (Manitoba, 2023).

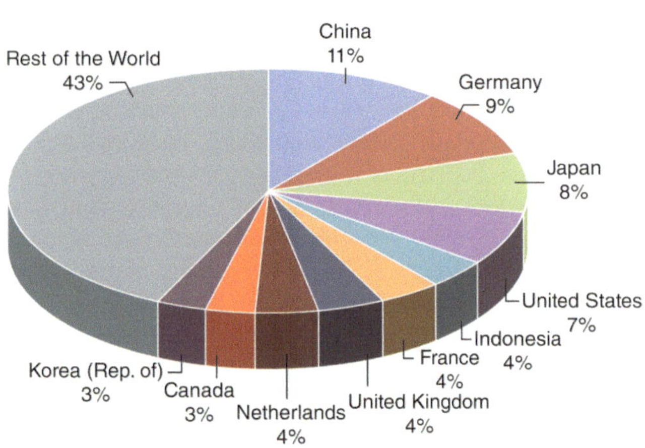

FIGURE 1.3 Top 10 countries importing modified starch in 2022.

1.4 RECENT TRENDS IN ADVANCEMENT OF STARCH MODIFICATION USING GREEN TECHNOLOGY

Starch is used as a natural polymer in numerous sectors of the food industry, including food product creation and packaging. Nevertheless, natural starch has inadequate physicochemical, functional, and structural qualities. To enhance the characteristics of native starch, many innovative and eco-friendly methods have been used. This section succinctly addresses modern and green technologies that increase the qualities of native starch, including cold plasma technology, ultrasonic technology, microwave technology, irradiation technology, and the use of nanotechnology in starch modification (Shrivastava et al., 2024).

1.4.1 Cold Plasma Technology

Cold plasma is emerging as a viable non-thermal solution for starch modification, providing substantial benefits over traditional chemical procedures. This environmentally sustainable method employs partly ionized gas to modify starch characteristics via three main mechanisms: cross-linking, depolymerization, and plasma etching (Gupta et al., 2023, 2024a). Exposure of starch to cold plasma results in the interaction of reactive species with starch molecules, leading to the disruption of the crystalline structure, increased water solubility, and alteration of functional characteristics. Research on different starch varieties has shown that cold plasma treatment may markedly alter morphology, generating holes on granule surfaces while facilitating both cross-linking and depolymerization of starch chains (Gupta et al., 2024c). The applications of plasma-modified starch are broad owing to the customizable characteristics of the treatment. Researchers may meticulously regulate alterations to get desired qualities by modifying factors such as plasma power (60–230 W), exposure duration (10–60 minutes), and gas type (argon, helium, nitrogen, or oxygen) (Kaur & Annapure, 2023; Okyere et al., 2022). These alterations lead to changed gelatinization properties, decreased viscosity, increased solubility, and better gel stability, making plasma-treated starches advantageous for both food and non-food sectors (Gupta et al., 2024e).

1.4.2 Ultrasonic Technology

Ultrasonic technology has evolved as an eco-friendly and effective approach for starch modification, using mechanical and cavitation effects to transform its structural and functional characteristics. Ultrasound applied to starch suspensions uses high-frequency sound waves to generate microbubbles, which collapse forcefully during cavitation, resulting in localized zones of elevated pressure and temperature (Dutta & Sit, 2024; Hu et al., 2015). This process disturbs starch granules by forming holes and fissures on their surfaces, dissolving amylopectin chains, and enhancing amylose content through selective depolymerization. A recent study suggested that when corn starch was modified by ultrasonic treatment, ultrasonication fragmented the starch granules, resulting in a reduction in particle size and a significant enhancement in the ζ potential (He et al., 2025). Another study also suggested that when buckwheat starch (BWS) was modified by ultrasonic treatment, the gel hardness and loss modulus of BWS were significantly decreased. The augmentation of water content during ultrasound significantly improved the swelling capacity of BWS and decreased the peak viscosity. Moreover, the elevation of water content and ultrasonic power resulted in a substantial decline in the crystallinity of BWS, inhibiting the creation of organized structures. Furthermore, subsequent to ultrasonic treatment, the particle size of BWS decreased, and the surface exhibited roughness and concavity (Yan et al., 2025).

1.4.3 Microwave Technology

Using dielectric heating, microwave technology has become an adaptable physical technique for starch modification that is free of chemical interventions and may produce structural and functional modifications. Upon exposure to microwave radiation (300–900 W), the fast oscillation of polar molecules, such as water, causes molecular friction to produce localized heat, breaking hydrogen bonds and changing crystalline areas in starch granules (Oyeyinka et al., 2021; Rostamabadi et al., 2023; Yi et al., 2024). This process alters the granule shape by forming internal holes and surface fissures. This was shown in potato starch treated at 300 W for 1, 3, and 5 minutes; the starch showed enhanced water absorption capacity, from 0.82 to 1.16 g/g, and decreased oil absorption capacity, from 0.63 to 0.53 g/g (Kumar et al., 2020). Another study also suggested that a cassava starch with 20% moisture content exposed to 900 W microwaves for 30 minutes demonstrated increased resistant starch content (up to 52.8%) as a result of amylopectin chain restructuring and the production of B+V-type crystalline polymorphs. The extent of modification varies depending on the starch's origin and other factors (Huong et al., 2021; Khurshida & Deka, 2022). Despite this, gelatinization behavior is also altered by microwave treatment. For example, lotus root starch underwent dual modification using 700 W microwaves and cold plasma, which increased pasting temperature while lowering peak viscosity by 38% (Sun et al., 2023).

1.4.4 Irradiation Technology

Irradiation technology has emerged as a novel technology for starch modification. By using ionizing radiation, such as gamma rays and electron beams, and non-ionizing ultraviolet (UV) radiation to change structural and functional characteristic of starch. Most research has suggested that gamma irradiation targets amylopectin over amylose and causes depolymerization and cross-linking by cleaving glycosidic bonds via free radicals (Annapure et al., 2025; Chung et al., 2015). One study suggests that gamma irradiation, with doses ranging from 1 to 50 kGy, promotes water solubility by breaking up crystalline areas while maintaining granule shape, reduces pasting viscosity by up to 90%, and lowers gelatinization enthalpy (Chung et al., 2015). A recent study suggests that when browntop millet starch was modified by gamma irradiation at 10 kGy, hydrophilicity was enhanced by increased functional characteristics, including an increase in water absorption capacity to 1.89 g/g at 10 kGy. Reduced crystallinity by 27.18%, especially in a 10-kGy treatment, was found by X-ray diffraction, indicating structural disturbance. Reduced thermal stability was shown by the gelatinization enthalpy (ΔH), which dropped from 184.18 J/g (native) to 148.65 J/g at 10 kGy. Granular breakdown was indicated by the peak viscosity decreasing from 3150 cP (native) to 409 cP at 10 kGy (Kaushik et al., 2025).

1.5 CONSUMPTION OF STARCH IN FOOD SYSTEMS

One of the main sources of energy in the human diet is starch, a basic carbohydrate. It is widely found in basic foods, including legumes, potatoes, taro, and cassava, as well as in grains such as rice, wheat, and maize. Beyond its nutritional value, starch plays a number of functions in food systems, including influencing the stability, texture, and viscosity of different food items. It is essential in processed foods, including baked products, sauces, soups, and confections, because of its functional qualities, which include gelatinization, retrogradation, and thickening. To enhance performance under industrial processing conditions such as heat, shear, and acidity, native starches are often altered physically, chemically, or enzymatically.

1.5.1 Food Packaging

The use of starch in food packaging is gradually growing as a result of increasing environmental concerns and regulatory pressure to cut down on plastic waste. A variety of products, including trays, wrapping films, and coatings for fruits, vegetables, and baked products, are made from starch-based materials. Furthermore, these materials' ability to work with active packaging technologies—such as adding antimicrobial agents like essential oils or nanoparticles—adds value by guaranteeing food safety and prolonging shelf life (Gupta et al., 2025a). In the industrial and commercial food packaging industries, starch's function as a renewable and biodegradable material is anticipated to rise as the need for sustainable packaging continues to escalate. A recent study exhibited that when packaging films based on taro peel starch were developed, the physicochemical and mechanical properties of the films were significantly enhanced (Gupta et al., 2024a). Another study suggested that for starch-based pH-sensitive packaging materials, the incorporation of blackberry extract at 0%, 5%, 10%, and 15% in starch-base film significantly improved swelling capacity (from 35.27% to 43.06%), water vapor permeance (WVP; from 1.99 to 3.69 × 10^{-10} g/(Pa × m × s)), water content (from 9.6% to 21.36%), and dissolution rate in water (from 288.05% to 459.89%) (Cvetković et al., 2025). Another recent study investigated the incorporation of zinc nanoparticle in starch-based nanocomposite film. The findings suggested that when nano-ZnO was added, the WVP rose dramatically from 1.16 g/m^2 day kPa (pure starch film) to 2.25 g/m^2 day kPa. Moreover, the tensile strength improved significantly with the introduction of ZnO nanoparticles at greater concentrations (≥ 1%); values increased from 6.2 MPa in pure starch films to 25.1 MPa at 3% ZnO nanoparticle inclusion (Dhiman et al., 2025).

1.5.2 Food Product Development

Starch is a fundamental carbohydrate extensively utilized in food product development due to its multifunctional properties, including its role as a thickening, gelling, and stabilizing agent (Gupta et al., 2025b). Naturally present in cereals, tubers, and legumes, starch contributes significantly to the texture, mouthfeel, and structural integrity of various food items, such as bakery products, snacks, sauces, and dairy formulations. Its ability to undergo gelatinization and retrogradation during processing makes it invaluable in achieving desirable rheological and sensory characteristics in finished food products. Moreover, starch serves as a key ingredient in the development of gluten-free and low-fat formulations, catering to the growing demands of health-conscious consumers. Modification of starches, through physical, enzymatic, or emerging non-thermal techniques such as cold plasma treatment, enhances functional attributes such as solubility, freeze–thaw stability, and resistance to shear or acidic conditions. These advancements make starch a vital component in the innovation of sustainable, clean-label, and functional foods.

1.6 SUMMARY

Starch, a polysaccharide composed of amylose and amylopectin, serves as a critical renewable resource in food and non-food industries due to its functional versatility and biodegradability. Derived from cereals (maize, wheat, rice), tubers (potato, cassava), legumes (peas, lentils), and palms (sago), its properties vary by botanical origin, influencing applications in food texture modification, bioplastics, pharmaceuticals, and textiles. Global starch production reached 134.5 million tonnes in 2022, driven by population growth and demand for sustainable materials, with the Asia-Pacific region, particularly China, dominating output. Conventional modification methods, such as chemical and enzymatic treatments, face criticism for their environmental impact, prompting a shift toward green technologies. Cold plasma (60–230 W) induces cross-linking and depolymerization, enhancing solubility and gel stability. Ultrasonication (20–40 kHz) leverages cavitation to reduce particle size and improve enzyme sensitivity, as seen in corn starch with 38% lower peak viscosity. Microwave treatment (300–900 W) disrupts hydrogen bonds, increasing resistant starch content up to 52.8% in cassava starch. Gamma irradiation (1–50 kGy) reduces viscosity by 90% through free radical–mediated glycosidic cleavage, while nanotechnology produces starch nanoparticles (20–600 nm) for applications such as drug encapsulation (287 mg/g adsorption capacity). In food systems, starch is vital for energy and functional roles in texture, viscosity, and shelf-life extension. Recent advances in food packaging include starch-based pH-sensitive films with blackberry extract, showing enhanced swelling capacity (43.06%) and water vapor permeability. Nanocomposites reinforced with zinc oxide nanoparticles achieve tensile strengths of 25.1 MPa, ideal for antibacterial packaging. Modified starches also enable clean-label product development, such as gluten-free formulations and low-glycemic-index foods. The starch industry is evolving through interdisciplinary innovation, focusing on hybrid approaches such as microwave–enzyme synergies and AI-driven modeling to optimize functionality. Sustainable modification technologies align with global trends toward biodegradable materials and circular economies, positioning starch as a cornerstone of future food security and material science.

REFERENCES

Annapure, U. S., Ranjitha Gracy, T. K., & Eazhumalaie, G. (2025). Interaction of reactive species with food molecules. In R. Mahendran, P. Birwal, & M. R. Goyal (Eds.), *Nonthermal plasma technology for food and food products* (pp. 73–107). Apple Academic Press. https://doi.org/10.1201/9781003543084-4

Butler, V., & Tetlow, I. J. (2024). Starch synthesis in plants. In L. Nilsson (Ed.), *Starch in food: Structure, function, and applications* (pp. 1–33). Woodhead Publishing. https://doi.org/10.1016/B978-0-323-96102-8.00009-7

Carvalho, H. J. M., Barcia, M. T., & Schmiele, M. (2024). Non-conventional starches: Properties and potential applications in food and non-food products. *Macromol*, *4*(4), 886–909. https://doi.org/10.3390/macromol4040052

Cheng, Y. T., Huang, P. H., Chan, Y. J., Chiang, P. Y., Lu, W. C., Hsieh, C. W., Liang, Z. C., Yan, B. W., Wang, C. C. R., & Li, P. H. (2024). Investigate the composition and physicochemical properties attributes of banana starch and flour during ripening. *Carbohydrate Polymer Technologies and Applications*, *7*, 100446. https://doi.org/10.1016/j.carpta.2024.100446

Chung, K. H., Othman, Z., & Lee, J. S. (2015). Gamma irradiation of corn starches with different amylose-to-amylopectin ratio. *Journal of Food Science and Technology*, *52*(10), 6218–6229. https://doi.org/10.1007/s13197-014-1700-4

Cvetković, K., Lazarević, A., Stamenković Stojanović, S., Dordevic, S., Dordevic, D., Karabegović, I., & Danilović, B. (2025). pH-sensitive starch-based packaging films enhanced with wild blackberry extract. *Processes*, *13*(4), 1148. https://doi.org/10.3390/pr13041148

Dhiman, S., Kumari, A., Kumari, S., & Sharma, R. (2025). Advanced biodegradable starch-based nanocomposite films incorporating zinc oxide nanoparticles: Synthesis, characterization, and efficacy in antibacterial food packaging applications. *Journal of Environmental Chemical Engineering*, *13*(3), 116296. https://doi.org/10.1016/j.jece.2025.116296

Dutta, D., & Sit, N. (2024). Ultrasound treatment. In S. P. Bangar, K.V. Sunooj, & A. K. Siroha (Eds.), *Starch: Structure, properties, and modifications for food applications* (pp. 132–164). CRC Press. https://doi.org/10.1201/9781032655598-6

Guo, L., Chen, H., Zhang, Y., Yan, S., Chen, X., & Gao, X. (2023). Starch granules and their size distribution in wheat: Biosynthesis, physicochemical properties and their effect on flour-based food systems. *Computational and Structural Biotechnology Journal*, *21*, 4172–4186. https://doi.org/10.1016/j.csbj.2023.08.019

Gupta, R. K., Guha, P., & Srivastav, P. P. (2022). Natural polymers in bio-degradable/edible film: A review on environmental concerns, cold plasma technology and nanotechnology application on food packaging - A recent trends. *Food Chemistry Advances*, *1*, 100135. https://doi.org/10.1016/j.focha.2022.100135

Gupta, R. K., Guha, P., & Srivastav, P. P. (2023). Effect of high voltage dielectric barrier discharge (DBD) atmospheric cold plasma treatment on physicochemical and functional properties of taro (*Colocasia esculenta*) starch. *International Journal of Biological Macromolecules*, *253*(P2), 126772. https://doi.org/10.1016/j.ijbiomac.2023.126772

Gupta, R. K., Guha, P., & Srivastav, P. P. (2024a). Dielectric barrier discharge plasma: A green and novel method to change the structure of taro peel starch and improve the physicochemical properties of taro peel starch films. *Plasma Processes and Polymers*, *21*(9), 2400047. https://doi.org/10.1002/ppap.202400047

Gupta, R. K., Guha, P., & Srivastav, P. P. (2024b). Exploring the potential of taro (Colocasia esculenta) starch: Recent developments in modification, health benefits, and food industry applications. *Food Bioengineering*, *3*(3), 365–379. https://doi.org/10.1002/fbe2.12103

Gupta, R. K., Guha, P., & Srivastav, P. P. (2024c). Insights into the physicochemical, crystalline, multistructural, thermal and digestibility behavior of taro starch (*Colocasia esculenta*) modified by cold plasma treatment with high voltage dielectric barrier discharge (HVDBD): An alternative technique for chemical modification of starch and environmental sustainability. *Journal of Cleaner Production*, *461*, 142710. https://doi.org/10.1016/j.jclepro.2024.142710

Gupta, R. K., Guha, P., & Srivastav, P. P. (2024d). Physical action of nonthermal cold plasma technology for starch modification. *Food Physics*, *1*, 100011. https://doi.org/10.1016/j.foodp.2024.100011

Gupta, R. K., Guha, P., & Srivastav, P. P. (2024e). Effect of cold plasma treatment and plasma-activated water on physicochemical and structural properties of starch: A green and novel approach for environmental sustainability. *Plasma Processes and Polymers*, *21*(4), 2300204. https://doi.org/10.1002/ppap.202300204

Gupta, R. K., Pipliya, S., Patel, J., Srivastav, P. P., Castro-Muñoz, R., Wang, C. K., & Nemţanu, M. R. (2025a). Essential oil-embedded starch based films for active packaging: Insights into migration, stability, and preservation efficiency. *Trends in Food Science & Technology*, *163*, 105149. https://doi.org/10.1016/j.tifs.2025.105149

Gupta, R. K., Gnana Moorthy Eswaran, U., Kumar, S., Pipliya, S., Kaur, B., & Srivastav, P. P. (2025b). Synergistic modification of starch polysaccharide via sequential treatments with ultrasonic and cold plasma technology. *Journal of Food Processing and Preservation*, *2025*(1), 9063007. https://doi.org/10.1155/jfpp/9063007

He, M., Chen, L., Liu, Y., Teng, F., & Li, Y. (2025). Effect of ultrasonic pretreatment on physicochemical, thermal, and rheological properties of chemically modified corn starch. *Food Chemistry*, *463*, 141061. https://doi.org/10.1016/j.foodchem.2024.141061

Hu, A., Jiao, S., Zheng, J., Li, L., Fan, Y., Chen, L., & Zhang, Z. (2015). Ultrasonic frequency effect on corn starch and its cavitation. *LWT - Food Science and Technology*, *60*(2), 941–947. https://doi.org/10.1016/j.lwt.2014.10.048

Huong, N. T. M., Hoa, P. N., & Van Hung, P. (2021). Effects of microwave treatments and retrogradation on molecular crystalline structure and in vitro digestibility of debranched mung-bean starches. *International Journal of Biological Macromolecules*, *190*, 904–910. https://doi.org/10.1016/j.ijbiomac.2021.09.032

Kaur, P., & Annapure, U. S. (2023). Effects of pin-to-plate atmospheric cold plasma for modification of pearl millet (*Pennisetum glaucum*) starch. *Food Research International*, *169*, 112930. https://doi.org/10.1016/j.foodres.2023.112930

Kaushik, A., Saxena, D. C., & Singh, S. (2025). Modification of browntop millet (*Brachiaria ramosa*) starch: Effects of pulse electric field and gamma-irradiation on morphological, thermal, rheological, and powder flow properties. *International Journal of Biological International Journal of Biological Macromolecules*, *309*, 143132. https://doi.org/10.1016/j.ijbiomac.2025.143132

Khurshida, S., & Deka, S. C. (2022). Application of microwave and hydrothermal treatments for modification of cassava starch of Manipur region, India and development of cookies. *Journal of Food Science and Technology*, *59*(1), 344–354. https://doi.org/10.1007/s13197-021-05020-9

Kumar, Y., Singh, L., Sharanagat, V. S., Patel, A., & Kumar, K. (2020). Effect of microwave treatment (low power and varying time) on potato starch: Microstructure, thermo-functional, pasting and rheological properties. *International Journal of Biological Macromolecules*, *155*, 27–35. https://doi.org/10.1016/j.ijbiomac.2020.03.174

Lal, M. A., & Bhatla, S. C. (2023). Metabolism of storage carbohydrates. In M. A. Lal & S. C. Bhatla (Eds.), *Plant physiology, development and metabolism* (pp. 231–259). Springer Nature Singapore. https://doi.org/10.1007/978-981-99-5736-1_9

Lv, Z., Yu, K., Jin, S., Ke, W., Fei, C., Cui, P., & Lu, G. (2019). Starch granules size distribution of sweet potato and their relationship with quality of dried and fried products. *Starch - Stärke*, *71*(5–6), 1800175. https://doi.org/10.1002/star.201800175

Manitoba. (2023). *Global market for starch and starch products.* 3–9. https://www.gov.mb.ca/agriculture/markets-and-statistics/trade-statistics/pubs/starch-world-market.pdf

Shrivastava, A., Gupta, R. K., & Srivastav, P. P. (2024). Exploring novel frontiers of advancements in purple yam (*Dioscorea alata* L.) starch extraction, modification, characterization, applications in food and other industries. *Measurement: Food*, *15*, 100196. https://doi.org/10.1016/j.meafoo.2024.100196

Nath, R., & Dutta, K. (2024). Sources of starch and its morphology: A review. In S. Chakrabarti, A. A. Sakib, S. Chattopadhyay, S. Poddar, A. Bhattacharya, & M. Gangopadhyaya (Eds.), *Interdisciplinary research in technology and management* (pp. 402–411). CRC Press. https://doi.org/10.1201/9781003430469-47

Obadi, M., & Xu, B. (2024). A review of the effects of physical processing techniques on the characteristics of legume starches and their application in low-glycemic index foods. *International Journal of Biological Macromolecules*, *279*, 135124. https://doi.org/10.1016/j.ijbiomac.2024.135124

Okyere, A. Y., Rajendran, S., & Annor, G. A. (2022). Cold plasma technologies: Their effect on starch properties and industrial scale-up for starch modification. *Current Research in Food Science*, *5*, 451–463. https://doi.org/10.1016/j.crfs.2022.02.007

Olawoye, B., Popoola-Akinola, O., Fagbohun, O. F., Adetola, D. B., Adeloye, A. E., Abdulwahab, A. A., Ogunjemilusi, M. A., & Olaoye, I. O. (2024). Starch modification using sustainable processing technology: Impacts and applications. *Starch - Stärke*, *77*(3), 2400115. https://doi.org/10.1002/star.202400115

Oyeyinka, S. A., Akintayo, O. A., Adebo, O. A., Kayitesi, E., & Njobeh, P. B. (2021). A review on the physicochemical properties of starches modified by microwave alone and in combination with other methods. *International Journal of Biological Macromolecules*, *176*, 87–95. https://doi.org/10.1016/j.ijbiomac.2021.02.066

Phukan, M., & Nongkhlaw, F. T. (2022). Starch: An overview of its conventional and unconventional sources, pharmaceutical and industrial applications. *Current Trends in Pharmaceutical Research*, *9*(1), 140–161.

Raman, R. (2023). *19 Foods that are high in starch.* Healthline. https://www.healthline.com/nutrition/high-starch-foods

Rostamabadi, H., Demirkesen, I., Hakgüder Taze, B., Can Karaca, A., Habib, M., Jan, K., Bashir, K., Nemţanu, M. R., Colussi, R., & Reza Falsafi, S. (2023). Ionizing and nonionizing radiations can change physicochemical, technofunctional, and nutritional attributes of starch. Food Chemistry: X, *19*, 100771. https://doi.org/10.1016/j.fochx.2023.100771

Shrivastava, A., Gupta, R. K., & Srivastav, P. P. (2024). Exploring novel frontiers of advancements in purple yam (*Dioscorea alata* L.) starch extraction, modification, characterization, applications in food and other industries. *Measurement: Food*, *15*, 100196. https://doi.org/10.1016/J.MEAFOO.2024.100196

Singh, N. (2020). Functional and physicochemical properties of pulse starch. In B. K. Tiwari, A. Gowen, & B. McKenna (Eds.), *Pulse foods: Processing, quality and nutraceutical applications* (pp. 87–112). Academic Press. https://doi.org/10.1016/B978-0-12-818184-3.00005-2

Starch Europe. (2024). *The European starch industry.* https://starch.eu/the-european-starch-industry/

Sun, X., Sun, Z., Saleh, A. S. M., Lu, Y., Zhang, X., Ge, X., Shen, H., Yu, X., & Li, W. (2023). Effects of various microwave intensities collaborated with different cold plasma duration time on structural, physicochemical, and digestive properties of lotus root starch. *Food Chemistry*, *405*(Pt A), 134837. https://doi.org/10.1016/j.foodchem.2022.134837

Susanto, B., Tosuli, Y. T., Adnan, Cahyadi, Nami, H., Surjosatyo, A., Alandro, D., Nugroho, A. D., Rashyid, M. I., & Muflikhun, M. A. (2024). Characterization of sago tree parts from Sentani, Papua, Indonesia for biomass energy utilization. *Heliyon*, *10*(1), e23993. https://doi.org/10.1016/j.heliyon.2024.e23993

Verma, V. C., Kumar, A., Zaidi, M. G. H., Verma, A. K., Jaiswal, J. P., Singh, D. K., Singh, A., & Agrawal, S. (2018). Starch isolation from different cereals with variable amylose/amylopectin ratio and its morphological study using SEM and FT-IR. *International Journal of Current Microbiology and Applied Sciences*, *7*(10): 211–228. https://doi.org/10.20546/ijcmas.2018.710.022

Vilpoux, O. F., & Junior, J. F. S.S. (2022). Global production and use of starch. In M. P. Cereda & O. F. Vilpoux (Eds.), *Starchy crops morphology, extraction, properties and applications: Vol 1. Underground starchy crops of South American origin: Production, processing, utilization and economic perspectives* (pp. 43–66). Academic Press. https://doi.org/10.1016/B978-0-323-90058-4.00014-1

WebMD. (2024). *5 High starch foods to avoid.* https://www.webmd.com/diet/foods-high-in-starch

Yan, Y., Jia, M., Zhou, Z., Xiao, S., Lin, P., Wang, Y., Fu, Y., & Wang, X. (2025). Effect of ultrasonic treatment on the physicochemical properties of buckwheat starch: Based on the ultrasonic power and moisture content. *Ultrasonics Sonochemistry*, *116*, 107333. https://doi.org/10.1016/j.ultsonch.2025.107333

Yi, M., Tang, X., Liang, S., He, R., Huang, T., Lin, Q., & Zhang, R. (2024). Effect of microwave alone and microwave-assisted modification on the physicochemical properties of starch and its application in food. *Food Chemistry*, 446, 138841. https://doi.org/10.1016/j.foodchem.2024.138841

Zarski, A., Bajer, K., & Kapuśniak, J. (2021). Review of the most important methods of improving the processing properties of starch toward non-food applications. *Polymers*, *13*(5), 832. https://doi.org/10.3390/polym13050832

2 Starch Structure and Functionality

Sachchidanand Pandey, Abhishek Kumar, Yaghuvendra Kumar, Shreyash Shirsat, P Srinivasa Rao, and Jayeeta Mitra
Agricultural and Food Engineering Department, Indian Institute of Technology Kharagpur, Kharagpur, West Bengal, India

2.1 INTRODUCTION

Starch is a primary energy source for human beings and is synthesized by plants as a carbohydrate reserve (Gupta et al., 2022). Starch can be found out in cereals, legumes, roots, and tubes. Some common sources of starch are maize, rice, rye, peas, potato, taro, and cassava (Wang & Guo 2020; Gupta et al., 2023). Starch is often utilized in the food business as anticaking, encapsulating, thickening, gelling, stabilizing, emulsifying, fat substitute, and resistant starch (Ai & Jane, 2024). Starch derived from cereal endosperms serves predominantly as food, supplying about 50% of the daily energy needs of the human being. Starch is an important food ingredient and a multifunctional biological material utilized worldwide, spanning multiple industrial sectors, such as food, health, textiles, chemicals, and engineering. The adaptable nature of starch in applications in industry is determined primarily by its physicochemical characteristics and functioning. Starch is synthesized in amyloplasts as tiny granules exhibiting varied morphology depending on source, with shapes including flat, round, oval, ogival, elongated, polyhedral, and lenticular and with diameters ranging from submicrons to more than 100 μm in diameter. It consists of two forms of glucan polymers, namely, amylose and amylopectin. A starch granule usually contains roughly 20% amylose and 80% amylopectin (Ohdan et al., 2006). Despite the lower amylose content in starch, amylose significantly affects the morphology, digestibility, adhesiveness, water absorption, cooking quality, and processing of the products (Wu et al., 2019). Amylose primarily consists of linear chains of glucose residues connected by α-1,4 links and exhibiting a low degree of branching at α-1,6 linkage, with each chain containing about 1000 residues (Martin & Smith, 1995). Amylopectin has extensive branching and features relatively short chains, making it a highly intricate structured molecule. The short chains of amylopectin are arranged in double helices that contribute to starch crystallinity (Gupta et al., 2024c). Studies suggests that the branching characteristics and molecular weight of starch, which significantly impact the morphology and dimensions of granules, may be utilized to foresee specific functionalities, including texture, pasting, and retrogradation. Amylose exhibits the greater proportional connection with gel and pasting textural qualities; however, amylopectin demonstrates a stronger relationship with stiffness (Mua & Jackson 1997). The native starch exhibits restricted functioning; for example, stability fluctuates adversely at varying pH levels and the starch is insoluble in water at ambient temperature. That's why it is need to be modified to enhance desired characteristics and functionality such as texture, solubility, viscosity, and heat stability. These properties are crucial for optimal industrial processing and product improvement. Using physical methods such as solubility, gelatinization, and retrogradation, characteristics of the starch can be modified without compromising chemical integrity (Egharevba, 2019; Khlestkin et al., 2018). Chemical methods such as etherification, esterification, acetylation, cationization, cross-linking hydrolysis, and oxidation destroy existing chemical bonds and create new bonds; as a consequence, properties of the products change. Also, biotechnological processes such as enzymatic and genetic alterations are being investigated (BeMiller & Huber 2015).

Starch possesses various applications in culinary products. It undergoes gelatinization upon heating, which is why it serves as a thickening agent in the preparation of soups, sauces, gravies, and custards; upon cooling, it produces a gel appropriate for the production of gummy candies and puddings. It possesses excellent water-binding properties, enabling it to absorb and hold moisture well, which is advantageous for maintaining softness in baked goods. Starch serves as a texturizing agent to improve mouthfeel in dairy confections and processed meals, as well as to provide creaminess in sauces and soups. It can serve as a fermentation substrate in the manufacture of bioethanol and beer. Additional roles include as adhesives, binders, and sweeteners. Another intriguing application is substitution for fat in food products to create low-fat options with lower calorie content, as modified starch can replicate the properties of fat. Starches with elevated amylose content exhibit exothermic properties and establish a stronger sustaining amylose–lipid complex. Starch–lipid complexes can serve as fat substitutes in certain food products, prolonging shelf life and enhancing freeze–thaw stability. Additionally, starch can regulate glucose levels in diabetic patients (Sengupta et al., 2023). Researchers are utilizing the film-forming characteristics of starch to create edible coatings and biodegradable packaging materials. Additionally, starch can encapsulate flavors, vitamins, and probiotics for controlled release.

DOI: 10.1201/9781003564287-2

2.2 STARCH STRUCTURE

Starch is synthesized in seeds, tubers, roots, and rhizomes as semi-crystalline granules, each exhibiting distinct features specific to the plant species. Although the size and appearance of the granules vary with their source, structurally they exhibit similar design (Bertoft, 2017). Starch granules possess multiple structural levels, encompassing their size and morphology, with dimensions ranging from 10^{-6} to 10^{-10}. The precise role of amylose is ambiguous, whereas amylopectin is widely recognized as the primary contributor to the semi-crystalline nature of the granules (Gupta et al., 2024a).

2.2.1 Starch Biosynthesis

Understanding the biosynthesis of amylose is crucial because it influences both the cooking and eating quality of the product and the usability of starch for industrial applications (Okpala et al., 2022). The biosynthesis of amylose and amylopectin encompasses a sequence of enzyme reactions that construct glucose molecules into elongated chains or branched configurations. Despite both amylose and amylopectin being composed of glucose, their structures differ considerably. Amylose and amylopectin are both synthesized by enzymes known as starch synthases. These enzymes incorporate the glucose monomer from adenosine diphosphate glucose (ADP-glucose) to the non-reducing terminus of an elongating glucan chain.

2.2.2 Biosynthesis of Amylose and Amylopectin

Amylopectin biosynthesis involves the coordinated action of starch synthase and the branching enzyme. Starch synthase facilitates the addition of glucose units from ADP-glucose (ADPG) to the non-reducing end of a growing glucan chain, creating α-1,4-glycosidic bonds and resulting in the formation of extended linear glucose chains (Gupta et al., 2024e). The branched architecture of amylopectin arises through the activity of the branching enzyme, also known as 1,4-α-glucan branching enzyme, which catalyzes the formation of α-1,6-glycosidic bonds. This is achieved by transferring a short segment, usually six or seven glucose units, from one part of the glucan chain to another, thereby introducing branch points. The branching enzyme hydrolyzes the α-1,4-glycosidic bond and subsequently reinstates the segment through an α-1,6-linkage, creating a branch point. Following branching, starch synthase persists in elongating the newly produced chains at the branch points. The process of elongation and branching persists, ending in a highly branched structure referred to as amylopectin.

The synthesis of amylose necessitated solely one isoform of starch synthase, namely, granule-bound starch synthase I (GBSSI). Numerous well-characterized mutants across various species are devoid of amylose, all of which specifically exhibit a deficiency in GBSSI action. Alternative isoforms of starch synthase are incapable of replacing GBSSI in amylose production. GBSSI is recognized for its distinct localization inside the plastid compared to other starch synthase isoforms. GBSSI is predominantly attached to starch granules, while other isoforms of starch synthase and starch-branching enzymes are situated within the granules and in the adjacent stroma. Compelling evidence indicates that GBSSI is dispersed within the granule interior, rather than solely affixed to the granule's surface, and that a portion of the encapsulated GBSSI is functional. GBSSI is encapsulated within the amylopectin matrix. The activity of GBSSI may be assessed in isolated intact granules, demonstrating that a portion of the concealed GBSSI remains active and that ADPG can access the GBSSI site within the granule. Nonetheless, instead of a uniform distribution of the small amount of amylose throughout the granule, iodine staining indicates that the majority of the amylose is localized in a central core region. This suggests that amylose is synthesized deep within the granule while the granule expands by the deposition of amylopectin at its surface. The production of amylose within the granule accounts for its lower degree of branching because glucans synthesized within the granule are likely less prone to branching than are those on the granule's surface. GBSSI binds tightly to the amylopectin matrix of starch granules. This area exhibits a strong affinity for maltooligosaccharides (MOSs). The MOSs infiltrate the granules to the locus of GBSSI activity, serving as a primer for amylose synthesis. GBSSI subsequently extends the MOSs to produce elongated glucan chains. The restricted permeability of the amylopectin matrix permits short MOSs to diffuse easily in and out of the granule, but longer MOSs synthesized by GBSSI are unable to escape the granule. The elongated glucans are contained inside the amylopectin matrix and are subsequently extended to create amylose (Denyer et al., 2001; Seung, 2020; Wang et al., 2015).

2.2.3 The Starch Granule

Once amylose and amylopectin are synthesized, they are packaged into starch granules in plastids within plant cells. Starch granules are semi-crystalline in nature. The dimensions, morphology, and molecular configuration of the granules are influenced by the species, plant variety, and gene–environment interactions. All starches, regardless of their plant source, consist of polymeric structures based on D-glucopyranose. Starch polymers consist of individual glucose units connected by α-1,4-glycosidic bonds and branched by α-1,6-glycosidic bonds (Himashree et al., 2022). Amylopectin forms the majority of the granule structure, while amylose is a minor component and typically forms a more amorphous, less ordered phase within the granule. Amylopectin short chains form amorphous region with lesser degree of polymerization than amylose. X-ray diffraction studies have shown that the double helices formed by neighboring branch chains of amylopectin can crystallize into one of three polymorphic forms: A-type, B-type, and C-type (Gupta et al., 2024b). In the A-type structure, the double helices are tightly packed within a

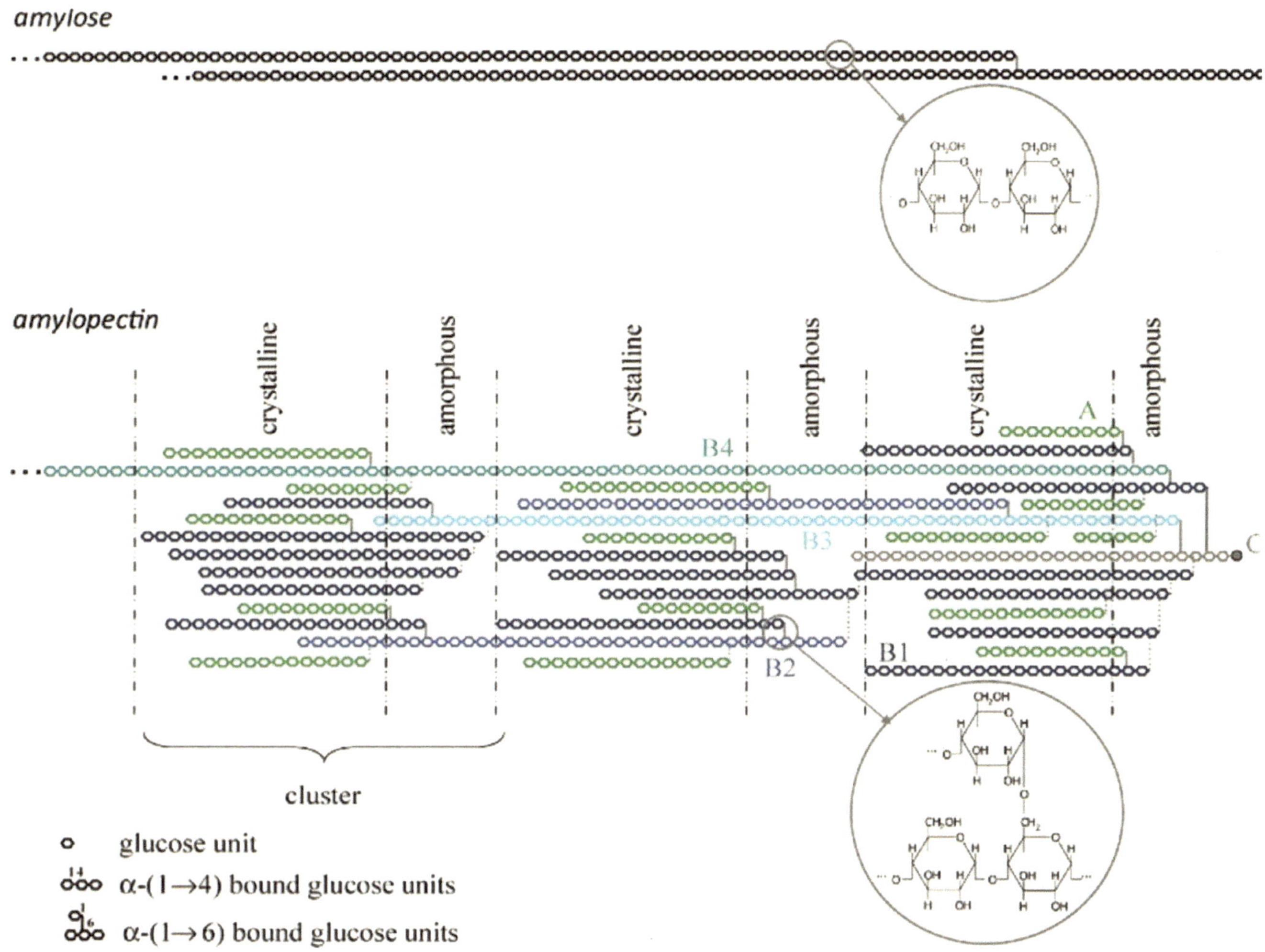

FIGURE 2.1 The cluster model representing the structure of the starch, alternate crystalline and amorphous region, proposed by Hizukuri (1986). Also illustrated are glucose chains linearly connected by α-1,4-glycosidic linkage and branching at α-1,6-glycosidic linkage.

monoclinic unit cell (a = 20.83 Å, b = 11.45 Å, c = 10.58 Å) belonging to the B2 space group and containing eight water molecules. In contrast, the B-type structure features a hexagonal unit cell ($a = b$ = 18.5 Å, c = 10.4 Å) with a P61 space group and accommodates 36 water molecules. The C-type starch represents a combination of A- and B-type polymorphs and displays an amylopectin branch-chain length distribution that falls between the two (Bertoft, 2017; Gupta et al., 2024c)).

2.2.4 Amylopectin and Amylose Structure

Amylopectin has a more intricate structure compared to amylose. It consists of glucose units arranged in linear chains through α-1,4-glycosidic bonds, with branching points formed by α-1,6-glycosidic linkages (Gupta et al., 2024d). The degree of polymerization in these chains is typically less than 100. The branches are typically spaced every 20 to 25 glucose units, creating a dense, tree-like structure (Koekuyt & Marangoni, 2024). The adjacent glucose chains of amylopectin form double helices making crystalline lamellae, while branching linkage joints form amorphous lamellae. Within a starch granule, crystalline and amorphous lamellae are arranged in an alternating pattern, resulting in a semi-crystalline structure. Amylose, which contains very few branches, is primarily located within the amorphous regions of the amylopectin matrix. It consists of long chains of glucose units connected by α-1,4-glycosidic bonds, with a degree of polymerization ranging from 100 to 10,000, and includes only a small number of α-1,6-linked branches (Figure 2.1). The structures of these molecules influence some functional properties of food products, such as mouthfeel and digestibility (Bertoft et al., 2024).

2.3 STARCH FUNCTIONALITY

2.3.1 Physicochemical Properties

2.3.1.1 Gelatinization: Temperature, Process, and Influencing Factors

"Gelatinization" refers to a permanent structural transformation that takes place at several product design scales, from micro to macro. These structural changes in starch are primarily influenced by a number of unknown variables, including thermal impact and subsequent manufacturing processes. Putting the contributing components into two categories—raw material qualities and process parameters—is essential to better understanding heat-induced changes (Schirmer et al., 2015).

The main structural changes of starch granules that take place when a system with a specific starch-to-water ratio is heated are known as gelatinization (Shrivastava et al., 2024). Prior to gelatinization, heating starch granules in excess water to temperatures higher than 52°C causes an irreversible glass transition in the amorphous region. The gelatinization temperature varies from 55°C to 80°C, contingent upon the type of starch (Mounir et al., 2024). Gelatinization begins as water penetrates the granule, hydrating the amorphous regions before moving to the crystalline areas, resulting in the breakdown of granules through water absorption. This results in a degradation of the crystalline structure and in birefringence, facilitating swelling and the leaching of amylose (Jiménez et al., 2012). Prolonged heating results in the melting of amylopectin crystallites, leading to an amorphous state, starch solubilization, and the leaching of starch polymers (amylose and amylopectin), thereby increasing the viscosity of the starch paste. Figure 2.2 depicts the phenomenon and the phase shift of starch when exposed to elevated temperatures in an aqueous environment (Schirmer et al., 2015). The characteristic temperature for swelling and gelatinization is specific to each starch, influenced by the native crystalline structures and by the amylose to amylopectin ratio in the sample. Amylose-free granules exhibit significantly greater swelling than those with amylose because elevated amylose concentrations inhibit granular swelling and broaden the temperature range for gelatinization. Increased amylose concentrations elevate the enthalpy of gelatinization and result in the formation of amorphous gelatinization thermogram peaks. Low-amylose (waxy) starches typically swell at approximately 55°C, while high-amylose starches primarily swell at temperatures exceeding 100°C (Koekuyt & Marangoni, 2024). A number of variables affect the progress of starch gelatinization: (1) the physicochemical characteristics of the starch granule, such as the ratio of amylose to amylopectin, the fine structures, and the molecular weight distribution of amylose and amylopectin; (2) the amount of water that is available; (3) processing parameters, such as temperature, time, and shear; and (4) the presence of other ingredients, such as proteins, lipids, and other polysaccharides (Mounir et al., 2024).

2.3.1.2 Retrogradation and Its Implications

"Retrogradation" of starch refers to the reorganization of amylopectin–amylose chains and molecules that occurs during pasting and persists into the energy-loss phase of gelation and storage (Dobosz et al., 2019). During retrogradation, segments of amylopectin and amylose recrystallize, resulting in a semi-crystalline structure distinct from the crystallinity of native starch (Gong et al., 2019). The starch paste viscosity increases upon cooling, resulting from gel

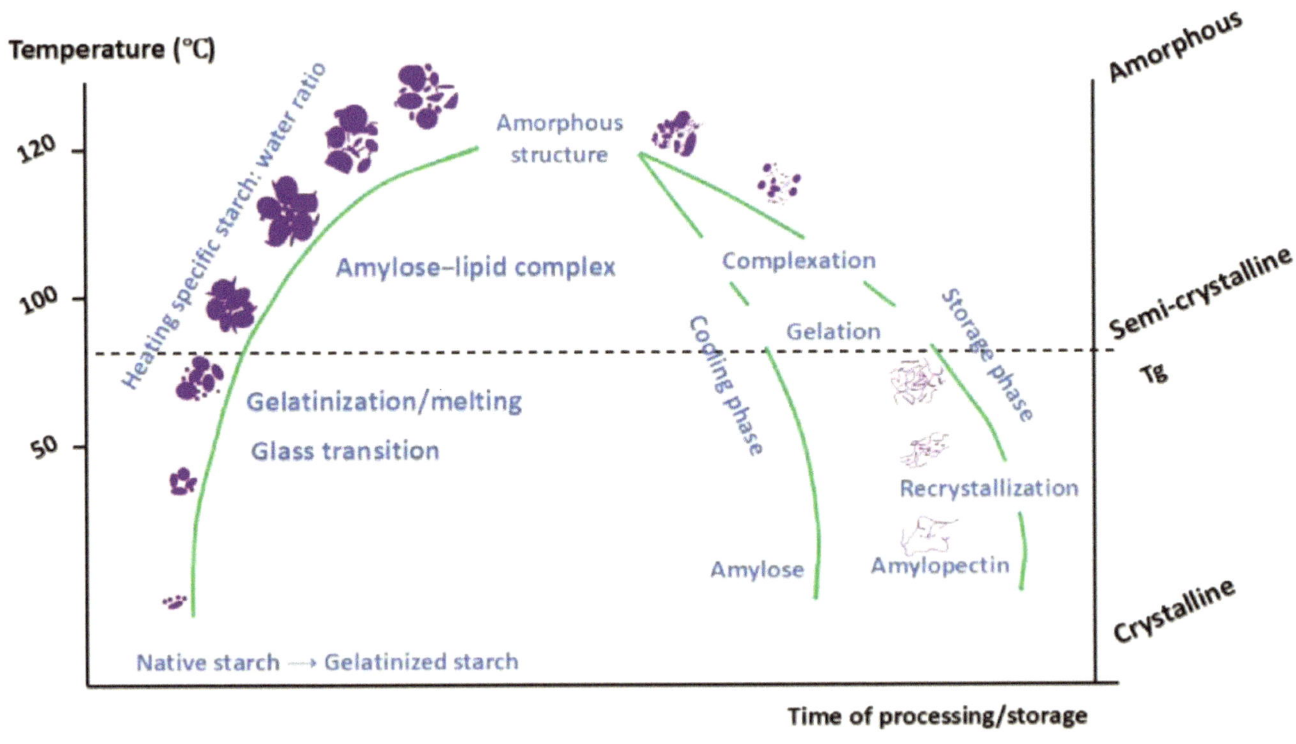

FIGURE 2.2 Phase diagram illustrating the states and phase transitions of starch under a temperature profile. Starch transitions from a crystalline to an amorphous structure upon heating, followed by recrystallization during cooling and storage (Tg, gelatinization temperature; AM, amylose; AP, amylopectin). (Schirmer et al., 2015)

formation through intermolecular interactions between amylopectin and amylose molecules. Upon standing, starch gels undergo retrogradation, resulting in the formation of insoluble B-type crystallites. The correlation between linear lengths of α-(1,4)-linked glucose units inside the polymers is responsible for this behavior (Copeland et al., 2009). While amylose retrogradation in processed foods has a significant impact on stickiness, water absorption, and digestibility, amylopectin retrogradation may be a more significant component in bread and cake staling. The texture and stickiness of rice are influenced by amylose, whereas amylopectin is associated mostly with pasting properties, gelatinization temperature, and heating (Balet et al., 2019).

The retrogradation of starch occurs in two phases, influenced by the proportions of amylopectin and amylose in the sample as well as by the cooling rate, cooling temperature, and storage temperature. The two stages are characterized as short-term and long-term retrogradation. Short-term retrogradation occurs swiftly during gelation and subsequent cooling, involving the rearrangement of amylose helices into double helical chains via inter- and intramolecular associations, as illustrated in Figure 2.3 (Tamaki et al., 2011). Long-term retrogradation, as indicated by its name, takes place during extended storage periods. This facilitates the reformation of helices in amylopectin chains, which subsequently associate to create helix-helix structures (Wang et al., 2021). Amylose, characterized by its linear structure and extended length, is capable of forming numerous stable intramolecular associations. Consequently, heat treatments following retrogradation generally fail to eliminate or diminish the amylose crystallinity observed (Dobosz et al., 2019). The associations between amylose and amylopectin during long-term retrogradation are significantly weaker than those between amylose molecules and exhibit a strength comparable to that of amylopectin–amylopectin associations (Vamadevan & Bertoft, 2018). The crystallinity of amylopectin generally decreases during heat treatment following retrogradation, while the amylose crystals persist. This facilitates the formation of amylose–amylopectin associations, potentially resulting in the further development of amylopectin crystals via secondary nucleation. Amylose in high-amylose starches and type 4 amylopectin starches, characterized by a high population of a higher degree of polymerization chains and double helices, can function as a crystalline stabilizer and re-nucleation site for the re-establishment of amylopectin crystals (Koekuyt & Marangoni, 2024).

2.3.1.3 Pasting Behavior

Pasting entails the swelling of starch granules as water penetrates, leading to the dissociation of inter- and intrahelical bonds and resulting in a loss of structural order within the granule. As the granule swells, pores and channels develop within the network, facilitating the leaching of amylose and smaller chain fragments from the granule. As the pasting process progresses, the swelling of the granule ultimately disrupts its integrity, leading to complete rupture of the granules (Kumar & Khatkar, 2017). The properties of starch granules, including swelling, degree of fragmentation, and retrogradation during processing and storage, determine the textural qualities and stability of food based on high-moisture starch. Consequently, understanding the mechanics of pasting and the creation of gels and pastes is crucial. Differentiating between a paste and a gel may be difficult; "paste" denotes the hot system produced from newly cooked starch, while "gel" signifies the system after it has cooled (Mounir et al., 2024). The two phases of a hot starch paste are (1) a continuous phase that contains partially and completely dissolved starch molecules, mainly low-molecular-weight amylopectin and amylose, and (2) granule fragments. Swollen granules and granule shadows constitute a discontinuous phase of swollen starch granules from which amylose and/or amylopectin have been leached (Mounir et al., 2024).

The pasting characteristics of flour–water and starch–water mixtures are often assessed using a Rapid Visco

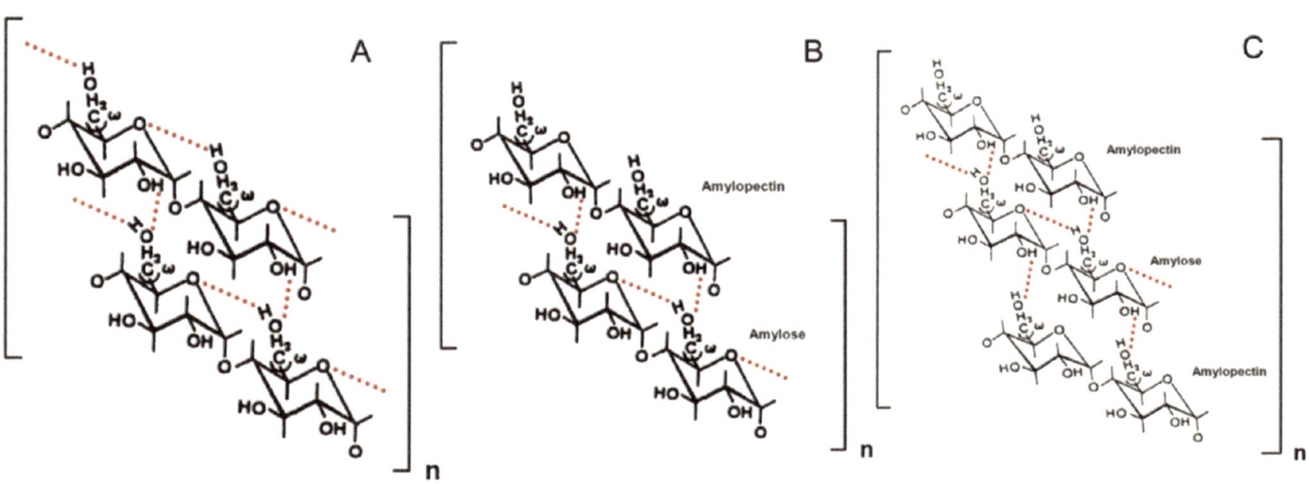

FIGURE 2.3 Illustrative representations of (A and B) intermolecular and intramolecular hydrogen bonding interactions seen in native starches in aqueous solution and (C) retrograded starch gels. The dotted red lines indicate hydrogen bonding (Tamaki et al., 2011).

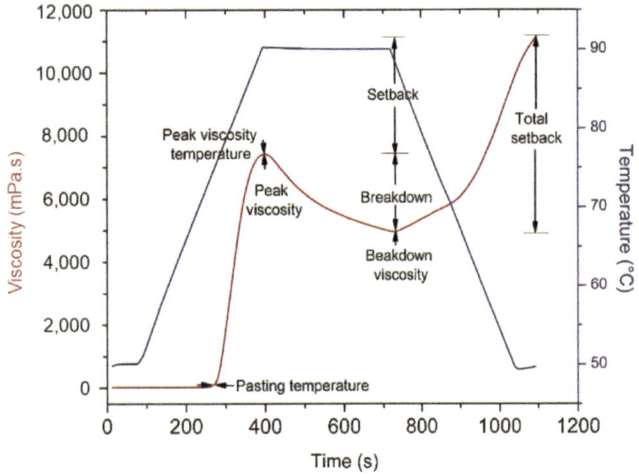

FIGURE 2.4 A standard RVA pasting profile of a cereal, highlighting the primary parameters assessed during analysis (Balet et al., 2019).

Analyzer (RVA), a viscometer that facilitates heating and cooling while measuring the sample's resistance to controlled shear. Functionality and structural features are correlated using the RVA, which mimics food processing (Balet et al., 2019). The viscosity parameters are assessed by varying the temperature and time across different intervals, as illustrated in Figure 2.4. The process consists of five distinct phases: (1) an initial phase involving the addition of water to starch; (2) a controlled heating phase in which the suspension temperature is raised from ambient to 95°C; (3) an isothermal phase during which the suspension is held at 95°C with shearing; (4) a cooling phase that reduces the suspension temperature to approximately 50°C; and (5) a final holding phase maintaining the temperature at 50°C (Balet et al., 2019).

2.3.2 Functional Properties in Food Systems

2.3.2.1 Thickening and Gelling

The thickening characteristics develop because even at very low concentrations, large molecular weight molecules in solution expand.

According to Torrens and Castellano (2018), long-chain polysaccharide molecules typically dwell in solution as conformationally disordered "random coils," with their structure changing continuously due to Brownian motion.

According to Copeland et al. (2009) and Torrens and Castellano (2018), thickeners exhibit their functional properties in emulsions through four mechanisms: (1) preserving an emulsion with a high internal phase volume, (2) increasing the external phase viscosity, (3) creating an elastic network inside the external phase, and (4) reducing the external phase volume. Among the factors influencing the gelatinization of starch, which enhances starch's thickening properties, are the size and shape of the starch granules. When native starch is hydrothermally treated, the granules grow up to 30 times their initial size before bursting, which releases the starch material and creates a colloidal dispersion in water (Wang et al., 2023).

2.3.2.2 Water Holding Capacity (WHC)

Water holding capacity (WHC) is the capacity of a specific flour or starch to retain its intrinsic and added water under conditions of force application, such as pressing, centrifuging, or heating. It plays a crucial role in swelling and gelatinization, contributing to the texture development of food products. It also influences sensorial characteristics, including mouthfeel, and enhances nutritional aspects by improving satiety and bulkiness (Chiranthika et al., 2022). Starch's water-holding capacity (WHC) influences viscosity, texture, mouthfeel, moisture management, and shelf life, among other functional attributes of food, and is a critical factor in formulation due to its role in retaining and binding water (Shen et al., 2022). It is necessary to modify native starches to improve their functioning because they are not appropriate for industrial uses. The fundamental structure of starch is altered by thermal processes, which impact the starch's solubility, digestibility, swelling, and water holding capacity. As a result, thermally processed starches are used in a variety of culinary products, including soups, jellies, noodles, pudding, custard, cakes, and biscuits (Lian et al., 2022).

2.3.2.3 Syneresis and Stability in Frozen Foods

The sensory appeal of frozen foods made with starches depends on their capacity to withstand freeze–thaw cycles. The matrix readily releases water when frozen food is thawed. The overall frozen food quality of items decreases as a result of syneresis caused by the separation of water from gel phases. Differential scanning calorimetry (DSC) and measuring the liquid being separated from the gel are two methods for evaluating the freeze–thaw stability of any starch gel (Vamadevan & Bertoft, 2018). Starches with higher amylose content exhibit a higher tendency for retrogradation, while amylopectin-free starches demonstrate a reduced tendency toward this phenomenon. As the frequency of freeze–thaw cycles rises, waxy starch gel's syneresis also increases. It has been shown that chemical modification improves starch's freeze–thaw stability (Vamadevan & Bertoft, 2015). Hydrocolloids have shown potential in lowering syneresis due to their ability to retain water, which improves starch freeze–thaw stability (BeMiller, 2021).

2.3.3 Functional Properties in Non-Food Applications

Starch biomass has the potential to be converted into various chemical products through chemical and biological processing, applicable in the food, paper, pharmaceutical, and adhesive industries (Marques et al., 2017).

2.3.3.1 Role in Bioplastics and Pharmaceuticals

The retrogradation process re-associates the dispersed amylose molecules to form flexible gels after the gelatinous starch melt cools from its heated state. The filler is evenly incorporated into the gels' three-dimensional matrix, which is made up of a continuous layer of amylose molecules. For industrial uses, especially in the food packaging industry, starch's diverse properties are beneficial (Lagarón et al., 2016). Starch-based biofilms exhibit favorable oxygen barrier properties, making them suitable for use as edible coatings for high-respiration-rate foods, such as fruits and vegetables. The films may inhibit respiration, leading to delayed oxidation. Additionally, these cost-effective edible starch-based films are typically tasteless, colorless, and odorless (Saha et al., 2020). Starch must be modified through plasticization or by blending with other polymeric materials to enhance its applicability. Chemical and physical modifications, as well as the linking of various approaches, can enhance the processability of starch. Various polymers, including poly(ethylene-co-vinyl alcohol), polycaprolactone, and polyvinyl alcohol, are mixed with starch to create starch-based products. Numerous industrial processes, including extrusion, blow molding, injection molding, film blowing, and shaping, use thermoplastics made from starch. Despite the changes, the expanding impact of water causes them to have poor resistance to water vapor. These films are appropriate only for hydrophilic dry products, and they are made using the starch component of wheat blends (Saha et al., 2020; Sapper et al., 2019). Starch is used to create biodegradable, controlled-life pharmaceutical packaging materials. It is also processed into sheets for the purpose of protecting glass dishes, trays, and similar items. The main application for starch-based bioplastics is as hard, flexible packaging, such as bags and sacks. To mitigate brittleness, plasticizers are used while making starch packing materials (Saha et al., 2020).

Regenerative medicine (RM) is an interdisciplinary field of study that combines concepts from engineering and the biological sciences to design products that can functionally or structurally replace tissues, facilitating the healing process (Gopinath et al., 2022). It aims to restore function to organs and tissues affected by damage, aging, and disease (Adewale et al., 2022). The biodegradability of RM applications for starch, facilitated by α-amylase produced by the pancreas, is a crucial factor in its development (Labelle et al., 2020). Wound healing and tissue engineering are two primary areas of focus in regenerative medicine. At the moment, RM is a low-volume, high-value industry for the use of non-food starch. High-purity starch, free from impurities such as residual proteins and fibers, is necessary for this application area, along with modifications for incorporation into bio-composites (Labelle et al., 2020). Cadexomer iodine is a commercial starch product that contains 0.9% (w/w) iodine physically immobilized inside the microsphere matrix. It is made of an epichlorohydrin cross-linked 3D hydrophilic starch. Exudate and particle materials from the wound surface can be absorbed by the product. Iodine is produced when the dressing absorbs moisture, aiding in wound washing and having a bactericidal impact on methicillin-resistant *Staphylococcus aureus* (Adewale et al., 2022).

2.3.3.2 Use in Paper, Adhesives, and Textile Industries

The paper industry is one of the largest sectors after the food industry for starch use. Industrial starch serves as a sizing agent and binder throughout the papermaking process. Sizing is the process of adding substances to paper to change its durability and absorption properties by acting as a protective coating or filler. By restricting the capillary action of fibers with the use of a sizing agent, liquid absorption of dry paper can be reduced, producing printing surfaces that are more accurate, consistent, and cost-effective (Han et al., 2021). In addition to reducing surface porosity and sizing, fuzzing affects smoothness, abrasiveness, printability, finish, reusability, and surface bond strength. However, at room temperature native starch's insolubility in different solvents, shear force changes, vulnerability to pH, temperature, and insufficient water barrier qualities limit its application in papermaking (Han et al., 2021).

Commercial starch adhesives utilized in corrugated boxes comprise three components: a mixture containing additional starch, borax, and water, to create the final adhesive referred to as a carrier. The presence of glucosidic and hydroxyl functional groups facilitates extensive hydrogen bonding in starch, which is a key factor for effective adhesion (Adewale et al., 2022). Native starch as an adhesive is suboptimal due to several factors: (1) its hydrophilic properties result in poor water resistance, (2) its low bonding strength limits its applicability in demanding adhesive for wood, and (3) high viscosity complicates its use due to its high molecular weight (Hemmilä et al., 2017).

After the petroleum industry, the textile sector is one of the most polluting in the world. A key tactic for attaining sustainability in the textile sector is the use of sustainable chemicals for textile printing, finishing, and sizing because wet textile manufacturing is very polluting (N. Singh & Sahu, 2019). The bond between the yarn body and starch film is enhanced by binding fibers together and minimizing yarn hair by reattaching it to the yarn body; adhesion is essential for boosting strength. However, native starch's brittleness results in poor adhesive qualities (W. Li et al., 2018). Research has focused on starch modification techniques such as cationization, double etherification, and grafting. Biopolymer coatings and nano-finishing are essential in the development of functional finishes. To achieve shrink resistance while preserving the fabric's natural properties, wool can be coated with biopolymers such as wheat starch, chitosan, and gum arabic. This is a sustainable alternative to energy-intensive techniques such as UV, plasma treatments, and ozone, as well as conventional treatments that harm the environment, such as the chlorine–Hercosett process (Rani et al., 2020). As useful finishing agents, zinc oxide–starch nanocomposites give textiles antibacterial and UV protection (Adewale et al., 2022; Mehrotra, 2020).

2.3.4 Factors Influencing Starch Functionality

2.3.4.1 Botanical Origin and Granule Size

Starches from various botanical sources exhibit unique gelatinization properties, which are indicative of the varying arrangement of double helical crystalline structures within starch granules and structures of starch molecules. Starch has distinct properties such as retrogradation rate (which is related to syneresis), gelling ability, paste viscosity and clarity, and gelatinization temperature and enthalpy change, depending on the botanical source. Comprehending the correlations between the functional and the structural attributes of starch is essential for the advancement of novel starch variations with preferred functionalities for use in food and other sectors (Vamadevan & Bertoft, 2018). Amylopectin extracted from starches of various botanical sources exhibits distinct distributions of branch-chain lengths. Compared to amylopectin of the B-type starch, amylopectin of the A-type polymorphic starch has a larger percentage of short branch chains (i.e., DP 6–12) that are restricted to a single cluster and a smaller percentage of long branch chains that cross many clusters. Granule size has a major impact on the usefulness and suitability of starch in culinary and other applications. In the size of the starch granules, there can be a seven-fold variation within one botanical species. Numerous parameters, such as the availability of nutrients, the activity and expression of distinct enzymes, the temperature at which the plant grows, the maturity of the storage organ, and the spatial position within the storage organ, all play a role in the biological conservation and regulation of the granule size. Variation in granule size may hold biological significance, such as improved cellular packing or a gradual energy release during germination. This variation also presents opportunities for broader applications of starch, particularly in modification and processing to attain specific functional properties in food and other sectors (M. Li et al., 2023).

2.3.4.2 Environmental Factors during Plant Growth

In plants, starch is a major storage substance that is produced by a convoluted biosynthesis process and is present in both storage and production organs. Glucose is produced by the quick digestion and absorption of these carbohydrates (Kim & Kim, 2021). Environmental and genetic factors affect the quality and content of starch. Compared to storing proteins, starch is more susceptible to drought and heat stress. The stressors affect starch synthesis negatively by having a substantial effect on the anthesis and grain-filling time. Enzymes involved in sucrose metabolism and starch synthesis are inhibited by drought and heat stress during the grain-filling phase. It is essential to elucidate the mechanisms of starch and sucrose during plant responses (Kim & Kim, 2021). Regarding heat tolerance in grain starch content, an extensive amount of genetic variation has been found among wheat species (Farooq et al., 2011). Particularly noteworthy are modifications to the deposition of amylose and amylopectin and the formation of starch granules (Thitisaksakul et al., 2012). During the wheat grain-filling process, high temperatures change the size distribution of the starch granules in the grains and reduce their starch concentration. This then changes the amylopectin chain length in endosperm starches, which leads to a less-than-ideal starch granule shape (Lu et al., 2014). Higher nighttime temperatures in emerging grains have been associated with a twofold increase in the starch-degrading enzymes alpha-amylase, beta-amylase, and isoamylase III and a decrease in the levels of the small subunit of adenosine diphosphate glucose pyrophosphorylase. Likewise, higher nighttime temperatures reduce grain-filling time and negatively impact grain structure more than do daytime temperatures (Impa et al., 2020).

2.4 Analytical Methods for Studying Starch Structure

2.4.1 Microscopy

Acquiring an in-depth understanding about structural and physical traits of raw materials is vital for achieving superior product quality and advancing new technologies. Microscopic techniques are integral in assessing the physical attributes and texture of food components. Such knowledge is instrumental in refining product quality and improving food processing methods. Three microscopy techniques are commonly utilized—light microscopy, atomic force microscopy, and electron microscopy—each of which is discussed in detail.

2.4.2 Light Microscopy

Microscopic examination of starch has a long history, beginning in the early 17th century, with Antonie van Leeuwenhoek's discoveries playing a key role. Van Leeuwenhoek was the first to observe and confirm the presence of starch in different plant tissues using microscopy (Seidemann, 1968). The transmission light microscope (TLM) is one of the earliest and most widely used microscopy techniques, in which light passes through a sample and is subsequently collected. However, TLM is limited by its low contrast. In contrast, polarized light microscopy (PLM) involves illuminating samples with polarized light, and the light transmitted through the sample is filtered by a perpendicular polarizer. This setup allows the detection of depolarized light from the sample, making PLM particularly suitable for studying anisotropic materials (Xiao et al., 2020). Light microscopy (LM) is crucial for studying microstructural changes in starch granules caused by hydrothermal treatment, processes like granule expansion, and amylose leakage and their impact on rheological behavior. It complements other techniques for analyzing structural and chemical alterations in processed starch systems (Błaszczak, 2020). Light microscopy plays a key role in differentiating between amylose and amylopectin by using iodine staining. This helps us to evaluate the dispersal of amylose and amylopectin, aiding in the understanding of starch pasting behavior within both dispersed and continuous media (Svegmark & Hermansson, 1991). These techniques play an important role in analyzing starch granule microstructures and their changes during processing.

2.4.3 Atomic Force Microscopy

Atomic force microscopy (AFM), invented by Gerd Binnig and his team in 1986, is an advanced form of scanning probe microscopy following scanning tunneling microscope (STM) principles (Binnig et al., 1986). It is extensively employed in surface science field to enable imaging at atomic-scale resolution, achieving dimensions on the order of 10^{-10} m (Jagtap & Ambre, 2006). It produces high-resolution photographs of samples not merely through optical observation but via the interaction of a sharp point that physically detects the surface of the sample (F. Zhu, 2017). It offers numerous advantages over other microscopic techniques. These include its exceptional imaging resolution, capacity for analyzing non-conductive samples, adaptability to diverse environments such as liquids or gases, and ability to measure nanoscale structures with precision (Wei et al., 2024). In the last 20 years, it has been increasingly utilized in the field of food systems (Morris et al., 2011). It has been utilized for taking image of starch granules and mixtures, providing detailed insights into structures such as blocklet formations, amylose chains, and changes during processes like gelatinization and retrogradation (Gupta et al., 2024a). Moreover, this instrument is employed to study starch interactions with iodine, enzymes, proteins, and fatty acids, connecting microscopic structures to larger-scale properties. AFM-based force spectroscopy further offers valuable information on elasticity, adhesion, and physicochemical characteristics, highlighting its potential for advanced starch analysis (F. Zhu, 2017). Park et al. (2011) have employed AFM to investigate the morphology of starch granules. Furthermore, when starch granules come in contact of iodine vapor in humid conditions, the hair-like structures formed can be easily identified by AFM. The ability of AFM to provide high-resolution imaging and to analyze various starch properties at the nano scale significantly gives us deep understanding about behavior of starch and interactions in food systems. Its versatility in different environments and its ability to link microscopic structure with macroscopic functionality make AFM a powerful tool in food science research.

2.4.4 Electron Microscopy

Electron microscopy was originally developed to detect inorganic substances, offering a resolution approximately 1000 times higher than that of light microscopy by utilizing electrons instead of light waves (Swanson & Wallace, 2014). Because it uses electron beams, it achieves a resolution of 0.5 nm and magnifications up to 10,000,000×. The main types of electron microscopy are scanning electron microscopy (SEM), transmission electron microscopy (TEM), and scanning transmission electron microscopy (STEM) (Kwiecińska et al., 2019). SEM uses a focused electron beam and a metal-coated sample to generate high-resolution images of surface topology and composition (Chakraborty et al., 2020). Due to its high resolution, it can be effectively applied to investigate the microstructure of starch, enabling the evaluation of shape and size. Govindaraju et al. (2022) analyzed rice starch using SEM to examine its polyhedral shape and to investigate how this morphology is altered during enzymatic hydrolysis. In another study, corn starch was treated in combination with cold plasma and enzymatic hydrolysis. The resulting modifications were evaluated using SEM, which revealed that the deeper holes formed over the surface of corn starch granules by cold plasma and enzymatic hydrolysis treatment (Davoudi et al., 2022). SEM analysis is essential for understanding the structural changes in starch granules following chemical modifications, as it provides detailed insights into surface morphology and particle structure. In the case of umbut palm stem starch, treatment with CH_3COOH caused irregular fractures, voids, and some fragmented particles that remained attached to larger granules. Treatment with HCl resulted in damaged, rough granules with flocs and an uneven surface, whereas treatment with H_2SO_4 produced smooth, intact granules with no visible voids, indicating a more uniform modification (Yunita et al., 2024). Electron microscopy, particularly SEM, is an effective tool for investigating the structural properties and modifications of starch granules at a microscopic level. Its ability to reveal detailed surface topology and changes in morphology provides critical insights into starch behavior during processing and chemical treatments.

2.4.5 X-ray Diffraction

X-ray analysis has been employed for over a century to visualize the internal structure of various objects. However, its application to evaluate the plant seeds began only in the early 1980s (Musaev & Beletskiy, 2021). X-ray diffraction (XRD) is a technique employed to examine how the molecules are arranged in the structure of crystalline materials by identifying their long-range order and to analyze non-crystalline materials by investigating their short-range order. This method facilitates the analysis of lattice parameters, material phases, grain size, crystallinity levels, and structural imperfections (Khan et al., 2020). In the context of starch–lipid complexes, it helps to identify the type of crystalline structures formed (e.g., A-type, V-type). XRD analysis revealed that native buckwheat starch exhibited A-type crystallinity, while buckwheat starch–myristic acid complexes formed V-type structures at higher temperatures (67.5°C–90°C). Relative crystallinity % increased with temperature and myristic acid concentration, but excessive myristic acid at lower temperatures (60°C–75°C) disrupted crystallinity (Oskaybaş-Emlek et al.,

2022). Ma et al. (2023) have studied the molecular structure of starch and apigenin. XRD analysis demonstrated that amylopectin and corn starch exhibit A-type crystalline structures, characterized by specific diffraction peaks. In contrast, apigenin displayed distinct, sharp diffraction peaks. Following complexation with apigenin, the crystalline structures of amylopectin and corn starch were disrupted, resulting in amorphous complexes while retaining residual characteristic peaks associated with apigenin. In another study, rice starch with varying amylose content was studied. Native rice starches displayed characteristic A-type polymorphic structures, with approximately 35%–47% crystallinity. Enzymatic hydrolysis reduced the crystallinity and the peaks of disintegration of crystalline, attributed to the preferential degradation of both crystalline and amorphous regions. Descreased crystallinity with increasing amylose content was observed, highlighting amylopectin as the primary contributor to the crystalline structure of starch (Govindaraju et al., 2022). The formation and behavior of resistant starch in non-pigmented rice starch subjected to repeated autoclaving were investigated. XRD analysis revealed that the autoclave-treated starches displayed a mixed crystalline pattern, combining A-type, B-type, and V-type characteristics, with less intense and broad peaks observed at angles 15°, 17°, 20°, and 23°. Structural damage to the crystalline regions was evident, as confirmed by SEM images. The crystallinity ranged from 29.63% to 33.86%, driven by the rearrangement of amylose and amylopectin chains during retrogradation (Anjali et al., 2024). XRD has proven essential for analyzing starch–lipid complexes and structural changes in starch, offering valuable insights into crystallinity, phase transitions, and complexation effects. The technique continues to enhance our understanding of starch behavior in various conditions, aiding in the development of novel starch-based materials.

2.4.6 Differential Scanning Calorimetry (DSC)

Thermal analysis, one of the earliest analytical techniques, saw its modern development begin in 1887 with Henri Le Chatelier's pioneering thermometric measurements, followed by William Roberts-Austen's introduction of differential thermal analysis in 1899. The commercialization of instruments in the 1960s marked the onset of rapid advancements, enabling extensive application across various industries. Today, the analysis of new materials necessitates specialized techniques, tailored measurement parameters, and precise sample preparation (Toledo, 2016). Differential scanning calorimetry (DSC) is a widely utilized technique to evaluate thermal characteristics that enables the characterization of material properties across diverse fields (e.g., polymers, foods, pharmaceuticals, and life sciences), providing essential insights into behavior, measurement methodologies, and data interpretation (Almoselhy, 2020).

The thermal properties of potato starch encapsulated within calcium alginate beads were investigated using DSC. The results demonstrated a significant increase in melting temperatures (T_p) and gelatinization temperatures (T_o, T_c) compared to unmodified potato starch, with an enhancement of 5.4°C–7.1°C. This improvement is attributed to the calcium alginate coating, which inhibits the swelling of starch granules and reduces water absorption. Further increasing the concentrations of sodium alginate increased T_o, T_c, and ΔH (Cui et al., 2022). DSC analysis showed that freeze–thawing treatment lowered the gelatinization temperatures and enthalpy (ΔH) of tuntas compared to potatoes, while peak temperature remained unaffected. The reduction in ΔH indicates a partial disruption of the crystalline structure intuntas, reducing the energy requirement for gelatinization. These results are consistent with previous studies and comparable to Andean and Loc'ka potato starches (Martínez et al., 2022). In another study, potato starch was improved using mild heating freeze–thawing (MHFT) and mild gelatinization freeze–thawing (MGFT) treatments and analyzed using DSC. The results showed that melting temperature, initially about 60°C, increased to about 67°C after treatment and reduced the enthalpy requirement from about 14 to about 10 J/g. These changes indicate enhanced thermal stability, attributed to the reassociation of soluble starch and partial disruption of the crystalline structure, highlighting the importance of pre-treatment in improving starch thermal properties (Wang et al., 2022). Montoya-Anaya et al. (2023) has studied the gelatinization properties of potato starch. The DSC thermograms showed the gelatinization process of potato starch, with an onset temperature (T_o) of 61.78°C, a peak temperature (T_p) of 64.45°C, and a final temperature (T_f) of 90°C, along with an enthalpy of gelatinization (ΔH) of 10.96 J/g. These values align with variations reported in the literature, which depend on the starch's botanical origin, crystallinity, and composition, particularly amylose and lipid content, as these factors influence both gelatinization temperatures and enthalpy. Additionally, an endothermic peak at 114.56°C was attributed to the dissociation of the amylose–lipid complex, emphasizing the impact of structural and compositional properties on the thermal behavior of starch. DSC provides valuable insights into the thermal properties and gelatinization behavior of starch, offering crucial information for optimizing starch modification processes. The technique's ability to characterize starch under various treatments highlights its importance in enhancing thermal stability and understanding starch behavior in different applications.

2.4.7 Rapid Visco Analyzer (RVA)

The Rapid Visco Analyzer (RVA) was introduced by the Bread Research Institute of Australia to measure sprout damage in wheat and barley. The earliest version lacked automated temperature control, relying on manual heating and thermometer monitoring. The RVA is recognized for its repeatability, consistency, and adaptability, making it an ideal tool for analyzing cereal starch (Balet et al., 2019).

The instrument has traditionally been used for cereals and starchy foods to analyze different factors involved in the starch pasting properties of a sample, such as viscosity and pasting temperature (Cozzolino, 2016). The findings indicate that, at higher temperatures, waxy and normal starches exhibited reductions in viscosities and gel adhesiveness due to thermal degradation, whereas high-amylose starches underwent complete gelatinization and viscosity development above 120°C (Liu et al., 2019). Through high-temperature Rapid Visco Analyzer (HTRVA) analysis, it was observed that protein content and glutenin abundance were inversely associated with setback and final viscosities, while the fine structural characteristics of amylose and amylopectin also played a significant role in influencing pasting behavior (Li et al., 2022). Tian et al. (2022), investigated the relationship between starch structure and high-pressure gelation properties using RVA 4800. Starches with higher amylose content, longer amylopectin chains (DP > 36), and lower amylopectin with short chains (DP ≤ 36) showed higher peak temperatures and lower viscosities within 95°C to 140°C. Within 130°C to 140°C, rheological and texture parameters were primarily influenced by amylose molecular structure. Guo et al. (2024), has demonstrated the use of the RVA to investigate structural changes in maize starches during pasting and gelation with varying amylose contents. Key transitions, including granule swelling, structural disruption, and gel network formation, were observed at 95°C, followed by gel reorganization. Analyzing starch thermal and viscosity behaviors, offering insights into structure–function relationships, in elucidating the molecular mechanisms governing the pasting properties of high amylose maize starch (HAMS) starches under high-temperature conditions, providing valuable insights for advancements in food processing and product formulation.

2.5 EMERGING TRENDS IN STARCH RESEARCH

2.5.1 Starch Modification Using Emerging Novel Processing Techniques

Starch is an affordable and highly adaptable industrial material, extensively utilized across various sectors, such as food, chemicals, petrochemicals, adhesives, paints, paper, textiles, pharmaceuticals, and animal feed. Its wide range of applications leads to diverse functional requirements. However, natural sources of starch and their inherent properties are limited. Therefore, advancing and optimizing starch modification techniques is crucial to obtain new functionalities and broaden the application potential of starch, thereby improving starch's performance and versatility (Ashogbon & Akintayo, 2014; Maniglia et al., 2021). Starches are often modified through physical, chemical, enzymatic, or genetic methods to improve and tailor their functional properties. Among these, physical modification involving the use of moisture, heat, shear forces, or radiation is gaining popularity, especially in food-related applications, because it avoids the use of chemical reagents, making it a safer and more natural option (Zia-ud-Din et al., 2017). While chemical modification remains widely used, physical methods are gaining importance for achieving desired functional properties across industries (Gałkowska et al., 2023). The ongoing development of new technologies and equipment in the starch industry, coupled with progress in mechanochemistry, has led to growing interest in non-thermal physical modifications of starch. These methods are valued for being environmentally friendly, straightforward, and easy to apply. They have a notable impact on the structure, functionality, and chemical reactivity of starch, thereby significantly broadening starch's potential applications (Wang et al., 2023). The incorporation of advanced technologies in starch modification represents a pivotal trend in the field. Cutting-edge methods such as ozone treatment (Castanha et al., 2020; Pandiselvam et al., 2019), cold plasma (Gupta et al., 2024), high-pressure processing (Colussi et al., 2020; Li et al., 2015), pulsed electric fields (Wu et al., 2019; Zeng et al., 2016), and both ultrasonic and ultrasound treatments (Zhu et al., 2012; Zhu, 2015) have shown great promise in altering the properties of starch granules while maintaining their structural integrity (Maniglia et al., 2021). These methods enhance the functional properties of starch, including gelatinization and pasting behaviors, while simultaneously promoting the formation of resistant starch. This type of starch is particularly beneficial for health, as it improves gut microbiota composition and aids in glycemic control, making it valuable for managing blood sugar levels in functional food applications (Dupuis et al., 2014). Traditionally, research has prioritized modifying starch and only subsequently exploring its potential uses, but this approach overlooks the benefits of tailoring modifications for specific applications from the outset. A reverse approach, starting with a specific application and determining the required starch modifications is more efficient (Maniglia et al., 2021). This requires a deeper understanding of the relationships between starch structure, properties, and processing methods. Identifying the desired functionality for various applications and the necessary structural or property changes in starch molecules is essential for effectively tailoring modifications (Compart et al., 2023; He et al., 2023). Despite the challenges in accessing detailed information about industrial requirements, such insights are vital for designing new modification approaches. Ultimately, achieving targeted functionalities in starch modification demands an innovative understanding of how structure, properties, and processing interact, enabling precise tailoring for specific applications. This involves not only screening synergy possibilities among technologies but also prioritizing application-driven research.

2.5.2 Starch-Based Biodegradable Material

Global forecasts indicate that plastic production could exceed 650 million tons by 2050, representing a staggering increase of over 22,000% since 1950—a trend that highlights the critical need for sustainable alternatives (Fayshal, 2024). Starch-based materials are widely used in edible food packaging coatings, often enhanced with functional additives such as antioxidants, antimicrobials, emulsifiers, and probiotics to improve their effectiveness and value (Miteluţ et al., 2021; Gupta et al., 2025a)). The development of biodegradable materials from starch has become a key focus of research, driven by the urgent need to address environmental issues such as plastic pollution. As a renewable and biodegradable material, starch presents a promising eco-friendly alternative to conventional plastics. Its integration into bioplastics and composites is garnering increasing attention as industries strive for eco-friendly solutions to minimize the environmental footprint of packaging (Singh et al., 2024; Zeng et al., 2016). Bioplastics produced from starch, notably thermoplastic starch (TPS), have shown great promise as substitutes for petroleum-derived plastics. TPS is formed by modifying the granular structure of starch using heat and plasticizers, resulting in a malleable material (Rahardiyan et al., 2023). However, unmodified TPS exhibits drawbacks such as limited mechanical strength and high sensitivity to moisture. To overcome these challenges, researchers are innovating bio-based materials that improve the functionality of conventional starch composites. Additives such as nanocellulose, lignin, and biopolymers like polylactic acid (PLA) are being incorporated to enhance tensile strength, thermal resistance, and moisture durability (Jacob et al., 2024; Jayarathna et al., 2022; Ng et al., 2024; Samir et al., 2022). Advanced processing methods, including reactive extrusion and 3D printing, are transforming the landscape of starch-based composites. Reactive extrusion facilitates in-process chemical modifications, enabling better integration of additives and improved material properties (Edeleva et al., 2023). Meanwhile, 3D printing provides a precise platform for fabricating intricate structures, opening up avenues for tailored applications across industries such as packaging and healthcare. These advancements are expanding the potential of starch-based materials and enhancing their ability to compete with traditional plastics (Siddiqui et al., 2024). A significant innovation in starch applications is the creation of active packaging systems focused on sustainability. Starch-based films and coatings are being designed to incorporate barrier properties and functional additives such as antimicrobials, antioxidants, and oxygen scavengers. These solutions not only protect food products from environmental influences but also help extend shelf life and improve safety (Khandeparkar et al., 2024). For example, films infused with essential oils or nanoparticles demonstrate strong antimicrobial effects, making them particularly suitable for packaging fresh produce and meats. The combination of biodegradability and enhanced functionality makes starch-based active packaging a valuable advancement for the food industry (García-Guzmán et al., 2022; Majeed et al., 2023; Gupta et al., 2025b).

2.6 CONCLUSION

Starch is one of the most versatile and affordable food components, can be extracted from almost any plant, and is renewable, colorless, odorless, nontoxic, and biodegradable. It has been utilized in multiple industrial products. The widespread use of starch has increased demand for it and the requirement of a specific starch in many of applications necessitating the need for its modification. Starch is a natural biopolymer; it represents a transformative resource in the quest for sustainable alternatives to plastics. From biodegradable bioplastics to multifunctional active packaging, starch-based innovations are driving progress toward a more sustainable future. Active food packaging is not only effective in preserving food products; it is also environmentally friendly. Active and antimicrobial agents added with packaging material interact with the food through the absorption or release of substances preventing growth of food-contaminating microorganisms, maintaining the food status and increasing shelf life. However, starch-based films have some challenges that limit large-scale application, such as moisture sensitivity, low mechanical strength, and poor barrier and preservative properties. Although significant progress has been made in starch modification technologies, the processes of modifying starch for the use in nanotechnology are severely lacking for the production of nanofibers and wall materials in nano-encapsulation and edible films. Starch also has great potential to produce healthy products that lower glucose levels in blood and reduce the danger of diabetes and obesity. Continuous efforts are required to develop modification techniques to meet the requirement of specific starch-modified products. Despite these advancements, several hurdles must be addressed to fully optimize starch-based materials for industrial applications. Key areas for improvement include balancing performance with cost-effectiveness, enhancing resistance to water, and improving mechanical robustness. Collaborative efforts between academia and industry are essential to tackle these challenges and to effectively scale up production.

REFERENCES

Adewale, P., Yancheshmeh, M. S., & Lam, E. (2022). Starch modification for non-food, industrial applications: Market intelligence and critical review. *Carbohydrate Polymers*, *291*, 119590. https://doi.org/10.1016/j.carbpol.2022.119590

Ai, Y., & Jane, J. lin. (2024). Understanding starch structure and functionality. In L. Nilsson (Ed.), *Starch in food: Structure, function, and applications*. Woodhead Publishing. https://doi.org/10.1016/B978-0-323-96102-8.00018-8

Almoselhy, R. I. M. (2020). Applications of differential scanning calorimetry (DSC) in oils and fats research: A review. *American Research Journal of Agriculture*, *6*(1), 1–9. https://doi.org/10.21694/2378-9018.20002

Anjali, K. U., Kamatchi, A. R., Sundaramoorthy, H., & Rayhana, J. S. (2024). Resistant starch formation and behavior changes of non-pigmented rice starches by repeated autoclaving. *Biomass Conversion and Biorefinery*, *15*, 16261–16271. https://doi.org/10.1007/s13399-024-06333-3

Ashogbon, A. O., & Akintayo, E. T. (2014). Recent trend in the physical and chemical modification of starches from different botanical sources: A review. *Starch-Stärke*, *66*(1–2), 41–57.

Balet, S., Guelpa, A., Fox, G., & Manley, M. (2019). Rapid Visco analyser (RVA) as a tool for measuring starch-related physiochemical properties in cereals: A review. *Food Analytical Methods*, *12*(10), 2344–2360. https://doi.org/10.1007/s12161-019-01581-w

BeMiller, J. N. (2021). Effect of hydrocolloids on normal and waxy maize starches cross-linked with epichlorohydrin. *Food Hydrocolloids*, *112*, 106260. https://doi.org/10.1016/j.foodhyd.2020.106260

Bertoft, E. (2017). Understanding starch structure: Recent progress. *Agronomy*, *7* (3), 56. https://doi.org/10.3390/AGRONOMY7030056

Bemiller, J. N., & Huber, K. C. (2015). Physical modification of food starch functionalities. *Annual Review of Food Science and Technology*, *6*, 19–69.

Binnig, G., Quate, C. F., & Gerber, C. (1986). Atomic force microscope. *Physical Review Letters*, *56*(9), 930.

Błaszczak, W. (2020). Light microscopy as a tool to evaluate the functionality of starch in food. *Foods*, *9*, 1–15.

Castanha, N., Miano, A. C., Jones, O. G., Reuhs, B. L., Campanella, O. H., & Augusto, P. E. D. (2020). Starch modification by ozone: Correlating molecular structure and gel properties in different starch sources. *Food Hydrocolloids*, *108*, 106027. https://doi.org/10.1016/j.foodhyd.2020.106027

Chakraborty, I., Pallen, S., Shetty, Y., Roy, N., & Mazumder, N. (2020). Advanced microscopy techniques for revealing molecular structure of starch granules. *Biophysical Reviews*, *12*(1), 105–122. https://doi.org/10.1007/s12551-020-00614-7

Chiranthika, N. N. G., Chandrasekara, A., & Gunathilake, K. D. P. P. (2022). Physicochemical characterization of flours and starches derived from selected underutilized roots and tuber crops grown in Sri Lanka. *Food Hydrocolloids*, *124*, 107272. https://doi.org/10.1016/j.foodhyd.2021.107272

Colussi, R., Kringel, D., Kaur, L., da Rosa Zavareze, E., Dias, A. R. G., & Singh, J. (2020). Dual modification of potato starch: Effects of heat-moisture and high pressure treatments on starch structure and functionalities. *Food Chemistry*, *318*, 126475. https://doi.org/10.1016/j.foodchem.2020.126475

Compart, J., Singh, A., Fettke, J., & Apriyanto, A. (2023). Customizing starch properties: A review of starch modifications and their applications. *Polymers*, *15*(16), 3491. https://doi.org/10.3390/polym15163491

Copeland, L., Blazek, J., Salman, H., & Tang, M. C. (2009). Form and functionality of starch. *Food Hydrocolloids*, *23*(6), 1527–1534. https://doi.org/10.1016/j.foodhyd.2008.09.016

Cozzolino, D. (2016). The use of the Rapid Visco Analyser (RVA) in breeding and selection of cereals. *Journal of Cereal Science*, *70*, 282–290. https://doi.org/10.1016/j.jcs.2016.07.003

Cui, C., Jiang, H., Guan, M., Ji, N., Xiong, L., & Sun, Q. (2022). Characterization and in vitro digestibility of potato starch encapsulated in calcium alginate beads. *Food Hydrocolloids*, *126*, 107458. https://doi.org/10.1016/j.foodhyd.2021.107458

Davoudi, Z., Azizi, M. H., & Barzegar, M. (2022). Porous corn starch obtained from combined cold plasma and enzymatic hydrolysis: Microstructure and physicochemical properties. *International Journal of Biological Macromolecules*, *223*, 790–797. https://doi.org/10.1016/j.ijbiomac.2022.11.058

Denyer, K. A. Y., Johnson, P., Zeeman, S., & Smith, A. M. (2001). The control of amylose synthesis. *Journal of Plant Physiology*, *158*(4), 479–487. https://doi.org/10.1078/0176-1617-00360

Dobosz, A., Sikora, M., Krystyjan, M., Tomasik, P., Lach, R., Borczak, B., Berski, W., & Lukasiewicz, M. (2019). Short- and long-term retrogradation of potato starches with varying amylose content. *Journal of the Science of Food and Agriculture*, *99*(5), 2393–2403. https://doi.org/10.1002/jsfa.9446

Dupuis, J. H., Liu, Q., & Yada, R. Y. (2014). Methodologies for increasing the resistant starch content of food starches: A review. *Comprehensive Reviews in Food Science and Food Safety*, *13*(6), 1219–1234. https://doi.org/10.1111/1541-4337.12104

Edeleva, M., De Smit, K., Debrie, S., Verberckmoes, A., Marien, Y. W., & D'hooge, D. R. (2023). Molecular scale-driven upgrading of extrusion technology for sustainable polymer processing and recycling. *Current Opinion in Green and Sustainable Chemistry*, *43*, 100848. https://doi.org/10.1016/j.cogsc.2023.100848

Farooq, M., Bramley, H., Palta, J. A., & Siddique, K. H. M. (2011). Heat stress in wheat during reproductive and grain-filling phases. *Critical Reviews in Plant Sciences*, *30*(6), 491–507. https://doi.org/10.1080/07352689.2011.615687

Fayshal, M. A. (2024). Current practices of plastic waste management, environmental impacts, and potential alternatives for reducing pollution and improving management. *Heliyon*, *10*(23), e40838. https://doi.org/10.1016/j.heliyon.2024.e40838

Gałkowska, D., Kapuśniak, K., & Juszczak, L. (2023). Chemically modified starches as food additives. *Molecules*, *28*(22), 1–19. https://doi.org/10.3390/molecules28227543

Gupta, R. K., Guha, P., & Srivastav, P. P. (2022). Natural polymers in bio-degradable/edible film: A review on environmental concerns, cold plasma technology and nanotechnology application on food packaging-A recent trends. *Food Chemistry Advances*, *1*, 100135. https://doi.org/10.1016/j.focha.2022.100135

Gupta, R. K., Guha, P., & Srivastav, P. P. (2023). Effect of high voltage dielectric barrier discharge (DBD) atmospheric cold plasma treatment on physicochemical and functional properties of taro (Colocasia esculenta) starch. *International Journal of Biological Macromolecules*, *253*, 126772. https://doi.org/10.1016/j.ijbiomac.2023.126772

Gupta, R. K., Guha, P., & Srivastav, P. P. (2024a). Dielectric barrier discharge plasma: A green and novel method to change the structure of taro peel starch and improve the physicochemical properties of taro peel starch films. *Plasma Processes and Polymers*, *21*(9), 2400047. https://doi.org/10.1002/ppap.202400047

Gupta, R. K., Guha, P., & Srivastav, P. P. (2024b). Exploring the potential of taro (*Colocasia esculenta*) starch: Recent developments in modification, health benefits, and food industry applications. *Food Bioengineering*, *3*(3), 365–379. https://doi.org/10.1002/fbe2.12103

Gupta, R. K., Guha, P., & Srivastav, P. P. (2024c). Insights into the physicochemical, crystalline, multistructural, thermal and digestibility behavior of taro starch (*Colocasia esculenta*) modified by cold plasma treatment with high voltage dielectric barrier discharge (HVDBD): An alternative technique for chemical modification of starch and environmental sustainability. *Journal of Cleaner Production*, *461*, 142710. https://doi.org/10.1016/j.jclepro.2024.142710

Gupta, R. K., Guha, P., & Srivastav, P. P. (2024d). Physical action of nonthermal cold plasma technology for starch modification. *Food Physics*, *1*, 100011. https://doi.org/10.1016/j.foodp.2024.100011

Gupta, R. K., Guha, P., & Srivastav, P. P. (2024e). Effect of cold plasma treatment and plasma-activated water on physicochemical and structural properties of starch: A green and novel approach for environmental sustainability. *Plasma*

Processes and Polymers, *21*(4), 2300204. https://doi.org/10.1002/ppap.202300204

Gupta, R. K., Pipliya, S., Patel, J., Srivastav, P. P., Castro-Muñoz, R., Wang, C. K., & Nemţanu, M. R. (2025a). Essential oil-embedded starch based films for active packaging: Insights into migration, stability, and preservation efficiency. *Trends in Food Science & Technology*, 163, 105149. https://doi.org/10.1016/j.tifs.2025.105149

Gupta, R. K., Gnana Moorthy Eswaran, U., Kumar, S., Pipliya, S., Kaur, B., & Srivastav, P. P. (2025b). Synergistic modification of starch polysaccharide via sequential treatments with ultrasonic and cold plasma technology. *Journal of Food Processing and Preservation*, *2025*(1), 9063007. https://doi.org/10.1155/jfpp/9063007

García-Guzmán, L., Cabrera-Barjas, G., Soria-Hernández, C. G., Castaño, J., Guadarrama-Lezama, A. Y., & Rodríguez Llamazares, S. (2022). Progress in starch-based materials for food packaging applications. *Polysaccharides*, *3*(1), 136–177. https://doi.org/10.3390/polysaccharides3010007

Gong, B., Cheng, L., Gilbert, R. G., & Li, C. (2019). Distribution of short to medium amylose chains are major controllers of in vitro digestion of retrograded rice starch. *Food Hydrocolloids*, *96*, 634–643. https://doi.org/10.1016/j.foodhyd.2019.06.003

Gopinath, V., Kamath, S. M., Priyadarshini, S., Chik, Z., Alarfaj, A. A., & Hirad, A. H. (2022). Multifunctional applications of natural polysaccharide starch and cellulose: An update on recent advances. *Biomedicine & Pharmacotherapy*, *146*, 112492. https://doi.org/10.1016/j.biopha.2021.112492

Govindaraju, I., Zhuo, G. Y., Chakraborty, I., Melanthota, S. K., Mal, S. S., Sarmah, B., Baruah, V. J., Mahato, K. K., & Mazumder, N. (2022b). Investigation of structural and physico-chemical properties of rice starch with varied amylose content: A combined microscopy, spectroscopy, and thermal study. *Food Hydrocolloids*, *122*(August 2021), 107093. https://doi.org/10.1016/j.foodhyd.2021.107093

Guo, K., Tian, Y., Podzimska-Sroka, D., Kirkensgaard, J. J. K., Herburger, K., Enemark-Rasmussen, K., Hassenkam, T., Petersen, B. L., Blennow, A., & Zhong, Y. (2024). Structural evolution of maize starches with different amylose content during pasting and gelation as evidenced by rapid visco analyser. *Food Chemistry*, *461*, 140817. https://doi.org/10.1016/j.foodchem.2024.140817

Gupta, R. K., Guha, P., & Srivastav, P. P. (2024). Physical action of nonthermal cold plasma technology for starch modification. *Food Physics*, *1*, 100011. https://doi.org/10.1016/j.foodp.2024.100011

Han, B., Niang, J., Rao, H., Lyu, N., Oda, H., Sakamoto, S., Yang, Y., & Sablier, M. (2021). Paper fragments from the Tibetan Samye Monastery: Clues for an unusual sizing recipe implying wheat starch and milk in early Tibetan papermaking. *Journal of Archaeological Science: Reports*, *36*, 102793. https://doi.org/10.1016/j.jasrep.2021.102793

He, R., Li, S., Zhao, G., Zhai, L., Qin, P., & Yang, L. (2023). Physicochemical characteristics, and industrial usability: starch modification with molecular transformation, physicochemical characteristics, and industrial usability: A state-of-the-art review. *Polymers*, *15*(13), 2935.

Hemmilä, V., Adamopoulos, S., Karlsson, O., & Kumar, A. (2017). Development of sustainable bio-adhesives for engineered wood panels: A review. *RSC Advances*, *7*(61), 38604–38630. https://doi.org/10.1039/C7RA06598A

Himashree, P., Sengar, A. S., & Sunil, C. K. (2022). Food thickening agents: Sources, chemistry, properties and applications: A review. *International Journal of Gastronomy and Food Science*, *27*, 100468. https://doi.org/10.1016/j.ijgfs.2022.100468

Hizukuri, S. (1986). Polymodal distribution of the chain lengths of amylopectins, and its significance. *Carbohydrate Research*, *147*(2), 342–347. https://doi.org/10.1016/S0008-6215(00)90643-8

Impa, S. M., Vennapusa, A. R., Bheemanahalli, R., Sabela, D., Boyle, D., Walia, H., & Jagadish, S. V. K. (2020). High night temperature induced changes in grain starch metabolism alters starch, protein, and lipid accumulation in winter wheat. *Plant, Cell & Environment*, *43*(2), 431–447. https://doi.org/10.1111/pce.13671

Jacob, J., Linson, N., Mavelil-Sam, R., Maria, H. J., Pothan, L. A., Thomas, S., Kabdrakhmanova, S., & Laroze, D. (2024). Poly(lactic acid)/nanocellulose biocomposites for sustainable food packaging. *Cellulose*, *31*(10), 5997–6042. https://doi.org/10.1007/s10570-024-05975-w

Jagtap, R. N., & Ambre, A. H. (2005). Atomic force microscopy (AFM): Basics and its important applications for polymer characterization: An overview. *Journal of Polymer Materials*, *22*(1), 1–26.

Jayarathna, S., Andersson, M., & Andersson, R. (2022). Recent advances in starch-based blends and composites for bioplastics applications. *Polymers*, *14*(21), 1–24. https://doi.org/10.3390/polym14214557

Jiménez, A., Fabra, M. J., Talens, P., & Chiralt, A. (2012). Edible and biodegradable starch films: A review. *Food and Bioprocess Technology*, *5*(6), 2058–2076. https://doi.org/10.1007/s11947-012-0835-4

Khan, H., Yerramilli, A. S., D'Oliveira, A., Alford, T. L., Boffito, D. C., & Patience, G. S. (2020). Experimental methods in chemical engineering: X-ray diffraction spectroscopy—XRD. *Canadian Journal of Chemical Engineering*, *98*(6), 1255–1266. https://doi.org/10.1002/cjce.23747

Khandeparkar, A. S., Paul, R., Sridhar, A., Lakshmaiah, V. V., & Nagella, P. (2024). Eco-friendly innovations in food packaging: A sustainable revolution. *Sustainable Chemistry and Pharmacy*, *39*, 101579. https://doi.org/10.1016/j.scp.2024.101579

Khlestkin, V. K., Peltek, S. E., & Kolchanov, N. A. (2018). Review of direct chemical and biochemical transformations of starch. *Carbohydrate Polymers*, *181*, 460–476. https://doi.org/10.1016/J.CARBPOL.2017.10.035

Kim, K.-H., & Kim, J. (2021). Understanding wheat starch metabolism in properties, environmental stress condition, and molecular approaches for value-added utilization. *Plants*, *10*(11), 2282. https://doi.org/10.3390/plants10112282

Koekuyt, H. A., & Marangoni, A. G. (2024). The solid state and nanostructure of starch: Effects on starch functionality. *Critical Reviews in Food Science and Nutrition*, *0*(0), 1–22. https://doi.org/10.1080/10408398.2024.2388279

Kumar, R., & Khatkar, B. S. (2017). Thermal, pasting and morphological properties of starch granules of wheat (*Triticum aestivum* L.) varieties. *Journal of Food Science and Technology*, *54*(8), 2403–2410. https://doi.org/10.1007/s13197-017-2681-x

Kwiecińska, B., Pusz, S., & Valentine, B. J. (2019). Application of electron microscopy TEM and SEM for analysis of coals, organic-rich shales and carbonaceous matter. *International Journal of Coal Geology*, *211*, 103203. https://doi.org/10.1016/j.coal.2019.05.010

Labelle, M., Ispas-Szabo, P., & Mateescu, M. A. (2020). Structure-functions relationship of modified starches for pharmaceutical and biomedical applications. *Starch - Stärke*, *72*(7–8), 2000002. https://doi.org/10.1002/star.202000002

Lagarón, J. M., López-Rubio, A., & José Fabra, M. (2016). Bio-based packaging. *Journal of Applied Polymer Science*, *133*(2), 42971. https://doi.org/10.1002/app.42971

Li, H. T., Kerr, E. D., Schulz, B. L., Gidley, M. J., & Dhital, S. (2022). Pasting properties of high-amylose wheat in conventional and high-temperature Rapid Visco Analyzer: Molecular contribution of starch and gluten proteins. *Food Hydrocolloids*, *131*, 107840. https://doi.org/10.1016/j.foodhyd.2022.107840

Li, M., Daygon, V. D., Solah, V., & Dhital, S. (2023). Starch granule size: Does it matter? *Critical Reviews in Food Science and Nutrition*, *63*(19), 3683–3703. https://doi.org/10.1080/10408398.2021.1992607

Li, W., Tian, X., Liu, L., Wang, P., Wu, G., Zheng, J., Ouyang, S., Luo, Q., & Zhang, G. (2015). High pressure induced gelatinization of red adzuki bean starch and its effects on starch physicochemical and structural properties. *Food Hydrocolloids*, *45*, 132–139. https://doi.org/10.1016/j.foodhyd.2014.11.013

Li, W., Xu, Z., Wang, Z., Liu, X., Li, C., & Ruan, F. (2018). Double etherification of corn starch to improve its adhesion to cotton and polyester fibers. *International Journal of Adhesion and Adhesives*, *84*, 101–107. https://doi.org/10.1016/j.ijadhadh.2018.03.001

Lian, F., Sun, D. W., Cheng, J. H., & Ma, J. (2022). Improving modification of structures and functionalities of food macromolecules by novel thermal technologies. *Trends in Food Science and Technology*, *129*, 327–338. https://doi.org/10.1016/j.tifs.2022.10.001

Liu, S., Yuan, T. Z., Wang, X., Reimer, M., Isaak, C., & Ai, Y. (2019). Behaviors of starches evaluated at high heating temperatures using a new model of Rapid Visco Analyzer—RVA 4800. *Food Hydrocolloids*, *94*, 217–228. https://doi.org/10.1016/j.foodhyd.2019.03.015

Lu, D., Shen, X., Cai, X., Yan, F., Lu, W., & Shi, Y.-C. (2014). Effects of heat stress during grain filling on the structure and thermal properties of waxy maize starch. *Food Chemistry*, *143*, 313–318. https://doi.org/10.1016/j.foodchem.2013.07.089

Ma, Y., Chen, Z., Wang, Z., Chen, R., & Zhang, S. (2023). Molecular interactions between apigenin and starch with different amylose/amylopectin ratios revealed by X-ray diffraction, FT-IR and solid-state NMR. *Carbohydrate Polymers*, *310*(1), 120737. https://doi.org/10.1016/j.carbpol.2023.120737

Majeed, T., Dar, A. H., Pandey, V. K., Dash, K. K., Srivastava, S., Shams, R., Jeevarathinam, G., Singh, P., Echegaray, N., & Pandiselvam, R. (2023). Role of additives in starch-based edible films and coating: A review with current knowledge. *Progress in Organic Coatings*, *181*, 107597. https://doi.org/10.1016/j.porgcoat.2023.107597

Martin, C., & Smith, A. M. (1995). Starch biosynthesis. *The Plant Cell*, *7*(7), 971–985. https://doi.org/10.1105/TPC.7.7.971

Mua, J. P., & Jackson, D. S. (1997). Relationships between functional attributes and molecular structures of amylose and amylopectin fractions from dorn starch. *Journal of Agricultural and Food Chemistry*, *45*(10), 3848–3854.

Maniglia, B. C., Castanha, N., Rojas, M. L., & Augusto, P. E. (2021). Emerging technologies to enhance starch performance. *Current Opinion in Food Science*, *37*, 26–36. https://doi.org/10.1016/j.cofs.2020.09.003

Marques, S., Moreno, A. D., Ballesteros, M., & Gírio, F. (2017). Starch biomass for biofuels, biomaterials, and chemicals. In: S. Vaz Jr. (Ed.), *Biomass and green chemistry: Building a renewable pathway* (pp. 69–94). Springer. https://doi.org/10.1007/978-3-319-66736-2_4

Martínez, P., Betalleluz-Pallardel, I., Cuba, A., Peña, F., Cervantes-Uc, J. M., Uribe-Calderón, J. A., & Velezmoro, C. (2022). Effects of natural freeze-thaw treatment on structural, functional, and rheological characteristics of starches isolated from three bitter potato cultivars from the Andean region. *Food Hydrocolloids*, *132*, 107860. https://doi.org/10.1016/j.foodhyd.2022.107860

Mehrotra, V. (2020). Contingent antidotes to civil conflict: A comparative case analysis of development projects as conflict drivers and mitigators. *Asian Journal of Research in Social Sciences and Humanities*, *10*(7), 1–7. https://doi.org/10.5958/2249-7315.2020.00009.X

Miteluț, A. C., Popa, E. E., Drăghici, M. C., Popescu, P. A., Popa, V. I., Bujor, O. C., Ion, V. A., & Popa, M. E. (2021). Latest developments in edible coatings on minimally processed fruits and vegetables: A review. *Foods*, *10*(11), 1–18. https://doi.org/10.3390/foods10112821

Montoya-Anaya, D. G., Madera-Santana, T. J., Aguirre-Mancilla, C. L., Grijalva-Verdugo, C., Gonzales-Garcia, G., Nuñez-Colín, C. A., Rodriguez-Nuñez, J. R., Montoya-Anaya, D. G., Madera-Santana, T. J., Aguirre-Mancilla, C. L., Grijalva-Verdugo, C., Gonzales-Garcia, G., Nuñez-Colín, C. A., & Rodriguez-Nuñez, J. R. (2023). Physicochemical characterization of residual potato (*Solanum tuberosum*) starch recovered from the potato chips industry in Mexico. *Biotecnia*, *25*(2), 60–72.

Morris, V. J., Woodward, N. C., & Gunning, A. P. (2011). Atomic force microscopy as a nanoscience tool in rational food design. *Journal of the Science of Food and Agriculture*, *91*(12), 2117–2125.

Mounir, S., Ghandour, A., Shatta, A., & Farid, E. (2024). Starch: Properties and functionality. In S. P. Bangar, K.V. Sunooj, & A. K. Siroha (Eds.), *Starch: Structure, properties, and modifications for food applications* (pp. 1–46). CRC Press. https://doi.org/10.1201/9781032655598-1

Musaev, F. B., & Beletskiy, S. L. (2021). History and prospects for the application of X-ray diffraction analysis in seed breeding and seed study. *Journal of the Russian Universities. Radioelectronics*, *24*(6), 6–15. https://doi.org/10.32603/1993-8985-2021-24-6-6-15

Ng, K. M. P., Tan, I. S., Foo, H. C. Y., Lau, J. S. Y., Lam, M. K., & Wong, M. K. (2024). Durable polylactic acid bionanocomposites with biomass-derived nanocellulose additives: Recent advances in production. *Journal of the Taiwan Institute of Chemical Engineers*, September, 105774. https://doi.org/10.1016/j.jtice.2024.105774

Oskaybaş-Emlek, B., Özbey, A., Aydemir, L. Y., & Kahraman, K. (2022). Production of buckwheat starch-myristic acid complexes and effect of reaction conditions on the physicochemical properties, X-ray pattern and FT-IR spectra. *International Journal of Biological Macromolecules*, *207*, 978–989. https://doi.org/10.1016/j.ijbiomac.2022.03.189

Ohdan, K., Fujii, K., Yanase, M., Takaha, T., & Kuriki, T. (2006). Enzymatic synthesis of amylose. *Biocatalysis and Biotransformation*, *24*(1–2), 77–81. https://doi.org/10.1080/10242420600598152

Okpala, N. E., Aloryi, K. D., An, T., He, L., & Tang, X. (2022). The roles of starch branching enzymes and starch synthase in the biosynthesis of amylose in rice. *Journal of Cereal Science*, *104*, 103393. https://doi.org/10.1016/J.JCS.2021.103393

Pandiselvam, R., Manikantan, M. R., Divya, V., Ashokkumar, C., Kaavya, R., Kothakota, A., & Ramesh, S. V. (2019). Ozone: An advanced oxidation technology for starch modification. *Ozone: Science and Engineering*, *41*(6), 491–507. https://doi.org/10.1080/01919512.2019.1577128

Park, H., Xu, S., & Seetharaman, K. (2011). A novel in situ atomic force microscopy imaging technique to probe surface morphological features of starch granules. *Carbohydrate Research*, *346*(6), 847–853.

Rahardiyan, D., Moko, E. M., Tan, J. S., & Lee, C. K. (2023). Thermoplastic starch (TPS) bioplastic, the green solution for single-use petroleum plastic food packaging: A review. *Enzyme and Microbial Technology*, *168*, 110260. https://doi.org/10.1016/j.enzmictec.2023.110260

Rani, S., Kadam, V., Rose, N. M., Jose, S., Yadav, S., & Shakyawar, D. B. (2020). Wheat starch, gum arabic and chitosan biopolymer treatment of wool fabric for improved shrink resistance finishing. *International Journal of Biological Macromolecules*, *163*, 1044–1052. https://doi.org/10.1016/j.ijbiomac.2020.07.061

Saha, T., Hoque, M. E., & Mahbub, T. (2020). Biopolymers for sustainable packaging in food, cosmetics, and pharmaceuticals. In F. M. Al-Oqla & S. M. Sapuan (Eds.), *Advanced processing, properties, and applications of starch and other bio-based polymers* (pp. 197–214). Elsevier Inc. https://doi.org/10.1016/B978-0-12-819661-8.00013-5

Samir, A., Ashour, F. H., Hakim, A. A. A., & Bassyouni, M. (2022). Recent advances in biodegradable polymers for sustainable applications. *NPJ Materials Degradation*, *6*(1). https://doi.org/10.1038/s41529-022-00277-7

Sapper, M., Talens, P., & Chiralt, A. (2019). Improving functional properties of cassava starch-based films by incorporating xanthan, gellan, or pullulan gums. *International Journal of Polymer Science*, *2019*, 1–8. https://doi.org/10.1155/2019/5367164

Schirmer, M., Jekle, M., & Becker, T. (2015). Starch gelatinization and its complexity for analysis. *Starch/Staerke*, *67*(1–2), 30–41. https://doi.org/10.1002/star.201400071

Seidemann, J. (1968). Beiträge zur Geschichte der Stärke-Mikroskopie. 1. Mitt. *Starch - Stärke*, *20*(5), 169–172.

Sengupta, A., Chakraborty, I., G. I., & Mazumder, N. (2023). An insight into the physicochemical characterisation of starch-lipid complex and its importance in food industry. *Taylor & Francis*, *39*(7), 4198–4212. https://doi.org/10.1080/87559129.2021.2021936

Seung, D. (2020). Amylose in starch: Towards an understanding of biosynthesis, structure and function. *New Phytologist*, *228*(5), 1490–1504.

Shen, Y., Hong, S., Singh, G., Koppel, K., & Li, Y. (2022). Improving functional properties of pea protein through "green" modifications using enzymes and polysaccharides. *Food Chemistry*, *385*, 132687. https://doi.org/10.1016/j.foodchem.2022.132687

Shrivastava, A., Gupta, R. K., & Srivastav, P. P. (2024). Exploring novel frontiers of advancements in purple yam (*Dioscorea alata* L.) starch extraction, modification, characterization, applications in food and other industries. *Measurement: Food*, *15*, 100196. https://doi.org/10.1016/j.meafoo.2024.100196

Siddiqui, M. A. S., Rabbi, M. S., Ahmed, R. U., & Billah, M. M. (2024). Biodegradable natural polymers and fibers for 3D printing: A holistic perspective on processing, characterization, and advanced applications. *Cleaner Materials*, *14*, 100275. https://doi.org/10.1016/j.clema.2024.100275

Singh, N., & Sahu, O. (2019). Sustainable cyclodextrin in textile applications. In Shahid-ul-Islam and B.S. Butola (Eds.), *The impact and prospects of green chemistry for textile technology* (pp. 83–105). Elsevier. https://doi.org/10.1016/B978-0-08-102491-1.00004-6

Singh, P., Pandey, V. K., Singh, R., Singh, K., Dash, K. K., & Malik, S. (2024). Unveiling the potential of starch-blended biodegradable polymers for substantializing the eco-friendly innovations. *Journal of Agriculture and Food Research*, *15*, 101065. https://doi.org/10.1016/j.jafr.2024.101065

Svegmark, K., & Hermansson, A.-M. (1991). Distribution of amylose and amylopectin in potato starch pastes: Effects of heating and shearing. *Food Structure*, *10*(2), 2.

Swanson, E., & Wallace, W. D. (2014). Handling and interpretation of heart transplant biopsies. In C. Day (Ed.), *Methods in molecular biology* (Vol. 1180, pp. 283–291). Springer. https://doi.org/10.1007/978-1-4939-1050-2_16

Tamaki, Y., Konishi, T., & Tako, M. (2011). Gelation and retrogradation mechanism of wheat amylose. *Materials*, *4*(10), 1763–1775. https://doi.org/10.3390/ma4101763

Thitisaksakul, M., Jiménez, R. C., Arias, M. C., & Beckles, D. M. (2012). Effects of environmental factors on cereal starch biosynthesis and composition. *Journal of Cereal Science*, *56*(1), 67–80. https://doi.org/10.1016/j.jcs.2012.04.002

Tian, Y., Qu, J., Zhou, Q., Ding, L., Cui, Y., Blennow, A., Zhong, Y., & Liu, X. (2022). High pressure/temperature pasting and gelling of starch related to multilevel structure-analyzed with RVA 4800. *Carbohydrate Polymers*, *295*, 119858. https://doi.org/10.1016/j.carbpol.2022.119858

Toledo, M. (2016). *Thermal analysis in practice: Tips and hints.* Mettler-Toledo GmbH.

Torrens, F., & Castellano, G. (2018). Brownian motion, random trajectory, diffusion, fractals, theory of chaos, and dialectics. In R.K. Haghi, E. Besalu, M. Jaroszewski, S. Thomas, & K.M. Praveen (Eds.), *Modern physical chemistry: Engineering models, materials, and methods with applications* (pp. 27–36). Apple Academic Press. https://doi.org/10.1201/9781315143118-11

Vamadevan, V., & Bertoft, E. (2015). Structure-function relationships of starch components. *Starch - Stärke*, *67*(1–2), 55–68. https://doi.org/10.1002/star.201400188

Vamadevan, V., & Bertoft, E. (2018). Impact of different structural types of amylopectin on retrogradation. *Food Hydrocolloids*, *80*, 88–96. https://doi.org/10.1016/j.foodhyd.2018.01.029

Wang, N., Li, C., Miao, D., Hou, H., Dai, Y., Zhang, Y., & Wang, B. (2023). The effect of non-thermal physical modification on the structure, properties and chemical activity of starch: A review. *International Journal of Biological Macromolecules*, *251*, 126200. https://doi.org/10.1016/j.ijbiomac.2023.126200

Wang, K., Hasjim, J., Wu, A. C., Li, E., Henry, R. J., & Gilbert, R. G. (2015). Roles of GBSSI and SSIIa in determining amylose fine structure. *Carbohydrate Polymers*, *127*, 264–274. https://doi.org/10.1016/J.CARBPOL.2015.03.081

Wang, S., & Guo, P. (2020). Botanical sources of starch. *Starch Structure, Functionality and Application in Foods*, 9–27.

Wang, S. Y., Zhang, C., Liu, Q. Q., Wang, Z. J., Wan, K. X., Qian, J. Y., Zhang, L., Wu, C., & Li, Q. (2022). Modification of potato starch by critical melting pretreatment combined with freeze-thawing: Preparation, morphology, structure, and functionality. *LWT*, *158*, 113109. https://doi.org/10.1016/j.lwt.2022.113109

Wang, X., Fang, J., Cheng, L., Gu, Z., & Hong, Y. (2023). Interaction of starch and non-starch polysaccharides in raw potato flour and their effects on thickening stability. *International Journal of Biological Macromolecules*, *242*(P2), 124702. https://doi.org/10.1016/j.ijbiomac.2023.124702

Wang, Y., Zhan, J., Lu, H., Chang, R., Qiu, L., & Tian, Y. (2021). Amylopectin crystal seeds: Characterization and their effect on amylopectin retrogradation. *Food Hydrocolloids*, *111*, 106409. https://doi.org/10.1016/j.foodhyd.2020.106409

Wei, M., Zhang, Y., Wang, Y., Liu, X., Li, X., & Zheng, X. (2024). Employing atomic force microscopy (AFM) for microscale investigation of interfaces and interactions in membrane fouling processes: New perspectives and prospects. *Membranes*, *14*(2), 35.

Wu, C., Wu, Q. Y., Wu, M., Jiang, W., Qian, J. Y., Rao, S. Q., Zhang, L., Li, Q., & Zhang, C. (2019). Effect of pulsed electric field on properties and multi-scale structure of japonica rice starch. *LWT, 116*, 108515. https://doi.org/10.1016/j.lwt.2019.108515

Xiao, H., Wang, S., Xu, W., Yin, Y., Xu, D., Zhang, L., Liu, G. Q., Luo, F., Sun, S., Lin, Q., & Xu, B. (2020). The study on starch granules by using darkfield and polarized light microscopy. *Journal of Food Composition and Analysis, 92*, 103576. https://doi.org/10.1016/j.jfca.2020.103576

Yunita, I., Prendika, W., & Mutia, R. (2024). Morphological analysis of chemically umbut palm stem starch granules using scanning electron microscope. *BIO Web of Conferences, 99*, 02005. https://doi.org/10.1051/bioconf/20249902005

Zeng, F., Gao, Q. Y., Han, Z., Zeng, X. A., & Yu, S. J. (2016). Structural properties and digestibility of pulsed electric field treated waxy rice starch. *Food Chemistry, 194*, 1313–1319. https://doi.org/10.1016/j.foodchem.2015.08.104

Zhu, F. (2015). Impact of ultrasound on structure, physicochemical properties, modifications, and applications of starch. *Trends in Food Science and Technology, 43*(1), 1–17. https://doi.org/10.1016/j.tifs.2014.12.008

Zhu, F. (2017). Atomic force microscopy of starch systems. *Critical Reviews in Food Science and Nutrition, 57*(14), 3127–3144. https://doi.org/10.1080/10408398.2015.1094650

Zhu, J., Li, L., Chen, L., & Li, X. (2012). Study on supramolecular structural changes of ultrasonic treated potato starch granules. *Food Hydrocolloids, 29*(1), 116–122. https://doi.org/10.1016/j.foodhyd.2012.02.004

Zia-ud-Din, Xiong, H., & Fei, P. (2017). Physical and chemical modification of starches: A review. *Critical Reviews in Food Science and Nutrition, 57*(12), 2691–2705.

3 Chemical Approaches to Starch Modification

Homendra Kumar Sahu and Pramod Kumar Omre
Department of Post-Harvest Process and Food Engineering, Govind Ballabh Pant University of Agriculture and Technology, Pantnagar, Uttarakhand

3.1 INTRODUCTION

Starch is a semi-crystalline polymer naturally derived from many plant sources and is known for its plentifulness, non-toxicity, biodegradability, and cost-effectiveness. Starch-based products are widely utilised in several sectors, such as food, paper, textiles, plastics, cosmetics, adhesives, and medicines. Starchy foods are the main source of dietary carbohydrates, greatly contributing to human caloric consumption and accounting for almost 80% of the worldwide average (Marta et al., 2022; Rashwan et al., 2024; Gupta et al., 2024a). Starch consists of two primary polymer compounds, amylose and amylopectin (as show in Figure 3.1), comprising glucose monomer units (Gutierrez-Montiel et al., 2024). Starch is the primary energy source for humans and is also a crucial component in food manufacturing. Natural starch possesses several deficiencies, including insufficient solubility, inadequate stability, pronounced hydrophilicity, and subpar mechanical characteristics (Averous & Boquillon, 2004; Ghanbarzadeh et al., 2011; Lv et al., 2022).

Native starch exhibits constrained functional properties; therefore, modification is employed to address its limitations, enhance or diminish its inherent characteristics, or impart specific properties to meet industrial requirements, broadening its applicability for food and non-food purposes (Wu et al., 2024; Gupta et al., 2022). Glycemic index and its character can also be reduced by modification of starch, which provides healthier diet (Chen et al., 2018). Physical, chemical, and enzymatic changes can improve starch. Each modification method confers unique characteristics to the altered starch (Marta et al., 2022; Gupta et al., 2023), as seen in Table 3.1. The most common chemical modifying techniques for starch include esterification, oxidation, etherification, and hydrolysis, with esterification becoming the most extensively utilised approach (Chen et al., 2018; Fang, 2002; Lv et al., 2022). Chemical modifications employ the incorporation of functional groups by particular reagents such as hydroxyls, phosphorus, ethers, esters, acids, hydrogen peroxide, and sodium hypochlorite into the backbone structure of native starch (Almeida, Monteiro, et al., 2024). The modifications influence the uniformity, related processes, gel stability of

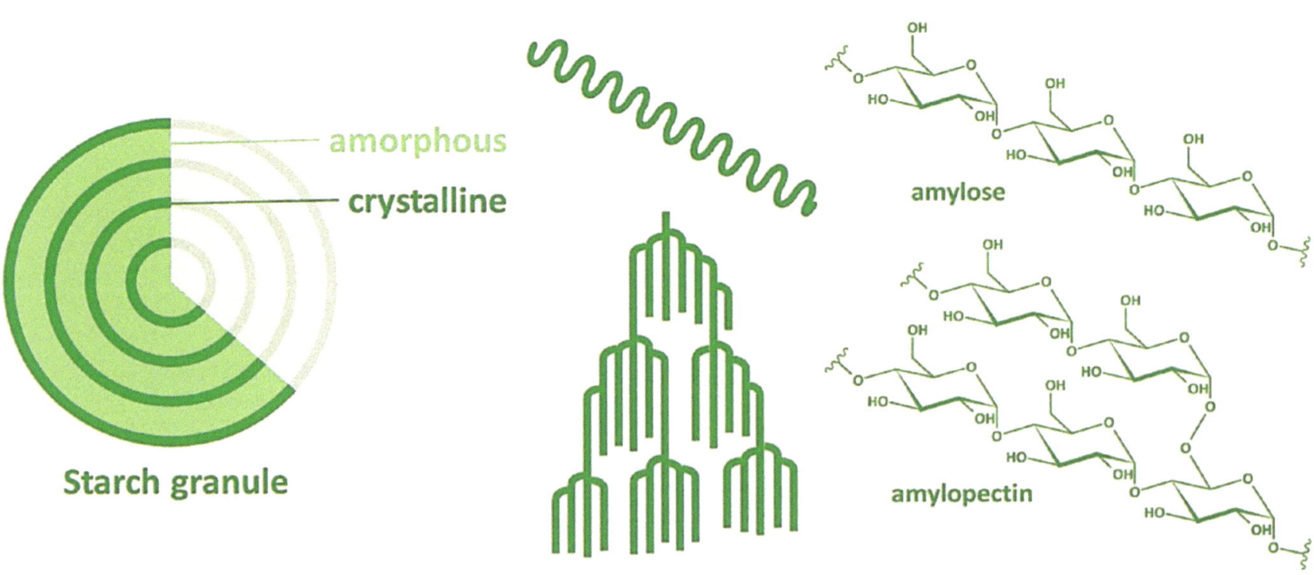

FIGURE 3.1 Native starch and its major components.

Source: **Imre & Vilaplana, 2020.**

DOI: 10.1201/9781003564287-3

TABLE 3.1
Chemical Uses for Modification of Starch

Starch Source	Chemical Use	Properties and Findings	References
Tiger nut starch	Citric acid	Enhanced the capability of the starch to stabilise Pickering emulsion	Lv et al., 2022
Corn starch	Citric acid and carboxymethyl cellulose	Enhanced the hydrophobic properties of maize starch, thereby impeding its retrogradation	Ghanbarzadeh et al., 2011
Potato starch, Rice starch	Citric acid	Enhanced transparency, thermal stability, dimensional stability, and water vapour barrier properties	Pooja et al., 2024
Sweet potato starch	Octenyl succinic anhydride (OSA)	Enhanced physical stability, elevated emulsification index, and reduced droplet size	Cui et al., 2024
Rice starch	Butyric acid	Increased resistant starch from 24.33% to 47.72%, with increased substitution and reduced crystallinity	Wu et al., 2024
Sorghum starches	Octenyl succinic anhydride	Increased swelling power and solubility while reducing gel strength and gelatinisation temperature (from 72.2°C to 61.5°C)	Zhang et al., 2021
Acid-converted corn starch	Trimethylammonium chloride and 3-chloro-2-hydroxy-1-propanesulfonic acid sodium salt	Increased the tensile strengths of starch, had positive effect on the desizability of starch, and enhanced paste stability and moisture regain	Li et al., 2018
Sweet potato starch	Hydroxyl propyl	Significantly improved gel stability, water solubility, digestibility, and cold storage stability of native starches	Senanayake et al., 2014
Sorghum starch	Epiclorohydrin (EPI)	Progressively decreased amylose content, swelling power, and solubility with increasing concentrations of EPI	Sandhu et al., 2021
Pearl millet starch	Epiclorohydrin (EPI)	Changes in structural properties, film-forming, and in vitro digestibility	Siroha et al., 2021
Potato starch	Sodium trimetaphosphate (STMP)/ sodium tripolyphosphate (STPP)	Improved rheological and pasting properties	Heo et al., 2017
Potato, millet, chickpea, tapioca, gram, finger millet, green maize, and wheat	Nitric acid (HNO_3), sulphuric acid (H_2SO_4), hydrochloric acid (HCl), and phosphoric acid (H_3PO_4)	Thinner slurried yielded from cereal and root starches than from pulse and tuber starches	Singh & Ali, 2008
Pearl millet starch	Stearic acid	Increased thermal stability, improved hydrophobicity, and improved barrier properties	A. Sultana et al., 2025

the final product, and the morphology and formation of hydrogen bonds of pure starch. Starch characteristics are modified at structural level, with minor changes in the external appearance of the starch molecule (Marta et al., 2022).

3.2 CHEMICAL MODIFICATION TECHNIQUES

3.2.1 Esterification

Esterification (Figure 3.2) is process frequently employing acids, acid chlorides, or anhydrides at different concentration. Starch can be made more hydrophobic if the number of longer nonpolar side chains is increases, but loss in crystalline granular form of starch will be observed (Imre & Vilaplana, 2020). Starch esterification is a prominent technique incorporating alkyl fatty acid or alkenyl succinic acid components into starch molecules. Esterification, a crucial chemical process for modifying the structure of starch granules and augmenting their usefulness, has substantially impacted food quality. The esterification of the hydroxyl group in glucose diminishes the hydrogen bonding capacity of amylose or amylopectin, reducing the starch's reactivity with moisture and improving its longevity (Gupta et al., 2024b). This structural modification in starch can diminish its reactivity with moisture, enhancing its longevity. Octenyl succinic anhydride (OSA), the sole legally permitted food emulsifier for the hydrophobic alteration of starch, has exceptional amphiphilic and interfacial characteristics in modified starch (Ji et al., 2025). An OSA content of 3% or below is acceptable for food production. The OSA-modified starch can increase viscosity and has diversified molecular weight distribution. Modified tapioca starch, corn starch, and rice starch paste have demonstrated increased viscosity and viscoelasticity when treated with OSA. This suggests that the use of OSA may interfere with the consistency of dough production, as observed by Gong et al. (2025) and Park et al. (2004). A significant modification approach is butyrylation, an esterification process that integrates butyric acid into the starch framework. This method alters several features of starch, including digestibility, thermal stability, and pasting ability, rendering it an optimal choice for diverse industrial applications.

3.2.2 Etherification

Etherification is the process of starch modification by substituting hydroxyl groups in starch with modifiable groups,

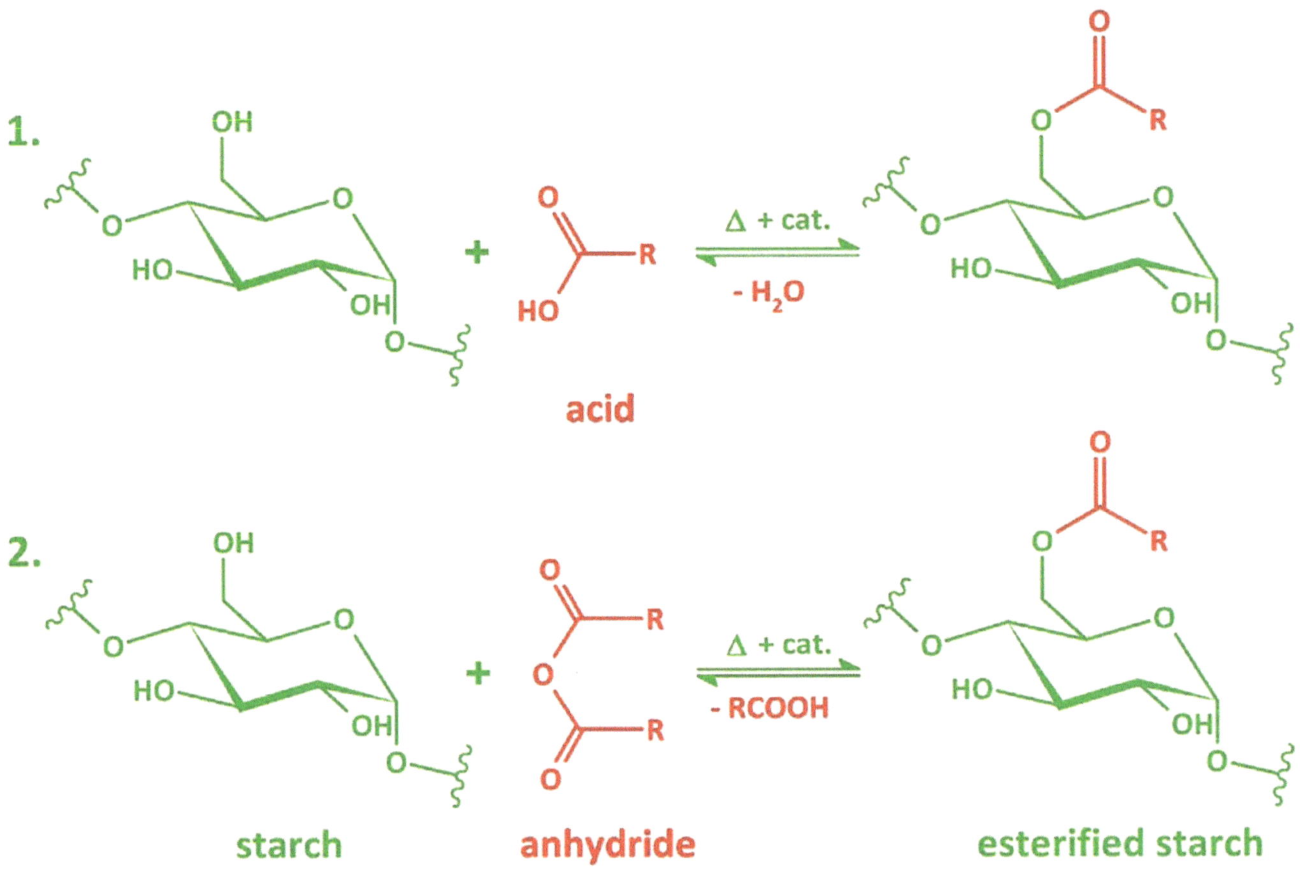

FIGURE 3.2 Esterification with acids or anhydrides.

Source: **Imre & Vilaplana, 2020.**

such as carboxymethyl and hydroxypropyl, creating an ether chain (Salimi et al., 2023). Hydroxypropylation (HP) is commonly utilised in the chemical modification of native starch because it markedly improves thermal properties during gelatinisation (Senanayake et al., 2014). Hydroxypropylated starch products play a crucial role in the food industry by reducing the tendency of starch pastes and gels to retrograde while maintaining the starch's ability to provide bulk, body, and other functional properties to food products (Fu et al., 2019). Carboxymethylation is a technique employed to increase the hydrophilicity of certain polysaccharides by substituting carboxymethyl groups (as shown in Figure 3.3) for the original hydroxyl groups to enhance bioactivity and foster anti-oxidative properties. Because of its simple process, low cost, and low toxicity, carboxymethylation is a popular chemical modification method for polysaccharides (Ahmad, 2021). The carboxymethylation process utilises the Williamson ether synthesis, allowing for the etherification of primary and/or secondary alcohol groups in polysaccharides with

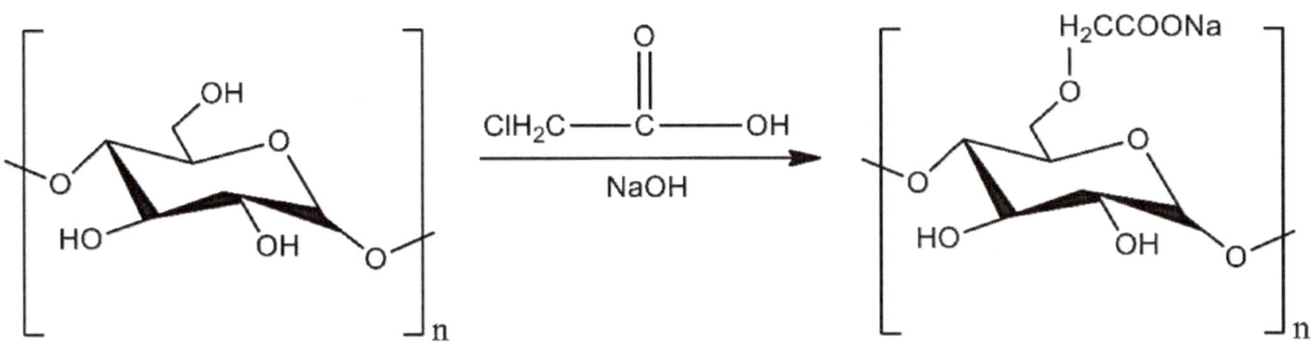

FIGURE 3.3 Etherification of starch by carboxymethylation process.

Source: **Ahmad, 2021.**

carboxymethyl groups (Chakka & Zhou, 2020). The structure of polysaccharides is altered following carboxymethylation, resulting in modifications to physicochemical and biological aspects. Bioactivities depend not only on the polysaccharide supply but also on chemical structural characteristics, such as the degree of substitution, which is affected by reagent concentration, reaction conditions, and solvent temperature during carboxymethylation. Carboxymethylation holds significant promise for application in the food industry to enhance the functions of polysaccharides (Xie et al., 2021). Etherification has been shown to enhance the tensile strengths of starch fibres and the swelling capacity of films. Etherification enhanced the stability and moisture retention of the film (Li et al., 2018).

3.2.3 Oxidation

Oxidised starch is derived through the oxidative alteration of native starch to address its deficiencies, including restricted hydrophilicity, low chemical reactivity, inadequate solubility, and suboptimal functional characteristics. Effective oxidation modification has made the starch abundantly available in nature renewable and biodegradable (Li et al., 2020). Oxidation is a prevalent technique for chemical change. The procedure entails oxidising primary or secondary hydroxyl groups inside the glucose molecules. Hydroxyl groups are initially transformed into carbonyl groups, then converted into carboxyl groups (Dimri et al., 2023). The oxidation reaction mostly transpires at the hydroxyl groups at the C-2, C-3, and C-6 locations (as shown in Figure 3.4), where hydroxyl groups are converted to carbonyl groups and carboxyl groups. Sodium hypochlorite, nitrogen dioxide, chromic acid, and permanganate are mostly used as oxidising agent (Chen et al., 2018). However, it is sodium hypochlorite (NaOCl) that serves as the primary reagent for the oxidation of starch, providing a clear understanding of the process. The oxidation of secondary alcohol groups leads to the hydrolysis of glycosidic bonds and the cleavage of monomeric rings, ultimately resulting in decreased polymer performance and lower viscosity (Liu et al., 2018). Oxidation can augment the functionality and reactivity of native starch by incorporating additional functional groups. These groups generally lack inert natural starch, and their inclusion can enhance the overall functioning and use of the modified starch.

3.2.4 Cross-Linking

Cross-linking is the process of starch modification in which the starch molecules are linked together by cross-linking agents (Gupta et al., 2024c). This is done by creating covalent or ionic bonds (as shown in Figure 3.5) between hydroxyl groups (–OH) on adjacent starch chains (Guo et al., 2015; Zhao et al., 2024). Various agents such as sodium tripolyphosphate ($Na_5P_3O_{10}$), monosodium phosphate (NaH_2PO_4), epichlorohydrin (C_3H_5ClO), sodium trimetaphosphate ($NA_3P_3O_9$), phosphorus oxychloride ($POCl_3$), and a mixture of adipic acid and acetic anhydride are commonly employed for the modification process, which improves the stability and durability of starch (Chen et al., 2018; Heo et al., 2017). Citric acid, a naturally occurring organic acid, serves as a cross-linking agent which creates a strong hydrogen bond in starch molecules (Shrivastava et al., 2024). Due to the effect of cross-linking process, changes observed include increased gelatinisation temperature, mechanical characteristics, shear stability, and resistance to pH, which improves the thermal and water stability of cross-linked starch films (Pooja et al., 2024; Wilpiszewska et al., 2019; Woggum et al., 2014). Cross-linked starch exhibits has been shown to enhance resistance to retrogradation, improve freeze–thaw stability, and augment stability during cooking, shearing, and processing at diverse temperatures and pH levels (Punia Bangar et al., 2024; Gupta et al., 2024d).

3.2.5 Grafting

Grafting of starch is a form of chemical modification in which synthetic monomers are covalently bonded to the

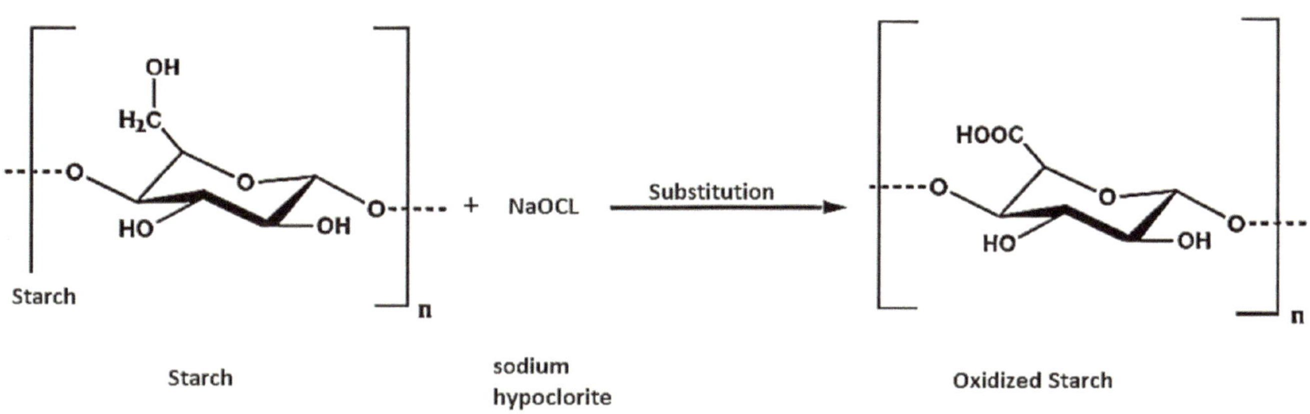

FIGURE 3.4 Oxidation method of starch modification.

Source: **Sukhija et al., 2016.**

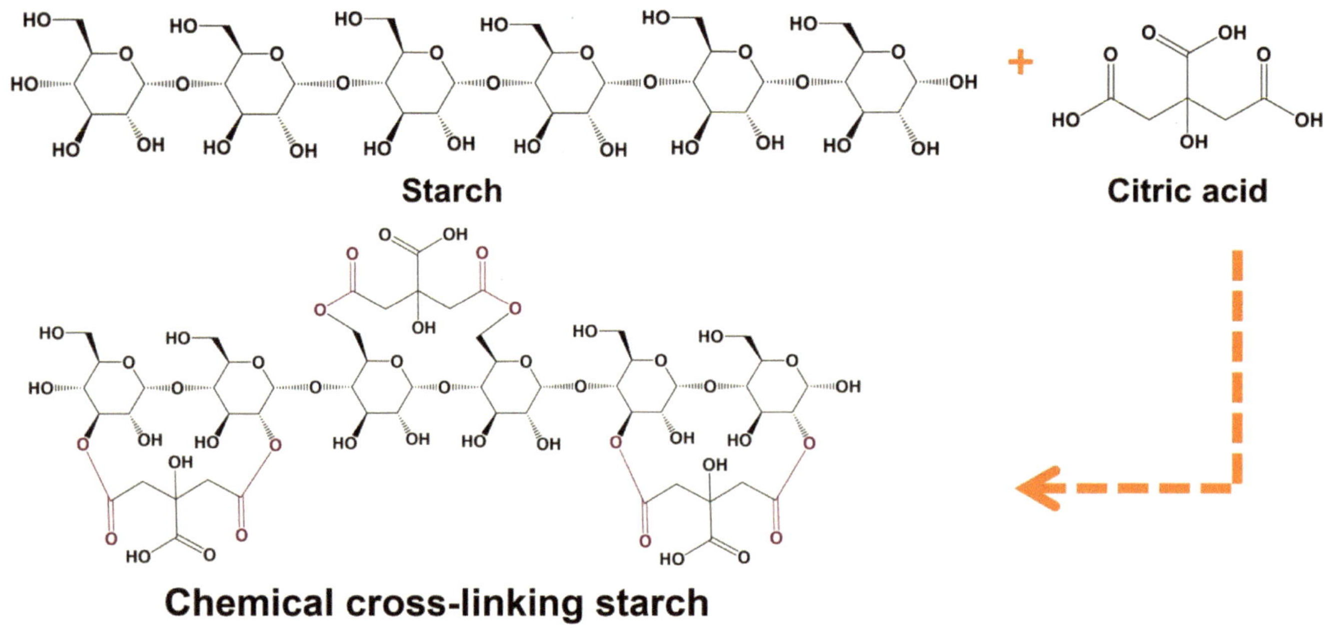

FIGURE 3.5 Cross-linking modification of starch by citric acid.

Source: Liu et al., 2024.

backbone of the starch polymer. This produces starch-based graft copolymers with enhanced or unique characteristics (Wang et al., 2020). By creating free radicals on the backbone of the starch and enabling these macro radicals to react with monomers, copolymers can be created (Meshram et al., 2009). Starch graft copolymers, with their wide-ranging applications in agriculture, medicine, and the food industry, are a subject of significant interest. A range of free-radical initiators including potassium persulfate ($K_2S_2O_8$), potassium permanganate ($KMnO_4$), ceric ammonium nitrate {$(NH_4)_2[Ce(NO_3)_6]$}, azobisisobutyronitrile ($C_8H_{12}N_4$), ammonium persulfate ($(NH_4)_2S_2O_8$), dibenzoyl peroxide [$(C_6H_5CO)_2O_2$], and Fenton's reagent are commonly used to generate free radicals (as shown in Figure 3.6) on the polymer backbone during the graft copolymerisation process (Choudhary et al., 2020). The two primary methods for producing starch-grafted copolymers, the "grafting from" and "grafting onto" approaches, as Meimoun et al. (2018) have demonstrated, further underscore the versatility of these copolymers and the synthesis methods.

3.3 CHARACTERISATION OF CHEMICALLY MODIFIED STARCH

Chemically modified starch is widely studied for its improved functional properties, making it useful in various industries. Structural, physicochemical, and functional changes caused by chemical treatments, including esterification, etherification, oxidation, and cross-linking, characterise chemically modified starch. These changes improve starch qualities such as thermal stability, solubility, viscosity, and resistance to enzymatic degradation, thus qualifying them for several industrial uses. The successful application of these modified starches in food, drugs, textiles, and biodegradable materials depends on a deep understanding of their molecular structure and functional

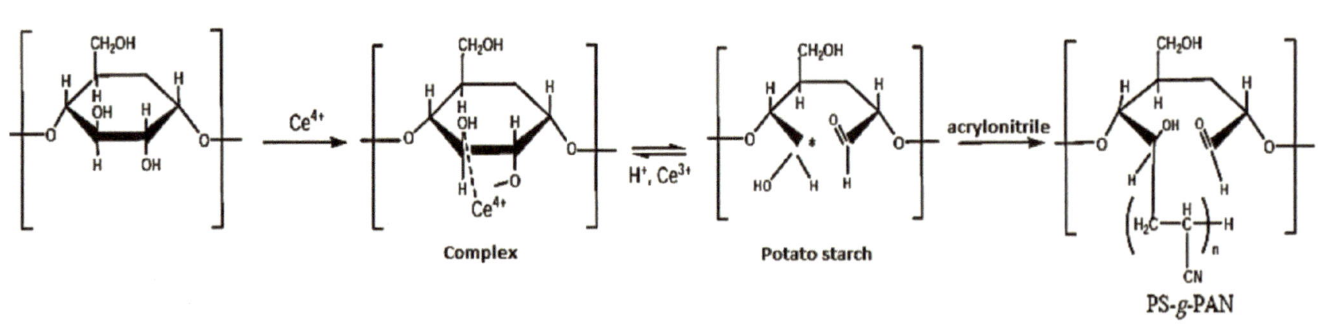

FIGURE 3.6 Mechanism of grafting acrylonitrile on potato starch with Ce_{4+} as free-radical initiator.

Source: Apopei et al., 2012.

behaviour (Gupta et al., 2024e). Techniques such as Fourier transform infrared spectroscopy (FTIR), X-ray diffraction (XRD), scanning electron microscopy (SEM), differential scanning calorimetry (DSC), and nuclear magnetic resonance (NMR) are instrumental in this understanding. Below are some key aspects of its characterisation.

3.3.1 Structural Analysis by X-ray Diffraction

The successful incorporation of butyryl acid into the starch molecules was confirmed through Fourier transform infrared spectroscopy. As the degree of substitution (DS) rises, the surface roughness of rice starch butyrate material progressively escalates, causing a decline in the smooth structure on specific surfaces, ultimately leading to cracks and collapses. This increase in surface roughness may lead to reduced aesthetics and structural weaknesses in the material. Furthermore, the crystallinity decreased as the degree of substitution increases (Wu et al., 2024). Gong et al. (2025) have documented the successful integration of starch molecules in octenyl succinic anhydride alongside the inhibition of glycosylase action. Incorporating modified starch significantly influences the development of the gluten network structure; however, the precise mechanisms governing gluten remain inadequately elucidated. Furthermore, the XRD graph indicated changes in relative crystallinity but not changes in the crystalline structure of starch.

The formation of starch–stearic acid complex was observed in pearl millet starch. As observed in SEM images, the presence of dents and holes in the starch granules suggests the leaching of amylose, which forms a complex with stearic acid. This study indicates that an increase in the concentration of stearic acid leads to the promotion of thermally stable complex synthesis when starch is treated with stearic acid (A. Sultana et al., 2025). Imre and Vilaplana (2020) investigated the modification of native corn starch treated with anhydride (Figure 3.7c, d, and e) or acid reagents (Figure 3.7f, g, and h). They found that, in the absence of a significant amount of water, even samples with a relatively high DS from anhydride treatment tend to

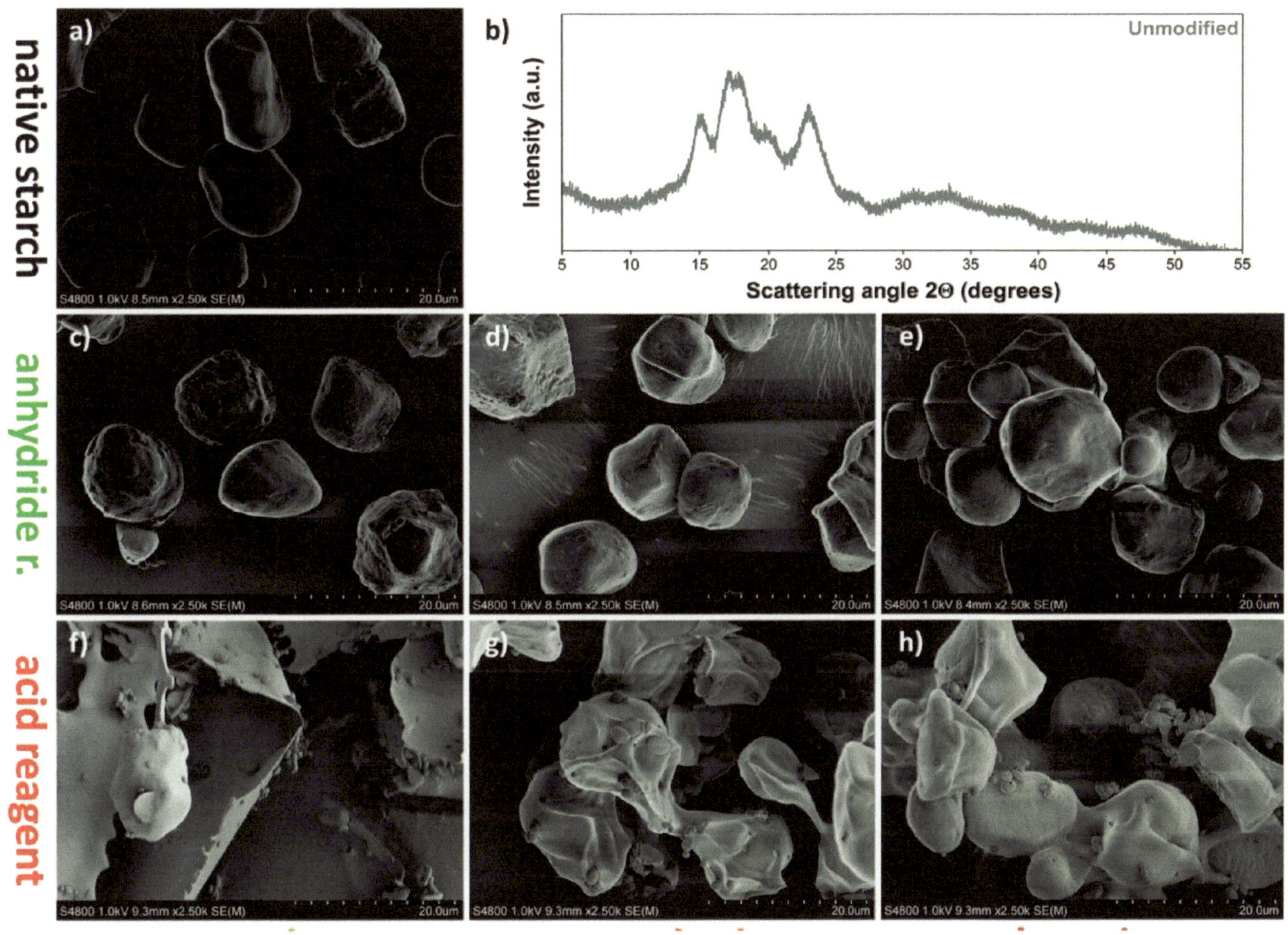

FIGURE 3.7 Morphology of acid- and anhydride-modified starches: Scanning electron micrographs (2500× magnification) of (a) unmodified, (c) acetic (d) propionic, and (e) butyric anhydride as well as (f) acetic (g) propionic, and (h) butyric acid–treated native corn, the respective XRD spectra of (b) unmodified.

Source: **Imre and Vilaplana, 2020.**

retain their distinct granular structure. Acylation leads to an increase in surface roughness, a phenomenon frequently noted in esterified starches. Treatment with acid reagents yields highly amorphous, elongated particles that progressively coalesce and fuse, potentially resulting in a fibrous structure when acetic acid is used.

3.3.2 Functional Property Analysis

Functional property analysis is the study of how modified starches behave during preparation, processing, storage, and consumption. This is critical for the texture, stability, and appearance of food products (Gupta et al., 2025b). Chemically modified sweet potato starch shows improvement in its functional characteristics, digestibility, storage stability, gel stability, and water solubility when treated with hydroxyl propyl. These improvements could significantly increase the application of native starch in the food industry (Senanayake et al., 2014). Oxidation modification decreases relative crystallinity, viscosity, swelling power, retrogradation tendency, and pasting temperature of starches (Lewicka et al., 2015). The process of etherification has successfully enhanced the tensile strengths of corn starch fibres, with a progressive increase observed as the levels of etherification were elevated. Etherification improved the stability of the paste and the moisture of the films (Q. Liu et al., 2018). The swelling power, water holding capacity, peak viscosity, and enthalpy of quinoa starch have been increased with modification. Ketchups formulated with the chosen modified starch exhibited significant enhancements in apparent viscosity, colour, and textural qualities, with the sample containing 100% modified quinoa starch demonstrating the highest degree of "shear-thinning." At the same time, the starch's physical stability matched that of commercial modified starch (Zakeri et al., 2025).

3.3.3 Rheological Studies

Starch modified through esterification exhibits a reduced pasting temperature and a notably higher peak viscosity. This phenomenon can be attributed to the tightly arranged internal starch granules held together by hydrogen bonding and other forces, limiting water entry. The OSA group introduced a significant spatial site resistance effect, that is, the resistance encountered by water molecules as they try to penetrate the starch granules. This resistance diminishes the influence of hydrogen bonding and facilitates easier water penetration, consequently enhancing viscosity (Liu et al., 2018). Applying STMP/STPP for cross-linking has been shown to lead to significant alterations in the physicochemical, structural, rheological, and pasting characteristics of potato starch. The use of STMP/STPP at concentrations ranging from 0.0125% to 0.05% resulted in a notable increase in both peak and final viscosity while simultaneously leading to a reduction in breakdown viscosity. Consequently, cross-linking with STMP/STPP (0.0125%–0.05%) could be appropriate for industrial applications in the roles of thickeners, stabilisers, and binders (Heo et al., 2017).

3.3.4 Thermal Analysis

Guo et al. (2015) and Pooja et al. (2024) indicated that potato starch and rice starch cross-linked with citric acid exhibit notable improvements in transparency, an increase in thermal stability of the polymer at 164°C, enhanced dimensional stability, and potential for customised water vapour barrier properties, resulting in significantly more hydrophobic surfaces, which are advantageous for packaging applications. Misman et al. (2015) documented the etherification process of sago starch by treating the starch with benzyl chloride in an ethanol solvent. Their findings indicated that this solvent-based etherification yields starch with a higher degree of substitution, enhanced thermal stability, and improved flow properties.

3.4 APPLICATIONS OF CHEMICALLY MODIFIED STARCH

Acid-modified (thin boiling) starches, with their broad applications in industries such as paper, textiles, and food, are typically modified using hydrochloric acid. The granular and molecular features of the starch, which vary significantly depending on the source, are crucial in determining the qualities of the end product, along with the acid strength, temperature, and treatment length (Singh & Ali, 2008). Analysis of chemically modified starches has indicated reduced gelatinisation temperatures, except for the S5 starch, which exhibited notable effects on pasting characteristics. The findings suggest significant promise for its application as a food ingredient within the industry (Senanayake et al., 2014). Cross-linked starches, single- or dual-modified, find extensive applications in the food industry. These substances are commonly utilised as fat substitutes in mayonnaise, soups, ice creams, and muffins to create low-fat versions of these items. Their high process tolerance and resistance to pH and shear make them a popular choice as texturisers and thickening agents (Mazumder & Rahman, 2024). Oxidised starches serve various functions, including as drug delivery agents, food additives, thickeners, emulsifiers, gelling agents, stabilisers, bread conditioners, and replacements for gum arabic. Oxidised starches are important components in various applications, including paper, textiles, laundry finishing, and binding materials. The unique characteristics of the oxidation process render them adaptable and in high demand (Dimri et al., 2023). A. Sultana et al. (2025) have modified mango kernel starch to manage and prevent hypercholesterolaemia, demonstrating that acid hydrolysis treatment can significantly enhance water holding and amylose content while maintaining oil-binding capacities. It was also discovered that modified mango kernel starch could serve as a promising alternative for therapeutic intervention in individuals with hypercholesterolaemia, showing potential as a fat replacer. A study

TABLE 3.2
Various Applications of Chemically Modified Starches

Starch Source	Application	References
Pearl millet starch	Food packaging	A. Sultana et al., 2025
Corn starch	Starch-based adsorbent for calcium ion adsorption	Cao & Ding, 2024
Sorghum starch	Emulsifier	Zhang et al., 2021
Mango kernel starch	Fat replacer	S. Sultana et al., 2025
Chickpea flour	Fillings, thickeners, and gelling agents	Sarıkaya et al., 2025
Hydroxypropyl distarch phosphate (HDSP)	Ice cream ink for 3D printing	Xu et al., 2024
Yam starch	Absorbent	Fang et al., 2024
White sorghum starch	Biodegradable films, frozen foods, and pasta and noodle formulations	Biduski et al., 2017; Magallanes-Cruz et al., 2023
Quinoa seed starch	Pickering emulsions, pharmaceutical formulations	Abdul Hadi et al., 2020; Magallanes-Cruz et al., 2023
Mango seed starch	Plastic films, gum and confectionary products	Magallanes-Cruz et al., 2023
Avocado seed starch	Cream soup	Cornelia & Christianti, 2018
Arrowroot	Edible coating	Marta et al., 2025
Cassava starch	Starch-based adhesives	Monroy et al., 2023

conducted by Xu et al. (2024) delves into the 3D printability of inks derived from modified starch. The authors put forward a potential mechanism for creating high-precision ice cream ink utilising modified starch to advance the use of starch-based materials in the realm of personalised ice cream processing. Table 3.2 presents a range of applications for chemically modified starch.

3.5 CHALLENGES AND ENVIRONMENTAL CONSIDERATIONS

Though chemically altered starches have great promise for many uses, certain issues and environmental problems must be resolved to guarantee sustainable growth. After using hazardous or toxic reagents during chemical modification operations such as esterification, etherification, or oxidation, if not used and disposed of correctly, these drugs might endanger the surroundings and human health. Many chemical changes also demand great energy input or produce waste and byproducts that need careful treatment. Environmentally friendly starch modification techniques still have limited scalability, which is restricted mostly by cost, regulatory compliance, and process efficiency. The end-of-life effects of chemically altered starch-based products are also rather important. Although natural starch is biodegradable, several chemical changes might prevent its degradability or generate persistent leftovers in the surroundings. This generates a dichotomy between improving performance and maintaining biodegradability, so careful design and life-cycle evaluation of modified starch products are necessary. Future research should concentrate on developing green chemistry techniques, including solvent-free methods, recyclable or benign reagents, and bio-based or enzymatic catalysts, to address these problems. Aligning starch modification technologies with circular economy ideas will depend mostly on stressing closed-loop systems and reducing environmental effects.

3.6 FUTURE PROSPECTS AND TRENDS

The primary approach for chemical modification involves using OSA; however, environmental and health concerns necessitate a revision of this methodology. Alternative straightforward methods, including thermal, non-thermal, and enzymatic treatments or combinations of these, may alter the starch structure and enhance its interfacial interactions (Almeida, Santos, et al., 2024). Further research is required to modify banana starch through alternative methods, including enzymatic and non-thermal physical techniques (such as ultrasonication, pulse electric field, and high-pressure treatment), to enhance our understanding and improve the characteristics of banana starch (Marta et al., 2022). Thermoplastic-modified starches are anticipated to be significant sustainable fillers in advancing polymer blends. Chemical modifications will further improve compatibility and dispersion in various polymeric resin matrices, creating new opportunities for high-performance and environmentally friendly materials. Recent applications of chemically modified starches include chelators, cryoprotectants, drying aids, fat replacers, flavour carriers, precursors for flavour and colour via Maillard reactions, and substrates in fermentation processes. Phosphorylated rice and tapioca starches are increasingly employed as fat replacers owing to their water-binding capabilities, which can mimic the sensory attributes of oil, including a slick mouthfeel and tenderness (Chen et al., 2018). High amylose starch, also known as resistant starch, is gaining popularity as an innovative biopolymer for food packaging due to its enhanced mechanical and barrier qualities relative to native starch films (Fatima et al., 2024; Gupta et al., 2025a).

3.7 REGULATORY AND SAFETY ASPECTS

Research indicates that the modification of starches, particularly phosphorylated/cross-linked starches, has significant nutritional benefits for humans. These starches, believed to be digested slowly, offer a unique advantage in the food industry (Singh et al., 2007). The maximum permissible replacement levels for starch acetates, phosphates, and hydroxypropylated starches are 2.5%, 0.4%, and 10%, respectively. Similarly, safe cross-linked dietary starches possess one substituent cross-linking group for every 1000 or more anhydroglucose units. In contrast to natural starches, which may be labelled simply as "starch," chemically modified starches must be distinctly identified by their specific names on food labels, according to Food Safety and Standards Authority of India regulations. This commitment to transparency ensures regulatory adherence and informs clients of the type of component utilised in the product. Food additives designated as "modified starch" are chemically changed starches with an E-number categorised under the International Numbering System for Food Additives (INS).

In contrast, native starches and starches that have been physically or enzymatically modified are classified as components rather than additives. Consequently, they may be designated only as "starch" from various sources without E-numbers (Park and Kim, 2021). Modified food starch is permissible in food according to the United States Code of Federal Regulations (21 CFR §172.892) within established safety parameters. Acetic anhydride can provide a maximum of 2.5% acetyl groups, whereas epichlorohydrin is limited to 0.3%, with residual propylene chlorohydrin capped at 5 ppm. If classified as GRAS (Generally Recognized as Safe), alpha-amylase and pullulanase are also authorised. The final product must be designated "food starch-modified" and utilised solely to achieve the intended technical outcome.

3.8 CONCLUSION

Chemical modification of starch provides a flexible and efficient method for changing starch's characteristics for many industrial and commercial uses. Native starch may be modified by esterification, etherification, oxidation, cross-linking, and grafting to improve its thermal stability, solubility, mechanical strength, emulsification capacity, and retrogradation resistance. These alterations rectify the constraints of native starch, facilitating its application in functional food systems, biodegradable packaging, medicinal formulations, and innovative technologies such as 3D food printing. Notwithstanding the myriad benefits, obstacles persist regarding environmental consequences, adherence to regulations, and preserving biodegradability. Future initiatives should prioritise green chemistry methodologies, sustainable reagent use, and process optimisation to conform to circular economy objectives. With ongoing study, chemically modified starch will remain integral to creating new, sustainable, and health-oriented goods across several sectors.

REFERENCES

Abdul Hadi, N., Marefati, A., Matos, M., Wiege, B., & Rayner, M. (2020). Characterization and stability of short-chain fatty acids modified starch Pickering emulsions. *Carbohydrate Polymers*, *240*, 116264. https://doi.org/10.1016/j.carbpol.2020.116264

Ahmad, M. M. (2021). Recent trends in chemical modification and antioxidant activities of plants-based polysaccharides: A review. *Carbohydrate Polymer Technologies and Applications*, *2*, 100045. https://doi.org/10.1016/j.carpta.2021.100045

Almeida, R. L. J., Monteiro, S. S., Santos, N. C., Rios, N. S., & Dos Santos, E. S. (2024). Starch modification and its application in Pickering emulsion stabilization: A review. *Journal of Food Measurement and Characterization*, *18*(6), 4984–5003. https://doi.org/10.1007/s11694-024-02550-6

Almeida, R. L. J., Santos, N. C., Monteiro, S. S., Rios, N. S., & Silvino Dos Santos, E. (2024). Exploring the potential of native and modified starch and starch nanocrystals in Pickering emulsions: Current advances, future perspectives, and challenges. *Food Bioscience*, *61*, 104675. https://doi.org/10.1016/j.fbio.2024.104675

Apopei, D. F., Dinu, M. V., & Drăgan, E. S. (2012). Graft copolymerization of acrylonitrile onto potatoes starch by ceric ion. *Digest Journal of Nanomaterials and Biostructures*, *7*(2), 707–716.

Averous, L., & Boquillon, N. (2004). Biocomposites based on plasticized starch: Thermal and mechanical behaviours. *Carbohydrate Polymers*, *56*(2), 111–122. https://doi.org/10.1016/j.carbpol.2003.11.015

Biduski, B., Silva, F. T. D., Silva, W. M. D., Halal, S. L. D. M. E., Pinto, V. Z., Dias, A. R. G., & Zavareze, E. D. R. (2017). Impact of acid and oxidative modifications, single or dual, of sorghum starch on biodegradable films. *Food Chemistry*, *214*, 53–60. https://doi.org/10.1016/j.foodchem.2016.07.039

Cao, Y., & Ding, S. (2024). Study on the enhanced calcium ion adsorption capacity of magnetic materials modified by the enzymatic and activated comodification esterification of starch (Fe_3O_4@EES). *Chemistry Africa*, *7*(8), 4617–4630. https://doi.org/10.1007/s42250-024-01044-8

Chakka, V. P., & Zhou, T. (2020). Carboxymethylation of polysaccharides: Synthesis and bioactivities. *International Journal of Biological Macromolecules*, *165*, 2425–2431. https://doi.org/10.1016/j.ijbiomac.2020.10.178

Chen, Y.-F., Kaur, L., & Singh, J. (2018). Chemical modification of starch. In *Starch in food* (pp. 283–321). Elsevier. https://doi.org/10.1016/B978-0-08-100868-3.00007-X

Cornelia, M., & Christianti, A. (2018). Utilization of modified starch from avocado (*Persea americana* Mill.) seed in cream soup production. *IOP Conference Series: Earth and Environmental Science*, *102*, 012074. https://doi.org/10.1088/1755-1315/102/1/012074

Cui, Q., Mahfouzi, M., Zhang, H., Gao, R., & Guo, Z. (2024). Characteristics of sweet potato starch particles modified by various methods for use in food-grade Pickering emulsion. *LWT*, *210*, 116831. https://doi.org/10.1016/j.lwt.2024.116831

Choudhary, S., Sharma, K., Sharma, V., & Kumar, V. (2020). Grafting polymers. In T. J. Gutiérrez (Ed.), *Reactive and functional polymers, Vol. 2* (pp. 199–243). Springer International Publishing. https://doi.org/10.1007/978-3-030-45135-6_8

Dimri, S., Aditi, Bist, Y., & Singh, S. (2023). Oxidation of starch. In V. S. Sharanagat, D. C. Saxena, K. Kumar, & Y. Kumar (Eds.), *Starch: Advances in modifications, technologies and applications* (pp. 55–82). Springer International Publishing. https://doi.org/10.1007/978-3-031-35843-2_3

Fang, J. (2002). The preparation and characterisation of a series of chemically modified potato starches. *Carbohydrate Polymers*, *47*(3), 245–252. https://doi.org/10.1016/S0144-8617(01)00187-4

Fang, J., Lu, J., Zhao, X., Zhang, T., Ye, H., Fang, T., Wang, Y., & Guo, N. (2024). Zearalenone removal using inactivated yeast embedded in porous modified yam starch aerogels and its application in corn silk tea. *Food Chemistry*, *460*, 140593. https://doi.org/10.1016/j.foodchem.2024.140593

Fatima, S., Khan, M. R., Ahmad, I., & Sadiq, M. B. (2024). Recent advances in modified starch based biodegradable food packaging: A review. *Heliyon*, *10*(6), e27453. https://doi.org/10.1016/j.heliyon.2024.e27453

Fu, Z., Zhang, L., Ren, M., & BeMiller, J. N. (2019). Developments in hydroxypropylation of starch: A review. *Starch - Stärke*, *71*(1–2), 1800167. https://doi.org/10.1002/star.201800167

Ghanbarzadeh, B., Almasi, H., & Entezami, A. A. (2011). Improving the barrier and mechanical properties of corn starch-based edible films: Effect of citric acid and carboxymethyl cellulose. *Industrial Crops and Products*, *33*(1), 229–235. https://doi.org/10.1016/j.indcrop.2010.10.016

Gong, J., Xu, W., Zhang, C., Zhu, Q., Qin, X., Zhang, H., & Liu, G. (2025). Effects of esterification and enzymatic modification on the properties of wheat starch and dough. *Food Hydrocolloids*, *158*, 110509. https://doi.org/10.1016/j.foodhyd.2024.110509

Guo, L., Liu, R., Li, X., Sun, Y., & Du, X. (2015). The physical and adsorption properties of different modified corn starches. *Starch - Stärke*, *67*(3–4), 237–246. https://doi.org/10.1002/star.201400200

Gutierrez-Montiel, E., Ávila-Orta, C. A., Cabrera-Canales, Z. E., Covarrubias-Gordillo, C. A., Reyes-Rodríguez, P. Y., Rodríguez-Velázquez, J. G., Cabello-Romero, J. N., & Fonseca-Florido, H. A. (2024). Effect of oxidized, maleate and dual chemical modification through extrusion on thermoplastic starch properties. *Polymer Bulletin*, *81*(3), 2525–2544. https://doi.org/10.1007/s00289-023-04857-8

Gupta, R. K., Guha, P., & Srivastav, P. P. (2022). Natural polymers in bio-degradable/edible film: A review on environmental concerns, cold plasma technology and nanotechnology application on food packaging-A recent trends. *Food Chemistry Advances*, *1*, 100135. https://doi.org/10.1016/j.focha.2022.100135

Gupta, R. K., Guha, P., & Srivastav, P. P. (2023). Effect of high voltage dielectric barrier discharge (DBD) atmospheric cold plasma treatment on physicochemical and functional properties of taro (*Colocasia esculenta*) starch. *International Journal of Biological Macromolecules*, *253*, 126772. https://doi.org/10.1016/j.ijbiomac.2023.126772

Gupta, R. K., Guha, P., & Srivastav, P. P. (2024a). Dielectric barrier discharge plasma: A green and novel method to change the structure of taro peel starch and improve the physicochemical properties of taro peel starch films. *Plasma Processes and Polymers*, *21*(9), 2400047. https://doi.org/10.1002/ppap.202400047

Gupta, R. K., Guha, P., & Srivastav, P. P. (2024b). Exploring the potential of taro (*Colocasia esculenta*) starch: Recent developments in modification, health benefits, and food industry applications. *Food Bioengineering*, *3*(3), 365–379. https://doi.org/10.1002/fbe2.12103

Gupta, R. K., Guha, P., & Srivastav, P. P. (2024c). Insights into the physicochemical, crystalline, multistructural, thermal and digestibility behavior of taro starch (*Colocasia esculenta*) modified by cold plasma treatment with high voltage dielectric barrier discharge (HVDBD): An alternative technique for chemical modification of starch and environmental sustainability. *Journal of Cleaner Production*, *461*, 142710. https://doi.org/10.1016/j.jclepro.2024.142710

Gupta, R. K., Guha, P., & Srivastav, P. P. (2024d). Physical action of nonthermal cold plasma technology for starch modification. *Food Physics*, *1*, 100011. https://doi.org/10.1016/j.foodp.2024.100011

Gupta, R. K., Guha, P., & Srivastav, P. P. (2024e). Effect of cold plasma treatment and plasma-activated water on physicochemical and structural properties of starch: A green and novel approach for environmental sustainability. *Plasma Processes and Polymers*, *21*(4), 2300204. https://doi.org/10.1002/ppap.202300204

Shrivastava, A., Gupta, R. K., & Srivastav, P. P. (2024). Exploring novel frontiers of advancements in purple yam (*Dioscorea alata* L.) starch extraction, modification, characterization, applications in food and other industries. *Measurement: Food*, *15*, 100196. https://doi.org/10.1016/j.meafoo.2024.100196

Gupta, R. K., Pipliya, S., Patel, J., Srivastav, P. P., Castro-Muñoz, R., Wang, C. K., & Nemţanu, M. R. (2025a). Essential oil-embedded starch based films for active packaging: Insights into migration, stability, and preservation efficiency. *Trends in Food Science & Technology*, 163, 105149. https://doi.org/10.1016/j.tifs.2025.105149

Gupta, R. K., Gnana Moorthy Eswaran, U., Kumar, S., Pipliya, S., Kaur, B., & Srivastav, P. P. (2025b). Synergistic modification of starch polysaccharide via sequential treatments with ultrasonic and cold plasma technology. *Journal of Food Processing and Preservation*, *2025*(1), 9063007. https://doi.org/10.1155/jfpp/9063007

Heo, H., Lee, Y.-K., & Chang, Y. H. (2017). Rheological, pasting, and structural properties of potato starch by cross-linking. *International Journal of Food Properties*, *20*, 2138–2150. https://doi.org/10.1080/10942912.2017.1368549

Imre, B., & Vilaplana, F. (2020). Organocatalytic esterification of corn starches towards enhanced thermal stability and moisture resistance. *Green Chemistry*, *22*(15), 5017–5031. https://doi.org/10.1039/D0GC00681E

Ji, R., Xu, J., Yu, Y., Song, S., Zhang, X., & Zhang, W. (2025). Improving the emulsification properties of corn starch by esterification combined with freeze-thawing and enzymatic treatment. *Food Chemistry*, *470*, 142664. https://doi.org/10.1016/j.foodchem.2024.142664

Lewicka, K., Siemion, P., & Kurcok, P. (2015). Chemical modifications of starch: Microwave effect. *International Journal of Polymer Science*, *2015*, 1–10. https://doi.org/10.1155/2015/867697

Li, J., Zhou, M., Cheng, F., Lin, Y., Shi, L., & Zhu, P.-X. (2020). Preparation of oxidized corn starch with high degree of oxidation by Fenton-like oxidation assisted with ball milling. *Materials Today Communications*, *22*, 100793. https://doi.org/10.1016/j.mtcomm.2019.100793

Li, W., Xu, Z., Wang, Z., Liu, X., Li, C., & Ruan, F. (2018). Double etherification of corn starch to improve its adhesion to cotton and polyester fibers. *International Journal of Adhesion and Adhesives*, *84*, 101–107. https://doi.org/10.1016/j.ijadhadh.2018.03.001

Liu, M., Yao, W., Shan, J., Zheng, H., Yang, Y., Cao, L., Qi, B., Tan, H., Sun, C., Zhang, X., & Zhang, Y. (2024). Preparation of high-performance antibacterial/antifungal citric acid–starch adhesives based on physical entanglement and chemical cross-linking. *International Journal of Biological Macromolecules*, *279*, 135560. https://doi.org/10.1016/j.ijbiomac.2024.135560

Liu, Q., Li, M., Xiong, L., Qiu, L., Bian, X., Sun, C., & Sun, Q. (2018). Oxidation modification of debranched starch for the preparation of starch nanoparticles with calcium ions. *Food Hydrocolloids*, *85*, 86–92. https://doi.org/10.1016/j.foodhyd.2018.07.004

Lv, X., Guo, C., Ma, Y., & Liu, B. (2022). Effect of citric acid esterification on the structure and physicochemical properties of tigernut starch. *International Journal of Biological Macromolecules*, *222*, 2833–2842. https://doi.org/10.1016/j.ijbiomac.2022.10.062

Magallanes-Cruz, P. A., Duque-Buitrago, L. F., & Del Rocío Martínez-Ruiz, N. (2023). Native and modified starches from underutilized seeds: Characteristics, functional properties and potential applications. *Food Research International*, *169*, 112875. https://doi.org/10.1016/j.foodres.2023.112875

Marta, H., Cahyana, Y., Djali, M., & Pramafisi, G. (2022). The properties, modification, and application of banana starch. *Polymers*, *14*(15), 3092. https://doi.org/10.3390/polym14153092

Marta, H., Chandra, S., Cahyana, Y., Sukri, N., Pangawikan, A. D., Yuliana, T., & Arifin, H. R. (2025). Arrowroot (*Maranta arundinaceae* L.) starch-based edible coating formulation and its application to shelf-life extension of tomato (*Solanum lycopersicum* L.). *Carbohydrate Polymer Technologies and Applications*, *9*, 100674. https://doi.org/10.1016/j.carpta.2025.100674

Mazumder, N., & Rahman, M. H. (Eds.). (2024). *Advanced research in starch*. Springer Nature Singapore. https://doi.org/10.1007/978-981-99-9527-1

Meimoun, J., Wiatz, V., Saint-Loup, R., Parcq, J., Favrelle, A., Bonnet, F., & Zinck, P. (2018). Modification of starch by graft copolymerization. *Starch - Stärke*, *70*(1–2), 1600351. https://doi.org/10.1002/star.201600351

Meshram, M. W., Patil, V. V., Mhaske, S. T., & Thorat, B. N. (2009). Graft copolymers of starch and its application in textiles. *Carbohydrate Polymers*, *75*(1), 71–78. https://doi.org/10.1016/j.carbpol.2008.06.012

Misman, M. A., Azura, A. R., & Hamid, Z. A. A. (2015). Physicochemical properties of solvent based etherification of sago starch. *Industrial Crops and Products*, *65*, 397–405. https://doi.org/10.1016/j.indcrop.2014.11.009

Monroy, Y., Hamet, M. F., Rivero, S., & García, M. A. (2023). Tailor-made starch-based adhesives chemically modified with NaOH:urea and their applications on a cellulosic substrate. *International Journal of Biological Macromolecules*, *247*, 125423. https://doi.org/10.1016/j.ijbiomac.2023.125423

Park, S., & Kim, Y.-R. (2021). Clean label starch: Production, physicochemical characteristics, and industrial applications. *Food Science and Biotechnology*, *30*(1), 1–17. https://doi.org/10.1007/s10068-020-00834-3

Park, S., Chung, M., & Yoo, B. (2004). Effect of octenylsuccinylation on rheological properties of corn starch pastes. *Starch - Stärke*, *56*(9), 399–406. https://doi.org/10.1002/star.200300274

Pooja, N., Shashank, S., Singh, B. N., & Mazumder, N. (2024). Advancing sustainable bioplastics: Chemical and physical modification of starch films for enhanced thermal and barrier properties. *RSC Advances*, *14*(33), 23943–23951. https://doi.org/10.1039/D4RA04263H

Punia Bangar, S., Sunooj, K. V., Navaf, M., Phimolsiripol, Y., & Whiteside, W. S. (2024). Recent advancements in cross-linked starches for food applications: A review. *International Journal of Food Properties*, *27*(1), 411–430. https://doi.org/10.1080/10942912.2024.2318427

Rashwan, A. K., Younis, H. A., Abdelshafy, A. M., Osman, A. I., Eletmany, M. R., Hafouda, M. A., & Chen, W. (2024). Plant starch extraction, modification, and green applications: A review. *Environmental Chemistry Letters*, *22*(5), 2483–2530. https://doi.org/10.1007/s10311-024-01753-z

Salimi, M., Channab, B., El Idrissi, A., Zahouily, M., & Motamedi, E. (2023). A comprehensive review on starch: Structure, modification, and applications in slow/controlled-release fertilizers in agriculture. *Carbohydrate Polymers*, *322*, 121326. https://doi.org/10.1016/j.carbpol.2023.121326

Sandhu, K. S., Siroha, A. K., Punia, S., Sangwan, L., Nehra, M., & Purewal, S. S. (2021). Effect of degree of cross linking on physicochemical, rheological and morphological properties of sorghum starch. *Carbohydrate Polymer Technologies and Applications*, *2*, 100073. https://doi.org/10.1016/j.carpta.2021.100073

Sarıkaya, S., İnanlar, B., Younis, H. G. R., Zhao, G., & Ye, F. (2025). Chickpea (*Cicer arietinum* Linn.) starch: Extraction, composition, structure, properties, modifications and food applications. *Grain & Oil Science and Technology*, *8*(2), 118–136. https://doi.org/10.1016/j.gaost.2025.03.003

Senanayake, S., Gunaratne, A., Ranaweera, K. K. D. S., & Bamunuarachchi, A. (2014). Effect of hydroxypropylation on functional properties of different cultivars of sweet potato starch in Sri Lanka. *International Journal of Food Science*, *2014*, 1–8. https://doi.org/10.1155/2014/148982

Singh, J., Kaur, L., & McCarthy, O. J. (2007). Factors influencing the physico-chemical, morphological, thermal and rheological properties of some chemically modified starches for food applications: A review. *Food Hydrocolloids*, *21*(1), 1–22. https://doi.org/10.1016/j.foodhyd.2006.02.006

Singh, V., & Ali, S. Z. (2008). Properties of starches modified by different acids. *International Journal of Food Properties*, *11*(3), 495–507. https://doi.org/10.1080/10942910802083774

Siroha, A. K., Bangar, S. P., Sandhu, K. S., Trif, M., Kumar, M., & Guleria, P. (2021). Effect of cross-linking modification on structural and film-forming characteristics of pearl millet (*Pennisetum glaucum* L.) starch. *Coatings*, *11*(10), 1163. https://doi.org/10.3390/coatings11101163

Sukhija, S., Singh, S., & Riar, C. S. (2016). Effect of oxidation, cross-linking and dual modification on physicochemical, crystallinity, morphological, pasting and thermal characteristics of elephant foot yam (*Amorphophallus paeoniifolius*) starch. *Food Hydrocolloids*, *55*, 56–64. https://doi.org/10.1016/j.foodhyd.2015.11.003

Sultana, A., Punia Bangar, S., & Whiteside, W. S. (2025). Effect of stearic acid modification on properties of pearl millet starch. *Biomass Conversion and Biorefinery*, *15*(6), 8745–8753. https://doi.org/10.1007/s13399-024-05623-0

Sultana, S., Das, S., Halder, P., Banerjee, S., Koley, H., & Sengupta, S. (2025). A novel approach in reducing hypercholesterolemia using modified mango kernel starch and its role as a potential fat replacer. *Food Research International*, *212*, 116420. https://doi.org/10.1016/j.foodres.2025.116420

Wang, X., Huang, L., Zhang, C., Deng, Y., Xie, P., Liu, L., & Cheng, J. (2020). Research advances in chemical modifications of starch for hydrophobicity and its applications: A review. *Carbohydrate Polymers*, *240*, 116292. https://doi.org/10.1016/j.carbpol.2020.116292

Wilpiszewska, K., Antosik, A. K., & Zdanowicz, M. (2019). The effect of citric acid on physicochemical properties of hydrophilic carboxymethyl starch-based films. *Journal of Polymers and the Environment*, *27*(6), 1379–1387. https://doi.org/10.1007/s10924-019-01436-9

Woggum, T., Sirivongpaisal, P., & Wittaya, T. (2014). Properties and characteristics of dual-modified rice starch based biodegradable films. *International Journal of Biological Macromolecules*, *67*, 490–502. https://doi.org/10.1016/j.ijbiomac.2014.03.029

Wu, Q., Yang, Y., Xu, Y., Wang, B., Liu, X., Wang, Y., Zhang, G., Bian, X., Ma, C., & Zhang, N. (2024). Impact of butyric acid modification on the structural and functional properties of rice starch. *Current Research in Food Science*, *9*, 100874. https://doi.org/10.1016/j.crfs.2024.100874

Xie, L., Shen, M., Wang, Z., & Xie, J. (2021). Structure, function and food applications of carboxymethylated polysaccharides: A comprehensive review. *Trends in Food Science & Technology*, *118*, 539–557. https://doi.org/10.1016/j.tifs.2021.09.016

Xu, M., Xu, Y., Ji, S., Amrouche, A. T., Li, Y., Zhou, Z., Shen, J., Li, K., & Lu, B. (2024). Investigation of the 3D printability of modified starch-based inks and their formation mechanism: Application in ice cream. *Food Hydrocolloids*, *154*, 110038. https://doi.org/10.1016/j.foodhyd.2024.110038

Zakeri, S. M., Alimi, M., Shokoohi, S., & Shahidi, S.-A. (2025). Acetylated distarch adipate quinoa starch in tomato ketchup: Modification process and functional charactristics. *Applied Food Research*, *5*(1), 100708. https://doi.org/10.1016/j.afres.2025.100708

Zhang, J., Ran, C., Jiang, X., & Dou, J. (2021). Impact of octenyl succinic anhydride (OSA) esterification on microstructure and physicochemical properties of sorghum starch. *LWT*, *152*, 112320. https://doi.org/10.1016/j.lwt.2021.112320

Zhao, Y., Zheng, Z., Zhao, Y., Chen, J., & Tang, S. (2024). Cross-linked modification of tapioca starch by sodium trimetaphosphate: An influence on its structure. *Food Chemistry: X*, *23*, 101670. https://doi.org/10.1016/j.fochx.2024.101670

4 Biotechnological and Enzymatic Advances in Starch Modification

Odangowei Inetiminebi Ogidi
Department of Biochemistry, Faculty of Basic Medical Sciences, Bayelsa Medical University, Yenagoa, Bayelsa State, Nigeria

Iniebiyo Felagha
Department of Chemical Sciences, Faculty of Basic and Applied Sciences, University of Africa, Toru-Orua Bayelsa State, Nigeria

4.1 INTRODUCTION

Starch, a complex carbohydrate primarily composed of amylose and amylopectin, is one of the most abundant and versatile natural polymers in the world. It is widely used in various industries, including food, pharmaceuticals, and textiles, due to its availability, biodegradability, and functional properties. However, native starch exhibits limitations in terms of solubility, viscosity, gelation behaviour, and digestibility, making it less suitable for certain applications (Singh et al., 2018d; Gupta et al., 2023). To overcome these limitations, starch can be modified chemically, physically, or enzymatically to enhance its functionality and broaden its application spectrum (Kaur et al., 2020; Gupta et al., 2024b).

While chemical and physical modification methods, such as acid hydrolysis, esterification, and oxidation, have been widely used for starch modification, they often present drawbacks, including environmental concerns, the use of toxic reagents, and the formation of byproducts (Ogidi & Emaikwu, 2024; Sharma et al., 2021b; Shrivastava et al., 2024). In contrast, biotechnological and enzymatic approaches provide more sustainable and eco-friendly alternatives by using natural catalysts to modify starch in a controlled manner (Li et al., 2019a). Enzymatic modification involves the use of specific enzymes such as α-amylase, pullulanase, isoamylase, and other glycoside hydrolases to target and modify the starch structure, thus improving the starch's properties and functionality (Kaur et al., 2020; Ogidi & Emaikwu, 2024).

Enzymatic modification offers several advantages over traditional methods, including high specificity, mild reaction conditions, and the ability to preserve the natural properties of starch while enhancing its desired characteristics. For instance, α-amylase can hydrolyse the α-1,4-glycosidic bonds in amylose and amylopectin, producing smaller oligosaccharides, while pullulanase and isoamylase specifically target α-1,6 linkages in amylopectin, resulting in debranching and an increase in amylose content (Singh et al., 2018d). These modifications can improve the solubility, gelatinisation temperature, swelling power, and pasting properties of starch, making it more suitable for various food formulations and industrial applications (Zhang et al., 2020a). Moreover, enzymatic processes can promote the formation of resistant starch, a type of starch that resists digestion in the small intestine and offers numerous health benefits, including improved glycaemic control and gut health (Li et al., 2019a).

Recent advances in enzyme engineering and biotechnology have significantly expanded the scope of enzymatic starch modification. Enzyme immobilisation, genetic engineering, and the development of thermostable enzymes have enhanced the efficiency, specificity, and scalability of these processes (Sharma et al., 2021b). As a result, enzymatically modified starches with tailored properties are increasingly being used in a variety of sectors, including the food industry, for low-calorie and health-oriented products; the pharmaceutical industry, for controlled drug delivery systems; and the bio-based material industry, for biodegradable plastics (Okereke et al., 2020; Zhang et al., 2020a). This chapter explores biotechnological and enzymatic advances in starch modification, highlighting the enzymatic mechanisms, the advantages of enzymatic over traditional methods, and the current and potential future applications of modified starch in various industries.

4.2 AN OVERVIEW OF STARCH AND ITS MODIFICATION

Granules of starch are present in several parts of verdant plants, including stems, tubers, roots, fruits, and grains. In addition to being an essential part of human diet, they are used as a main ingredient in making various sucrose derivatives, ethanol, bio-plastics and paper, among other things (Fan & Picchioni, 2020; Preiss, 2018; Gupta et al., 2024c). Starch is composed of two fundamental molecular components: amylose (AM) chains, which are typically linear and joined by α-1,4 bonds, and amylopectin (AP) chains, which also have α-1,4 connections but possess several α-1,6-linked branches. Starch crops such as wheat, rice, rye,

DOI: 10.1201/9781003564287-4

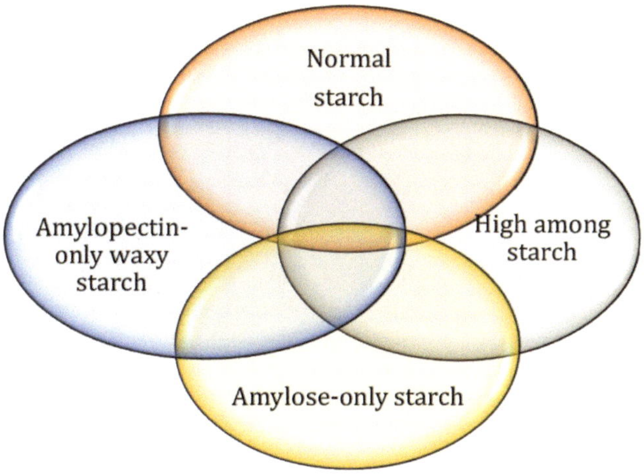

FIGURE 4.1 Classification of starch.

potato, corn, and cassava often have starch with an amylose content ranging from 15% to 30% (Bertoft, 2013).

In addition, like AM, starch may possess an isomeric composition characterised by a branching arrangement and a dispersion of molecular weights (Bertoft, 2013). A comprehensive classification approach that encompasses all aspects of amylose content has just been implemented. According to Zhong et al. (2020) and Carciofi et al. (2012), there are four types of starch that fall into this category (Figure 4.1): normal starch, high among starch, AM-only starch, and AP-only waxy starch. Scientists and manufacturers in industrial contexts frequently alter starch to make it more useful and to get beyond its restrictions (Maniglia et al., 2021). Starch can be modified to meet specific requirements, allowing for the production of new derivatives or products with improved properties. Aesthetic improvements, easier digestion for dietary benefits, clearer product labels, and lower production and recipe development costs are among potential outcomes of food innovation (Maniglia et al., 2021). The starch modification business is expected to experience tremendous expansion, with the industrial starch market surpassing billions in US dollars by 2028 (Maniglia et al., 2021). In the production process, starch-based goods are regularly improved with new functional and value-added qualities (Maniglia et al., 2021).

Starch has evolved considerably throughout the years to meet the specific needs of many businesses and individuals. Enzymatic, chemical, and physical changes are the three main types of modifications (Maniglia et al., 2021). The most frequently used industrial method in improving starch functionality is chemical modification, according to Maniglia et al. (2021), because it is both efficient and cheap. There are three main chemical reactions that affect the three -OH groups at carbon numbers 2, 3, and 6, and the breakdown of glucosidic bonds: oxidation, esterification, and etherification (Masina et al., 2017). Nevertheless, concerns arise regarding the potential impact on consumer well-being and the ecological harm resulting from chemical alteration. Given the limitations of chemical treatments, it is imperative to explore other approaches that employ physical and enzymatic techniques for modification (Maniglia et al., 2021). Physical alteration has several advantages over chemical modification, including enhanced consumer acceptance, heightened safety, and reduced environmental contamination. Cold plasma treatment, high pressure, milling, ultrasonic treatment, heat-moisture treatment, annealing, and high-speed shearing are among the physical treatments that have been developed in recent decades (Schmiele et al., 2019).

Nevertheless, physical alteration is not sufficiently precise to alter specific starch properties and is significantly less efficient than the majority of chemical approaches. Porous starch particles have been generated by physical methods, such as microwave and ultrasonic treatments. Undoubtedly, enzymatic approaches exert a far bigger influence (Schmiele et al., 2019). They assert that enzyme-based techniques are more precise and structurally specific than physical procedures when it comes to altering starch structures and their physicochemical qualities. Using enzyme methods offers two main advantages. Producing goods with exact and specialised structures is made possible by these enzymes' great selectivity and substrate specificity. Furthermore, unlike physical and chemical treatments, which can generate undesired byproducts under harsh conditions, these methods operate in mild reaction settings (Karaki et al., 2016). Indeed, it is well acknowledged that enzymatic methods for altering starch are crucial and possess significant potential for further development.

4.3 STRUCTURE OF STARCH

The morphology of the granular surface has an impact on the enzymatic activity involved in catalysing interfacial amylolytic reactions on granular starch. The ability to break down fully gelatinised starch is influenced mostly by the structure of the starch itself. The hierarchical structure of native granular starch can be represented at several sizes. At the molecular scale, there exist AM and AP molecules, with a dimension of 0.1 nm. The lamellar structure consists of both crystalline and amorphous areas, with a size ranging from 8 to 11 nm (Zhong et al., 2021; Gupta et al., 2024d). The growth rings, with a dimension of 0.1 μm, alternate between regions that lack a definite crystalline structure and regions that contain a partially ordered arrangement of atoms (Zhong et al., 2021). Sizes of the overall granular structure range from 1 to 100 μm, according to Bertoft (2013). Even while the granular condition of native granular starch is unaffected by modifications, its levels can be changed to a certain degree. Some studies have shown that subjecting native starch granules to microwave treatment or high hydrostatic pressure can partially degrade the starch molecules, which can disturb the granules' relatively ordered structure (Liu et al., 2022; Zhong et al., 2019). In Figure 4.2, the chemical structure of starch is illustrated.

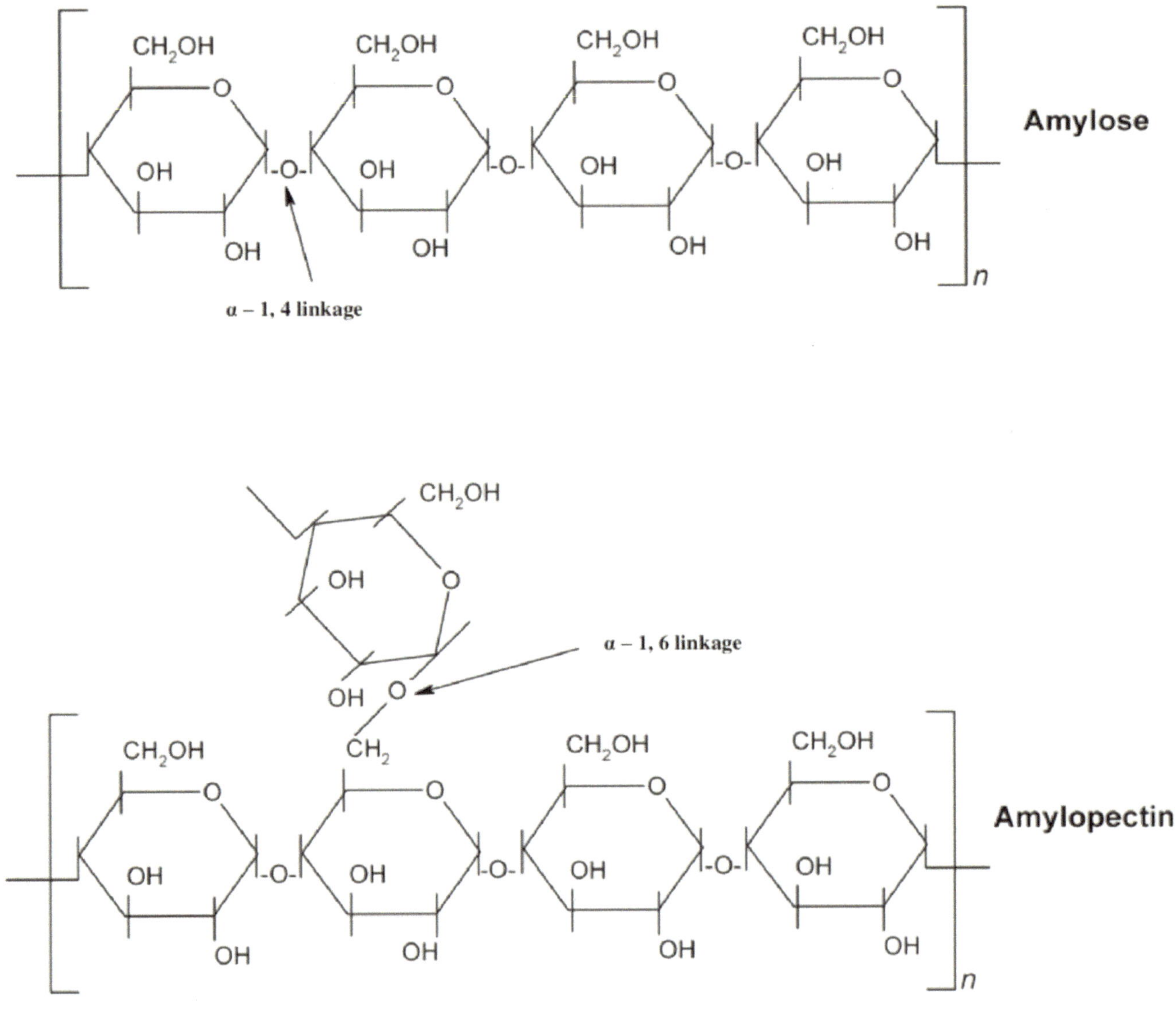

FIGURE 4.2 Chemical structure of starch.

Basic taxonomy divides granular starch into two varieties: native granular starch, which remains untreated, and pre-treated granular starch, which has been artificially modified to have a certain structure. Producing gelatinised starch requires soaking starch granules in a lot of water and then heating them to a specific temperature that is appropriate for that kind of starch (Ai & Jane, 2018). The starch molecules lose their crystalline structure and turn into a liquid after this treatment. This treatment disrupts their organised organisation and leads to the absorption of a significant quantity of water (Ai & Jane, 2018).

4.4 ENZYMES THAT MODIFY STARCH

Enzymes are extensively utilised in diverse industries because of their exceptional ability to function as catalysts. Enzymes such as xylanase, glucoamylase, α-amylase, fungal acid protease, β-amylase, pullulanase, and glucose isomerase are crucial in the starch business for processing and removing starch from various raw materials (Vitolo, 2020). One type of biocatalyst is the enzyme, which can come from either animals, bacteria, or plants. These are GRAS-approved (generally recognized as safe) and utilised extensively to enhance several food processing areas (Obadi & Xu, 2021). In the biological transformation of starch, several enzymes are used, including glycosyltransferase, α-amylase, debranching enzymes, β-amylase, cyclodextrin, and glucoamylase (Bangar et al., 2022; Wang et al., 2020).

The α-1,4-glycosidic bonds in starch molecules, especially at the non-reducing end, can be broken by the enzyme α-amylase. Because of this procedure, the starch paste solution's molecular size and viscosity are reduced. Some α-1,4-glycosidic linkages can be hydrolysed by the enzyme β-amylase. This process shortens the side chain, increases branch density, and eliminates maltose residue. The α-1,6-glycosidic linkages are degraded by glycoamylase as time goes on, while the α-1,4-glycosidic bonds are cleaved at

TABLE 4.1
Starch-Modifying Enzymes and Their Applications

Enzyme	Organisms	Reaction	Application	References
α-Amylase	*Bacillus stearothermophilus*	Maltose production from amylose	Spoilage prevention of bakery products	Ao et al. (2007), Kelly et al. (2009), Lee et al. (2008)
β-Amylase	Barley (malt)	External side chain of amylopectin shortening	Improvement of emulsification and prevention of starch	Miao et al. (2015), Wursch & Gummy (1994), Xu et al. (2015), Yao et al. (2003)
Amylosucrase	*Neisseria polysaccharea*	Transglycosylation of glucosyl unit	Slow digestible starch	Jane et al. (1999), Kim et al. (2016), Shin et al. (2010)
Branching enzyme	*Deinococcus radiodurans*	Formation of α-1,6 linkages and hydrolysis of α-1,4 linkages	Slow digestible starch and thermostable glycogen	Palomo et al. (2009), Shinohara et al. (2001), Takata et al. (2003), van der Maarel et al. (2003)
α-Glucanotransferase	*Thermus aquatieus*	Molecular transglycosylation	Prevention of spoilage of rice	He et al. (2017), Kim et al. (2013), Mun et al. (2009), Nguyen et al. (2014)
Cyclodetrin glucanotransferase	Alkalophilic *Baccillus*	Formation of α-1,4-glycosidic linkage	Retard of bread	Nguyen et al. (2014), Shim et al. (2007)

the non-reducing end. An enzyme called pullulanase helps break down oligosaccharides, amylopectin, and pullulan into their individual 1,6 bonds. To enzymatically degrade starch, pullulanase is usually employed alongside glucoamylase or β-amylase during the saccharification process. Amylopectin has α-1,6-glycosidic connections that can be broken down by isoamylase, pullulanase, and glycogen 6-glucanohydrolase. Amylose and oligosaccharides are the end products of this process (Bangar et al., 2022). Enzymes that modify starch and their applications are shown in Table 4.1.

According to Li et al. (2017) and Hutabarat and Stevensen (2023), the linear chains of amylose and amylopectin have their α-1,4-glycosidic links cleaved by the branching enzyme. Afterwards, new α-1,6-glycosidic linkages are formed by initiating a transglycosylation process. Pelulanase, isoamylase, and oligo-1,6-glucosidase are enzymes that break down starch. Enzymes mainly target the α-1,6-glycosidic linkages in starch, which leads to the primary synthesis of linear amylose. According to Amaraweera et al. (2021), enzymes including amylosucrase, α-amylase, and amylomaltase have the ability to break both the 1,4- and 1,6-glycosidic linkages, resulting in the formation of shorter chains of amylose or amylopectin.

4.4.1 Schematic Mechanism of Enzymatic Modification of Starch

Enzymatic modification of starch involves specific enzymes acting on the molecular structure of starch, targeting glycosidic bonds in amylose and amylopectin. This process modifies the physicochemical and functional properties of starch, such as solubility, pasting behaviour, and digestibility. Below is a detailed explanation of the mechanisms.

4.4.1.1 Structure of Native Starch

Native starch consists of two polysaccharides: amylose, consisting of linear chains of glucose linked by α-1,4-glycosidic bonds, and amylopectin, consisting of branched chains with α-1,4-glycosidic bonds in the linear regions and α-1,6-glycosidic bonds at the branch points (Gupta et al., 2024a).

4.4.1.2 Enzymes Used in Starch Modification

1. α-Amylase
    - Hydrolyses internal α-1,4-glycosidic bonds in amylose and amylopectin.
    - Produces oligosaccharides, maltose, and glucose.
    - Reduces molecular weight and viscosity of starch solutions.
    - Applications: Production of syrups, improved solubility.
2. Pullulanase
    - Specifically hydrolyses α-1,6-glycosidic bonds in amylopectin (debranching enzyme).
    - Converts branched amylopectin into linear chains, increasing apparent amylose content.
    - Applications: Improved gelation properties, reduced retrogradation.
3. Isoamylase
    - Similar to pullulanase, targets α-1,6-glycosidic linkages.
    - Produces linear polysaccharides with a focus on disrupting branch points.
    - Applications: Enhanced crystallinity, resistant starch production.
4. β-Amylase
    - Hydrolyses α-1,4-glycosidic bonds from the non-reducing ends, releasing maltose units.
    - Leaves a β-limit dextrin resistant to further hydrolysis.
5. Glucose Isomerase
    - Converts glucose from hydrolysed starch into fructose.
    - Applications: High-fructose corn syrup production.

4.4.1.3 Mechanism of Enzymatic Modification

- Step 1: Granule Surface Interaction
    - Enzymes bind to the surface of starch granules. The amorphous regions, which are less ordered, are hydrolysed first.
- Step 2: Hydrolysis or Debranching
    - α-Amylase cleaves internal α-1,4 bonds, reducing the size of polysaccharides.
    - Pullulanase and isoamylase hydrolyse α-1,6 bonds, breaking branches in amylopectin.
    - β-Amylase trims chains from the ends, releasing maltose.
- Step 3: Structural Transformation: The starch structure becomes fragmented.
    - Short-chain oligosaccharides are produced.
    - Amylopectin transforms into linear molecules.
    - Crystalline regions may remain intact initially but degrade with extended enzymatic action.
- Step 4: Altered Properties: Enzymatic action modifies the starch properties.
    - Increased solubility.
    - Reduced viscosity.
    - Lower gelatinisation temperature.
    - Enhanced digestibility or resistant starch formation.

4.4.2 Starch Modification via Enzymatic Processes

In recent years, physical and chemical approaches have been more commonly used to produce modified starch, with enzyme changes playing a supporting role. Enzymes are preferred over chemical treatment due to their perceived safety and environmental benefits (Park et al., 2018a). Modifying the digestibility of starch through enzymatic alteration is a precise and environmentally friendly procedure. Enzymes such as pullulanase, α-amylase, and branching enzymes are capable of achieving this. Because of its enhanced semi-crystalline structure and increased presence of linear short chains, enzyme-modified starch displays a higher amount of resistant starch, according to Zhang and Bao (2023). Because of this, it becomes less susceptible to enzymatic hydrolysis that occurs during digestion. Enzymes can raise the proportion of resistant starch in various starches, according to various reports (Harder et al., 2015; Kim et al., 2015; Sorndech et al., 2016).

Researchers have used these enzymes in malt production (Chu et al., 2014) and to explore how starch structure and type affect starch breakdown (Naguleswaran et al., 2014). Starch undergoes hydrolysis, resulting in the cleavage of glycosidic linkages and the formation of smaller units. Starch forms either polymeric or oligomeric molecules when hydrolytic enzymes break the glycosidic linkages. There are essentially three main classes of these enzymes: Amylose and amylopectin include α-1,4-glycosidic connections that can be cleaved by α-amylase and other endo-acting enzymes. Glycosidic linkages at the substrate's non-reducing ends are sequentially cleaved by exo-acting enzymes to form monomeric or oligomeric structures. In contrast to the exo-acting enzyme β-amylase, which breaks down α-1,4 bonds on amylose and amylopectin gradually, a debranching enzyme known as pullulanase preferentially cleaves α-1,6-glycosidic connections (Bertoft, 2013; Hutabarat & Stevensen, 2023).

The utilisation of microcapsules made from enzyme-treated maize starch with hydrolysis durations of 16 hours (ETCS 16h-GA) and 20 hours (ETCS 20h-GA), followed by coating with gum arabic, facilitated the encapsulation of ascorbic acid (AA). The study found that ETCS 16h-GA microcapsules exhibit reduced degradation of AA during storage, improved protection of AA, and the capacity to release the molecule in a controlled manner in the gastrointestinal tract. According to Leyva-López et al. (2019), this characteristic renders them an effective method for enclosing functional food components. Enzyme hydrolysis by pullulanase can change the granule shape of potato starch. Subsequently, nanoprecipitation will be employed on the hydrolysates to generate starch nanoparticles with dimensions ranging from 20 to 50 nm. Nanoparticles often measure less than 10 μm in size, which is significantly smaller than traditional starch granules. The pullulanase enzyme worked best when placed in a pH 4 phosphate buffer solution with 160 NPUN/g (pullulanase units per gram) of starch and maintained at 60 °C.

Consequently, Chorfa et al. (2022) created hydrolysates consisting of starch polymers with reduced molecular weights. By utilising β-amylase and transglucosidase for dual-enzymatic modification, the molecular weight of maize starch was decreased while the quantity of short chains and α-1,6 links was enhanced. Achieving a maximum concentration of 33.5% SDS (slowly digestible starch), compared to the original starch, was possible by using dual-enzyme hydrolysis for a duration of half a day (Chorfa et al., 2022). The findings indicated that the application of β-amylase and transglucosidase on starches resulted in an increased production of short chains and a decreased production of long chains compared to the untreated starches. The medication moreover enhanced the quantity of α-1,6 connections, resulting in a deceleration of digestion (Miao et al., 2015). Cereal crops (such as maize) and tuber crops (such as potatoes and cassava) undergo acid hydrolysis and enzymatic hydrolysis to produce nanostarch. The enzymatic procedure used to make maize nanostarch results in a smaller particle size than cassava and potato nanostarch. Nanostarch comprises a substantial proportion of both corn starch (29% by weight) and potato starch (41% by weight) (Chorfa et al., 2022). In addition, nanostarches exhibit reduced melting enthalpies in comparison to their larger counterparts. The nanostarch underwent acid hydrolysis, resulting in a reduction in size to 18 ± 3 nm. Conversely, enzyme hydrolysis caused an increase in size to 162 ± 23 nm (Miao et al., 2015).

Additionally, the enzymatic technique enhances its adaptability while ensuring environmental safety (Dukare et al., 2021). An industrial α-amylase was employed to enzymatically alter the cassava starch (*Manihot esculenta* Corpoica M-Tai) near its pasting temperature. The technique resulted in a range of yields and in dextrose equivalents from 0% to 45.5%. Granules underwent exo-corrosion as shown by scanning electron microscopy (SEM) examination, which was based on the degree of hydrolysis. Moreover, after the change, there was an increase in the level of crystallinity. Hydrolysed starches exhibited reduced setback and amylose values compared to the unmodified starch. This indicates that the hydrolysed starch is more susceptible to retrogradation and water loss (syneresis) (Dukare et al., 2021).

In order to enhance its ability to stabilise ice cream, taro starch undergoes enzymatic processing with α-amylase. This change improves the overrun, foam stability, and viscosity of the ice cream. The application of α-amylase to taro starch resulted in enhanced functional characteristics. The second sample had a greater swelling capacity of 18.28 g/g (grams absorbed liquid per gram absorbent), while the extracted starch had a capacity of 17.43 g/g. Similarly, the starch that was extracted had an enzyme digestibility of 59.32%, which was lower than the digestibility of the other sample, which was 62.7%. The addition of 0.5% enzyme-modified taro starch resulted in greater viscosity (110.21Cp) and overrun (65.67%), which revealed the improved stability and quality of the ice cream. The enhanced performance of taro starch, which has been changed by α-amylase, can be attributed to its significant capacity to swell or retain gas. Substituting enzyme-modified sweet potato starch for fat is common in ice cream recipes. An inverse relationship was seen between the concentration of enzyme-modified starch and the fat level, with a drop from 9.85% to 5.79%.

A study carried out by Kale et al. (2020) found that sweet potato starch that has been enzymatically transformed is an excellent fat alternative when making ice cream. Thermostable α-amylase and pullulanase enzymes were utilised to alter corn starch, leading to a notable enhancement in resistant starch. Two enzymes, thermostable α-amylase and pullulanase, were used to convert corn starch into resistant starch (RS). The maximum amount of resistant starch that could be obtained was 10.75%. Enzymatic hydrolysis for 35 minutes, 8 U/g of pullulanase, and a liquid containing 15% starch were used to achieve this. According to Liu et al. (2022), this value was much larger than for native starch or starch treated with just one enzyme. Amylosubtilin and amylase produced by *Bacillus licheniformis* are incorporated into enzymatically modified potato starch to replace fat. The results showed that yoghurt made with modified potato starch (i.e., with amylosubtilin and amylase) was similar to yoghurt made with unmodified potato starch. This suggests that this starch product can be used as a replacement for regular yoghurt in low-fat varieties. Sopawong et al. (2022) used partial gelatinisation (PG), heat-moisture treatment (HMT), and pullulanase treatment among the methods used for alternative lotus seed flour (Sopawong et al., 2022).

Resistant starch levels fell and rapidly digested starch levels rose in the PG case. On the other hand, with HMT, RS levels reached 86.49% of the total starch, while amylose levels reached 34.57%. Pasting time was significantly reduced and viscosity differences were significantly larger in pullulanase-treated flour compared to unmodified wheat and other modified flours. Lotus seed flour was able to be thickened more effectively while retaining its low glycaemic index. By heating granular potato starch to 35 °C and adding the debranching enzyme isoamylase to a solution containing 40% water by weight, Ulbrich et al. (2021) modified the starch. A constant concentration of 200 UmL^{-1} of enzyme solution was maintained when the volume was progressively increased to 100, 250, and 400 μL/50 g of starch. The starch polysaccharides undergo partial breakdown, leading to a little decrease in the amylose portion's weight average molar mass. That debranching enzymes such as **isoamylase** and polyurethane may one day replace acid-based treatments is further supported by this.

Zhang et al. (2020a) showed that pullulanase and amylosucrase could gradually modify amylopectin. Amylosucrase was used to lengthen the amylopectin side chains, and pullulanase was used to eliminate branching from the resultant structure. As a final product, microparticles were generated via the self-assembly of linear chains. Aggregated nanoparticles ranging in size from 200 to 400 nm were present within these microparticles. It is crucial to note that these microscopic particles can float in water without creating a gel. The microparticles had up to 67.2% resistant starch, which is significantly more than the 9.1% found in native amylopectin. To create digestibility-reducing functional starch microparticles, enzyme treatments were employed in their manufacture. The modified starch underwent substantial molecular and crystalline structure changes as a result of these treatments. More than ever before, this technology facilitates the synthesis of novel modified starches while also improving their environmental impact.

4.4.3 Benefits of Enzymatically Modifying Starch

Physical approaches are cost-effective and environmentally friendly. However, because they have the ability to alter the starch structure, their industrial use is restricted. 3D printing, packaging, the food and non-food industries, and physically modified starches all reap benefits. Altering starch's structure can be achieved chemically by processes such as grafting, acid hydrolysis, etherification, esterification, cross-linking, and dual modification. After these changes are done to achieve certain product features, the end result is chemically modified starch or derivatives of it. Chemically modified starches find use in various industries, such as agriculture, medications, bio-based adhesives, packaging, superabsorbents, and wastewater treatment (Amaraweera et al., 2021; Iyiola et al., 2023; Ogidi & Akpan, 2023; Ogidi et al., 2023, 2024).

Consumers do not appreciate chemical methods of starch modification for several reasons, including worries about safety (Onyango, 2016) and the release of multiple chemicals that are harmful to the environment and need to be recycled, such as phosphates, hypochlorites, and acetates. Enzyme reactions exhibit high sensitivity and precision, resulting in reduced production of byproducts when applied to starches compared to physical or chemical alterations (Park et al., 2018b). The clean label approach of enzymatically altering starches has several advantages. The bundle includes enhancements in purity, reliability in product quality, absence of undesired components, cost, and beneficial characteristics (Park & Kim, 2021). Enzymatic starch modification enhances the freeze–thaw capabilities of starch by increasing the amount of resistant starch and producing shorter, straight amylopectin chains. Furthermore, this alteration not only impacts the digestion of the starch paste but also reduces its viscosity and enhances its elastic properties (Park & Kim, 2021).

4.5 EFFECT OF ENZYMATIC MODIFICATION ON PHYSICOCHEMICAL AND TECHNO-FUNCTIONAL PROPERTIES OF STARCH

Enzymatic modification of starch enhances its functionality by altering its physicochemical and techno-functional properties, making it more suitable for specific food and industrial applications. Enzymes such as α-amylase, pullulanase, and isoamylase selectively target glycosidic bonds in starch molecules, leading to changes in properties such as colour, solubility, amylose content, turbidity, pH, swelling power, and thermal, pasting, and rheological behaviours (Table 4.2).

4.5.1 Colour

Enzymatic modification can influence the colour of starch, primarily due to changes in molecular structure and reaction byproducts.

- *Lighter appearance:* Hydrolysis can lead to a reduction in molecular aggregates, enhancing the clarity and brightness of starch dispersions. This is especially beneficial in applications such as clear soups and sauces (Sharma et al., 2021a).
- *Enzyme-specific effects:* The type of enzyme used influences colour. For example, pullulanase-treated starches often appear clearer than native starches due to reduced retrogradation (Zhang et al., 2020c).

4.5.2 Solubility

Enzymatic treatment significantly increases the solubility of starch by breaking down the glycosidic bonds, reducing molecular weight, and increasing the proportion of shorter chains.

- *Increased solubility:* α-Amylase hydrolysis enhances water solubility by producing smaller oligosaccharides, which dissolve more readily in water (Li et al., 2019a).

4.5.3 Amylose Content

Enzymatic modification alters the amylose-to-amylopectin ratio, which is crucial for functionality.

- *Debranching effects:* Pullulanase and isoamylase increase apparent amylose content by hydrolysing amylopectin branches into linear chains, mimicking amylose (Singh et al., 2018a).
- *Functional benefits:* Higher amylose content improves gelling properties and film-forming capacity, making the starch suitable for edible coatings and biodegradable films.

4.5.4 Turbidity

Turbidity is influenced by the structural organisation of starch molecules.

- *Reduction in turbidity:* Enzymatic modification reduces turbidity in starch dispersions by breaking down insoluble fractions and retrograded chains (Sharma et al., 2021a).

4.5.5 pH Stability

The pH of starch dispersions can be affected by enzymatic treatment.

- *Neutral pH stability:* Enzymatic hydrolysis generally does not significantly alter the pH of starch, making it stable in a wide range of food applications (Li et al., 2019b).

4.5.6 Swelling Power

Swelling power is reduced after enzymatic modification, as hydrolysis decreases the ability of starch granules to retain water.

- *Reduced swelling:* Amylase-treated starches exhibit lower swelling power due to the partial breakdown of granular structure (Singh et al., 2018a).

4.5.7 Thermal Properties

Enzymatic modification affects the gelatinisation temperature and enthalpy of starch.

- *Lower gelatinisation temperature:* Starch treated with debranching enzymes, such as pullulanase, exhibits reduced gelatinisation temperatures due to the breakdown of amylopectin branches (Zhang et al., 2020b).
- *Improved stability:* Modified starches often demonstrate enhanced thermal stability, making them suitable for high-temperature applications.

TABLE 4.2
Effect of Enzymatic Modification on the Physicochemical and Functional Properties of Starch

Treatment	Enzyme Used	Temperature (°C)	Time (min)	pH	Moisture Content (%)	Starch Damage (%)	Gelatinisation Temperature (°C)	Viscosity (mPa.s)	Swelling Power (%)	Solubility (%)	Water Absorption Capacity (%)	Emulsification Capacity (%)	Reference
Control	–	–	–	6.5	10.2 ± 0.5	2.1 ± 0.2	65.4 ± 0.8	450 ± 10	12.5 ± 0.8	5.2 ± 0.5	150 ± 10	30 ± 5	Kumar et al. (2019)
Amylase	α-Amylase	50	30	6.5	9.5 ± 0.3	5.5 ± 0.5	62.1 ± 0.9	350 ± 15	10.8 ± 0.6	7.5 ± 0.8	120 ± 8	40 ± 6	Singh et al. (2017)
Glucoamylase	Glucoamylase	60	60	5.5	8.8 ± 0.4	8.2 ± 0.8	58.5 ± 1.1	250 ± 20	9.2 ± 0.7	10.2 ± 1.0	100 ± 5	50 ± 8	Park et al. (2018a)
Pullulanase	Pullulanase	55	45	6.0	9.2 ± 0.5	6.8 ± 0.7	60.8 ± 0.7	320 ± 18	11.5 ± 0.9	6.5 ± 0.6	130 ± 9	35 ± 5	Lee et al. (2020)
Combination	α-Amylase + glucoamylase	55	60	6.0	8.5 ± 0.4	9.5 ± 1.0	56.2 ± 1.3	200 ± 25	8.5 ± 0.8	12.1 ± 1.2	90 ± 6	55 ± 9	Singh et al. (2017)
β-Amylase	β-Amylase	50	30	6.5	9.8 ± 0.4	4.2 ± 0.4	63.5 ± 0.9	380 ± 12	10.2 ± 0.7	6.8 ± 0.7	110 ± 7	38 ± 5	Wang et al. (2020)
Debranching Enzyme	Isoamylase	55	45	6.0	9.5 ± 0.5	5.8 ± 0.6	61.2 ± 0.8	300 ± 15	11.2 ± 0.9	7.2 ± 0.8	120 ± 8	42 ± 6	Li et al. (2019a)
Transglucosidase	Cyclodextrin glycosyltransferase	60	60	5.5	8.2 ± 0.4	9.0 ± 1.0	57.5 ± 1.2	220 ± 20	9.0 ± 0.8	11.5 ± 1.2	90 ± 6	58 ± 9	Zhang et al. (2020a)
α-Glucosidase	α-Glucosidase	55	45	6.0	9.0 ± 0.5	6.5 ± 0.7	60.5 ± 0.8	280 ± 12	10.8 ± 0.9	8.0 ± 0.9	110 ± 7	45 ± 6	Liu et al. (2022)

4.5.8 Pasting Properties

Enzymatically modified starch shows altered pasting profiles due to changes in molecular weight and amylopectin branching.

- *Lower peak viscosity:* Hydrolysed starches have lower peak viscosities because smaller molecules reduce the viscosity of the paste (Sharma et al., 2021c).
- *Improved stability:* Pullulanase-treated starches exhibit stable pasting profiles, making them suitable for industrial processes requiring consistency (Li et al., 2019b).

4.5.9 Rheological Properties

The rheological behaviour of starch, including its viscoelastic properties, is significantly influenced by enzymatic modification.

- *Reduced viscosity:* Amylase-treated starch dispersions exhibit lower viscosity, beneficial for applications requiring fluidity, such as beverages (Singh et al., 2018b).
- *Enhanced gelling:* Isoamylase-treated starches form firmer gels due to an increased amylose content and reduced branching, which enhances cross-linking during gelatinisation.

4.6 EFFECT OF ENZYMATIC MODIFICATION ON STRUCTURAL PROPERTIES OF STARCH

Enzymatic modification of starch is a precise method for tailoring its structural and functional properties. This modification involves enzymes such as α-amylase, pullulanase, and isoamylase, which selectively target glycosidic bonds. These enzymatic actions lead to significant alterations in the starch granule's morphology, crystallinity, chemical structure, and molecular weight distribution. Advanced analytical techniques, including SEM, atomic force microscopy (AFM), X-ray diffraction (XRD), Fourier transform infrared spectroscopy (FTIR), and molecular analyses, provide comprehensive insights into these changes.

4.6.1 Scanning Electron Microscopy (SEM)

SEM captures the morphological changes in starch granules induced by enzymatic hydrolysis. Enzymatic hydrolysis, especially by α-amylase, creates visible pits, cracks, and rough surfaces. Pullulanase treatment often results in smoother erosion patterns due to the selective hydrolysis of amylopectin branches (Zhang et al., 2020b). Prolonged enzymatic exposure leads to the breakdown of granules into smaller fragments or complete solubilisation, which is also refer to as granule disintegration, as shown in SEM images of amylase-treated starch (Li et al., 2019b).

4.6.2 Atomic Force Microscopy (AFM)

AFM provides nanoscale visualisation of the granule surface and topography changes that involve increased surface roughness. AFM reveals a progressive increase in surface roughness after enzymatic treatment, with clear depressions and fragmented structures (Sharma et al., 2021c). Crystalline-amorphous transitions – the disruption of crystalline regions and the generation of amorphous zones – are detectable at the nano scale. Pullulanase and isoamylase treatments exhibit a more fragmented crystalline network (Singh et al., 2018b).

4.6.3 X-ray Diffraction (XRD)

XRD quantifies changes in crystallinity and crystalline structure in enzymatically modified starch. Enzymatic hydrolysis decreases relative crystallinity by selectively breaking down first amorphous regions, followed by crystalline structures during extended treatment which is commonly termed reduced crystallinity (Zhang et al., 2020d). The enzymatic action may induce shifts in crystalline polymorphic transformations. Pullulanase treatment can lead to a transition from A-type or B-type patterns to V-type crystallinity due to the reorganisation of linear chains (Li et al., 2019c).

4.6.4 Fourier Transform Infrared (FTIR) Spectroscopy

FTIR spectroscopy highlights chemical bond changes and molecular rearrangements (Gupta et al., 2025b). A decrease in the intensity of peaks associated with glycosidic bond cleavage (e.g., 1155 cm^{-1} for C–O–C stretching) confirms enzymatic hydrolysis. The crystalline to amorphous ratio of the 1047 cm^{-1} (crystalline) to 1022 cm^{-1} (amorphous) bands decreases after enzymatic treatment, indicating reduced crystallinity (Singh et al., 2018c). FTIR spectra may show new peaks indicating the generation of resistant starch fractions after debranching enzyme treatments (Sharma et al., 2021b).

4.6.5 Molecular Weight Distribution

Enzymatic hydrolysis affects the molecular weight of starch components, particularly amylopectin. Hydrolysis by α-amylase produces dextrins and oligosaccharides, leading to a broadening of molecular weight distribution. Gel permeation chromatography studies confirm significant reductions in weight average molecular weight (Zhang et al., 2020d). Enzymatic debranching enzymes such as pullulanase and isoamylase selectively cleave amylopectin branch points, producing shorter chains and linear fragments (Li et al., 2019c).

4.6.6 Amylopectin Branch Length Distribution

Branch length distribution, a key determinant of starch functionality, is significantly altered by enzymatic treatment. Pullulanase and isoamylase specifically hydrolyse α-1,6-glycosidic linkages in amylopectin, increasing the proportion of shorter branch chains. This shift enhances starch solubility and reduces retrogradation tendencies (Singh et al., 2018c). Short linear chains formed during debranching can retrograde into a tightly packed structure, increasing the content of resistant starch. These changes are essential for dietary fibre enhancement in food applications (Sharma et al., 2021d).

4.7 BIOTECHNOLOGICAL MANIPULATION OF STARCH METABOLISM

Starch is the primary carbohydrate that plays a crucial function in various sectors and offers significant nutritional advantages for humans (Gupta et al., 2022). Though maize accounts for the vast majority of industrial starch production, other plants, such as arrowroot (*Maranta arundinacea*), sago palm (*Metroxylon sagu*), cassava, potato, rice, and wheat, all play important roles. The physicochemical qualities of starches produced from different plants can vary significantly due to their distinct polymer compositions and structures. These features include gelatinisation temperature, cooked paste viscosity, and gel viscosity. The diverse applications of starch are dictated by its inherent characteristics, which subsequently affect its practicality. When selecting crops for either grazing or keeping harvested crop organs, it is crucial to take into account the starch content (Wang et al., 2007).

For instance, whereas starch is converted into sugars during the ripening of fruits, this process has a contrary impact on potato tubers, leading to their deterioration. Modifying crop starch metabolism offers several advantages, such as enhancing starch accumulation in harvested organs, restricting starch breakdown (depending on the crop and goal), or enhancing the usability of starch in food and industry by altering its structure. The second component, concerning starch, is intriguing because most specialised applications of starch necessitate subsequent physical or chemical alterations to enhance their effectiveness. The vast potential for bioengineering a wide range of starches by manipulating the expression of specific genes is derived from the extensive variety of enzymes that control the formation of starch in a particular plant organ. An effective approach to alter the starch metabolism genes of sexually propagated diploid species is to crossbreed them with targeted mutations, potentially employing targeting induced local lesions in genomes (TILLING) methods as an additional tool. When working with polyploid and/or asexually reproducing crops, it may be more beneficial or necessary to use biotechnological approaches to increase or partially decrease the activity of the target gene, or to achieve the desired trait by expressing a gene from a different species (Smith, 2008).

4.7.1 Enhancing Starch Biosynthesis Using Biotechnological Methods

The primary method for enhancing starch accumulation has predominantly involved manipulating the plant enzyme ADP-glucose pyrophosphorylase (AGPase). This enzyme undergoes strict control through allosteric mechanisms; it serves as the substrate for starch synthesis. As a result, plants have successfully overexpressed an enzyme with a unique regulatory mechanism by utilising a mutant AGPase gene from *Escherichia coli* (*glgC16*). Starch concentration increased by 60% on a fresh-weight basis after inserting the *glgC16* gene into the plastids of specific transgenic potato lines as compared to control plants (Stark et al., 1992). Nevertheless, there have been no significant advancements observed in experiments involving alternative types of potato (Sweetlove et al., 1996), cassava (Ihemere et al., 2006), maize (Wang et al., 2007), and other plant species. Hence, enhancing starch levels in plant storage organs may not be most effectively achieved only by altering AGPase (Smith, 2008).

In order to accelerate the manufacture of starch, it is highly recommended to increase the availability of ATP to the plastid, as this would promote the synthesis of ADP-glucose. When a particular protein called plastidial envelope adenylate translocator from *Arabidopsis* was increased in potatoes, the levels of ADP-glucose were twice as high in untreated tubers and the starch content was 16%–36% higher (Geigenberger et al., 2001; Tjaden et al., 1998). Despite a doubling of the starch content in potato tubers, ADP-glucose levels increased tenfold due to a plastidial adenylate kinase's reduced activity. In order to produce ATP and AMP, this enzyme converts two molecules of ADP. This impact was validated by both controlled laboratory settings and outdoor studies (Geigenberger et al., 2001). If adenylate kinase is downregulated, plastids may get a greater amount of ATP instead of ADP. When ADP is liberated from ADP-glucose during glucan polymerisation, this exchange occurs (Geigenberger et al., 2001).

4.7.2 Manipulation of Starch Breakdown to Enhance Crop Quality

In crops where starch turnover is high, slowing down the breakdown rate is another way to boost starch content. Adding phosphate groups to amylopectin by α-glucan water dikinase (GWD) is crucial for starch hydrolysis, as previously stated. The starch-synthesis genes of transgenic clover (*Trifolium repens*), ryegrass (*Lolium perenne*), transgenic *Medicago sativa*, and silage maize cultivars have been reduced. Starch is a readily digested carbohydrate. The digestibility of the fodder crops was improved in all of these situations due to a large increase in leaf starch levels resulting from the reduction in GWD (Mottram et al., 2002). In order to improve the quality of potato tubers, researchers have also utilised it to reduce GWD activity, hence inhibiting starch degradation. Cold sweetening occurs when potatoes are refrigerated; this happens because the tubers absorb sugars such as fructose and glucose after breaking down starch. The Maillard reaction, which darkens food unappealingly and creates carcinogenic acrylamide, is more likely to occur when foods with reducing sugars are fried (Mottram et al., 2002).

It is possible to prevent acrylamide synthesis while frying (Rommens et al., 2007) and cold sweetening (Lorberth et al., 1998) by blocking GWD activity in potato tubers. It is surprising that the exact enzymes that break down potato tuber starch when it is refrigerated are still a mystery. The latest advancements in *Arabidopsis* research will aid in their identification and offer further possibilities for mitigating cold-induced sweetening. Enhancing the breakdown capabilities of starch crops can increase their potential for producing more bioethanol. To enhance the economic feasibility of converting starch into ethanol, self-processing plants have been developed by introducing a heat-resistant α-amylase (Lanahan et al., 2006) into maize (*Zea mays*). Amylase is found in parts of cells that aren't responsible for making or storing starch. Therefore, starch buildup during crop growth is unaffected by amylase. The starch contained in the kernels undergoes gelatinisation when they are crushed and boiled in water. In this process, the thermophilic amylase enzyme initiates the decomposition of starch into sugars that can be fermented (Lanahan et al., 2006).

4.7.3 Manipulation of Starch Structure to Diversify Its Applications

Plant starch structural modification mainly aims to change amylopectin to amylose ratios, amylopectin chain length distribution, or starch phosphate concentration (Gupta et al., 2024e). Industrial starches are typically more advantageous when they predominantly consist of amylopectin or amylose, as these two constituents possess distinct physicochemical characteristics. Several mutants with an unusually high amylose content or starches consisting solely of amylopectin have been discovered in different legume and cereal species (Kossmann & Lloyd, 2000). Transgenic technologies have been used to transfer these features to potatoes and cassava, two crops known for their challenging breeding process. Several factors have led to this. Initially, it is not universally true that mutations in specific species consistently result in reduced crop productivity. However, this may not be the situation for potatoes and cassava. Furthermore, these starches possess unique characteristics that make them highly valuable in various industrial applications. An example of the value of potatoes lies in their elevated phosphate content. A decrease in granule-bound starch synthase (GBSS) expression led to modified starch and the elimination of amylose in the first transgenic potato and cassava plants (Raemakers et al., 2005; Tallberg et al., 1998; Visser et al., 1991). Commercially viable plants were introduced some 20 years ago, and no adverse effects on starch content were observed. The absence of approval has prevented the commercialisation of transgenic crops in Europe (Raemakers et al., 2005).

The alternative involves the introduction of a genetically engineered GBSS enzyme, initially derived from a diploid potato variety, into a commercially cultivated potato cultivar. This enzyme is derived from a potato variety that inherently lacks amylose. The genetically modified potatoes are used to extract amylose-free starch, which is then marketed to the public. A recent study demonstrated that a non-transgenic variant of cassava lacks amylose (Ceballos et al., 2007). This discovery presents the opportunity to cheaply produce amylose-free cassava starch, without the need to navigate the regulatory challenges associated with transgenic varieties. Producing high-amylose potato starch is more challenging than originally anticipated. The main approach employed has been to decrease the number of branching enzyme (BE) isoforms. Despite inhibiting the main form (BEI), which resulted in a loss of over 95% of measured activity, minimal changes were seen in the starch structure, and the amylose concentration remained unchanged (Kossmann, 1992; Safford et al., 1998). The result was a starch with an observable amylose content of 38%, in contrast to the 30% seen in the starch from control plants (Jobling et al., 1999), because of the decrease in the minor tuber isoform (BEII). Simultaneous downregulation of both isoforms (Schwall et al., 2000) resulted in a substantial increase in amylose content, reaching as high as 75% of the granule. Curiously, there appears to be a correlation between starch phosphorylation and its division in potato tubers during the process of starch production. The simultaneously suppression of GWD and BEs resulted in a 90% increase in amylose levels compared to the suppression of BEs (Lorberth, 1996; Uwer et al., 2006). An appreciable rise in the phosphate concentration of the starch was noted when the activity of the BE was suppressed (Uwer et al., 2006).

4.8 GENETIC MODIFICATION

An effective strategy involves utilising transgenic technology to specifically target the enzymes responsible for starch production. After the crop is harvested, it may undergo chemical or enzymatic modifications that have the potential to negatively impact the ecosystem. Johnson et al. (1999) and Neelam et al. (2012) both acknowledge that both conventional plant breeding approaches and biotechnology can be employed to induce genetic alterations. Starch synthetase I (SSI), GBSS, and starch synthetase III (SSIII) were simultaneously downregulated to synthesise starch without amylose. Consequently, the utilisation of waxy starch led to the production of shorter amylopectin chains. A genetically modified type of maize, wheat, sorghum, or amaranth is commonly known as a mutant maize. This starchy substance possesses numerous applications in the food sector due to its rapid gelatinisation ability, enabling the creation of transparent pastes that do not solidify (Neelam et al., 2012).

Cereals with elevated amylose starch content result from a genetic alteration in the starch branching enzyme (SBE) IIB, also referred to as amylose extender (ae). Scientists employed antisense technology to identify and suppress the specific potato gene responsible for tuber disintegration. This discovery enabled the creation of starch with a somewhat higher amylose concentration (Jobling et al., 1999). However, the presence of SBEI and SBEII enzymes is necessary to restrict the formation of starches that have an amylose concentration over 60% (Schwall et al., 2000). A more advanced method was employed to produce starches with increased amylose content by inhibiting the function of a gene using SBII antibodies. The derivative is a crucial precursor for the creation of enzyme-resistant polymers, which are utilised for the specific transportation of medicines to the colon. Another valuable type of high-amylose starch is known as resistant starch (Bird et al., 2000; Ogidi & Felagha, 2024).

Different enzymes, such as starch synthetase, ramifying enzymes, and debranching enzymes, each possessing a unique shape, control the arrangement of modified amylopectin. Due to the structural alteration of amylopectin, the activity of a certain enzyme is hindered, resulting in the inability to generate any novel features of amylopectin. Suppressing the SSII and SSIII isoforms in certain rice plants leads to a reduction in the gelatinisation of functional derivatives at temperatures lower than 50°C (Neelam et al., 2012).

4.9 APPLICATION OF ENZYMATIC MODIFIED STARCH IN FOOD SYSTEMS

Enzymatic modification of starch is a significant technological advancement in the food industry. Starch, a polysaccharide commonly derived from corn, potato, wheat, or tapioca, is widely used in food systems for its functional properties, such as thickening, gelling, and stabilising packaging (Gupta et al., 2025a). However, native starches often have limitations, such as susceptibility to retrogradation, poor solubility, and instability under extreme processing conditions. Enzymatic modification offers a targeted approach to overcome these challenges by altering the structure and properties of starch, thereby enhancing its functionality in food applications. Some of the functional properties and applications are discussed below.

4.9.1 Improved Textural Properties

Enzymatically modified starches are commonly used as thickeners and stabilisers in products such as soups, sauces, and dressings. By enzymatically hydrolysing starch, smaller molecular weight fragments are produced, leading to improved solubility and a smoother texture. For instance, amylase-treated starches are particularly effective in creating clear and stable sauces without cloudiness (Zhang et al., 2020c).

4.9.2 Enhanced Stability

Enzymatic treatments can enhance the thermal and freeze–thaw stability of starches, making them suitable for frozen and refrigerated food products. For example, pullulanase and isoamylase are used to debranch amylopectin, which reduces retrogradation tendencies and prevents syneresis in frozen desserts and pie fillings (Li et al., 2019c).

4.9.3 Calorie Reduction in Foods

Resistant starch, a product of enzymatic modification, is increasingly used in low-calorie foods due to its limited digestibility. It functions as a dietary fibre and is incorporated into baked goods, snacks, and breakfast cereals to improve fibre content and reduce glycaemic response (Sharma et al., 2021b).

4.9.4 Improved Gelling Properties

Enzymatically modified starch can also be used to improve the gelling properties of food products. For example, starch treated with glucosyltransferases can form gels with higher clarity and elasticity, making them ideal for confectionery applications (Singh et al., 2018c).

4.9.5 Clean-Label Products

Enzymatic modifications are considered a "clean-label" alternative to chemical modification, addressing consumer demand for natural and minimally processed ingredients. These starches are used in organic or clean-label food products, such as baby foods and dairy alternatives (Zhang et al., 2020d).

4.10 PROSPECTS FOR THE FUTURE

When determining the optimal methodology for modifying starch, it is crucial to take into account not only the outcomes achieved through different methods, but also the advantages and disadvantages associated with each treatment. The treatments and outcomes documented in the literature were derived from limited-scale trials conducted in controlled settings. Hence, it is imperative to examine the viability of expanding or implementing a continuous process utilising these approaches, disregarding the advantages of these approaches for starch modification over conventional procedures. It is vital to gather data on energy consumption and large-scale construction methods, as well as other factors related to cost estimation, in order to facilitate future planning. Assessing the technological feasibility is crucial as well. The starch sector is consistently growing, and its adaptability is being improved through modification techniques. Furthermore, there is an increasing demand for ecologically sustainable approaches that have minimal negative impact on the environment. Thus, the starch industry will undoubtedly gain advantages from the previously described environmentally friendly initiatives, which are appealing substitutes (Neelam et al., 2012).

4.11 CONCLUSION

In recent years, the starch business has experienced significant growth, leading to an abundance of modification techniques that have enhanced the versatility of the product. Starch can be altered chemically or physically to meet a wide range of needs in the food and non-food industries. It is possible to enhance the utilisation of non-traditional starches and vegetable byproducts that contain starch in various industrial applications by modifying the starches using physical, chemical, genetic, or enzymatic methods. Unconventional starches can be substituted for regular starch in industrial operations, either completely or partially, depending on their availability and cost. In order to choose the most suitable starch modification technique, it is crucial to identify the specific property needs of each operation. When making this decision, it is important to consider the following factors: structural qualities, availability, market trends, and application requirements. Continuous advancements in starch modification are constantly occurring. The value of starch can be enhanced by altering its structure through various methods, resulting in a functional product that possesses the necessary properties for specific industrial applications.

Derivation processes, such as esterification, etherification, cross-linking, and cationisation, are commonly employed to chemically alter starch. These approaches can modify the hydroxyl groups present in starch. Dual modification is an emerging strategy that combines physical and chemical processes. The molecular mass of the material is decreased by employing enzyme treatment. In addition, enzymes are used to eliminate amylopectin branches. Plant genetic modification can be achieved by the application of biotechnology or conventional plant breeding techniques. Starch can undergo a physical alteration due to this process, resulting in a reduction in its granule size and an increase in its solubility in water. Native grains can be subjected to various combinations of pressure, shear, temperature/moisture, and irradiation for physical treatment. Altering any physical object is a simple, cost-effective, and secure process. These methods of transformation are favoured since chemical or biological substances can pose a risk to one's health.

REFERENCES

Ai, Y., & Jane, J. I. (2018). Understanding starch structure and functionality. In M., Sjöö, and L. Nilsson (Eds.), *Starch in food* (2nd ed.) (pp. 151–178). Woodhead Publishing.

Amaraweera, S. M., Gunathilake, C., & Gunawardene, O. H. (2021). Development of starch-based materials using current modification techniques and their applications: A review. *Molecules*, *26*(22), 6880.

Ao, Z., Sinsek, S., Zhang, G., & Venkatachalam, M. (2007). Starch with a slow digestion property produced by altering its chain length, branch chain density, and crystalline structure. *Journal of Agricultural and Food Chemistry*, *55*, 4540–4547.

Bangar, S. P., Ashogbon, A. O., Singh, A., Chaudhary, V., & Whiteside, W. S. (2022). Enzymatic modification of starch: A green approach for starch applications. *Carbohydrate Polymers*, *287*, 119265.

Bertoft, E. (2013). On the building block and backbone concepts of amylopectin structure. *Cereal Chemistry*, *90*(4), 294–311.

Bird, A. R., Brown, I. L., & Topping, D. L. (2000). Starches, resistant starches, the gut microflora and human health. *Current Issues in Intestinal Microbiology*, *1*(1), 25–37.

Carciofi, M., Blennow, A., Jensen, S. L., Shaik, S. S., Henriksen, A., Buléon, A., Holm, P. B., & Hebelstrup, K. H. (2012). Concerted suppression of all starch branching enzyme genes in barley produces amylose-only starch granules. *BMC Plant Biology*, *12*, 223. https://doi.org/10.1186/1471-2229-12-223

Ceballos, H., Sánchez, T., Morante, N., Fregene, M., Dufour, D., Smith, A. M., Denyer, K., Pérez, J. C., Calle, F., & Mestres, C. (2007). Discovery of an amylose-free starch mutant in cassava (*Manihot esculenta* Crantz). *Journal of Agricultural and Food Chemistry*, *55*(18), 7469–7476. https://doi.org/10.1021/jf070633y

Chorfa, N., Nlandu, H., Belkacemi, K., & Hamoudi, S. (2022). Physical and enzymatic hydrolysis modifications of potato starch granules. *Polymers*, *14*(10), 2027. https://doi.org/10.3390/polym14102027

Chu, S., Hasjim, J., Hickey, L. T., Fox, G., & Gilbert, R. G. (2014). Structural changes of starch molecules in barley grains during germination. *Cereal Chemistry*, *91*(5), 431–437.

Dukare, A. S., Arputharaj, A., Bharimalla, A. K., Saxena, S., & Vigneshwaran, N. (2021). Nanostarch production by enzymatic hydrolysis of cereal and tuber starches. *Carbohydrate Polymer Technologies and Applications*, *2*, 100121.

Fan, Y., & Picchioni, F. (2020). Modification of starch: A review on the application of "green" solvents and controlled functionalization. *Carbohydrate Polymers*, *241*, 116350.

Geigenberger, P., Stamme, C., Tjaden, J., Schulz, A., Quick, P. W., Betsche, T., Kersting, H. J., & Neuhaus, H. E. (2001). Tuber physiology and properties of starch from tubers of transgenic potato plants with altered plastidic adenylate transporter activity. *Plant Physiology*, *125*(4), 1667–1678. https://doi.org/10.1104/pp.125.4.1667

Gupta, R. K., Guha, P., & Srivastav, P. P. (2022). Natural polymers in bio-degradable/edible film: A review on environmental concerns, cold plasma technology and nanotechnology application on food packaging-A recent trends. *Food Chemistry Advances*, *1*, 100135. https://doi.org/10.1016/j.focha.2022.100135

Gupta, R. K., Guha, P., & Srivastav, P. P. (2023). Effect of high voltage dielectric barrier discharge (DBD) atmospheric cold plasma treatment on physicochemical and functional properties of taro (Colocasia esculenta) starch. *International Journal of Biological Macromolecules*, *253*, 126772. https://doi.org/10.1016/j.ijbiomac.2023.126772

Gupta, R. K., Guha, P., & Srivastav, P. P. (2024a). Dielectric barrier discharge plasma: A green and novel method to change the structure of taro peel starch and improve the physicochemical properties of taro peel starch films. *Plasma Processes and Polymers*, *21*(9), 2400047. https://doi.org/10.1002/ppap.202400047

Gupta, R. K., Guha, P., & Srivastav, P. P. (2024b). Exploring the potential of taro (Colocasia esculenta) starch: Recent developments in modification, health benefits, and food industry applications. *Food Bioengineering*, *3*(3), 365–379. https://doi.org/10.1002/fbe2.12103

Gupta, R. K., Guha, P., & Srivastav, P. P. (2024c). Insights into the physicochemical, crystalline, multistructural, thermal and digestibility behavior of taro starch (Colocasia esculenta) modified by cold plasma treatment with high voltage dielectric barrier discharge (HVDBD): An alternative technique for chemical modification of starch and environmental sustainability. *Journal of Cleaner Production*, *461*, 142710. https://doi.org/10.1016/j.jclepro.2024.142710

Gupta, R. K., Guha, P., & Srivastav, P. P. (2024d). Physical action of nonthermal cold plasma technology for starch modification. *Food Physics*, *1*, 100011. https://doi.org/10.1016/j.foodp.2024.100011

Gupta, R. K., Guha, P., & Srivastav, P. P. (2024e). Effect of cold plasma treatment and plasma-activated water on physicochemical and structural properties of starch: A green and novel approach for environmental sustainability. *Plasma Processes and Polymers*, *21*(4), 2300204. https://doi.org/10.1002/ppap.202300204

Shrivastava, A., Gupta, R. K., & Srivastav, P. P. (2024). Exploring novel frontiers of advancements in purple yam (Dioscorea alata L.) starch extraction, modification, characterization, applications in food and other industries. *Measurement: Food*, *15*, 100196. https://doi.org/10.1016/j.meafoo.2024.100196

Gupta, R. K., Pipliya, S., Patel, J., Srivastav, P. P., Castro-Muñoz, R., Wang, C. K., & Nemţanu, M. R. (2025a). Essential Oil-Embedded Starch based Films for Active Packaging: Insights into Migration, Stability, and Preservation Efficiency. *Trends in Food Science & Technology*, 163, 105149. https://doi.org/10.1016/j.tifs.2025.105149

Gupta, R. K., Gnana Moorthy Eswaran, U., Kumar, S., Pipliya, S., Kaur, B., & Srivastav, P. P. (2025b). Synergistic Modification of Starch Polysaccharide via Sequential Treatments with Ultrasonic and Cold Plasma Technology. *Journal of Food Processing and Preservation*, *2025*(1), 9063007. https://doi.org/10.1155/jfpp/9063007

Harder, H., Khol-Parisini, A., & Zebeli, Q. (2015). Treatments with organic acids and pullulanase differently affect resistant starch and fiber composition in flour of various barley genotypes (*Hordeum vulgare* L). *Starch - Stärke*, *67*(5-6), 512–520.

He, L., Park, S. H., Hai Dang, N. D., Duong, H. X., Duong, T. P. C., Tran, P. L., Park, J. T., Ni, L., & Park, K. H. (2017). Characterization and thermal inactivation kinetics of highly thermostable ramie leaf β-amylase. *Enzyme and Microbial Technology*, *101*, 17–23. https://doi.org/10.1016/j.enzmictec.2017.02.011

Hutabarat, D. J. C., & Stevensen, J. (2023). Physicochemical properties of enzymatically modified starch: A review. *IOP Conference Series: Earth and Environmental Science*, *1169*(1), 012093.

Ihemere, U., Arias-Garzon, D., Lawrence, S., & Sayre, R. (2006). Genetic modification of cassava for enhanced starch production. *Plant Biotechnology Journal*, *4*(4), 453–465. https://doi.org/10.1111/j.1467-7652.2006.00195.x

Iyiola, A. O., Kolawole, A. S., Ajayi, F. O., Ogidi, O. I., & Ogwu, M. C. (2023). Sustainable water use and management for agricultural transformation in Africa. In Madhav, S., Srivastav, A. L., Izah, S. C., van Hullebusch, E. D. (eds) *Water resources management for rural development: Challenges and mitigation* (pp. 287–300). Elsevier. https://doi.org/10.1016/B978-0-443-18778-0.00001-5

Jane, J., Chen, Y. Y., & Lee, L. F. (1999). Effects of amylopectin branch chain length and amylose content on the gelatinization and pasting properties of starch. *Cereal Chemistry*, *76*, 629–637.

Jobling, S. A., Schwall, G. P., Westcott, R. J., Sidebottom, C. M., Debet, M., Gidley, M. J., Jeffcoat, R., & Safford, R. (1999). A minor form of starch branching enzyme in potato (*Solanum tuberosum* L.) tubers has a major effect on starch structure: Cloning and characterisation of multiple forms of SBE A. *The Plant Journal*, *18*(2), 163–171. https://doi.org/10.1046/j.1365-313x.1999.00441.x

Johnson, L. A., Baumel, C. P., Hardy, C. L., & White, P. J. (1999). *Identifying valuable corn quality traits for starch production. A project of the Iowa Grain Quality Initiative Traits Task Team.* Center for Crops Utilization Research, Iowa Agriculture & Home Economics Experiment Station, Iowa State University.

Kale, R. V., Sontakke, M. D., Raut, G. S., & Chavan, V. R. (2020). Use of enzyme modified sweet potato starch in formulation of ice cream. *International Journal of Chemical Studies*, *8*(4), 3002–3008.

Karaki, N., Aljawish, A., Humeau, C., Muniglia, L., & Jasniewski, J. (2016). Enzymatic modification of polysaccharides: Mechanisms, properties, and potential applications: A review. *Enzyme and Microbial Technology*, *90*, 1–18.

Kaur, L., Singh, J., & Sodhi, N. S. (2020). Enzymatic treatments and their impact on the physicochemical properties of starch. *Food Hydrocolloids*, *109*, 106122.

Kelly, R. M., Dijkhuizen, L., & Leemhuis, H. (2009). Starch and α-glucan acting enzymes, modulating their properties by directed evolution. *Journal of Biotechnology*, *140*, 184–193.

Kim, B. S., Kim, H. S., & Yoo, S. H. (2015). Characterization of enzymatically modified rice and barley starches with amylosucrase at scale-up production. *Carbohydrate Polymers*, *125*, 61–68.

Kim, H., Choi, S., Park, C. S., & Moon, T. W. (2016). Low digestion property of amylosucrase-modified waxy adlay starch. *Food Science and Biotechnology*, *25*, 457–460.

Kim, Y. L., Mun, S., Park, K. H., Shim, J. Y., & Kim, Y. R. (2013). Physicochemical functionality of 4-α-glucanotransferase-treated rice flour in food application. *International Journal of Biological Macromolecules*, *60*, 422–426.

Kossmann, J. (1992). Klonierung und funktionelle Analyze kodierend fuer am Kohlenhydratstoffwechsel der Kartoffel beteiligte Proteine. PhD thesis. Technical University of Berlin. 58 pp.

Kossmann, J., & Lloyd, J. (2000). Understanding and influencing starch biochemistry. *Critical Reviews in Biochemistry and Molecular Biology*, *19*, 171–226.

Kumar, P., Kumar, V., & Sharma, S. (2019). Enzymatic modification of starch: A review. *Journal of Food Science and Technology*, *56*(2), 531–542.

Lanahan, M. B., Basu, S. S., Batie, C. J., Chen, W., Craig, J., & Kinkema, M. (2006). *Self-processing plants and plant parts.* U.S. Patent No. 7,102,057 B2. Assigned to Syngenta Participations AG; issued September 5, 2006. Priority filed August 27, 2001.

Lee, C. K., Le, Q. T., Kim, Y. H., Shim, J. H., Lee, S. J., Park, J. H., Lee, K. P., Song, S. H., Auh, J. H., Lee, S. J., & Park, K. H. (2008). Enzymatic synthesis and properties of highly branched rice starch amylose and amylopectin cluster. *Journal of Agricultural and Food Chemistry*, *56*(1), 126–131. https://doi.org/10.1021/jf072508s

Lee, S. M., Lee, J., & Kim, B. Y. (2020). Effects of pullulanase on the physicochemical properties of starch. *Starch - Stärke*, *72*(1-2), 1900095. https://doi.org/10.1002/star.201900095

Leyva-López, R., Palma-Rodríguez, H. M., LópezTorres, A., Capataz-Tafur, J., Bello-Pérez, L. A., & Vargas-Torres, A. (2019). Use of enzymatically modified starch in the microencapsulation of ascorbic acid: Microcapsule characterization, release behavior and in vitro digestion. *Food Hydrocolloids*, *96*, 259–266.

Li, J., Zhang, H., & Chen, Y. (2019a). Enzymatic modification of starch and its application in food systems: A review. *Journal of Food Science and Technology*, *56*(5), 2005–2016.

Li, K., Zhu, L., Li, H. Zhu, Y., Pan, C., Gao, X., & Wei, L. (2019b). Structural characterization and rheological properties of a pectin with anti-constipation activitiy from the roots of

Arctium lappa L. *Carbohydrate Polymers*, *215*, 119–129. https://doi.org/10.1016/j.carbpol.2019.03.051

Li, Y., Xu, J., Zhang, L., Ding, Z., Gu, Z., & Shi, G. (2017). Investigation of debranching pattern of a thermostable isoamylase and its application for the production of resistant starch. *Carbohydrate Research*, *446-447*, 93–100. https://doi.org/10.1016/j.carres.2017.05.016

Liu, Y., Jiang, F., Du, C., Li, M., Leng, Z., Yu, X., & Du, S. K. (2022). Optimization of corn resistant starch preparation by dual enzymatic modification using response surface methodology and its physicochemical characterization. *Foods (Basel, Switzerland)*, *11*(15), 2223.

Lorberth, R. (1996). Charakterisierung von RL1:ein neues Enzym des Staerkemetabolismus. PhD thesis. Free University Berlin. 86 pp.

Lorberth, R., Ritte, G., Willmitzer, L., & Kossmann, J. (1998). Inhibition of a starch-granule-bound protein leads to modified starch and repression of cold sweetening. *Nature Biotechnology*, *16*(5), 473–477. https://doi.org/10.1038/nbt0598-473

Maniglia, B. C., Castanha, N., Le-Bail, P., Le-Bail, A., & Augusto, P. E. D. (2021). Starch modification through environmentally friendly alternatives: A review. *Critical Reviews in Food Science and Nutrition*, *61*(15), 2482–2505. https://doi.org/10.1080/10408398.2020.1778633

Masina, N., Choonara, Y. E., Kumar, P., du Toit, L. C., Govender, M., Indermun, S., & Pillay, V. (2017). A review of the chemical modification techniques of starch. *Carbohydrate Polymers*, *157*, 1226–1236. https://doi.org/10.1016/j.carbpol.2016.09.094

Miao, M., Li, R., Huang, C., Jiang, B., & Zhang, T. (2015). Impact of β-amylase degradation on properties of sugary maize soluble starch particles. *Food Chemistry*, *177*, 1–7. https://doi.org/10.1016/j.foodchem.2014.12.101

Mottram, D. S., Wedzicha, B. L., & Dodson, A. T. (2002). Acrylamide is formed in the Maillard reaction. *Nature*, *419*(6906), 448–449. https://doi.org/10.1038/419448a

Mun, S., Kim, Y. L., Kang, C. G., Park, K. H., Shim, J. Y., & Kim, Y. R. (2009). Development of reduced-fat mayonnaise using 4αGTase-modified rice starch and xanthan gum. *International Journal of Biological Macromolecules*, *44*(5), 400–407. https://doi.org/10.1016/j.ijbiomac.2009.02.008

Naguleswaran, S., Vasanthan, T., Hoover, R., & Bressler, D. (2014). Amylolysis of amylopectin and amylose isolated from wheat, triticale, corn and barley starches. *Food Hydrocolloids*, *35*, 686–693.

Neelam, K., Vijay, S., & Lalit, S. (2012). Various techniques for the modification of starch and the applications of its derivatives. *International Research Journal of Pharmacy*, *3*(5), 25–31.

Nguyen, D. H. D., Tran, P. L., & Li, D. (2014). Modification of rice grain starch for lump-free cooked rice using thermostable disproportionating enzymes. *Food Research International*, *63*, 55–56.

Obadi, M., & Xu, B. (2021). Review on the physicochemical properties, modifications, and applications of starches and its common modified forms used in noodle products. *Food Hydrocolloids*, *112*, 106286.

Ogidi, O. I., & Akpan, U. M. (2023). Impacts of chemical use in agricultural practices: Perspectives of soil microorganism and vegetation. In M. C. Ogwu & S. C. Izah (Eds.),*One health implications of agrochemicals and their sustainable alternative*. Springer Nature Singapore.

Ogidi, O. I., & Emaikwu, N. G. (2024). Plant phenolic compound isolation and its bioinformatics approaches of molecular mechanisms in antimicrobial activities and resistance. In P. D. Pathak, R. Raut, S. Jaramillo-Isaza, P. Borkar, & R. H. Jhaveri (Eds.), *Computational approaches in biomaterials, bioinformatics and biomedical engineering applications*. CRC Press, Taylor and Francis Group.

Ogidi, O. I., & Felagha, I. (2024). Gut microbiome and diet interaction. In In S. Patnaik, A. M. Hamad, D. Paul, P. K. Dutta, & M. Shafiq (Eds.), *Nutrition controversies and advances in autoimmune disease*. IGI Global. https://doi.org/10.4018/979-8-3693-5528-2.ch012

Ogidi, O. I., Izah, S. C., & Akpan, U. M. (2023). Food quality and agrochemical use: Integrated monitoring, assessment, and management policies. In M. C. Ogwu & S. C. Izah (Eds.), *One Health implications of agrochemicals and their sustainable alternative*. Springer Nature Singapore Pte Ltd.

Ogidi, O. I., Izah, S. C., & Tobia, P. S. (2024). Biosurfactant production from cassava processing wastes. In M. C. Ogwu, S. C. Izah, A. A. C. Alves, & S. C. Babu (Eds.), *Sustainable cassava: Strategies from production through waste management*. Elsevier. pp. 419–438.

Okereke, J. N., Nduka, J. N., Adanma, U. A., & Ogidi, O. I. (2020). Heavy metals in cassava (*Manihot esculenta* Crantz) harvested from farmlands along highways in Owerri, Nigeria. *Turkish Journal of Agriculture—Food Science and Technology*, *8*(4), 800–806. https://doi.org/10.24925/turjaf.v8i4.800-806.2737

Onyango, C. (2016). Starch and modified starch in bread making: A review. *African Journal of Food Science*, *10*(12), 344–351.

Palomo, M., Kralj, S., van der Maarel, M. J., & Dijkhuizen, L. (2009). The unique branching patterns of *Deinococcus* glycogen branching enzymes are determined by their N-terminal domains. *Applied and Environmental Microbiology*, *75*(5), 1355–1362. https://doi.org/10.1128/AEM.02141-08

Park, E. Y., Ma, J. G., & Kim, J. (2018a). Effect of dual modification of HMT and crosslinking on physicochemical properties and digestibility of waxy maize starch. *Food Hydrocolloids*, *75*, 33–40.

Park, J. H., Kim, J. H., & Lee, Y. B. (2018b). Enzymatic modification of starch using glucoamylase. *Journal of Food Engineering*, *223*, 137–145.

Park, S., & Kim, Y. R. (2021). Clean label starch: Production, physicochemical characteristics, and industrial applications. *Food Science and Biotechnology*, *30*, 1–17.

Preiss, J. (2018). Plant starch synthesis. In M. Sjöö & L. Nilsson (Eds.), *Starch in food: Structure, function and applications* (2nd ed.) (pp. 3–95). Woodhead Publishing.

Raemakers, K., Schreuder, M., & Suurs, L. (2005). Improved cassava starch by antisense inhibition of granule-bound starch synthase I. *Molecular Breeding*, *16*, 163–72.

Rommens, C. M., Haring, M. A., Swords, K., Davies, H. V., & Belknap, W. R. (2007). The intragenic approach as a new extension to traditional plant breeding. *Trends in Plant Science*, *12*(9), 397–403. https://doi.org/10.1016/j.tplants.2007.08.001

Safford,R.,Jobling,S.A.,&Sidebottom,C.M.(1998).Consequences of antisense RNA inhibition of starch branching enzyme activity on properties of potato starch. *Carbohydrate Polymers*, *35*, 55–68.

Schmiele, M., Sampaio, U. M., Gomes, P. T. G., & Clerici, M. T. P. S. (2019). *Physical modifications of starch*. In M. T. P. S. Clerici & M. Schmiele (Eds.), *Starches for food application* (pp. 223–269). Academic Press.

Schwall, G. P., Safford, R., Westcott, R. J., Jeffcoat, R., Tayal, A., Shi, Y. C., Gidley, M. J., & Jobling, S. A. (2000). Production of very-high-amylose potato starch by inhibition of SBE A and B. *Nature Biotechnology*, *18*(5), 551–554. https://doi.org/10.1038/75427

Sharma, A., Yadav, B. S., & Ritika, R. (2021a). Enzyme-modified starch and its functionality: A systematic review. *International Journal of Biological Macromolecules*, 178, 251–259. https://doi.org/10.1016/j.ijbiomac.2021.04.103

Sharma, A., Yadav, B. S., & Ritika, R. (2021b). Nanoscale insights into enzymatic hydrolysis of starch: Application of AFM and FTIR. *International Journal of Biological Macromolecules*, *178*, 152–159. https://doi.org/10.1016/j.ijbiomac.2021.05.104

Sharma, A., Yadav, B. S., & Ritika, R. (2021c). Physicochemical and functional properties of enzymatically modified starches: A comprehensive review. *International Journal of Biological Macromolecules*, 178, 152–159. https://doi.org/10.1016/j.ijbiomac.2021.05.104

Sharma, A., Yadav, B. S., & Ritika, R. (2021d). Resistant starch: Physiological roles and food applications. *Journal of Functional Foods*, 83, 104557. https://doi.org/10.1016/j.jff.2021.104557

Shim, J. H., Seo, N. S., Roh, S. A., Kim, J. W., Cha, H. J., & Park, K. H. (2007). Improved bread-baking process using *Saccharomyces cerevisiae* displayed with engineered cyclodextrin glucanotransferase. *Journal of Agricultural and Food Chemistry*, *55*, 4735–4740.

Shin, H. J., Choi, S. J., Park, C. S., & Moon, T. W. (2010). Preparation of starches with low glycaemic response using amylosucrase and their physicochemical properties. *Carbohydrate Polymers*, *82*, 489–497.

Shinohara, M. L., Ihara, M., Abo, M., Hashida, M., Takagi, S., & Beck, T. C. (2001). A novel thermostable branching enzyme from an extremely thermophilic bacterial species, *Rhodothermus obamensis*. *Applied Microbiology and Biotechnology*, *57*, 653–659.

Singh, N., Kaur, L., & Sodhi, N. S. (2018a). Enhancing gelling properties of starches using enzymatic modification: Current trends and future directions. *Food Hydrocolloids*, *81*, 449–456. https://doi.org/10.1016/j.foodhyd.2017.12.025

Singh, N., Kaur, L., & Sodhi, N. S. (2018b). Functional and physicochemical characterization of starch modified enzymatically. *Food Hydrocolloids*, *81*, 449–456. https://doi.org/10.1016/j.foodhyd.2017.12.025

Singh, N., Kaur, L., & Sodhi, N. S. (2018c). Functional characteristics of starch enzymatically modified for food applications. *Starch - Stärke*, *70*(5–6), 1800154.

Singh, N., Kaur, L., & Sodhi, N. S. (2018d). Structural characterization of starch granules modified enzymatically: Insights from XRD and FTIR. *Food Hydrocolloids*, *81*, 449–456. https://doi.org/10.1016/j.foodhyd.2017.12.025

Singh, S., Singh, N., & Ezekiel, R. (2017). Enzymatic modification of starch: A review. *J. Food Sci. Techno.*, *54*(4), 931–943. https://doi.org/

Smith, A. M. (2008). Prospects for increasing starch and sucrose yields for bioethanol production. *The Plant Journal*, *54*, 546–58.

Sopawong, P., Warodomwichit, D., Srichamnong, W., Methacanon, P., & Tangsuphoom, N. (2022). Effect of physical and enzymatic modifications on composition, properties and in vitro starch digestibility of sacred lotus (*Nelumbo nucifera*) seed flour. *Foods*, *11*(16), 2473.

Sorndech, W., Sagnelli, D., & Meier, S. (2016). Structure of branching enzyme- and amylomaltase modified starch produced from well-defined amylose to amylopectin substrates. *Carbohydrate Polymers*, *152*, 51–61.

Stark, D. M., Timmerman, K. P., Barry, G. F., Preiss, J., & Kishore, G. M. (1992). Regulation of the amount of starch in plant tissues by ADP-glucose pyrophosphorylase. *Science*, *258*, 7–92.

Sweetlove, L. J., Burrell, M. M., & Rees, T. (1996). Starch metabolism in tubers of transgenic potato (*Solanum tuberosum*) with increased ADPglucose pyrophosphorylase. *Biochemical Journal*, *320*, 493–98.

Takata, H., Ohdan, K., Takaha, T., Kuriki, T., & Okada, S. (2003). Properties of branching enzyme from hyperthermophilic bacterium, *Aquifex aeolicus*, and its potential for production of highly-branched cyclic dextrin. *Journal of Applied Glycoscience*, *50*, 15–20.

Tallberg, A., Hofvander, P., Persson, P. T., & Wikstrom, O. (1998). Genetically engineered modification of potato to obtain amylopectin-type starch. US Patent application no. US 5824798.

Tjaden, J., Mohlmann, T., Kampfenkel, K., Henrichs, G., & Neuhaus, H. E. (1998). Altered plastidic ATP/ADP-transporter activity influences potato (*Solanum tuberosum* L.) tuber morphology, yield and composition of tuber starch. *The Plant Journal*, *16*, 531–40.

Ulbrich, M., Asiri, S. A., Bussert, R., & Flöter, E. (2021). Enzymatic modification of granular potato starch using isoamylase: Investigation of morphological, physicochemical, molecular, and techno-functional properties. *Starch - Stärke*, *73*(1-2), 2000080.

Uwer, U., Frohberg, C., Pilling, J., & Landschuetze, V. (2006). Transgenic plants synthesizing high amylose starch. US Patent application no. US 7112718 B2.

van der Maarel, M. J. E. C., Vos, A., Sanders, P., & Dijkhuizen, L. (2003). Properties of the glucan branching enzyme of the hyperthermophilic bacterium *Aquifex aeolicus*. *Biocatalysis and Biotransformation*, *21*, 199–207.

Visser, R. G. F., Somhorst, I., Kuipers, G. J., Ruys, N. J., Feenstra, W. J., & Jacobsen, E. (1991). Inhibition of the expression of the gene for granule-bound starch synthase in potato by antisense constructs. *Molecular and General Genetics MGG*, *225*, 289–96.

Vitolo, M. (2020). Enzymatic modification of starch. *World Journal of Pharmacy and Pharmaceutical Sciences*, *9*(4), 1341.

Wang, S., Wang, J., Liu, Y., & Liu, X. (2020). Starch modification and application. In S. Wang (Ed.), *Starch structure, functionality and application in foods* (pp. 131–149). Springer Singapore.

Wang, Z. Y., Chen, X. P., & Wang, J. H. (2007). Increasing maize seed weight by enhancing the cytoplasmic ADP-glucose pyrophosphorylase activity in transgenic maize plants. *Plant Cell, Tissue and Organ Culture*, *88*, 83–92.

Wursch, P., & Gummy, D. (1994). Inhibition of amylopectin retrogradation by partial beta-amylolysis. *Carbohydrate Research*, *256*, 129–137.

Xu, Y., Huang, Q., Fu, X., & Jane, J. (2015). Modification of starch octenylsuccinate by β-amylase hydrolysis in order to increase its emulsification properties. *Food Hydrocolloids*, *48*, 55–61.

Yao, Y., Zhang, J., & Ding, X. (2003). Partial β-amylolysis retards starch retrogradation in rice products. *Journal of Agricultural and Food Chemistry*, *51*, 4066–4071.

Zhang, B., Chen, Y., & Li, Z. (2020a). Clean-label starches: Enzymatic modification and applications in food systems. *Trends in Food Science & Technology*, *96*, 1–10. https://doi.org/10.1016/j.tifs.2019.12.004

Zhang, B., Chen, Y., & Li, Z. (2020b). Influence of enzymatic hydrolysis on the techno-functional properties of starch: A systematic review. *Trends in Food Science & Technology*, *96*, 1–10. https://doi.org/10.1016/j.tifs.2019.12.004

Zhang, B., Chen, Y., & Li, Z. (2020c). Morphological and structural changes in starch granules subjected to enzymatic hydrolysis: SEM, XRD, and FTIR studies. *Trends in Food Science & Technology*, *96*, 1–10. https://doi.org/10.1016/j.tifs.2019.12.004

Zhang, B., Chen, Y., & Li, Z. (2020d). The role of enzymatic modification in enhancing starch functionality: A comprehensive review. *Trends in Food Science & Technology*, *96*, 50–68. https://doi.org/10.1016/j.tifs.2019.11.012

Zhang, Z., & Bao, J. (2023). Recent advances in modification approaches, health benefits, and food applications of resistant starch. *Starch - Stärke*, *75*(9–10), 2100141.

Zhong, Y., Bertoft, E., Li, Z., Blennow, A., & Liu, X. (2020). Amylopectin starch granule lamellar structure as deduced from unit chain length data. *Food Hydrocolloids*, *108*, 106053.

Zhong, Y., Liang, W., Pu, H., Blennow, A., Liu, X., & Guo, D. (2019). Short-time microwave treatment affects the multi-scale structure and digestive properties of high-amylose maize starch. *International Journal of Biological Macromolecules*, *137*, 870–877.

Zhong, Y., Tian, Y., & Liu, X. (2021). Influence of microwave treatment on the structure and functionality of pure amylose and amylopectin systems. *Food Hydrocolloids*, *119*, 106856.

5 Nanotechnology in Starch Modification

Naveen Chandrakar
Bio-Research Laboratory, Rajendra Mishra School of Engineering Entrepreneurship, Indian Institute of Technology, Kharagpur, India

Shubhra Shekhar
School of Chemical Engineering, University of New South Wales, Sydney, NSW, Australia

Debapam Saha
Calcutta Institute of Technology, Uluberia, Howrah, West Bengal, India

Ayan Sarkar
Department of Agricultural and Food Engineering, Indian Institute of Technology Kharagpur, India

Debjyoti Acharya
Department of Microbiology, Vidyasagar University, Paschim Medinipur, West Bengal, India

Mamoni Banerjee
Bioresearch Laboratory, Rajendra Mishra School of Engineering Entrepreneurship, Indian Institute of Technology, Kharagpur, India

5.1 INTRODUCTION

5.1.1 Starch: A Versatile Biopolymer

Starch is the most abundant carbohydrate in plants and is widely used by animals; it is synthesized primarily through photosynthesis. It is the prime energy providing polysaccharides in human nutrition, as well as an effective raw material for several foods, pharmaceuticals, and packaging industries. Starch is mostly produced from tuber plants (e.g., potato, cassava), cereals (e.g., wheat, paddy, maize), pseudocereals (e.g., buckwheat, amaranth, chia), and millets (e.g., bajra, ragi, jowar) (Compart et al., 2021; Gupta et al., 2023; Saha et al., 2024; Sarkar et al., 2024). Structurally, starch consists of several simpler branching units of amylose (20%–25%) and amylopectin (75%–80%), which are polymers of D-glucopyranose, that are interconnected by α-1,4- and α-1,6-glycosidic linkages (Compart et al., 2021). Starch granules consist of alternating amorphous and crystalline layers, formed by amylose interacting with large amylopectin branches and by linked short-branched amylopectin, respectively. Their crystallinity typically ranges from 20% to 45%, depending on the amylose-to-amylopectin ratio, which varies with the starch's source and age (Gonçalves et al., 2024). Starch has become a highly versatile biopolymer owing to its biodegradability, low cost, and wide availability. In the food industry, starch plays a central role as a thickening agent, stabilizer, and fat replacer, enhancing texture and shelf life in numerous processed products (Baranwal et al., 2022). Its application extends into pharmaceuticals. Due to its biocompatibility and non-toxicity, it serves as a binder, disintegrant, and filler in tablet formulations (Gamage, Liyanapathiranage, et al., 2022; Rodrigues & Emeje, 2012). Also, in cosmetics, starch is favored for its absorbency, soft feel, and ability to deliver active compounds, making it a key component in powders, creams, and facial masks (Gupta et al., 2022). Apart from its uses in food and nutrition, starch-based biodegradable films and composites are extensively used in the production of packaging materials (Gupta et al., 2025a). They offer sustainable alternatives to synthetic chemical and plastic materials with acceptable barrier properties and compostability (Cheng et al., 2021).

All authors have contributed equally to this chapter.

DOI: 10.1201/9781003564287-5

5.1.2 Why Do You Mean by Starch Modification?

Starch modification is the physical, chemical, enzymatic, or genetic alteration of native starch to improve or tailor its functional properties, such as solubility, stability, viscosity, and resistance to retrogradation, for specific industrial applications (Amaraweera et al., 2021; Singh et al., 2007; Gupta et al., 2024c). Modification of starch is one of the crucial step after its extraction because the native starch possesses many limitations, as described by Wang and Ren (2020) and Zhang et al. (2014); these include the following:

- *Poor thermal stability:* Native starch undergoes degradation at high temperatures, limiting its use in heat-processed foods.
- *High retrogradation tendency:* Native starch easily recrystallizes during storage, causing undesirable texture changes in foods.
- *Low shear resistance:* Native starch breaks down under high shear, making it unsuitable for mechanical food processing.
- *Tendency to syneresis:* Water separates from gels upon storage, reducing shelf life and product quality.
- *Inconsistent pasting behavior:* Native starch presents difficulty in processing and incorporating in functional or novel food products.
- *Low processability in non-food industries:* Native starch lacks compatibility with hydrophobic polymers in packaging.

5.1.3 What Do You Mean by Nanotechnology? How Can It Modify Starch?

Food nanotechnology is an emerging discipline that presents exciting new prospects for the food sector. It focuses on the manipulation and application of materials at the nanoscale (1–100 nm), where unique properties develop due to increased surface area and quantum effects (Sekhon, 2010). In the context of starch, nanotechnology enables the development of starch nanoparticles (SNPs) or nanocomposites through processes such as acid hydrolysis, ultrasonication, high-pressure homogenization, or enzymatic treatment. These nano-sized starch materials show improved mechanical strength, water resistance, controlled release behavior, and improved functional properties, making them ideal for advanced applications in food packaging, pharmaceuticals, drug delivery, and biodegradable films. For example, SNPs have been shown to improve film-forming ability, thermal stability, and barrier properties when incorporated into bioplastics or used as nanocarriers for active compounds (Gamage, Thiviya, et al., 2022; Torres & De-la-Torre, 2022). Moreover, starch nanocomposites can be engineered to have antimicrobial, antioxidant, or controlled release properties, expanding their utility in active and intelligent packaging systems (Torres & De-la-Torre, 2022).

5.2 KEY HIGHLIGHTS OF THE CHAPTER

- Explains the fundamentals of nanotechnology and its importance in modifying starch
- Introduces starch-based nanomaterials and composites: their unique, ground-breaking features, properties, and uses
- Explains the various cutting-edge technological advances in their synthesis pathways
- Discusses the potential applications of these modified starches in industries
- Discusses current challenges, research prospects, and future opportunities in this emerging domain.

5.3 FUNDAMENTALS OF NANOTECHNOLOGY AND STARCH

Nanotechnology is the science and engineering of designing and manipulating of materials at the nanoscale level, typically ranging from 1 to 100 nm. This is the size where the matters exhibit enhanced properties and distinct properties compared to the bulk material (Khan et al., 2019). These distinct properties are governed by various factors, such as an increase in surface-area-to-volume ratios, quantum confinement effect, and dominance of material boundary properties. These properties in turn affect various optical, electrical, magnetic, thermal and catalytic behaviors (Joudeh & Linke, 2022). For instance, the optical property known as localized surface plasmon resonance (LSPR) is different in metallic nanoparticles from the bulk material. In this phenomenon, the light that interacts with the oscillating electron at the interface of the metal nanoparticle causes the excitation wavelength to be altered (Shekhar et al., 2023). Gold nanoparticles exhibit a deep ruby red color in solution, which is due to the phenomenon of surface plasmon. However, this is absent for the bulk gold (Altammar, 2023). In addition, nanomaterials have significant applications in domains such as electronics, medicine, food science, energy, and remediation due to their higher reactivity, strength, and conductivity. Furthermore, nanoscale size also enhances interactions in the formation of matrix materials or composites. Figure 5.1 illustrates the classification of nanohybrids into organic and inorganic types, highlighting their key applications in the biomedical and food sectors.

The synthesis of nanomaterials generally follows two distinct strategies: top-down and bottom-up.

1. *Top-down:* The top-down approach is a common one that uses mechanical force to break down larger bulk materials into the nanoscale range; methods include milling (ball milling), high-pressure homogenization, and ultrasonication. Some chemical and enzymatic hydrolysis methods are also used. The top-down approach has the advantages of being generally straightforward and cost-effective. However, a significant downside of

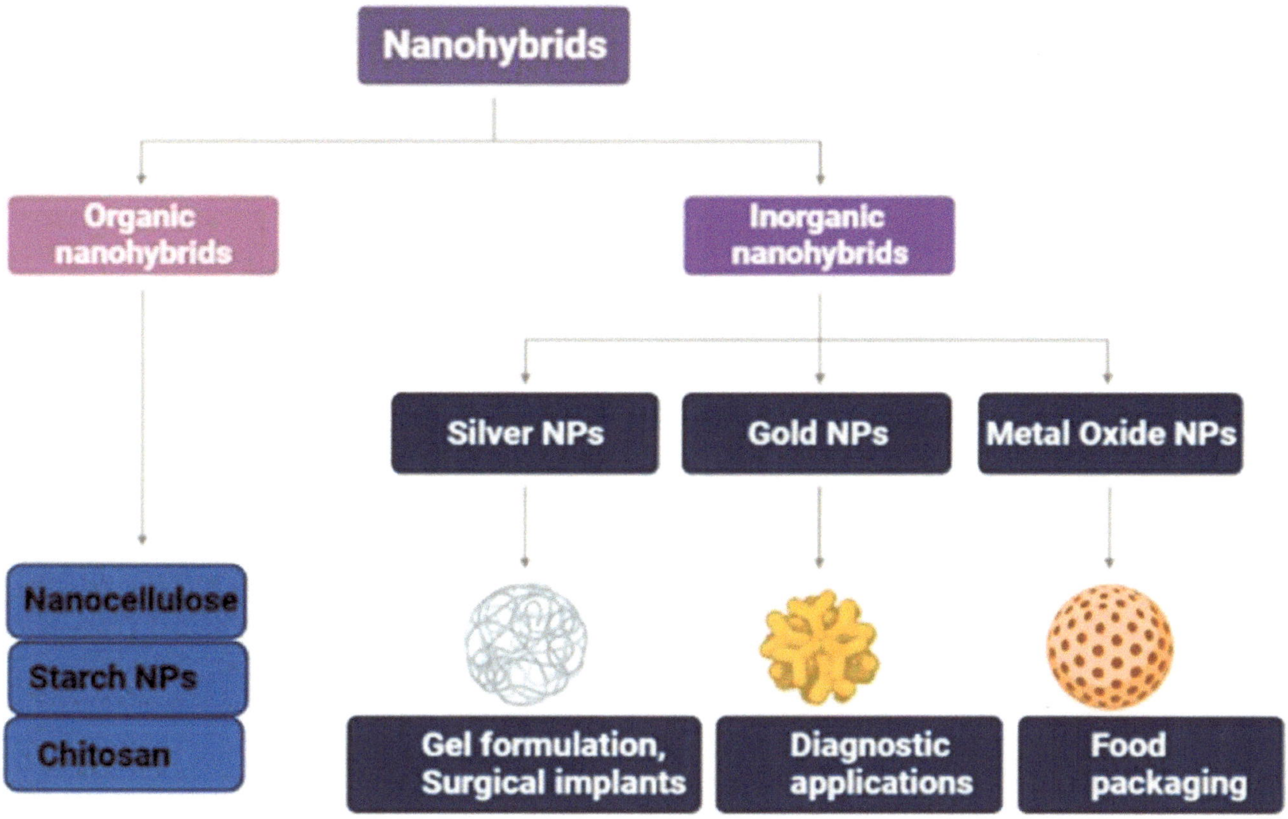

FIGURE 5.1 Flowchart of nanohybrids categorized as organic (nanocellulose, starch nanoparticles [NPs], chitosan) and inorganic (silver, gold, metal oxide NPs), with related applications (Das et al., 2025).

this approach is irregular shape, size, and surface structure control, which leads to lower efficiency and specificity (Phanthong et al., 2018).

2. *Bottom-up:* The bottom-up approach, as its name suggests, consists of constructing the material from the smaller unit of atoms, molecules, or oligomers into a nanostructure. This approach consists of methods such as nanoprecipitation (also known as antisolvent precipitation or solvent displacement), self-assembly, ionic gelation, and electrospinning. This approach is considered better than the top-down approach due to its controlled morphology. These benefits help researchers use nanoparticles of a specified shape and size for various applications (Jha et al., 2009).

5.4 STARCH-BASED NANOMATERIALS: TYPES AND SYNTHESIS

After cellulose, starch is the second most abundant plant-based biopolymer. Starch is one of the primary carbohydrate sources of energy storage in plants and is also a vital component of the human diet. Starch is a long chain of saccharides composed of a number of D-glucose units (Gupta et al., 2022; Shrivastava et al., 2024). These D-glucose units join using different linkages to make amylose and amylopectin. Amylose are linear chained polymer molecule made up of $\alpha(1\rightarrow4)$-linked glucose units. Amylose constitutes 20% to 30% of the common starch; however, the proportion of amylose varies significantly based on the plant source. Among the starches obtained from waxy maize starch, the amylose content is little to none, whereas in the case of high-amylose varieties, the content can be over 70% (Pfister & Zeeman, 2016). Amylose chains often show a twisted/spiral shape/helical conformation, are found in the amorphous region of starch granules, and are relatively resistant to enzymatic digestion (Kang et al., 2023; Gupta et al., 2024a). Another component of the starch is amylopectin. Amylopectin is a branched chain polymer constituting 70%–80% of most starches (depending significantly on the source) (Martens et al., 2018). It consists of short $\alpha(1\rightarrow4)$-linked glucose chains interconnected by $\alpha(1\rightarrow6)$-glycosidic bonds, forming a large, branched structure. Amylopectin, being branched and composed of a larger number of glucose units, has a much higher molecular weight than amylose. Due to the branched structure, amylopectin forms a double helical structure that is packed together to form a semi-crystalline region within the granules. Amylopectin is more digestible than amylose. However, amylose and amylopectin are incompatible for dispersal in aqueous solution during film formation under certain conditions (Cornejo-Ramírez et al., 2018; Gupta et al., 2025b). Table 5.1 compares synthesis methods for starch nanomaterials.

TABLE 5.1
Comparison of Synthesis Methods for Starch Nanomaterials

Starch Sample	Approach Type	Synthesis Method	Process Conditions	Key Findings	Figures, Illustrations, SEM Images	Ref.
Native corn starch	Top-down	US + HPH	US reduced size to ~340 nm; HPH to < 60 nm	Improved swelling, solubility, thermal stability; stable carotenoid nanoemulsion	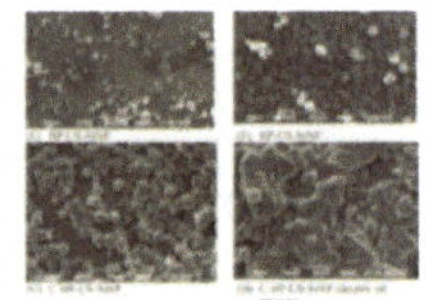	(Chutia & Mahanta, 2021)
Debranched starch (DBS) + β-cyclodextrin	Self-assembly	Ultrasonication + recrystallization	4 °C, varied ultrasound durations, different DBS polymerization degrees	Particle size control, high tangeretin loading (83.2%), 99.21% retention after 20 days, stable in simulated human microenvironments, non-toxic	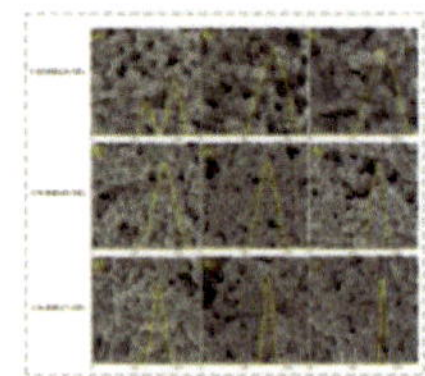	(Hu et al., 2020)
Quinoa, maize, waxy maize starch	Self-assembly	Enzymatic hydrolysis	Formation of SNPs via debranching; encapsulated chrysin and rutin	Encapsulation efficiency: 74.3%–79.1%; Quinoa starch nanoparticles (QSNPs) best; > 90% chrysin and 85% rutin retained after 15 days; sustained release; enhanced antioxidant activity		(Remanan & Zhu, 2024)
Starch–glycerol matrix	Nanocomposite formulation	ZnO nanoparticle incorporation (1%–3%)	Casting method; ZnO enhances structure, UV resistance, antibacterial action	Tensile strength increased (6.2 → 25.1 MPa); water vapor permeance (WVP) increased; UV shielding; reactive oxygen species (ROS)-mediated antibacterial activity; shelf-life extension of grapes and tomatoes	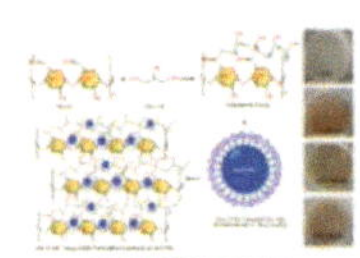	(Dhiman et al., 2025)
Starch (ST) + Carboxymethyl-cellulose (CMC)	Nanocomposite hydrogel	Green synthesis with CuO nanoparticles and borax cross-linking	UV/visible light exposure; tested under varied pHs and temperatures	High catalytic degradation of methyl red and methylene blue dyes; stable, elastic (~87.5%), water-retentive hydrogel; superior to pure CuO or CMC/ST hydrogels; potential for drug delivery	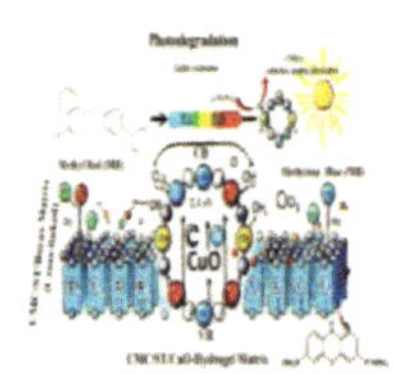	(Khan et al., 2025)
Starch + PVA	Electrospun nanofiber formulation	Enzymatic synthesis of sodium lactobionate and electrospinning	Nanofiber diameter: ~297–368 nm; tested up to 1000 μg/mL for 24 h	Biocompatible; low hemolysis (0.08%–3.18%); antioxidant activity (62.59% ABTS); safe profile (no ROS, NO, dsDNA effects); potential in pharma and food	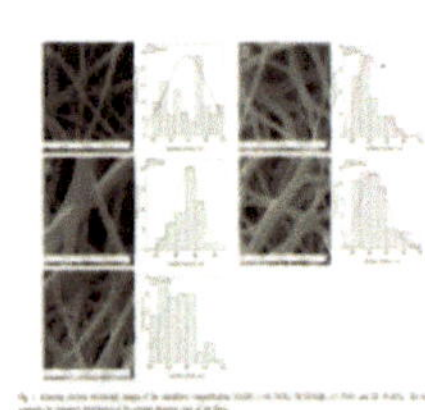	(Vencato et al., 2025)
Jamun seed starch (JaSS) + tamarind kernel xyloglucan (XG)	Bio-nanocomposite film formation	Blending JaSS and XG; incorporation of chitosan nanoparticles (ChNPs) at various concentrations (up to 3% w/w)	3% ChNPs: tensile strength 20.42 MPa, contact angle 89°, WVP 13.26 g/h·m^2	ChNPs improved film crystallinity, thermal stability, mechanical properties, and antimicrobial activity; JaSS/XG/ChNP coating extended sapota shelf life up to 12 days		(Santhosh & Sarkar, 2024)

(Continued)

TABLE 5.1 *(Continued)*
Comparison of Synthesis Methods for Starch Nanomaterials

Starch Sample	Approach Type	Synthesis Method	Process Conditions	Key Findings	Figures, Illustrations, SEM Images	Ref.
Carboxymethyl starch sodium (CMS-Na) + ovalbumin (OVA)	Protein/ polysaccharide nanoparticle-stabilized high-internal-phase Pickering emulsion (HIPE)	Fabrication of OVA/CMS-Na complexed nanoparticles (OCCNPs) by modulating OVA/CMS-Na interactions	Particle size 378.6 nm; stable up to 120 °C; contact angle ~88.9°	High-internal-phase emulsions (HIPE) reduced 5-Hydroxymethylfurfural (5-HMF) (75.7%), lipid hydroperoxide (LH) (46.5%), malondialdehyde (MDA) (27.4%) in biscuits; enabled 52% salt reduction without compromising sensory quality	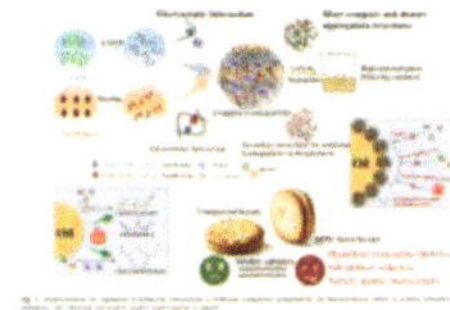	(Xie et al., 2025)
Waxy maize starch	Acid hydrolysis + mechanical treatment	Cold sulfuric acid hydrolysis (3.16 M, 14.7% solids) followed by ultrasonication (60% amplitude, 3 min)	Hydrolysis: 2–6 days at 4 °C, 40 °C, or cycled 4/40 °C; centrifugation: 10,000 rpm, 10 min	High nanoparticle yield (78%) and crystallinity using 4 °C for 6 days; particles were 50–90 nm; more resistance to ultrasonication at longer hydrolysis duration	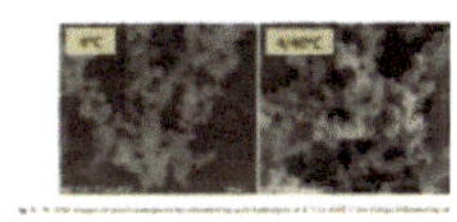	(Kim et al., 2013)
Sorghum, foxtail, and pearl millet starch	Acid hydrolysis + ultrasonication + encapsulation	Acid hydrolysis to form starch nanoparticles, ultrasonication-assisted encapsulation of β-carotene	Particle size: 416–587 nm; zeta potential: −17.98 to −22.31 mV; encapsulation efficiency: 78.32–89.65%	Sustained release of β-carotene in intestinal conditions; improved antioxidant activity vs. free β-carotene; no adverse interactions during encapsulation	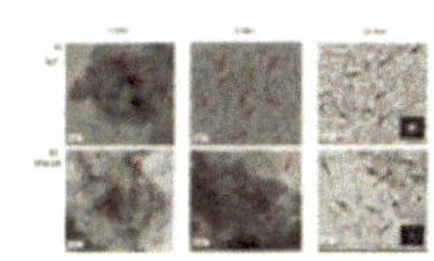	(Cuthbert et al., 2017)
Native starch	Nanoprecipitation	Varying polymer concentration, NaOH, solvent ratio, stirring speed, flow rate, non-solvent phase	Stirring speed: 500–900 rpm; starch concentration: 0.5–2%; pH 4–9; cryoprotectants for freeze-drying stability	Particle size reduced from 161.5 nm to 136.0 nm as stirring speed increased; zeta potential from −1.97 mV to −3.9 mV; good stability at pH 7.4; cryoprotectants retained nanoparticle size post freeze-drying	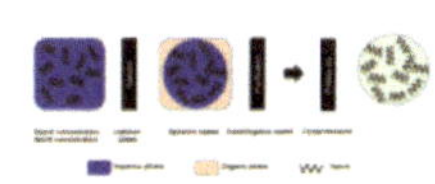	(Hernández-Giottonini et al., 2024)
Native starch	Enzymatic, acidic, and alkaline hydrolysis	Hydrolysis (enzymatic/ acidic/alkaline) + ultrasonication + precipitation + freeze-drying	Freeze-drying at −70 °C and 100 Pa; starch concentration: 10% (w/v); ultrasonication post-hydrolysis	Enzymatically prepared SNPs had the smallest size (~225 nm, < 100 nm by TEM), highest zeta potential (−26.3 mV), highest antioxidant activity (3.36%), and good emulsifying properties (oil separation time: 41 s); amorphous or low-crystallinity structure depending on hydrolysis method	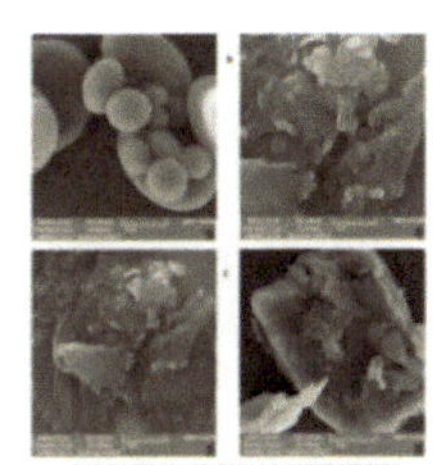	(Rastmanesh et al., 2024)
Corn Starch	Oil–water interface technique	Static interface and stirring conditions tested	-Temp: 25 °C vs 60 °C - Static vs stirred - Aggregation rate varied	V-type starch nanoparticles (144–377 nm; smaller at 25 °C) showed more ordered helices under static conditions, while stirring reduced order and crystallinity.	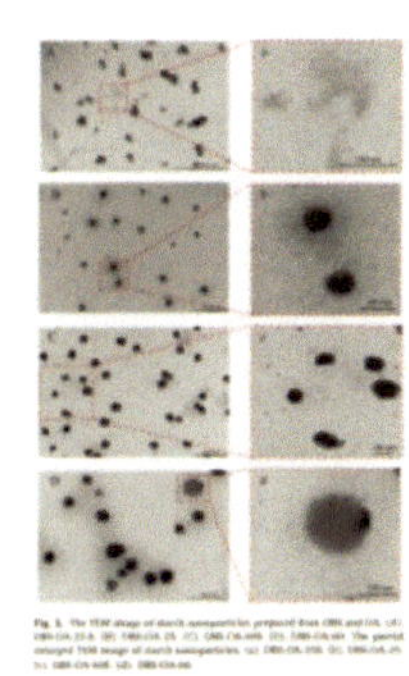	(Lu et al., 2023)

(Continued)

TABLE 5.1 *(Continued)*
Comparison of Synthesis Methods for Starch Nanomaterials

Starch Sample	Approach Type	Synthesis Method	Process Conditions	Key Findings	Figures, Illustrations, SEM Images	Ref.
Chestnut starch	Acid and enzymatic hydrolysis	Ultrasonic + acid hydrolysis to prepare SNPs, followed by α-amylase and glucoamylase to create porosity	SNP size: ~173.9 nm; PSNP size: ~286.3 nm; synergistic hydrolysis created porous structure; adsorption experiments performed	Porous starch nanoparticle (PSNPs) showed enhanced water, oil, and methylene blue (MB) adsorption; higher specific surface area (SSA); enzymatic + acid treatments decreased total starch and amylose content; increased solubility and relative crystallinity (RC); decreased short-range order and swelling at high temps; SSA correlated positively with solubility, WAC, MB adsorption, and RC	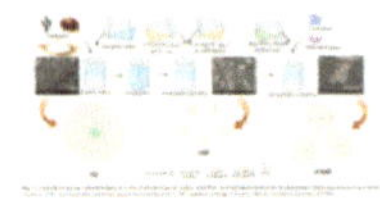	(Liu et al., 2024)
Corn, wheat, faba bean, field pea	Physical–chemical combination	Ultrasonic-assisted dissolution + rapid nanoprecipitation	Varied starch types, starch concentration, solvent:antisolvent ratio	Pulse starch SNPs had smaller, uniform size; higher amylose led to better crystallinity and thermal stability	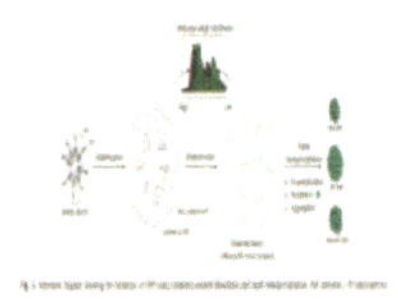	(Dong et al., 2022)
Guinea starch	Physical–chemical combination	Ultrasonication + nanoprecipitation	50% amplitude, 30 min, starch:ethanol = 1:4, 85% desirability	SNPs (206–391.5 nm) with VH-type structure; increased solubility, oil/water absorption, gelatinization temp; stable viscosity; good shear-thinning behavior	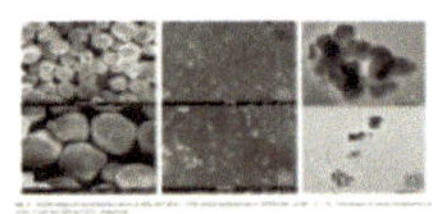	(Bist et al., 2024)

Starch are generally stored intracellularly in the form of discrete, semi-crystalline granules inside the plant (Gupta et al., 2024e). The morphological structure of the starch granules can be spherical, oval, lenticular, or polygonal, and granule size ranges from submicron level to 100 μm, depending on the botanical source (Bertoft, 2017). Native starch granules are insoluble or semi-soluble in cold water, but when heated in the presence of sufficient water, they undergo an irreversible process called gelatinization. Gelatinization is the process in which the granules swell and lose their crystallinity, leading to a significant increase in the viscosity of the starch. When cooled, the starch tends to retrograde, a process involving re-association and crystallization of starch chains, which leads to gel formation, increased firmness, and syneresis (Agama-Acevedo et al., 2019). Although these properties exist in native starches, starch-based nanoparticles are often a topic of interest. With the properties and applications being explored, starch can be transformed into various nanostructured materials, called starch nanomaterials (SNMts). These SNMts possess unique characteristics such as enhanced solubility, altered physicochemical properties, and high surface area compared to native starch (Ma et al., 2025).

5.4.1 Classification of Starch Nanomaterials

SNMts are a recent growing area of research and application and have gained significant interest due to their diverse structures and versatile applications. Starch nanoparticles (SNPs) are often amorphous, exhibiting a low degree of crystallinity. In contrast to it, there are starch nanocrystals (SNCs) that have highly crystalline fractions of starch granules, obtained by selective removal of the amorphous regions by acid or enzymatic hydrolysis. SNCs exhibit distinct mechanical and thermal properties due to their high crystallinity and platelet-like shape and have faster digestion rates than native starch granules (Ma et al., 2025). In contrast, starch nanofibers (SNFs) are one-dimensional, cylindrical nanostructures prepared via electrospinning. These nanofibers have a high aspect ratio and a diameter in the nanoscale range, providing them with unique functional and structural advantages. SNFs are often fabricated with chemically modified starch and are often blended with other polymers to facilitate fiber formation. These nanofibers show great potential in bioplastic formation due to their high reinforcing capability (Liu et al., 2017). Beyond these main categories, other starch-based nanostructures have also been explored, such as starch nanovesicles (SNVs) and starch nano-micelles (SNMs). SNVs are lipid vesicles coated with starch nanomaterials. SNVs, due to the presence of a bilayer structure, are capable of encapsulating both hydrophilic and hydrophobic compounds. SNMs typically have a size below 200 nm. They have a core–shell structure with colloidal dispersions generally formed from amphiphilically modified starch. Both SNVs and SNMs are often used as nanocarriers for delivering drugs (Rayo-Mendez & Rodriguez-Llanos, 2024).

The major synthesis approaches for starch nanoparticles are either top-down and bottom-up. In top-down approaches, the native starch granules are broken into the nanoscale level. Acid hydrolysis is one of the most widely used methods for producing SNCs. In this method, sulfuric, hydrochloric, or citric acid is used to selectively hydrolyze amorphous regions while preserving crystalline domains. This process reduces molecular weight, amylose content, and viscosity while increasing crystallinity (Dai et al., 2019; Hu et al., 2024). Another selective method, enzymatic hydrolysis, uses enzymes. In this method, enzymes such as amylases or pullulanase are used to break down starch, singly or in combination with other techniques, such as self-assembly (Oh et al., 2020). High-pressure homogenization (HPH) is a physical approach in which the starch is forced through a narrow gap at high pressure, creating an intense hydrodynamic force. This force results in granule structure destruction and the formation of stable, uniform dispersions. HPH has also been successfully applied to modify other matrices, enhancing their nanocarrier potential (Ding & Kan, 2017). Ultrasonication (UT) is another top-down method. This method works through the application of high-intensity sound waves to the starch suspension, causing acoustic cavitation. Cavitation causes the rapid formation and collapse of microbubbles, in turn causing intense localized heat and shear forces, which in turn cause the polymer chain breakage and granule fragmentation. UT improves dispersion, solubility, and colloidal stability and is often combined with acid hydrolysis or nanoprecipitation for synergistic effects (Ahmad et al., 2020). Gamma irradiation, while less commonly used, is also a physical modification method in the context of cross-linking or film enhancement (Naikwadi et al., 2022). Milling techniques such as ball milling also fall into the top-down category. Milling works on mechanical principles, applying impacting and shear forces to grind starch granules to the nano scale. Milling is frequently combined with chemical treatments such as acid hydrolysis to improve outcomes (Joy et al., 2022).

The other broad category of synthesis methods is bottom-up; these approaches synthesize nanostructures from molecular or oligomeric precursors. Antisolvent precipitation, or the solvent displacement method, known more simply as nanoprecipitation, is one of the most widely adopted bottom-up techniques for SNP production. This method generally involves dissolving the gelatinized starch in a solvent such as water and adding an antisolvent such as ethanol, for which starch is insoluble. This induces the rapid aggregation and precipitation of starch molecules into nanoparticles. Nanoprecipitation is a relatively fast, easy, and generally surfactant-free process. However, this method is highly sensitive to process conditions such as concentration, solvent/antisolvent ratio, temperature, and stirring speed (Wu & Wang, 2022). Another bottom-up technique is self-assembly. This technique relies on spontaneous molecular organization driven by hydrogen bonding, electrostatic interactions, and hydrophobic interactions. This technique is especially relevant for amphiphilically modified starches that form nano-micelles or nanovesicles. The self-assembly process is triggered by direct dissolution, solvent evaporation followed by hydration, or dialysis, with parameters such as pH, temperature, and critical micelle concentration (CMC) needing precise control (Pei et al., 2023).

5.5 INCORPORATING NANOPARTICLES INTO STARCH MATRICES: TECHNIQUES AND MECHANISMS

i. *Solution blending and solvent casting:* This is the most common and straightforward method. Starch is first gelatinized or dissolved in water or another suitable solvent, and nanoparticles (such as silver, ZnO, or nanoclays) are dispersed into the solution using ultrasonication or mechanical stirring. The resulting homogeneous mixture is cast onto a flat surface, and the solvent is evaporated under controlled conditions to form films, as shown in Figure 5.2. This method ensures good nanoparticle dispersion and matrix interaction, which improves film barrier, mechanical, and antimicrobial properties (Borbolla-Jiménez et al., 2023). However, complete solvent removal and nanoparticle agglomeration can be concerns.
ii. *In situ nanoparticle synthesis:* In this method, nanoparticles are synthesized directly within the starch matrix, often by adding metal precursors (e.g., $AgNO_3$ or $Ag(NH_3)_2$ for silver) to a gelatinized starch solution and then triggering a chemical reduction (e.g., with heat, UV, or reducing agents) (Chen et al., 2019), as shown in Figure 5.3. This results in fine and uniformly distributed nanoparticles, improving matrix interaction. In situ synthesis also reduces the risk of nanoparticle aggregation seen in premade nanoparticle dispersion (Harish et al., 2023).
iii. *Surface modification of starch nanoparticles:* This involves chemically or physically altering starch nanoparticles to improve compatibility with other hydrophobic polymers or fillers (Figure 5.4). Techniques include grafting of functional groups

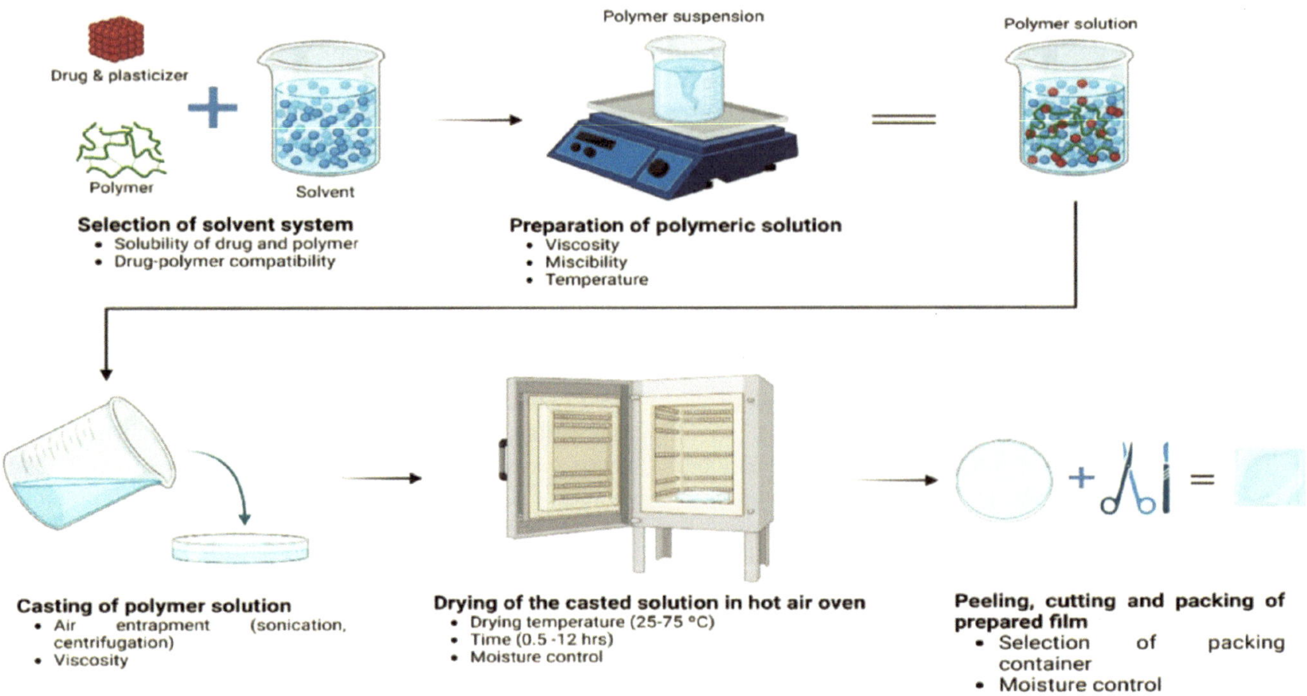

FIGURE 5.2 Process of nanoparticle incorporation by solvent casting technique (Borbolla-Jiménez et al., 2023).

(e.g., acetylation, silanization) and coating with surfactants or ligands (Singh et al., 2021). Modified starch shows better dispersion in matrices and stronger interactions with embedded nanoparticles or polymer blends, leading to enhanced thermal stability, mechanical strength, and film flexibility.

iv. *Electrospinning techniques:* Electrospinning uses a high-voltage electric field to draw a starch–polymer–nanoparticle solution into ultrafine fibers, which are collected as nonwoven mats, as shown in Figure 5.5. This method enables the creation of nanostructured composites with a high

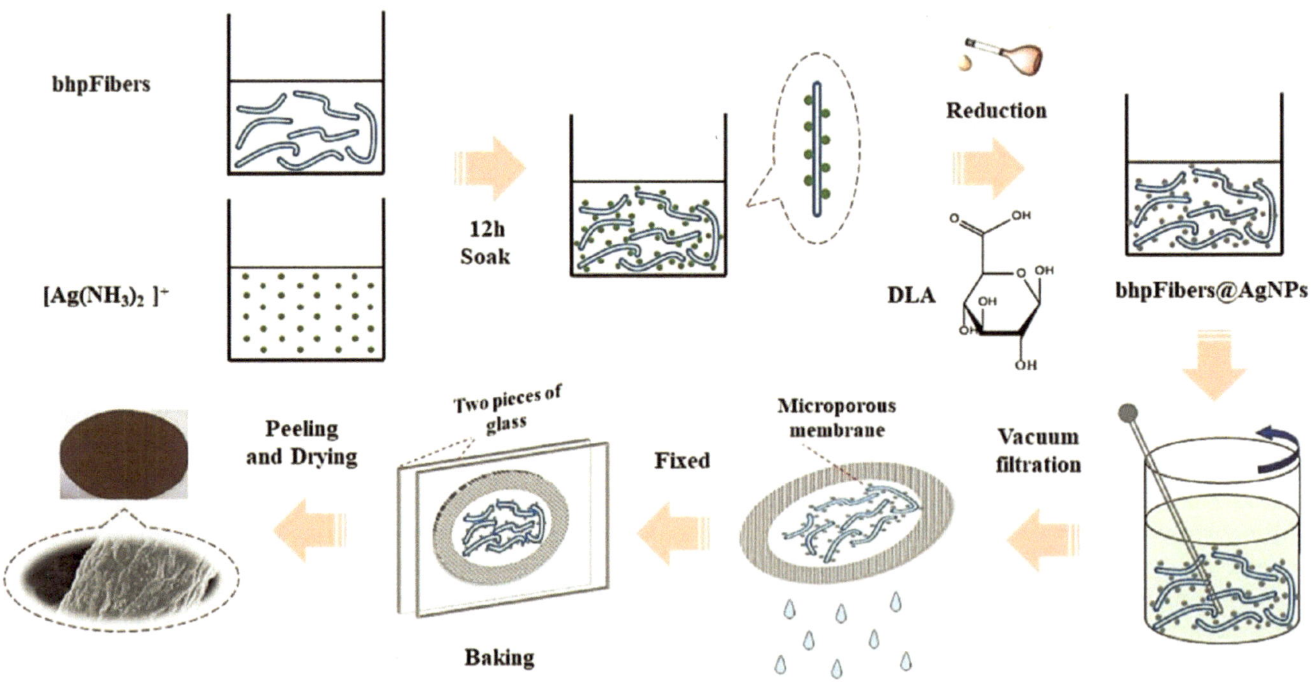

FIGURE 5.3 Procedure for in situ synthesis of silver nanoparticles. Here, bhp stands for bleached hardwood pulp which was pretreated and defibred to extract bhpfibres using a pulping machine with a beating degree of 48 °SR and 25000 rpm; DLA: D-glucuronic acid (Chen et al., 2019).

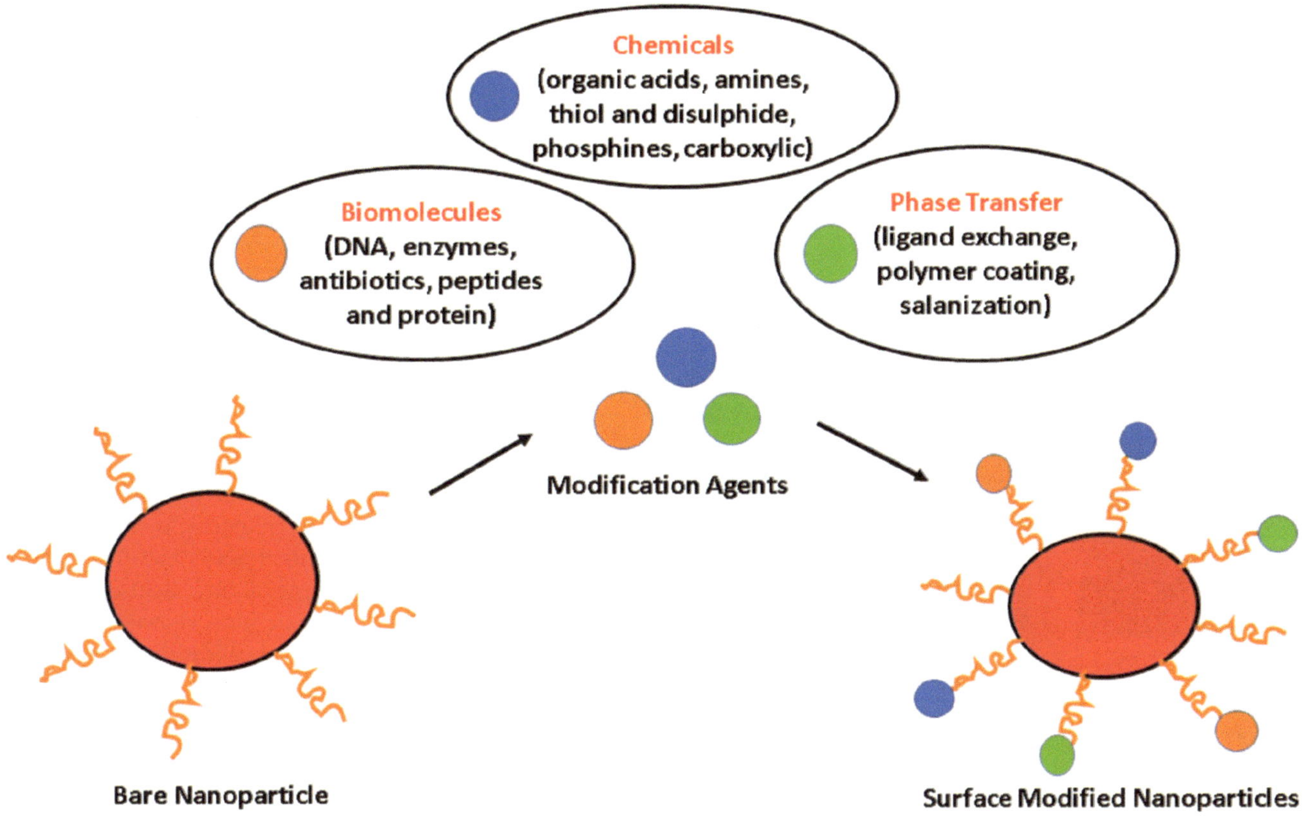

FIGURE 5.4 Surface modification of nanoparticles (Singh et al., 2021).

surface-area-to-volume ratio (Borbolla-Jiménez et al., 2023). Nanoparticles (e.g., TiO_2, ZnO, or SiO_2) can be embedded in the fibers or coated afterward. These nanofibers are ideal for active food packaging, wound dressings, or filtration membranes but require precise control over solution viscosity and voltage (García-Guzmán et al., 2022).

v. *Layer-by-layer assembly (LbL):* LbL involves alternating deposition of positively and negatively charged layers (e.g., starch with chitosan or polyanions) and nanoparticles on a substrate (Figure 5.6). This controlled build-up enables the design of multifunctional films with specific thickness, permeability, or bioactivity. LbL is particularly useful for encapsulation, controlled drug release, and smart packaging (e.g., pH-sensitive layers) (Darwish et al., 2022; Lengert et al., 2020). The key advantage is precision and modularity, though it is time-intensive.

vi. *3D Printing for customized starch nanocomposites:* This modern method allows for additive manufacturing of starch-based materials integrated

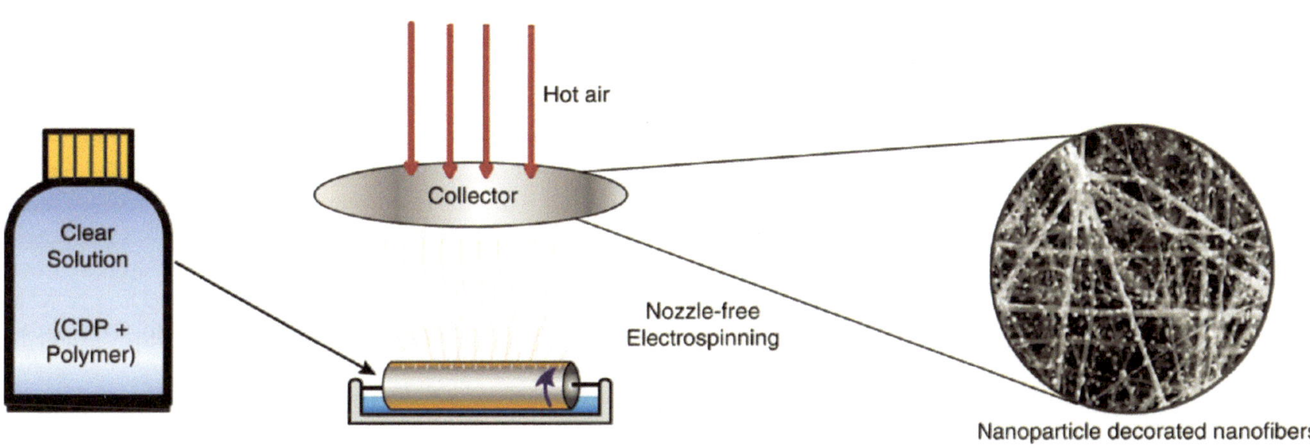

FIGURE 5.5 Schematic illustration of the electrospinning process (Borbolla-Jiménez et al., 2023).

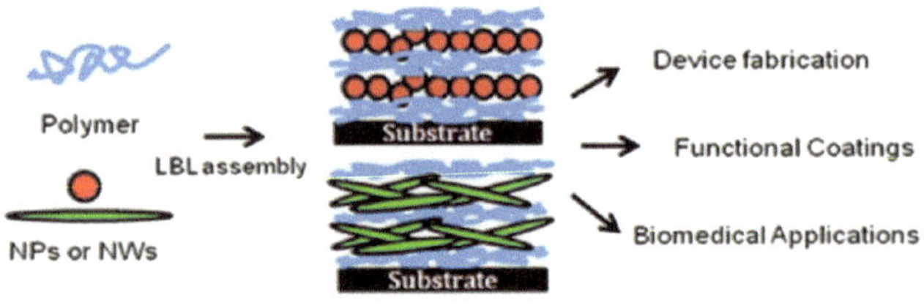

FIGURE 5.6 Schematics of incorporation of inorganic nanoparticles in layer-by-layer (LbL) assembly; NPs: nanoparticles, NWs: nanowires (Lengert et al., 2020).

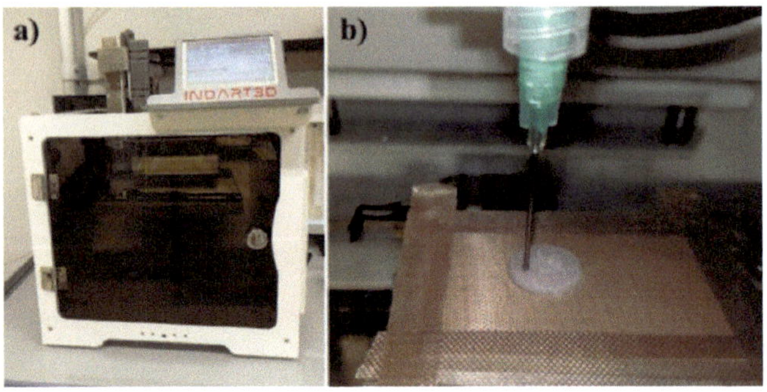

FIGURE 5.7 3D printing process with (a) Voladora 3D printer (Tumaker); (b) printing of customized starch composite tablets (González et al., 2022).

with nanoparticles by formulating a printable starch bio-ink (Figure 5.7). The starch nanoparticle mixture is extruded layer by layer to create tailored structures for biomedical scaffolds, packaging, or sensors (González et al., 2022; Pradhan et al., 2024; Saha et al., 2025). It offers high design flexibility and precision in architecture, although it requires advanced control of viscosity, nozzle pressure, and curing conditions. Integration with AI can also assist in process optimization (Saha et al., 2025; Shahbazi et al., 2024).

5.6 MULTIFUNCTIONAL PROPERTIES OF NANOTECHNOLOGY-MODIFIED STARCH AND ITS APPLICATION

5.6.1 Enhancement of the Mechanical Strength and Barrier Performance of Nanocomposites

Integrating starch nanocrystals (SNCs) into polymer matrices has been shown to enhance the mechanical properties of composites (Xu et al., 2010). Nevertheless, one major challenge remains the limited compatibility between SNPs and the polymer matrices, which impairs the effective development of nanocomposites. For instance, incorporating pea-derived SNCs as fillers in polyvinyl alcohol composites resulted in diminished tensile strength and reduced elongation at break when the SNC content increased up to 40 wt%. This decline is attributed to the poor interfacial adhesion between the filler and the polymer matrix (Chen et al., 2008). To address these compatibility issues, surface modification of the nanofillers has been proposed as an effective approach. Chemical modification using small molecules such as acetic anhydride or isocyanates can enhance the polarity or solubility of SNPs, altering both filler–filler and filler–matrix interactions. This modification can promote better dispersion within the matrix and facilitate the formation of covalent bonds at the interface, thereby improving the overall mechanical performance of the nanocomposite. Moreover, producing highly cross-linked starch structures limits the mobility of the starch chains, which helps prevent gelatinization of SNPs even under high-temperature conditions (up to 95 °C). In one study, corn-derived SNPs were cross-linked using citric acid following ethanol precipitation. When these citric acid–modified SNPs were incorporated into glycerol-plasticized pea starch composites, they dispersed uniformly without aggregation, owing to strong interactions between the modified SNPs and the polymer matrix. By adjusting the content of citric acid-crosslinked SNPs from 0 to 4%, the composite films showed significant enhancements: water vapor permeability dropped from 4.76×10^{-10} to 2.72×10^{-10} $g{\cdot}m^{-1}{\cdot}s^{-1}{\cdot}Pa^{-1}$, Young's modulus increased from 49.8 MPa to 125.1 MPa, and the glass transition temperature rose from 34.7 °C to 41.5 °C (Ma et al., 2008). In another approach, poly(ε-caprolactone) (PCL) chains were grafted onto the surface of SNCs via a "grafting from" method using microwave-assisted ring-opening polymerization. These functionalized nanoparticles were then incorporated into a poly(lactic acid) (PLA) matrix to fabricate nanocomposites. The resulting materials, containing 5% (w/w) PCL-grafted SNCs, demonstrated notable improvements in tensile strength and elongation at break due to the formation of a compatible interfacial region created by the long-grafted chains. However, at higher SNC loadings, the mechanical enhancements were offset by the tendency of the modified nanocrystals to aggregate with the PCL, reducing performance benefits (Yu et al., 2008).

5.6.2 Morphological Traits and Dimensions of SNPs

Various analytical methods have been employed to study the size distribution and morphology of starch nanoparticles (SNPs). These include transmission electron microscopy (TEM) (Kim et al., 2012; Kim et al., 2013), field emission scanning electron microscopy (FE-SEM) (Chin et al., 2014; Lamanna et al., 2013; Oladebeye et al., 2013; Sun et al., 2014), and atomic force microscopy (AFM) (Neethirajan et al., 2008). Additionally, techniques such as small-angle neutron scattering and dynamic light scattering have been utilized to further explore SNP morphology (Ding & Kan, 2016; Lamanna et al., 2013; Liu et al., 2016). The physical characteristics of SNPs are highly influenced by the botanical source and by the synthesis method, resulting in significant variability in the nanoparticles' shape and size (Torres et al., 2018). In terms of specific morphology, SNPs derived from potato starch (Chen et al., 2008), sago starch

(Bouthhegourd et al., 2011; Chin et al., 2011), and buckwheat starch (Li et al., 2023) predominantly exhibit spherical or oval shapes. In contrast, waxy maize starch corn starch (Song et al., 2008), and pea starch (Zheng et al., 2009) often form platelet-like nanostructures. These platelet-like SNPs typically range from 20 to 100 nm in length, 25 to 30 nm in width, and 6 to 8 nm in thickness. In contrast, SNPs with round or spherical morphologies show a broad size range, from as small as 8 nm to as large as 400 nm, depending on the starch origin and the processing method. Preparation methods also significantly influence particle size even when using the same starch source. For example, Liu et al. (2016) synthesized spherical SNPs measuring 30 to 40 nm by self-assembling short glucan chains from waxy maize starch at 50 °C. Pullulanase enzyme–treated waxy maize starch resulted in SNPs sized between 60 and 120 nm (Sun et al., 2014), while taro starch–derived SNPs exhibited sizes of 40 to 60 nm (Jiang et al., 2016). Similarly, proso millet starch that underwent pullulanase debranching and retrogradation at 4 °C produced SNPs within the 20 to 100 nm range (Sun et al., 2014). Moreover, sulfuric acid hydrolysis has been employed to produce SNPs from various starches, including mung bean, potato, normal maize, waxy maize, and high-amylose maize, which resulted in particle sizes of 53.7, 43.2, 41.0, 41.4, and 69.7 nm, respectively (Kim et al., 2012). A combined process of cold acid hydrolysis and ultrasonication yielded waxy maize SNPs with a globular morphology and sizes ranging from 50 to 90 nm (Kim et al., 2013). Additionally, Mukurumbira et al. (2017) observed squarish platelet-like SNPs, prepared by acid hydrolysis from the purple and white varieties of amadumbe corm, exhibiting average sizes of 260 nm and 220 nm when measured using SEM and DLS, respectively.

5.6.3 Crystallographic Properties of SNPs

Based on X-ray diffraction analyses, starch is classified into three primary crystalline structures: A, B, and C types. Imberty et al. (1987) introduced a model explaining the arrangement of double helices, highlighting the key distinctions between A- and B-type starch granules (Gupta el al., 2024d). In A-type crystallinity, the double helices are compactly organized, with water molecules positioned between these helices. Conversely, B-type structures display a more open arrangement, wherein water molecules reside in the central cavity formed by six double helices. The C-type configuration is generally regarded as a hybrid of the A and B forms (Gupta et al., 2024b). The degree of crystallinity is determined by the ratio of crystalline mass relative to the total mass of the nanocrystals. Ideally, this should approach 100% for polysaccharide nanocrystals; however, incomplete removal of amorphous areas often leads to lower crystallinity levels. Starches rich in amylopectin, when subjected to hydrolysis, typically exhibit higher crystallinity due to the limited presence of amorphous components, indicating that starches with inherently high crystallinity yield nanocrystals with correspondingly higher crystallinity. It has been observed that starches from various plant sources, despite having similar amylose levels, show no significant variation in the crystallinity of their native and nanocrystalline forms, suggesting that amylopectin plays a critical role in determining the crystallinity of starch nanocrystals (Duan et al., 2011). Furthermore, the method used for SNP preparation and the specific reaction conditions significantly influence the resulting crystallinity. Research indicates that acid hydrolysis tends to preserve the crystalline structure, while alkaline treatments disrupt the granules and alter their crystalline type (Torres et al., 2018). Generally, higher incubation temperatures promote the formation of A-type starch, while lower temperatures favor B-type structures (Kiatponglarp et al., 2015). For example, SNPs obtained through enzymatic hydrolysis of waxy maize starch using pullulanase followed by recrystallization at 4 °C displayed a combined B+V-type crystalline structure, characterized by peaks at 2θ values of 5.9°, 17.1°, 22.5°, and 24.3°. This pattern contrasts with the native starch, which shows an A-type pattern with diffraction peaks at 15.3°, 17.1°, 18.5°, and 23.5° (Sun et al., 2014). However, in separate research, SNPs produced from waxy maize starch at a moderate temperature of 50 °C maintained the A-type crystalline pattern even after recrystallization of the short glucan chains. Similarly, starch nanoparticles produced via acid hydrolysis (conducted over seven days at 40 °C) from waxy and normal maize (both A-type), high-amylose maize and potato starch (both B-type), and mung bean starch (C-type) retained their original crystalline forms, albeit with variations in diffraction intensity (Kim et al., 2012). Additionally, A-type starches such as waxy and normal maize were found to be more susceptible to acid hydrolysis than B-type starches such as high-amylose maize.

5.6.4 Biomedical and Drug Delivery Applications

Drug delivery systems are specialized formulations or technologies designed to transport therapeutic agents within the body, ensuring their release occurs at a controlled rate, specific time, and targeted location. Among these, nanotechnology-based delivery platforms—such as polymer-based nanoparticles, dendrimers modified with saccharides, and various nanocomposite structures—have emerged as effective alternatives, offering advantageous pharmacokinetic and pharmacodynamic profiles (Pooresmaeil & Namazi, 2021). Numerous investigations have focused on assessing the encapsulation processes, efficiency, and release dynamics of different drugs, antioxidants, and bioactive agents when incorporated into functionalized starch nanoparticles (SNPs). For instance, Najafi et al. (2016) explored the use of acetylated SNPs as carriers for ciprofloxacin, a broad-spectrum antimicrobial agent. The study demonstrated that SNPs with a high degree of acetylation (2.66) achieved superior drug encapsulation efficiencies (EEs), ranging from 50.2% to 89.7%, comparable to the EE of lipid-core nanocapsules (86.9%) (Venturini et al, 2011). This enhancement is believed to result from hydrogen bonding interactions between ciprofloxacin and the oxygen-bearing functional groups (carboxyl and hydroxyl) on the SNP surfaces. Similarly, bionanoparticles such as poly(DL-lactide-co-glycolide) have shown encapsulation efficiencies near 82.7% (Guenday et al., 2020). Additionally, succinylation, a method that introduces ester groups onto hydroxyl sites, has been used to increase SNP hydrophobicity and introduce negative surface charges, thereby enhancing the binding of cationic drugs through hydrophobic and electrostatic interactions (Puentes et al., 2020). These approaches broaden the range of poorly soluble compounds—including steroids, hormones, and polyphenols—that can be effectively encapsulated and delivered via SNPs. Furthermore, SNPs modified through cold plasma(CP)-induced oxidation exhibited an increase in surface negative charges, facilitating the adsorption of tea-derived polyphenols (Chang et al., 2009).

5.6.5 Role of SNPs in Microencapsulation and Targeted Release of Biofunctional Agents

The growing demand for functional foods with enhanced nutrition, minimal preservatives, and better sensory qualities has accelerated the use of nanoencapsulation technologies in the food industry (Fathi & Dadkhah, 2014). Nanoencapsulation, a key application of nanotechnology, involves enclosing bioactive compounds within nanoscale carriers. Compared to native starch, starch nanoparticles offer advantages such as larger surface area, reduced viscosity at higher concentrations, and improved encapsulation of active compounds (Chen et al., 2006). Encapsulation of polyphenols in SNPs helps mitigate stability issues and mask undesirable tastes (Fang & Bhandari, 2010). SNPs have been widely explored as carriers for the controlled delivery of various bioactives. Curcumin, a potent polyphenol from *Curcuma longa*, suffers from poor water solubility and stability, limiting its use. Encapsulation in SNPs enhances its stability, solubility, and sustained release (Rajeswari et al., 2011). For instance, cassava SNPs prepared via sulfuric acid hydrolysis successfully encapsulated curcumin, confirmed through fluorescence and Fourier transform infrared spectroscopy (FTIR) studies, showing potential as pharmaceutical carriers (Rajeswari et al., 2011). Similarly, sago SNPs improved the aqueous solubility and gradual release of curcumin due to their hydrophilic nature and swelling capacity (Chin et al., 2014). Acevedo-Guevara et al. (2018) also reported that both native and acetylated banana SNPs enhanced the gastric release of curcumin, achieving over 80% loading efficiency via hydrogen bonding interactions, confirmed by FTIR and X-ray diffraction analysis. Other studies have utilized SNPs for polyphenol-rich extracts, such as date palm pit phenolics, using cross-linked potato SNPs with encapsulation efficiencies of 62% (Jivan et al., 2014). SNPs from various sources, including horse chestnut and water chestnut, have been used for catechin encapsulation, preserving its bioactivity and stability during digestion (Ahmada et al., 2019). Additionally, hydroxyethyl starch–deoxycholic acid (HES-DOCA) SNPs have been developed for morin, improving its release and bioavailability for hyperuricemia treatment (Li et al., 2018). The release of salvianolic acid B was also improved through composites (Li et al., 2017). SNPs have been used to enhance the bioavailability of fat-soluble vitamin D3, where larger potato SNPs (99.2 nm) outperformed smaller high-amylose corn SNPs (32.04 nm) in encapsulation efficiency (Hasanvand et al., 2018). Additionally, arrowroot SNPs provided an effective matrix for andrographolide, a hydrophobic herbal compound, producing spherical nanoparticles under 300 nm suitable for controlled release (Winarti et al., 2014). Similarly, cassava SNPs were used to encapsulate anthocyanins and carotenoids from sweet potato leaves, enhancing their bioavailability, as confirmed by ATR spectroscopy. These studies collectively demonstrate the promising role of SNPs in improving the delivery, stability, and bioactivity of various hydrophobic and bioactive compounds for food, pharmaceutical, and nutraceutical applications. Table 5.2 lists applications of modified starch in different sectors.

TABLE 5.2
Applications of Modified Starch in Different Sectors

Modified Starch Sample	Nanotech Modification Method	Application	Characteristic Properties and Features	Limitations and Disadvantages	Figures, Illustrations, Images	Ref.
Decanoyl chloride/ curcumin-modified potato starch	Esterification + pH-driven curcumin loading (pH 8 & 12)	Pickering emulsion stabilizer (flaxseed oil)	Smaller size, high encapsulation (86.7%), high zeta potential (−41.2 mV), V-type crystallinity, contact angle up to 130.9°	pH-sensitive possible instability in extremes	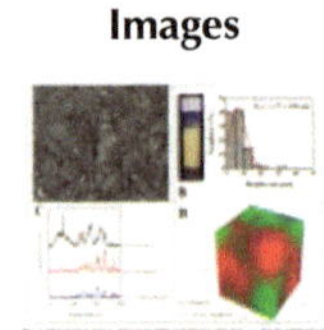	(Nobakht-Nia et al., 2025)
Tef and maize starch	Pasting with stearic acid + α-amylase hydrolysis	Potential nanofillers in food and non-food uses	Amylose–lipid (V-type) complexes; nanoparticle size ~2.4–10 nm; melting temp > 104 °C; yield ~24–30%	Long processing time (pasting 130 min); low yield	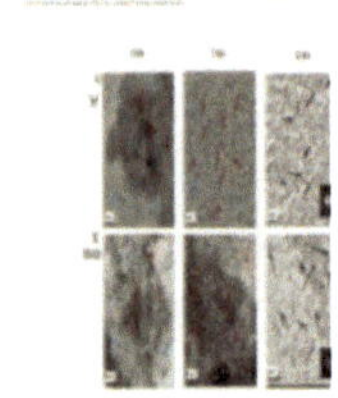	(Cuthbert et al., 2017)

(Continued)

TABLE 5.2 ***(Continued)***
Applications of Modified Starch in Different Sectors

Modified Starch Sample	Nanotech Modification Method	Application	Characteristic Properties and Features	Limitations and Disadvantages	Figures, Illustrations, Images	Ref.
Cyperus esculentus starch	Nanoprecipitation + OSA (octenyl succinic anhydride) modification	Functional food, emulsifier systems	SNP size: ~125.25 nm; OSA-modified SNPs: ~411–442 nm; V-type XRD; network morphology; increased solubility (from 70% to 90%), increased viscosity	Particle deformation after OSA; no measurable gelatinizati	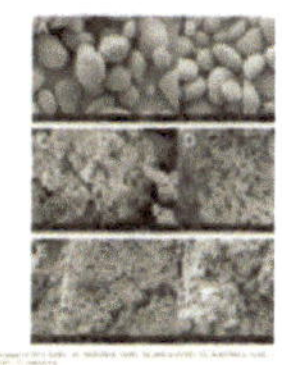	(Meng et al., 2025)
Zein-modified starch nanoparticles (Z-SNPs)	Zein coating via hydrogen bonding and hydrophobic interaction	Nanocarriers in nutraceuticals and pharmaceuticals	Irregular, worm-like, spherical shapes; contact angles: 91.6°, 86.9°, 84.8°; V-crystalline structure; shape controlled by temperature & zein concentration	Shape-dependent properties; sensitive to processing parameters	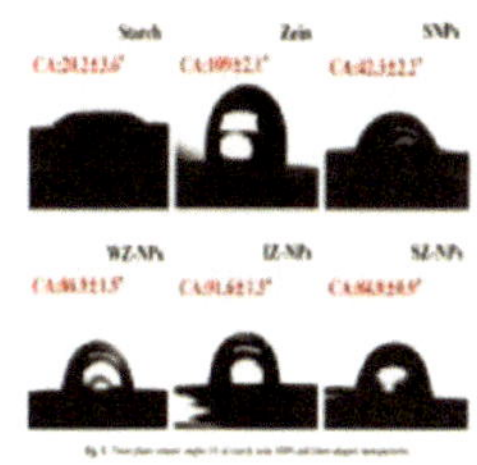 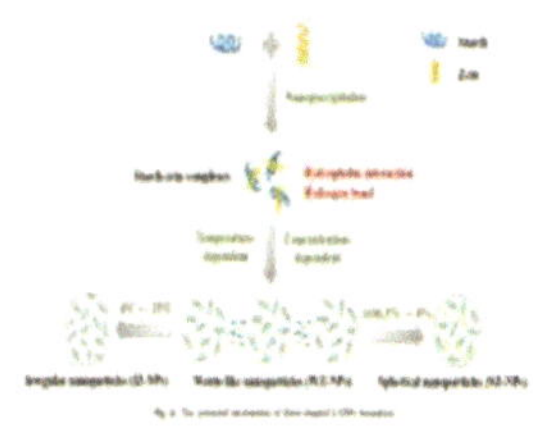	(Yang et al., 2025)
Dual-modified waxy maize starch	Cold plasma (0.8 and 1.2 A) + Heat-moisture treatment (HMT) (20%–30% moisture)	Functional starch with enhanced digestibility	Disrupted crystallinity (plasma), restored order (HMT), lower molecular weight, increased digestibility, biphasic thermal profile, improved thermal stability	Structural disruption from plasma may affect texture; optimization of HMT moisture level required	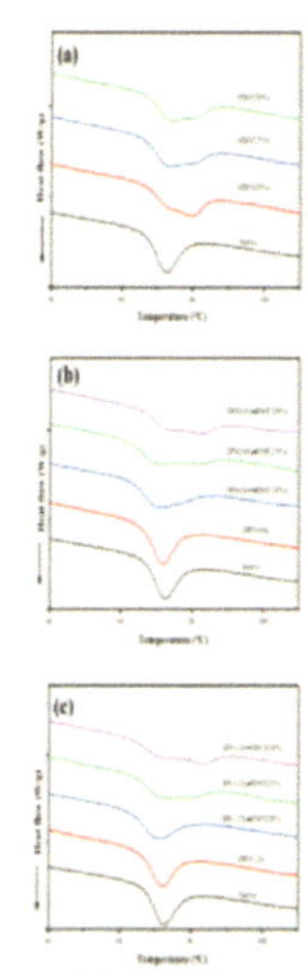	(Wang et al., 2025)
OSA-modified taro starch nanoparticles (OSA-SNPs)	Esterification with octenyl succinic anhydride (OSA) after self-assembly from debranched amylose	Stabilizer for oil–water systems	Increased particle size, improved dispersibility in nonpolar solvents, ester bond confirmed (FTIR 1727 cm^{-1}), tunable hydrophobicity via degree of substitution(DS)	Limited compatibility without OSA; synthesis dependent on DS control	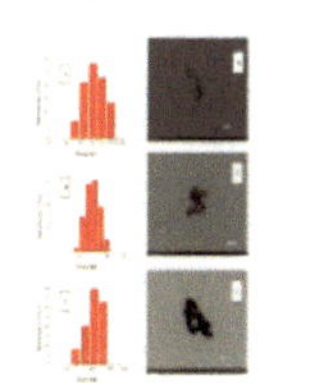	(Jiang et al., 2016)
Green banana and maize starches (Dgb-s, Hgb-s, Ym-s, Pm-s)	Heat-humidity treatment in an acid medium (denoted as CAHMT) (acid + heat-moisture) and OSA	Functional food ingredient (prebiotic, fat replacer)	High Resistant Starch (RS) (up to 90.52 g/100 g), improved Slowly Digestible Starch (SDS), enhanced viscosity & gelatinization (OSA), CAHMT increased amylose, diabetic-friendly potential	Lower starch yield in banana than maize; process-intensive	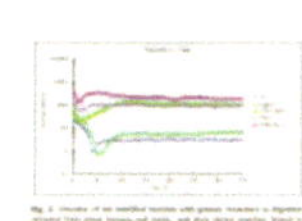	(Pacheco et al., 2025)

5.7 SPECIFIC NANOMATERIALS FOR STARCH MODIFICATION

i. *Metal nanoparticles:* Metal nanoparticles such as silver (AgNPs), gold (AuNPs), zinc oxide (ZnO-NPs), and titanium dioxide (TiO_2-NPs) have gained significant attention in starch modification due to their unique physicochemical properties and multifunctional benefits. When incorporated into starch matrices, these nanoparticles enhance thermal stability, mechanical strength, antimicrobial activity, and UV resistance, making starch-based films and composites more suitable for active food packaging, biomedical materials, and drug delivery. Silver nanoparticles, for instance, confer potent antimicrobial effects, thereby improving the shelf life of perishable goods (Yuan et al., 2021). ZnO and TiO_2 nanoparticles not only offer antibacterial and antioxidant properties but also improve water barrier characteristics, essential for packaging applications (Goudarzi et al., 2017). Gold nanoparticles, although more expensive, are prized for their biocompatibility and surface plasmon resonance, facilitating smart packaging and biosensor development (Ahari et al., 2022). These metal nanoparticles interact with starch primarily through hydrogen bonding or electrostatic interactions, resulting in well-integrated nanocomposites with enhanced performance and functionality.

ii. *Carbon nanotubes and graphene derivative:* Carbon nanotubes (CNTs) and materials such as graphene oxide (GO) are emerging as powerful nanofillers for enhancing the physicochemical and functional properties of starch-based materials. Due to their exceptional mechanical strength, thermal conductivity, electrical properties, and high aspect ratio, these carbon-based nanomaterials significantly improve the barrier, tensile, and thermal stability of starch composites (Bhadane & Mishra, 2024). When incorporated into starch matrices via blending or in situ methods, CNTs and GO form strong interfacial interactions—mainly through hydrogen bonding and π–π stacking—with starch chains, enhancing film flexibility, durability, and even imparting antimicrobial or electroactive properties suitable for intelligent packaging and biosensors (Islam & Mollik, 2020). Additionally, GO improves water resistance and reduces water vapor permeability, making starch–GO films promising for sustainable packaging applications (Islam & Mollik, 2020).

iii. *Nanoclays:* Nanoclays such as montmorillonite (MMT) and halloysite nanotubes (HNTs) are widely utilized in starch modification due to their layered structure, high aspect ratio, and excellent compatibility with hydrophilic polymers such as starch (Rouf & Kokini, 2017). These nanoclays enhance the mechanical strength, thermal stability, and water vapor barrier properties of starch-based films by forming intercalated or exfoliated structures within the starch matrix. MMT, in particular, improves film integrity and reduces oxygen and moisture permeability, making it ideal for eco-friendly food packaging applications. HNTs, with their tubular morphology and natural biocompatibility, offer controlled release functionalities, useful in drug delivery and active packaging. These nanoclays interact through hydrogen bonding and ionic interactions with starch chains, leading to structurally reinforced composites (Deshmukh et al., 2023).

iv. *Bio-based nanoparticles:* Bio-based nanoparticles such as chitosan nanoparticles and cellulose nanocrystals (CNCs) offer sustainable and biodegradable options for starch modification, significantly enhancing its functional performance. These nanoparticles improve the mechanical strength, thermal stability, and water barrier properties of starch-based films due to their high crystallinity and strong interfacial hydrogen bonding with starch chains (Noorbakhsh-Soltani et al., 2018). Chitosan nanoparticles also impart antimicrobial and antioxidant properties, making them ideal for active food packaging, while CNCs contribute to film rigidity and transparency, beneficial for biodegradable plastics (Ahmed et al., 2021). Their renewable origin, non-toxicity, and biodegradability make them especially attractive for environmentally friendly applications.

5.8 SUSTAINABILITY, SAFETY, AND REGULATORY ASPECTS

Nanotechnology has opened new horizons in enhancing starch-based materials for sustainable packaging. Starch nanocomposites present a biodegradable, renewable, and functional alternative to conventional plastics, offering improved barrier properties, mechanical strength, and shelf life for food products (Ghosh et al., 2025; Singh et al., 2024). Their development aligns with global sustainability goals and circular economy principles, significantly reducing environmental impact compared to fossil-based polymers (Dimkpa et al., 2012). The incorporation of nanomaterials enables the design of intelligent packaging solutions that actively protect food, reduce spoilage, and extend freshness. As these materials move toward commercialization, ensuring safety and compliance with global standards is key. International regulatory agencies such as the US Food and Drug Administration (FDA) and the European Food Safety Authority (EFSA) have established clear guidelines that support the responsible use of nanotechnology in food systems while encouraging innovation (Chaudhry et al., 2012).

5.8.1 Biodegradability and Environmental Impact of Starch Nanocomposites

Starch-based nanocomposites are inherently biodegradable, making them ideal for eco-friendly applications. The addition of nanomaterials enhances film strength and environmental performance without compromising biodegradability (Gamage, Thiviya, et al., 2022; Joshi, 2008). These materials degrade naturally via water absorption and enzymatic or microbial action, aligning with soil health and waste management goals (Dutta & Sit, 2023; Julinová et al., 2019). Chemical modification such as esterification, acetylation, and oxidation using green agents such as hydrogen peroxide improves the functional properties of starch nanoparticles, making them more stable and hydrophobic for food applications (Torres & De-la-Torre, 2022). This versatility allows tailored packaging solutions that balance durability with environmental responsibility (Su et al., 2022). Advanced testing methods such as SEM, TEM, AFM, thermal gravimetric analysis, FTIR, and Biochemical Oxygen Demand (BOD) provide detailed insights into degradation mechanisms and environmental impact (López et al., 2015; Sen & Das, 2018) (López et al., 2015; Sen & Das, 2018. The wide use of starch in agriculture, food, and pharmaceuticals combined with its low carbon footprint demonstrates its viability as a sustainable material platform (Singh et al., 2024).

5.8.2 Food Safety and Human Health Risks

Starch nanocomposites play a vital role in active and smart food packaging, enhancing safety by incorporating antimicrobial and antioxidant properties (Kumar & Thakur, 2020). Innovative uses include yam starch films for meat and cassava-based coatings for fruits, which contribute to food preservation and biodegradability (Cheng et al., 2021). While nanoparticles such as ZnO, TiO_2, and Ag offer powerful antimicrobial activity, their safety profiles are well studied to ensure responsible use. Ongoing research helps understand their behavior in the human body and food systems, contributing to the development of safe packaging materials (Chaudhry et al., 2012; Xuan et al., 2023). Cytotoxicity assays and in vitro studies are standard in evaluating biocompatibility, and procedures such as the use of polymyxin B help ensure reliable results by reducing testing interference (Nassarawa et al., 2024). The semicrystalline structure of starch allows the creation of various nanoscale forms SNPs, SNFs, and SNCs that are adaptable and increasingly safe. With robust safety assessment protocols, starch-based nanomaterials are well positioned for further expansion in food and biomedical sectors.

5.8.3 Regulatory Frameworks for Nano-Enabled Food Materials

Global regulatory frameworks are evolving to support safe and innovative applications of nanotechnology in food packaging. Agencies such as the FDA and the EFSA have established clear protocols for food contact materials that contain nanomaterials, focusing on parameters such as particle migration, toxicity, and labeling (Chaudhry et al., 2012). In the European Union, regulations such as EU 1169/2011 and EU 10/2011 require identification and risk assessments for nano-enabled materials. In the United States, the Federal Food, Drug, and Cosmetic Act mandates Food Contact Notifications (FCNs) for new materials unless they are generally recognized as safe (GRAS). Countries such as Canada, Australia, and New Zealand are developing case-by-case regulatory models that encourage responsible innovation. For starch nanocomposites, clarity around particle size, migration potential, and environmental behavior ensures compliance. Migration limits for key nanoparticles (e.g., 0.05 mg/kg for silver) and mandatory toxicological testing have been adopted to support safe commercialization (Herrera-Rivera et al., 2024). These standards provide a stable and forward-looking framework for accelerating the use of nanotechnology in sustainable food packaging.

5.9 FUTURE TRENDS AND CHALLENGES

The application of nanotechnology in starch modification is steadily advancing toward more sustainable, scalable, and functional materials for food packaging and environmental applications. Recent developments emphasize environmentally friendly and cost-effective modification methods such as enzymatic hydrolysis, ultrasonication, and electrospinning to enhance mechanical strength, thermal resistance, and barrier properties of starch-based films (Hassan et al., 2022; Sreejit et al., 2025). Incorporating functional nanofillers such as nanocellulose, ZnO, and TiO_2 has further improved the antimicrobial efficacy and durability of these films (Ramani et al., 2024). Despite this progress, challenges such as nanoparticle aggregation, migration, toxicity, and regulatory uncertainties persist. Addressing these through optimized dispersion techniques, comprehensive toxicological testing, and standardized manufacturing practices is essential for the safe and widespread adoption of starch nanocomposites in food packaging and environmental remediation.

5.9.1 Emerging Trends in Nanotechnology for Starch Modification for Scale-up

Current innovations favor physical and biological modification techniques over chemical methods, which often involve strong reagents and can impact the environment (Torres & De-la-Torre, 2022). Methods such as ultrasonication, milling, and high-pressure homogenization are being prioritized for their safety, efficiency, and scalability (Hassan et al., 2022; Sreejit et al., 2025). Enzymatic hydrolysis, in particular, provides a sustainable, high-yield alternative with minimal environmental waste (Kumari et al., 2020). New processing techniques including casting, extrusion, blowing, electrospinning, and electrospraying are being refined to improve the structural and barrier characteristics of high-amylose starch-based films (Sreejit et al., 2025; Wang et al., 2020). Green synthesis strategies such as nanoprecipitation, especially when combined with ultrasonication and environmentally friendly solvents, are showing promise in producing stable and well-dispersed starch nanoparticle solutions (Saha et al., 2025). The integration of clay minerals and bioactive agents (antioxidants, antimicrobials) continues to enhance film functionality (Bruna et al., 2023; García-Guzmán et al., 2022). Biopolymer blending and cross-linking, especially with nanocellulose (non-toxic under ISO 10993-5), are widely used to improve flexibility, moisture resistance, and mechanical strength (Camarero-Espinosa et al., 2016). Additionally, incorporating silver, copper, and metal oxide nanoparticles improves oxidative and antimicrobial properties, key for active food packaging (Ramani et al., 2024). Safer synthesis frameworks for nanoparticle production are also being explored to minimize exposure risks for workers and end users. Looking ahead, innovations in nanofiber-reinforced starch matrices and nanoencapsulation of bioactives (e.g., curcumin) are poised to significantly influence sustainable packaging and preservation, especially as safe and effective nanotechnologies become a core strategy in building resilient food systems by 2050.

5.9.2 Challenges in Scaling Up Production and Industrial Applications

Scaling up starch nanocomposite production for industrial applications requires addressing technical, environmental, and economic challenges. Green chemistry principles focusing on energy efficiency, reduced waste, and non-hazardous reagents are central to the development of sustainable processes (Torres & De-la-Torre, 2022). However, native starch suffers from inherent limitations such as hydrophilicity, brittleness, and poor mechanical resistance, which must be overcome through plasticization, blending with hydrophobic polymers (e.g., PLA, PHB), or reinforcement with natural fibers (Mittu et al., 2024). Nanotechnology offers promising solutions to these challenges. Nanocatalysts and nanosorbents are being used in environmental restoration, including wastewater treatment and sludge remediation (Yu et al., 2020). For example, starch-based magnetic nanocomposites have achieved over 90% removal of mercury ions in wastewater, and photocatalytic nanomaterials are used to degrade persistent organic pollutants (Wang et al., 2020; Xuan et al., 2023). Despite these advancements, nanotoxicity remains a key concern. Free radical generation, oxidative stress, and nanoparticle–protein interactions (protein coronas) in biological systems require deeper investigation (Neme et al., 2021). Regulatory clarity, cost-effectiveness, and consumer awareness are crucial for successful commercialization (Ramani et al., 2024). Functional enhancements using ZnO, TiO_2, graphene, and poly(methyl methacrylate) (PMMA)-based nanocomposites are improving the UV resistance and barrier properties of starch films (Flores et al., 2007; Goudarzi et al., 2017). Curcumin-loaded and acetylated SNPs have demonstrated antimicrobial effects and reduced water vapor permeability, offering additional functionality for active packaging (Torres & De-la-Torre, 2022). Furthermore, SNPs show potential in agricultural applications by enhancing seed vigor, root growth, and disease resistance (Dutta & Sit, 2024). While biocomposites show immense promise, scaling up requires attention to factors such as nanoparticle dispersion, migration control, and consumer safety. Techniques such as blending and reinforcement with nanocellulose can enhance mechanical and thermal stability (Tibolla et al., 2020), though nanoparticle aggregation remains a technical hurdle (Kumari et al., 2022). Retrogradation and moisture sensitivity still limit the long-term durability of native starch films (Cornejo-Ramírez et al., 2018; Halley & Avérous, 2014). To reduce competition with food crops, researchers are now exploring underutilized starch sources that provide unique functional properties and promote agricultural diversity (Dutta & Sit, 2024). Nevertheless, bio-based packaging currently represents less than 1% of the global market, largely due to water sensitivity and limited barrier function (Kumari et al., 2022). Continued efforts to enhance competitiveness and functionality under diverse conditions are essential.

5.10 CONCLUSION

Nanotechnology is revolutionizing the modification of starch into advanced, sustainable materials suitable for modern food packaging and biomedical applications. Starch stands out as a biodegradable, renewable biopolymer with immense potential. By incorporating nanomaterials such as nanoparticles, nanofibers, and nanogels, starch-based films can be engineered to offer enhanced strength, flexibility, barrier performance, and bioactivity. With innovations in green synthesis, surface modification, and safe nanofiller integration, starch nanocomposites are moving closer to commercialization. Challenges such as scalability, nanotoxicity, and consumer safety are actively being addressed through advanced research and supportive regulatory frameworks. Furthermore, the use of starch from underutilized crops offers new possibilities for sustainable agriculture and resource diversification. As the field progresses, starch-based nanomaterials are expected to play a transformative role in building circular, eco-efficient food systems and safer, smarter packaging solutions. By 2050, these biocomposites may not only replace petroleum-based plastics but also contribute to climate goals, food security, and public health.

REFERENCES

Acevedo-Guevara, L., Nieto-Suaza, L., Sanchez, L. T., Pinzon, M. I., & Villa, C. C. (2018). Development of native and modified banana SNP's as vehicles for curcumin. *International Journal of Biological Macromolecules*, *111*, 498–504.

Agama-Acevedo, E., Flores-Silva, P. C., & Bello-Perez, L. A. (2019). Cereal starch production for food applications. In Clerici, M.T.P.S. & Schmiele, M. *(Eds.) Starches for food application*. (pp. 71–102). Academic Press. Elsevier. https://doi.org/10.1016/B978-0-12-809440-2.00003-4.

Ahari, H., Fakhrabadipour, M., Paidari, S., Goksen, G., & Xu, B. (2022). Role of AuNPs in active food packaging improvement: A review. *Molecules*, *27*, 8027.

Ahmad, M., Gani, A., Hassan, I., Huang, Q., & Shabbir, H. (2020). Production and characterization of starch nanoparticles by mild alkali hydrolysis and ultra-sonication process. *Scientific Reports*, *10*, 3533.

Ahmada, M., Mudgil, P., Gani, A., Hamed, F., Masoodi, F. A., & Maqsood, S. (2019). Nanoencapsulation of catechin in SNP's: Characterization, release behaviour and bioactivity retention during simulated in-vitro digestion. *Food Chemistry*, *270*, 95–104.

Ahmed, E. M., Saber, D., Abd Elaziz, K., Alghtani, A. H., Felemban, B. F., Ali, H. T., & Megahed, M. (2021). Chitosan-based nanocomposites: Preparation and characterization for food packing industry. *Materials Research Express*, *8*, 025017.

Altammar, K. A. (2023). A review on nanoparticles: Characteristics, synthesis, applications, and challenges. *Frontiers in Microbiology*, *14*, 1155622.

Amaraweera, S. M., Gunathilake, C., Gunawardene, O. H., Fernando, N. M., Wanninayaka, D. B., Dassanayake, R. S., Rajapaksha, S., Manamperi, M., Fernando, A., & Kulatunga, C. A. (2021). Development of starch-based materials using current modification techniques and their applications: A review. *Molecules*, *26*, 6880.

Baranwal, J., Barse, B., Fais, A., Delogu, G. L., & Kumar, A. (2022). Biopolymer: A sustainable material for food and medical applications. *Polymers*, *14*, 983.

Bertoft, E. (2017). Understanding starch structure: Recent progress. *Agronomy*, *7*, 56.

Bhadane, P., & Mishra, A. (2024). Versatile, flexible rice starch–graphene oxide bio-nanocomposites. *Environmental Science: Water Research & Technology*, *10*, 2432–2441.

Bist, Y., Sharanagat, V. S., & Saxena, D. C. (2024). Synthesis, optimization, and characterization of precipitation derived starch nanoparticles from guinea seeds. *International Journal of Biological Macromolecules*, *265*, 131010.

Borbolla-Jiménez, F. V., Peña-Corona, S. I., Farah, S. J., Jiménez-Valdés, M. T., Pineda-Pérez, E., Romero-Montero, A., Del Prado-Audelo, M. L., Bernal-Chávez, S. A., Magaña, J. J., & Leyva-Gómez, G. (2023). Films for wound healing fabricated using a solvent casting technique. *Pharmaceutics*, *15*, 1914.

Bouthhegourd, E., Rajisha, K. R., Thomas, S., & Saiter, J. M. (2011). Natural rubber latex/ potato starch nanocrystals nanocomposites: Correlation morphology/electrical properties. *Materials Letters*, 65, 3615–3617.

Bruna, S. T., Elizabeth, H. N., Christiane, Q. M., Danielle, I., & Nelida, L. D. M. (2023). Physicochemical properties of potato starch films containing natural extracts of turmeric and hibiscus. *Magna Scientia Advanced Research and Reviews*, *8*, 1–8.

Camarero-Espinosa, S., Endes, C., Mueller, S., Petri-Fink, A., Rothen-Rutishauser, B., Weder, C., Clift, M., & Foster, E. (2016). Elucidating the potential biological impact of cellulose nanocrystals. *Fibers*, *4*, 21.

Chang, P. R., Ai, F., Chen, Y., Dufresne, A., & Huang, J. (2009). Effect of SNC's–graft polycaprolacton on mechanical properties of waterborne polyurethane- based nanocomposites. *Journal of Applied Polymer Science*, *111*(2), 619–627.

Chaudhry, Q., Gergely, A., & Bowman, D. (2012). Regulatory frameworks for food nanotechnologies. In Huang, Q. (Ed.) *Nanotechnology in the Food, Beverage and Nutraceutical Industries*. (pp. 82–98). Woodhead Publishing. Elsevier. https://doi.org/10.1533/9780857095657.1.82

Chen, H., Weiss, J., & Shahidi, F. (2006). Nanotechnology in nutraceuticals and functional foods. *Food Technology*, *60*, 30–36.

Chen, Y., Cao, X., Chang, P. R., & Huneault, M. A. (2008). Comparative study on the films of poly (vinyl alcohol)/ pea starch nanocrystals and poly (vinyl alcohol)/native pea starch. *Carbohydrate Polymers*, *73*(1), 8–17.

Chen, G., Yan, L., Wan, X., Zhang, Q., & Wang, Q. (2019). In situ synthesis of silver nanoparticles on cellulose fibers using d-glucuronic acid and its antibacterial application. *Materials*, *12*, 3101.

Cheng, H., Chen, L., Mcclements, D. J., Yang, T., Zhang, Z., Ren, F., Miao, M., Tian, Y., & Jin, Z. (2021). Starch-based biodegradable packaging materials: A review of their preparation, characterization and diverse applications in the food industry. *Trends in Food Science & Technology*, *114*, 70–82.

Chin, S. F., Pang, S. C., & Tay, S. H. (2011). Size-controlled synthesis of starch nanoparticles by a simple nanoprecipitation method. *Journal of Nanomaterials*, *86*, 1817–1819.

Chin, S. F., Mohd Yazid, S. N. A., & Pang, S. C. (2014). Preparation and characterization of starch nanoparticles for controlled release of curcumin. *International Journal of Polymer Science*, 1–8.

Chutia, H., & Mahanta, C. L. (2021). Properties of starch nanoparticle obtained by ultrasonication and high pressure homogenization for developing carotenoids-enriched powder and Pickering nanoemulsion. *Innovative Food Science & Emerging Technologies*, *74*, 102822.

Compart, J., Li, X., & Fettke, J. (2021). Starch: A complex and undeciphered biopolymer. *Journal of Plant Physiology*, *258*, 153389.

Cornejo-Ramírez, Y. I., Martínez-Cruz, O., Del Toro-Sánchez, C. L., Wong-Corral, F. J., Borboa-Flores, J., & Cinco-Moroyoqui, F. J. (2018). The structural characteristics of starches and their functional properties. *CyTA - Journal of Food*, *16*, 1003–1017.

Cuthbert, W. O., Ray, S. S., & Emmambux, N. M. (2017). Isolation and characterisation of nanoparticles from tef and maize starch modified with stearic acid. *Carbohydrate Polymers*, *168*, 86–93.

Dai, L., Zhang, J., & Cheng, F. (2019). Succeeded starch nanocrystals preparation combining heat-moisture treatment with acid hydrolysis. *Food Chemistry*, *278*, 350–356.

Darwish, M. S., Mostafa, M. H., & Al-Harbi, L. M. (2022). Polymeric nanocomposites for environmental and industrial applications. *International Journal of Molecular Sciences*, *23*, 1023.

Das, R., Kumar, A., Singh, C., & Kayastha, A. M. (2025). Innovative synthesis approaches and health implications of organic-inorganic nanohybrids for food industry applications. *Food Chemistry*, 464, 141821.

Deshmukh, R. K., Kumar, L., & Gaikwad, K. K. (2023). Halloysite nanotubes for food packaging application: A review. *Applied Clay Science*, *234*, 106856.

Dhiman, S., Kumari, A., Kumari, S., & Sharma, R. (2025). Advanced biodegradable starch-based nanocomposite films incorporating zinc oxide nanoparticles: Synthesis, characterization, and efficacy in antibacterial food packaging applications. *Journal of Environmental Chemical Engineering*, 13.

Dimkpa, C. O., Mclean, J. E., Britt, D. W., & Anderson, A. J. (2012). Bioactivity and biomodification of Ag, ZnO, and CuO nanoparticles with relevance to plant performance in agriculture. *Industrial Biotechnology*, *8*, 344–357.

Ding, Y., & Kan, J. (2016). Characterization of nanoscale retrograded starch prepared by a sonochemical method. *Starch/Starke*, *68*, 264–273.

Ding, Y., & Kan, J. (2017). Optimization and characterization of high pressure homogenization produced chemically modified starch nanoparticles. *Journal of Food Science and Technology*, *54*, 4501–4509.

Dong, H., Zhang, Q., Gao, J., Chen, L., & Vasanthan, T. (2022). Preparation and characterization of nanoparticles from cereal and pulse starches by ultrasonic-assisted dissolution and rapid nanoprecipitation. *Food Hydrocolloids*, 122, 107081.

Duan, B., Sun, P., Wang, X., & Yang, C. (2011). Preparation and properties of SNC's/ carboxymethyl chitosan nanocomposite films. *Starch/Starke*, *63*, 528–535.

Dutta, D., & Sit, N. (2023). Comparison of properties of films prepared from casein modified by ultrasound and autoclave treatment. *Journal of Food Measurement and Characterization*, *17*, 5426–5439.

Dutta, D., & Sit, N. (2024). Comprehensive review on developments in starch-based films along with active ingredients for sustainable food packaging. *Sustainable Chemistry and Pharmacy*, 39, 101534.

Fang, Z., & Bhandari, B. (2010). Encapsulation of polyphenol-A review. *Trend in Food Science and Technology*, *21*, 510–523.

Fathi, F., & Dadkhah, A. (2014). Characterisation and surface chemical modification of starch nanoparticles with lactid through ring opening polymerization. *International Journal of Nanoparticles*, *7*, 37–48.

Flores, S., Famá, L., Rojas, A. M., Goyanes, S., & Gerschenson, L. (2007). Physical properties of tapioca-starch edible films: Influence of filmmaking and potassium sorbate. *Food Research International*, *40*, 257–265.

Gamage, A., Liyanapathiranage, A., Manamperi, A., Gunathilake, C., Mani, S., Merah, O., & Madhujith, T. (2022). Applications of starch biopolymers for a sustainable modern agriculture. *Sustainability*, *14*, 6085.

Gamage, A., Thiviya, P., Mani, S., Ponnusamy, P. G., Manamperi, A., Evon, P., Merah, O., & Madhujith, T. (2022). Environmental properties and applications of biodegradable starch-based nanocomposites. *Polymers*, *14*, 4578.

García-Guzmán, L., Cabrera-Barjas, G., Soria-Hernández, C. G., Castaño, J., Guadarrama-Lezama, A. Y., & Rodríguez Llamazares, S. (2022). Progress in starch-based materials for food packaging applications. *Polysaccharides*, *3*, 136–177.

Ghosh, S., Mandal, R. K., Mukherjee, A., & Roy, S. (2025). Nanotechnology in the manufacturing of sustainable food packaging: A review. *Discover Nano*, *20*, 36.

Gonçalves, E. M., Silva, M., Andrade, L., & Pinheiro, J. (2024). From fields to films: Exploring starch from agriculture raw materials for biopolymers in sustainable food packaging. *Agriculture*, *14*, 453.

González, K., Larraza, I., Berra, G., Eceiza, A., & Gabilondo, N. (2022). 3D printing of customized all-starch tablets with combined release kinetics. *International Journal of Pharmaceutics*, *622*, 121872.

Goudarzi, V., Shahabi-Ghahfarrokhi, I., & Babaei-Ghazvini, A. (2017). Preparation of ecofriendly UV-protective food packaging material by starch/TiO2 bio-nanocomposite: Characterization. *International Journal of Biological Macromolecules*, *95*, 306–313.

Guenday, C., Anand, S., Gencer, H.B., Munafo, S., Moroni, L., Fusco, A., Donnarumma, G., Ricci, C., Hatir, P.C., Tuereli, N.G. and Tuereli, A.E., 2020. Ciprofloxacin-loaded polymeric nanoparticles incorporated electrospun fibers for drug delivery in tissue engineering applications. *Drug Delivery and Translational Research*, *10*, 706–720.

Gupta, S., Sharma, S., Nadda, A. K., Husain, M. S. B., & Gupta, A. (2022). Biopolymers from waste biomass and its applications in the cosmetic industry: A review. *Materials Today: Proceedings*, *68*, 873–879.

Gupta, R. K., Guha, P., & Srivastav, P. P. (2022). Natural polymers in bio-degradable/edible film: A review on environmental concerns, cold plasma technology and nanotechnology application on food packaging-A recent trends. *Food Chemistry Advances*, *1*, 100135. https://doi.org/10.1016/j.focha.2022.100135

Gupta, R. K., Guha, P., & Srivastav, P. P. (2023). Effect of high voltage dielectric barrier discharge (DBD) atmospheric cold plasma treatment on physicochemical and functional properties of taro (*Colocasia esculenta*) starch. *International Journal of Biological Macromolecules*, *253*, 126772. https://doi.org/10.1016/j.ijbiomac.2023.126772

Gupta, R. K., Guha, P., & Srivastav, P. P. (2024a). Dielectric barrier discharge plasma: A green and novel method to change the structure of taro peel starch and improve the physicochemical properties of taro peel starch films. *Plasma Processes and Polymers*, *21*(9), 2400047. https://doi.org/10.1002/ppap.202400047

Gupta, R. K., Guha, P., & Srivastav, P. P. (2024b). Exploring the potential of taro (Colocasia esculenta) starch: Recent developments in modification, health benefits, and food industry applications. *Food Bioengineering*, *3*(3), 365–379. https://doi.org/10.1002/fbe2.12103

Gupta, R. K., Guha, P., & Srivastav, P. P. (2024c). Insights into the physicochemical, crystalline, multistructural, thermal and digestibility behavior of taro starch (*Colocasia esculenta*) modified by cold plasma treatment with high voltage dielectric barrier discharge (HVDBD): An alternative technique for chemical modification of starch and environmental sustainability. *Journal of Cleaner Production*, *461*, 142710. https://doi.org/10.1016/j.jclepro.2024.142710

Gupta, R. K., Guha, P., & Srivastav, P. P. (2024d). Physical action of nonthermal cold plasma technology for starch modification. *Food Physics*, *1*, 100011. https://doi.org/10.1016/j.foodp.2024.100011

Gupta, R. K., Guha, P., & Srivastav, P. P. (2024e). Effect of cold plasma treatment and plasma-activated water on physicochemical and structural properties of starch: A green and novel approach for environmental sustainability. *Plasma Processes and Polymers*, *21*(4), 2300204. https://doi.org/10.1002/ppap.202300204

Gupta, R. K., Pipliya, S., Patel, J., Srivastav, P. P., Castro-Muñoz, R., Wang, C. K., & Nemţanu, M. R. (2025a). Essential oil-embedded starch based films for active packaging: Insights into migration, stability, and preservation efficiency. *Trends in Food Science & Technology*, 163, 105149. https://doi.org/10.1016/j.tifs.2025.105149

Gupta, R. K., Gnana Moorthy Eswaran, U., Kumar, S., Pipliya, S., Kaur, B., & Srivastav, P. P. (2025b). Synergistic modification of starch polysaccharide via sequential treatments with ultrasonic and cold plasma technology. *Journal of Food Processing and Preservation*, *2025*(1), 9063007. https://doi.org/10.1155/jfpp/9063007

Halley, P. J., & Avérous, L. (2014). *Starch polymers: From genetic engineering to green applications*. Elsevier.

Harish, V., Ansari, M., Tewari, D., Yadav, A. B., Sharma, N., Bawarig, S., García-Betancourt, M.-L., Karatutlu, A., Bechelany, M., & Barhoum, A. (2023). Cutting-edge advances in tailoring size, shape, and functionality of nanoparticles and nanostructures: A review. *Journal of the Taiwan Institute of Chemical Engineers*, *149*, 105010.

Hasanvand, E., Fathi, M., & Bassiri, A. (2018). Production and characterization of vitamin D3 loaded SNP's: Effect of amylopectin ratio and sonication parameters. *Journal of Food Science and Technology*, *55*, 1314–1324.

Hassan, N. A., Darwesh, O. M., Smuda, S. S., Altemimi, A. B., Hu, A., Cacciola, F., Haoujar, I., & Abedelmaksoud, T. G. (2022). Recent trends in the preparation of nano-starch particles. *Molecules*, *27*, 5497.

Hernández-Giottonini, K. Y., Quiñones-Rabago, J. A., Peñuñuri-Miranda, O., Rodríguez-Córdova, R. J., Zavala-Rivera, P., & Lucero-Acuña, A. (2024). Starch nanoparticle preparation by the nanoprecipitation technique: Effects of formulation parameters. *Colloids and Surfaces A: Physicochemical and Engineering Aspects*, *702*, 135022.

Herrera-Rivera, M. D. R., Torres-Arellanes, S. P., Cortés-Martínez, C. I., Navarro-Ibarra, D. C., Hernández-Sánchez, L., Solis-Pomar, F., Pérez-Tijerina, E., & Román-Doval, R. (2024). Nanotechnology in food packaging materials: Role and application of nanoparticles. *RSC Advances*, *14*, 21832–21858.

Hu, M., Lv, X., Wang, Y., Zhang, Y., & Dai, H. (2024). Guideline for the extraction of nanocellulose from lignocellulosic feedstocks. *Food Biomacromolecules*, *1*, 9–17.

Hu, Y., Qin, Y., Qiu, C., Xu, X., Jin, Z., & Wang, J. (2020). Ultrasound-assisted self-assembly of β-cyclodextrin/debranched starch nanoparticles as promising carriers of tangeretin. *Food Hydrocolloids*, *108*, 106021.

Imberty, A., Chanzy, H., Perez, S., Buleon, A., & Tran, V. (1987). New three–dimensional structure for A–type starch. *Macromolecules*, *20*, 2634–2636.

Islam, M. R., & Mollik, S. I. (2020). Enhanced electrochemical performance of flexible and eco-friendly starch/graphene oxide nanocomposite. *Heliyon*, *6*, e05292.

Jha, A. K., Prasad, K., Prasad, K., & Kulkarni, A. R. (2009). Plant system: Nature's nanofactory. *Colloids and Surfaces B: Biointerfaces*, *73*, 219–223.

Jiang, S., Dai, L., Qin, Y., Xiong, L., & Sun, Q. J. P. O. (2016). Preparation and characterization of octenyl succinic anhydride modified taro starch nanoparticles. *PLOS One*, *11*, e0150043.

Jivan, J. M., Yarmand, M., & Madadlou, A. (2014). Encapsulation of date palm pit extract via particulation of starch nanocrystals in a microemulsion. *International Journal of Food Science and Technology*, *49*, 920–923.

Joshi, S. (2008). Can nanotechnology improve the sustainability of biobased products? *Journal of Industrial Ecology*, *12*, 474–489.

Joudeh, N., & Linke, D. (2022). Nanoparticle classification, physicochemical properties, characterization, and applications: A comprehensive review for biologists. *Journal of Nanobiotechnology*, *20*, 262.

Joy, J., Krishnamoorthy, A., Tanna, A., Kamathe, V., Nagar, R., & Srinivasan, S. (2022). Recent developments on the synthesis of nanocomposite materials via ball milling approach for energy storage applications. *Applied Sciences*, *12*, 9312.

Julinová, M., Vaňharová, L., Jurča, M., Minařík, A., Duchek, P., Kavečková, J., Rouchalová, D., & Skácelík, P. (2019). Effect of different fillers on the biodegradation rate of thermoplastic starch in water and soil environments. *Journal of Polymers and the Environment*, *28*, 566–583.

Kang, X., Gao, W., Cui, B., & Abd El-Aty, A. M. (2023). Structure and genetic regulation of starch formation in sorghum (*Sorghum bicolor* (L.) *Moench*) endosperm: A review. *International Journal of Biological Macromolecules*, *239*, 124315.

Khan, I., Saeed, K., & Khan, I. (2019). Nanoparticles: Properties, applications and toxicities. *Arabian Journal of Chemistry*, *12*, 908–931.

Khan, W. U., Khan, S. U., Khan, S. U., Gul, A., Zaman, H., Khan, W. U., Ali, N., Aziz, A., Khan, M. I., Khan, D., & Gouda, M. M. (2025). Carboxymethyl-cellulose/starch/copper-oxide nanocomposite hydrogel green synthesis for organic pollutants photocatalytic degradation that supports health applications. *Colloids and Surfaces A: Physicochemical and Engineering Aspects*, *718*, 136919.

Kiatponglarp, W., Tongta, S., Rolland-Sabaté, A., & Buléon, A. (2015). Crystallization and chain reorganization of debranched rice starches in relation to resistant starch formation. *Carbohydrates Polymer*, *122*, 108–114.

Kim, H. Y., Lee, J. H., Kim, J. Y., Lim, W. J., & Lim, S. T. (2012). Characterization of nanoparticles prepared by acid hydrolysis of various starches. *Starch/Starke*, *64*, 367–373.

Kim, H.-Y., Park, D. J., Kim, J.-Y., & Lim, S.-T. (2013). Preparation of crystalline starch nanoparticles using cold acid hydrolysis and ultrasonication. *Carbohydrate Polymers*, *98*, 295–301.

Kumar, S., & Thakur, K. S. (2020). Active packaging technology to retain storage quality of pear cv. "Bartlett" during shelf-life periods under ambient holding after periodic cold storage. *Packaging Technology and Science*, *33*, 239–254.

Kumari, S., Yadav, B. S., & Yadav, R. B. (2020). Synthesis and modification approaches for starch nanoparticles for their emerging food industrial applications: A review. *Food Research International*, *128*, 108765.

Kumari, S. V. G., Pakshirajan, K., & Pugazhenthi, G. (2022). Recent advances and future prospects of cellulose, starch, chitosan, polylactic acid and polyhydroxyalkanoates for sustainable food packaging applications. *International Journal of Biological Macromolecules*, *221*, 163–182.

Lamanna, M., Morales, N. J., García, N. L., & Goyanes, S. (2013). Development and characterization of starch nanoparticles by gamma radiation: Potential application as starch matrix filler. *Carbohydrates Polymers*, *97*(1), 90–97.

Lengert, E. V., Koltsov, S. I., Li, J., Ermakov, A. V., Parakhonskiy, B. V., Skorb, E. V., & Skirtach, A. G. (2020). Nanoparticles in polyelectrolyte multilayer layer-by-layer (LbL) films and capsules—key enabling components of hybrid coatings. *Coatings*, *10*, 1131.

Li, X., Ge, S., Yang, J., Chang, R., Liang, C., Xiong, L., *et al.* (2017). Synthesis and study the properties of StNPs/gum nanoparticles for salvianolic acid B-oral delivery system. *Food Chemistry*, *229*, 111–119.

Li, J., Yang, Y., Lu, L., Ma, Q., & Zhang, J. (2018). Preparation, characterization and systemic application of self-assembled hydroxyethyl SNP's-loaded flavonoid Morin for hyperuricemia therapy. *International Journal of Nanomedicine*, 2129–2141.

Li, D., Cao, G., Yao, X., Yang, Y., Yang, D., Liu, N., ... & Yang, X. (2023). Tartary buckwheat-derived exosome-like nanovesicles against starch digestion and their interaction mechanism. *Food Hydrocolloids*, *141*, 108739.

Liu, C., Jiang, S., Han, Z., Xiong, L., & Sun (2016). In vitro digestion of nanoscale starch particles and evolution of thermal, morphological, and structural characteristics. *Food Hydrocolloids*, *61*, 344–350.

Liu, C., Qin, Y., Li, X., Sun, Q., Xiong, L., & Liu, Z. (2016). Preparation and characterization of starch nanoparticles via self-assembly at moderate temperature. *International Journal of Biological Macromolecules*, 84, 354–360.

Liu, G., Gu, Z., Hong, Y., Cheng, L., & Li, C. (2017). Electrospun starch nanofibers: Recent advances, challenges, and strategies for potential pharmaceutical applications. *Journal of Controlled Release*, *252*, 95–107.

Liu, M., Guo, X., Ma, X., Xie, Z., Wu, Y., & Ouyang, J. (2024). Physicochemical properties of a novel chestnut porous starch nanoparticle. *International Journal of Biological Macromolecules*, *261*, 129920.

López, O. V., Castillo, L. A., García, M. A., Villar, M. A., & Barbosa, S. E. (2015). Food packaging bags based on thermoplastic corn starch reinforced with talc nanoparticles. *Food Hydrocolloids*, *43*, 18–24.

Lu, H., He, X., Qin, Y., Ji, N., Dai, L., Xiong, L., Shi, R., Wang, T., & Sun, Q. (2023). Preparation and characterization of V-type starch nanoparticles by an oil-water interface method. *Food Hydrocolloids*, *138*, 108455.

Ma, X., Jian, R., Chang, P. R., & Yu, J. (2008). Fabrication and characterisation of citric acid modified starch nanoparticles/plasticized starch composites. *Biomacromolecules*, 9, 3314–3320.

Ma, S., He, J., Chen, Q., Zhou, J., Xie, F., & Cai, J. (2025). Advancements in starch-based nanomaterials for functional delivery and food packaging: A comprehensive review and future directions. *Journal of Future Foods*, *5*, 443–454.

Martens, B. M. J., Gerrits, W. J. J., Bruininx, E. M. A. M., & Schols, H. A. (2018). Amylopectin structure and crystallinity explains variation in digestion kinetics of starches across botanic sources in an in vitro pig model. *Journal of Animal Science and Biotechnology*, *9*, 91.

Meng, F., Tian, S., Chen, Y., & Liu, Z. (2025). Preparation and physicochemical properties of OSA modified *Cyperus esculentus* starch nanoparticles. *International Journal of Biological Macromolecules*, *292*, 140045.

Mittu, B., Lily, M. K., Begum, Z., Bala, R., Singh, M., & Kaur, M. (2024). Starch-based nanomaterials, their properties and sources. In Nayik, G.A. & Dar, A.H. (Eds.) *Starch Based Nanomaterials for Food Packaging.* (pp. 1–22). Academic Press. Elsevier. https://doi.org/10.1016/B978-0-443-18967-8.00008-6

Mukurumbira, A., Mariano, M., Dufresne, A., Mellem, J. J., & Amonsou, E. O. (2017). Microstructure, thermal properties and crysatllinity of amadube starch nanocrystals. *International Journal of Biological Macromolecules*, *102*, 241–247.

Naikwadi, A. T., Sharma, B. K., Bhatt, K. D., & Mahanwar, P. A. (2022). Gamma radiation processed polymeric materials for high performance applications: A review. *Frontiers in Chemistry*, *10*. https://doi.org/10.3389/fchem.2022.837111

Najafi, S. H. M., Baghaie, M., & Ashori, A. (2016). Preparation and characterization of acetylated starch nanoparticles as drug carrier: Ciprofloxacin as a model. *International Journal of Biological Macromolecules*, *87*, 48–54.

Nassarawa, S. S., Abdelshafy, A. M., Nasiru, M. M., Gambo, A., Dandago, M. A., Yusuf, H. L., Maulud, A. D. J., Sulaiman, S. A., & Mohammed, M. (2024). Starch-based nanomaterials safety aspects: Perspectives and future trends. In Nayik, G.A. & Dar, A.H. (Eds.) *Starch Based Nanomaterials for Food Packaging.* (pp. 269–284). Academic Press. Elsevier. https://doi.org/10.1016/B978-0-443-18967-8.00002-5

Neethirajan, S., Thomson, D., Jayas, D., & White, N. (2008). Characterization of the surface morphology of durum wheat starch granules using atomic force microscopy. *Microscopy Research and Technique*, *71*, 125–132.

Neme, K., Nafady, A., Uddin, S., & Tola, Y. B. (2021). Application of nanotechnology in agriculture, postharvest loss reduction and food processing: Food security implication and challenges. *Heliyon*, *7*, e08539.

Nobakht-Nia, M., Niakousari, M., Eskandari, M. H., Golmakani, M.-T., & Hosseini, S. M. H. (2025). Fabrication and characterization of decanoyl chloride/curcumin-modified potato starch nanoparticles and the potential application in the stabilization of flaxseed oil-in-water Pickering emulsions. *International Journal of Biological Macromolecules*, *307*, 141888.

Noorbakhsh-Soltani, S., Zerafat, M., & Sabbaghi, S. (2018). A comparative study of gelatin and starch-based nanocomposite films modified by nano-cellulose and chitosan for food packaging applications. *Carbohydrate Polymers*, *189*, 48–55.

Oh, S.-M., Lee, B.-H., Seo, D.-H., Choi, H.-W., Kim, B.-Y., & Baik, M.-Y. (2020). Starch nanoparticles prepared by enzymatic hydrolysis and self-assembly of short-chain glucans. *Food Science and Biotechnology*, *29*, 585–598.

Oladebeye, A. O., Oshodi, A. A., Amoo, I. A., & Karim, A. A. (2013). Morphology, X–ray diffraction and solubility of underutilized legume starch nanocrystals. *International Journal of Scientific Research*, *2*(3), 497–503.

Pacheco, M. T., Rojas, B. A., Burbano, K. Y., Zevallos, K. N., Mendoza, A. F., Rodríguez, D. H., Díaz, R. G., & Moreno, F. J. (2025). Modification of green banana and maize starches: A comparative study to select an ingredient with functional and technological potential. *LWT*, *220*, 117511.

Pei, Q., Jiang, B., Hao, D., & Xie, Z. (2023). Self-assembled nanoformulations of paclitaxel for enhanced cancer theranostics. *Acta Pharmaceutica Sinica B*, *13*, 3252–3276.

Pfister, B., & Zeeman, S. C. (2016). Formation of starch in plant cells. *Cellular and Molecular Life Sciences*, *73*, 2781–2807.

Phanthong, P., Reubroycharoen, P., Hao, X., Xu, G., Abudula, A., & Guan, G. (2018). Nanocellulose: Extraction and application. *Carbon Resources Conversion*, *1*, 32–43.

Pooresmaeil, M., & Namazi, H. (2021). Advances in development of the dendrimers having natural saccharides in their structure for efficient and controlled drug delivery applications. *European Polymer Journal*, *148*, 110356.

Pradhan, A., Saha, D., & Tripathy, P. P. (2025). A multivariate study on the effect of 3D printing parameters on the printability of gluten-free composite dough. *Journal of the Science of Food and Agriculture*, *105*(4), 2578–2593.gr https://doi.org/10.1002/jsfa.14031

Puentes, A. A. E., García-Gurrola, A., Rincón, S., Zepeda, A., & Martínez-Bustos, F. (2020). Effect of amylose/amylopectin content and succinylation on properties of corn starch nanoparticles as encapsulants of anthocyanins. *Carbohydrate polymers*, *250*, 116972.

Rajeswari, L. S., Moorthy, S. N., & Rajasekharan, K. N. (2011). Preparation of cassava starch nanoparticles and their application as a carrier system for curcumin delivery. *International Journal of Nanotechnology and Application*, *5*, 193–201.

Ramani, A., Taherabbas, S., Saji, R., Bumbadiya, M., Gandhi, K., & Seth, R. (2024). Nanotechnology: An emerging trend in the dairy industry—Applications and future challenges. *Food and Humanity*, *3*, 100409.

Rastmanesh, S., Jafarizadeh-Malmiri, H., Javadi, A., & Anarjan, N. (2024). Enzymatically and chemically starch nanoparticles preparation using ultrasonication, precipitation and lyophilization post-treatments: Screening and characterization. *International Journal of Biological Macromolecules*, *277*, 134506.

Rayo-Mendez, L. M., & Rodriguez-Llanos, J. (2024). In G.A. Valencia (Ed.), Starch nanovesicles and nanomicelles. In Valencia, G. A. (Ed.) *Starch nanomaterials and food applications.* (pp. 81–113). Cham: Springer Nature Switzerland. https://doi.org/10.1007/978-3-031-60086-9_5

Remanan, M. K., & Zhu, F. (2024). Encapsulation of chrysin and rutin using self-assembled nanoparticles of debranched quinoa, maize, and waxy maize starches. *Carbohydrate Polymers*, *337*, 122118.

Rodrigues, A., & Emeje, M. (2012). Recent applications of starch derivatives in nanodrug delivery. *Carbohydrate Polymers*, *87*, 987–994.

Rouf, T. B., & Kokini, J. L. (2017). Natural biopolymer-based nanocomposite films for packaging applications. In Jawaid, M. & Swain, S. K. (Eds.) *Bionanocomposites for packaging applications*. (pp. 149–177). Cham: Springer International Publishing. https://doi.org/10.1007/978-3-319-67319-6_8

Saha, D., Pradhan, A., Sarkar, A., & Tripathy, P. P. (2024). Recent trends in pseudocereal-based foods. In Dhull, S. B., Bains, A., Chawla, P., & Kaur, S. (Eds.) *Pseudocereals* (pp. 264–291). CRC Press.

Saha, D., Padhiary, M., Hoque, A., & Prasad, G. (2025). 3D printing technology for valorization of food processing wastes and byproducts: A systematic review. *Waste Management Bulletin*, *3*(4), 100192. https://doi.org/10.1016/j.wmb.2025.100192

Saha, D., Padhiary, M., & Chandrakar, N. (2025). AI vision and machine learning for enhanced automation in food industry: A systematic review. *Food and Humanity*, 100587. https://doi.org/10.1016/j.foohum.2025.100587

Saha, D., Panme, F. A., & Pradhan, A. (2025). Green technologies for sustainable agricultural management. In Kuanar, A., Kar, D., Das, A. P., & Bhanja, D. (Eds.) *Recent advances in bioremediation and phytoremediation* (pp. 245–271). Cham: Springer Nature Switzerland. https://doi.org/10.1007/978-3-031-77884-1_11

Santhosh, R., & Sarkar, P. (2024). Fabrication of jamun seed starch/tamarind kernel xyloglucan bio-nanocomposite films incorporated with chitosan nanopsarticles and their application watermelon sapota (*Manilkara zapota*) fruits. *International Journal of Biological Macromolecules*, *260*, 129625.

Sarkar, A., Saha, D., Pradhan, A., & Tripathy, P. P. (2024). Innovative processing technologies for pseudocereals. In Dhull, S.B., Bains, A., Chawla, P. & Kaur, S. (Eds.) *Pseudocereals* (pp. 191–215). CRC Press.

Sekhon, B. S. (2010). Food nanotechnology—an overview. *Nanotechnology, Science and Applications*, *4*(3), 1–15.

Sen, C., & Das, M. (2018). Biodegradability of starch based self-supporting antimicrobial film and its effect on soil quality. *Journal of Polymers and the Environment*, *26*, 4331–4337.

Shahbazi, M., Jäger, H., Ettelaie, R., Chen, J., Kashi, P. A., & Mohammadi, A. (2024). Dispersion strategies of nanomaterials in polymeric inks for efficient 3D printing of soft and smart 3D structures: A systematic review. *Advances in Colloid and Interface Science*, *333*, 103285.

Shekhar, S., Prakash, P., Shekhar, S., Singh, S. K., & Prasad, K. (2023). Ultrasound-assisted green synthesis of silver nanoparticles from *Allium sativum*, its characterization, antimicrobial capabilities and thermo-plasmonic studies. *Journal of Physics: Conference Series*, *2663*, 012020.

Shrivastava, A., Gupta, R. K., & Srivastav, P. P. (2024). Exploring novel frontiers of advancements in purple yam (*Dioscorea alata* L.) starch extraction, modification, characterization, applications in food and other industries. *Measurement: Food*, *15*, 100196. https://doi.org/10.1016/j.meafoo.2024.100196

Singh, A. K., Lee, M., Jang, D., & Lee, Y. S. (2024). Non-conventional starch nanoparticles: Novel avenues towards improving sustainability of the food packaging sector. *Trends in Food Science & Technology*, *143*, 104273.

Singh, J., Kaur, L., & Mccarthy, O. (2007). Factors influencing the physico-chemical, morphological, thermal and rheological properties of some chemically modified starches for food applications: A review. *Food Hydrocolloids*, *21*, 1–22.

Singh, R., Thakur, P., Thakur, A., Kumar, H., Chawla, P., Rohit, J., Kaushik, R., & Kumar, N. (2021). Colorimetric sensing approaches of surface-modified gold and silver nanoparticles for detection of residual pesticides: A review. *International Journal of Environmental Analytical Chemistry*, *101*, 3006–3022.

Song, S., Wang, C., Pan, Z., & Wang, X. (2008). Preparation and characterization of amphiphilic starch nanocrystals. *Journal of Applied Polymer Science*, *107*, 418–422.

Sreejit, V., Preetha, R., Mubeena, S. A., & Dhananjay, S. (2025). Green synthesized nano starch: Preparation and emerging avenues in food industry. *Journal of Nanoparticle Research*, *27*, 105.

Su, C.-Y., Li, D., Wang, L.-J., & Wang, Y. (2022). Biodegradation behavior and digestive properties of starch-based film for food packaging: A review. *Critical Reviews in Food Science and Nutrition*, *63*, 6923–6945.

Sun, Q., Fan, H., & Xiong, L. (2014). preparation and characterization of SNP's through ultrasonic- assisted oxidation methods. *Carbohydrates Polymer*, *106*, 359–436.

Sun, Q., Li, G., Dai, L., Ji, N., & Xiong, L. (2014). Green preparation and characterisation of waxy maize SNP's through enzymolysis and re crystallisation. *Food Chemistry*, *162*, 223–228.

Tibolla, H., Czaikoski, A., Pelissari, F. M., Menegalli, F. C., & Cunha, R. L. (2020). Starch-based nanocomposites with cellulose nanofibers obtained from chemical and mechanical treatments. *International Journal of Biological Macromolecules*, *161*, 132–146.

Torres, F. G., Arroyo, J., Tineo, C., & Trancoso, O. (2018). Tailoring the properties of native Andean potato starch nanoparticles using acid and alkaline treatment. *Starch*, 1–8.

Torres, F. G., & De-la-Torre, G. E. (2022). Synthesis, characteristics, and applications of modified starch nanoparticles: A review. *International Journal of Biological Macromolecules*, *194*, 289–305.

Vencato, A. A., Clerici, N. J., Rigotti, M., Branco, C. S., Carra, S., Malvessi, E., & Brandelli, A. (2025). Biocompatible PVA/gelatin and PVA/starch nanofibers incorporating sodium lactobionate obtained from enzymatic synthesis. *International Journal of Biological Macromolecules*, *309*, 143112.

Venturini, C. G., Jäger, E., Oliveira, C. P., Bernardi, A., Battastini, A. M., Guterres, S. S., & Pohlmann, A. R. (2011). Formulation of lipid core nanocapsules. *Colloids and Surfaces A: Physicochemical and Engineering Aspects*, *375*(1–3), 200–208.

Wang, J., Xue, X., Han, S., Fu, M., Feng, Y., Wang, Y., Dong, S., & Guo, P. (2025). Effects of dual modification by cold plasma and heat-moisture treatment on the structure and in vitro digestibility of waxy maize starch. *Innovative Food Science & Emerging Technologies*, *102*, 103991.

Wang, L., Sang, W., Xu, R., & Cao, J. (2020). Alteration of flesh color and enhancement of bioactive substances via the stimulation of anthocyanin biosynthesis in 'Friar' plum fruit by low temperature and the removal. *Food Chemistry*, *310*, 125862.

Wang, S., & Ren, F. (2020). Rheological, pasting, and textural properties of starch. In Wang, S. (Ed.) *Starch structure, functionality and application in foods* (pp. 121–129). Singapore: Springer Singapore. https://doi.org/10.1007/978-981-15-0622-2_7

Winarti, C., Sunarti, T. C., Mangunwidjaja, D., & Richana, N. (2014). Preparation of arrowroot starch nanoparticles by butanol complex precipitation and its application as bioactive encapsulation matrix. *International Food Research Journal*, *21*(6), 2207–2213.

Wu, C. Y., & Wang, W. (2022). Application of antisolvent precipitation method for formulating excipient-free nanoparticles of psychotropic drugs. *Pharmaceutics*, *14*, 819.

Xie, L., He, Z., Shen, S., Chen, L., Wang, T., & Li, X. (2025). Fabrication of thermotolerant ovalbumin and carboxymethyl starch sodium complexed nanoparticles and high-internal-phase Pickering emulsion for food chemical hazards inhibition. *International Journal of Biological Macromolecules*, *309*, 143072.

Xu, Y., Ding, W. Q., Liu, J., Li, Y., Kennedy, J. F., Gu, Q., et al. (2010). Preparation and characterisation of organic–soluble acetylated starch nanocrystals. *Carbohydrate Polymers*, *80*, 1078–1084.

Xuan, L., Ju, Z., Skonieczna, M., Zhou, P. K., & Huang, R. (2023). Nanoparticles-induced potential toxicity on human health: Applications, toxicity mechanisms, and evaluation models. *MedComm*, *4*, e327.

Yang, J., Zhang, C., Xian, M., Chen, Y., Zhang, L., & Qian, J.-Y. (2025). Formation mechanism of anisotropic zein-modified starch nanoparticles. *Food Chemistry*, *482*, 144177.

Yu, J. H., Ai, F., Dufresne, A., Gao, S., Huang, J., & Chang, P. R. (2008). Structure and mechanical properties of poly (lactic acid) filled with (starch nanocrystal)-graft-poly (ε-caprolactone). *Macromolecular Materials and Engineering*, *293*, 763–770.

Yu, M., Ji, N., Wang, Y., Dai, L., Xiong, L., & Sun, Q. (2020). Starch-based nanoparticles: Stimuli responsiveness, toxicity, and interactions with food components. *Comprehensive Reviews in Food Science and Food Safety*, *20*, 1075–1100.

Yuan, T., Zeng, J., Wang, B., Cheng, Z., Gao, W., Xu, J., & Chen, K. (2021). Silver nanoparticles immobilized on cellulose nanofibrils for starch-based nanocomposites with high antibacterial, biocompatible, and mechanical properties. *Cellulose*, *28*, 855–869.

Zhang, Y., Rempel, C., & Liu, Q. (2014). Thermoplastic starch processing and characteristics—a review. *Critical Reviews in Food Science and Nutrition*, *54*, 1353–1370.

Zheng, H., Ai, F., Chang, P. R., Huang, J., & Dufresne, A. (2009). Structure and properties of starch nanocrystals reinforced soy protein plastics. *Polymer Composites*, *30*, 474–480.

6 Responsive and Smart Starches

Anjali Tiwari, Arti Malviya, Shraddha Sharma, Pratibha Saxena, and Lisha Kurup
Lakshmi Narain College of Technology, Bhopal
Maulana Azad National Institute of Technology, Bhopal

6.1 INTRODUCTION

Starches are polysaccharides that serve as a major energy source in the human diet and have diverse industrial applications. The development of responsive and smart starches has opened new avenues for their use in advanced technological applications. When exposed to changes in environmental stimuli such as pH, temperature, and light, these novel materials show changes in their properties or possess enhanced functionalities tailored for specific uses. The chapter explores different types, mechanisms, and uses of responsive and smart starches. The quest for innovative materials with adaptive functionalities has led to the emergence of responsive and smart starches, a subset of biomaterials that combine the intrinsic properties of traditional starch with advanced, dynamic responses to external stimuli. These intelligent biomaterials have garnered significant attention due to their potential to revolutionize various industries, including food science, pharmaceuticals, and environmental technologies. Starch is a naturally occurring, biodegradable polymer that is widely accessible and reasonably priced. The repeating units of anhydroglucose (AGU) are grouped in linear (amylose, with α-1,4-glycosidic linkages) and branching (amylopectin, with α-1,6- and α-1,4-glycosidic linkages) fractions to form the comparatively basic structure of starch (Dominguez-Ayala et al., 2022; Gupta et al., 2022; Shrivastava et al., 2024). Starch is commonly chemically modified to change its inherent physicochemical properties by emphasizing the numerous hydroxyl groups that are present in its structure (Fan & Picchioni, 2020; Gupta et al., 2023). Grafting, etherification, and polymerization are used for the majority of alterations (Gilet et al., 2018; Meimoun et al., 2018; Wang et al., 2020). However, because of their immiscibility, mechanical properties, processability, and insufficient oxygen and vapour barrier properties, starch-based materials are challenging to manufacture (de Freitas et al., 2021; Jiang et al., 2020; Majeed et al., 2021; Nasiri et al., 2020). Similarly, when employing starch and synthetic polymers, biocompatibility, biodegradability, and cytotoxicity are important considerations (Lv et al., 2022; Sarwan et al., 2020). Starch, one of nature's most prevalent natural polymers, is widely used in the food industry because it is inexpensive, abundant, biodegradable, and edible (Ji et al., 2019). Amylose and amylopectin make up starch (Figure 6.1). Amylose has excellent film-forming properties and can form colourless and transparent films (Jiménez et al., 2012).

Plant essential oils, plant extracts, and nanoparticles can be added to starch films to improve their antioxidant, antibacterial, and barrier properties (Gupta et al., 2025a). Starch films can become pH-responsive by adding anthocyanins (C. Cui et al., 2021). Smart starches, also known as responsive starches, are chemically or physically modified starches designed to react to specific external triggers such as changes in temperature, pH, light, or moisture. Unlike conventional starches, which primarily serve as thickeners and stabilizers, smart starches possess the ability to alter their physical or chemical state in response to environmental changes (Gupta et al., 2025b). This responsiveness is achieved through the incorporation of functional groups or the integration of polymers that interact with the external stimuli in a controlled manner (Zhang et al., 2019). The development of smart starches is driven by the need for materials that not only perform traditional roles but also offer additional functionalities tailored to specific applications. For instance, in the food industry, smart starches can be engineered to change texture or release flavours based on temperature changes, enhancing consumer experiences and product shelf life. In the pharmaceutical field, responsive starches can enable controlled drug release, improving the efficacy and reducing side effects of medications. Furthermore, the environmental impact of conventional materials has spurred

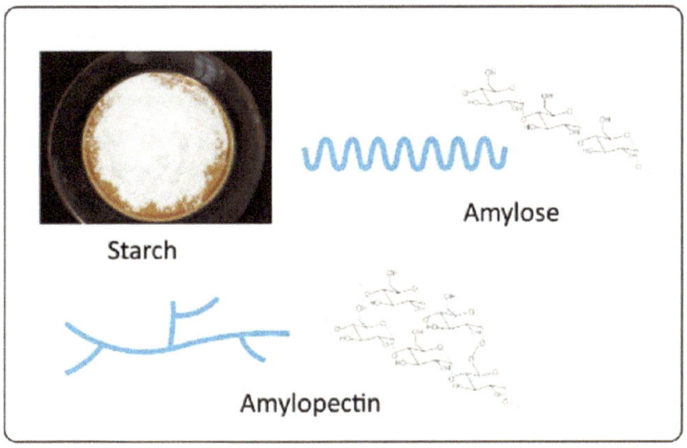

FIGURE 6.1 Chemical structure of amylose and amylopectin.

DOI: 10.1201/9781003564287-6

interest in biodegradable smart starches for use in sustainable packaging solutions and agricultural applications.

Responsive starch materials can have advantages in environmental and agricultural applications. Starch-based products are used in water treatment procedures to eliminate heavy metals and dyes owing to their absorbent and biodegradable properties (C. Cui et al., 2021). Starch-based solutions in agriculture provide slow-release fertilizers and enhance water retention, enhancing efficiency and mitigating environmental effect. Chemically modified starch derivatives have shown efficacy in increased oil recovery, providing environmentally benign alternatives to synthetic polymers while ensuring stability under harsh circumstances (Chen et al., 2022). Starch-based hydrogels and emulsifiers are used in the food sector to improve product quality and stability. They are especially beneficial for noodle production, encapsulating bioactive compounds, and stabilizing food emulsions, thus enhancing texture and prolonging shelf life. The adaptability of starch is further shown in industrial applications, where starch is incorporated into optical sensors for environmental monitoring and diagnostics. Responsive starch composites are used in these sensors to detect fluctuations in temperature, pH, and other factors, facilitating real-time data acquisition (Sun et al., 2019). The amalgamation of starch with nanotechnology has enhanced the use of starch across several areas. Starch-based materials augmented with nanoparticles have superior mechanical characteristics and responsiveness, enabling novel applications in biological, environmental, and industrial sectors (Chen et al., 2020). Nonetheless, obstacles such as scalability, longevity, and compatibility with new technology persist. Resolving these difficulties via continuous research and development would realize the full potential of responsive starch-based materials, facilitating their application across many sectors while enhancing sustainability and efficiency.

Responsive starches represent a significant advancement in materials science, displaying a range of behaviours in response to various external stimuli. These properties make them valuable in numerous fields, including food technology, pharmaceuticals, and materials science. Responsive and smart starch-based biomaterials represent an emerging class of materials that can adapt and react to environmental stimuli. These materials have advanced applications in drug delivery, smart packaging, tissue engineering, and environmental protection. This chapter will focus on understanding smart and responsive starches, their types and preparations, and evolving applications in various industries. Responsive and smart starches are innovative materials that respond to external stimuli or possess enhanced functionalities.

6.2 TYPES OF SMART AND RESPONSIVE STARCHES

Starch, a natural carbohydrate found abundantly in plants, is widely used due to its versatility and eco-friendly nature (Gupta et al., 2024a). Recent developments have led to the creation of smart and responsive starches, which can adapt to changes in their environment. Responsive starches are materials that can adapt their physical or chemical properties based on environmental stimuli, including temperature, pH, light, or chemicals. Similar to smart starches, they undergo transformations, such as swelling, shrinking, gelation, or dissolution, when exposed to particular conditions. These starches are designed to provide controlled responses in various applications, offering the possibility of creating self-regulating systems. A responsive starch shows great potential in acting as eco-friendly sustainable resource for industrial production and daily utility.

6.2.1 Thermo/Temperature-Sensitive Starches

Thermo-sensitive starches are advanced materials that respond to temperature variations, resulting in notable changes in their physical or chemical properties. These starches are widely studied for their ability to undergo reversible transitions between sol (liquid) and gel (semi-solid) states at specific temperatures. Such behaviour is particularly valuable in fields such as food science, pharmaceuticals, and biomedicine, where controlled material behaviour is essential. For instance, starches modified with thermo-responsive polymers such as poly(*N*-isopropylacrylamide) (PNIPAm) exhibit a lower critical solution temperature (LCST). Above this threshold, the material transitions from a liquid to a gel, driven by interactions between the polymer chains and the aqueous medium. This unique feature is harnessed to design temperature-responsive gels for food processing and targeted drug delivery systems (Yang et al., 2021).

The thermo-sensitive properties of these starches are attributed to their structural modifications, which include grafting functional polymers onto the starch backbone. These modifications enable phase transitions in response to heat, allowing the starch to alternate between solubility states. Below the LCST, the material remains dispersed as a solution, while above this temperature, it aggregates into a gel due to the collapse of polymer chains. This phase behaviour is critical in applications requiring precision in thermal responsiveness, such as sustained drug release and thermal gelation in food products.

These products are employed as smart packaging materials that can indicate temperature changes to ensure product integrity and in controlled release systems in pharmaceuticals and agrochemicals, where the release rate can be adjusted with temperature. Thermo-responsive starches are crucial in controlling the texture and stability of food products like sauces and puddings (Fennema, 1996). These starches are employed in drug delivery systems where temperature sensitivity is required to control release rates (Kwon et al., 2011).

6.2.1.1 Preparation of Thermo-Sensitive Starches

Thermo-sensitive starches are synthesized through chemical processes such as free radical polymerization or atom transfer radical polymerization (ATRP). These techniques enable the grafting of thermo-sensitive polymers such as PNIPAm onto the starch backbone. The addition of cross-linking agents, such as divinyl sulphone, further enhances the material's thermal stability and responsiveness. The resulting starches exhibit reversible sol–gel transitions, providing versatility for diverse applications (Chen et al., 2019).

The preparation process generally involves the following processes:

- *Starch activation:* The native starch is chemically or enzymatically activated to create functional groups that facilitate polymer attachment.
- *Polymer grafting:* Thermo-responsive polymers are covalently bonded to the starch backbone using grafting agents or initiators.
- *Cross-linking:* Cross-linking agents reinforce the structure, enabling more defined phase transitions.

6.2.1.2 Structural Representation

A structural representation of starch and divinyl sulphone is shown in Figure 6.2.

6.2.1.3 Key Components

- *Starch backbone:* A polysaccharide chain providing the structural framework.
- *Thermo-sensitive polymer (e.g., PNIPAm):* Responsible for the temperature-induced phase transitions.
- *Cross-linking agent:* Enhances mechanical strength and maintains polymer–starch integrity.

Recent advances in the preparation of thermo-sensitive starches include novel processing methods such as the following:

- *High hydrostatic pressure (HHP):* HHP involves subjecting starch to pressures of up to 600 MPa, which facilitates gelatinization and enhances properties such as swelling, solubility, and film-forming capacity. Unlike thermal processing, HHP preserves molecular integrity while reducing gelatinization temperature and viscosity, resulting in improved physical and functional properties. Studies on waxy wheat and mung bean starches have demonstrated improved digestibility and reduced crystallinity through this method.
- *Plasma processing:* Dielectric barrier discharge plasma induces structural changes such as oxidation and degradation of starch molecules, enhancing their responsiveness and functional attributes. For instance, plasma-treated potato and corn starches exhibit have been shown to improve stability and functional diversity (Gupta et al., 2024b).
- *Enzymatic modifications:* Enzyme treatments have been explored to functionalize starches with thermo-responsive polymers, providing eco-friendly alternatives to chemical methods.

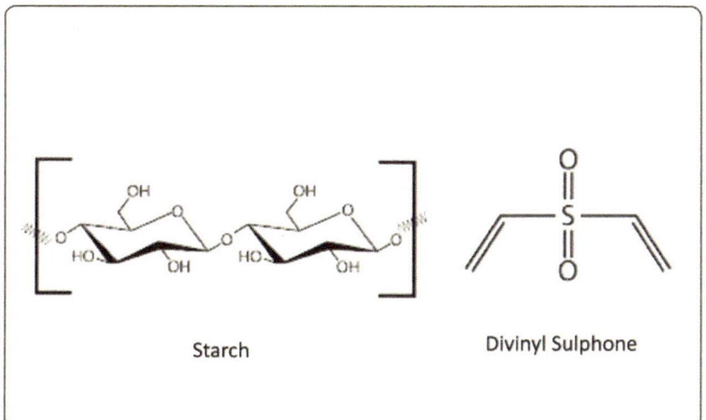

FIGURE 6.2 Chemical structure of starch and divinyl sulphone.

These innovative preparation methods are broadening the application spectrum of thermo-sensitive starches in sustainable technologies (Chen et al., 2019; Wang et al., 2022).

6.2.2 pH-Sensitive Starches

pH-sensitive starches are modified biopolymers that respond to environmental pH fluctuations by altering their structural properties, such as solubility, swelling behaviour, and degradation rates. These materials have become increasingly important in pharmaceutical formulations, especially in controlled-release drug delivery systems, where the goal is to achieve a sustained and targeted release of therapeutic agents at specific sites in the gastrointestinal (GI) tract. Their ability to adapt to changes in pH—especially the significant variation observed from the acidic stomach (pH 1.5–3.5) to the neutral/alkaline small intestine (pH 6–8)—has made them an attractive choice in the design of site-specific drug delivery systems (Zhao et al., 2022).

pH-responsive starches are employed in drug delivery systems for targeted release in specific areas of the GI tract. Properties such as solubility, degradation, and degree of swelling vary with changes in pH. These starches help create products with consistent texture and stability even when pH levels fluctuate (Haug & Lantzsch, 1983). They are used in controlled-release fertilizers that dissolve under certain pH conditions (Gaur et al., 2011). When starch is mixed with water, it forms gel networks called hydrogels. These hydrogels, when mixed with alginate and pectin, form pH-responsive hydrogels. An alginate pectin layer protects the starch core in acidic environment but opens up as pH increases. This mechanism is helpful in controlled release of any encapsulated compound in bio-inks (Matthias et al., 2024). In an investigation, starch citrate nanoparticles with pH responsiveness characteristics were synthesized using a green esterification process. The cytotoxicity study of human skin cells indicated that the prepared starch is non-toxic and suitable as a drug carrier (Chin et al., 2019). Copolymerization of starch with acrylate comonomers in

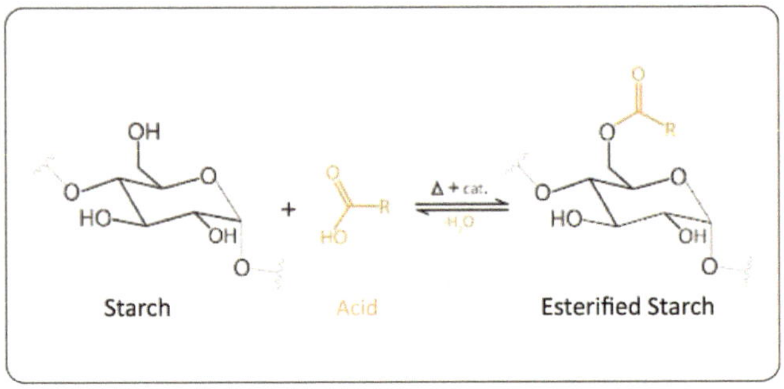

FIGURE 6.3 Chemical structure of starch and esterified starch.

the presence of potassium persulphate as a free radical initiator resulted in the formation of composite hydrogels which were found to be sensitive to pH. They showed quick drug release tendency at neutral pH than acidic conditions. This is attributed to the ionization of anionic groups on starch hydrogels increasing the electrostatic force of repulsion between them leading to swelling of polymer and ease in drug delivery (Quintanilla et al., 2020).

6.2.2.1 Mechanism of Action

The functionality of pH-sensitive starches relies on the incorporation of acid- or base-sensitive functional groups, such as carboxyl (–COOH) or amine (–NH_2) groups, into the starch structure as depicted in Figures 6.3 and 6.4. These groups play a key role in the material's ability to respond to changes in pH. Depending on the pH of the surrounding environment, these functional groups can do the following:

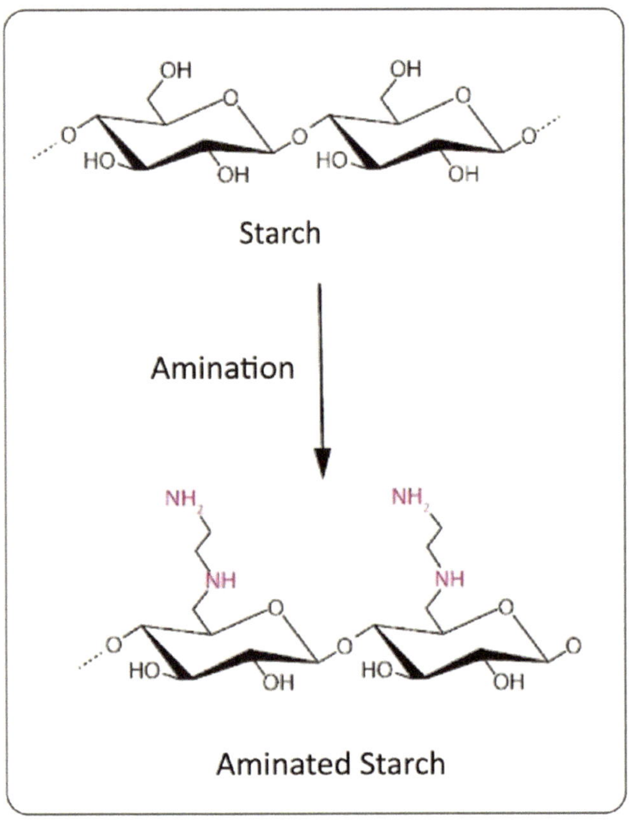

FIGURE 6.4 Amination of starch.

- Ionize (in alkaline or acidic conditions), leading to changes in solubility, swelling, and viscosity.
- Form or break cross-links, influencing the material's mechanical properties and degradation rate.

For instance, starch modified with carboxymethyl groups (–COOH) swells or dissolves in alkaline conditions due to the ionization of the carboxyl group, whereas starches functionalized with amine groups (–NH_2) are more soluble in acidic environments (Cui et al., 2021).

6.2.2.2 Preparation Methods for pH-Sensitive Starches

pH-sensitive starches are typically produced by chemical modification methods such as esterification, graft copolymerization, and cross-linking. These modifications allow for the introduction of functional groups that confer pH sensitivity to the starch.

- *Esterification:* This is a common method to introduce carboxymethyl groups into starch molecules. The process involves reacting starch with monochloroacetic acid or other reagents, creating a starch derivative that is capable of ionizing in basic conditions (Liu et al., 2021).
- *Graft copolymerization:* In this method, pH-sensitive polymers, such as polyacrylic acid (PAA) or polyethylene glycol (PEG), are grafted onto the starch backbone. These grafted chains enhance the material's responsiveness to pH changes and can further control the swelling and degradation behaviours of the starch.
- *Cross-linking:* Cross-linking starches through reagents like epichlorohydrin or glutaraldehyde allows for the creation of networks that swell in response to pH changes. Cross-linked starches typically show reduced solubility, but they can exhibit sustained release properties suitable for drug delivery applications (Vijaykumar et al., 2022).

6.2.2.3 Structural Representation

The introduction of pH-sensitive groups alters the structural characteristics of starch, which, in turn, influence its swelling behaviour, solubility, and degradation rate in different pH environments.

6.2.2.4 Key Components

The structure of a pH-sensitive starch may include the following:

- *Starch backbone:* This forms the structural core of the polymer.
- *Carboxyl group (–COOH):* Upon ionization in basic conditions (pH > 7), the starch swells due to

the electrostatic repulsion between the negatively charged carboxyl groups.

- *Amine group (–NH_2):* Under acidic conditions (pH < 6), amine groups protonate, increasing the solubility of the starch.
- *Responsive polymer chains:* These chains modify the interaction of starch with water and other solvents, controlling the starch's solubility and degradation in response to pH changes.

Recent research has focused on optimizing the properties of pH-sensitive starches for drug delivery and biomedical applications. Some recent advancements include the following:

- *Smart starch nanoparticles:* Researchers have successfully developed pH-sensitive starch-based nanoparticles for drug encapsulation. These nanoparticles can efficiently deliver active pharmaceutical ingredients (APIs) to the small intestine, where the pH is neutral or slightly alkaline. The nanoparticles release the API in a controlled manner, preventing premature drug release in the stomach (Bajpai et al., 2022).
- *Dual-responsive starches:* Recent studies have introduced starches that respond to both pH and temperature. These dual-responsive starches can be used in environments where both pH and temperature fluctuate, such as the human body or industrial processes. By incorporating both thermo-responsive and pH-sensitive groups, researchers can achieve even more precise control over drug release rates (Zhao et al., 2022).
- *Advanced grafting techniques:* The use of advanced grafting techniques has enhanced the efficiency of starch-based materials for bioactive compound delivery. These modifications improve the stability, biodegradability, and drug release profiles of starch-based systems. In addition, grafted starch polymers have been used in wound healing applications due to their ability to respond to local pH variations at the injury site. While starch is a natural biopolymer, modifications to create pH-sensitive versions must be carefully evaluated for biocompatibility and toxicity. Recent studies have assessed the safety of pH-sensitive starch derivatives in in vivo models. These studies confirm the low toxicity and excellent biocompatibility of modified starches, making them suitable for pharmaceutical and medical applications (Vijaykumar et al., 2022).

6.2.3 Light-Responsive Starches

Light-responsive materials, including light-responsive starches, have garnered significant attention due to their unique ability to change properties such as colour, solubility, or structure upon exposure to light. These materials, particularly those incorporating photochromic molecules such as azobenzene or spiropyran, offer a level of control and responsiveness that is particularly valuable in smart packaging, optical sensors, and environmental monitoring. The concept of light-sensitive starches is built on the incorporation of photoactive compounds that undergo reversible structural changes when exposed to specific wavelengths of light, typically ultraviolet (UV) or visible light.

The use of light-responsive starches in various applications, including drug delivery and environmental sensors, continues to expand, thanks to advancements in material science and nanotechnology. These advancements allow for better integration of starch-based materials with modern technologies, increasing their potential for sustainability, customizability, and multifunctionality.

Structural changes in starch molecules occur upon exposure to specific wavelengths of light. Starches that undergo structural changes when exposed to light are called light-responsive starches. This innovative category is still emerging but holds promise for advanced applications. Light-responsive starches are of two types:

- *Photo-cross-linked starches:* These starches form cross-links when exposed to light, which modifies their gelation or swelling properties.
- *Photo-sensitive starches:* Designed to change their solubility or other properties under light exposure, these starches offer new possibilities for responsive materials.

UV-responsive packaging that protects products from light-induced degradation is one of the most desirable applications of light-responsive starches. These materials can be utilized in packaging that changes properties under different lighting conditions. UV-responsive packaging protects products from light-induced degradation. These are employed in smart coatings and textiles with photo-responsive properties. They are used also in biotechnical industries to create smart materials that respond to light for controlled release or other functional purposes (Zhang et al., 2015).

6.2.3.1 Mechanism of Light Responsiveness in Starches

Light-responsive starches operate based on the incorporation of photochromic molecules into the starch matrix. These molecules undergo reversible changes in structure that affect their physical properties such as solubility, colour, and viscosity when exposed to light. Some of the most commonly used photochromic molecules are azobenzenes and spiropyrans, which are well known for their ability to undergo trans–cis isomerization (for azobenzenes) or ring-opening and ring-closing reactions (for spiropyrans) in response to light exposure.

- *Azobenzene derivatives:* Azobenzenes are perhaps the most widely studied photochromic compounds in light-responsive systems. Upon exposure to UV light, the azobenzene undergoes a cis–trans isomerization, changing its molecular conformation from a trans to a cis form, which can lead to a change in solubility, swelling behaviour, or mechanical properties. This transformation is reversible, meaning that exposure to visible light or heat can return the molecule to its original trans form, completing the cycle of responsiveness (Shen et al., 2021).

- *Spiropyran:* Spiropyrans are another class of photochromic molecules that can undergo reversible ring-opening and ring-closing reactions when exposed to UV or visible light. The change in structure is often accompanied by a significant change in colour, making spiropyrans particularly useful in optical sensing and smart packaging applications (Zhang et al., 2020.

These photochromic molecules are typically incorporated into the starch matrix via chemical bonding techniques or by physical mixing during the fabrication process. The photoactive properties of these molecules can significantly enhance the starch's responsiveness to light, enabling the material to undergo reversible transformations when exposed to light sources of specific wavelengths.

6.2.3.2 Preparation of Light-Responsive Starches

The preparation of light-sensitive starches involves the integration of photochromic molecules such as azobenzene and spiropyran into the starch polymer network. Several methods are employed to achieve this:

- *Covalent bonding:* The photochromic molecules can be covalently bonded to the starch molecules through reactions such as grafting or cross-linking. For example, azobenzene derivatives can be grafted onto the hydroxyl groups of starch through esterification or etherification reactions. This approach ensures a strong integration of the photoactive molecule within the starch matrix, leading to a stable light-responsive system (Chen et al., 2022).
- *Physical incorporation:* Alternatively, photochromic molecules can be physically mixed with starch during the polymerization process or by using solvent casting techniques. In this method, the molecules are dispersed within the starch matrix, allowing them to retain their light responsiveness properties while interacting with the starch structure. This method is often simpler but may result in less control over the spatial distribution of the photochromic molecules.
- *Nanocomposites:* Another promising method involves the preparation of starch-based nanocomposites, where photochromic molecules are embedded into starch nanoparticles or nanofibres. This method can improve the mechanical strength, stability, and light responsiveness of the starch material, making it suitable for applications in more demanding environments (Shen et al., 2021).
- *Cross-linking:* In some cases, starch is cross-linked with a photo-responsive polymer or monomer to increase the stability of the light-sensitive material and to control the network formation. Cross-linking can also improve the durability and control over response time to light exposure (Zhang et al., 2022).

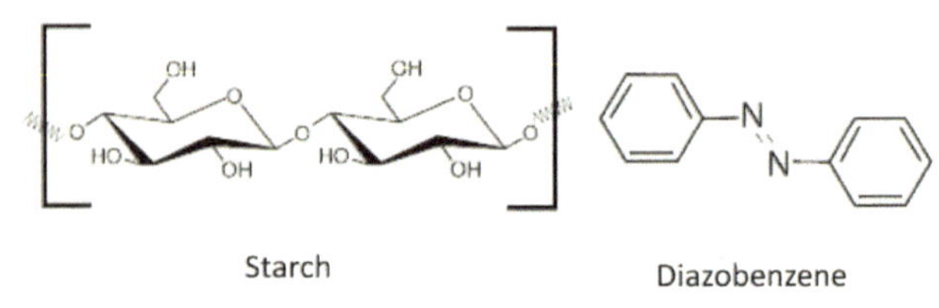

FIGURE 6.5 Chemical structure of starch and diazobenzene.

6.2.3.3 Structural Representation

Figure 6.5 shows the structure of starch and diazobenzene.

6.2.3.4 Key Components

The key components which are required for the formation of the light-responsive starches are as follows:

- *Starch backbone:* The core polymer that forms the matrix of the material, providing the structure and stability.
- *Photochromic molecule (spiropyran):* A molecule that undergoes reversible structural changes (e.g., ring opening and closing) upon exposure to light, leading to changes in colour or solubility.
- *Azobenzene derivative:* A photochromic molecule that undergoes cis–trans isomerization upon light exposure, affecting the material's physical properties such as solubility and swelling behaviour.
- *Cross-linked structure:* This network, formed by either the starch polymer or an additional cross-linking agent, maintains the stability of the material while allowing it to change properties in response to light.

6.2.4 Moisture/Humidity-Sensitive Starches

Moisture-sensitive starches have gained significant attention in recent years due to their ability to change their physical properties in response to variations in environmental moisture or humidity. These materials are especially valuable in applications requiring precise control over moisture levels, such as biodegradable films, agricultural coatings, and food packaging. The moisture-absorbing ability of these starches, combined with their biodegradability, makes them an excellent candidate for sustainable materials that can replace petroleum-based polymers. In addition to their use in packaging, these materials also hold promise in biomedical applications, such as controlled drug delivery and wound healing.

6.2.4.1 Mechanisms of Moisture Sensitivity in Starches

Moisture-sensitive starches derive their functionality from their ability to interact with water molecules, leading to changes in their physical properties. The ratio of amylose to amylopectin, the presence of hydrophilic polymers, and the kind of starch–water interaction are the main variables affecting this behaviour.

- *Amylose–amylopectin complex:* The two polysaccharides that make up starch are the linear polymer amylose and the branching polymer amylopectin (Gupta et al. 2024c). The amylose-to-amylopectin ratio significantly influences a starch's moisture sensitivity. Amylose is less hydrophilic than amylopectin, and starches with a higher amylopectin content are generally more hydrophilic, meaning they absorb more moisture from the environment (Gupta et al., 2024d). This increased moisture absorption results in swelling, softening, and increased flexibility of the starch material. In contrast, amylose-rich starches are more resistant to water absorption, making them suitable for applications where moisture resistance is needed (Zhou et al., 2022).
- *Hydrophilic polymers:* The introduction of hydrophilic polymers such as polyvinyl alcohol (PVA), glycerol, or PEG to starch matrices enhances their ability to absorb moisture. These polymers are highly hygroscopic and can form hydrogen bonds with water molecules, allowing them to absorb and retain moisture effectively. When incorporated into starch materials, hydrophilic polymers modify the material's mechanical properties, such as flexibility, permeability, and tensile strength, in response to ambient humidity (Gupta et al., 2024e). The presence of these polymers allows starch-based materials to adapt to changes in humidity by absorbing moisture in humid conditions or releasing moisture when the environment becomes drier (Zhang et al., 2019).
- *Water-induced changes:* The interaction between starch and water leads to changes in the material's structure. Upon moisture absorption, starch molecules swell and the hydrogen bonds between starch chains break, causing the polymer to become more plasticized and flexible. Conversely, when the humidity decreases, water molecules desorb, leading to the reformation of these bonds, which causes the starch material to become more rigid and less flexible. This dynamic change in mechanical properties allows moisture-sensitive starches to be used in a wide variety of applications, from packaging to agriculture (Abdollahi et al., 2020).

6.2.4.2 Preparation of Moisture-Sensitive Starches

The preparation of moisture-sensitive starches typically involves modifying the starch structure to enhance its interaction with water molecules. Several techniques have been developed to incorporate hydrophilic polymers or alter the amylose-to-amylopectin ratio to control moisture response.

- *Incorporation of hydrophilic polymers:* One of the most common approaches to preparing moisture-sensitive starches is by blending starch with hygroscopic polymers such as PVA, glycerol, or PEG. These polymers enhance the starch's ability to absorb moisture and adjust its mechanical properties in response to humidity changes. The hydrophilic nature of these polymers allows them to form strong hydrogen bonds with water, which increases the starch's moisture retention capacity and provides greater flexibility in humid environments (Sang et al., 2023). The addition of plasticizers like glycerol further enhances the flexibility of the starch matrix, making it more adaptable to changing environmental conditions.
- *Amylose-to-amylopectin ratio adjustment:* By altering the ratio of amylose to amylopectin in the starch source, researchers can control the moisture sensitivity of the material. Starches with a higher amylopectin content tend to absorb more water and exhibit greater swelling behaviour, making them ideal for applications that require a higher degree of moisture sensitivity. Conversely, starches with a higher amylose content are less moisture-sensitive and provide more resistance to water absorption, which is beneficial for applications that demand moisture resistance (Zhou et al., 2022).
- *Cross-linking:* Cross-linking is another technique used to modify starch for enhanced moisture sensitivity. In this process, chemical cross-links are introduced between starch molecules, which helps to reduce solubility in water while still allowing the material to absorb moisture. Cross-linked starch materials often exhibit improved mechanical properties and moisture retention, making them suitable for long-term applications, including biodegradable packaging and agricultural films (Sang et al., 2023).
- *Gelatinization and plasticization:* Starch can be gelatinized by heating it with water, which causes the starch granules to swell and release amylose into the solution. During this process, plasticizers such as glycerol or sorbitol are added to improve the flexibility and moisture absorption capacity of the

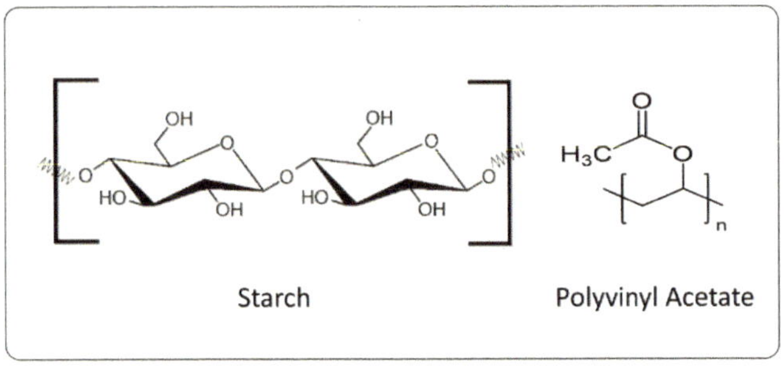

FIGURE 6.6 Chemical structure of starch and polyvinyl acetate.

starch. This process leads to the formation of gel-like structures that can be tailored to exhibit specific moisture sensitivity. The plasticizers prevent the starch from becoming too rigid or brittle and ensure that it can respond dynamically to changes in environmental humidity (Zhang et al., 2019).

6.2.4.3 Structural Representation

Figure 6.6 depicts the structure of starch and polyvinyl acetate.

6.2.4.4 Key Components

- *Starch backbone:* The core structure of the material, composed of starch polymers.
- *Hydrophilic polymer (PVA):* A hygroscopic polymer grafted onto the starch to enhance moisture absorption and modify the material's mechanical properties.
- *Amylose–amylopectin complex:* The amylose-to-amylopectin ratio is tailored to control the material's moisture response.
- *Humidity-sensitive polymer matrix:* A composite matrix formed by starch and hydrophilic polymers, which adjusts its properties in response to changes in moisture content.

6.2.5 Salt-Responsive Starches

Salt-responsive starches react to changes in salt concentration, affecting their swelling, viscosity, and gel strength. This responsiveness is essential for maintaining product stability under varying ionic conditions. They are also of two types:

- *Salinity-adjusted gelling starches:* These starches change their gelation properties based on salt concentrations, which helps in ensuring stability in different ionic environments.
- *Electrolyte-sensitive starches:* These starches are engineered to respond to specific ionic strengths, making them ideal for formulations requiring precise ionic control.

Salt influences the dielectric constant and viscosity property of starch slurries. It delays the gelatinization of starch; this process is dependent on the structure of starch and the concentration of the salt. In an investigation, the swelling power of starch decreased on increasing the salt concentration from 0% to 2%. Increased salt concentration also altered the gelatinization temperature of starch. Salt-responsive starch is loaded with properties useful for specific industrial applications. These starches ensure consistent product quality even with varying salt content. They are used in responsive hydrogels for applications such as drug delivery and tissue engineering (Gupta et al., 2013)

6.2.6 Enzyme-Responsive Starches

Enzyme-responsive starches are designed to interact with specific enzymes, leading to changes in the starch's properties. This interaction can be utilized for controlled release or targeted functions. These are of two types.

- *Digestible starches:* Engineered to be broken down by digestive enzymes, these starches provide controlled nutrient release.
- *Enzyme-cleavable starches:* These starches have linkages cleaved by specific enzymes, altering their viscosity or solubility in response to enzymatic activity.

Researchers have reported that α-amylase can hydrolyze glycosidic linkages present in starch into oligosaccharides, indicating starch's enzyme-responsive behaviour. Chin et al. (2019) successfully prepared an efficient starch-based hydrogel for investigating drug release. They fabricated a starch responsive to α-amylase by cross-linking carboxymethyl starch and ferric chloride solution to understand the drug release behaviour. They found that in presence of enzyme, the drug doxorubicin was released in an acceptable amount (0.5 mg/L). Enzyme-responsive starches have found their application in controlling digestion rates and nutrient release in food products. They are successfully employed in drug delivery systems that utilize enzyme-triggered release mechanisms (Thompson et al., 2014).

6.2.7 Nanocomposite Starches

The incorporation of nanoparticles into starch matrices has become an exciting approach to improving the material's mechanical, thermal, and barrier properties. Nanocomposite starches are typically reinforced with nanomaterials such as nanoclay, cellulose nanocrystals (CNCs), nanocellulose, and silica nanoparticles. These nanoparticles enhance starch's physical properties by improving its strength, moisture resistance, and thermal stability.

In food packaging applications, the addition of nanoparticles improves the barrier properties of starch-based films, which are crucial for preserving food quality by preventing moisture and oxygen from penetrating the packaging. Nanocomposite starch films are biodegradable alternatives to petroleum-based plastics, making them an eco-friendly choice for reducing plastic waste. For example, montmorillonite (MMT) nanoclay has been found to significantly improve the mechanical strength and water barrier properties of starch films, making the films suitable for applications in both food and non-food packaging (Silva et al., 2020). In the biomedical field, nanocomposite starches are also being used in wound dressings and tissue scaffolds, where their enhanced mechanical properties and biodegradability make them ideal for supporting tissue regeneration and promoting wound healing.

6.2.8 Bioactive Starches

Bioactive starches are starch-based materials integrated with bioactive compounds such as antioxidants, antimicrobials, or probiotics. These bioactive compounds confer additional functional properties to the starch, making it useful in a variety of health-related applications. For example, antioxidants such as polyphenols or vitamin C are often incorporated into starch matrices to provide free radical scavenging properties, which can help preserve the nutritional quality of food and reduce oxidative stress in the body (Zhang et al., 2023). Additionally, probiotics can be encapsulated within starch to deliver gut health benefits through prebiotic effects, enhancing the digestive health of individuals when used in food products.

In food packaging, bioactive starches are gaining attention for their ability to provide active packaging that extends the shelf life of perishable goods by reducing microbial growth and oxidation. Starch-based films incorporating antimicrobial agents such as chitosan or essential oils offer a natural alternative to synthetic preservatives, providing a safer and more sustainable option for food preservation (Patel et al., 2021). These bioactive starches are also gaining popularity in the creation of functional foods that promote health, such as foods enriched with probiotics to enhance gut microbiota or antioxidants to help prevent chronic diseases.

6.2.9 Modified Starches

Modified starches are starches that have undergone chemical, physical, or enzymatic modifications to impart specific functionalities, such as increased resistance to digestion, enhanced emulsification properties, and improved film-forming ability. Resistant starch, which is not digested in the small intestine, is one of the most well-known examples of modified starches. Resistant starch acts as a dietary fibre, providing health benefits such as improved gut health, blood sugar regulation, and potential weight management (González et al., 2022). Additionally, resistant starch can be used as a fat replacer in low-calorie foods, where it mimics the texture and mouthfeel of fat without contributing to calories.

Cross-linking is another common modification used to improve the structural stability and viscosity of starch. This process involves linking starch molecules together through chemical bonds, enhancing its resistance to water and heat. Cross-linked starches are widely used in the food industry for applications such as sauces, puddings, and gravies, where stable textures and consistency are crucial. In biodegradable packaging, modified starches are often used to create films that are not only strong and flexible but also break down in the environment over time, providing a sustainable alternative to traditional plastics (Xie et al., 2022).

6.3 APPLICATIONS OF SMART AND RESPONSIVE STARCHES

Responsive and intelligent starch-based materials have attracted considerable interest because of their versatility and sustainability in many industrial and biological applications. These materials use the intrinsic characteristics of starch—biocompatibility, renewability, and biodegradability—while integrating functional improvements via chemical or physical alterations. The sensitivity of starch derivatives to external stimuli, including temperature, pH, and light, renders them appropriate for many applications, such as medicine administration, environmental cleanup, and food technology (Sun et al., 2019). In the biomedical field, dual-responsive starch hydrogels that respond to both temperature and pH have been created to enable targeted medication release, therefore improving therapeutic efficacy and reducing adverse effects (Chen et al., 2020). Moreover, biodegradable starch nanoparticles function as carriers for the targeted delivery of pharmaceuticals and bioactive substances, presenting intriguing solutions for difficulties like cancer therapy (Wei et al., 2019). Starch-based scaffolds for tissue engineering provide versatile support for cellular proliferation and regeneration, dynamically adapting to physiological circumstances (Jung et al., 2013).

6.3.1 Drug Delivery: Adaptive and Intelligent Starches in Biomedical Applications

Drug delivery systems seek to enhance treatment efficacy by accurate drug release, tailored administration, and minimized adverse effects. Starch-based materials, especially those with reactive characteristics, have attracted considerable interest in this field. Their intrinsic biocompatibility, biodegradability, and capacity for customization make them appropriate for the development of sophisticated drug delivery systems.

6.3.1.1 Responsive Starch Hydrogels in Pharmaceutical Delivery

Responsive starch hydrogels, especially those enhanced with synthetic polymers, have surfaced as viable options for regulated and sustained drug delivery systems. Hybrid microgels composed of starch nanoparticles (SNPs) and poly(*N*-isopropylacrylamide) (p(NIPAM)) have thermo-responsive characteristics, making them suitable for temperature-sensitive drug release applications. These polymers may encapsulate pharmaceuticals such as methylene blue and release them in a regulated way at body temperature, enhancing therapeutic effectiveness while reducing negative effects. Ribeiro et al. conducted research revealing that SNPs-*co*-p(NIPAM) microgels efficiently contained methylene blue and exhibited notable swelling and deswelling behaviour in response to temperature fluctuations. Below the LCST, the microgels expanded, facilitating drug release in a hydrated form; conversely, above the LCST, the polymer contracted, diminishing the release rate. This adjustable characteristic facilitates accurate regulation of medication dose and release kinetics (da Costa Ribeiro et al. 2024).

6.3.1.2 Biodegradability and Biocompatibility

Starch-derived materials possess intrinsic biodegradability and biocompatibility, making them suitable for biomedical applications. In contrast to synthetic polymers, starch hydrogels decompose into non-toxic by-products, guaranteeing compatibility with human cells. These attributes are essential for drug delivery systems because they reduce the danger of inflammatory reactions and chronic toxicity (Ashogbon 2021).

6.3.1.3 Starch-Derived Nanoparticles for Pharmaceutical Delivery

Starch-derived nanoparticles provide an alternative method for medication delivery. Starch nanoparticles may be modified to transport various therapeutic agents, including as small-molecule medicines, proteins, and genetic material. Their diminutive size enhances cellular absorption, while surface alterations provide targeted distribution to certain tissues or organs. Cui et al. (2022) investigated starch-based hydrogels as transporters for food bioactive compounds and suggested their possible use in pharmaceuticals. Their research demonstrated that starch hydrogels may encapsulate bioactive chemicals, safeguard them from degradation, and release them in response to environmental stimuli such as pH or temperature fluctuations.

6.3.1.4 Dual-Responsive Starch Systems

Dual-responsive starch-based systems, which react to several stimuli (e.g., pH and temperature), have been explored for intricate drug delivery requirements. These systems are especially beneficial for tackling the problems of cancer treatment because the microenvironment differs markedly between healthy and malignant tissues. Starch derivatives modified with polymers such as PEG or pH-sensitive groups improve targeting specificity and minimize premature medication release.

6.3.1.5 Targeted Pharmacological Delivery

The incorporation of responsive starch materials with targeting agents, such as antibodies or ligands, has facilitated the creation of sophisticated tailored drug delivery systems. These technologies provide targeted medication delivery, minimizing systemic toxicity. Starch nanoparticles modified with folic acid may specifically target cancer cells that overexpress folate receptors, facilitating the direct delivery of chemotherapeutic drugs to the tumour location (Cui et al. 2022).

6.3.1.6 Clinical Applications and Future Outlook

Although starch-based drug delivery systems remain in the experimental phase, their potential for clinical applications is substantial. Future research aims to enhance their stability, increase manufacturing scale, and perform in vivo investigations to confirm their effectiveness and safety. Progress in nanotechnology and polymer research is expected to enhance the functionality of starch-based drug delivery systems, making them essential for next-generation therapies (da Costa Ribeiro et al. 2024).

6.3.1.7 Case Analyses

- Ribeiro et al. revealed that hybrid starch–p(NIPAM) microgels exhibited a 70% decrease in drug release rates at temperatures above the LCST, underscoring the efficacy of these microgels in controlled drug delivery (Jung et al., 2013).
- Cui et al. (2022) demonstrated that starch-based hydrogels cross-linked with polyacrylic acid efficiently released encapsulated bioactives at pH levels resembling intestinal circumstances, making them suitable for oral medication administration.
- Aaliya et al. (2022) conducted research using plasma-activated water (PAW) for the modification of starch, therefore improving the starch's functional characteristics and capacity for encapsulating and delivering sensitive molecules such as peptides and proteins.

6.3.2 Enhanced Oil Recovery: The Function of Adaptive and Intelligent Starches

Enhanced oil recovery (EOR) is an essential method in the petroleum sector, designed to augment the volume of crude oil retrieved from reservoirs. The use of biopolymers, particularly starch-based derivatives, in EOR has increased in popularity owing to biopolymers' environmental sustainability, economic efficiency, and adaptability (Ashogbon, 2021). Starch derivatives, especially those altered chemically or physically, provide superior viscosity augmentation, stability under harsh circumstances, and emulsification capabilities, making them appropriate for enhanced oil recovery applications.

6.3.2.1 Function of Starch Derivatives in Enhanced Oil Recovery

Modified starches, particularly those undergoing multiple modification methods such as succinylation and acetylation, have shown efficacy in enhancing oil recovery. These compounds exhibit elevated molecular weights, superior thickening properties, and considerable resistance to high-temperature and high-salinity environments characteristic of oil reservoirs. Their capacity to preserve viscosity and stability under extreme circumstances guarantees effective displacement of entrapped oil, resulting in enhanced recovery rates (Ashogbon, 2021).

6.3.2.2 Enhancement of Viscosity and Stability

In EOR, polymeric solutions are injected into reservoirs to enhance the mobility ratio between oil and the displacing water. Starch-based polymers, owing to their elevated viscosity, mitigate water fingering and improve the sweep efficiency of the flooding process. Research demonstrates that starch derivatives, including hydroxypropylated and cross-linked starches, maintain their viscosity at high temperatures and salinities, providing a notable benefit over synthetic counterparts (Ashogbon 2021).

6.3.2.3 Eco-Friendly Alternative

The biodegradability and renewability of starch make it a sustainable substitute for synthetic polymers conventionally used in EOR. In contrast to synthetic polymers that last in the environment and provide ecological hazards, starch-based materials decompose into non-toxic by-products, hence minimizing their environmental impact. Moreover, the plentiful supply and affordability of starch from agricultural sources enhance its feasibility in extensive EOR operations (Aaliya et al., 2022).

6.3.2.4 Prospective Trajectories

The development of responsive starch-based polymers for EOR signifies a promising prospect. The research aims to develop starch derivatives that react to reservoir conditions, including temperature and pH, to enhance dynamic oil displacement. Advancements in nanotechnology may facilitate the manufacture of starch-based nanoparticles for specific uses in porous reservoir medium (Aaliya et al., 2022).

6.3.3 Optical Sensors: The Utilization of Adaptive and Intelligent Starches

Optical sensors are essential instruments in several domains, such as environmental monitoring, biomedical diagnostics, and industrial applications. These sensors use materials that interact with light to monitor variations in environmental factors such as temperature, pH, or the presence of certain analytes. Responsive starch-based materials have recently emerged as interesting elements in optical sensor technology owing to their capacity to undergo structural and optical alterations in reaction to external stimuli.

6.3.3.1 Responsive Starch Materials for Optical Detection

Starch-based hydrogels and composites are attracting interest due to their adjustable optical characteristics. Temperature-responsive starch–lignin blends demonstrate changes in their structural structure in response to temperature fluctuations, modifying their refractive index. This characteristic is crucial for the construction of sensors that detect real-time temperature variations (Diaz-Baca & Fatehi, 2023).

Research by Ashogbon (2021) revealed that starch-based films infused with nanoparticles such as silver or gold exhibited improved plasmonic characteristics, facilitating its use in extremely sensitive surface plasmon resonance (SPR) sensors. These films demonstrated consistent performance across many environmental conditions, making them appropriate for long-term monitoring applications (Ashogbon 2021).

6.3.3.2 Mechanisms of Action

The method by which starch materials function in optical sensors is predicated on their capacity to swell, contract, or alter opacity in response to external stimuli such as the following:

- *Temperature sensitivity:* Cross-linked starch hydrogels exhibit reversible swelling and deswelling, which alters their light transmission and scattering characteristics, facilitating precise temperature measurement (Aaliya et al., 2022).
- *pH sensitivity:* pH-responsive starch hydrogels may modify their physical state, resulting in quantifiable alterations in optical characteristics. These devices are especially beneficial in biosensors for identifying pH variations linked to metabolic activities (Ashogbon, 2021).

6.3.3.3 Benefits of Starch-Based Optical Sensors

- *Biodegradability:* Starch-based materials spontaneously disintegrate, making them eco-friendly substitutes for synthetic polymers often used in optical sensors.
- *Cost-effectiveness:* Starch, sourced from renewable resources, is both economical and plentiful, lowering the production costs of optical sensors.
- *Customization:* Starch derivatives may undergo chemical or physical modifications to get certain responsiveness and optical characteristics, expanding their application spectrum (Ashogbon, 2021).

6.3.3.4 Utilizations

Following are the utilization of starch-based optical sensors:

- *Environmental monitoring:* Starch-based optical sensors detect temperature and pH fluctuations in aquatic environments, facilitating the evaluation of ecological health (Aaliya et al., 2022).
- *Biomedical diagnostics:* These materials are integrated into sensors for the assessment of physiological parameters, including the detection of pH variations in tissues or fluids (Aaliya et al., 2022).
- *Industrial application:* Starch-based sensors facilitate companies' process control by monitoring parameters such as temperature and chemical concentrations (Ashogbon, 2021).

6.3.3.5 Prospective Outlooks

Investigations into the amalgamation of starch-based materials with cutting-edge nanotechnologies are anticipated to improve the sensitivity and specificity of optical sensors. Combining starch hydrogels with quantum dots or graphene may enhance their optical performance, facilitating the detection of subtle variations in environmental conditions. Furthermore, the development of multi-responsive starch-based sensors may provide new opportunities for real-time multifunctional sensing systems (da Costa Ribeiro et al., 2024).

6.3.4 Emulsion Stabilizers: The Function of Responsive and Intelligent Starches

Emulsions are vital in several sectors, such as food, medicines, and cosmetics, where they function as delivery mechanisms for active compounds, taste enhancers, or functional agents. Emulsion stabilizers are essential for preserving the dispersion of one liquid inside another, so averting phase separation (da Costa Ribeiro et al., 2024; Hsieh et al., 2019). Starch-derived compounds, especially modified starches, have shown efficacy as emulsion stabilizers owing to their amphiphilic characteristics, surfactant qualities, and eco-friendliness.

6.3.4.1 Starch as an Emulsifying Agent

Native starches have restricted emulsifying capabilities owing to their hydrophilic characteristics. Chemical or physical alterations may substantially improve their amphiphilic characteristics, making them effective for emulsion stabilization. Modified starches such as octenyl succinate starch and hydroxypropyl starch include hydrophobic groups that engage with oil droplets and hydrophilic sections that interact with water, therefore forming stable emulsions (Cui et al., 2022).

6.3.4.2 Stabilization Mechanism

The stabilization of emulsions with starch-based materials depends on their capacity to adsorb at the oil–water interface, thereby decreasing interfacial tension and creating a protective barrier surrounding scattered droplets. This barrier inhibits coalescence and guarantees long-term stability. Furthermore,

Hydrophobic changes enhance the compatibility of starch molecules with oil phases, hence enhancing emulsification (Ashogbon, 2021). Cross-linked starches improve the mechanical stability of emulsions, providing resistance to shear and temperature fluctuations (Ashogbon 2021; Cui et al., 2022).

6.3.4.3 Industrial Applications

The varied applications of modified starches are given below:

- *Food sector:* Modified starches are extensively used as emulsion stabilizers in goods such as mayonnaise, salad dressings, and whipped toppings. They provide the necessary viscosity and texture while inhibiting phase separation during storage and processing (Hsieh et al., 2019).
- *Pharmaceuticals:* Starch-based stabilizers in drug delivery maintain the uniformity of emulsions used as carriers for hydrophobic medicines, enhancing bioavailability and stability (Aaliya et al., 2022).
- *Cosmetics:* In cosmetic compositions, starch stabilizers preserve emulsion consistency, facilitating easy application and extending shelf life (Ashogbon, 2021).

6.3.4.4 Benefits of Starch-Based Emulsion Stabilizers

The benefits of starch-based emulsion stabilizers include:

- *Biodegradability:* Starch-based stabilizers undergo natural decomposition, reducing environmental effect in comparison to synthetic stabilizers. Their safety profile makes them appropriate for food and medicinal uses.
- *Customizability:* Modifications provide specialized functioning for certain emulsion systems, including adjustments to oil–water ratios or viscosity requirements (Ashogbon, 2021).

6.3.4.5 Recent Developments

Recent research has investigated the use of dual-modified starches, including hydroxypropylated and cross-linked starches, to improve emulsion stability. These starches have enhanced heat and shear resistance, making them suitable for demanding applications in the food and petroleum sectors. The integration of nanoparticles into starch-based stabilizers has facilitated the development of Pickering emulsions, which are stabilized by solid particles rather than surfactants (Aaliya et al., 2022).

6.3.4.6 Prospective Outlooks

The incorporation of sensitive starch materials with cutting-edge technologies such as nanostructuring has the potential to transform their function as emulsion stabilizers. Intelligent starch-derived stabilizers that react to environmental stimuli such as pH or temperature variations may provide dynamic regulation of emulsion characteristics, broadening their use across several domains (Cui et al., 2022; da Costa Ribeiro et al., 2024).

6.3.5 Intelligent Polymers: Adaptive and Innovative Starches in Sophisticated Applications

Smart polymers are materials that dynamically adapt to environmental stimuli, such as temperature, pH, light, or ionic strength, resulting in substantial changes to their characteristics. Responsive and intelligent starch-based polymers constitute a sustainable and adaptable category of smart materials, combining the regenerative characteristics of starch with enhanced functionality achieved by chemical or physical alterations. These materials are used in medication delivery, tissue engineering, environmental sensing, and controlled release systems.

6.3.5.1 Responsive Starch-Derived Polymers

Starch-based polymers possess intrinsic hydrophilicity, biodegradability, and biocompatibility. Alterations such as grafting with synthetic polymers, cross-linking, or the inclusion of functional groups facilitate their response to certain stimuli:

- *Temperature sensitivity:* Starch–p(NIPAM) composites have thermo-responsive characteristics, expanding below the LCST and contracting above it. This feature has significant value in medication delivery and thermal sensing applications (da Costa Ribeiro et al., 2024).
- *pH responsiveness:* Cross-linked starch hydrogels containing polyacrylic acid or analogous pH-sensitive moieties exhibit swelling or shrinking behaviour in reaction to pH fluctuations, making them suitable for applications including targeted medication delivery in the gastrointestinal system (Cui et al., 2022).
- *Ionic responsiveness:* Starch derivatives with ionic capabilities react to variations in ionic strength, facilitating their use in water purification and selective ion adsorption (Cui et al., 2022).
- *Utilization of smart starch polymers in drug delivery systems:* Intelligent starch-derived hydrogels and nanoparticles provide regulated and precise medication delivery. Thermo-responsive and pH-sensitive devices provide accurate distribution to target locations, reducing systemic adverse effects (da Costa Ribeiro et al., 2024).
- *Utilization in tissue engineering:* Responsive starch-based scaffolds provide a dynamic extracellular matrix that facilitates cell growth and proliferation while adjusting to physiological circumstances.
- *Utilization in environmental sensing:* Smart starch polymers used in sensors identify environmental variations, such as pH, temperature, or pollutant concentrations, facilitating real-time monitoring and control systems. Starch-based smart polymers are used in slow-release fertilizers and water retention systems, enhancing resource efficiency and sustainability in agriculture (da Costa Ribeiro et al., 2024; Hsieh et al., 2019).

6.3.5.2 Benefits of Starch-Based Intelligent Polymers

The benefits of starch-based intelligent polymers are as follows;

- *Biocompatibility:* Starch is intrinsically safe for use in biomedical and environmental applications.
- *Customizability:* Chemical alterations enable exact regulation of responsiveness and mechanical characteristics.
- *Eco-friendliness:* Starch-based polymers, which are biodegradable and sourced from renewable materials, serve as a sustainable substitute for synthetic smart polymers.
- *Cost-effectiveness:* The plentiful supply of starch lowers manufacturing expenses relative to synthetic alternatives.

6.3.5.3 Recent Advancements

Recent improvements emphasize the integration of starch-based polymers with nanomaterials, including graphene oxide, quantum dots, and metallic nanoparticles, to augment their responsiveness and usefulness. Starch composites infused with graphene oxide demonstrate enhanced mechanical strength and electrical conductivity, expanding the applicability of starch to wearable electronics and smart coatings (Aaliya et al., 2022). Furthermore, dual- and multi-responsive starch-based polymers are being engineered to react to many stimuli for intricate biomedical and industrial applications.

6.3.5.4 Obstacles and Prospective Pathways

Although starch-based smart polymers have significant potential, issues such as scalability, durability during extended usage, and compatibility with sophisticated technology persist. Future research should address these limits by investigating innovative modification methods, hybrid materials, and biomimetic designs (da Costa Ribeiro et al., 2024). The amalgamation of artificial intelligence with smart polymer systems may provide autonomous reactions to intricate surroundings, hence broadening their applicability.

Responsive and intelligent starch-based materials signify a potential advancement in science and technology, propelled by their inherent sustainability, versatility, and use. The intrinsic characteristics of starch—biocompatibility, biodegradability, and availability—alongside sophisticated chemical and physical alterations, have facilitated its use across several domains, including biomedical, industrial, and environmental sectors. In medication delivery, these advanced materials provide accurate and regulated therapeutic administration, minimizing systemic adverse effects and improving patient outcomes. Their dual-responsive characteristics, including sensitivity to temperature and pH, provide tailored administration, especially in complex settings such as cancer microenvironments. In improved oil recovery, starch derivatives function as environmentally sustainable substitutes for synthetic polymers, providing strong performance in elevated temperature and high salt environments. The use of responsive starch materials in optical sensors has significantly transformed environmental monitoring and biological diagnostics by enabling real-time, precise detection of physical and chemical alterations. Simultaneously, their function as emulsion stabilizers in the food, pharmaceutical, and cosmetic sectors underscores their flexibility and significance in developing stable, efficient, and sustainable systems. Smart starch-based polymers have transformed opportunities in tissue engineering, environmental monitoring, and agriculture, demonstrating their versatility in response to advancing technological requirements. Their incorporation of nanotechnology and artificial intelligence expands their potential, creating opportunities for multifunctional, adaptive systems that tackle intricate problems. Nonetheless, obstacles persist. Challenges such as scalability, durability, and integration with sophisticated technologies must be resolved to fully realize the promise of these materials. Future studies should concentrate on enhancing their features, minimizing manufacturing costs, and investigating novel modification approaches to broaden their application scope. In summary, responsive and intelligent starch-based materials illustrate the intersection of sustainability and innovation. Their capability to fulfill the requirements of contemporary businesses while mitigating environmental issues establishes them as pivotal players in the future of materials science and technology. As research and development progress, these materials are poised to assume a progressively crucial role in promoting a more sustainable and technologically advanced future.

6.4 FUTURE PROSPECTS OF SMART AND RESPONSIVE STARCHES

Starch-based intelligent materials will have a significant role in the future due to their unique properties and benefits. Starch is a complex carbohydrate made of many units of glucose joined by glycosidic linkages. It contains many hydroxyl groups, which makes it hydrophilic in nature, thus decreasing its mechanical strength. To resolve the shortcomings of this macromolecule, some modification in its structure is done along with the addition of additives; these processes make starch more stable and durable (Alizadeh et al., 2021; Liu et al., 2022). In smart starches, the linear amylose molecules are amplified to give more mechanical strength to the starch, while branched amylopectin molecules are enhanced to increase its reactive tendency. Smart and responsive starches can be used in food packaging, drug delivery system, controlled release of fertilizer, humidity-responsive starches, smart windows, and so on. In food packaging, the application of such starches is increasing by the day. The application of integrated technology in food processing is seeing significant progress. The primary issue facing food processing companies is the storage of food items. Increasing the shelf life and preserving the quality of fresh food is made possible by the new technology of clever starches. Researchers are aiming to create starch films that will guarantee food safety and allow processors and consumers to see the current state of the food supply chain (Vilas et al., 2020). Even efforts to extend the shelf life of fish and poultry are ongoing (Kanatt, 2020). According to Tarsila et al. (2022) and Menzel et al. (2020), active packaging is the addition of specific substances, such as organic acids and natural plant extracts, that will be released either in the surrounding environment or in the packaged food at a specific time, extending the food's shelf life while maintaining its freshness and quality. Smart packaging also includes the use of pH indicators and temperature and time sensors, which will be helpful in monitoring the status of packaged food.

The usage of starch for drug delivery system is getting attention for emerging medical science technologies. Work on developing pH-responsive starch is giving a new turn to green chemistry. Studies are underway for the production of pH-responsive hydrogel from Eronga triticale starch. A starch-based delivery system for nutraceuticals is being used to effectively deliver bioactive materials (Cui et al., 2022; Cruz-Amaya et al., 2023; Meng et al., 2022). The structure of starch is modified by cross-linking to form smart microgels that are very helpful for drug delivery system. These microgels are smart materials responsive to external stimuli such as temperature, pH, and light (Y. Li et al., 2009). Moreover, the stability of starch-based Pickering emulsions is being taken advantage of for delivery of bioactive compounds. Studies have shown that an intelligent delivery system of starch-based Pickering emulsion containing thymol is helpful in defending against foodborne pathogens (Wang et al., 2023). Work on preparing smart starches for the controlled release of fertilizer is gaining increasing recognition. Fertilizer release systems optimize nutrient uptake and thereby reduce nutrient loss and promote plant growth. These systems not only produces good crop yield but also help in reducing the overuse of fertilizers, thus making sustainable agriculture more feasible (Salimi et al., 2023). This work needs more refinement for successful implementation at a large scale. Responsive starches use the hydrophilic nature of starch to develop responsive smart polymers. These starches undergo shape changes as a response to external stimuli such as pH, temperature, or humidity. For example, humidity-responsive starches (Willett, 2008) display shrinkages with increased relative humidity. The functionality of starch shows significant changes during hydrothermal treatments. This treatment brings about rearrangement of amylose and amylopectin, which gives rise to increased rigidity and crystallinity and decreased entropy. Such modification gives good thermal and shear stability to starch, which makes starch suitable for canned and frozen product preparation. Research is also underway for developing smart windows (Zhang et al., 2018), which will be a step towards energy conservation. These windows will be able to change their transmittance of light radiation with differences in solar irradiation. The solar transmittance of such windows could be regulated by applying heat, voltage, or light.

Although many methods have been adopted to form smart starches, these are still operational at the laboratory level. There remains a long way to go for the acceptance of starch-based materials by consumers at a large scale, and for this the starch products will have to pass through different manufacturing levels, meet safety regulations, and demonstrate environmental feasibility. Stability of these products depends on various factors such as type of starch, techniques used during preparation, and storage conditions, which may result in decreased homogeneity and thus in poor product quality. There are many drawbacks to starch-based materials, and studies are ongoing to overcome these shortcomings to make these materials practically viable. For future prospects, more exploratory research is required to enhance mechanical strength and durability. New applications must be explored to develop new opportunities for the usage of starch-based substances that will be helpful in making production scalable. In the near future, these starch-based intelligent materials will completely replace petroleum-based materials, leading to a sustainable and clean environment.

6.5 CONCLUSION

Starch, a polysaccharide made of α-amylose and amylopectin, forms the major reserve of carbohydrate in plants. This chapter deals with the way these starches are modified to form smart and responsive starches. Different types of smart and responsive starches, along with their structural representation and preparation, have been discussed: thermo-sensitive starches; pH-sensitive, light-responsive, moisture-sensitive starches; nanocomposite starches; bioactive starches; and modified starches. The thermo-sensitive starches are prepared by grafting thermo-sensitive polymer onto the starch matrix. To make thermally stable starch, the starch is cross-linked through divinyl sulphone. pH sensitivity can be induced by carboxylation or amination of starch. Incorporating photochromic materials such as azobenzene that undergo confirmation changes on light irradiation gives rise to light-responsive starches.

The chapter has covered different applications of smart and responsive starches. It has shown how these modified starches are being used in agriculture, pharmaceutical, food, medical, and material industries. Starch hydrogels are being used in drug delivery systems with much effect. Rebero et al. have shown that when solution temperature is below critical level, expansion of microgels takes place with the release of a drug in the hydrated form, whereas release decreases with contraction. In biomedical applications, hydrogels show much compatibility with human cells and are safer than synthetic polymers, which cause chronic toxicity. With the advancement of nanotechnology, starch nanoparticles are being used to deliver therapeutic agents to target cells. Starch nanoparticles loaded with folic acid are also helpful in delivering chemotherapeutic drugs to tumour locations.

These materials can also be used efficiently in oil recovery system. The compounds, due to their high molecular weight, show high thickening properties and resistance to heat and salinity, which are the appropriate properties for oil recovery. Starch-based films modify plasmonic character and thus, find use as optical sensors. As these intelligent polymers are not only cost-effective but also compatible with humans and our environment, they can be widely used as emulsion stabilizers; as sensors of temperature, pH, and ions; and in tissue engineering.

These intelligent materials, due to their unique properties, will play a significant role in the near future. We can see their usage in varied fields. More studies are needed to overcome their shortcomings and pave the way for their major role at a large scale on the ground level.

ACKNOWLEDGEMENT

The Indian Institute of Science Education and Research, Bhopal, is acknowledged by the authors for supporting this investigation. The authors also appreciate the support given by the administration of Lakshmi Narain College of Technology, Bhopal.

REFERENCES

Aaliya, B., Sunooj, K. V., Navaf, M., Akhila, P. P., Sudheesh, C., Sabu, S., Sasidharan, A., Sinha, S. K., & George, J. (2022). Influence of plasma-activated water on the morphological, functional, and digestibility characteristics of hydrothermally modified non-conventional talipot starch. *Food Hydrocolloids*, *130*, 107709. https://doi.org/10.1016/j.foodhyd.2022.107709

Abdollahi, M., Ranjbar, M., & Rashid, S. (2020). Moisture-responsive starch-based materials for food and pharmaceutical applications: A review. *Carbohydrate Polymers*, *249*, 116865.

Alizadeh, S. M., Maryam, A., Milad, T., Keyhan, M., & Julian, M. D. (2021). Recent advances in the development of smart and active biodegradable packaging materials. *Nanomaterials (Basel)*, *11*(5), 1331. https://doi.org/10.3390/nano11051331

Ashogbon, O. (2021). Dual modification of various starches: Synthesis, properties and applications. *Food Chemistry*, *342*, 128325. https://doi.org/10.1016/j.foodchem.2020.128325

Bajpai, A. K., Giri, P. S., & Gupta, R. (2022). pH-sensitive starch-based nanoparticles for controlled drug release. *International Journal of Pharmaceutics*, *623*, 121761.

Chen, H., Zhao, X., & Zhang, Q. (2019). Advances in the synthesis and applications of thermo-sensitive starch derivatives. *Carbohydrate Polymers*, *218*, 100–110.

Chen, Z., Li, S., & Wang, Y. (2022). Recent progress in starch-based nanocomposites for sustainable packaging applications. *International Journal of Biological Macromolecules*, *194*, 757–769.

Chin, S. F., Ain, N. B. R., Pang, S. C., Lee, B. K., & Hwang, S. S. (2019). pH-Responsive starch-citrate nanoparticles for controlled release of paracetamol. *Starch - Stärke*, *71*(9–10), 1800336. https://doi.org/10.1002/star.201800336

Cruz-Amaya, K. S., Del-Toro-Sánchez, C. L., Carvajal-Millan, E., Anda-Flores, Y. D., Hernandez-Martínez, D., Martinez-Robinson, K. G., & Cornejo-Ramirez, Y. I. (2023). Triticale starch-based pH-responsive hydrogel: Synthesis, characterization, diffusion, and new perspectives (version 3). *ChemRxiv*. Preprint.

Cui, C., Jia, Y., Sun, Q., Yu, M., Ji, N., Dai, L., Wang, Y., Qin, Y., Xiong, L., & Sun, Q. (2022). Review recent advances in the preparation, characterization, and food application of starch-based hydrogels, *Carbohydrate Polymers*, *291*, 119624. https://doi.org/10.1016/j.carbpol.2022.119624.

Cui, C., Wang, N. J. Y., Xiong, L., & Sunnn, Q. (2021). Bioactive and intelligent starch-based films: A review. *Trends in Food Science & Technology*, *116*, 854–869. https://doi.org/10.1016/j.tifs.2021.08.024

Cui, L., Zhang, L., & Li, Q. (2021). Graft copolymerization of pH-sensitive polymers onto starch for drug delivery applications. *Carbohydrate Polymers*, *264*, 118039.

da Costa Ribeiro, A., Tominaga, T. T., Bonadio, T. G. M., da Silveira, N. P., & Leite, D. C. (2024). A study on the behavior of smart starch-*co*-poly(*N*-isopropylacrylamide) hybrid microgels for encapsulation of methylene blue. *ACS Omega*, *9*(25), 27349–27357. https://doi.org/10.1021/acsomega.4c01947

de Freitas, A. d. S. M., Rodrigues, J. S., Maciel, C. C., Pires, A. A. F., Lemes, A. P., Ferreira, M., & Botaro, V. R. (2021). Improvements in thermal and mechanical properties of composites based on thermoplastic starch and kraft lignin. *International Journal of Biological Macromolecules*, *184*, 863–873. https://doi.org/10.1016/j.ijbiomac.2021.06.153

Diaz-Baca, J. A., & Fatehi, P. (2023). Temperature responsive crosslinked starch-Kraft lignin macromolecule. *Carbohydrate Polymers*, *313*, 120846. https://doi.org/10.1016/j.carbpol.2023.120846

Dominguez-Ayala, J. E., Soler, A., Mendez-Montealvo, G., & Velazquez, G. (2022). Supramolecular structure and technofunctional properties of starch modified by high hydrostatic pressure (HHP): A review. *Carbohydrate Polymers*, *291*, 119609. https://doi.org/10.1016/j.carbpol.2022.119609

Fan, Y., & Picchioni, F. (2020). Modification of starch: A review on the application of "green" solvents and controlled functionalization. *Carbohydrate Polymers*, *241*, 116350. https://doi.org/10.1016/j.carbpol.2020.116350.

Fennema, O. R. (1996). *Food chemistry*. Marcel Dekker.

Gaur, A., Kumar, P., & Singh, B. (2011). Development of pH-responsive starch-based hydrogels for controlled fertilizer release. *Bioresource Technology*, *102*(7), 4482–4488.

Gupta, R. K., Guha, P., & Srivastav, P. P. (2022). Natural polymers in bio-degradable/edible film: A review on environmental concerns, cold plasma technology and nanotechnology application on food packaging-A recent trends. *Food Chemistry Advances*, *1*, 100135. https://doi.org/10.1016/j.focha.2022.100135

Gupta, R. K., Guha, P., & Srivastav, P. P. (2023). Effect of high voltage dielectric barrier discharge (DBD) atmospheric cold plasma treatment on physicochemical and functional properties of taro (*Colocasia esculenta*) starch. *International Journal of Biological Macromolecules*, *253*, 126772. https://doi.org/10.1016/j.ijbiomac.2023.126772

Gupta, R. K., Guha, P., & Srivastav, P. P. (2024a). Dielectric barrier discharge plasma: A green and novel method to change the structure of taro peel starch and improve the physicochemical properties of taro peel starch films. *Plasma Processes and Polymers*, *21*(9), 2400047. https://doi.org/10.1002/ppap.202400047

Gupta, R. K., Guha, P., & Srivastav, P. P. (2024b). Exploring the potential of taro (*Colocasia esculenta*) starch: Recent developments in modification, health benefits, and food industry applications. *Food Bioengineering*, *3*(3), 365–379. https://doi.org/10.1002/fbe2.12103

Gupta, R. K., Guha, P., & Srivastav, P. P. (2024c). Insights into the physicochemical, crystalline, multistructural, thermal and digestibility behavior of taro starch (*Colocasia esculenta*) modified by cold plasma treatment with high voltage dielectric barrier discharge (HVDBD): An alternative technique for chemical modification of starch and environmental sustainability. *Journal of Cleaner Production*, *461*, 142710. https://doi.org/10.1016/j.jclepro.2024.142710

Gupta, R. K., Guha, P., & Srivastav, P. P. (2024d). Physical action of nonthermal cold plasma technology for starch modification. *Food Physics*, *1*, 100011. https://doi.org/10.1016/j.foodp.2024.100011

Gupta, R. K., Guha, P., & Srivastav, P. P. (2024e). Effect of cold plasma treatment and plasma-activated water on physicochemical and structural properties of starch: A green and novel approach for environmental sustainability. *Plasma Processes and Polymers*, *21*(4), 2300204. https://doi.org/10.1002/ppap.202300204

Gupta, R. K., Pipliya, S., Patel, J., Srivastav, P. P., Castro-Muñoz, R., Wang, C. K., & Nemţanu, M. R. (2025a). Essential oil-embedded starch based films for active packaging: Insights into migration, stability, and preservation efficiency. *Trends in Food Science & Technology*, *163*, 105149. https://doi.org/10.1016/j.tifs.2025.105149

Gupta, R. K., Gnana Moorthy Eswaran, U., Kumar, S., Pipliya, S., Kaur, B., & Srivastav, P. P. (2025b). Synergistic modification of starch polysaccharide via sequential treatments with ultrasonic and cold plasma technology. *Journal of Food Processing and Preservation*, *2025*(1), 9063007. https://doi.org/10.1155/jfpp/9063007

Gilet, A., Quettier, C., Wiatz, V., Bricout, H., Ferreira, M., Rousseau, C., Monflier, E., & Tilloy, S. (2018). Unconventional media and technologies for starch etherification and esterification. *Green Chemistry*, *20*(6), 1152–1168. https://doi.org/10.1039/C7GC03135A

González, M. A., Rodríguez, L. E., & Pérez, G. (2022). Modified starches as fat replacers in reduced-calorie foods: Applications and health benefits. *Food Science and Technology International*, *28*(5), 407–418.

Gupta, S., Patel, M., & Yadav, A. (2013). Electrolyte-sensitive starch hydrogels for controlled drug release. *Journal of Biomedical Materials Research Part A*, *101*(8), 2344–2354.

Haug, W., & Lantzsch, H. J. (1983). A method for rapid determination of phytate in cereals and cereal products. *Journal of the Science of Food and Agriculture*, *34*(12), 1423–1426. https://doi.org/10.1002/jsfa.2740341217

Ji, N., Ge, S., Li, M., Wang, Y., Xiong, L., & Qiu, L. (2019). Effect of annealing on the structural and physicochemical properties of waxy rice starch nanoparticles. *Food Chemistry*, *286*, 17–21. https://doi.org/10.1016/j.foodchem.2019.01.205

Jiang, T., Duan, Q., Zhu, J., Liu, H., & Yu, L. (2020). Starch-based biodegradable materials: Challenges and opportunities. *Advanced Industrial and Engineering Polymer Research*, *3*(1), 8–18. https://doi.org/10.1016/j.aiepr.2019.11.003

Jiménez, A., Fabra, M. J., Talens, P., & Chiralt, A. (2012). Edible and biodegradable starch films: A review. *Food and Bioprocess Technology*, *5*, 2058–2076. https://doi.org/10.1007/s11947-012-0835-4

Kanatt, S. R. (2020). Development of active/intelligent food packaging film containing Amaranthus leaf extract for shelf life extension of chicken/fish during chilled storage. *Food Packaging and Shelf Life*, *24*, 100506. https://doi.org/10.1016/j.fpsl.2020.100506

Kwon, I. K., Lee, Y. K., & Liu, S. (2011). Temperature-sensitive starch-based hydrogels for controlled drug release. *Journal of Controlled Release*, *151*(2), 187–193. https://doi.org/10.1016/j.jconrel.2011.01.002

Li, Y., De Vries, R., Slaghek, T., Timmermans, J., Cohen Stuart, M. A., & Norde, W. (2009). Preparation and characterization of oxidized starch polymer microgels for encapsulation and controlled release of functional ingredients. *Biomacromolecules*, *10*(7), 1931–1938. https://doi.org/10.1021/bm900337n

Liu, D., Zhao, P., Chen, J., Yan, Y., & Wu, Z. (2022). Recent advances and applications in starch for intelligent active food packaging: A review. *Foods*, *11*(18), 2879. https://doi.org/10.3390/foods11182879

Liu, Y., Yang, Z., & Yu, L. (2021). Smart starch-based hydrogels for biomedical applications. *International Journal of Biological Macromolecules*, *183*, 1000–1010. https://doi.org/10.1016/j.ijbiomac.2021.05.002

Lv, M., Du, Y., Zhang, T., Du, X., & Yin, X. (2022). Cassava starch-based thermo-responsive Pb(II)-imprinted material:

Preparation and adsorption performance on Pb(II). *Polymers*, *14*(4), 828. https://doi.org/10.3390/polym14040828

Majeed, Z., Ajab, Z., Guan, Q., Abbasi, A. Z., Mahmood, Q., Mahnashi, M. H., Alyami, B. A. A., Alqarni, A. O., Alqahtani, Y. S., & Mansor, N. (2021). Reduction in lignin peroxidase activity revealed by effects of lignin content in urea crosslinked starch under aerobic biodegradation in soil. *BioResources*, *16*(1), 1940–1948.

Matthias, D. R., Safoura, A., Bockstaele, F. V., & Ubeyitogullari, A. (2024). Development of a pH-responsive system based on starch and alginate-pectin hydrogels using coaxial 3D food printing. *Food Hydrocolloids*, *153*, 109989. https://doi.org/10.1016/j.foodhyd.2024.109989

Meimoun, J., Wiatz, V., Saint-Loup, R., Parcq, J., Favrelle, A., Bonnet, F., & Zinck, P. (2018). Modification of starch by graft copolymerization. *Starch/Staerke*, *70*(1–2), 1600351. https://doi.org/10.1002/star.201600351

Meng, C., Yingjun, C., Xiaoran, Y., Rongfei, Z., Juan, W., & Xiangyou, W. (2022). Effect of dual-modified cassava starches on intelligent packaging films containing red cabbage extracts. *Food Hydrocolloids*, *124*, 107225. https://doi.org/10.1016/j.foodhyd.2021.107225

Menzel, C., González-Martínez, C., Vilaplana, F., Diretto, G., & Chiralt, A. (2020). Incorporation of natural antioxidants from rice straw into renewable starch films. *International Journal of Biological Macromolecules*, *146*, 976–986. https://doi.org/10.1016/j.ijbiomac.2019.12.270

Nasiri, A., Wearing, J., & Dubé, M. A. (2020). Using lignin to modify starch-based adhesive performance. *ChemEngineering*, *4*(1), 3. https://doi.org/10.3390/chemengineering4010003

Patel, A., Patel, D., & Patel, P. (2021). Antimicrobial bioactive starch-based films for food packaging: Current trends and future perspectives. *Food Packaging and Shelf Life*, *29*, 100711.

Quintanilla, S. J. C., Abundis-Correa, V., Herrera-Flores, S. D., & Alvarez, A. J. (2020). pH-Sensitive starch-based hydrogels: Synthesis and effect of molecular components on drug release behavior. *Polymers*, *12*(9), 1974.

Salimi, M., Channab, B. E., El Idrissi, A., Zahouily, M., & Motamedi, E. (2023). A comprehensive review on starch: Structure, modification, and applications in slow/controlled-release fertilizers in agriculture. *Carbohydrate Polymers*, *322*, 121326. https://doi.org/10.1016/j.carbpol.2023.121326

Sang, L., Yang, Y., & Zhang, Y. (2023). Design of moisture-sensitive starch materials for sustainable packaging and agricultural applications. *International Journal of Biological Macromolecules*, *240*, 124351.

Sarwan, T., Kumar, P., Choonara, Y. E., & Pillay, V. (2020). Hybrid thermo-responsive polymer systems and their biomedical applications. *Frontiers in Materials*, *7*, 73.

Shen, L., Zhang, H., & Li, Y. (2021). Development of starch-based nanocomposite materials for food packaging applications: Challenges and opportunities. *International Journal of Biological Macromolecules*, *183*, 1422–1433.

Silva, M. A., Souza, L. C., & Gomes, F. O. (2020). Nanocomposite starch-based materials: An overview of applications in food packaging and biomedical fields. *Food Research International*, *138*, 109825.

Shrivastava, A., Gupta, R. K., & Srivastav, P. P. (2024). Exploring novel frontiers of advancements in purple yam (*Dioscorea alata* L.) starch extraction, modification, characterization, applications in food and other industries. *Measurement: Food*, *15*, 100196. https://doi.org/10.1016/j.meafoo.2024.100196

Rodrigues, A. T., Campos, B.P., Rabledo, F.E.M.A., & de Fatima, F.S.N. (2022). Natural bioactives in perspective: The future of active packaging based on essential oils and plant extracts themselves and those complexed by cyclodextrins. *Food Research International*, *156*, 111160.

Thompson, D. R., Reed, R., & Matthews, C. (2014). Enzyme-sensitive starch-based systems for pharmaceutical applications. *International Journal of Pharmaceutics*, *466*(1–2), 88–96.

Vijaykumar, S., Kumar, R., & Sharma, M. (2022). Recent advances in the development of starch-based nanocomposites: Applications in food packaging and biomedical fields. *Carbohydrate Polymers*, *273*, 118568.

Vilas, C., Mauricio-Iglesias, M., & García, M. R. (2020). Model-based design of smart active packaging systems with antimicrobial activity. *Food Packaging and Shelf Life*, *24*, 100446.

Wang, R., Yao, L., Peng, S., Liu, Z., Zhu, X., Li, H., Xu, D., Zhang, J., Mo, H., & Hu, L. (2023). An "intelligent -responsive" bactericidal system based on OSA-starch Pickering emulsion. *International Journal of Biological Macromolecules*, *235*, 123808.

Wang, S., Zhou, Y., & Zhu, C. (2022). Sustainable strategies for the synthesis of multi-responsive starch derivatives. *Green Chemistry*, *24*(7), 1550–1570.

Wang, X., Huang, L., Zhang, C., Deng, Y., Xie, P., Liu, L., & Cheng, J. (2020). Research advances in chemical modifications of starch for hydrophobicity and its applications: A review. *Carbohydrate Polymers*, *240*, Article 116292.

Willett, J. L. (2008). Humidity-responsive starch-poly(methyl acrylate) films. *Macromolecular Chemistry and Physics*, *209*, 764–772. https://doi.org/10.1002/macp.200700495

Xie, X., Zhang, X., & Liu, Y. (2022). Starch-based biodegradable packaging materials: A review of recent developments and future prospects. *Journal of Cleaner Production*, *366*, 132731.

Yang, X., Gao, Y., & Hu, W. (2021). Phase transition behavior of starch-based materials: A review of mechanisms and applications. *Macromolecular Chemistry and Physics*, *222*(12), 2100317.

Zhang, K., Shi, Y., Wu, L., Chen, L., Wei, T., Jia, X., Chen, Z., Li, M., Xu, Y., Wang, Y., Gao, Y., & Guo, X. (2018). Thermo- and pH-responsive starch derivatives for smart window. *Carbohydrate Polymers*, *196*, 209–216. https://doi.org/10.1016/j.carbpol.2018.05.039

Zhang, Q., Chen, M., & Zhou, J. (2015). Photo-responsive starch-based materials for advanced applications. *Advanced Functional Materials*, *25*(16), 2341–2349.

Zhang, Y., Chen, C. Chen,Y., Chen, Y. (2019). Effect of rice protein on the water mobility, water migration and microstructure of rice starch during retrogradation, *Food Hydrocolloids*, *91*, 136–142. https://doi.org/10.1016/j.foodhyd.2019.01.015

Zhang, Y., Liu, Z., & Wang, Z. (2023). Bioactive starches for functional foods and active packaging: A review. *Food Control*, *142*, 109297.

Zhang, Y., Ng, M., Hong, E. Y., Chan, A. K., Wu, N. M., Chan, M. H., Wu, L., & Yam, V. W. (2020). Synthesis and photoswitchable amphiphilicity and self-assembly properties of photochromic spiropyran derivatives. *Journal of Material Chemistry C*, *8*, 13676–13685. https://doi.org/10.1039/D0TC03301D

Zhao, X., Zhang, Y., & Li, J. (2022). Light-responsive starch materials for smart packaging applications: Recent advancements and future directions. *Journal of Materials Science*, *57*(12), 5123–5135.

Zhou, L., Wang, X., & Liu, Z. (2022). Functionalized starch-based materials for moisture-responsive applications in packaging and agriculture. *Food and Bioproducts Processing*, *131*, 235–245.

7 Microwave Technology for Starch Modification

Honey Gupta[a], *Nitamani Choudhury*[a], *Prudhvi Pasumarthi*[b], and *Sankar Chandra Deka*[a]
[a]Food Engineering and Technology Department, Tezpur University, Assam, India

Prudhvi Pasumarthi
[b]School of Engineering, University of Guelph, ON, Canada

7.1 INTRODUCTION

Starch is the primary energy source for people all over the world and is created by plants as a reserve carbohydrate. It is made in plant amyloplast cells as discrete granules that range in shape from round, oval, ogival, or elongated to flat, lenticular, or polyhedral and in diameter from submicrons to over 100 μm in various plants (Bertoft, 2017; Gupta et al., 2023). Major sources of starch include cereals (wheat, rice, barley, maize, sorghum), tubers (taro, potato, cassava), and fruits (banana, plantain, dates) (Bangar et al., 2022; Shrivastava et al., 2024). There are also some non-conventional sources that are emerging as new sources of starch, such as jackfruits, fava beans, kiwi, mango seed, and lychee (Morales-Trejo et al., 2021; Tagliapietra et al., 2020; Wang et al., 2021). Amylose and amylopectin are the two primary polymer types that make up starch. They are connected by α-1,4- and α-1,6-glycosidic linkages (Gupta et al., 2022). The amount of amylose and amylopectin in starch influences the starch's size, crystallinity, chemical makeup, functional qualities, and pace of digestion, among other characteristics (Nawaz et al., 2020). In the food industry, starch can be used in many ways, for example, as a binder, stabilizer, additive, thickener, gelling agent, or encapsulating agent (Kumari & Sit, 2023). As starch is easily accessible, affordable, biodegradable, and non-toxic, it is also widely employed in the pharmaceutical and other industries. Its primary usage in the pharmaceutical sector is as a drug delivery medium. (Masina et al., 2016). However, low shear and thermal stabilities, strong retrogradation behaviour, low fluidity, low paste clarity, high gel turbidity, and gelatinization temperature limit the use of native starch. These issues can be resolved via starch modification, which will increase the starch's overall utility. A variety of techniques, including physical, chemical, enzymatic, and genetic ones, can be used to alter the characteristics of starch. Moreover, several attributes, including resistant starch content, film-forming ability, film strength, adhesiveness, and techno-functional qualities, can be enhanced with the aid of these modifications (Kumari & Sit, 2023; Maniglia et al., 2020).

For starch modification, a novel and developing non-thermal physical technique is the use of microwaves. High-frequency electromagnetic waves having wavelengths between 1 mm and 1 m with frequencies between 300 MHz and 300 GHz are referred to as microwaves. Because of surface erosion, microwaves have a bigger effect on the food material's morphological characteristics. The use of microwaves is less time-consuming and is environmentally friendly, with great efficiency, uniform energy radiation, and good penetrability (Yi et al., 2024). This chapter will explore the history of microwaves, the variables that influence microwave modification, the impact of microwave treatment (MT) on the different properties of starch, and the ultimate application of MT in diverse industries.

7.2 HISTORICAL OVERVIEW OF MICROWAVE TECHNOLOGY

Microwave technology has a long history that began during World War II and witnessed significant advancements up to the 1960s. The cavity magnetron, created by British physicists John T. Randall and Henry A. Boot, was the key to the revelation of microwave heating (Osepchuk, 1984). The low-energy microwave electromagnetic radiation was inadvertently discovered by Raytheon scientists in the 1940s. It was Percy Spencer at Raytheon, who patented numerous microwave heating applications, and made the primary contribution to the invention of the microwave heating system. The Radarange, a freestanding water-cooled device that Raytheon introduced, was modified in the early 1950s as air-cooled floor and countertop models. The following elements make up a basic microwave system: the sample to be processed, the oven cavity, the step-up transformer, the magnetron, the waveguide, and the applicator. The magnetron is powered by the step-up transformer, which accounts for 50% of the weight of home microwave ovens. At 4000 volts, the magnetron transforms electrical energy into microwave energy. The most often used frequency for consumer-based microwaves is

DOI: 10.1201/9781003564287-7

2450 MHz, while 915 MHz is typically utilized on an industrial scale (Orsat et al., 2016).

7.3 STARCH MODIFICATION USING MICROWAVE

Microwaves are alternating electromagnetic fields that interact with polar molecules within a material. These molecules, possessing a permanent dipole moment, attempt to align themselves with the oscillating electric field generated by microwaves. As the field alternates at extremely high frequencies by changing direction billions of times per second, the polar molecules undergo rapid rotational movements to match the field's orientation. This phenomenon is known as dipole rotation. Subsequently, the rotating molecules collide with neighbouring molecules, generating intermolecular friction. The combination of dipole rotation and molecular friction rapidly increases the temperature of the system in a volumetric manner (Prudhvi et al., 2021).

Starch is a natural polymer with a radial arrangement of alternating crystalline and amorphous regions formed by glucose chains. The crystalline regions primarily consist of branched glucose chains linked at the 1,6 carbon atoms, arranged in a double-helical structure known as amylopectin (Gupta et al., 2024a). In contrast, the amorphous regions are predominantly composed of linear glucose chains connected at the 1,4 carbon atoms, forming single helices called amylose (Wang et al., 2015). When starch is exposed to microwaves, the molecular movements and heat generated within the system disrupt the double- and single-helical arrangements within the starch granule. These disruptions cause the glucose chains to reorganize into a less ordered and weaker crystalline structure than that of their native state. This structural alteration significantly modifies the functional properties of starch, making it suitable for specialized applications. Water, a common polar molecule present in starch, also contributes to this process. The heat generated by the rotation of water molecules and their intermolecular collisions can lead to moisture loss at elevated temperatures. This loss creates vapour pressure within the starch granule, further promoting the disruption of its structure (Oyeyinka et al., 2021).

The modification of starch using microwaves is highly dependent on moisture content, as water molecules play a critical role in molecular movement and heat generation. In low-moisture systems, the limited availability of water results in concentrated energy dissipation, causing rapid temperature increases in localized areas. Conversely, in high-moisture systems, the abundance of water molecules distributes the energy more evenly, creating a buffering effect that moderates the overall temperature rise (Braşoveanu & Nemtanu, 2013). Microwave power and treatment duration are additional process parameters that directly influence temperature. Studies suggested that microwave power of a minimum 2 W/g of starch is needed to cause a significant modification in starch characteristics (Xu et al., 2018). The treatment time depends on the microwave power density (W/g of starch). Lower microwave powers often require longer durations and vice versa (Kumar et al., 2020). Microwave treatment under high-moisture conditions (typically > 60%), at elevated power levels, and for extended durations can generate sufficient heat to induce starch gelatinization. However, intense microwave treatments at controlled temperatures and moisture contents could result in depolymerized starch. During this process, water penetrates the starch granules, leaching out amylose and disrupting the granular structure of the starch. Upon cooling, the gelatinized starch may undergo retrogradation, forming resistant starch, characterized by low swelling capacity and increased resistance to digestion.

Microwave treatment is a physical modification technique that can replicate the effects of conventional physical methods such as heat-moisture treatment (HMT) and annealing, both of which result in structural disruption and reorganization of starch. In HMT, starch is treated with 25%–30% moisture content at temperatures above its gelatinization point but with insufficient moisture to achieve complete gelatinization. In contrast, annealing involves heating starch at moisture levels exceeding 40% to a temperature between the glass transition and gelatinization points. Microwave treatment can be optimized by adjusting parameters such as power and duration to achieve similar temperature conditions, enabling rapid HMT or annealing processes (Pinto et al., 2018). There are also other modification methods that modify starch in similar mechanisms. For instance, high hydrostatic pressure (HHP) is an emerging modification technique that induces non-thermal gelatinization of starch at low temperatures (Shen et al., 2017). This process, like microwave treatment, promotes structural reorganization (retrogradation), offering a non-thermal alternative for starch modification.

7.4 PHYSICOCHEMICAL AND FUNCTIONAL PROPERTIES OF MICROWAVE-MODIFIED STARCH

7.4.1 Effect of Microwave Treatment on Structural Properties of Starch

7.4.1.1 Surface Morphology

Depending on its botanical source, starch, a semi-crystalline biological molecule, can have a variety of sizes and forms. Common shapes of native starch range in size from 1 to 175 μm and include polygonal, spherical, flaky, ovoid, and irregular morphologies (Gupta et al., 224b). According to several studies, the morphology and birefringence of starches are affected differently by different combinations of microwave power and time (Table 7.1) (Oyeyinka et al., 2021; Tao et al., 2020). According to a study on potato starch by Kumar et al. (2020), granules remained unchanged after 1 minute of treatment; however, when the duration was increased to 3 and 5 minutes at the same power level of 300 W, the granules began to lose their forms and surface

TABLE 7.1
Effect of Microwave Treatment on Structural and Thermal Properties of Starch

Starch Source	Microwave Treatment	Surface Morphology	Relative Crystallinity	Thermal Properties	References
Cassava roots	30% moisture; 600 and 700 W; 5, 15, 30, and 60 s	The changes were severe over time. Pores and indentations developed on the surface.	The crystalline pattern remained unchanged, although peak intensities were lowered when compared to native starch. There was also a drop in RC.	Increase in T_o, T_p, and T_c. Reduction in ΔH.	(Oyeyinka et al., 2020)
White finger millet	20% moisture; 510 W; 10 min	Granules were more irregular and agglomerated.	Additional peak was found at 19.95°, indicating crystalline reorientation. Reduction in relative crystallinity.	T_o, T_p, and T_c values increased.	(Balakumaran et al., 2023)
Mango kernel (Tommy and Sugar)	900 W; 1 min	Both kinds saw an increase in particle sizes. Despite the granules' unchanged shape, their surface hardened, and a few tiny pores appeared.	MT reduced the crystallinity.	Increase in T_o, T_p, and T_c by 1.1 and 1.33 °C, 0.41 and 0.55 °C, and 1.42 and 0.58 °C. Decrease in ΔH.	(Ramírez-Brewer et al., 2023)
Maize (amylopectin and amylose granules)	30% moisture; 400 W/g; 1, 2, 3, 4, 6, and 8 min	–	AP, 0–3 min: decrease in peak intensities; disruption of B-type structure and formation of V_h-type. During 2–8 min, intensity of V_h-type reduced. Further increase in treatment to 4–8 min enhanced the granular peaks' intensity. AM, 0–2 min: increase in RC. 2–8 min: Decrease in RC.	–	(Zhong et al., 2021)
Proso millet	500 W; 10 min	Pore sizes are uniform, and the structure is tidy and consistent.	No peaks in MT starch. RC found to be reduced to zero.	–	(Zheng et al., 2020)
Waxy maize	30% moisture; 1600 W; 5, 10, and 20 min	–	RC reduced to 10.47%, 6.42%, and 2.91% in the treated starch from 19.58% in NS.	T_o, T_p, and T_c increased with increase in treatment time. ΔH reduced.	(Yang et al., 2017)
Potato	30%, 50%, and 70% moisture; 800 W; 5 min	Small fractures at the starch surface, with rough texture.	RC increased from 31.72 in NS to 37.57 in MW-70.	T_o and T_p increased with moisture content. Reduction in conclusion temperature. ΔH reduced.	(Xia et al., 2018)
Chestnut	30 W 90 s Dual modification: UC (500 W, 60 min) + MC (UMC) And MC+UC (MUC)	Single MT caused wrinkles and roughness. UMC and MUC had more roughness at the surface and pits. MUC particles were more aggregated.	Peak intensity found to be weak for all the modified starch. RC of NS was 27.85%. it decreased for modified starches MUC (23.75%), UMC (23.98%), and MT (24.75%).	UMC showed highest T_o (61.8 °C) and T_p (64.3 °C). ΔH was lowest for UMC (0.97 J/g) < MUC (1.18 J/g). For NS, ΔH was 2.91 J/g.	(Wang et al., 2020)
Mung bean	Debranching + MT (20% moisture; 350 W 3 min)	NS: Oval or elliptical and round. DS: Aggregation and formation of crystalline structure. MDS: No further effect.	NS: A-type crystalline structure. DS: B+V. MDS: No changes in XRD pattern after DS; had B+V type pattern.	–	(Huong et al., 2021)
Breadfruit	30% moisture; 1 W/g; 20 min	Disappearance of granular structure; small holes appeared.	B-type NS to A+B type in modified starch. RS decreased from 26.07% to 14.99%.	–	(Marta et al., 2019

Note: ΔH, gelatinization enthalpy; DS, debranched; MC, microwave modified starch; MDS, microwave-assisted debranched; MT, microwave treatment; NS, native starch; RC, relative crystallinity; T_c, conclusion gelatinization temperature; T_o, onset gelatinization temperature; T_p, peak gelatinization temperature; UC, ultrasonic modified starch; XRD, X-ray diffraction; MUC, microwave-ultrasonic modified starch; UMC, ultrasonic microwave modified starch.

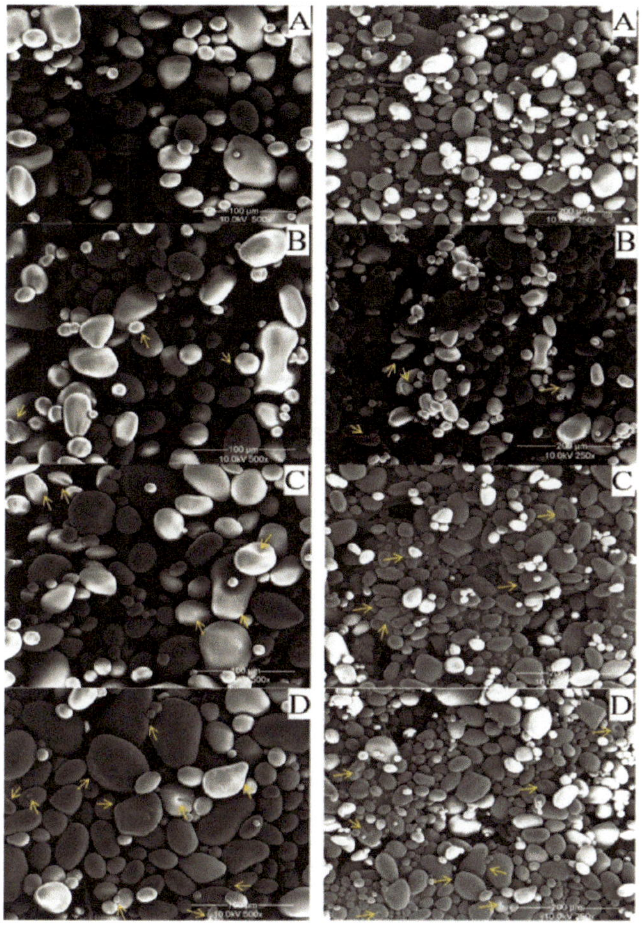

FIGURE 7.1 SEM images of (A) control and (B–D) microwave-treated potato starch (B, 1 min; C, 3 min; D, 5 min) (Kumar et al., 2020).

fractures appeared (Figure 7.1). Because the outside pressure is low and the moisture inside the granules turns into steam during MT, high pressure is created, which causes the existing cracks to widen as the steam seeks to escape. Additionally, some new cracks are created as a result of rising temperatures. The MT occasionally increases the size of starch granules, as demonstrated in two different kinds of mango kernel starch. After MT treatment, the Sugar (Sg) variety's size increased from 5.62–22.14 μm to 8.05–24.14 μm, while the Tommy (To) variety's size increased from 5.89–20.81 μm to 7.37–24.3 μm. This kind of modifications influence the rheological properties of starch (Ramírez-Brewer et al., 2023). The length of treatment has a major impact on how MT affects the size and form of starch; the alterations are more noticeable for longer the treatments (Navaf et al., 2024). The morphology of cassava starch treated at 600 and 700 W for varied durations of 5, 15, 30, and 60 seconds revealed no impacts on the structural integrity of the starch at 5 or 15 seconds, but the effects became more noticeable at 30 or 60 seconds. There were surface indentations at 600 W for 30 seconds, but the sample treated at 700 W had more pores. As the period rose to 60 seconds, the pores grew larger and the indentations grew greater (Oyeyinka et al., 2020). After 10 minutes of MT of proso millet starch at 500 W, the stromal wall of the starch gel thinned, exhibiting consistent pore size and a tidy, regular network topology (Zheng et al., 2020).

7.4.1.2 Crystallinity

Starch is composed of multilayer structures with alternating amorphous and semi-crystalline growth rings. The crystalline area is made up of double helices generated by the aggregation and entwining of chains longer than ten glucose units. Lamellar and crystalline forms have a direct impact on starch characteristics such as gelatinization and digestibility (Tao et al., 2020). Starch is classified into four types of crystal patterns (A, B, C, and V) based on its botanical origin. Using X-ray diffraction (XRD) techniques, the various patterns of starch granules are analyzed (Gupta et al., 2024c). Starch granules' crystallinity and structure are not significantly impacted by microwave treatment (Table 7.1). However, B-type starch can change into A or C types following MT, while A-type starch's crystallinity can occasionally decrease (Yi et al., 2024). For instance, after MT, the modified mango kernel starch's crystallinity value decreases from 26.70% to 24.02% for the Sg variety and from 24.50% to 22.30% for the To variety. However, all the samples had the identical XRD patterns (Figure 7.2) (Ramírez-Brewer et al., 2023). Similarly, when cassava starch was subjected to MT at 600 and 700 W for 5, 15, 30, and 60 seconds, the XRD results revealed that at higher power and a maximum length of 60 seconds, the reduction in relative crystallinity (RC) was greater than when the samples were microwaved at 600 W for 60 seconds. Also, the peak intensities were lower for microwaved starch. These alterations may be related to damage in the crystalline area and the breaking of hydrogen bonds in the starch granules (Oyeyinka et al., 2020). MT may also increase the crystallinity of starches

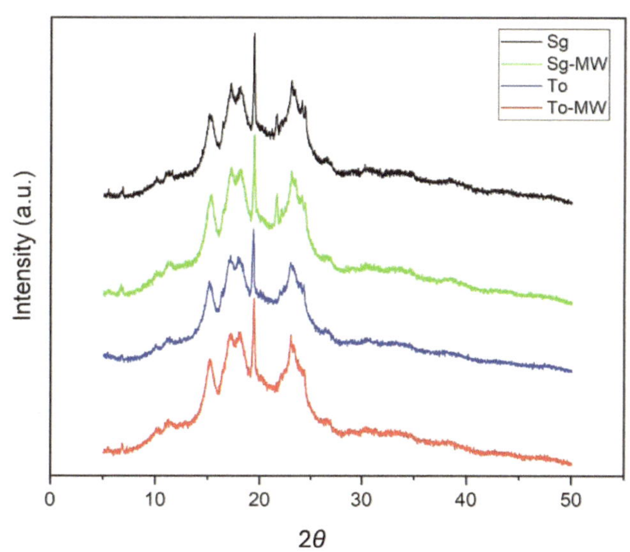

FIGURE 7.2 X-ray diffraction patterns of native and microwave-modified mango kernel starches from Sugar (Sg, Sg-MW) and Tommy (To, To-MW) varieties.

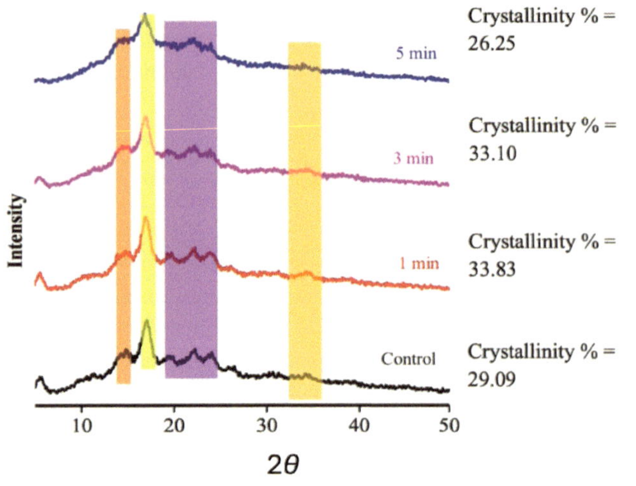

FIGURE 7.3 X-ray diffraction patterns for control and microwave-treated potato starch, with changes in relative crystallinity over different treatment times.

(Tao et al., 2020). For example, in a study on pure amylopectin (AP) and amylose (AM) in maize, Zhong et al. (2021) discovered that MT initially reduced the peak intensities of AP molecules during a period of 0–3 minutes, whereas increasing the treatment duration to 4–8 minutes resulted in an increase in granular peaks. The intensity was augmented by the degraded AP molecules' significant reaggregation as the treatment increased. Conversely, when treatment was administered for 2–8 minutes, the RC of AM granules dropped, while it increased during a period of 0–2 minutes. At first, AM granules showed a mix of the V_h-type allomorph (at $2\theta = 8°$, 13°, 15°, and 20°) and B-type allomorph (at $2\theta = 17°$, 22°, and 26°). However, when treatment was administered for 0–2 minutes, the strength of the peaks at 2θ of 22° and 26° rose, and the peak became more intense at 2θ of 20° increased. This indicates that amylose and endogenous uncomplexed lipids disrupted the B-type allomorph and concurrently formed the V_h-type allomorph. Nevertheless, additional treatment (2–8 min) reduced the intensity of the V_h-type crystalline allomorph, indicating its depletion, maybe because of amylose or lipid degradation, which causes the V-type allomorph to become unstable. Figure 7.3 shows the XRD patterns of control and microwave-treated potato starch, along with the changes in relative crystallinity values over different treatment times (Kumar et al., 2020).

7.4.2 Effect of Microwave Treatment on Various Thermal Properties

7.4.2.1 Pasting Properties

The term "pasting" describes the changes that take place in starch when it is heated under constant rotating stress. This causes the starch granules to inflate and eventually burst, which allows amylose to leach out and produce a thick paste. These characteristics include pasting temperature (PT), breakdown viscosity (BV), setback viscosity (SV), peak viscosity (PV), and final viscosity (FV) (Gupta et al., 2024e). A Rapid Visco Analyzer (RVA) or rheometer can be used to determine these qualities. Numerous food and non-food industries depend on these qualities (Bashir & Aggarwal, 2019). Research has indicated that microwave heating primarily diminishes the starch's pasting qualities; the longer the treatment period, the more noticeable the effect (Yi et al., 2024). The viscograms of microwave-treated waxy maize starch (wms) revealed a reduction in PV, FV, BV, and SV. The decline was higher as treatment time increased (5, 10, and 20 min). This drop is related to the low MW of starches following treatment. However, the PT increased as treatment progressed. This demonstrates that modified wms were very resistant to swelling and rupture (Yang et al., 2017). On the other hand, the MT of taro starch (25% moisture content raised pasting characteristics such as PV, BV, SV, and FV, while decreasing pasting temperature (Deka & Sit, 2016).

In another study, microwave heating (540 W for 0–30 min) was used as a pretreatment to increase the effect of annealing (ANN) on rice starch. It was found that the combined effect of microwave and annealing (MW-ANN) treatment increased the peak and breakdown viscosities, with MW20-ANN having the highest effect of all treatments. Similar results were obtained for final and setback viscosities, demonstrating that microwave pretreatment improves pasting properties (Y. Zhong et al., 2019).

7.4.2.2 Gelatinization Properties

The process of heating starch with excess water (atleast 60%) is called gelatinization. In the process, the starch granules start to disintegrate. First, the water penetrates the amylopectin in amorphous regions (less ordered), causing granule swelling due to the disruption of weaker hydrogen bonds. This is followed by the breaking down of crystalline regions (double helical arrangement), expansion, and birefringnce that cause amylose to leach into the surrounding water.. Beginning at the hilum(centre of the starch granule), gelatinization progressively moves outward until the entire granule structure is destabilized, forming a viscous paste or gel. Amylopectin's molecular structure and the ratio of amylopectin content to that of amylose determine the starch's ability to gelatinize. In the food sector, this characteristic is essential for it affects on stability, viscosity, and texture. A popular method for elucidating the gelatinization characteristics of starch is differential scanning calorimetry (DSC) (Chakraborty et al., 2022). Enthalpy of gelatinization (ΔH), onset gelatinization temperature (T_o), peak gelatinization temperature (T_p), and conclusion gelatinization temperature (T_c) are of the vital gelatinization parameters. The MT of starch includes the effects of electromagnetic polarization and the dielectric heating impact of microwaves. During MT, friction in the polar molecules produces a lot of heat, which raises the starch grain temperatures and alters the starch's properties (Table 7.1) (Zheng et al., 2020). When waxy maize starch was microwaved at 1600W for varying periods, the T_o, T_p, and T_c values increased. The molecular structure of the crystalline region, which is linked to the distribution of short amylopectin chains, influences the gelatinization temperatures. The gelatinization enthalpy dropped after treatment, showing that the starch structure had lost its double-helical order (Yang et al., 2017). Typically, starches with less amylose than amylopectin exhibit a high gelatinization temperature (Oyeyinka et al., 2021). However, when microwave heating is employed in conjunction with other treatments or when starch is dual modified, the impact on gelatinization properties may differ (Yi et al., 2024). In a study on corn starch, Zhou et al. (2022) discovered that when starch was treated with lysine (0.5%–4%, dry starch base) and then microwaved for a specific amount of time (2–6 min), the T_o, T_p, and T_c were high in MC-Lys compared to MC-CS and CS-lys. This could be because lysine binds to corn starch when it is heated, inhibiting the gelation of starch granules and enhancing the starch's thermal stability.

When potato starch was microwave-treated at varying moisture contents (30%, MW-30; 50%, MW-50; 70%, MW-70) for 5 minutes at 800 W, the gelatinization temperature of the starch samples rose in comparison to native potato starch; however, the changes differed depending on the moisture content. The values in MW-30 and MW-50 increased from 60.12 ± 0.19 to 61.92 ± 0.74, 60.95 ± 0.22, respectively, and to 60.09 ± 0.15 in MW-70, which was a little drop from NS. T_p followed a similar pattern, with the highest values for 30% MC being 65.74 ±0.21. T_c decreased in starches that were microwave-treated. The lowest was 80.49 ± 1.12 for MW-30. However, NS had the largest ΔH, at 19.74 ± 0.23, which decreased to 15.12 ± 0.10 for MW-30, 17.08 ± 0.26 for MW-50, and 18.53 ± 0.72 for MW-70. The MT lengthens double helices and raises the amylopectin concentration of starch, which improves crystal stability (Xia et al., 2018).

7.4.3 Changes in Functional Properties

7.4.3.1 Amylose Contents

Starch is composed of two types of polymers amylose (20%–30%) and amylopectin (70%–80%). Based on amylose concentrations, starch is further divided into high amylose starch (> 40%), normal starch (20%–30%), and waxy starch (< 10%). The molecular weight and structure of amylose and amylopectin are also different (Gupta et al., 2024d). Amylose is a low-molecular-weight polymer (1.03–4.89 × 105) that mostly works as a non-branched molecule and is composed of linear chains joined by α-1,4-glycoside linkages. With a cluster-like shape that suggests heavy branching, amylopectin has a greater molecular weight (7.08–9.88 × 107) (Cornejo-Ramírez et al., 2018). It is believed that MT breaks down the α-1,6-glycosidic linkage more efficiently than the α-1,4-glycosidic linkage, increasing the amount of amylose (Table 7.2) (Zailani et al., 2021). The amylose

TABLE 7.2
Effect of Microwave Treatment on Functional Properties of Starch

Starch Source	Microwave Treatment	Functional Properties	References
Sago	30% MC; 900 W; 5–20 min	↑ Amylose content, SP, %S and WBC. ↓ OBC.	(Zailani et al., 2021)
Rice	~30% MC; 500 W, 30 and 90 s	↑ Amylose content, SP, and %S. MT significantly enhanced the molecular weight of amylose and amylopectin. RS increased from 43.53% to 48.68%. In dual modification with CP, highest RS was found for M90-P2 (58.77%).	(Sun et al., 2021)
Breadfruit	30% MC; 1 W/g; 20 min	No significant difference changes in SP and solubility. ↑ WAC, from 1.75 ± 0.05 to 2.63 ± 0.46 g/g in the modified starch. ↑ % syneresis from 1.51 ± 0.37 in NS to 25.99 ± 1.65 in modified starch.	(Marta et al., 2019)
Buckwheat	30% MC, 720 W, 30–120 s	↑ Amylose content from 25.5 ± 0.2 to 35.1 ± 0.3. ↑ WHC from 1.21 ± 0.06 to 1.41 ± 0.03. ↓ OHC from 1.13 ± 0.02 to 1.02 ± 0.01. ↓ SP and % S.	(Negi et al., 2025)
Lotus seed flour	1–5 min	↑ WHC, OHC, emulsifying activity, emulsifying stability, and least gelation concentration (LGC).	(Nawaz et al., 2018)
Tibetan qingke (TQ)	700 W, 60 s	↑ RDS, 31.2 ± 0.4 (NS) to 55.0 ± 0.9 in microwave-treated starch. ↓ SDS, 39.0 ± 1.1 (NS) to 13.7 ± 1.7 in treated starch. ↑ RS, 31.3 % in the treated starch.	(Bai et al., 2022)
Cyperus esculentus	10% MC, 700 W, 90 s	↑ RS content for both regelatinized and non-regelatinized starch from 20.27 ± 1.78 in NS to 27.65 ± 0.91 and 29.70 ± 0.510. ↑ Freeze–thaw stability of microwave-treated starch.	(Wang et al., 2023)

Note: MC, moisture content; MT, microwave treatment; OBC, oil binding capacity; OHC, oil holding capacity; %S, solubility; RDS, rapid-digesting starch; RS, resistant starch; SDS, slow-digesting starch; SP, swelling power; WAC, water absorption capacity; WBC, water binding capacity; WHC, water holding capacity.

content of modified rice starch increased from 13% in NS to 15.15% after microwave treatment. It is possible that the polymerization of amylose chains during microwave heating alters the helix shape and increases the quantity of ordered short-range double helices, which increases amylose's capacity to form a complex with iodine (Sun et al., 2021). Similar results were found for taro starch when treated with microwave alone and in combination with other heat treatment, the amylose content was enhanced from 13.90% in NS to 15.12% in microwave-treated starch (Deka & Sit, 2016).

7.4.3.2 Water and Oil Absorption Capacity

The quantity of water absorbed per gram of starch is known as WAC (water absorption capacity). It is a crucial functional characteristic that affects food items' consistency, texture, and ability to retain moisture. Most of the time, MT raises the WAC because it breaks up the starch granules or increases their amylose content, which improves their ability to absorb or hold water (Table 7.2). Additionally, during MT, the crystalline structure of the starch granules changes to an amorphous state, allowing for greater reactivity with water (Negi et al., 2025). In certain instances, however, it is discovered that MT lowers the WAC. Oyeyinka et al. (2020), for instance, discovered that following MT, the WAC capacity of cassava starch decreased to 3%–32% for 700 W treated starch and 2%–14% for 600 W treated starch. This decrease may be the result of amorphous region rearrangement, or the drop in RC may also impair starch WAC.

Oil absorption capacity (OAC) is an important property of starch as it influences the flavour retention and overall mouth feel of food products. According to Nawaz et al. (2020) the oil holding capacity (OHC) of starch extracted from microwave-treated flour was increased with increasing treatment time. But in another study, a negligible decrease in OAC was seen with MT white finger millet starch. A decrease in the number of hydrophobic amino acids, decreased water activity, or a decreased propensity of starch to interact with oil because of modification could all contribute to the decrease in OAC (Table 7.2) (Balakumaran et al., 2023).

7.4.3.3 Swelling Power (SP) and Solubility (%S)

The starch undergoes a number of changes during food processing and heating, including swelling, dissolution of the starch granule, and binding with water. The swelling power (SP) provides information about the changes in the crystalline and amorphous regions of starch, whereas solubility (%S) describes how the starch particles dissolve after being heated in water for 30 minutes at a specific temperature (Gupta et al., 2025b). MT causes the crystalline portion of starch to reorganize, resulting in disordered molecules that may lessen the starch's ability to expand (Table 7.2) (Yi et al., 2024). According to Ma et al. (2020), power and time also have a significant impact on the SP and solubility values of starch. Using varying power (640, 720, and 800 W) and time conditions (60, 120, and 180 seconds), they administered MT to waxy hull-less barley starch (WS) and non-waxy hull-less barley starch (NWS). According to the study, SP declined to varying degrees as power and time increased. However, after applying MT to the in-kernel WS and in-kernel NWS at different powers and time, the solubility values were generously increased. For WS that was microwave-treated, it improved over time, but the solubility value decreased as power increased. Breadfruit starch samples yielded similar swelling power values. The increase in swelling power did not differ significantly from native breadfruit starch. In the same study, other treatments showed a considerable improvement in solubility but MT showed no increase. The drop in crystallinity index suggests a rise in the amorphous region of treated starch, which could also induce an increase in solubility (Marta et al., 2019). A recent work by Negi et al. (2025) examined buckwheat starch that was microwave-treated at 720 W for varied durations (30–120 s). The researchers discovered a substantial drop in both SP and %S ($p < 0.05$) as treatment duration increased. The decrease in swelling and solubility is justified by the rise in amylose content because amylose inhibits the mobility of starch particles in the external environment.

7.4.3.4 Syneresis

Syneresis is the term used to describe the liquid that spontaneously exudes from the gel. In a gel or colloidal solution, the water content is physically trapped. Syneresis is caused when this liquid separates due to contraction of the gel or due to phase separation in the colloidal solution.. Because it is a poor predictor of stability and preservation quality, syneresis is regarded as an undesirable feature of starch paste and food applications (Table 7.2). The greater amylose content of legume starches contributes to their increased rate of syneresis compared to cereal starches. This effect mostly impacts the stability of foods such as yogurt and jellies that are stored at low temperatures (Wani et al., 2016; Zailani et al., 2021). The syneresis in microwave-treated buckwheat starch dropped dramatically from 45.38% to 43.92% as the microwave treatment time increased from 0 to 120 seconds. Nevertheless, as the gel was stored, the syneresis percentage rose because of the amylopectin chain and amylose leaching interacting to form junction zones and release water (Negi et al., 2025). The rise in syneresis shows that starch granules' capacity to connect to water molecules diminishes with cooling. Increased syneresis correlates with reduced freeze–thaw stability and accelerated starch aging. When creating frozen dough and speed dumplings, using starch with strong freeze–thaw resilience would reduce quality loss and mouthfeel changes (Cai et al., 2024).

7.4.3.5 Digestibility Properties

Following high-temperature processing, native starch becomes gelatinized and some of its molecules are completely dissolved, making it easier for glucose to be digested. In the small intestine, this glucose is then absorbed by glucose transporters and transported throughout the body by the blood circulation system. Numerous diseases, including obesity, diabetes, cardiovascular disease, and other lifestyle disorders, are brought on by elevated blood glucose levels.

Starch is classified as resistant starch (RS), slow-digesting starch (SDS), and rapid-digesting starch (RDS) based on the rate at which glucose is released and absorbed in the gastrointestinal tract. The large intestine ferments RS to create short-chain fatty acids (SCFAs) and their breakdown products, which are favourable to human health (Xu et al., 2021). Starch digestibility is influenced by a number of parameters, such as the starch's source, amylose concentration, crystallinity index, and amylopectin molecular structure. In addition, the type of alteration influences digestibility. Previous research suggests that microwave treatment slows digestion. Starch digestibility is also affected by starch moisture content, power, and microwave treatment time (Table 7.2). The digestibility of starch reduces when moisture content is in the range of 20%–35%, but when moisture level reaches 40% or higher, the rate of digestion increases (Oyeyinka et al., 2021). Huong et al. (2021) observed increased RS in debranched mung bean starches (DS) followed by microwave treatment with 20% mc, at 350 W for 3 minutes. The debranched starch was retrograded at different temperatures (28 °C, 4 °C, and −18 °C). The RS content increased from 15.6 ± 0.9 in NS to 52.8 ± 0.9 in modified starch. On the contrary, Zhou et al. (2022) postulated that MT alone reduced the SDS and RS content of corn starch to 51.49% while raising the RDS content. These changes in digestibility properties likely resulted from the disruption of granules and internal structure, which leads to reduction in branching, exposing more sites for enzymatic attack. However, when MT is given in combination with lysine addition, the SDS content increased, reaching highest at optimal conditions of 4 minutes of MT and 3% lysine addition, suggesting better modification that resist enzymatic hydrolysis.

7.5 MICROWAVE-ASSISTED CHEMICAL MODIFICATION OF STARCH

Various chemical methods are used for modifying starches from various sources in order to improve the functional attributes of the starch and broaden its usage. Chemical modification of starch mainly involves the introduction of a functional group in the structure of a starch molecule, which alters the starch's many physicochemical characteristics and, consequently, its functional qualities. Oxidation, esterification, hydrolysis and cross-linking are some of the chemical methods used in modifying starch. However, research on utilization of microwave energy for improving the chemical modification of starches is currently the focus of many researchers because microwave heating leads to reaction enhancement, consumes less energy, and produces starch with a high degree of substitution. The following section describes the microwave-assisted chemical modification of starches from various sources.

7.5.1 Microwave-Assisted Oxidation

For starch modification, oxidation is one of the most popular techniques. Aldehyde or carboxyl groups are created during this process by oxidizing the glucose units of starch's main or secondary hydroxyl groups. Oxidation of starches can be carried out in various ways, including the use of oxidizing agents such as air or oxygen with such catalysts as transition metal ions, using inorganic peroxides (H_2O_2), organic peroxides such as NaClO or $NaIO_4$, nitrogen compounds such as HNO_3 or N_2O_4, and metal compounds such as chromates and bromates. Oxidation of starch, irrespective of the method used, can be improved using microwave treatment. Komulainen et al. (2013) carried out bromate oxidation of native wheat starch under microwave treatment. The oxidation process was improved in terms of reduced need of oxidant for the reaction, shortened reaction time, and high yield of oxidized product. Microwave-assisted and convectional heat-assisted zinc derivatives of potato starch were prepared by Staroszczyk and Janas (2010); monoesterified and cross-linked starch was formed, and microwave-assisted process was found to be faster than convectional heating, reducing the reaction time from 120 to 30 minutes. Oxidized products showed reduction in melting temperature and melting enthalpy, with increased water solubility and water binding capacity. Thus, such studies demonstrate that oxidation process can be more improved with microwave heating than with simple convectional heating.

7.5.2 Microwave-Assisted Esterification

In esterification of starches, hydroxyl groups are converted to hydrophobic ester groups, hence imparting new functional properties to the starch. Esterification is carried out in the presence of organic acids and their derivatives, such as anhydrides, chlorides and oxychlorides. Use of microwave heating to aid chemical modification speeds the conversion process, results in better conversion of the starches, and leads to the production of thermally stable products. Zhao et al. (2018) carried out acetylation of potato starches with acetic anhydride and glacial acetic acid (1:1 v/v) by pretreating the starch with 600 W microwave. Remarkable differences were observed in both microwave pre-treated and non-microwave treated acetylated starches. However, the microwave pretreatment resulted in higher degree of substitution, infrared absorption peak intensity, solubility, while the swelling power, relative crystallinity and peak viscosity were reduced. Wang et al. (2022b) prepared maleic acid sweet potato starch using microwave treatment and achieved a degree of substitution of 0.0712.This modification enhanced the solubility, gel transparency, freeze–thaw stability, and resistant starch content.

7.5.3 Microwave-Assisted Hydrolysis

Starch modified by acid hydrolysis alters the granular structure of the starch. Starch's low viscosity, improved water solubility, and strong film-forming ability are all altered as a result of these structural alterations. Acid hydrolysis of starches can be effectively carried out under microwave treatment and is found to be more effective than traditional heating. Physicochemical changes in the faba bean starch due to hydrochloric acid hydrolysis assisted by microwave irradiation was studied by González-Mendoza et al. (2022). Acid hydrolysis under microwave treatment produced starches with low viscosity, high solubility, high amylose content, increased resistant starch, and desirable thermal properties, with a potential to be used in different industries. In a study by Wang et al. (2022a), sweet potato starch was modified by microwave alone and by microwave-assisted L-malic acid treatment, and physicochemical and digestive properties were compared. The results of this study indicated that both treatments decreased the swelling power and gelatinization enthalpy of starch while the resistant starch content was increased after microwave-assisted malic acid hydrolysis of sweet potato starch and decreased with microwave treatment alone. Depending on the desirable properties of the starches with varying uses, such modifications in starches from different sources can be carried out.

7.5.4 Microwave-Assisted Cross-linking

Cross-linking of starches involves addition of intra- and intermolecular bonds at different locations in the starch molecule. Various cross-linking agents such as epichlorohydrin, sodium trimetaphosphate, sodium tripolyphosphate, and phosphorus chloride are used in developing cross-linked starch. Introduction of various covalent and non-covalent linkages in the starch molecule alters various properties of starch, such as increased thermal stability, improved resistance to retrogradation, increased freeze–thaw stability, and increased stability to low pH (Bangar et al., 2024). Microwave-assisted cross-linking of talipot starch reduced the swelling capacity and in vitro digestibility while increasing solubility when compared to conventionally cross-linked starch (Aaliya et al., 2022). In a study by Otemuyiwa and Aina (2021), breadfruit starch underwent chemical alteration with microwave assistance, and its physicochemical characteristics and in vitro starch hydrolysis were assessed. In comparison to native starch, phosphorylated starch demonstrated a lower rate of hydrolysis and a lower glycemic index, suggesting potential for application in the creation of functional meals.

7.6 APPLICATIONS OF MICROWAVE-MODIFIED STARCH

Starch is one of the most commonly utilized carbohydrates, finding application in food as well as non-food sectors such as pharmaceuticals and packaging. Modification of starches using microwave heating alone or in combination with other methods may lead to development of starches with customizable properties, widening their range of applications.

7.6.1 Food Industry

Microwave-modified starches have various properties, such as reduced swelling power, decreased viscosity, increased thermal stability, increased freeze–thaw stability, increased resistance to retrogradation, and decreased in vitro digestibility due to increase in resistant starch. Starches with low viscosity can be used to develop edible coatings and to replace fat in high-fat products. Modified starches can be used to stabilize emulsion and in frozen pie fillings. Khurshida and Deka (2021) worked on application of microwave-modified cassava starch in development of cookies. Microwave modification increased the resistant content of cassava starch, so this study was focused on using calorie-rich cassava starch with high resistant starch content as a functional ingredient in developing cookies. In another study by Garske et al. (2023), in order to create gluten-free bread, chickpea flour and microwave-treated cassava starch were utilized as alternatives. Microwave treatment altered the structural and functional characteristics of starch such that it could be effectively used as a substitute for additives such as carboxymethyl cellulose and xanthan gum for developing gluten-free bread. Thus, studies show that microwave-modified starches can be used as a functional ingredient in developing products with desirable characteristics.

7.6.2 Pharmaceuticals

In recent years much research has been done regarding usage of modified starches in development of drugs. Microwave-treated starches, due to their increased porosity, can be successfully employed in tablet formulations where controlled release is desired. Singh et al. (2013) evaluated the use of microwave-assisted acrylamide-grafted sago starch as a controlled release polymer for lamivudine tablets. The results of this study showed that optimized formulation of the polymer exhibited good stability and release profile. Zheng et al. (2023) studied the structural and functional properties of microwave-treated pregelatinized maize starch and its effect on enalapril maleate release tablets, the results of which indicated that pregelatinized starch can be used effectively as an excipient in drug delivery.

7.6.3 Packaging

Starch, a biopolymer with many desirable properties such as good foamability, processability, ana compostable nature, is promising for use in developing various kinds of packaging materials (Gupta et al., 2025a). However, native starches have limitations such as brittleness and rigidity which limit their application. Attempts are therefore being made to modify starches to widen their applications, and microwave modification is one such promising technology. Amaraweera et al. (2022) prepared and characterized biodegradable foams from dual-modified cassava starch combining microwave treatment and freeze drying. Those biodegradable foams developed exhibited density of 124 to 245 kg/m^3 and show great potential to be used as a sustainable alternative to petroleum-based packaging materials. Pearl millet starch–based films were developed and characterized by Bangar et al. (2021). Here, pearl millet starch was modified using microwave treatment before developing the films. Films prepared from modified starch showed high solubility, opacity, and low moisture when compared to native starch.

7.7 CONCLUSION

Poor stability and hence limited application of native starches from various sources rationalizes the modification of starch. Various physical, chemical, and enzymatic methods are used for modification of starches. Microwave radiation is one such novel physical technique used to alter various characteristics of starch. This process has a number of advantages over other approaches, including uniform sample heating, reduced time requirements, and non-toxicity. This chapter delved into the technology and mechanism of starch modification using microwave radiation and its impact on various physicochemical, structural, thermal, and functional properties of starch. Properties of starch such as digestibility, amylose concentration, and relative crystallinity were all found to be highly affected by microwave treatment as ascertained and justified through various studies summarized in this chapter. Moreover, microwave technology seems promising for use in conjunction with other modification techniques, especially chemical modification techniques such as oxidation, cross-linking, and esterification. This leads to enhancement of the reactions involved and to the production of starches with better stability and functional characteristics. Overall, it can be concluded that microwave modification of starches alone or in combination with other techniques widens the range of application of starches in different fields such as food processing, pharmaceuticals, and packaging due to improved properties. Hence, there is immense scope for exploring such applications of microwave-modified starches in these and related fields.

REFERENCES

Aaliya, B., Sunooj, K. V., John, N. E., Navaf, M., Akhila, P. P., Sudheesh, C., Sabu, S., Sasidharan, A., Mir, S. A., & George, J. (2022). Impact of microwave irradiation on chemically modified talipot starches: A characterization study on heterogeneous dual modifications. *International Journal of Biological Macromolecules*, *209*, 1943–1955. https://doi.org/10.1016/j.ijbiomac.2022.04.172

Amaraweera, S. M., Gunathilake, C., Gunawardene, O. H. P., Dassanayake, R. S., Fernando, N. M., Wanninayaka, D. B., Rajapaksha, S. M., Manamperi, A., Gangoda, M., Manchanda, A., Fernando, C. A., Kulatunga, A. K., & Manipura, A. (2022). Preparation and characterization of dual-modified cassava starch-based biodegradable foams for sustainable packaging applications. *ACS Omega*, *7*(23), 19579–19590. https://doi.org/10.1021/acsomega.2c01292

Bai, Y., Zhou, H., Guo, X., & Zhu, K. (2022). Structural changes and components' interactions alter the digestion property of in-kernel starch from thermally processed Tibetan qingke. *Food Research International*, *162*, 111950. https://doi.org/10.1016/j.foodres.2022.111950

Balakumaran, M., Nath, K. G., Giridharan, B., Dhinesh, K., Dharunbalaji, A., Malini, B., & Sunil, C. (2023). White finger millet starch: Hydrothermal and microwave modification and its characterisation. *International Journal of Biological Macromolecules*, *242*, 124619. https://doi.org/10.1016/j.ijbiomac.2023.124619

Bangar, S. P., Ashogbon, A. O., Singh, A., Chaudhary, V., & Whiteside, W. S. (2022). Enzymatic modification of starch: A green approach for starch applications. *Carbohydrate Polymers*, *287*, 119265. https://doi.org/10.1016/j.carbpol.2022.119265

Bangar, S. P., Nehra, M., Siroha, A. K., Petrů, M., Ilyas, R. A., Devi, U., & Devi, P. (2021). Development and characterization of physical modified pearl millet starch-based films. *Foods*, *10*(7), 1609. https://doi.org/10.3390/foods10071609

Bangar, S. P., Sunooj, K., Navaf, M., Phimolsiripol, Y., & Whiteside, W. S. (2024). Recent advancements in cross-linked starches for food applications: A review. *International Journal of Food Properties*, *27*(1), 411–430. https://doi.org/10.1080/10942912.2024.2318427

Bashir, K., & Aggarwal, M. (2019). Physicochemical, structural and functional properties of native and irradiated starch: A review. *Journal of Food Science and Technology*, *56*(2), 513–523. https://doi.org/10.1007/s13197-018-3530-2

Bertoft, E. (2017). Understanding starch structure: Recent progress. *Agronomy*, *7*(3), 56. https://doi.org/10.3390/agronomy7030056

Braşoveanu, M., & Nemţanu, M. R. (2013). Behaviour of starch exposed to microwave radiation treatment. *Starch - Stärke*, *66*(1–2), 3–14. https://doi.org/10.1002/star.201200191

Cai, M., Zhang, Y., Cao, H., Li, S., Zhang, Y., Huang, K., Song, H., & Guan, X. (2024). Exploring the remarkable effects of microwave treatment on starch modification: From structural evolution to changed physicochemical and digestive properties. *Carbohydrate Polymers*, *343*, 122412. https://doi.org/10.1016/j.carbpol.2024.122412

Chakraborty, I., N, P., Mal, S. S., Paul, U. C., Rahman, M. H., & Mazumder, N. (2022). An insight into the gelatinization properties influencing the modified starches used in food industry: A review. *Food and Bioprocess Technology*, *15*(6), 1195–1223. https://doi.org/10.1007/s11947-022-02761-z

Cornejo-Ramírez, Y. I., Martínez-Cruz, O., Del Toro-Sánchez, C. L., Wong-Corral, F. J., Borboa-Flores, J., & Cinco-Moroyoqui, F. J. (2018). The structural characteristics of starches and their functional properties. *CyTA – Journal of Food*, *16*(1), 1003–1017. https://doi.org/10.1080/19476337.2018.1518343

Deka, D., & Sit, N. (2016). Dual modification of taro starch by microwave and other heat moisture treatments. *International Journal of Biological Macromolecules*, *92*, 416–422. https://doi.org/10.1016/j.ijbiomac.2016.07.040

Garske, R. P., Mercali, G. D., Thys, R. C. S., & Cladera-Olivera, F. (2023). Cassava starch and chickpea flour pre-treated by microwave as a substitute for gluten-free bread additives. *Journal of Food Science and Technology*, *60*(1), 53–63. https://doi.org/10.1007/s13197-022-05586-y

González-Mendoza, M. E., Martínez-Bustos, F., Castaño-Tostado, E., & Amaya-Llano, S. L. (2022). Effect of microwave irradiation on acid hydrolysis of faba bean starch: Physicochemical changes of the starch granules. *Molecules*, *27*(11), 3528. https://doi.org/10.3390/molecules27113528

Gupta, R. K., Guha, P., & Srivastav, P. P. (2022). Natural polymers in bio-degradable/edible film: A review on environmental concerns, cold plasma technology and nanotechnology application on food packaging-A recent trends. *Food Chemistry Advances*, *1*, 100135. https://doi.org/10.1016/j.focha.2022.100135

Gupta, R. K., Guha, P., & Srivastav, P. P. (2023). Effect of high voltage dielectric barrier discharge (DBD) atmospheric cold plasma treatment on physicochemical and functional properties of taro (*Colocasia esculenta*) starch. *International Journal of Biological Macromolecules*, *253*, 126772. https://doi.org/10.1016/j.ijbiomac.2023.126772

Gupta, R. K., Guha, P., & Srivastav, P. P. (2024a). Dielectric barrier discharge plasma: A green and novel method to change the structure of taro peel starch and improve the physicochemical properties of taro peel starch films. *Plasma Processes and Polymers*, *21*(9), 2400047. https://doi.org/10.1002/ppap.202400047

Gupta, R. K., Guha, P., & Srivastav, P. P. (2024b). Exploring the potential of taro (*Colocasia esculenta*) starch: Recent developments in modification, health benefits, and food industry applications. *Food Bioengineering*, *3*(3), 365–379. https://doi.org/10.1002/fbe2.12103

Gupta, R. K., Guha, P., & Srivastav, P. P. (2024c). Insights into the physicochemical, crystalline, multistructural, thermal and digestibility behavior of taro starch (*Colocasia esculenta*) modified by cold plasma treatment with high voltage dielectric barrier discharge (HVDBD): An alternative technique for chemical modification of starch and environmental sustainability. *Journal of Cleaner Production*, *461*, 142710. https://doi.org/10.1016/j.jclepro.2024.142710

Gupta, R. K., Guha, P., & Srivastav, P. P. (2024d). Physical action of nonthermal cold plasma technology for starch modification. *Food Physics*, *1*, 100011. https://doi.org/10.1016/j.foodp.2024.100011

Gupta, R. K., Guha, P., & Srivastav, P. P. (2024e). Effect of cold plasma treatment and plasma-activated water on physicochemical and structural properties of starch: A green and novel approach for environmental sustainability. *Plasma Processes and Polymers*, *21*(4), 2300204. https://doi.org/10.1002/ppap.202300204

Gupta, R. K., Pipliya, S., Patel, J., Srivastav, P. P., Castro-Muñoz, R., Wang, C. K., & Nemţanu, M. R. (2025a). Essential oil-embedded starch based films for active packaging: Insights into migration, stability, and preservation efficiency. *Trends in Food Science & Technology*, *163*, 105149. https://doi.org/10.1016/j.tifs.2025.105149

Gupta, R. K., Gnana Moorthy Eswaran, U., Kumar, S., Pipliya, S., Kaur, B., & Srivastav, P. P. (2025b). Synergistic modification of starch polysaccharide via sequential treatments with ultrasonic and cold plasma technology. *Journal of Food Processing and Preservation*, *2025*(1), 9063007. https://doi.org/10.1155/jfpp/9063007

Huong, N. T. M., Hoa, P. N., & Van Hung, P. (2021). Effects of microwave treatments and retrogradation on molecular crystalline structure and in vitro digestibility of debranched mung-bean starches. *International Journal of Biological Macromolecules*, *190*, 904–910. https://doi.org/10.1016/j.ijbiomac.2021.09.032

Khurshida, S., & Deka, S. C. (2021). Application of microwave and hydrothermal treatments for modification of cassava starch of Manipur region, India and development of cookies. *Journal of Food Science and Technology*, *59*(1), 344–354. https://doi.org/10.1007/s13197-021-05020-9

Komulainen, S., Verlackt, C., Pursiainen, J., & Lajunen, M. (2013). Oxidation and degradation of native wheat starch by acidic bromate in water at room temperature. *Carbohydrate Polymers*, *93*(1), 73–80. https://doi.org/10.1016/j.carbpol.2012.06.001

Kumar, Y., Singh, L., Sharanagat, V. S., Patel, A., & Kumar, K. (2020). Effect of microwave treatment (low power and varying time) on potato starch: Microstructure, thermo-functional, pasting and rheological properties. *International Journal of Biological Macromolecules*, *155*, 27–35. https://doi.org/10.1016/j.ijbiomac.2020.03.174

Kumari, B., & Sit, N. (2023). Comprehensive review on single and dual modification of starch: Methods, properties and applications. *International Journal of Biological Macromolecules*, *253*, 126952. https://doi.org/10.1016/j.ijbiomac.2023.126952

Ma, M., Zhang, Y., Chen, X., Li, H., Sui, Z., & Corke, H. (2020). Microwave irradiation differentially affect the physicochemical properties of waxy and non-waxy hull-less barley starch. *Journal of Cereal Science*, *95*, 103072. https://doi.org/10.1016/j.jcs.2020.103072

Maniglia, B. C., Castanha, N., Le-Bail, P., Le-Bail, A., & Augusto, P. E. D. (2020). Starch modification through environmentally friendly alternatives: A review. *Critical Reviews in Food Science and Nutrition*, *61*(15), 2482–2505. https://doi.org/10.1080/10408398.2020.1778633

Marta, H., Cahyana, Y., Arifin, H. R., & Khairani, L. (2019). Comparing the effect of four different thermal modifications on physicochemical and pasting properties of breadfruit (*Artocarpus altilis*) starch. *International Food Research Journal, 26*(1), 269–276.

Masina, N., Choonara, Y. E., Kumar, P., Du Toit, L. C., Govender, M., Indermun, S., & Pillay, V. (2016). A review of the chemical modification techniques of starch. *Carbohydrate Polymers*, *157*, 1226–1236. https://doi.org/10.1016/j.carbpol.2016.09.094

Morales-Trejo, F., Trujillo-Ramírez, D., Aguirre-Mandujano, E., Lobato-Calleros, C., Vernon-Carter, E. J., & Alvarez-Ramirez, J. (2021). Ultrasound-assisted extraction of lychee (*Litchi chinensis* Sonn.) seed starch: Physicochemical and functional properties. *Starch - Stärke*, *74*(1–2), 2100092. https://doi.org/10.1002/star.202100092

Navaf, M., Bangar, S. P., & Sunooj, K. (2024). Microwave treatment. In S. P. Bangar, K.V. Sunooj, & A. K. Siroha (Eds.), *Starch* (pp. 65–84). CRC Press. https://doi.org/10.1201/9781032655598-3

Nawaz, H., Shad, M. A., Saleem, S., Khan, M. U. A., Nishan, U., Rasheed, T., Bilal, M., & Iqbal, H. M. (2018). Characteristics of starch isolated from microwave heat treated Lotus (*Nelumbo nucifera*) seed flour. *International Journal of Biological Macromolecules*, *113*, 219–226. https://doi.org/10.1016/j.ijbiomac.2018.02.125

Nawaz, H., Waheed, R., Nawaz, M., & Shahwar, D. (2020). *Physical and chemical modifications in starch structure and reactivity*. IntechOpen eBooks. https://doi.org/10.5772/intechopen.88870

Negi, P., Joia, S., Siddiqi, R. A., Pasrija, D., & Singh, T. P. (2025). Impact of microwave treatment on the physicochemical, structural, and rheological properties of buckwheat (*Fagopyrum esculentum* L.) Starch. *Starch - Stärke*, *77*(4), e202400162. https://doi.org/10.1002/star.202400162

Orsat, V., Raghavan, G., & Krishnaswamy, K. (2016). Microwave technology for food processing. In M. Regier, K. Knoerzer, & H. Schubert (Eds.), *The microwave processing of foods* (2nd ed., pp. 100–116). Elsevier. https://doi.org/10.1016/b978-0-08-100528-6.00005-x

Osepchuk, J. (1984). A history of microwave heating applications. *IEEE Transactions on Microwave Theory and Techniques*, *32*(9), 1200–1224. https://doi.org/10.1109/tmtt.1984.1132831

Otemuyiwa, I. O., & Aina, A. F. (2021). Physicochemical properties and in-vitro digestibility studies of microwave assisted chemically modified breadfruit (*Artocarpus altilis*) starch. *International Journal of Food Properties*, *24*(1), 140–151. https://doi.org/10.1080/10942912.2020.1861007

Oyeyinka, S. A., Akintayo, O. A., Adebo, O. A., Kayitesi, E., & Njobeh, P. B. (2021). A review on the physicochemical properties of starches modified by microwave alone and in combination with other methods. *International Journal of Biological Macromolecules*, *176*, 87–95. https://doi.org/10.1016/j.ijbiomac.2021.02.066

Oyeyinka, S. A., Akinware, R. O., Bankole, A. T., Njobeh, P. B., & Kayitesi, E. (2020). Influence of microwave heating and time on functional, pasting and thermal properties of cassava starch. *International Journal of Food Science & Technology*, *56*(1), 215–223. https://doi.org/10.1111/ijfs.14621

Pinto, V. Z., Deon, V. G., Moomand, K., Vanier, N. L., Pilatti-Riccio, D., Da Rosa Zavareze, E., Lim, L., & Dias, A. R. G. (2018). Characteristics of modified carioca bean starch upon single and dual annealing, heat-moisture-treatment, and sonication. *Starch - Stärke*, *71*(3–4), 1800173. https://doi.org/10.1002/star.201800173

Prudhvi, P., Deepika, S., & Sutar, P. (2021). Modeling moisture and solids transfer kinetics during a novel microwave assisted water absorption-desorption process of dry red gram (*Cajanus cajan* L.) Splits. *Journal of Food Engineering*, *318*, 110891. https://doi.org/10.1016/j.jfoodeng.2021.110891

Ramírez-Brewer, D., Quintana, S. E., & García-Zapateiro, L. A. (2023). Effect of microwave treatment on technological, physicochemical, rheological and microstructural properties of mango (*Mangifera indica*) kernel starch variety Tommy and Sugar. *LWT*, *187*, 115311. https://doi.org/10.1016/j.lwt.2023.115311

Shen, X., Shang, W., Strappe, P., Chen, L., Li, X., Zhou, Z., & Blanchard, C. (2017). Manipulation of the internal structure of high amylose maize starch by high pressure treatment and its diverse influence on digestion. *Food Hydrocolloids*, *77*, 40–48. https://doi.org/10.1016/j.foodhyd.2017.09.015

Shrivastava, A., Gupta, R. K., & Srivastav, P. P. (2024). Exploring novel frontiers of advancements in purple yam (*Dioscorea alata* L.) starch extraction, modification, characterization, applications in food and other industries. *Measurement: Food*, *15*, 100196. https://doi.org/10.1016/j.meafoo.2024.100196

Singh, A. V., & Nath, L. K. (2013). Evaluation of microwave assisted grafted sago starch as controlled release polymeric carrier. *International Journal of Biological Macromolecules*, *60*, 62–68.

Staroszczyk, H., & Janas, P. (2010). Microwave-assisted preparation of potato starch silicated with silicic acid. *Carbohydrate Polymers*, *81*(3), 599–606. https://doi.org/10.1016/j.carbpol.2010.03.017

Sun, X., Saleh, A. S., Sun, Z., Ge, X., Shen, H., Zhang, Q., Yu, X., Yuan, L., & Li, W. (2021). Modification of multi-scale structure, physicochemical properties, and digestibility of rice starch via microwave and cold plasma treatments. *LWT*, *153*, 112483. https://doi.org/10.1016/j.lwt.2021.112483

Tagliapietra, B. L., Felisberto, M. H. F., Sanches, E. A., Campelo, P. H., & Clerici, M. T. P. S. (2020). Non-conventional starch sources. *Current Opinion in Food Science*, *39*, 93–102. https://doi.org/10.1016/j.cofs.2020.11.011

Tao, Y., Yan, B., Fan, D., Zhang, N., Ma, S., Wang, L., Wu, Y., Wang, M., Zhao, J., & Zhang, H. (2020). Structural changes of starch subjected to microwave heating: A review from the perspective of dielectric properties. *Trends in Food Science & Technology*, *99*, 593–607. https://doi.org/10.1016/j.tifs.2020.02.020

Wang, G., Li, C., Zhang, X., Wang, Q., Cao, R., Liu, X., Yang, X., & Sun, L. (2023). The changed multiscale structures of tight nut (*Cyperus esculentus*) starch decide its modified physicochemical properties: The effects of non-thermal and thermal treatments. *International Journal of Biological Macromolecules*, *253*, 126626. https://doi.org/10.1016/j.ijbiomac.2023.126626

Wang, J., Lan, T., Lei, Y., Suo, J., Zhao, Q., Wang, H., Lei, J., Sun, X., & Ma, T. (2021). Optimization of ultrasonic-assisted enzymatic extraction of kiwi starch and evaluation of its structural, physicochemical, and functional characteristics. *Ultrasonics Sonochemistry*, *81*, 105866. https://doi.org/10.1016/j.ultsonch.2021.105866

Wang, M., Wu, Y., Liu, Y., & Ouyang, J. (2020). Effect of ultrasonic and microwave dual-treatment on the physicochemical properties of chestnut starch. *Polymers*, *12*(8), 1718. https://doi.org/10.3390/polym12081718

Wang, S., Li, C., Copeland, L., Niu, Q., & Wang, S. (2015). Starch retrogradation: A comprehensive review. *Comprehensive Reviews in Food Science and Food Safety*, *14*(5), 568–585. https://doi.org/10.1111/1541-4337.12143

Wang, W., Hu, A., Li, J., Liu, G., Wang, M., & Zheng, J. (2022a). Comparison of physicochemical properties and digestibility of sweet potato starch after two modifications of microwave alone and microwave-assisted L-malic acid. *International Journal of Biological Macromolecules*, *210*, 614–621. https://doi.org/10.1016/j.ijbiomac.2022.04.215

Wang, W., Liu, T., Zhi, W., Zhou, Y., Hu, A., & Zheng, J. (2022b). Study on the preparation and digestibility of malic acid sweet potato starch ester under microwave assistance. *Starch - Stärke*, *74*(9–10), 2200011. https://doi.org/10.1002/star.202200011

Wani, I. A., Sogi, D. S., Hamdani, A. M., Gani, A., Bhat, N. A., & Shah, A. (2016). Isolation, composition, and physicochemical properties of starch from legumes: A review. *Starch - Stärke*, *68*(9–10), 834–845. https://doi.org/10.1002/star.201600007

Xia, T., Gou, M., Zhang, G., Li, W., & Jiang, H. (2018). Physical and structural properties of potato starch modified by dielectric treatment with different moisture content. *International Journal of Biological Macromolecules*, *118*, 1455–1462. https://doi.org/10.1016/j.ijbiomac.2018.06.149

Xu, T., Li, X., Ji, S., Zhong, Y., Simal-Gandara, J., Capanoglu, E., Xiao, J., & Lu, B. (2021). Starch modification with phenolics: Methods, physicochemical property alteration, and mechanisms of glycaemic control. *Trends in Food Science & Technology*, *111*, 12–26. https://doi.org/10.1016/j.tifs.2021.02.023

Xu, X., Chen, Y., Luo, Z., & Lu, X. (2018). Different variations in structures of A- and B-type starches subjected to microwave treatment and their relationships with digestibility. *LWT*, *99*, 179–187. https://doi.org/10.1016/j.lwt.2018.09.072

Yang, Q., Qi, L., Luo, Z., Kong, X., Xiao, Z., Wang, P., & Peng, X. (2017). Effect of microwave irradiation on internal molecular structure and physical properties of waxy maize starch. *Food Hydrocolloids*, *69*, 473–482. https://doi.org/10.1016/j.foodhyd.2017.03.011

Yi, M., Tang, X., Liang, S., He, R., Huang, T., Lin, Q., & Zhang, R. (2024). Effect of microwave alone and microwave-assisted modification on the physicochemical properties of starch and its application in food. *Food Chemistry*, *446*, 138841. https://doi.org/10.1016/j.foodchem.2024.138841

Zailani, M. A., Kamilah, H., Husaini, A., & Sarbini, S. R. (2021). Physicochemical properties of microwave heated sago (*Metroxylon sagu*) starch. *CyTA – Journal of Food*, *19*(1), 596–605. https://doi.org/10.1080/19476337.2021.1934550

Zhao, K., Li, B., Xu, M., Jing, L., Gou, M., Yu, Z., Zheng, J., & Li, W. (2018). Microwave pretreated esterification improved the substitution degree, structural and physicochemical properties of potato starch esters. *LWT*, *90*, 116–123. https://doi.org/10.1016/j.lwt.2017.12.021

Zheng, M., Xiao, Y., Yang, S., Liu, H., Liu, M., Yaqoob, S., Xu, X., & Liu, J. (2020). Effects of heat–moisture, autoclaving, and microwave treatments on physicochemical properties of proso millet starch. *Food Science & Nutrition*, *8*(2), 735–743. https://doi.org/10.1002/fsn3.1295

Zheng, Y., Liu, R., Hou, X., Zhuang, X., Wu, H., Yin, D., & Yang, Y. (2023). Structural and physicochemical properties of microwave-processing pregelatinized maize starch, and its influence on drug release from tablets. *Journal of Drug Delivery Science and Technology*, *84*, 104452. https://doi.org/10.1016/j.jddst.2023.104452

Zhong, Y., Tian, Y., Liu, X., Ding, L., Kirkensgaard, J. J. K., Hebelstrup, K., Putaux, J., & Blennow, A. (2021). Influence of microwave treatment on the structure and functionality of pure amylose and amylopectin systems. *Food Hydrocolloids*, *119*, 106856. https://doi.org/10.1016/j.foodhyd.2021.106856

Zhong, Y., Xiang, X., Zhao, J., Wang, X., Chen, R., Xu, J., Luo, S., Wu, J., & Liu, C. (2019). Microwave pretreatment promotes the annealing modification of rice starch. *Food Chemistry*, *304*, 125432. https://doi.org/10.1016/j.foodchem.2019.125432

Zhou, Y., Wang, M., Li, W., Liu, G., Wang, W., Zhi, W., Wang, M., Wang, R., Hu, A., & Zheng, J. (2022). Effects of dual modification of lysine and microwave on corn starch: In vitro digestibility and physicochemical properties. *International Journal of Biological Macromolecules*, *220*, 426–434. https://doi.org/10.1016/j.ijbiomac.2022.08.075

8 Gamma Irradiation and Its Impact on the Physicochemical Properties of Starch

Naveen Chandrakar and Mamoni Banerjee
Bio-Research Laboratory, Rajendra Mishra School of Engineering Entrepreneurship, Indian Institute of Technology, Kharagpur, India

Debapam Saha
Calcutta Institute of Technology, Uluberia, Howrah, West Bengal, India

Ayan Sarkar
Department of Agricultural and Food Engineering, Indian Institute of Technology Kharagpur, India

8.1 INTRODUCTION

Starch is a vital, renewable biopolymer naturally synthesized and stored in plants as an energy reserve. It is essential in the food sector because of its affordability, biodegradability, and ability to modify its physicochemical properties. As a key polysaccharide, starch serves as a primary dietary energy source for humans and a fundamental food ingredient (Goswami & Mahanta, 2021; HerCeg et al., 2010; Shi et al., 2019). Starch is present in a wide range of foods, including grains, roots, tubers, seeds, legumes, and leaves (Gupta et al., 2023; Shrivastava et al., 2024). It is found in various plant tissues as granules, and each plant source has its own unique granule structure, shape, size, and makeup (Öztürk & Mutlu, 2019). Two polysaccharides make up starch: amylopectin, an extensively branched molecule that contains α-1,4 and α-1,6 linkages, and amylose, a straight molecule with α-1,4-glycosidic bonds. Amylose makes up 20%–30% of starch, with 500–2000 glucose units, while amylopectin comprises 70%–85%, consisting of over a million glucose units. Amylose forms helices and is less easily hydrolyzed, whereas amylopectin's branching occurs every 24–30 glucose units. Starch also contains small amounts of phosphorus, lipids, and proteins, with variations such as waxy starches being nearly all amylopectin and others having high amylose content (Chakraborty et al., 2020; Egharevba, 2019; Öztürk & Mutlu, 2019). In food companies, starch is often used as a thickening, bulking, gelling, and water-retaining ingredient. It is essential for improving the texture of a variety of dishes and can alter viscosity, texture, adhesion, moisture retention, and the formation of gels and films (Shetty et al., 2024; Gupta et al., 2022). Common examples include bread, pasta, cereals, and starchy vegetables such as potatoes and corn.

8.1.1 The Need for Starch Modifications

All foods experience compositional changes after harvesting, including fruit ripening, bacterial or enzymatic destruction of cellular or chemical structures, and bacterial generation of toxins. Due to enzymatic or microbiological degradation, food quality will decrease if it isn't eaten shortly after harvest. Approximately 25% to 33% of the world's crop is lost to things such as bacterial or fungal infections, excessively quick ripening, growth linked to storage, and insect infestation (El-Ramady et al., 2015). Amylose is prone to retrogradation, leading to the formation of firm gels and robust films, whereas amylopectin, which is more stable in water, produces softer gels and less durable films (Milani & Golkar, 2021). Moreover, native starch has several limitations, such as insoluble nature in organic solvents, brittleness, decreased shear strength, limited resistance to digestion, and susceptibility to acids, enzymes, heat, and retrogradation. These issues restrict its use in some industrial applications, leading food processors to seek starches with improved functional and performance characteristics (Goswami & Mahanta, 2021). These restrictions are linked to natural starch's amphiphilic character and complex macromolecular structure, which comprises methylene (–CH) groups, hydrophobic α-glycosidic linkages, and a variety of hydroxyl groups on the homoglucans (Chimonyo et al., 2020). Despite their ubiquity and importance, native starches can pose challenges in food systems, particularly in terms of nutrient absorption and overall health. Therefore, modifying starches is essential to

DOI: 10.1201/9781003564287-8

improve their functional properties, improving their nutritional profile and making them better suited for a range of dietary uses (Chiu & Solarek, 2009; Gupta et al., 2025b).

8.1.2 Food Irradiation: The State of the Art

Food irradiation is a method to reduce or eliminate microbes and insects in order to improve food safety and prolong shelf life. This method has been in development since the early 20th century but has seen varied acceptance globally. Below is an overview of its current state, applications, safety, and regulatory aspects of food irradiation (Ihsanullah & Rashid, 2017). Food irradiation entails subjecting food to a certain amount of ionizing radiation, usually from electron beams, X-rays, or gamma rays (Y-rays; created by isotopes such as cesium-137, ^{137}Cs, and cobalt-60, ^{60}Co). These radiation types have enough energy to damage the DNA of bacteria such as *Salmonella* and *E. coli*, effectively killing them or inhibiting their growth without making the food itself radioactive (Farkas, 1998; Sruthi et al., 2023). Additionally, irradiation serves as a preservation method, prolonging the shelf life of perishable food products by eliminating spoiling microbes. It is particularly beneficial in the control of insects, especially for treating tropical fruits, as it eliminates pests without the need for harmful chemicals. Moreover, irradiation can delay processes such as ripening and sprouting in vegetables like potatoes, allowing for longer storage and minimizing waste. This makes food irradiation a valuable tool in enhancing food safety, quality, and longevity across the supply chain (Andrews et al., 1998). In the European Union (EU), modified starches are defined and regulated under EC No. 1333/20083. Some are considered food additives and must be labeled accordingly, while others are not (Kaur et al., 2024). Food irradiation safety is regulated by the Food and Drug Administration (FDA) within the United States. The agency has approved various foods for irradiation, including meats, fruits, vegetables, and spices. The World Health Organization (WHO) and other international bodies have also endorsed the safety of food irradiation, concluding that it poses no toxicological, microbiological, or nutritional risks when applied correctly (Roberts, 2003). Despite the benefits of food irradiation, consumer acceptance of irradiated foods has been slow compared to its use in medical and industrial applications. This hesitance is partly rooted in historical anti-nuclear sentiments and misconceptions about the technology. However, studies have shown that irradiated foods do not significantly alter taste, texture, or nutritional value, and the FDA requires labeling to inform consumers of irradiation.

8.1.3 Gamma Irradiation as a Potential Non-thermal Technology for Starch Modification

To overcome these shortcomings and achieve the desired characteristics for food applications, various modification techniques, including physical, chemical, enzymatic, and genetic engineering techniques, have been used (Kaur et al., 2024). These procedures are time-consuming, expensive, complicated, and reagent-intensive. They also pollute the environment. Novel techniques for starch modification have been developed as a consequence. Ultrasonic treatment, high-pressure water treatment, microwave radiation, ozonation, ohmic radiation, plasma, and radiation from gamma rays are examples of nonconventional procedures, which are mostly physical processes. These methods have been offered to be quick, economical, and ecologically neutral since they don't need catalysts or substantial sample preparation, they don't create unwanted residues, they don't utilize polluting chemicals, and they stop harmful compounds from infiltrating into treated items (Milani & Golkar, 2021; Nemţanu & Braşoveanu, 2010b). Gamma irradiation (Y-irradiation) stands out as a promising non-thermal technology for starch modification. By breaking down starch molecules and altering their structure, gamma irradiation can improve starch physical and chemical characteristics, such as gelatinization, pasting, and starch recrystallization behavior. This method offers a controlled way to enhance starch functionality without the need for chemical additives, making it a clean-label technology with broad applications in the food industry.

8.1.4 Objectives of the Chapter

The objectives of this chapter are to examine how gamma irradiation affects starch's physicochemical characteristics, talk about the underlying processes that underlie these changes, and assess the possible industrial uses for gamma-irradiated starches. The various non-destructive techniques and strategies used for irradiated starch characterizations such as in vitro analysis, optical microscopy, X-ray diffraction (XRD), Fourier transform infrared spectroscopy (FTIR), scanning electron microscopy (SEM), and digital scanning calorimetry (DSC), are also addressed (see Figure 8.1). Additionally, the chapter aims to provide insights into the advantages and limitations of using gamma irradiation as a starch modification technique in enhancing starch functionality for diverse food applications.

8.2 WORKING PRINCIPLES

8.2.1 Mechanism of Starch Irradiation

In the realm of food chemistry, it is necessary to understand the effect of irradiation on the primary constituents of foods, particularly carbohydrates (such as starch), in order to make inferences about the physicochemical properties of irradiated foodstuffs. The mechanism of reaction in carbohydrates is independent of their structural complexity. The effects of irradiation in simple, in both bound sugars and in relatively complex polymeric materials comprising these subunits (Fernandes et al., 2017).

Free radicals produced by ionizing radiation, such as electron beams and gamma rays, can cause chemical

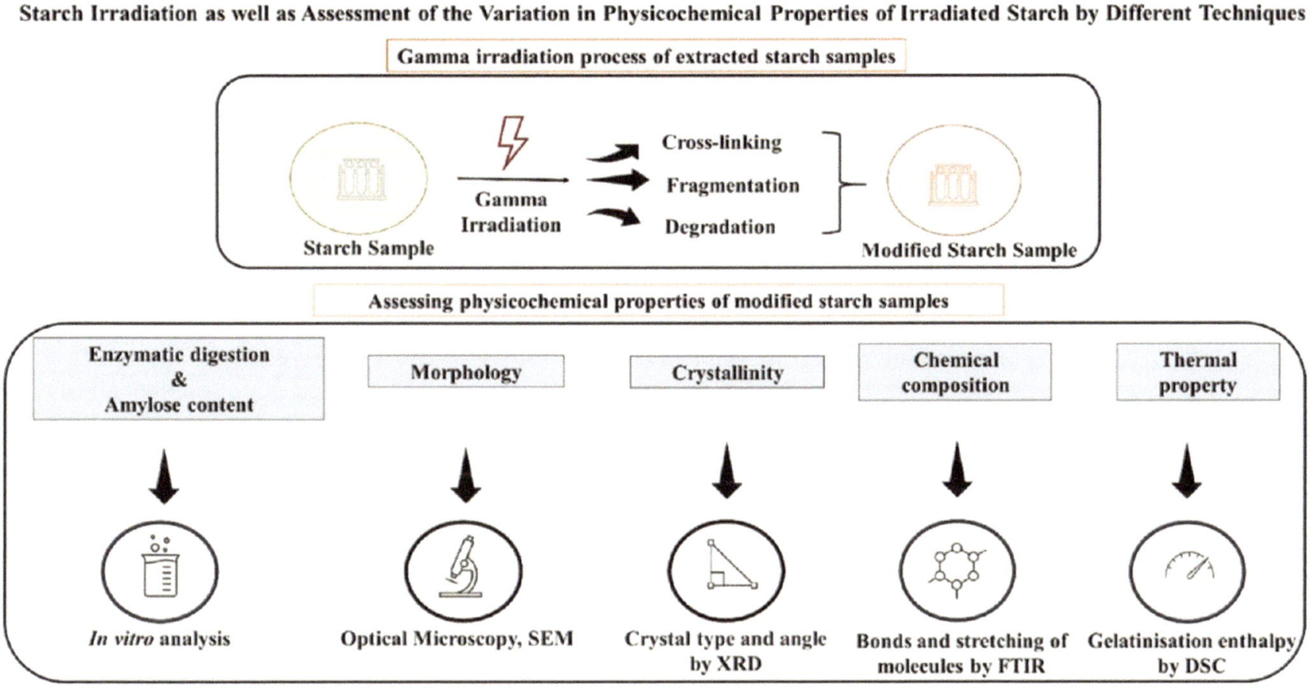

FIGURE 8.1 Overview of various non-destructive assessment and characterization techniques for irradiated starches (Sunder et al., 2022).

changes and starch fragmentation. It is believed that the mechanism for this plays a major role in the physicochemical alterations in starchy meals, including in decreased viscosity and improved water solubility (Bhat & Karim, 2009). During radiation exposure (e.g., by gamma rays), glycosidic bonds at the chain ends of starch granules are broken, resulting in the breakdown of large macromolecules into smaller chains (Bhat & Karim, 2009). Research indicates that radiation also results in a decrease in the crystalline phase of starch, along with a disruption in the ordered amylose and amylopectin distributed within the granules (Polesi et al., 2016). Furthermore, radiation can promote starch cross-linking in the presence of oxygen. It is well established that radiation processing produces radiolytic degradation products (RDPs). Three important radioactive byproducts of carbohydrates are formaldehyde, acetaldehyde, and formic acid (Kong, 2023). Additionally, it has been shown that, regardless of their source (cereals, pulses, vegetables, or millets), all radiolytic end products produced by starch irradiation are identical (Kong, 2023). Figure 8.2 illustrates the production of free radicals in the radiation-exposed starch molecule amylopectin. It is well known that radiation processing causes cross-linking in the starch matrix, which is a very important phenomenon. Nowadays, many cross-linking techniques utilized in the food business depend on substances that are obtained chemically, such vinyl chloride, phosphoryl chloride, epichlorohydrin, sodium tripolyphosphate, and monosodium phosphate (Nagasawa et al., 2004; Zhiguang et al., 2023). This radiation-induced cross-linking presents a valuable alternative.

8.2.2 Radiation Sources and Dosages

Fundamentally, there are three types of radiation applied on food: X-rays, gamma rays (γ-rays), and electron beams (EBs). Based on their properties and the sources (source of radiation) to which they relate, these ionizing radiations are absorbed into food accordingly for particular purposes, as shown in Table 8.1. Typically, isotopes such as cobalt-60 or cesium-137 are used for gamma radiation because they have a deep penetration and thus effectively treat bulk foods. Electron beams, generated by accelerators, have lower penetration than gamma rays but are suitable for treating thin layers of food. X-rays, produced by converting electron beams, offer

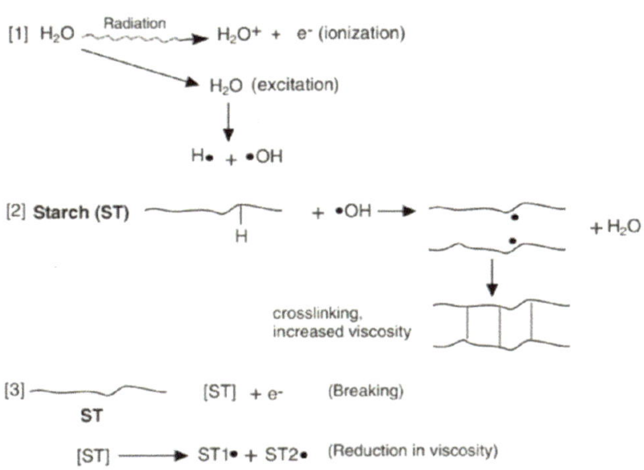

FIGURE 8.2 Mechanism of irradiation induced chemical modifications on structures of food starches (Bhat & Karim, 2009).

TABLE 8.1
Type and Sources of Radiation Used in Food Processing

Types of Radiation	Sources of Radiation	Dose	Characteristics	Applications
Gamma rays	Radioactive variants of the elements cesium (cesium-137) and cobalt (cobalt-60) release gamma rays.	0.02–10 kGy (For sterilization purposes, doses can go up to 25 kGy.)	• High penetrating power • High efficiency • Low throughput	• Insect disinfestation • Sterilization
X-rays	A high-energy electron stream is reflected off a target object, often one that contains heavy metals, into food, producing X-rays.	0.1–7.5 kGy for pathogen elimination and shelf-life extension	• High penetrating power • Low efficiency • High throughput	• Pathogen elimination • Shelf-life extension
Electron beams	An electron accelerator shoots a stream of fast-moving electrons into food, creating an electron beam, also known as an e-beam, which is comparable to X-rays.	< 1–10 kGy	• Low penetrating power • High efficiency • High throughput	• Sprout inhibition • Sterilization

Sources: **Saha et al. (2024)**; **Sarkar et al. (2024)**

a balance between penetration and efficiency (Indiarto et al., 2020; Mshelia et al., 2023; Nemtanu & Brasoveanu 2010a).

Dose and dose rate are the preliminary variables for the irradiation of foods. The dose is fundamentally a measure to gauge and quantify the treatment. The amount of energy from ionizing radiation that a mass of food material (solid, liquid, or semi-solid) absorbs under irradiation treatment is termed "absorbed dose" or dose (D). Irradiation doses are measured in Grays (Gy) (Ganguly et al., 2012). One Gray is equivalent to the reception of 1 J/kg of energy. The dosage of irradiation mainly determines the extent of radiolytic changes in the constitutional properties of a food system. A food material subjected to lower doses exhibits an increase in the formation of radiolytic products with increasing dosages but low doses do not affect the nature of the food material (Mshelia et al., 2023). However, at higher doses, some alteration can be seen, as in meat irradiation (Indiarto et al., 2023). Radiation dose governs the effect of the radiation on texture, rheology, color, and other organoleptic properties of irradiated food. The quantity of radiation energy received per unit of time is the primary measure of dose rate for foods (Kong, 2023). When a radiolytic food item undergoes irradiation and the dose rate is such that the free radicals and other active molecules in the system are uniformly dispersed and clashes between reaction orders take place with the same intermediate species (particularly first- and second-order reactions), the yield of the product is significantly affected. However, pseudo-first-order reactions mostly occurring in food are somewhat less likely to undergo any changes with the dose rates (Kong, 2023). In foods, minor losses of nutrients and other essentials during irradiation are believed to be nullified with comparatively higher dose rates of application, particularly with electron beams from accelerators, rather than from lower dose rates of typical gamma irradiation (e.g., ^{60}Co) (Sahoo et al., 2023). The reason for this could be that in such treatments, the end products favor recombination over reactions with other species, resulting in reduced indirect mechanism. Thus, in commercial applications, dosages of irradiation vary based on the purpose: low doses (up to 1 kGy) are used for delaying ripening and inhibiting sprouting (Akamine & Moy, 2018; Peng et al., 2024; Yoon et al., 2023), medium doses (1–10 kGy) are effective for controlling pathogens and pests, and high doses (above 10 kGy) are used for sterilizing foods (Yang et al., 2023). Regulatory agencies such as the FDA and WHO have established guidelines to ensure that irradiation dosages used in food processing are safe, are effective, and do not compromise the nutritional or sensory qualities of the food.

8.2.3 Gamma Irradiation of Food Starches: Advantages and Disadvantages

The advantages of irradiation on food starches include the following:

- Irradiation enhances the water solubility of starch, making it easier to incorporate starch into various food formulations.
- Radiation treatment reduces the viscosity of starch, which can be beneficial for specific food applications where a thinner consistency is desired.
- It breaks down starch macromolecules into smaller chains, resulting in modified starches with unique functional properties.
- Irradiated starches can show lower gelatinization temperatures, which are useful in products requiring lower cooking temperatures.
- In the presence of oxygen, irradiation promotes cross-linking in starch molecules, enhancing their stability and functionality in food systems.
- Radiation reduces microbial contamination and enzymatic degradation, prolonging the shelf life of starch-based foods.
- Unlike traditional chemical methods, irradiation provides a clean, chemical-free way to modify starches without the need for additives.

- Irradiation decreases the starch granules' crystalline structure, which results in improved digestibility and altered texture in food products.

The disadvantages of irradiation on food starches include the following:

- Irradiation may lead to minor degradation of some essential nutrients, such as vitamins, in starch-rich foods.
- Irradiation can produce RDPs such as formic acid, acetaldehyde, and formaldehyde, which may raise concerns about safety and quality.
- Overexposure to radiation can cause excessive molecular breakdown, leading to undesirable changes in texture and consistency of starch-based products.
- Excessive irradiation may disrupt the useful qualities of starch, such its capacity to thicken, gel, or bind, which could impact the performance in specific food applications.
- Setting up and maintaining irradiation facilities requires significant investment, making the process costlier than other starch modification methods.
- A shortcoming in consumer consciousness exists and, at times, skepticism regarding the safety of irradiated foods, which can affect marketability.
- In certain cases, irradiation can affect the sensory characteristics of starch-rich foods, leading to slight changes in flavor or aroma.
- Electron beam irradiation has limited penetration depth, restricting its use to thin layers of starch or small batches of starch-based products.

8.3 INTERACTION BETWEEN GAMMA IRRADIATION AND STARCH MOLECULES

8.3.1 Effect on Structural and Morphological Properties of Starches

A-type and B-type starch crystals can be identified by wide-angle X-ray diffraction (WAXD). C-type starch combines A and B polymorphs (Gupta et al., 2024c). Gamma irradiation alters the structure and crystalline arrangement of starch granules, which WAXD can detect. At dosages over 2 kGy, γ-irradiation reduces starch granule crystallite content (Chung & Liu, 2010; Sofi et al., 2013; Wani et al., 2014; Zhu, 2016). Crystallinity decrease relies on polymorph type, with the compact A type being less susceptible to irradiation than the hydrated B type (Pérez & Bertoft, 2010). Increased water enhances starch's vulnerability to degradation caused by γ-irradiation (Kamal et al., 2007). Small-angle X-ray diffraction investigation confirmed the fragmentation of crystalline domains inside granules (Ciesla et al., 1991; Gupta et al., 2024b) At the lower doses (< 2 kGy), which affect crystallinity differently, one research found that 1 kGy enhanced rice starch crystallinity (Bao et al., 2005). Others observed a decrease in potato starch at 0.1–0.5 kGy (Lu et al., 2012; Singh et al., 2011). More research is needed to clarify the effects of lower doses. Lower dose rates can increase crystallinity by encouraging structural reorganization in corn starch:

$$\text{Crystallinity index CI (\%)} = \frac{S_c}{S_t} \times 100,$$

where S_c is the area of the crystalline peaks and S_t is the total area under the XRD curve, which includes both crystalline and amorphous regions.

SEM and light microscopy were used to study morphological changes in starches from different sources after γ-irradiation (Abu et al., 2006; Chung & Liu, 2010; Ezekiel et al., 2007; Gani et al., 2013; Wani et al., 2014; Gupta et al., 2024a). Higher dosages cause cracks and damage to isolated starch granules, whereas lower amounts do not (Sofi et al., 2013; Wani et al., 2014). Bean starch (C-type) is more vulnerable than potato starch (B-type) to visual degradation (Chung & Liu, 2010). Polarized light microscopy demonstrates granules' anisotropic structure, and at greater irradiation levels (such as 50 kGy), the maltese cross pattern in certain granules lessens, indicating crystallite rupture, especially in the center (Chung & Liu, 2009).

8.3.2 Effect on Physicochemical Properties of Starches

γ-Irradiation caused varying degrees of color changes in starch. Broad bean starch lost color, gained yellowness, and exhibited little lightness after 15 kGy radiation (Sofi et al., 2013). Similar increases in yellowness were observed in wheat, corn, Indian horse chestnut, rice, and potato starches (Lee et al., 2003; Wani et al., 2014). The color change could be vital to the Maillard reaction, which occurs between proteins and sugars (Lee et al., 2003). Earlier reviews reported that γ-irradiation also degraded starch molecules, reducing their molecular size (Tomasik & Zaranyika, 1995). Recent studies using chromatographic analysis and measuring reducing sugars have provided more insight into this degradation (Duarte & Rupnow, 1994; Habib et al., 2023; Rayas-Duarte & Rupnow, 1993; Yoon et al., 2010). Molecular starch breakdown was more significant in water than in dry form, as shown by a reduction in molecular weight and gyration radius (Kamal et al., 2007).

In the case of white rice, γ-irradiation doses ranging from 0.5 to 9 kGy led to a decrease in amylose's molecular size, as denoted by the smaller size of the amylose and amylopectin mixture observed using efficient size exclusion chromatography (Bao et al., 2005). However, direct numerical assessment of amylose dimensions is limited, as degraded amylopectin fractions overlap with the amylose peak in the chromatograms. Amylose appears to be less affected by irradiation than amylopectin, likely because of its smaller size (Sokhey & Chinnaswamy, 1993).

Several studies have shown that γ-irradiation typically leads to the degradation of amylopectin, resulting in a decrease in differences in starch dimension and gyration radius from natural sources (Bao et al., 2005; Chung & Liu, 2009). Waxy corn starch exhibited greater sensitivity to radiolytic degradation than corn starch that contains 25%–70% amylose, indicating that amylose is less susceptible to damage. An increase in amylopectin molecules in white and black pepper starches, possibly because of free radical–induced starch chain cross-linking, as reported by Sokhey & Chinnaswamy (1993). High-performance anion exchange chromatography demonstrated that applying lower doses of γ-irradiation, such as 10 kGy, resulted in an increased proportion of unit chains with a degree of polymerization ranging from 6 to 12 in corn, potato, white, and corn bean starches (Chung & Liu, 2009; Chung & Liu, 2010). Irradiation, such as at 9 kGy, is believed to more efficiently cleave amylopectin chains in the amorphous regions while leaving the crystalline areas largely intact (Bao et al., 2005). Some of the crucial modifications and alterations in physicochemical properties of irradiated food starches have been listed in Table 8.2. Cross-linking through free radical reactions may also lead to new structures forming in the starch (Esteves et al., 1997).

TABLE 8.2
Effect of Gamma Irradiation on Physicochemical Properties of Food Starches

Sl. No.	Food Starch Type	Gamma Irradiation Dosages (kGy)	Treatment Conditions (Temp., RH, pH etc.)	Effect on Physiochemical Properties	Ref.
1.	White japonica rice starch	0.5, 1.0, 3.0, 5.0, 7.0 and 9.0 kGy	11.7% MC, room temp.; 0.5 kGy/h dosage rate; ^{60}Co irradiator	• Mw decreased from 1.48×10^9 to 2.36×10^8 and Rz from 384.1 nm to 236.8 nm. • The amylopectin chains broke at the amorphous areas below 9 kGy as a result of Y-irradiation.	(Bao et al., 2005)
2.	Corn starch	2, 10, and 50 kGy	9% MC, room temp.; 0.40, 0.67, and 2 kGy/h dose rates; ^{60}Co gamma source	• SDS revealed the opposite pattern, with the RS content showing a minor drop at 2 kGy and an increase up to 50 kGy. • After being exposed to 50 kGy of radiation, the apparent amylose concentration progressively dropped from 28.7% for native starch to 20.9%. • Melting of amylose–lipid complex at 50 kGy was observed.	(Chung & Liu, 2009)
3.	Corn starch	1, 5, 10, 25 and 50 kGy	1 kGy/h dose rate; room temp.	• Waxy samples showed an increased amylose-like fractions at 10 kGy. • At 25 and 50 kGy, low-amylose starches lost their pasting ability.	(Chung et al., 2015)
4.	Whole wheat starch	0.5, 1, 2.5, 5, and 10 kGy	^{60}Co as source irradiator; 23 ± 2 °C temp.	• Decrease with an increase in dose in the peak, final, and setback viscosities. • Dose caused the amylose content to rise from 25.33% to 36.03%. • Swelling, solubility, syneresis, and freeze–thaw stability improved.	(Bashir & Aggarwal, 2017; Bashir et al., 2017)
5.	Acetylated wheat starch	1, 3, 5, 7, and 9 kGy	12% MC; room temp.; ^{60}Co gamma-ray source; 1 kGy dose rate	• The viscosities of the setback, hot, cool, and peak pastes decreased with dose. • Breakdown viscosity increased. • Up to a dose of 9 kGy, the amylose–lipid complex's V-type crystallinity remained unaffected.	(Kong et al., 2016)
6.	Potato starch	0.01, 0.05, 0.1, and 0.5 kGy	^{60}Co source	• At 0.5 kGy, a complete disorder of the crystalline structure and carboxyl content of 0.09%–0.11% were noted. • Gelatinization temperature increased. • Reduction in peak, trough, breakdown, final, and gel hardness viscosities.	(Singh et al., 2011)
7.	Buckwheat and potato starch	5, 10, 15, and 20 kGy	2.5 kGy/h dose rate using ^{60}Co	• In potatoes, the apparent amylose content dropped from 27.01% to 16.11%, and in buckwheat, from 26.84% to 22.12%. • pH, swelling power and syneresis decreased. • Carboxyl content (0% to 23%), freeze–thaw stability, WAC, OAC, and transmittance increased.	(Verma et al., 2018)

(Continued)

TABLE 8.2 *(Continued)*
Effect of Gamma Irradiation on Physicochemical Properties of Food Starches

Sl. No.	Food Starch Type	Gamma Irradiation Dosages (kGy)	Treatment Conditions (Temp., RH, pH etc.)	Effect on Physiochemical Properties	Ref.
8.	Buckwheat and oat starch	5, 10, 15, and 20 kGy	23 ± 2 °C temp.; ^{60}Co	• Starches' percentage crystallinity was decreased by Y-irradiation. • A reduction in the swelling index and apparent amylose content, but solubility index increased.	(Dar et al., 2018)
9.	Sonicated pea and vetch starch	5 kGy	Sonication (33 kHz for 45 min); 12 MC; ^{60}Co as γ source; room temp.; 83 Gy/min dose rate	• Apparent amylose content varied from 31.08 to 27.58 g/100 g. • Synersis decreased significantly after dual treatment from 47.82 to 35.22 g/100 g (pea) and from 46.36 to 26.98 g/100 g (vetch) when stored for 120 h.	(Majeed et al., 2018)
10.	Broad bean starch	0, 5, 10, or 15 kGy	^{60}Co source irradiator; 20 ± 2 °C temp.; 83 Gy/min dose rate; 20–40 RH; 0.1 MPa pressure.	• C-type pattern with a lower peak density was seen in the X-ray diffraction pattern. • Shapes of the starch granules ranged from 6.8 to 14.9 mm lengthwise and from 3.1 to 10.5 mm in breadth. They were round, oval, irregular, or elliptical.	(Sofi et al., 2013)
11.	Chickpea starch	0.5, 1, 2.5, 5, and 10 kGy	^{60}Co as source irradiator, 23 ± 2°C temp., 0.5 kGy/h dose rate	• As per dosage, there was a reduction in peak, final, setback, trough, and pasting temperatures. • Gelatinization temperatures T_o, T_p, and T_c decreased. • Slight surface fissures observed at 5 and 10 kGy.	(Bashir & Aggarwal, 2017)
12.	Cowpea starch	2, 10, and 50 kGy.	^{60}Co source; samples were stored at −18 °C	• Pasting and swelling properties decreased. • Rise in peak gelatinization temp. • Starch granule surface order (crystallinity) not affected.	(Abu et al., 2006)
13.	Ultrasonicated Indian barnyard millet starch	5 and 10 kGy	ultrasonication (20 kHz; 30 and 60 min); room temp.	• The protein content decreased slightly, by 2.5%–3.8%. • The starches exhibited a greater value for G′ than for G″ => elastic nature	(Dimri & Singh, 2024)
14.	Sorghum starch	10 and 25 kGy	^{60}Co gamma-ray source; ambient temp.	• A 22% and 32% decrease in the activity of α- and β-amylase, respectively. • Hydration and gelatinization; the irradiated granules readily burst and dissolved.	(Mukisa et al., 2012)
15.	Black rice starch	300 Gy	^{60}Co gamma rays; room temp.	• The starch content decreased after irradiation to 73.77% (M5 variety) and 62.20% (M6 variety). • Irradiated starches had lower amylopectin levels.	(Bachtari & Listyawati, 2020)
16.	Brown rice starch	5, 10, 15, and 20 kGy	^{60}Co as source irradiator; 2.5 kGy/h dose rate; 23 ± 2 °C temp.	• The distinctive A-type pattern and reduction in crystallinity percentage (22.53%–20.41%) were shown by XRD. • Pasting characteristics declined with dosage, although FRAP values and DPPH% inhibition rose.	(Kumar et al., 2017)
17.	Sonicated lentil starch	5 kGy	Sonication (33 kHz, 45 min); 12% MC; ^{60}Co as gamma source; room temp.	• Amylose content of starch was 31.16 ± 1.80 g/100 g, which decreased after dual modification. • Significant decrease in syneresis observed after 120 h storage. • FTIR study showed that the CH_2 bending and O–H, C–H, and O=C stretch intensities had decreased.	(Liang et al., 2024; Majeed et al., 2017)
18.	Oat starch	0 (control), 5, 10, 15, and 20 KGy	12 MC; ^{60}Co source irradiator; 2 KGy/min dose rate; 20 ± 0.5 °C temp.	• A significant drop in pasting temperature was seen in RVA profiles together with a reduction in peak, trough, final, and set back viscosities. • ATR-FTIR confirmed hydroxyl group exposure in starch chains. • SEM revealed surface ridge formations.	(Mukhtar et al., 2017)
19.	Alkali-extracted rice starch (K-322 and K-448 rice varities)	0 (control), 5, 10, and 20 kGy	12% MC; ^{60}Co gamma source; 20 ± 0.5 °C temp; 2 kGy/h dose rate	• The pasting qualities, pH, swelling strength, syneresis, and amylose concentration were all reduced. • The granules had a distorted or irregular polygonal shape. XRD showed A-type pattern.	(Ashwar et al., 2014)

(Continued)

TABLE 8.2 *(Continued)*
Effect of Gamma Irradiation on Physicochemical Properties of Food Starches

Sl. No.	Food Starch Type	Gamma Irradiation Dosages (kGy)	Treatment Conditions (Temp., RH, pH etc.)	Effect on Physiochemical Properties	Ref.
20.	Quinoa starch	0 (non-irradiation), 1, 2, 4, and 8 kGy	Room temp.; EBI, 5 sec; 2 kGy/h dose rate	• Swelling ability not affected between 0 and 4 kGy. • Solubility increased at > 4 kGy. • Starch crystallinity decreased. • SDS and RS content increased.	(Du et al., 2020)
21.	Amaranth starch	0 kGy (control, non-irradiated), 2, 4, 6, 8, and 10 kGy	Room temp.; 1 kGy/h dosage rate	• Pasting viscosities decreased • Degree of crystallinity and thermal parameters decreased.	(Kong et al., 2009)
22.	Flaxseed starch	5, 10, 15, and 20 kGy	10 MeV ER; 9.06 Gy/min dose rate	• γ-Tocopherol content decreased. • No impact on proximate composition, cyanide concentration, or fatty acid profiles.	(Beheshti Moghadam et al., 2019)
23.	Tapioca starch	0, 5, 10, and 20 kGy	LDPE bag; gamma cell 5000 with ^{60}Co source; 1.8 kGy/h dose rate	• Solubility and light transmittance improved. • Syneresis and swelling capacity reduced.	(Kanatt, 2020)
24.	Sweet potato starch	0.0, 0.20, 0.30, and 0.40 kGy	^{60}Co source; 409 Gy/h	• Starch MC varied between 10.02% and 11.88%. • Starch pH varied from 6.26 to 7.07. • Maximum viscosity dropped from 1008.2 to 937.0 BU.	(Ocloo et al., 2011)
25.	Banana starch	5, 10, 15, 20, and 25 kGy	^{60}Co gamma source; 25 ± 1.0 °C temp.; 2 kGy/h dose rate	• The B-type X-ray pattern of starches remains unaffected. • Pasting properties improved. • Carboxyl content, starch's in vitro digestibility, solubility, and WAC increased. • Reduced retrogradation, syneresis and low thermal decomposition with reduced amylose content.	(Koteswara Reddy et al., 2015; Reddy et al., 2015)
26.	Cassava starch	1 to 10 kGy	^{60}Co gamma source; room temp.	• Starch granules deformed and fragmented. • Breakage of glycosidic bonds in starch chains. • Crystallinity decreased at > kGy dosage.	(Janthanasakulwong & Yoksan, 2024; Tran et al., 2022)

* ^{60}Co, cobalt-60; ATR, attenuated total reflectance; DPPH%, 2, 2-Diphenyl-1-picrylhydrazyl; EBI, electron beam irradiation; ER, electron radiation; FRAP, Ferric reducing antioxidant power; FTIR, Fourier transform infrared spectroscopy; G′, storage modulus; G″, loss modulus; LDPE, low density polyethylene; MC, moisture content; Mw, weight-average molecular weight; OAC, oil absorption capacity; RH, relative humidity; RS, resistant starch; RVA: Rapid Visco Analyzer; Rz, gyration radius; SDS, slowly digestible starch; SEM, scanning electron microscopy; WAC, water absorption capacity; XRD: X-ray diffraction.

8.3.3 Effect on Functional Properties of Food Starches

A variety of γ-irradiation doses and rates have been shown to reduce granule swelling, increase water solubility and amylose leaching in starches derived from diverse plant sources. A favorable association had been noted between irradiation dosage (0–500 kGy) and limited disintegration of starch granules, and the generation of lower-molecular-weight fragments that promote solubility may enhance maize starch solubility at 30 °C and 90 °C (Liu et al., 2012). γ-Irradiation generally enhances the water absorption capacity of starches by disrupting granule structures and exposing hydrophilic groups, allowing more water binding. It also weakens hydrogen bonds within starch molecules, increasing the surface area for water interaction.

The impact on oil absorption capacity (OAC) varies with irradiation dose. At lower doses, OAC increases as starch granules degrade, allowing more oil interaction. However, at higher doses, excessive degradation can reduce OAC, as the structure necessary for oil retention is lost. The breakdown of amylose and amylopectin chains initially boosts oil holding capacity, but too much degradation weakens this capacity. Additionally, syneresis increases after gamma irradiation, as weaker gel structures formed by irradiation lose water more easily during storage, as listed in Table 8.2. The disruption of the starch gel network reduces its water holding capacity, leading to faster water release (syneresis) during freezing and thawing cycles.

8.3.4 Effect on Thermal Properties of Starches

γ-Irradiation significantly impacted the gelatinization characteristics of the starch, as observed through DSC analysis. A lower rate (0.4–2 kGy/h, 10 kGy) increased the degree of gelatinization (DH) in corn starch (Chung & Liu, 2009). The impact of γ-irradiation on gelatinization characteristics is contingent upon variables such as moisture level, starch type, and irradiation settings. When the dosage is above 50 kGy, both gelatinization temperatures and degree of hydrolysis diminished, presumably owing to the breakdown of crystalline domains inside the starch granules. (Liu et al., 2012). γ-Irradiation degrades starch by breaking amylose and amylopectin chains, weakening the granules and decreasing the gelatinization temperature. Irradiated starches tend to gelatinize more quickly because water molecules penetrate the granules more easily (Gupta et al., 2024e).

Gamma irradiation generally lowers gelatinization enthalpy because the radiation-induced degradation reduces the energy required for decomposition crystalline structure of starch. This decrease in enthalpy is more significant at elevated levels of gamma irradiation. Starches that have been gamma-irradiated may also show altered retrogradation behavior; the degradation of starch chains can diminish their tendency to retrograde, resulting in less firmness in gels formed after cooking. The lower molecular weight of irradiated starches can slow down the recrystallization process, impacting texture and thermal stability during storage (Gupta et al., 2024d). Additionally, gamma irradiation typically decreases the thermal stability of starch, as the shorter polymer chains created by irradiation are less heat-resistant, leading to quicker breakdown at elevated temperatures. This reduction in stability could affect the processing conditions for foods that contain starch.

8.3.5 Effect on Pasting and Rheological Properties of Starches

The most often observed rheological characteristic of the irradiation starch is pasting, or the viscosity development of a starch and moisture combination under regulated heating and cooling cycles with continuous shear. Most studies indicate that γ-irradiation reduces starch viscosity during pasting, regardless of the starch source. Peak viscosity (PV), pasting time (PT), trough viscosity (TV), final viscosity (FV), and setback (SB) have been shown to negatively correlate with dosage throughout a broad range of doses (0–500 kGy). This decrease in viscosity is also associated with reduced swelling capacity, increased solubility, and higher amylose leaching, along with lowered intrinsic and apparent viscosity (as detailed in Table 8.2). Additionally, starch type and composition significantly influence the pasting properties impacted by radiation. For example, the susceptibilities of two cultivars of amaranth starch to the effects of the irradiation (2–10 kGy) on pasting varied, most likely because of differences in the composition and organization of crystalline and amorphous areas inside the granules.

A dynamic oscillatory assessment has been used to evaluate the irradiation starch's rheological characteristics. Amaranth and potato starch gels' storage modulus and loss modulus were observed to decrease with radiation (up to 10 kGy), (Kong et al., 2009; Singh et al., 2011), indicating that irradiated starch forms weaker gels. At this step, starch paste gelation is largely regulated by the reformation of amylose released during granule expansion, which fills the amylose network (Brunnschweiler et al., 2005). Consequently, the dynamic rheological findings may be ascribed to the disintegration of granules and the breaking down of starch molecules induced by irradiation.

8.4 INDUSTRIAL APPLICATIONS OF GAMMA-IRRADIATED STARCHES

Gamma-irradiated starches have emerged as versatile and valuable ingredients across various industries due to their enhanced physicochemical properties. The modifications induced by gamma irradiation, such as altered gelatinization behavior, improved solubility, and increased binding capacity, open up new possibilities for the applications of starches in different sectors.

8.4.1 Food Industry

Gamma irradiation can alter starch's swelling power, solubility, and pasting properties, which are important for modifying food texture. For example, irradiated waxy corn starch exhibits reduced swelling power and increased solubility, making it suitable for improving the texture of instant noodles and snack foods (Chung et al., 2015). Irradiated potato starch shows decreased viscosity and increased transparency, enhancing the texture of sauces, gravies, and puddings (Bashir & Aggarwal, 2019). Compared to native starches, cross-linked starches have better emulsifying power, are less digestible, and have less calories, making them useful fat substitutes. They are often used in the production of ice cream, milk products, mayonnaise, and low-fat butter spreads (Chen et al., 2018). Under specific processing conditions, gamma irradiation can stabilize starch-based foods by enhancing their resilience and functionality. For instance, in high-temperature processes like extrusion cooking, irradiated maize starches exhibit improved thermal and shear stability, making them more effective. Additionally, irradiated starches reduce retrogradation, helping to extend shelf life and maintain texture stability in baked and frozen foods by preventing staling (in baked goods) and preserving the desired consistency (Bashir & Aggarwal, 2019; Braşoveanu & Nemţanu, 2018; Chung et al., 2015a). Gamma irradiation increases resistant starch, which may be healthy at 50 kGy, waxy maize starch had the greatest resistant starch concentration. Normal, high-amylose, and waxy maize starches demonstrate greater resistance. Irradiation-resistant starch may be utilized to make functional meals with prebiotic and glycemic benefits (Bashir & Aggarwal, 2019). Gamma-irradiated starches improve the clarity and stability of fruit juices and dairy products, while also enhancing the gelling properties in starch-based custards and desserts. Additionally, they modify the viscosity and flow characteristics of starch-based sauces and dressings, contributing to better texture and consistency in these products (Braşoveanu & Nemţanu, 2018; Sunder et al., 2022). In conclusion, gamma irradiation proves to be a powerful tool in enhancing the functional properties of starches across various food applications.

8.4.2 Pharmaceutical Industry

Gamma irradiation enhances the functional characteristics of starch, such as by reducing viscosity and increasing solubility, both qualities that are essential for excipient performance. This treatment also helps in reducing microbial contamination, thus ensuring quality as well as safety of pharmaceutical products. The irradiation process is recognized for being environmentally friendly compared to traditional chemical modifications, reducing the risk of residual chemicals in the final product (Braşoveanu & Nemţanu, 2018). Gamma irradiation modifies starches to improve their performance in drug delivery. The technique can modify the physicochemical characteristics of starch, including solubility, viscosity, and gelatinization temperature. For instance, studies have shown that gamma irradiation may enhance the resistant starch content, which is beneficial for controlled drug release. This modification allows for the development of formulations that can provide a sustained release of active pharmaceutical ingredients, enhancing therapeutic efficacy and patient compliance (Bashir & Aggarwal, 2019; Chung et al., 2015a). Studies demonstrate that the impact of gamma irradiation on starch is contingent upon the dose and the specific kind of starch used. Corn starches with differing amylose-to-amylopectin ratios demonstrate varying reactions to irradiation. Higher doses of gamma radiation can lead to significant changes in pasting properties and thermal behavior, which can be tailored for specific pharmaceutical applications (Braşoveanu & Nemţanu, 2018). The modifications brought about by gamma irradiation can enhance the disintegration and dissolution properties of starches, making them suitable for use as binders and fillers in tablets and capsules. Furthermore, in controlled-release drug delivery, the gamma-irradiated starches altered molecular structure and reduced crystallinity of irradiated starches enable them to encapsulate active pharmaceutical ingredients and release them gradually over time. This property is particularly valuable in developing sustained-release formulations that improve patient compliance and therapeutic outcomes. It enhances safety and environmental impact by avoiding harmful chemicals, effectively controls microbial contamination to extend shelf life and ensure safety in pharmaceuticals, and improves starch functionality by increasing solubility and stability, making it a more effective excipient in various formulations.

8.4.3 Packaging Industry

Gamma-irradiated starches provide a vital function in the packaging industry by enhancing biodegradable materials and biomaterials. They efficiently diminish moisture migration, exchange of gases, and oxidative reactions; mitigate physiological illnesses; and postpone textural alterations. Edible films may be composed of polysaccharides, proteins, and lipids, with protein-based films distinguished by their enhanced gas barrier qualities relative to films derived from lipids or polysaccharides. Nonetheless, the use of protein films in food packaging is constrained by their comparatively inadequate resistance to water vapor and diminished mechanical strength compared to synthetic polymers (Hassan et al., 2018). Gamma irradiation facilitates the cross-linking of starch with other biodegradable polymers, such as polyvinyl alcohol (PVA). This process enhances the overall performance of the packaging materials (Bashir & Aggarwal, 2019; Kanatt, 2020). The impact of irradiation on edible films varies significantly depending on the type of substrate used. When cross-linking occurs, the polymer network becomes more robust, enhancing both tensile and puncture strength, though it results in reduced elongation. Cross-linking also leads to lower gas and water permeability. Conversely, in the absence of cross-linking, irradiation causes the polymer network to become looser, leading to a notable decrease in both mechanical strength and permeability (Arvanitoyannis & Dionisopoulou, 2010). Gamma-irradiated starch can be used in the creation of biomaterials for active packaging films that include antibacterial agents or antioxidants. This application prolongs the shelf life of food goods by inhibiting microbial proliferation and oxidation. The irradiation method modifies the physicochemical characteristics of starch, enhancing solubility and reducing viscosity. These changes are beneficial for creating films that maintain their integrity and functionality under various environmental conditions. The application of gamma-irradiated starch is consistent with the trend toward sustainable materials in packaging, as shown in Figure 8.3. As a renewable resource, starch can be sourced from various agricultural products, making it an eco-friendly option for packaging solutions as an effective alternative to traditional plastics (Bashir & Aggarwal, 2019; Braşoveanu & Nemţanu, 2018; Chung et al., 2015a; Kanatt, 2020). Gamma-irradiated starches can also be blended with other biopolymers to create composite materials with enhanced barrier properties, making them suitable for packaging perishable foods and other sensitive products. Moreover, the biodegradability of starch-based packaging materials ensures that they decompose naturally, minimizing the ecological footprint of packaging trash (Gupta et al., 2025a). In addition to packaging, gamma-irradiated starches are being explored for use in other biomaterials, such as medical implants, wound dressings, and agricultural films. Their biocompatibility and biodegradability make them ideal promising possibilities for these applications, further expanding their industrial utility.

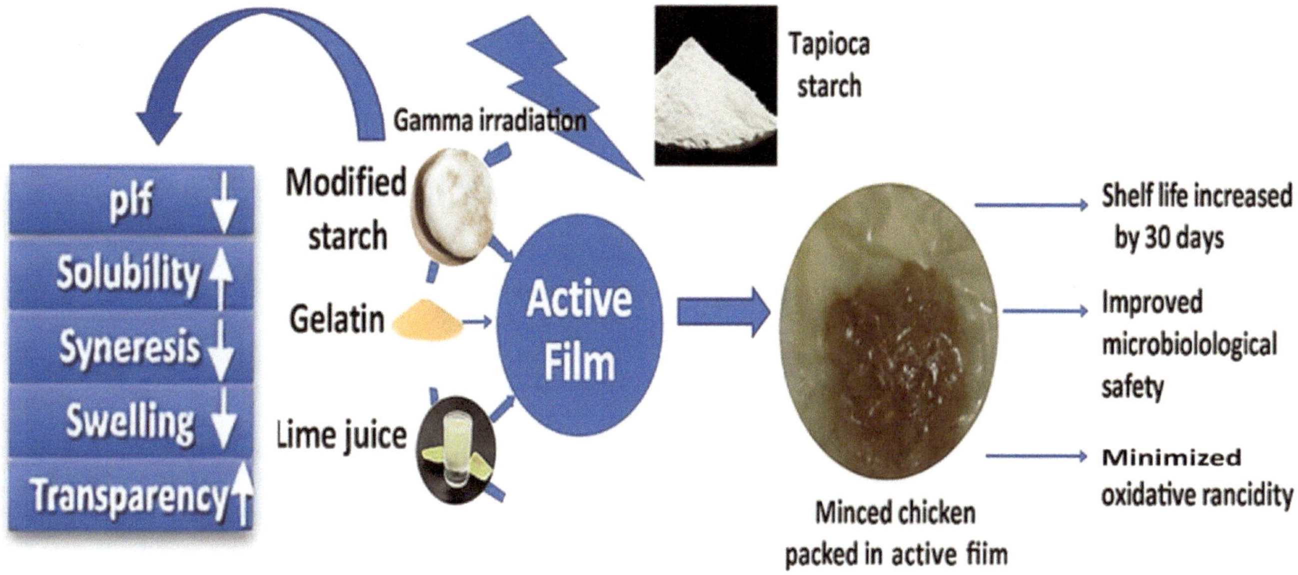

FIGURE 8.3 Properties of irradiated starches and their applications in food processing and packaging (Kanatt, 2020).

8.5 CHALLENGES TO TECHNOLOGY ADOPTION

Despite the promising potential of gamma-irradiated starches across various industries, the widespread adoption of this technology faces several challenges. These challenges stem from concerns related to radiation safety, regulatory hurdles, industrial scalability, and consumer acceptance. Addressing these barriers is crucial for the successful integration of gamma irradiation as a mainstream technology for starch modification.

8.5.1 Radiation Safety and Regulatory Aspects for Irradiated Starches

The adoption of gamma irradiation for starch modification faces significant challenges related to radiation safety and regulatory frameworks, which vary across nations. Worker safety is another critical concern, requiring strict adherence to safety protocols and the implementation of robust shielding and monitoring systems to protect those involved in the irradiation process. Additionally, the environmental effects of gamma irradiation, including the supervision of radioactive waste and the prevention of contamination, are closely scrutinized, often necessitating detailed environmental assessments as part of the regulatory approval process. The complexity of navigating diverse regulatory landscapes further complicates the global adoption of irradiated starches, as different jurisdictions impose different standards and approval processes. For example, while the EU enforces stringent guidelines about the application of irradiation in food items, the United States and Australia have more lenient regulations, leading to disparities in international trade and market access. Moreover, the lack of standardized practices for irradiation dosages and treatment conditions contributes to inconsistencies in product quality, underlining the need for harmonized global standards. Companies looking to adopt this technology must also contend with extensive testing, documentation, and labeling requirements to meet regulatory demands, which can delay market entry and add significant costs. Despite these challenges, gamma irradiation remains a promising method for starch modification, provided that these safety and regulatory hurdles are effectively addressed.

8.5.2 Industrial Adoption and Challenges

The global adoption of gamma irradiation technology for starch modification encounters several significant challenges that affect its industrial implementation. A major obstacle is the high initial investment required for setting up irradiation facilities, which entails substantial costs for equipment, safety measures, and regulatory compliance. This financial burden can be particularly onerous for small and medium enterprises, potentially hindering market entry and slowing technological growth. Additionally, the complexity of gamma irradiation systems necessitates specialized training and expertise, which can deter companies from adopting the technology. The reliance on radioactive isotopes such cobalt-60 introduces supply chain risks, as disruptions in isotope availability can affect operational stability and increase costs. Regulatory hurdles further complicate adoption, with diverse and often stringent regulations across jurisdictions requiring extensive documentation and testing. The lengthy approval processes can delay market entry and limit the scope of applications for irradiated starches. Public perception of radiation also poses a challenge, with concerns about safety and health risks potentially leading to consumer resistance. Furthermore, competition from potentially faster-processing alternative technologies, such as X-ray and electron beam irradiation, and simpler regulatory requirements can divert interest and investment away from gamma irradiation. Overcoming these barriers is crucial for fully realizing the potential of gamma-irradiated starches and advancing their use in food safety, pharmaceutical applications, and sustainable packaging (Bashir & Aggarwal, 2019; Braşoveanu & Nemţanu, 2018; Chung et al., 2015; Kong, 2023).

8.5.3 Consumer Acceptance and Global Market Trends for Irradiated Starches

Consumer acceptance is a pivotal factor influencing the market potential of gamma-irradiated starches, despite the technology's clear benefits for food safety and preservation. Negative perceptions of irradiation, driven by misconceptions about radiation contamination and potential nutritional loss, often deter consumers from embracing these products. Effective education and transparent labeling are essential for overcoming these concerns and building trust. Regional acceptance varies, with some areas, such as North America and parts of Asia, showing more openness to irradiation technology than regions with stricter regulations or greater skepticism. As global market trends shift toward safer and longer-lasting food products, gamma-irradiated starches could see increased adoption, especially if positioned as a clean-label and environmentally friendly alternative to chemical modifications. Advances in irradiation technology and the rise of e-commerce also present opportunities to enhance consumer awareness and acceptance, potentially driving growth in markets where regulatory and cultural barriers can be addressed. To fully leverage the potential of gamma-irradiated starches, a concerted effort involving scientific research, regulatory navigation, industry collaboration, and consumer education is necessary. Consumer acceptance of irradiated starches is shaped by factors such as safety perceptions, knowledge of the technology, and market trends. Although gamma irradiation is recognized for its benefits in increasing shelf life and enhancing food safety, challenges remain in gaining widespread acceptance. Safety concerns persist among consumers, with many fearing that irradiated foods are radioactive or contain harmful compounds, often due to misconceptions and a lack of understanding of the irradiation process. Effective education and communication are key to addressing these fears, emphasizing the safety, nutritional equivalence, and benefits of irradiated foods. Clear labeling practices that provide transparency and highlight the advantages of irradiation can further build consumer trust. Additionally, cultural factors play a significant role, as acceptance varies across different regions, requiring targeted marketing strategies that consider these cultural nuances (Castell-Perez & Moreira, 2021; Maherani et al., 2016). Moreover, gamma-irradiated starches face market challenges due to misconceptions about safety and nutrition, despite benefits for food safety and preservation. Consumer education, transparent labelling, and collaboration among stakeholders are key to building trust. Furthermore, growing demand for clean-label, sustainable products offers an opportunity to position them as a natural alternative to chemically modified starches.

8.6 CONCLUSION AND FUTURE SCOPE

Gamma irradiation has shown promise as a method of altering starch, offering transformative benefits over conventional methods. This non-thermal approach significantly enhances starch's structural, physical, functional, thermal, and rheological properties, enhancing its performance across diverse industry applications. Through gamma irradiation, starch granules undergo critical morphological changes, including in size, shape, and surface texture, that enhance their functionality in improving food texture, stability, and shelf life. Additionally, these modifications support advancements in drug delivery systems and the development of eco-friendly packaging solutions. Despite their potential, the broader adoption of gamma-irradiated starches faces hurdles such as concerns about radiation safety, complex regulatory landscapes, and consumer acceptance issues. Overcoming these challenges requires a concerted education effort, transparent communication, and addressing regulatory and economic barriers.

Looking forward, several key areas present opportunities for further research and development. Optimizing irradiation conditions, including dosage levels and treatment environments, could lead to more precise and beneficial modifications of starch. Expanding the scope of industrial applications beyond the traditional realms of food, pharmaceuticals, and packaging to emerging fields like biomedicine, nanotechnology, and environmental engineering could unlock new uses for irradiated starches. Additionally, efforts to harmonize global regulatory frameworks are essential to facilitate the widespread adoption of this technology. Research into the long-term effectiveness of radioactively treated starches and associated environmental impact, particularly in biodegradable packaging, will be crucial in addressing regulatory and consumer concerns. Enhancing consumer awareness and trust through targeted educational initiatives and transparent labeling will also significantly promote market acceptance. By addressing these challenges and harnessing ongoing innovations, gamma irradiation can revolutionize starch modification, contributing to advancements in food science, material engineering, and sustainable practices.

REFERENCES

Abu, J. O., Duodu, K. G., & Minnaar, A. (2006). Effect of γ-irradiation on some physicochemical and thermal properties of cowpea (*Vigna unguiculata* L. Walp) starch. *Food Chemistry*, *95*(3), 386–393.

Akamine, E. K., & Moy, J. H. (2018). Delay in postharvest ripening and senescence of fruits. In Josephson, E.S. & Peterson, M.S. (Eds.) *Preservation of food by ionizing radiation* (pp. 129–158). CRC Press.

Andrews, L. S., Ahmedna, M., Grodner, R. M., Liuzzo, J. A., Murano, P. S., Murano, E. A., Rao, R. M., Shane, S., & Wilson, P. W. (1998). Food preservation using ionizing radiation. In Cavieres, M. F., Knaak, J. B., Wezel, A. P. V., Tjeerdema, R. S. & Vighi, M. (Eds.) *Reviews of environmental contamination and toxicology* (pp. 1–53). Springer. https://doi.org/10.1007/978-1-4612-2208-8_1

Arvanitoyannis, I. S., & Dionisopoulou, N. K. (2010). Irradiation of edible films of plant and animal origin. In Arvanitoyannis, I. S. (Ed.) *Irradiation of food commodities* (pp. 609–634). Elsevier. https://doi.org/10.1016/b978-0-12-374718-1.10015-x

Ashwar, B. A., Shah, A., Gani, A., Rather, S. A., Wani, S. M., Wani, I. A., Masoodi, F. A., & Gani, A. (2014). Effect of gamma irradiation on the physicochemical properties of alkali-extracted rice starch. *Radiation Physics and Chemistry*, *99*, 37–44.

Bachtari, R., & Listyawati, S. (2020). Starch, amylose and amylopectin levels of M5 and M6 generations of black rice irradiated by gamma Co^{60} ray. *Journal of Physics: Conference Series*. (Vol. 1436, No. 1, p. 012116). IOP Publishing.

Bao, J., Ao, Z., & Jane, J. (2005). Characterization of physical properties of flour and starch obtained from gamma-irradiated white rice. *Starch - Stärke*, *57*(10), 480–487.

Bashir, K., & Aggarwal, M. (2017). Physicochemical, thermal and functional properties of gamma irradiated chickpea starch. *International Journal of Biological Macromolecules*, *97*, 426–433.

Bashir, K., & Aggarwal, M. (2019). Physicochemical, structural and functional properties of native and irradiated starch: A review. *Journal of Food Science and Technology*, *56*(2), 513–523. https://doi.org/10.1007/s13197-018-3530-2

Bashir, K., Swer, T. L., Prakash, K. S., & Aggarwal, M. (2017). Physico-chemical and functional properties of gamma irradiated whole wheat flour and starch. *LWT*, *76*, 131–139.

Beheshti Moghadam, M., Rezaei, M., Behgar, M., & Kermanshahi, H. (2019). Effects of gamma and electron radiation on chemical composition and some phyto-chemical properties of whole flaxseed. *Journal of Radioanalytical and Nuclear Chemistry*, *321*, 1019–1025.

Bhat, R., & Karim, A. (2009). Impact of radiation processing on starch. *Comprehensive Reviews in Food Science and Food Safety*, *8*(2), 44–58.

Braşoveanu, M., & Nemţanu, M.-R. (2018). Aspects on starches modified by ionizing radiation processing. In Huicochea, E. F. & Villalobos, R. R. (Eds.) *Applications of modified starches*. IntechOpen. https://doi.org/10.5772/intechopen.71626

Brunnschweiler, J., Luethi, D., Handschin, S., Farah, Z., Escher, F., & Conde-Petit, B. (2005). Isolation, physicochemical characterization and application of yam (*Dioscorea* spp.) starch as thickening and gelling agent. *Starch - Stärke*, *57*(3–4), 107–117.

Castell-Perez, M. E., & Moreira, R. G. (2021). Irradiation and consumers acceptance. In Knoerzer, K. & Muthukumarappan, K. *(Eds.) Innovative food processing technologies* (pp. 122–135). Elsevier. https://doi.org/10.1016/b978-0-12-815781-7.00015-9

Chakraborty, I., Pallen, S., Shetty, Y., Roy, N., & Mazumder, N. (2020). Advanced microscopy techniques for revealing molecular structure of starch granules. *Biophysical Reviews*,*12*, 105–122.

Chen, Y.-F., Kaur, L., & Singh, J. (2018). Chemical modification of starch. In Sjöö, M. & Nilsson, L. (Eds.) *Starch in food* (pp. 283–321). Elsevier. https://doi.org/10.1016/b978-0-08-100868-3.00007-x

Chimonyo, W., Fletcher, B., & Peng, Y. (2020). Starch chemical modification for selective flotation of copper sulphide minerals from carbonaceous material: A critical review. *Minerals Engineering*, *156*, 106522.

Chiu, C.-W., & Solarek, D. (2009). Modification of starches. In BeMiller, J. & Whistler, R. (Eds.) *Starch* (pp. 629–655). Academic Press. Elsevier

Chung, H. J., & Liu, Q. (2009). Effect of gamma irradiation on molecular structure and physicochemical properties of corn starch. *Journal of Food Science*, *74*(5), C353–C361.

Chung, H.-J., & Liu, Q. (2010). Molecular structure and physicochemical properties of potato and bean starches as affected by gamma-irradiation. *International Journal of Biological Macromolecules*, *47*(2), 214–222.

Chung, K.-H., Othman, Z., & Lee, J.-S. (2015). Gamma irradiation of corn starches with different amylose-to-amylopectin ratio. *Journal of Food Science and Technology*, *52*(10), 6218–6229. https://doi.org/10.1007/s13197-014-1700-4

Ciesla, K., Zoltowski, T., & Mogilevsky, L. Y. (1991). Detection of starch transformation under γ-irradiation by small-angle X-ray scattering. *Starch - Stärke*, *43*(1), 11–12.

Dar, M. Z., Deepika, K., Jan, K., Swer, T. L., Kumar, P., Verma, R., Verma, K., Prakash, K. S., Jan, S., & Bashir, K. (2018). Modification of structure and physicochemical properties of buckwheat and oat starch by γ-irradiation. *International Journal of Biological Macromolecules*, *108*, 1348–1356.

Dimri, S., & Singh, S. (2024). Transformations in properties of Indian barnyard millet starch as induced by ultra-sonication and gamma irradiation. *Starch - Stärke*, 76, 2300218.

Du, Z., Li, Y., Luo, X., Xing, J., Zhang, Q., Wang, R., Wang, L., & Chen, Z. (2020). Effects of electron beam irradiation on the physicochemical properties of quinoa and starch microstructure. *Starch-Stärke*, *72*(11–12), 1900178.

Duarte, P. R., & Rupnow, J. (1994). Gamma-irradiated dry bean (*Phaseolus vulgaris*) starch: Physicochemical properties. *Journal of Food Science*, *59*(4), 839–843.

Egharevba, H. O. (2019). Chemical properties of starch and its application in the food industry. In M. Emeje (Ed.), *Chemical properties of starch* (ch. 5). IntechOpen. http://dx.doi.org/10.5772/intechopen.87777

El-Ramady, H. R., Domokos-Szabolcs, É., Abdalla, N. A., Taha, H. S., & Fári, M. (2015). Postharvest management of fruits and vegetables storage. In Lichtfouse, E. (Ed.) *Sustainable agriculture reviews* (pp. 65–152). Springer. https://doi.org/10.1007/978-3-319-09132-7_2

Esteves, M.-P., Girio, F., Amaral-Collaço, M., Andrade, M., & Empis, J. (1997). Characterization of starch from white and black pepper treated by ionizing radiation. *Sciences des Aliments*, *17*(3), 289–298.

Ezekiel, R., Rana, G., Singh, N., & Singh, S. (2007). Physicochemical, thermal and pasting properties of starch separated from γ-irradiated and stored potatoes. *Food Chemistry*, *105*(4), 1420–1429.

Farkas, J. (1998). Irradiation as a method for decontaminating food. *International Journal of Food Microbiology*, *44*(3), 189–204. https://doi.org/10.1016/s0168-1605(98)00132-9

Fernandes, A., Pereira, C., Antonio, A. L., & Ferreira, I. C. F. R. (2017). Food irradiation chemistry. In I. C. F. R. Ferreira, A. L. Antonio, & S. Cabo Verde (Eds.), *Food irradiation technologies* (pp. 210–236). Royal Society of Chemistry.

Ganguly, S., Mukhopadhayay, S. K., & Biswas, S. (2012). Preservation of food items by irradiation process. *International Journal of Chemical and Biochemical Sciences*, *1*(2012), 11–13.

Gani, A., Gazanfar, T., Jan, R., Wani, S., & Masoodi, F. (2013). Effect of gamma irradiation on the physicochemical and morphological properties of starch extracted from lotus stem harvested from Dal Lake of Jammu and Kashmir,

India. *Journal of the Saudi Society of Agricultural Sciences*, *12*(2), 109–115.

Gupta, R. K., Guha, P., & Srivastav, P. P. (2022). Natural polymers in bio-degradable/edible film: A review on environmental concerns, cold plasma technology and nanotechnology application on food packaging-A recent trends. *Food Chemistry Advances*, *1*, 100135. https://doi.org/10.1016/j.focha.2022.100135

Gupta, R. K., Guha, P., & Srivastav, P. P. (2023). Effect of high voltage dielectric barrier discharge (DBD) atmospheric cold plasma treatment on physicochemical and functional properties of taro (*Colocasia esculenta*) starch. *International Journal of Biological Macromolecules*, *253*, 126772. https://doi.org/10.1016/j.ijbiomac.2023.126772

Gupta, R. K., Guha, P., & Srivastav, P. P. (2024a). Dielectric barrier discharge plasma: A green and novel method to change the structure of taro peel starch and improve the physicochemical properties of taro peel starch films. *Plasma Processes and Polymers*, *21*(9), 2400047. https://doi.org/10.1002/ppap.202400047

Gupta, R. K., Guha, P., & Srivastav, P. P. (2024b). Exploring the potential of taro (*Colocasia esculenta*) starch: Recent developments in modification, health benefits, and food industry applications. *Food Bioengineering*, *3*(3), 365–379. https://doi.org/10.1002/fbe2.12103

Gupta, R. K., Guha, P., & Srivastav, P. P. (2024c). Insights into the physicochemical, crystalline, multistructural, thermal and digestibility behavior of taro starch (*Colocasia esculenta*) modified by cold plasma treatment with high voltage dielectric barrier discharge (HVDBD): An alternative technique for chemical modification of starch and environmental sustainability. *Journal of Cleaner Production*, *461*, 142710. https://doi.org/10.1016/j.jclepro.2024.142710

Gupta, R. K., Guha, P., & Srivastav, P. P. (2024d). Physical action of nonthermal cold plasma technology for starch modification. *Food Physics*, *1*, 100011. https://doi.org/10.1016/j.foodp.2024.100011

Gupta, R. K., Guha, P., & Srivastav, P. P. (2024e). Effect of cold plasma treatment and plasma-activated water on physicochemical and structural properties of starch: A green and novel approach for environmental sustainability. *Plasma Processes and Polymers*, *21*(4), 2300204. https://doi.org/10.1002/ppap.202300204

Gupta, R. K., Pipliya, S., Patel, J., Srivastav, P. P., Castro-Muñoz, R., Wang, C. K., & Nemţanu, M. R. (2025a). Essential Oil-Embedded Starch based Films for Active Packaging: Insights into Migration, Stability, and Preservation Efficiency. *Trends in Food Science & Technology*, 163, 105149. https://doi.org/10.1016/j.tifs.2025.105149

Gupta, R. K., Gnana Moorthy Eswaran, U., Kumar, S., Pipliya, S., Kaur, B., & Srivastav, P. P. (2025b). Synergistic Modification of Starch Polysaccharide via Sequential Treatments with Ultrasonic and Cold Plasma Technology. *Journal of Food Processing and Preservation*, *2025*(1), 9063007. https://doi.org/10.1155/jfpp/9063007

Goswami, B., & Mahanta, D. (2021). Starch and its derivatives: Properties and applications. In M. I. A. Inamuddin, R. Boddula, & T. Altalhi (Eds.), *Polysaccharides: Properties and applications* (pp. 253–281). Scrivener.

Habib, M., Jan, K., Qureshi, I., Rani, S., & Bashir, K. (2023). Gamma irradiation of starch. In *Starch: Advances in modifications, technologies and applications* (pp. 385–407). Springer.

Hassan, B., Chatha, S. A. S., Hussain, A. I., Zia, K. M., & Akhtar, N. (2018). Recent advances on polysaccharides, lipids and protein based edible films and coatings: A review. *International Journal of Biological Macromolecules*, *109*, 1095–1107. https://doi.org/10.1016/j.ijbiomac.2017.11.097

HerCeg, I. L., Jambrak, A. R., ŠubArIć, D., Brnčić, M., Brnčić, S. R., Badanjak, M., Tripalo, B., Ježek, D., Novotni, D., & Herceg, Z. (2010). Texture and pasting properties of ultrasonically treated corn starch. *Czech Journal of Food Sciences*, *28*(2), 83–93.

Ihsanullah, I., & Rashid, A. (2017). Current activities in food irradiation as a sanitary and phytosanitary treatment in the Asia and the Pacific Region and a comparison with advanced countries. *Food Control*, *72*, 345–359. https://doi.org/10.1016/j.foodcont.2016.03.011

Indiarto, R., Irawan, A. N., & Subroto, E. (2023). Meat irradiation: A comprehensive review of its impact on food quality and safety. *Foods*, *12*(9), 1845.

Indiarto, R., Pratama, A. W., Sari, T. I., & Theodora, H. C. (2020). Food irradiation technology: A review of the uses and their capabilities. *International Journal of Engineering Technology and Trends*, *68*(12), 91–98.

Janthanasakulwong, P., & Yoksan, R. (2024). Effect of gamma ray dose on granular and molecular structures of gamma ray-irradiated cassava starch and its application in bioplastics. *International Journal of Biological Macromolecules*, 279, 135330.

Kamal, H., Sabry, G. M., Lotfy, S., Abdallah, N. M., Ulanski, P., Rosiak, J., & Hegazy, E. S. A. (2007). Controlling of degradation effects in radiation processing of starch. *Journal of Macromolecular Science, Part A: Pure and Applied Chemistry*, *44*(8), 865–875.

Kanatt, S. R. (2020). Irradiation as a tool for modifying tapioca starch and development of an active food packaging film with irradiated starch. *Radiation Physics and Chemistry*, *173*, 108873.

Kaur, L., Kaur, R., & Singh, J. (2024). Chemical modification of starch. In Nilsson, L. (Ed.) *Starch in food* (pp. 97–117). Elsevier.

Kong, X. (2023). Gamma irradiation of starch. In Sui, Z. & Kong, X. (Eds.) *Physical modifications of starch* (pp. 103–143). Springer.

Kong, X., Kasapis, S., Bao, J., & Corke, H. (2009). Effect of gamma irradiation on the thermal and rheological properties of grain amaranth starch. *Radiation Physics and Chemistry*, *78*(11), 954–960.

Kong, X., Zhou, X., Sui, Z., & Bao, J. (2016). Effects of gamma irradiation on physicochemical properties of native and acetylated wheat starches. *International Journal of Biological Macromolecules*, *91*, 1141–1150.

Koteswara Reddy, C., Vidya, P. V., Vijina, K., & Haripriya, S. (2015). Modification of poovan banana (*Musa* AAB) starch by γ-irradiation: Effect on in vitro digestibility, molecular structure and physico-chemical properties. *International Journal of Food Science & Technology*, *50*(8), 1778–1784.

Kumar, P., Prakash, K. S., Jan, K., Swer, T. L., Jan, S., Verma, R., Deepika, K., Dar, M. Z., Verma, K., & Bashir, K. (2017). Effects of gamma irradiation on starch granule structure and physicochemical properties of brown rice starch. *Journal of Cereal Science*, *77*, 194–200.

Lee, Y. S., Oh, S. H., Lee, J. W., Kim, J. H., Kim, D. S., & Byun, M. W. (2003). Effects of gamma irradiation on physicochemical and textural properties of starches. *Food Science and Biotechnology*, *12*(5), 508–512.

Liang, D., Liu, Q., Luo, H., Luo, L., Temirlan, K., & Li, W. (2024). The effect of maltose on structural, physicochemical, and digestive properties of lentil starch under electron beam irradiation. *Foods*, *13*(16), 2544.

Liu, T., Ma, Y., Xue, S., & Shi, J. (2012). Modifications of structure and physicochemical properties of maize starch by γ-irradiation treatments. *LWT, 46*(1), 156–163.

Lu, Z.-H., Donner, E., Yada, R. Y., & Liu, Q. (2012). Impact of γ-irradiation, CIPC treatment, and storage conditions on physicochemical and nutritional properties of potato starches. *Food Chemistry, 133*(4), 1188–1195.

Maherani, B., Hossain, F., Criado, P., Ben-Fadhel, Y., Salmieri, S., & Lacroix, M. (2016). World market development and consumer acceptance of irradiation technology. *Foods, 5*(4). https://doi.org/10.3390/foods5040079

Majeed, T., Wani, I. A., Hamdani, A. M., & Bhat, N. A. (2018). Effect of sonication and γ-irradiation on the properties of pea (*Pisum sativum*) and vetch (*Vicia villosa*) starches: A comparative study. *International Journal of Biological Macromolecules, 114*, 1144–1150.

Majeed, T., Wani, I. A., & Hussain, P. R. (2017). Effect of dual modification of sonication and γ-irradiation on physicochemical and functional properties of lentil (*Lens culinaris* L.) starch. *International Journal of Biological Macromolecules, 101*, 358–365.

Milani, J. M., & Golkar, A. (2021). Starch modification by novel technologies and their functionality. In Galanakis, C. M. (Ed.) *Food structure and functionality* (pp. 157–179). Elsevier.

Mshelia, R. D.-z, Dibal, N. I., & Chiroma, S. M. (2023). Food irradiation: An effective but under-utilized technique for food preservations. *Journal of Food Science and Technology, 60*(10), 2517–2525.

Mukhtar, R., Shah, A., Noor, N., Gani, A., Wani, I. A., Ashwar, B. A., & Masoodi, F. (2017). γ-Irradiation of oat grain–effect on physico-chemical, structural, thermal, and antioxidant properties of extracted starch. *International Journal of Biological Macromolecules, 104*, 1313–1320.

Mukisa, I. M., Muyanja, C. M., Byaruhanga, Y. B., Schüller, R. B., Langsrud, T., & Narvhus, J. A. (2012). Gamma irradiation of sorghum flour: Effects on microbial inactivation, amylase activity, fermentability, viscosity and starch granule structure. *Radiation Physics and Chemistry, 81*(3), 345–351.

Nagasawa, N., Yagi, T., Kume, T., & Yoshii, F. (2004). Radiation crosslinking of carboxymethyl starch. *Carbohydrate Polymers, 58*(2), 109–113.

Nemţanu, M., & Braşoveanu, M. (2010a). Aspects regarding the rheological behavior of wheat starch treated with accelerated electron beam. *Romanian Journal of Physics, 55*, 111–117.

Nemţanu, M., & Braşoveanu, M. (2010b). Functional properties of some non-conventional treated starches. In *Biopolymers*. IntechOpen.

Ocloo, F., Otoo, G., Nkansah, R., Mahami, T., Odonkor, S., & Quayson, E. (2011). Functional and physicochemical characteristics of starch obtained from gamma-irradiated sweet potato (*Ipomea batatas* L.). *Journal of Agriculture and Food Technology, 1*(7), 116–122.

Öztürk, S., & Mutlu, S. (2019). Physicochemical properties, modifications, and applications of resistant starches. In Clerici, M. T. P. S. & Schmiele, M. (Eds.) *Starches for food application* (pp. 297–332). Elsevier.

Peng, X., Li, Y., Wang, S., Gao, Y., Lei, X., & Ren, Y. (2024). Electron beam irradiation inhibited potato sprouting by regulating the metabolism of membrane lipid peroxidation and antioxidant. *Potato Research*, 68(1), 391–407.

Pérez, S., & Bertoft, E. (2010). The molecular structures of starch components and their contribution to the architecture of starch granules: A comprehensive review. *Starch - Stärke, 62*(8), 389–420.

Polesi, L. F., Sarmento, S. B. S., de Moraes, J., Franco, C. M. L., & Canniatti-Brazaca, S. G. (2016). Physicochemical and structural characteristics of rice starch modified by irradiation. *Food Chemistry, 191*, 59–66.

Rayas-Duarte, P., & Rupnow, J. (1993). Gamma-irradiation affects some physical properties of dry bean (*Phaseolus vulgaris*) starch. *Journal of Food Science, 58*(2), 389–394.

Reddy, C. K., Vidya, P. V., Vijina, K., & Haripriya, S. (2015). Modification of poovan banana (*Musa* AAB) starch by γ-irradiation: Effect on in vitro digestibility, molecular structure and physico-chemical properties. *International Journal of Food Science and Technology, 50*, 1778–1784.

Roberts, P. B. (2003). Irradiation of foods: Legal and consumer aspects. In Caballero, B. (Ed.) *Encyclopedia of food sciences and nutrition* (pp. 3396–3401). Elsevier. https://doi.org/10.1016/b0-12-227055-x/00656-8

Saha, D., Pradhan, A., Sarkar, A., & Tripathy, P. P. (2024). Recent trends in pseudocereal-based foods. In Dhull, S. B., Bains, A., Chawla, P., & Kaur, S. (Eds.) *Pseudocereals* (pp. 264–291). CRC Press.

Sahoo, M., Aradwad, P., Panigrahi, C., Kumar, V., & Naik, S. (2023). Irradiation of food. In Chhikara, N., Panghal, A. & Chaudhary, G. (Eds.) *Novel Technologies in Food Science*, 333–373. Wiley. https://doi.org/10.1002/9781119776376.ch9

Sarkar, A., Saha, D., Pradhan, A., & Tripathy, P. P. (2024). Innovative processing technologies for pseudocereals. In Dhull, S.B., Bains, A., Chawla, P., & Kaur, S. (Eds.) *Pseudocereals* (pp. 191–215). CRC Press.

Shetty, S., Govindaraju, I., Rebello, A. S., Chatterjee, D., Rahman, M. H., & Mazumder, N. (2024). Physicochemical modification and characterization of starch used in the food industry: A review. In Mazumder, N. (Ed.) *Advanced biophysical techniques for polysaccharides characterization* (pp. 1–46). Elsevier.

Shi, M., Jing, Y., Yang, L., Huang, X., Wang, H., Yan, Y., & Liu, Y. J. P. (2019). Structure and physicochemical properties of malate starches from corn, potato, and wrinkled pea starches. *Polymers,11*(9), 1523.

Shrivastava, A., Gupta, R. K., & Srivastav, P. P. (2024). Exploring novel frontiers of advancements in purple yam (*Dioscorea alata* L.) starch extraction, modification, characterization, applications in food and other industries. *Measurement: Food, 15*, 100196. https://doi.org/10.1016/j.meafoo.2024.100196

Singh, S., Singh, N., Ezekiel, R., & Kaur, A. (2011). Effects of gamma-irradiation on the morphological, structural, thermal and rheological properties of potato starches. *Carbohydrate Polymers, 83*(4), 1521–1528.

Sofi, B. A., Wani, I. A., Masoodi, F. A., Saba, I., & Muzaffar, S. (2013). Effect of gamma irradiation on physicochemical properties of broad bean (*Vicia faba* L.) starch. *LWT, 54*(1), 63–72.

Sokhey, A., & Chinnaswamy, R. (1993). Chemical and molecular properties of irradiated starch extrudates. *Cereal Chemistry, 70*(3), 260–268.

Sokhey, A., & Hanna, M. (1993). Properties of irradiated starches. *Food Structure, 12*(4), 2.

Sruthi, P., Srinivasa, U. M., Rastogi, N. K., & Naidu, M. M. (2023). Principles of irradiation and its equipment. In Jafari, S.M. & Therdthai, N. (Eds.) *Non-thermal food processing operations* (pp. 205–220). Elsevier. https://doi.org/10.1016/b978-0-12-818717-3.00002-0

Sunder, M., Mumbrekar, K. D., & Mazumder, N. (2022). Gamma radiation as a modifier of starch: Physicochemical perspective. *Current Research in Food Science*, *5*, 141–149. https://doi.org/10.1016/j.crfs.2022.01.001

Tomasik, P., & Zaranyika, M. F. (1995). Nonconventional methods of modification of starch. *Advances in Carbohydrate Chemistry and Biochemistry*, *51*, 243–318.

Tran, M. Q., Nguyen, V. B., & Tran, X. A. (2022). Gamma radiation modification of cassava starch and its characterization. *Polymer Engineering & Science*, *62*(4), 1197–1204.

Verma, R., Jan, S., Rani, S., Jan, K., Swer, T. L., Prakash, K. S., Dar, M., & Bashir, K. (2018). Physicochemical and functional properties of gamma irradiated buckwheat and potato starch. *Radiation Physics and Chemistry*, *144*, 37–42.

Wani, I. A., Jabeen, M., Geelani, H., Masoodi, F. A., Saba, I., & Muzaffar, S. (2014). Effect of gamma irradiation on physicochemical properties of Indian horse chestnut (*Aesculus indica* Colebr.) starch. *Food Hydrocolloids*, *35*, 253–263.

Yang, J., Pan, M., Han, R., Yang, X., Liu, X., Yuan, S., & Wang, S. (2023). Food irradiation: An emerging processing technology to improve the quality and safety of foods. *Food Reviews International*, 40(8): 2321–2343.

Yoon, H.-S., Yoo, J.-Y., Kim, J.-H., Lee, J.-W., Byun, M.-W., Baik, B.-K., & Lim, S.-T. (2010). In vitro digestibility of gamma-irradiated corn starches. *Carbohydrate Polymers*, *81*(4), 961–963.

Yoon, K. N., Yoon, Y. S., Hong, H. J., Park, J. H., Song, B. S., Eun, J. B., & Kim, J. K. (2023). Gamma irradiation delays tomato (*Solanum lycopersicum*) ripening by inducing transcriptional changes. *Journal of the Science of Food and Agriculture*, *103*(13), 6640–6653.

Zhiguang, C., Rui, Z., Qi, Y., & Haixia, Z. (2023). The effects of gamma irradiation treatment on starch structure and properties: A review. *International Journal of Food Science & Technology*, *58*(9), 4519–4528.

Zhu, F. (2016). Impact of γ-irradiation on structure, physicochemical properties, and applications of starch. *Food Hydrocolloids*, *52*, 201–212.

9 Effect of Ultrasound on Physicochemical and Functional Properties of Starch

Honey Gupta and Sankar Chandra Deka
Tezpur University, Assam, India

Prudhvi Pasumarthi
University of Guelph, ON, Canada

Payel Dhar
Chennai Institute of Technology, Kundrathur, Tamil Nadu, India

9.1 INTRODUCTION

Starch serves multiple functions in the processed food industry, such as thickening, stabilizing, gelling, and acting as a precursor for sweeteners and polyols. Beyond food, starch is crucial in non-food sectors such as paper manufacturing, textile sizing, and fermentation for ethanol production (Kennedy & White, 1987). Native starch (NS) exists as granules, with each plant source producing starch that differs in appearance, particle size, polymer structure, and physical properties. These granules have semi-crystalline regions and are primarily composed of amylose (AM) and amylopectin (AMP). The linear (1,4)-α-D-glucose chain is referred to as AM, while AMP features a branched structure where, the sixth carbon of one of the 1,4 linked D-glucose molecules bonds with the anomeric carbon (C_1) of another glycosidic chain (Figure 9.1). Typically, AMP constitutes 70%–75% of normal starches by weight (Gupta et al., 2023). The varying structures, sizes, and AM-to-AMP ratios give each starch its unique characteristics (Apriyanto et al., 2022).

Key sources of industrial starch include corn, cassava, potato, and wheat, with smaller amounts derived from plants such as arrowroot, rice, and legumes (Gupta et al., 2022). Starch is most often used after cooking in water, which breaks down the granules and releases their polymers (Bertolini, 2009). The global starch market has grown consistently, reaching 134.5 million tonnes in 2022, and is

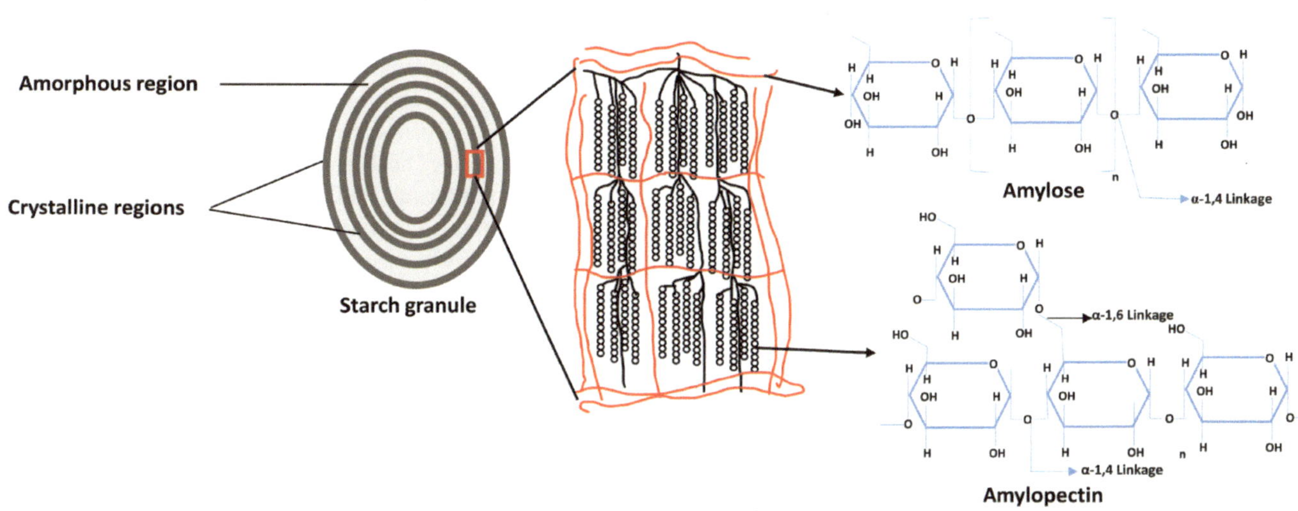

FIGURE 9.1 Pictorial representation of starch granule: Alternating amorphous, crystalline regions; amylose and amylopectin chains (conceptualized based on Kumar et al., 2023).

DOI: 10.1201/9781003564287-9

projected to grow at a compound annual rate of 5.1%, hitting 199.8 million tonnes by 2030. The food and beverage sector accounts for the majority (55%) of starch revenue, with additional demand from the feed, pharmaceutical, and food service sectors. China leads the global starch market, anticipating 5.9% annual growth, to reach 74.4 million tonnes by 2030. The U.S. starch market stood at 30.6 million tonnes in 2022, with further growth expected, along with 3.7% growth in Canada and 3.3% in Germany from 2022 to 2030 (Government of Manitoba, 2023).

9.1.1 Characteristics of Native Starches

The size, shape, and proportion of starch granules vary among starch sources. Studies on cereal grains consistently show that A- and B-type granules are more abundant than other types (Gupta et al., 2024a; Shrivastava et al., 2024). In wheat, A-type granules, which are larger than 15 μm, are disk-shaped, while spherical B- and C-type granules are 5–15 and < 5 μm, respectively (Sandeep et al., 2010). Wheat starch contains a higher proportion of B-type granules (20%) than does rye, which consists of 10%–15% B-type granules but up to 85% A-type granules. In barley starch, A-type granules dominate, at 74.7% in normal state, compared to waxy (66.4%) and null (43.9%) barley, while high-AM barley has the least (19.4%) (Ao & Jane, 2007). Potato starch granules tend to be large and oval, whereas maize starch granules are smaller and either round or polygonal (Swinkels, 1985).

All starches consist of linear AM and highly branched AMP chains, which largely influence the starches' properties. Cereal starches contain higher AM (20%–28%) than do root and tuber starches (17%–21%). AM chains tend to reorganize quickly post-gelatinization, a process known as retrogradation. Longer AM molecules retrograde at a slower rate, while smaller AM molecules associate more quickly (Seung, 2020). Ultrasound (US) modification affects starch polymerization, and the tendency of AM to polymerize varies based on the type of starch. AM molecules of root and tuber starches have relatively higher susceptibility to polymerization than those found in cereal starches (Zhu, 2015).

Pasting temperature is a threshold level beyond which the starch gains viscosity upon stirring. Cereal starches have significantly higher pasting temperatures than roots, tubers, and non-amylose starches (waxy). A higher peak viscosity indicates greater thickening ability (Jane et al., 1999). Root and tuber starches exhibit higher peak viscosities than cereal starches. The negatively charged phosphate groups that aid in granule expansion of root and tuber starches enhance their swelling power. In contrast, maize and wheat starches have lower swelling power, partly because of lipids complexing with their AM chains. The presence of lipids in cereal starches also reduces starch solubility compared to tuber and root starches (Li & Yeh, 2001). Tuber and waxy starches generally produce pastes with less stringiness and cohesiveness, while cereal starch pastes show contrasting textural properties to tubersand waxy starches (Sterling, 1978). Pastes made from root, tuber, and waxy starches are clear, translucent, and transparent, unlike those from cereal starches. Mechanical shear typically increases the flowability of starch pastes. Waxy and root starches have limited stability to shear compared to cereal starches, while potato starch pastes show moderate stability. Cereal starches retrograde faster than root and tuber starches, while waxy starches show the slowest retrogradation rate due to their higher AMP content (Ai & Jane, 2015).

9.1.2 Starch Modification Methods

Extracted or isolated starches without modification cannot be used for a variety of applications due to the limitations in their characteristic behavior. For instance, low water solubility hinders their application in liquid and semi-solid foods; low shear and thermal stability restrict intensive processing; low paste transparency challenges their application in gels and pastes; high syneresis, retrogradation tendency, and limited freeze–thaw stability make them unsuitable for application in refrigerated commodities; high gel turbidity affects the visual quality of gels; and quick digestion results in higher glycemic index that adversely affects health (Kumari & Sit, 2023; Gupta et al., 2024b).

To address these challenges, the starch characteristics are modulated using chemical, enzymatic, or physical methods. Chemical modification involves attaching different functional groups to hydroxyl (OH) groups of starch through oxidation, esterification, cross-linking, acid thinning, etherification, or acetylation (Park et al., 2018; Gupta et al., 2024c). The resultant characteristic behavior is dependent on the type of bonds attached to the starch hydroxyl group. In enzyme-assisted modifications, glycosidic bonds are debranched and the starches are hydrolyzed to alter molecular structure and contents of both AM and AMP (Shin et al., 2010).

Physical modification methods, increasingly popular due to being "green-label" (referring to very little to no use of chemicals or biological agents), are sustainable in both economic and environmental aspects. These methods involve thermal treatments such as heat-moisture, annealing, dry heat, hydrothermal, and microwave treatments, which have the potential to negatively affect minor nutrients, appearance, and texture. These challenges can be addressed by alternative non-thermal methods such as mechanical milling, cold plasma, US, irradiation, pulsed electric fields, and high-hydrostatic-pressure processing to modify starch properties based on their unique mechanisms (Ren et al., 2021).

9.2 ULTRASOUND TECHNOLOGY

US is a green and environmentally friendly technology that has gained appeal in the food processing industries because of its numerous applications, such as surface sterilization, enzyme deactivation, and extraction, preservation, and alteration of food components (Fu et al., 2020). US has a long history, dating back to hundreds of years, however the breakthrough in the field of US occurred when the Curie brothers studied the piezoelectric effect, which has been the major source for producing ultrasonic waves since the 1880s (Ravikumar, 2017). US, or ultrasonic, waves have a frequency between 20 Hz and 20 KHz. US is divided into three categories based on the instrument used: ultrasonic bath type, ultrasonic probe type, and sonicator mixing those types. The main components of bath-mode US instruments are a container, a transducer, a heater, and a timer knob; probe-type US instruments consist of temperature controller, a container, a sonotrode, and sometimes a stirrer (Fu et al., 2020). Low-energy US, characterized by frequencies below 100 KHz and power levels under 1 W/cm^2, is employed for the non-destructive characterization and monitoring of food materials. This technique can evaluate the quality of fresh produce to identify adulteration and track alterations in food properties throughout processing. Low-energy US is particularly effective for applications related to quality control and non-destructive testing. High-energy US, operating at frequencies between 20 and 500 KHz with intensities exceeding 1 W/cm^2, is predominantly utilized in the food industry for modifying food structures, disrupting cell walls, and reducing particle sizes. This approach is highly disruptive and can profoundly change the mechanical, chemical, and physical characteristics of food. High-power US is utilized in processes such as emulsification, homogenization, extraction, and microbial inactivation. The cavitation effect produced by high-power US generates intense localized heat and pressure, which enhances processes such as mixing and extraction (Arvanitoyannis et al., 2015).

9.2.1 Working Principle of Ultrasound Technology

The workings of US are based on generating sound waves with frequencies higher than what humans can hear, i.e., 20 KHz. Transducers are the main part of the device used and convert one form of energy into another. They are employed in US to transform electrical energy into acoustic waves, or sound. Two main types of transducers are used to produce US waves: piezoelectric and magnetostrictive (Bhargava et al., 2021). The materials used in piezoelectric transducers display the piezoelectric effect, which is the process by which mechanical stress applied to a material produces an electrical charge. Applying an alternating voltage causes the material to vibrate, producing US waves. Common piezoelectric materials include lead zirconate titanate (PZT), barium titanate, and quartz. Magnetostrictive transducers operate based on the magnetostrictive effect, wherein a magnetic field causes certain materials (often ferromagnetic) to change shape or dimensions. This change generates ultrasonic waves. The transducer typically consists of a magnetostrictive rod and a coil that generates a magnetic field (Arvanitoyannis et al., 2015). When ultrasonic vibrations go through a material containing food product, they induce high-energy cavitation. Microscopic bubbles are constantly forming, expanding, and collapsing in this phenomenon. Significant amounts of mechanical and chemical energy are released when these bubbles burst, and this changes the food system's chemical and physical characteristics. This process can modulate the texture, conformation, and stability of the food product, leading to modifications throughout the food matrix (Ravikumar, 2017). Figure 9.2 is a schematic representation of the working mechanism of ultrasound. When used for starch treatment, US can help in both extracting and modifying the starch's physical and biochemical properties.

9.2.2 Factors Affecting Ultrasound Treatment

Factors that affect US treatment include the medium of the food matrix, amplitude, time, temperature of the system,

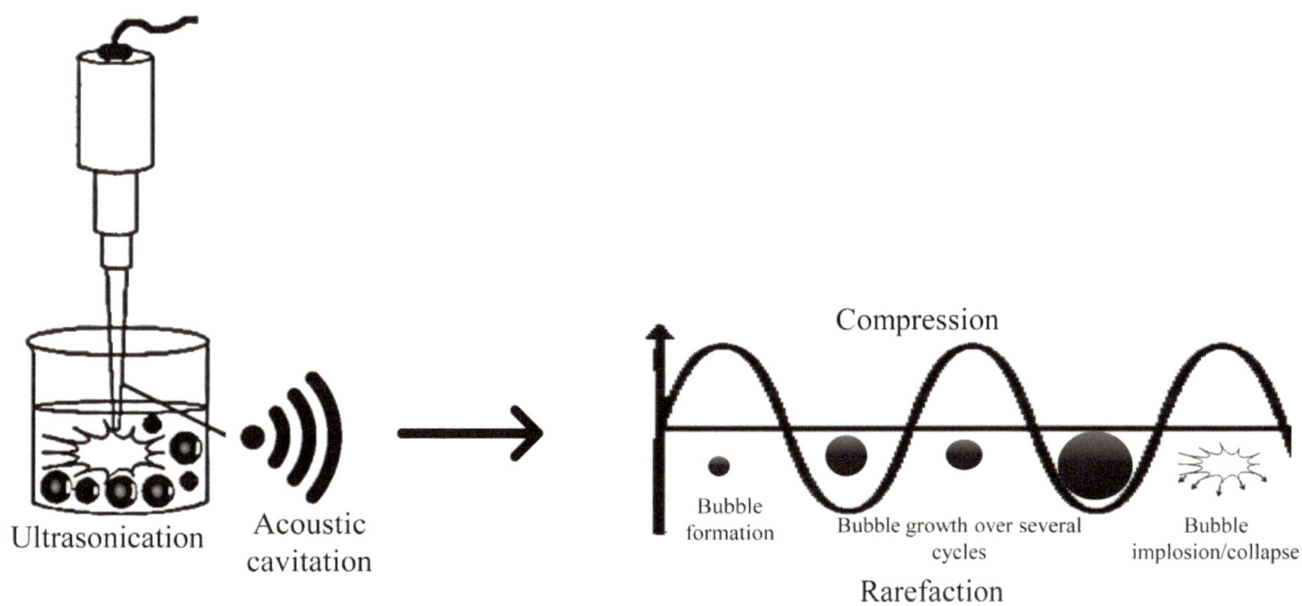

FIGURE 9.2 Schematic representation of the working mechanism of ultrasound.

concentration, and frequency of the US (Zhu, 2015). In the context of US, amplitude and frequency are the two primary parameters that significantly affect the characteristics of food materials. Lower frequencies from 20 to 100 kHz are primarily employed for food processing applications, and higher frequencies (> 100 kHz) are more suitable for analytical purposes. Lower frequencies are said to generate stronger cavitation effects, which assist in food processing, while higher frequencies are more suitable for the qualitative or quantitative analysis of food. Similarly, amplitude, the strength of the US waves, has different effects based on its degree. Higher amplitude has more pronounced cavitation effects, which easily disrupt food microstructures, enhance mass transfer, and improve extraction efficiency. It also changes texture and viscosity, reducing the hardness and chewiness of meat products with minimum impact on other quality parameters (Arvanitoyannis et al., 2015).

Temperature and time also greatly influence the properties of food material during ultrasonication (Ranjha et al., 2021). Increasing either parameter generally leads to higher extraction or to changes in properties. However, prolonged exposure beyond an optimal threshold can result in damage to the food material. To maintain the temperature longer, the use of pulse mode is advised. In pulse mode, the US runs for certain time and then goes off for a set period. This cycling helps to prevent excessive heat buildup and protect the integrity of the food material (Shen et al., 2023).

In US treatment, the medium plays an important role during wave transmission. Generally, solid and liquid media are used in US treatment. Out of these two, liquid media are the more efficient one due to their ability to effectively transmit sound waves. In liquid media, cavitation is easily achieved, which is beneficial for processes such as emulsification, extraction, and microbial inactivation. Solid media are utilized mostly in meat products for tenderization and in the extraction of bioactive components from plant tissues. The viscosity of the medium also plays an important role, as higher viscosity makes the formation of bubbles more difficult, and for better results, more power is required (Astráin-Redín et al., 2020). The solute/solvent concentration also needs to be carefully managed during extraction and modification processes. An optimal ratio ensures efficient cavitation and better starch granule disruption. During modification processes, if the solid/starch concentration is too high, viscosity will increase, which hinders cavitation, and if the proportion of solvent is too high, the energy may not be effectively transferred to the starch particles, lowering modification efficiency (Medina-Torres et al., 2017).

9.3 ULTRASOUND FOR STARCH MODIFICATION

Ultrasound is a novel and sustainable technique for modifying the functional and physicochemical properties of starch. It offers several advantages, including economic viability, reduced processing time, higher selectivity, minimal chemical usage, and superior product quality (Rahaman et al., 2021). Subjecting starch to ultrasonic waves results in severe surface damage and breakage of starch glycosidic linkages (both 1–4 and 1–6), causing a significant increase in linear amylose chains (Figure 9.3). This increase in linear chains can cause significant modulations in crystallinity, swelling power, solubility, pasting behavior, thermal stability, gel strength, retrogradation tendency, digestibility and so on, depending on the molecular weight and degree of polymerization (DP) of the starch chains. The breakdown of amylose and amylopectin branch chains can result in low DP (< 30) linear chains, which might not be assessable by analytical methods and could result in reduced amylose content. It might also increase water solubility, digestibility (especially rapidly digestible starch) with restricted swelling, pasting viscosities, retrogradation, crystallinity, thermal stability, and gel strength. However, debranching at certain branch points may produce long linear amylose chains, potentially leading to characteristics that are contradictory to those just mentioned. In this section, the effects of US on various starch sources, including cereals, roots, tubers, legumes, and unconventional sources, along with their modification outcomes are discussed.

The food and nutrition of humans consist mostly of cereal grains. These grains are primarily divided into two groups: widely consumed/cultivated (rice, wheat and corn)

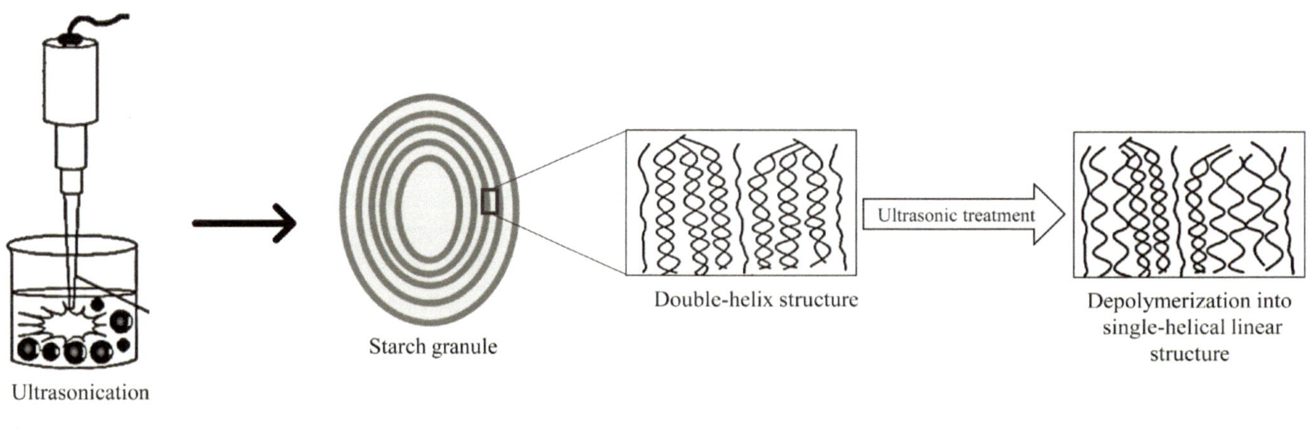

FIGURE 9.3 Effects of ultrasound treatment on starch granules.

and minor grains that are underutilized but gaining interest. The major components of these grains are starch, which makes up between 60% and 70% of their weight. But in native state, these starches often require modification to address functional limitations (Wang et al., 2021). Kaur and Gill (2019) studied US treatment on cereal starches using 100 W power at 100% amplitude for 15 and 30 minutes. The treatment increased solubility and swelling capacity while decreasing resistant starch (RS) levels. Surface morphology analysis revealed the formation of holes and depressions, particularly in wheat and maize starch, attributed to cavitation-induced bubble collapse. A 15-minute treatment resulted in higher storage (G′) and loss (G″) moduli, enhancing viscosity and elastic behavior. However, prolonged exposure (30 minutes) caused significant structural damage, reducing moduli and viscosity while shifting the starch gel toward fluid-like behavior. This was offset by a decrease in tan δ, indicating a firmer gel suitable for applications requiring solid-like consistency. For rice starch, exposure to different power levels (150, 300, 450, and 600 W) at 25°C for 20 minutes caused a reduction in relative crystallinity from 34.60% in native starch to 26.24% at 600 W. This decrease in crystallinity results from the crystalline regions being more prone to damage than the amorphous regions. The particle size was also decreased due to the disruption of larger aggregates. Rapid Visco Analyzer (RVA) results showed increased peak viscosity, reduced gelatinization enthalpy, and lower pasting temperature, reflecting structural modifications. The peak viscosity, which measures the starch's expanding capacity, rose because of the internal structural damage caused by gelatinization. However, despite the general trend of reduced viscosity due to the scission of long starch chains, which decreases molecular interactions, peak viscosity increased in rice starch treated with US. This anomaly is likely due to variations in starch varieties and types, which influence how the starch responds to US treatment (Yang et al., 2019).

Thermal–US treatment, combining heat and ultrasound, further enhances starch functionality. Golkar et al. (2021) investigated its effects on corn starch at varying temperatures (25°C, 45°C, 65°C) and durations (30 and 60 minutes). The results indicated a shear-thinning behavior in starch pastes, with reduced viscosity due to severe structural damage. Herschel–Bulkley modeling revealed lower consistency coefficients (κ′) and yield stress (τ0) values. G′ and G″ moduli also decreased, except for starch treated at 65°C for 60 minutes, which exhibited enhanced solid-like behavior due to retrogradation. These findings highlight the potential of thermal–US treatment for producing starches with diverse rheological profiles.

The market for starch is growing quickly, and to satisfy the growing demand, it is essential to have an abundant supply of this raw material, particularly for the commodity market. Root and tuber starches such as cassava, potato, and sweet potato play significant roles in food production. These starches are regionally significant, with sweet potatoes predominant in China and yams and taro vital in Africa. Despite their lesser production scale, these starch-rich crops play vital roles in their respective regions and contribute significantly to the global starch market (Vilpoux et al., 2019). Monroy et al. (2018) studied cassava starch subjected to US treatment, observing rougher granule surfaces and a bimodal size distribution. X-ray diffraction (XRD) and Fourier transform infrared spectroscopy (FTIR) confirmed reduced crystallinity and increased amorphous regions, respectively, suggesting molecular disarray. Rheological analysis showed improved paste stability under refrigeration, attributed to enhanced water interactions and gel stability. The granules' rougher texture and larger surface area are probably what caused this improvement. It can enhance interactions with water and other components in food matrices, leading to better gel formation and stability. Wang et al. (2020) investigated how a combination of ultrasonic and freeze–thaw processes affected the properties of potato starch, focusing on enhancing its lipophilicity and functional properties. The findings revealed a substantial increase in oil absorption capacity (OAC), which elevated from 59.62% for NS to 80.2% following seven cycles of US–freeze–thaw treatment. This significant enhancement in OAC can be attributed to the structural alterations induced by the treatments, which caused the starch granules' specific surface area to rise and micropores to develop. The combined US and freeze–thaw treatments significantly increased the resistant starch content of modified potato starch. The structural modifications induced by these treatments, including the formation of micropores and enhanced porosity, contributed to a reduction in digestibility. Furthermore, the study indicated that while the crystalline structure and chemical groups of the modified starch stayed the same, relative crystallinity, gelatinization enthalpy, and pasting viscosity significantly decreased (Gupta et al., 2024d). These changes suggest a transformation in the starch's physicochemical characteristics, enhancing its functionality.

Pulses, the edible seeds of legume crops, are vital for both food and feed, gaining recognition for their nutritional value and role in food security. As part of the Food and Agriculture Organization's 2030 agenda for sustainable development, legumes have seen significant growth in cultivation and production. Between 2012 and 2021, the area of legumes harvested worldwide grew from 229.5 to 269.4 million hectares, while production rose from 421.5 million to 578.8 million tons (Rostamabadi et al., 2024). Legume starches, derived from pulses such as lentils, chickpeas, and mung beans, are valued for their nutritional benefits. Ashogbon et al. (2020) note that legume starches typically contain lipid (0.06%–0.87%), ash (0.01%–0.65%), AM (18.42%–51.69%), and nitrogen (0.00%–0.20%). High AM content improves nutritional quality but can lead to poor stability and increased syneresis, necessitating modification (Dhull et al., 2024). Ultrasonication significantly impacts legume starch solubility. For instance, chickpea starch showed reduced solubility but increased swelling power after US treatment due to cavitation effects. The disruption of molecular bonds during ultrasonication leads to a rearrangement of AM and AMP, which in turn modifies the swelling properties of the starch granules (Bitik et al., 2019). Shen et al. (2021) observed surface corrosion on mung bean starch granules with reduced crystallinity and enhanced gelatinization temperatures, indicating improved structural stability. US treatment also increased pasting viscosity and gel formation, whereas plasma treatment yielded contrasting effects. Additionally, US combined with bamboo leaf flavonoids (BLFs) altered the rheological properties of pea starch, enhancing viscoelasticity and solid-like behavior (Hu et al., 2023).

Unconventional starch sources, such as unripe mango, banana, jackfruit, and rhizomes like ginger and turmeric, offer unique properties for industrial applications. These sources are increasingly valued for their low glycemic index and functional benefits. Many of these unconventional sources, especially bio-wastes from industrial processing, are rich in starch and bioactive compounds such as dietary fibers and phenolics. Valorizing these bio-wastes may result in novel applications within food and non-food sectors as research increasingly emphasizes the extraction of starch from these materials (Makroo et al., 2021). US-assisted extraction from unconventional sources improves yield and functional properties. For instance, mango kernel starch yield increased to 54% compared to 42% using conventional methods. Mieles-Gómez et al. (2023) observed elevated AM content, enhanced water and oil holding capacities, and increased solubility and swelling power, along with antioxidant properties. High-power US treatment of kiwi starch produced fine particles with a porous structure due to cavitation bubble collapse, which creates high temperatures and pressures. This increases surface area, improving interactions with water and enhancing functional properties. The optimal condition for maximizing resistant starch was 600 W for 30 minutes, but treatment conditions can be adjusted based on application needs (Wang et al., 2022).

Hence, we can say that ultrasound is a versatile and sustainable method for starch modification across various sources. It induces structural and functional changes that enhance the applicability of starch in food and non-food industries. With the tailoring of treatment parameters such as power, amplitude, and duration, US can produce starches with desired characteristics, making it a valuable tool for addressing the limitations of native starches.

9.4 DUAL MODIFICATION BY ULTRASOUND

The hydroxyl groups and simple glycosidic connections in starch enable modification by physical, chemical, enzymatic, or biotechnological means, alone or in combination (Ashogbon, 2018a). Changes are usually made to add certain functional characteristics, improving things such as adhesion, texture, gel clarity, gloss, and film formation and decreasing things such as syneresis, retrogradation, and gelling tendencies. These changes can also alter cooking qualities and enhance the stability of freeze–thaw cycles (BeMiller, 2015 Mehfooz et al., 2019). Starches can acquire favorable properties through modification, including enhanced mechanical strength, resistance to retrogradation, hydrophobicity, thermal stability, and amphiphilicity (BeMiller, 2015; Winkler et al., 2014). Due to the intrinsic limitations of NS, which include their inability to dissolve in water, poor dissociation at high temperatures, and susceptibility to syneresis and retrogradation, multiple modification procedures are justified (Ashogbon & Akintayo, 2014; Ashogbon, 2018b; Gupta et al., 2024e). Furthermore, Sudheesh et al. (2019) found that NS pastes and gels tended to become unstable when subjected to high temperatures, stress, or pH fluctuations during processing. Consequently, it becomes necessary to modify these NSs.

Certain food and industrial applications have found single-modification approaches for starches to be insufficient in certain circumstances. For instance, the process of succinylation offers a number of advantageous qualities, including better freeze–thaw stability, higher viscosity, improved thickening power, enhanced paste clarity, delayed retrogradation, and enhanced solubility in cold water (Moin et al., 2016). To overcome these restrictions, an extra modification is required because the succinylation process is unstable when subjected to high temperatures (Ačkar et al., 2015). Dual-modification techniques have been created with the goal of optimizing the functionality of starches that have undergone a solo modification, to further boost their performance and usability. The functional applications of starch are largely determined by its physical, chemical, and functional characteristics. In the food sector, starch utilization is influenced by factors such as starches' swelling capacity, water solubility index, pasting behavior, thermal stability, and in vitro digestibility (Sudheesh et al., 2020). Dual modification of starch has a higher level of complexity due to the interplay of various single-modification techniques which result in alterations to the AM and AMP structures. When preparing dual-modified starches, several factors must be carefully considered, including the sequence of modifications, the choice of modifying agents, the medium, and reaction conditions, as well as the starch's botanical source. It is essential to recognize the distinct characteristics of each type of starch, as their properties can vary significantly even under identical treatment conditions. Variations exist within the more general categories of starch sources, which include cereals, legumes, roots and tubers, and green fruits, even though it may be possible to generalize certain qualities within these groupings. Accurately predicting the functional and physicochemical features of dual modification of starches across multiple plant origins is difficult, if not impossible, due to these inherent variations. When starch is modified twice, the first modification seems to be done mostly to prepare for the second modification. This encompasses the granular surfaces as well as additional starch molecular components, such as the strength of the links inside the starch chains. The first modification is essential to improving the efficacy of this secondary treatment, and the secondary modification has a substantial effect on the final qualities of the dual-modified starch.

Modifying starch using US is a favorable approach for improving the performance of the drug delivery systems using starch-based material. US-assisted techniques have gained recognition as effective methods for starch modification due to their capacity to induce significant physical and chemical changes in starch molecules.

9.4.1 Ultrasound-Aided Physical Methods

Physical modifications of starch are changes made to starch properties without altering the free hydroxyl groups located at the C-2, C-3, and C-6 positions on the anhydroglucose units (AGUs) that make up starch molecules. These hydroxyl groups remain intact, and no organic substances or external chemical compounds, such as enzymes, are initiated into the starch granules. The modification process relies purely on physical processes, including sound waves, pulse electric, cold plasma, heat, vapor or hydraulic pressure, and steam, though it may occasionally result in the breaking of glycosidic bonds, leading to depolymerization and a subsequent decrease in the molecular weight of starch molecules.

Ultrasonic modification is an innovative and promising technology. This method has gathered significant interest due to its safe nature, environmentally friendly attributes, and high efficiency. Ultrasonic sinusoidal waves, with wavelengths ranging from 15 km to 30 cm, are employed for energy generation that agitates particles dispersed in organic solution. As these sound waves propagate, they cause pressure fluctuations that lead to the formation of cavitation bubbles. The collapse of these bubbles generates high-pressure and high-temperature conditions (Han et al., 2021). Ultrasonication has found extensive use in synthetic chemistry for minimizing waste, designing environmentally friendly compounds and methods, and optimizing the utilization of components, particularly renewable resources derived from organic matters (Kardos & Luche, 2001). Its application to starch aligns with the principles of green chemistry and green technology. Typically, NS undergoes physical, chemical, or enzymatic modifications to develop a variety of different applications. Physical modification significantly enhances the effectiveness and reactivity of NS, especially ultrasonication within dual-modification systems.

Physically modified starches have gained increased interest from both researchers and consumers due to their eco-friendly and health-conscious attributes when compared to chemically and enzymatically modified starches. High-AM corn starch (consisting of 70% AM) has been treated through melting with glycerol as the plasticizer, facilitated by ultrasonication at a frequency of 20 kHz. Findings indicated that ultrasonication significantly enhanced the efficiency of the melting process by reducing the thickness and improving the solubility index of the treated high-AM corn starch (Lima & Andrade, 2010). Ultrasonication treatment generally contributes to a better solubility index of starch on physicochemical properties. Future investigations could explore the effects of ultrasonic modification on the physical treatments of starch, such as extrusion technology, electric field, and hydrothermal treatments. These experiments would necessitate the development of advanced instrumental configurations.

The polymeric chains close to the microbubbles quickly collapse due to ultrasonic wavelengths leading to the rupture of intramolecular bonds (C–C, –OH, etc.) and the development of new bonds with other free molecules (Rahaman et al., 2021). The starches from tubers such as cassava were modified utilizing ultrasonication at 40 kHz. The findings indicated that modified starches experienced a reduction in crystallinity, and cracks appeared on the starch surfaces, though the starch granule size remained unchanged. The starch surface becomes porous and cracked as a result of the US treatment, making the granules more prone to penetration and making the crystalline regions easier to break up. In other research, natural starches in both potatoes and maize were altered using US. The treated starches' relative crystallinity decreased. The cavitation effect caused by US, which breaks internal links and jeopardizes the structural integrity of the starch granules, is most likely reason for this result (Rahaman et al., 2021). In research by Ouyang et al. (2021), ultrasonication produced varying impacts on starch. The findings indicated that ultrasonicated B-type starch developed more cracks and pores and rougher starch surfaces than did A- and C-type starches. A-type starch exhibited increased resistant starch content, a greater tendency for retrogradation, and enhanced thermal stability. In contrast, B- and C-type starches exhibited reductions in the same properties but a rise in total AM content. The investigation showed that the nature of starch (amorphous and crystallinity) plays a critical role in the physicochemical and microstructural changes induced by ultrasonication. The structure of A-type starch became more organized, creating double helices that promote the development of resistant starch. This modification could be optimized for utilization in bakery products. Ultrasonication predominantly affected the amorphous regions of starch, resulting in fragmentation of AM molecules into smaller ones. This will form additional surface cracks. These modifications are suitable for producing clear, smooth gels and developing films. According to Raza et al. (2021), modification of tuber crop starch demonstrated increased solubility index, water and oil holding capacities, swelling power, and gelatinization temperature, along with surface cracks. Additionally, ultrasonicated taro starch showed enhanced clarity, pasting properties, and stability for freeze–thaw products, indicating its possible application in the development of frozen products. According to Sit et al. (2014), starch subjected to ultrasonication treatment exhibited enhanced properties of improved functionality, higher AM content, better paste clarity, and more surface cracks, whereas crystallinity and gel hardness decreased. Modified oat starch has been used in several food industries, such as coatings, dairy products, and baking, because of its enhanced porosity, decreased crystallinity, and paste clarity. These suggest that starch's functional qualities are improved by ultrasonication but that its crystallinity is decreased. The intensity and power of the ultrasonication, as well as the origin of starch, are highly connected with the microstructural variations in starch. Ultrasonication can also be utilized as a pretreatment before physical modifications to increase surface fissures, cracks, and pores, allowing for increasing the efficiency of the NS. This kind of alteration is used in frozen goods to maintain the emulsions in mayonnaise and salad dressings. Additionally, it serves as a carrier in drug delivery systems and as a coating material in paper manufacturing to enhance printability.

In research by Majeed et al. (2017), lentil starch was modified using physical methods, starting with low-power US followed by gamma irradiation. The findings indicated that the pasting properties of lentil starch remained unaffected by ultrasonication alone but decreased significantly when subjected to dual modification. The total AM content of NS also exhibited a reduction after ultrasonication and further declined with dual modification. Both ultrasonication and combined modification led to a decrease in the *L* value, while the *a* and *b* values increased. Syneresis of any starchy product exhibited a slight decrease after ultrasonication, but a substantial reduction was observed following 120 hours of storage after dual modification. Ultrasonication did not significantly impact the functional properties of lentil starch, whereas dual treatments caused a notable reduction in these properties. FTIR analysis revealed that ultrasonication and dual modification led to decreased intensities in the OH, CH, and OC stretching vibrations, as well as CH_2 bending. The physical, chemical, and functional characteristics of legume starch were not appreciably changed by the sonication procedure on its own. Dual alteration, however, caused these traits to vary significantly. After radiation, there was an increase in transmittance, solubility, water and oil absorption capacities, and solubility. Both ultrasonicated and gamma-irradiated starch showed a decrease in swelling and syneresis, the latter possibly improving texture in cooking. In order to preserve quality qualities during storage, syneresis must decrease and transmittance must rise. The combined impacts of irradiation (5 kGy) and ultrasonication significantly lowered pasting qualities, which may be advantageous for applications needing lower viscosity. Thus, changing the physical characteristics of legume starches using different physical techniques could give them new qualities and increase their range of uses.

Majeed et al. (2018) investigated the modification of pea seed starch using physical methods, beginning with low-power ultrasonication followed by gamma irradiation. Both ultrasonication and the combined treatments led to a reduction in AM content. Syneresis remained relatively unchanged after ultrasonication but decreased significantly during storage following dual modification. The combination of ultrasonication and irradiation substantially altered the functionality and pasting properties of the starches. The physicochemical, functional, and pasting properties of starches treated with ultrasonication alone did not differ significantly from those of NSs. Dual treatments, however, significantly altered these characteristics. Following the radiation treatment, some functional qualities, including water holding capacity, oil holding capacity, solubility index, and transmittance, increased, while other properties, including swelling properties, pasting properties, and syneresis, decreased. The reduction in the swelling properties may be advantageous for improving texture during cooking. The alterations in syneresis and transmittance due to dual modification have a positive impact on food preparations, as reduced syneresis and increased transmittance during storage are crucial for maintaining the quality attributes of starch. The substantial decrease in pasting properties resulting from the combined effects of ultrasonication and irradiation treatment could be useful in applications where lower viscosity is desired. Therefore, modifying the physical, chemical, and functional properties of pulse starches through various physical methods could impart novel characteristics, broadening their potential applications.

The effects of heat-moisture treatment (HMT) and US treatment on the starch of mung beans (MBs) were studied by Dhull et al. (2024). The results showed that after HMT and the combination of both modifications, the AM content of MB starch dropped, whereas it increased after US treatment. The water absorption capacity increased as a result of the alterations, with the US-modified starch showing the largest increase. In contrast, oil absorption capacity and light transmittance were generally reduced by the adjustments, with the exception of HMT and combined HMT-US treatment. The modified MB starch had less swelling power and solubility than NS. In comparison to the modified starches, NS showed the highest peak viscosity, based on the pasting properties. The modified and native starches did not differ significantly in FTIR spectra or X-ray diffraction patterns. Nevertheless, morphological examination of the changed starch granules showed significant alterations, such as the existence of surface cracks and holes. These results imply that physical treatments such as HMT and US treatment have a substantial effect on the varied characteristics of MB starch, which may increase its suitability for use in a variety of food formulations. MB starch's physical alterations can dramatically change its functional qualities (e.g., solubility and swelling capacity), thermal qualities (e.g., pasting, gelatinization, and thermal stability), and structural qualities (e.g., crystallinity and surface features). Granule size increased noticeably as a result of the adjustments, and morphological changes such cavitation, aggregation, grooves, notches, and cracks were seen. After US treatment, the amount of AM was increased; however, when HMT and US were combined, the amount of AM decreased. The native MB starch's pasting capabilities showed a higher peak viscosity than the modified starches. According to the results, HMT and US, either separately or together, have a great deal of potential to improve the uses of MB starch in a variety of food products. Furthermore, investigating cutting-edge methods such as ultrasound may result in the creation of starches that are more environmentally friendly. The findings thus have important ramifications for the use of modified starches in the pharmaceutical and food industries. In general, the degree, duration, and source of the starch all affect how ultrasonic treatments work. According to earlier research, starch granules treated with US may disintegrate, have their molecular size reduced, or have surface flaws (Sujka, 2017). Ozonated starch is first produced by ozonating native corn starch, which is subsequently treated with US waves to create ozonated–ultrasonicated (O_3-US) starch or ultrasonicated–ozonated (US-O_3) starch. In contrast, US-O_3 starch is produced by ozonating the starch after it has been ultrasonically sonicated (Castanha et al., 2019). Ozone treatment and dual modifications (O_3-US or US-O_3) induced significant changes in the physicochemical and functional properties of the starches, including increased apparent AM content, the presence of carbonyl and carboxyl groups, and decreased pH and molecular size distribution of starch. US treatment alone had no appreciable effect on the physical properties of maize starch. The two modification treatments had no effect on these alterations in the starch's molecular size distribution, structure, or crystallinity. There is a discernible difference in the paste clarity between ultrasonic–ozonated starch and the former; the latter shows more clarity.

HMT and US treatment (UT) were used to modify the starch of cowpeas (*Vigna unguiculata*) (Acevedo et al., 2022). The XRD patterns and granule shapes were unaffected by the treatments. On the other hand, ^{1}H NMR (Proton Nuclear Magnetic Resonance)analysis showed that AMP's branching degree decreased after the individual treatments of both HMT and UT. The treated starch showed a greater improvement in thermal stability, but the treatments had no effect on the gelatinization enthalpy. The combined alterations led to a rise in resistant starch content of up to 11%, while HMT-UT samples saw an increase in slowly digested starch (SDS) content of up to 30%. With longer heating times, a progressive decrease in pasting viscosity was seen. These results imply that two different modification processes improve cowpea starch's functional qualities and increase its potential use as a food additive. The physicochemical characteristics of cowpea starch underwent considerable changes while receiving dual-modification (HMT-UT) treatments. UT and short-term (2-hour) HMT broke α-1,6 bonds in AMP, reducing its branching degree, according to NMR spectroscopy; α-1,4 bonds appeared to be affected by longer-duration HMT (5 and 12 hours). Cowpea starch's relative crystallinity was enhanced by the treatments, but the X-ray diffraction pattern and granule shape remained unchanged. Every procedure used improved the starch pasting profile's heat stability and decreased its viscosity. By varying the treatment time, HMT was shown to be the most successful in modifying the pasting qualities of cowpea starch. While HMT-UT increased the amount of SDS, HMT increased the fraction of RS. The results show that cowpea starch can be modified using HMT and dual HMT-UT modification, which will increase the number of applications for the starch in different food products.

Because modified starches are more functional, they have gained a lot of attention. Bitik et al. (2019) reported that ultrasonication or microfluidization techniques modified the starches fromlegumes and chickpeas. Starches showed greater swelling power after the treatments; however, either treatments considerably reduced their solubility. As the shear rate rose, the apparent viscosity decreased in both samples. Furthermore, the starches' gelatinization temperatures were decreased by the treatments. When compared to NS, both methods significantly decreased the mean diameter, particle-size distribution, and span values. Images from scanning electron microscopy showed significant morphological changes in starch. The changed starches showed longer relaxation periods, which suggested changes in water mobility, according to time-domain NMR studies. Through the use of Fourier transform infrared spectroscopy, structural alterations were further verified. Modifying the starches of chickpeas and lentils was found to be possible through ultrasonication or microfluidizationmethods. Both techniques individually y lowered the apparent viscosity of native chickpea starch and the size of NS particles. The gelatinization enthalpy of both legume starches was notably decreased by microfluidization. According to NMR data, changed starches had longer relaxation times, which suggested that their free water content had increased. The structural alterations in the NS caused by both treatments were validated by FTIR data, which were in line with SEM observations. The study found that depending on the targeted qualities, ultrasonication and microfluidization are both useful techniques for modifying starch. The synthesis of readily gelatinized starches was found to be more successfully achieved using microfluidization, although the practical applications of ultrasonication were enhanced by its higher isolation efficiency and reduced production costs.

Hydrothermal treatment (Yang et al., 2021) and autoclaving (Ratnaningsih et al., 2020) are physical modification techniques used to improve the functionality of starch, such as gelatinization, light transmittance, pasting properties, freeze–thaw stability, and firmness. But these techniques can also result in a decrease in paste viscosity and swelling power, especially when annealing is involved (Wang et al., 2014). These methods of modification have an effect on the morphological and microstructural aspects of starch in addition to its functional capabilities. For example, after undergoing treatments such as microwave (Kumar et al., 2020), US (Sit et al., 2014), X-rays (De Kerf et al., 2001), ball milling (Kumar et al., 2017), and γ-irradiation (Lima et al., 2021), the crystallinity of starch obtained from different sources diminishes. Furthermore, the starch's porous surfaces and crack count tend to increase as a result of these treatments (Lima et al., 2021). These changes have the potential to upset molecular balance, change AMP into AM, weaken linkages, and reduce relative crystallinity. The origin of the starch being treated, together with variables such as power, intensity, duration, and pressure, all affect the results of non-thermal physical modification operations (Ashogbon & Akintayo, 2014).

9.4.2 Ultrasound-Aided Chemical Methods

The process of chemically modifying starch entails adding functional groups to the granules' natural polymer structure, which significantly changes starch's physicochemical characteristics. Important characteristics such as pasting behavior, proximate composition, gelatinization, and starch retrogradation are all impacted by this alteration. It is possible to stabilize intra- and intermolecular linkages at different points within the starch granules by chemically altering starch (Zia-ud-Din & Fei, 2017). The method usually targets the hydroxyl groups found in starches, making it possible to modify them specifically for use in culinary or industrial purposes. Chemical modifications are especially common in the food industry, where the functional and chemical properties of the modified starches are determined by a variety of factors, including the type and distribution of the substituting agent, the degree of substitution, reaction conditions, and the source of the starch (Haroon et al., 2016).

Derivatization procedures including acetylation, cationization, acid hydrolysis, oxidation, and cross-linking are examples of common chemical modification techniques. These techniques do have some drawbacks, though, as issues with consumer safety and environmental impact have been brought up. Combining several chemical modification techniques to create novel and inventive alterations is becoming increasingly popular as a way to address these issues. Furthermore, in order to produce starches with extremely particular functional qualities, combining chemical alteration with physical processes including radiation, extrusion, and microwave processing has been investigated (Moad, 2011). By combining strategies, starches may become more adaptable and useful across a number of sectors, increasing increased suitability for a variety of specialized applications.

In a number of starch modification processes, including oxidation, cationization, cross-linking, and esterification, US has been shown to shorten reaction durations and increase the degree of substitution. Starches from plants such as yams, potatoes, cassava, and maize have been shown to have these benefits (Ashogbon & Akintayo, 2014). For instance, the time needed for the oxidation or cationization of potato starch was drastically shortened by using ultrasonication, from up to a day to only a few minutes. Additionally, this method made the process more environmentally friendly by doing away with the requirement for *p*-toluenesulfonic acid (Hong et al., 2016). Furthermore, the degree of substitution rose to 38.7% when ultrasonic treatment was used before alkalization in the carboxymethylation of maize starch (Shi & Hu, 2013). Similarly, simultaneous ultrasonic treatment during the oxidation of corn starch sped up the reaction and improved efficiency, causing higher concentrations of carbonyl (C=O) and carboxyl (COOH) groups in the modified starch (Chong et al., 2013).

Exploring the impact of ultrasonication treatment on chemical modifications, such as phosphorylation, could be highly beneficial. As previously mentioned in the context of physicochemical properties and physical structure, ultrasonic treatment can increase starch's swelling power and solubility while creating pores, cracks, and holes in the granules. This porous structure facilitates the diffusion of chemical reagents and catalysts, thereby enhancing the likelihood of chemical reactions (Zhang et al., 2012). Additionally, free groups, such as hydroxyl and hydrogen, generated from solvent (water) through cavitation and sonolysis, may further enhance such chemical reactions as oxidation (Chong et al., 2013). Further research is needed to understand the role of these free radicals in dual modification (ultrasonication and chemical modification). Optimizing ultrasonication conditions is also crucial, as the sequence of ultrasonication in carboxymethylation can significantly influence the degree of substitution (Shi & Hu, 2013).

Acid hydrolysis effectively reduced starch particle size but resulted in irregularly shaped modified starch. When combined with US treatment, acid hydrolysis facilitated the formation of uniformly shaped starch nanoparticles. The synergistic effect of US and acid hydrolysis significantly influenced the production of starch nanoparticles with well-defined sizes. According to Zhou et al. (2021), acid hydrolysis using hydrochloric acid (HCl) decreased starch particle diameter to the micrometer range. However, when acid hydrolysis was followed by US-assisted modification, the starch particle diameter was further reduced to the nanometer range, yielding nanoparticles with spherical and ellipsoidal shapes. The dual-modification process also led to reduced total enthalpic energy and crystallinity in the starch nanoparticles. This combination proves beneficial for achieving higher yields of starch nanoparticles, which have applications as fat replacers and drug carriers in various industries. Cassava starch was modified using two different sequences: US followed by acetylation, and acetylation followed by US. The results indicated that both the sequence and type of modification significantly affected the properties of the dual-modified starch. When US was applied prior to acetylation, the starch exhibited increased solubility, enhanced pasting properties, and a higher content of acetyl groups, while showing a reduction in swelling power. US cavitation breaks both intra- and intermolecular connections in starch, allowing molecules to create new bonds with the altered starch structure. This is most likely the origin of the observed property changes (Khurshida et al., 2021). Alterations in the order of changes had no discernible impact on the starch's digestibility or granular structure. In contrast to single alterations, dual modifications resulted in more surface holes and cracks as well as increased SDS and RS content. Potential uses for these modified starches include textiles, paper, soups and other food products, and pharmaceuticals.

Zimmikand starch underwent two separate modifications: ultrasonication and citric acid. The outcomes showed that although the glass transition temperature was lowered by citric acid esterification alone, the dual modification of starch had a significantly lower gelatinization temperature than starch modification using a single technique. Whereas single alterations produced less noticeable surface fissures and cracks, the dual-modification method caused the breakup of starch clumps. Higher quantities of citric acid also resulted in a rise in overall crystallinity, most likely as a result of the cross-linking reagents engaging with accessible molecular sites and strengthening the more crystalline arrangement. Applications for the twofold modification of starch include thickening agents, extruded items, and bread and bakery products (Tappiban et al., 2022). The combined effects of esterification and the subsequent ultrasonic procedure on tiger nut starch nanoparticles and their possible usage as fillers in chitosan-based films were examined by Yan et al. (2021). According to the study, esterification followed by US treatment decreased the starch's A-type crystallinity and nanoparticle size while increasing the suspension's diffusion and consistency because of the ultrasonic cavitation. With the addition of these starch nanoparticles, the chitosan-based films' material strength increased, probably as a result of their smaller size. These modified starch nanoparticles can be used in the creation of packaging films, biodegradable films, paper, and nanotechnology.

Chemical modification of starch involves altering its structure through processes such as cross-linking, etherification, oxidation, and esterification. These methods are widely utilized in both industrial and laboratory settings, largely due to the availability and low cost of chemicals. However, there is an increasing focus on using organic reagents, such as acetic anhydride, succinic anhydride, and propylene oxide, to ensure consumer safety. Chemical modifications lead to significant changes in the molecular and techno-functional properties of starch, including increased AM content and higher levels of resistant starch (Lee & Chang, 2019). These modifications can also enhance relative crystallinity (Zhang et al., 2019), improve emulsion stability (Zhang et al., 2021), and result in reduced swelling and solubility (Colussi et al., 2015; Sandhu et al., 2021) as well as in changes in pasting properties (Mei et al., 2015).

The process of chemical modification introduces new functional groups into the starch, which disrupts its granular structure (Zhou et al., 2016). Techniques such as acetylation, esterification, and cross-linking improve properties such as gelation, retrogradation tendency, hydrophobicity, and the ability to develop packaging films. This approach offers numerous advantages, making it suitable for producing a wide range of goods, comprising starch-based gels, starch nanocrystals, films (Sharma et al., 2021; Zhang et al., 2019), and Pickering emulsions (Zhang et al., 2021), as well as ready-to-eat foods, frozen goods, and canned products. However, it is important to consider the environmental impact of chemical modification, as it generates significant amounts of wastewater and the use of concentrated chemicals poses potential health risks due to their toxicity or allergenic properties.

9.4.3 Ultrasound-Aided Enzymatic Methods

To investigate the impact of dual modification using ultrasonication and enzymatic treatment, pea starch was simultaneously modified with pullulanase and subjected to various US amplitudes on a pre-cooked substrate (Lu et al., 2018). Ultrasonication has been frequently used to improve the enzymatic susceptibility of starches from a range of plant-based sources, including maize, mung beans, rice, and wheat (Rahman & Lamsal, 2021). For example, glucoamylase catalysis significantly raised the productivity of ultrasonicated microporous maize starch (Wu et al., 2011). Hu et al. (2013) also found that the capacity of dual-frequency ultrasonication—25 kHz and 40 kHz—to increase the α-AM concentration in mung bean starch was greater than that of single-frequency ultrasonication—25 kHz or 40 kHz. However, in certain instances, it has been shown that ultrasonication either reduces the stability and activity of specific enzymes (Ozbek & Ulgen, 2000) or has minimal impact on the activity of β-amylase in sweet potatoes and the hydrolysis of specific starches by α-amylases from different *Bacillus* strains (Apar et al., 2006). The enhanced enzyme penetration into starch granules, which is made possible by the formation of pores and structural damage brought on by ultrasonication, is probably the reason for the enhanced efficacy of enzymatic starch modification through ultrasonication. The interaction between the enzyme and the substrate is improved by this damage, which increases the interfacial area between them. However, it is important to carefully optimize the experimental conditions to maximize the benefits of ultrasonication in enzymatic starch modifications while avoiding excessive damage to the starch granules when necessary.

The physical and functional characteristics of resistant starch made by ultrasonication–enzymatic hydrolysis (UE) and native corn starch were investigated by Wang et al. (2023). Along with a decrease in swelling power, the use of ultrasonication and enzyme treatments increased the amount of resistant starch, AM, and solubility. At 41.71% RS, this twofold alteration produced the least digestibility. Particle-size distribution analyses and scanning electron microscopy analyses showed that the RS samples had more irregularly shaped, rough surfaces and larger granule sizes. Analysis using FTIR and XRD revealed that no new functional groups were created throughout the alteration procedure. All of the RS samples' crystal structures, however, changed from A-type to B+V-type, and the RS-UE's relative crystallinity and short-range order both rose. In particular, RS-UE showed the highest relative crystallinity at 45.54% and the highest molecular order at R1047/1022, with a value of 0.8769. These results imply that the dual-modification method, which includes enzyme treatment and

US, is useful for increasing RS content and has potential for use in the production of low-glycemic-index and resistant starch meals. This approach has the potential to improve the nutritional quality of starch-based products based on the changes it induces in their structural and physicochemical qualities.

Enzymatic modification is a pricey yet environmentally friendly process that produces novel starches with the potential to improve porosity and raise RS content. In order to encapsulate bioactive ingredients in food products, the bread industry frequently uses this healthier starch. Li et al. (2022) assert that enzyme modification is an effective means of producing starch with a greater resistant starch content, thermal stability, and gelling ability (Li et al., 2017).

Starch is ideal for a wide range of industrial applications because it can be changed using two different processes to increase certain qualities. These changes, for instance, make it possible to create intelligent and sensitive materials, such as self-assembling or self-healing materials, in the field of 3D printing. Dual alterations have significant potential benefits, including increased resilience, efficiency, and environmental effect. A deeper understanding of the microstructure of starch may offer vital insights into the ways in which modifications at the molecular level impact its characteristics, opening the door to more targeted and effective modification techniques. In the future, depending on end users' requirements, threefold modification schemes might be created to target even more specific qualities. These modifications, whether dual, threefold, or fourfold, could be achieved by adjusting various parameters, such as the chemical concentration, the duration and intensity of power, or the enzyme concentration and incubation time. Additionally, compounds derived from plants might be used to modify NSs, potentially increasing the acceptance of modified starches among consumers. These advanced techniques promise to further improve the functional properties and thermal stability of starches with dual modification, producing products with precisely regulated attributes to satisfy various industry demands.

9.5 INDUSTRIAL IMPORTANCE

Starch, a highly versatile biomaterial, finds extensive applications across industries such as foods, pharmaceuticals, and textiles. In recent years, the adoption of ultrasound treatment for starch modification has grown significantly. This technique is gaining popularity due to its ability to alter starch properties without the use of chemicals, making it a safer and more sustainable option for applications in food products and other industrial sectors. This section explores the diverse industrial uses of ultrasound-treated starch and highlights its functional advantages.

9.5.1 Food Industry

Ultrasound-treated starch serves a variety of purposes in the food industry, including as a texture modifier, stabilizer, emulsifier, and agent for encapsulation and delivery (Park & Kim, 2020). Additionally, ultrasound treatment is employed to lower the antinutrient content in grains, enhancing their nutritional value and suitability for consumption (Yadav et al., 2021). A study by Hu et al. (2023) on modified pea starch found the most notable changes in rheological properties, mainly in viscosity and elasticity of the starch gels. US treatment increased viscosity, enhancing the use of the starch in products such as soups, gravies, and sauces. The elasticity of the starch gel was improved with UT, making it suitable to use in gluten-free products, such as noodles and bread. The combination of US and plasma treatment significantly impacts the molecular weight and gelatinization properties of starch. Incorporating the modified starch into products such as noodles enhances texture and cooking quality due to improved gelatinization and stability (Shen et al., 2021). When *Lactobacillus acidophilus* was encapsulated with a 10% concentration of porous starch, the maximum survival rate of the bacteria in gastrointestinal fluid was observed. This was the result of an evaluation of the encapsulation efficiency using US-treated porous starch (Xing et al., 2014).

9.5.2 Packaging Industry

Modified starch has applications that extend beyond the food industry, particularly in packaging and 3D printing, owing to its improved functional properties and the enhanced strength of packaging materials (Adewale et al., 2022; Gupta et al., 2025a). Ultrasound treatment further boosts the film-forming capabilities of starch. Gelatinized starch subjected to ultrasound treatment exhibits superior transparency, structural integrity, and moisture resistance, making it highly suitable for film production (Yudhistira et al., 2024). US extraction of starch from sago pith waste produced a bioplastic film with a tensile strength of 0.9 MPa and water vapor permeability of 1.11×10^{-8} g m^{-1} s^{-1} Pa^{-1}. Compared to control starch, the US-treated starch showed higher swelling power (14.5 g gel/g) and solubility (29.3%). Additionally, the films were darker, with significant color differences (ΔE of 15.6) from control films (Tan et al., 2021).

Ultrasound treatment is also employed to produce starch nanocrystals (SNCs) that exhibit superior functionalities and properties. For instance, a study by Hakke et al. (2020) demonstrated that using ultrasound for SNC preparation, as opposed to traditional acid hydrolysis, resulted in SNCs with enhanced surface charge and increased crystallinity. These improvements are attributed to the cavitation effect generated during ultrasound treatment. Moreover, the incorporation of these SNCs into films significantly enhanced the mechanical properties of polyurethane nanocomposite films, highlighting their potential in advanced material applications. Ultrasound treatment at power levels of 200–750 W increased the opacity and water barrier performance of cassava starch–acetic acid films and decreased the water adsorption of the films (Wu et al., 2020). The surface of films were also smoother and uniform after US

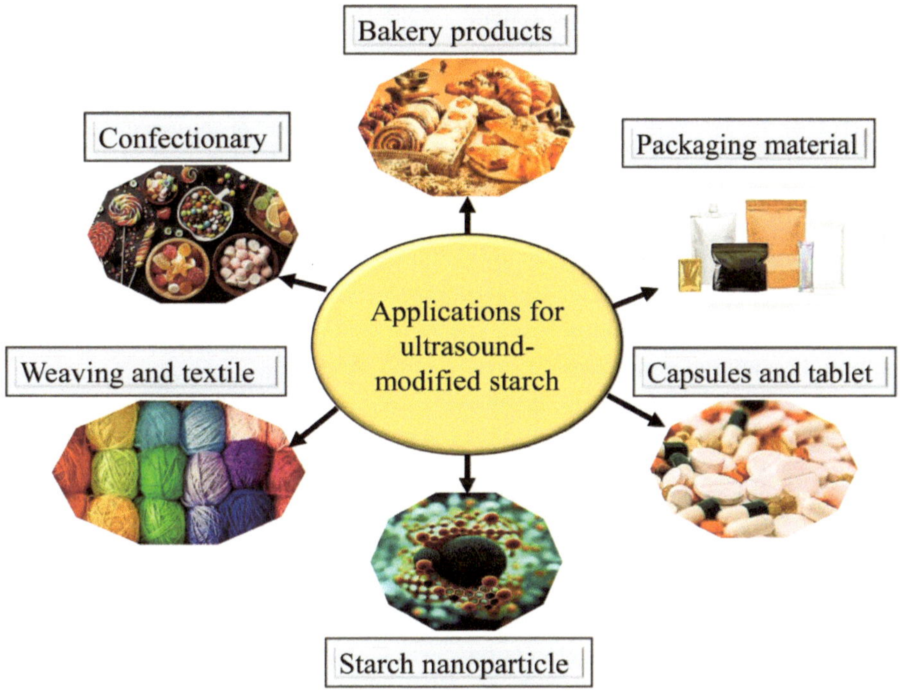

FIGURE 9.4 Applications of ultrasound-modified starch.

treatment due to the breakup of the agglomerated starch clumps (Gupta et al., 2025b).

9.5.3 Biomedical Industry

Conventional drug delivery materials are often associated with toxic effects and other limitations that make them less suitable for use. As a result, the pharmaceutical industry has shifted its focus toward biodegradable and non-toxic alternatives such as starch. Starch is abundant in nature, cost-effective, and biocompatible and exhibits high encapsulation efficiency, making it an attractive candidate for drug delivery systems. Furthermore, starch can be easily modified to enhance its functionality. In contrast, metal- and metal oxide–based delivery systems have demonstrated significant toxicity concerns. These systems can trigger the generation of reactive oxygen species (ROS), leading to oxidative stress, which in turn causes DNA damage, lipid peroxidation, and protein oxidation. Such cellular damage can escalate to organ dysfunction, underscoring the need for safer and more sustainable drug delivery materials such as starch (Sivamaruthi et al., 2022). Abbasian and Khayyatalimohammadi (2023) used ultrasound-assisted modification of starch to prepare a advanced drug delivery system. A novel nanocomposite, starch/iron oxide nanoparticles (Fe_3O_4)/metal-organic framework MIL-88(Fe), was synthesized using an ultrasound-assisted method, enhancing the integration of starch with a MIL-88(Fe) framework. This carrier exhibited controlled and sustained antibiotic release, biocompatibility, and improved antibacterial activity, highlighting the potential of ultrasound in creating functional starch-based materials for biomedical applications.

Figure 9.4 shows the various applications of ultrasound-modified starch in food and other industries.

9.6 CHALLENGES, POTENTIAL PROSPECTS, AND CONCLUSIONS

US is a starch modification technique that creates distinctive characteristics of starch through the cavitation mechanism. It alters surface morphology and interactions between linear and branched glucose chains in starch. Despite significant modification potential, the technology is associated with several drawbacks that must be addressed for effective large-scale implementation. A key issue is the lack of optimized standard protocols for US techniques, leading to inconsistent results. For example, the generation of tiny air bubbles in a US-treated system can pose a significant impedance mismatch and could reflect US waves instead of penetrating the food matrix. Addressing this issue through improved methods that minimize reflections would enhance process design and efficiency. Furthermore, the US is less effective in dense, complex food matrices, such as those rich in proteins, where it may not penetrate deep to the intended areas. This effect can vary depending on the type of food and whether the proteins are concentrated or distributed in the overall food system. Additionally, free radicals generated by US could oxidize active compounds in certain solid foods, potentially changing the properties of the matrix or rendering it unsafe for consumption. Although US is often considered a "green" technology, its economic feasibility is critical for transitioning from laboratory-scale to industrial applications. Unfortunately, there is a scarcity of studies evaluating the economic sustainability of

US processes. The need for a solid–liquid biphasic system for efficient US application may necessitate added drying steps, increasing both effluent production and operational expenses. However, the potential benefits of improved modification efficiency, shorter treatment times, and reduced maintenance costs could cover these additional expenses. Highlighting these benefits is vital for promoting the commercial adoption of US-based methods.

The US has shown significant promise in the food industry, with expectations for expanded use in the future. It can be integrated with other modification techniques, such as chemical, enzymatic, and thermal and non-thermal physical methods, to enhance its effects. US modification overcomes the limitations of NS by enhancing thermal stability, pasting characteristics, freeze–thaw stability, and gel clarity while also reducing retrogradation and digestibility. Additionally, the future of US in food processing is likely to see increased automation and miniaturization, with the development of portable and handheld devices, making the technology more accessible across different stages of production. As we gain deeper insights into its effects on various starches, the application of US will become more precise and targeted, particularly in terms of its influence on physicochemical properties. US technology is also a sustainable alternative in food processing, potentially reducing the need for chemical preservatives and extending product shelf life, thus helping to reduce food waste. However, the high initial cost of US equipment and the need for skilled operators may limit its adoption by smaller food producers or those in low-income regions. Further research is necessary to refine these technologies, making them safer, more efficient, and cost-effective for the food industry.

REFERENCES

Abbasian, M., & Khayyatalimohammadi, M. (2023). Ultrasound-assisted synthesis of MIL-88(Fe) conjugated starch-Fe_3O_4 nanocomposite: A safe antibacterial carrier for controlled release of tetracycline. *International Journal of Biological Macromolecules*, *234*, 123665.

Acevedo, B. A., Villanueva, M., Chaves, M. G., Avanza, M. V., & Ronda, F. (2022). Modification of structural and physicochemical properties of cowpea (*Vigna unguiculata*) starch by hydrothermal and ultrasoundtreatments. *Food Hydrocolloids*, *124*, 107266.

Ačkar, Đ., Babić, J., Jozinović, A., Miličević, B., Jokić, S., Miličević, R., Rajič, M., & Šubarić, D. (2015). Starch modification by organic acids and their derivatives: A review. *Molecules*, *20*(10), 19554–19570.

Adewale, P., Shokrolllahi Yancheshmeh, M., & Lam, E. (2022). Starch modification for non-food, industrial applications: Market intelligence and critical review. *Carbohydrate Polymers*, *291*, 119590.

Ai, Y., & Jane, J. L. (2015). Gelatinization and rheological properties of starch. *Starch - Stärke*, *67*, 213–224.

Ao, Z., & Jane, J. L. (2007). Characterization and modeling of the A- and B-granule starches of wheat, triticale, and barley. *Carbohydrate Polymers*, *67*(1), 46–55.

Apriyanto, A., Compart, J., & Fettke, J. (2022). A review of starch, a unique biopolymer: Structure, metabolism and in planta modifications. *Plant Science*, *321*, 111223.

Apar, D. K., Turhan, M., & Özbek, B. (2006). Enzymatic hydrolysis of starch by using a sonifier. *Chemical Engineering Communications*, *193*(9), 1117–1126.

Arvanitoyannis, I. S., Kotsanopoulos, K. V., & Savva, A. G. (2015). Use of ultrasounds in the food industry-methods and effects on quality, safety, and organoleptic characteristics of foods: A review. *Critical Reviews in Food Science and Nutrition*, *57*(1), 109–128.

Ashogbon, A. O. (2018a). Contradictions in the study of some compositional and physicochemical properties of starches from various botanical sources. *Starch - Stärke*, *70*(1–2), 1600372.

Ashogbon, A. O. (2018b). Current research addressing physical modification of starch from various botanical sources. *Global Nutrition and Dietetics*, *1*(1), 1–7.

Ashogbon, A. O. (2021). Dual modification of various starches: Synthesis, properties and applications. *Food Chemistry*, *342*, 128325.

Ashogbon, A. O., Akintayo, E. T., Oladebeye, A. O., Oluwafemi, A. D., Akinsola, A. F., & Imanah, O. E. (2020). Developments in the isolation, composition, and physicochemical properties of legume starches. *Critical Reviews in Food Science and Nutrition*, *61*(17), 2938–2959.

Astráin-Redín, L., Ciudad-Hidalgo, S., Raso, J., Condón, S., Cebrián, G., & Álvarez, I. (2020). *Application of high-power ultrasound in the food industry*. IntechOpen eBooks.

Bertolini, A. (Ed.). (2009). *Starches: Characterization, properties, and applications*. CRC Press.

BeMiller, J. N., & Huber, K. C. (2015). Physical modification of food starch functionalities. *Annual Review of Food Science and Technology*, *6*(1), 19–69.

Bhargava, N., Mor, R. S., Kumar, K., & Sharanagat, V. S. (2021). Advances in application of ultrasound in food processing: A review. *Ultrasonics Sonochemistry*, *70*, 105293.

Bitik, A., Sumnu, G., & Oztop, M. (2019). Physicochemical and structural characterization of microfluidized and sonicated legume starches. *Food and Bioprocess Technology*, *12*(7), 1144–1156.

Castanha, N., Lima, D. C., Junior, M. D. M., Campanella, O. H., & Augusto, P. E. D. (2019). Combining ozone and ultrasound technologies to modify maize starch. *International Journal of Biological Macromolecules*, *139*, 63–74.

Chong, W. T., Uthumporn, U., Karim, A. A., & Cheng, L. H. (2013). The influence of ultrasound on the degree of oxidation of hypochlorite-oxidized corn starch. *LWT*, *50*(2), 439–443.

Colussi, R., Lisie Mello El Halal, S., Pinto, V. Z., Bartz, J., Gutkoski, L. C., da Rosa Zavareze, E., & Dias, A. R. G. (2015). Acetylation of rice starch in an aqueous medium for use in food. *LWT*, *62*, 1076–1082.

De Kerf, M., Mondelaers, W., Lahorte, P., Vervaet, C., & Remon, J. P. (2001). Characterisation and disintegration properties of irradiated starch. *International Journal of Pharmaceutics*, *221*(1–2), 69–76.

Dhull, S. B., Tanwar, M., Khatkar, S. K., Chandak, A., Chawla, P., & Goksen, G. (2024). Exploring the effects of thermal and non-thermal modification methods on morphological, functional, and pasting properties of mung bean starch. *Innovative Food Science & Emerging Technologies*, *92*, 103581.

Fu, X., Belwal, T., Cravotto, G., & Luo, Z. (2020). Sono-physical and sono-chemical effects of ultrasound: Primary applications in extraction and freezing operations and influence on food components. *Ultrasonics Sonochemistry*, *60*, 104726.

Golkar, A., Milani, J. M., Motamedzadeghan, A., & Esmaeilzadeh Kenari, R. (2021). Physicochemical, structural, and rheological characteristics of corn starch after thermal-ultrasound processing. *Food Science and Technology International*, *29*(2), 168–180.

Gupta, R. K., Guha, P., & Srivastav, P. P. (2022). Natural polymers in bio-degradable/edible film: A review on environmental concerns, cold plasma technology and nanotechnology application on food packaging-A recent trends. *Food Chemistry Advances*, *1*, 100135. https://doi.org/10.1016/j.focha.2022.100135

Gupta, R. K., Guha, P., & Srivastav, P. P. (2023). Effect of high voltage dielectric barrier discharge (DBD) atmospheric cold plasma treatment on physicochemical and functional properties of taro (Colocasia esculenta) starch. *International Journal of Biological Macromolecules*, *253*, 126772. https://doi.org/10.1016/j.ijbiomac.2023.126772

Gupta, R. K., Guha, P., & Srivastav, P. P. (2024a). Insights into the physicochemical, crystalline, multistructural, thermal and digestibility behavior of taro starch (Colocasia esculenta) modified by cold plasma treatment with high voltage dielectric barrier discharge (HVDBD): An alternative technique for chemical modification of starch and environmental sustainability. *Journal of Cleaner Production*, *461*, 142710. https://doi.org/10.1016/j.jclepro.2024.142710

Gupta, R. K., Guha, P., & Srivastav, P. P. (2024b). Dielectric barrier discharge plasma: A green and novel method to change the structure of taro peel starch and improve the physicochemical properties of taro peel starch films. *Plasma Processes and Polymers*, *21*(9), 2400047. https://doi.org/10.1002/ppap.202400047

Gupta, R. K., Guha, P., & Srivastav, P. P. (2024c). Exploring the potential of taro (Colocasia esculenta) starch: Recent developments in modification, health benefits, and food industry applications. *Food Bioengineering*, *3*(3), 365–379. https://doi.org/10.1002/fbe2.12103

Gupta, R. K., Guha, P., & Srivastav, P. P. (2024d). Physical action of nonthermal cold plasma technology for starch modification. *Food Physics*, *1*, 100011. https://doi.org/10.1016/j.foodp.2024.100011

Gupta, R. K., Guha, P., & Srivastav, P. P. (2024e). Effect of cold plasma treatment and plasma-activated water on physicochemical and structural properties of starch: A green and novel approach for environmental sustainability. *Plasma Processes and Polymers*, *21*(4), 2300204. https://doi.org/10.1002/ppap.202300204

Gupta, R. K., Pipliya, S., Patel, J., Srivastav, P. P., Castro-Muñoz, R., Wang, C. K., & Nemţanu, M. R. (2025a). Essential oil-embedded starch based films for active packaging: Insights into migration, stability, and preservation efficiency. *Trends in Food Science & Technology*, 163, 105149. https://doi.org/10.1016/j.tifs.2025.105149

Gupta, R. K., Gnana Moorthy Eswaran, U., Kumar, S., Pipliya, S., Kaur, B., & Srivastav, P. P. (2025b). Synergistic modification of starch polysaccharide via sequential treatments with ultrasonic and cold plasma technology. *Journal of Food Processing and Preservation*, *2025*(1), 9063007. https://doi.org/10.1155/jfpp/9063007

Government of Manitoba. (2023). *Global market for starch and starch products*. https://www.gov.mb.ca/agriculture/markets-and-statistics/trade-statistics/pubs/starch-world-market.pdf

Hakke, V. S., Bagale, U. D., Boufi, S., Babu, G. U. B., & Sonawane, S. H. (2020). Ultrasound assisted synthesis of starch nanocrystals and its applications with polyurethane for packaging film. *Journal of Renewable Materials*, *8(3)*, 239–250.

Han, L., Cao, S., Yu, Y., Xu, X., Cao, X., & Chen, W. (2021). Modification in physicochemical, structural and digestive properties of pea starch during heat-moisture process assisted by pre-and post-treatment of ultrasound. *Food Chemistry*, *360*, 129929.

Haroon, M., Wang, L., Yu, H., Abbasi, N. M., Saleem, M., Khan, R. U., Ullah, R. S., Chen, Q., & Wu, J. (2016). Chemical modification of starch and its application as an adsorbent material. *RSC Advances*, *6*(82), 78264–78285.

Hong, J., Zeng, X.-A., Brennan, C. S., Brennan, M., & Han, Z. (2016). Recent advances in techniques for starch esters and the applications: A review. *Foods*, *5*(3), 50.

Hu, A., Lu, J., Zheng, J., Sun, J., Yang, L., Zhang, X., ... & Lin, Q. (2013). Ultrasonically aided enzymatical effects on the properties and structure of mung bean starch. *Innovative Food Science & Emerging Technologies*, *20*, 146–151.

Hu, R., Wu, L., Liao, X., Zhang, F., & Zheng, J. (2023). Synergistic modification of ultrasound and bamboo leaf flavonoid on the rheological properties, multi-scale structure, and in vitro digestibility of pea starch. *Food Chemistry*, *429*, 136959.

Jane, J., Chen, Y. Y., Lee, L. F., McPherson, A. E., Wong, K. S., Radosavljevic, M., & Kasemsuwan, T. (1999). Effects of amylopectin branch chain length and amylose content on the gelatinization and pasting properties of starch. *Cereal Chemistry*, *76*(5), 629–637.

Kardos, N., & Luche, J. L. (2001). Sonochemistry of carbohydrate compounds. *Carbohydrate Research*, *332*(2), 115–131.

Kaur, H., & Gill, B. S. (2019). Effect of high-intensity ultrasound treatment on nutritional, rheological and structural properties of starches obtained from different cereals. *International Journal of Biological Macromolecules*, *126*, 367–375.

Kennedy, J. F., & White, C. A. (1987). Starch chemistry and technology. *Carbohydrate Polymers*, *7*(2) 161–162.

Khurshida, S., Das, M. J., Deka, S. C., & Sit, N. (2021). Effect of dual modification sequence on physicochemical, pasting, rheological and digestibility properties of cassava starch modified by acetic acid and ultrasound. *International Journal of Biological Macromolecules*, *188*, 649–656.

Kumar, G., Le, D. T., Durco, J., Cianciosi, S., Devkota, L., & Dhital, S. (2023). Innovations in legume processing: Ultrasound-based strategies for enhanced legume hydration and processing. *Trends in Food Science and Technology*, *139*, 779.

Kumar, P., Prakash, K. S., Jan, K., Swer, T. L., Jan, S., Verma, R., Deepika, K., Dar, M. Z., Verma, K., & Bashir, K. (2017). Effects of gamma irradiation on starch granule structure and physicochemical properties of brown rice starch. *Journal of Cereal Science*, *77*, 194–200.

Kumar, Y., Singh, L., Sharanagat, V. S., Patel, A., & Kumar, K. (2020). Effect of microwave treatment (low power and varying time) on potato starch: Microstructure, thermo-functional, pasting and rheological properties. *International Journal of Biological Macromolecules*, *155*, 27–35.

Kumari, B., & Sit, N. (2023). Comprehensive review on single and dual modification of starch: Methods, properties and applications. *International Journal of Biological Macromolecules*, *253*, 126952.

Lee, Y.-K., & Chang, Y. H. (2019). Structural and in vitro digestibility properties of esterified maca starch with citric acid and its application as an oil-in-water (O/W) pickering emulsion stabilizer. *International Journal of Biological Macromolecules*, *134*, 798–806.

Li, J., Kong, X., & Ai, Y. (2022). Modification of granular waxy, normal and high-AM maize starches by maltogenic α-amylase to improve functionality. *Carbohydrate Polymers*, *290*, 119503.

Li, J. Y., & Yeh, A. I. (2001). Relationships between thermal, rheological characteristics and swelling power for various starches. *Journal of Food Engineering*, *50*(3), 141–148.

Li, Y., Li, C., Gu, Z., Hong, Y., Cheng, L., & Li, Z. (2017). Effect of modification with 1, 4-α-glucan branching enzyme on the rheological properties of cassava starch. *International Journal of Biological Macromolecules*, *103*, 630–639.

Lima, F. F., & Andrade, C. T. (2010). Effect of melt-processing and ultrasonic treatment on physical properties of high-amylose maize starch. *Ultrasonics Sonochemistry*, *17*(4), 637–641.

Lima, D. C., Maniglia, B. C., Matta Junior, M. D., Bail, P. L.-, Bail, A. L.-, & Augusto, P. E. D. (2021). Dual-process of starch modification: Combining ozone and dry heating treatments to modify cassava starch structure and functionality. *International Journal of Biological Macromolecules*, *167*, 894–905.

Lu, Z.-H., Belanger, N., Donner, E., & Liu, Q. (2018). Debranching of pea starch using pullulanase and ultrasonication synergistically to enhance slowly digestible and resistant starch. *Food Chemistry*, *268*, 533–541.

Makroo, H. A., Naqash, S., Saxena, J., Sharma, S., Majid, D., & Dar, B. N. (2021). Recovery and characteristics of starches from unconventional sources and their potential applications: A review. *Applied Food Research*, *1*(1), 100001.

Majeed, T., Wani, I. A., & Hussain, P. R. (2017). Effect of dual modification of sonication and γ-irradiation on physicochemical and functional properties of lentil (*Lens culinaris* L.) starch. *International Journal of Biological Macromolecules*, *101*, 358–365.

Majeed, T., Wani, I. A., Hamdani, A. M., & Bhat, N. A. (2018). Effect of sonication and γ-irradiation on the properties of pea (*Pisum sativum*) and vetch (*Vicia villosa*) starches: a comparative study. *International Journal of Biological Macromolecules*, *114*, 1144–1150.

Medina-Torres, N., Ayora-Talavera, T., Espinosa-Andrews, H., Sánchez-Contreras, A., & Pacheco, N. (2017). Ultrasound assisted extraction for the recovery of phenolic compounds from vegetable sources. *Agronomy*, *7*(3), 47.

Mehfooz, T., Ali, T. M., & Hasnain, A. (2019). Effect of cross-linking on characteristics of succinylated and oxidized barley starch. *Journal of Food Measurement and Characterization*, *13*, 1058–1069.

Mei, J. Q., Zhou, D. N., Jin, Z. Y., Xu, X. M., & Chen, H. Q. (2015). Effects of citric acid esterification on digestibility, structural and physicochemical properties of cassava starch. *Food Chemistry*, *187*, 378–384.

Mieles-Gómez, L., Quintana, S. E., & García-Zapateiro, L. A. (2023). Ultrasound-assisted extraction of Mango (*Mangifera indica*) kernel starch: Chemical, techno-functional, and pasting properties. *Gels*, *9*(2), 136.

Moad, G. (2011). Chemical modification of starch by reactive extrusion. *Progress in Polymer Science*, *36*, 218–237.

Moin, A., Ali, T. M., & Hasnain, A. (2016). Effect of succinylation on functional and morphological properties of starches from broken kernels of Pakistani Basmati and Irri rice cultivars. *Food Chemistry*, *191*, 52–58.

Monroy, Y., Rivero, S., & García, M. A. (2018). Microstructural and techno-functional properties of cassava starch modified by ultrasound. *Ultrasonics Sonochemistry*, *42*, 795–804.

Ouyang, Q., Wang, X., Xiao, Y., Luo, F., Lin, Q., & Ding, Y. (2021). Structural changes of a-, b- and c-type starches of corn, potato and pea as influenced by sonication temperature and their relationships with digestibility. *Food Chemistry*, *358*, 129858.

Özbek, B., & Ülgen, K. Ö. (2000). The stability of enzymes after sonication. *Process Biochemistry*, *35*(9), 1037–1043.

Park, S., & Kim, Y.-R. (2020). Clean label starch: Production, physicochemical characteristics, and industrial applications. *Food Science and Biotechnology*, *30*(1), 1–17.

Park, S. H., Na, Y., Kim, J., Kang, S. D., & Park, K. H. (2018). Properties and applications of starch modifying enzymes for use in the baking industry. *Food Science and Biotechnology*, *27*, 299–312.

Rahaman, A., Kumari, A., Zeng, X.-A., Farooq, M. A., Siddique, R., Khalifa, I., Siddeeg, A., Ali, M., & Manzoor, M. F. (2021). Ultrasound based modification and structural-functional analysis of corn and cassava starch. *Ultrasonics Sonochemistry*, *80*, 105795.

Rahman, M. M., & Lamsal, B. P. (2021). Ultrasound-assisted extraction and modification of plant-based proteins: Impact on physicochemical, functional, and nutritional properties. *Comprehensive Reviews in Food Science and Food Safety*, *20*(2), 1457–1480.

Ranjha, M. M. N. A., Irfan, S., Lorenzo, J. M., Shafique, B., Kanwal, R., Pateiro, M., Arshad, R. N., Wang, L., Nayik, G. A., Roobab, U., & Aadil, R. M. (2021). Sonication, a potential technique for extraction of phytoconstituents: A systematic review. *Processes*, *9*(8), 1406.

Ratnaningsih, N., Harmayani, E., & Marsono, Y. (2020). Physicochemical properties, in vitro starch digestibility, and estimated glycemic index of resistant starch from cowpea (*Vigna unguiculata*) starch by autoclaving-cooling cycles. *International Journal of Biological Macromolecules*, *142*, 191–200.

Ravikumar, M. (2017). Ultrasonication: An advanced technology for food preservation. *International Journal of Pure & Applied Bioscience*, *5*(6), 363–371.

Raza, H., Ameer, K., Ma, H., Liang, Q., & Ren, X. (2021). Structural and physicochemical characterization of modified starch from arrowhead tuber (*Sagittaria sagittifolia* L.) using tri-frequency power ultrasound. *Ultrasonics Sonochemistry*, *80*, 105826.

Ren, Y., Yuan, T. Z., Chigwedere, C. M., & Ai, Y. (2021). A current review of structure, functional properties, and industrial applications of pulse starches for value-added utilization. *Comprehensive Reviews in Food Science and Food Safety*, *20*(3), 2765–2804.

Rostamabadi, H., Nowacka, M., Kumar, Y., Xu, S., Colussi, R., Frasson, S. F., Singh, S. K., & Falsafi, S. R. (2024). Green modification techniques: Sustainable approaches to induce novel physicochemical and technofunctional attributes in legume starches. *Trends in Food Science & Technology*, *146*, 104389.

Sandeep, S., Singh, N., Isono, N., & Noda, T. (2010). Relationship of granule size distribution and AMP structure with pasting, thermal, and retrogradation properties in wheat starch. *Journal of Agricultural and Food Chemistry*, *58*(2), 1180–1188.

Sandhu, K. S., Siroha, A. K., Punia, S., Sangwan, L., Nehra, M., & Purewal, S. S. (2021). Effect of degree of cross-linking on physicochemical, rheological and morphological properties of sorghum starch. *Carbohydrate Polymer Technologies and Applications*, *2*, 100073.

Seung, D. (2020). AM in starch: Towards an understanding of biosynthesis, structure and function. *New Phytologist*, *225*(6), 2356–2371.

Sharma, V., Kaur, M., Sandhu, K. S., Kaur, S., & Nehra, M. (2021). Barnyard millet starch cross-linked at varying levels by sodium trimetaphosphate (STMP): Film forming, physicochemical, pasting and thermal properties. *Carbohydrate Polymer Technologies and Applications*, *2*, 100161.

Shen, H., Guo, Y., Zhao, J., Zhao, J., Ge, X., Zhang, Q., & Yan, W. (2021). The multi-scale structure and physicochemical properties of mung bean starch modified by ultrasound combined with plasma treatment. *International Journal of Biological Macromolecules*, *191*, 821–831.

Shen, L., Pang, S., Zhong, M., Sun, Y., Qayum, A., Liu, Y., Rashid, A., Xu, B., Liang, Q., Ma, H., & Ren, X. (2023). A comprehensive review of ultrasonic assisted extraction (UAE) for bioactive components: Principles, advantages, equipment, and combined technologies. *Ultrasonics Sonochemistry*, *101*, 106646.

Shi, H., & Hu, X. (2013). Preparation and structure characterization of carboxymethyl corn starch under ultrasonic irradiation. *Cereal Chemistry*, *90*(1), 24–28.

Shin, H. J., Choi, S. J., Park, C. S., & Moon, T. W. (2010). Preparation of starches with low glycaemic response using amylosucrase and their physicochemical properties. *Carbohydrate Polymers*, *82*(2), 489–497.

Shrivastava, A., Gupta, R. K., & Srivastav, P. P. (2024). Exploring novel frontiers of advancements in purple yam (*Dioscorea alata* L.) starch extraction, modification, characterization, applications in food and other industries. *Measurement: Food*, *15*, 100196. https://doi.org/10.1016/j.meafoo.2024.100196

Sit, N., Misra, S., & Deka, S. C. (2014). Yield and functional properties of taro starch as affected by ultrasound. *Food and Bioprocess Technology*, *7*, 1950–1958.

Sivamaruthi, B. S., Nallasamy, P. K., Suganthy, N., Kesika, P., & Chaiyasut, C. (2022). Pharmaceutical and biomedical applications of starch-based drug delivery system: A review. *Journal of Drug Delivery Science and Technology*, *77*, 103890. https://doi.org/10.1016/j.jddst.2022.103890

Sterling, C. (1978). Textural qualities and molecular structure of starch products. *Journal of Texture Studies*, *9*(3), 225–255.

Sudheesh, C., Sunooj, K. V., Navaf, M., Bhasha, S. A., George, J., Mounir, S., ... & Sajeevkumar, V. A. (2020). Hydrothermal modifications of nonconventional kithul (*Caryota urens*) starch: Physico-chemical, rheological properties and in vitro digestibility. *Journal of Food Science and Technology*, *57*(8), 2916–2925.

Sujka, M. (2017). Ultrasonic modification of starch: Impact on granules porosity. *Ultrasonics Sonochemistry*, *37*, 424–429.

Swinkels, J. J. M. 1985. Composition and properties of commercial NSes. *Starch - Stärke*, *37* (1), 1–5. https://doi.org/10.1002/star.19850370102

Tan, S. X., Andriyana, A., Lim, S., Ong, H. C., Pang, Y. L., & Ngoh, G. C. (2021). Rapid ultrasound-assisted starch extraction from sago pith waste (SPW) for the fabrication of sustainable bioplastic film. *Polymers*, *13*(24), 4398. https://doi.org/10.3390/polym13244398

Tappiban, P., Zhao, J., Zhang, Y., Gao, Y., Zhang, L., & Bao, J. (2022). Effects of single and dual modifications through electron beam irradiation and hydroxypropylation on physicochemical properties of potato and corn starches. *International Journal of Biological Macromolecules*, *220*, 1579–1588.

Vilpoux, O. F., Brito, V. H., & Cereda, M. P. 2019. Starch extracted from corms, roots, rhizomes, and tubers for food application. In M. T. P. Silva Clerici & M. Schmiele (Eds.), *Starches for food application* (pp. 103–165). Academic Press. https://doi.org/10.1016/b978-0-12-809440-2.00004-6

Wang, J., Lv, X., Lan, T., Lei, Y., Suo, J., Zhao, Q., Lei, J., Sun, X., & Ma, T. (2022). Modification in structural, physicochemical, functional, and in vitro digestive properties of kiwi starch by high-power ultrasound treatment. *Ultrasonics Sonochemistry*, *86*, 106004.

Wang, M., Liu, G., Li, J., Wang, W., Hu, A., & Zheng, J. (2023). Structural and physicochemical properties of resistant starch under combined treatments of ultrasound, microwave, and enzyme. *International Journal of Biological Macromolecules*, *232*, 123331.

Wang, Q., Li, L., & Zheng, X. (2021). Recent advances in heat-moisture modified cereal starch: Structure, functionality and its applications in starchy food systems. *Food Chemistry*, *344*, 128700.

Wang, S., Hu, X., Wang, Z., Bao, Q., Zhou, B., Li, T., & Li, S. (2020). Preparation and characterization of highly lipophilic modified potato starch by ultrasound and freeze-thaw treatments. *Ultrasonics Sonochemistry*, *64*, 105054.

Wang, S., Wang, J., Yu, J., & Wang, S. (2014). A comparative study of annealing of waxy, normal and high-AM maize starches: The role of AM molecules. *Food Chemistry*, *164*, 332–338.

Winkler, H., Vorwerg, W., & Rihm, R. (2014). Thermal and mechanical properties of fatty acid starch esters. *Carbohydrate Polymers*, *102*, 941–949.

Wu, F., Zhou, Z., Liang, M., Zhong, L., & Xie, F. (2020). Ultrasonication improves the structures and physicochemical properties of cassava starch films containing acetic acid. *Starch – Stärke*, *73* (1–2), 2000094.

Wu, Y., Du, X., Ge, H., & Lv, Z. (2011). Preparation of microporous starch by glucoamylase and ultrasound. *Starch-Stärke*, *63*(4), 217–225.

Xing, Y., Xu, Q., Ma, Y., Che, Z., Cai, Y., & Jiang, L. (2014). Effect of porous starch concentrations on the microbiological characteristics of microencapsulated *Lactobacillus acidophilus*. *Food & Function*, *5*(5), 972.

Yadav, S., Mishra, S., & Chandra Pradhan, R. (2021). Ultrasound-assisted hydration of finger millet (*Eleusine coracana*) and its effects on starch isolates and antinutrients. *Ultrasonics Sonochemistry*, *73*, 105542.

Yan, X., Diao, M., Yu, Y., Gao, F., Wang, E., Wang, Z., & Zhang, T. (2021). Influence of esterification and ultrasound treatment on formation and properties of starch nanoparticles and their impact as a filler on chitosan based films characteristics. *International Journal of Biological Macromolecules*, *179*, 154–160.

Yang, W., Kong, X., Zheng, Y., Sun, W., Chen, S., Liu, D., Zhang, H., Fang, H., Tian, J., & Ye, X. (2019). Controlled ultrasound treatments modify the morphology and physical properties of rice starch rather than the fine structure. *Ultrasonics Sonochemistry*, *59*, 104709.

Yang, Z., Hao, H., Wu, Y., Liu, Y., & Ouyang, J. (2021). Influence of moisture and AM on the physicochemical properties of rice starch during heat treatment. *International Journal of Biological Macromolecules*, *168*, 656–662.

Yudhistira, B., Husnayain, N., Punthi, F., Gavahian, M., Chang, C.-K., & Hsieh, C.-W. (2024). Progress in the application of emerging technology for the improvement of starch-based active packaging properties: A review. *ACS Food Science & Technology*, *4*(9), 1997–2012.

Zhang, H., Hou, H., Liu, P., Wang, W., & Dong, H. (2019). Effects of acid hydrolysis on the physicochemical properties of pea starch and its film forming capacity. *Food Hydrocolloids*, *87*, 173–179.

Zhang, J., Ran, C., Jiang, X., & Dou, J. (2021). Impact of octenyl succinic anhydride (OSA) esterification on microstructure and physicochemical properties of sorghum starch. *LWT*, *152*, 112320.

Zhang, L., Zuo, B., Wu, P., Wang, Y., & Gao, W. (2012). Ultrasound effects on the acetylation of Dioscorea starch isolated from *Dioscorea zingiberensis* C. H. Wright. *Chemical Engineering and Processing: Process Intensification*, 54, 29–36.

Zhang, W., Cheng, B., Li, J., Shu, Z., Wang, P., & Zeng, X. (2021). Structure and properties of octenyl succinic anhydride-modified high-AM japonica rice starches. *Polymers*, *13*(8), 1325.

Zhou, F., Liu, Q., Zhang, H., Chen, Q., & Kong, B. (2016). Potato starch oxidation induced by sodium hypochlorite and its effect on functional properties and digestibility. *International Journal of Biological Macromolecules*, *84*, 410–417.

Zhou, L., He, X., Ji, N., Dai, L., Li, Y., Yang, J., Xiong, L., & Sun, Q. (2021). Preparation and characterization of waxy maize starch nanoparticles via hydrochloric acid vapor hydrolysis combined with ultrasonication treatment. *Ultrasonics Sonochemistry*, *80*, 105836.

Zhu, F. (2015). Impact of ultrasound on structure, physicochemical properties, modifications, and applications of starch. *Trends in Food Science & Technology*, *43*(1), 1–17.

Zia-ud-Din, H. X., & Fei, P. (2017). Physical and chemical modification of starches: A review. *Critical Reviews in Food Science and Nutrition*, *57*(12), 2691–2705.

10 Pulsed Electric Field and Microwave-Assisted Dual Modification Techniques

Pulsed Electric Field (PEF) and Microwave-Assisted Methods

Khumbaron Kiranbala Kabui
Ph.D Scholar, NIFTEM-T

Swetha Priya Gali
Ph.D Scholar, NIFTEM-T

R. Meenatchi
Associate Professor, Department of Primary Processing Storage and Handling, NIFTEM-T

K. A. Athmaselvi
Associate Professor, Centre of Excellence for Grain Sciences, NIFTEM- Thanjavur, Tamil Nadu, India

10.1 INTRODUCTION

10.1.1 Overview of Starch and Its Industrial Relevance

Starch is a natural polysaccharide made of two glucose polymers: amylose and amylopectin that acts as the reserve food of a plant. It is present in wheat, corn, rice, cassava, and potatoes, which makes it one of the most popular biopolymers in the global market (Bertoft, 2017). Starch is industrially useful because it has significant functional characteristics such as stabilizing, binding, film-forming, thickening, and gelling properties (Gupta et al., 2022). For instance, in the food industry, starch finds application as a thickening agent in sauces, soup, and gravies, while it is used as a stabilizer in pudding, yogurt, and baked foods. It plays an important role in food formulation and processing because it is involved in changing the texture as well as the viscosity of the food products that are important in the development of mouthfeel, consistency, and shelf life of the products (Cornejo-Ramírez et al., 2018).

Apart from the food industry, starch has a vital role to play in various other industries, such as paper, textiles, adhesives and pharmaceuticals. In the paper industry, starch increases the stiffness, surface strength, and brightness of printing on paper. It binds the fibers higher, and because of this tensile power can be larger but the paper will remain as smooth (Li et al., 2019a). In textiles, during the sizing of yarns in weaving, sizing is done to give strength to the yarn so that breakage will not happen and productivity is increased. They are naturally degradable than other synthetic sub-institutional glue requirements (Tesfaye et al., 2018). In adhesives, starch is generally used for binding books, packaging, and corrugated boards. In addition, in the bioethanol industry, starch also acts as feedstock that can be converted into fermentable sugars, ultimately resulting in the production of a renewable source of energy (Maia et al., 2020).

10.1.2 The Need for Starch Modification

Starch modification is vital because native starches cannot meet the requirements of many advanced industries. The native starches are cheap and multifunctional but have low solubility and instability in continuous high/low temperatures, pH range, or retrogradation (recrystallization of starch molecules), which impacts both the texture and the shelf life of finished products. These weaknesses prevent native starches from being used in many industrial processes that require stability, consistency, and performance (Witczak et al., 2016). To solve these problems, much research has been done to change the characteristics of starch based on specific requirements through physical, chemical, or enzymatic modification processes leading to modified starch. Modified starches have been specifically engineered for improved attributes such as greater gel strength, higher viscosity, heat resistance, acid resistance, and shear resistance as well as better water retention and freeze–thaw stability (Punia, 2020; Gupta et al., 2023). Moreover, the

DOI: 10.1201/9781003564287-10

modification of starch creates opportunities in new fields such as biodegradable plastics and sustainable packaging, where modified starch affords better film-forming capabilities, strength, and flexibility than its unmodified form. Such improvements are important for designing environmentally friendly materials that can replace conventional plastic materials. The relentless advancement of modified starch addresses the increasing need for efficient, performance-optimized, sustainable, affordable solutions across various sectors (Amaraweera et al., 2021).

10.1.3 Traditional vs. Modern Starch Modification Techniques

Traditional starch modification techniques are based on physical methods (e.g., heat treatments) and chemical methods (e.g., oxidation, acetylation) to improve texture and stability, among other properties. However, these approaches often require a lot of energy, and chemical residues are left behind. Modern non-thermal techniques, such as ultrasound and high-pressure processing (Punia, 2020; Gupta et al., 2025b), pulsed electric field, and microwaves, cold plasma technique, modify starch without heat while preserving its nutritional value (Thirumdas et al., 2015; Gupta et al., 2024a). With nanotechnology, modifications are made at molecular levels which improve starch function and bioavailability (Wang & Zhang, 2021). Through enzymatic modifications and genetic engineering biotechnology, specific traits such as improved digestibility or functionality can be offered with tailored starches (Masina et al., 2017). Thus, these advanced methods tend to be more sustainable and efficient, as well as meeting the expectations of health-conscious consumers.

10.1.4 Introduction to Dual Modification Approaches

Modifying starches in two ways using pulsed electric field (PEF) and microwave (MW) technologies is a new method for improving starches' properties without thermal processing techniques. In PEF, short bursts of very high-voltage electrical pulses are applied to starch suspensions to disrupt the arrangement of granules, leading to altered physical and chemical properties such as increased swelling ability, enhanced solubility, and better gelatinization efficiency. PEF has an edge over other modulating agents because it modifies starches at lower temperatures, thus retaining their nutritional value while enhancing their functionality (Soltanzadeh et al., 2022). In contrast, MW treatment depends on electromagnetic radiation to create heat within the granules themselves, causing simultaneous agitation of water molecules contained therein. This process results in structural changes such as the breakdown of crystalline regions followed by greater molecular mobility, which translates into improved pasting properties, reduced retrogradation tendencies, and increased digestibility (Ekezie et al., 2017).

The combination of PEF and MW technologies has synergistic effects that cannot be achieved by using either technology alone to modify starches. PEF pretreatment disrupts granule structure, making the starch more amenable to subsequent MW processing, and is a tool to control subsequent modifications to its functional properties (Leong & Oey, 2022). These modifying strategies for starches afford advantages relating to efficiency (processing time and energy) and therefore the fine-tuning of starch to meet specific industrial characteristics while at the same time allowing the complementary nature of the changes (Mannozzi et al., 2018).

Also, PEF and MW technologies can work together to minimize the use of chemical reagents in starch processing recipes, appealing to the increasing demand for clean-label, sustainable food ingredients. Upon considering these factors, investigating dual modification of starch with PEF and MW technologies brings improvements in functionality and environmental efficiencies and is therefore a relevant approach to modifying starch in today's food and non-food marketplaces (Bakr, 2020).

10.2 HIGHLIGHTS

- According to research, dual modification with microwave and PEF processes enhances starch features such as solubility, gelatinization, and viscosity, leading to better functional properties for food application.
- Both are non-thermal processes that help save energy and minimize chemical input from traditional starch modification processes, making them greener.
- When used together, both PEF and MW produce synergistic results, improving the efficiency of starch modification while yielding high-quality starch with better texture, stability, and performance across several industrial applications.

10.3 MECHANISMS AND EFFECTS OF PULSED ELECTRIC FIELD TECHNOLOGY ON STARCH

10.3.1 Basic Principles of PEF Technology

PEF technology processes foods without heat by using bursts of high-voltage pulses to modfy cells in food, leading to various beneficial effects. The fundamental concept of PEF includes applying fields (ranging from 1 to 50 kV/cm) for brief durations (microseconds to milliseconds) to a food product positioned between two electrodes (Soltanzadeh et al., 2022). When a voltage is applied in this scenario, it generates a field that triggers a change in potential across the cell membranes known as electroporation. While the intensity and duration of these pulses vary, it can create permanent pores that could damage the structure of cell membranes, leading to electroporation. This technique enables targeted modifications to be carried out on cells while maintaining their sensory properties, making PEF technology particularly advantageous for heat-treatedfoods (Arshad et al., 2021).

The efficiency of PEF treatment is determined by several key parameters, including electric field strength, pulse duration and frequency, treatment time, and food conductivity. Larger electric field strength and a longer duration of treatment normally widen the amount of electroporation, while higher intensity widens its efficacy (Versteeg, 2016). Nonetheless, the optimization of such parameters should be judicious so as not to inflict too much damage on cells or to create undesired modification that would compromise food quality (Maloney & Harrison, 2016). As PEF can produce homogenous electric field treatment in a short time, it is a highly operative method of sterilization or pasteurization for liquid and semi-liquid foods such as juices, milk, and smoothies.

One benefit of PEF is the ability to inactivate microbes at lower temperatures and thus do less damage to certain properties of the food matrix, such as color, taste, texture, and nutritional quality. It also has the potential to improve mass transfer processes, which creates opportunities for utilization in applications such as the extraction of bioactive compounds from food matrices, enhancement of drying and freezing rates, and increasing juice yield from fruits and vegetables due to higher cell permeability leading to the outflow of certain intracellular components (Pataro et al., 2018).

10.3.2 Mechanism of PEF-Induced Starch Modification

The application of PEF targets the modification of the starch structure through pulses of high-voltage electricity, which alters the granule content. This process creates a field that causes rapid changes in the starch granules by disrupting the hydrogen bonding and crystalline regions of amylose and amylopectin (Han et al., 2020). Due to this, the permeability of starch granules increases, and granules can swell partially or completely, enhancing their solubilization and gelatinization properties. The degree of modification is a function of such PEF parameters as electric field strength and pulse duration and also depends on specific characteristics of the starch itself such as its botanical origin and moisture content (Castro et al., 2023).

Electrical pulses highly excites water molecules inside the starch grains, causing localized high heating through rapid molecular movement. These local energy sources distort the crystalline structure that had been semi-crystalline, and it becomes disordered. The breakdown of these crystalline regions increases the starch's water absorption capacity and hence its swelling propensity as well as viscosity at high temperatures (Li et al., 2019a). This alteration has broad implications, especially in food industries where improved thickening properties for produces such as soups, sauces, and many processed foods are sought after. Furthermore, lower crystalline orientation reduces the gelatinization point, which would allow gelation at temperatures lower than those needed for the untreated starch (Maniglia et al., 2021a).

PEF-induced electroporation of starch granules also causes both amylose and amylopectin to leach out, which can be exploited for modifying the functionality of starch for diverse industrial applications. In this respect, these molecules enhance the pasting properties of the starch to make it even more suited to various industrial needs, where there is a need for quick viscosity development (Duque et al., 2020). The modified starch structure is more bioavailable because the parts of the starch that have been crystallized have been destroyed and are therefore more prone to enzymatic hydrolysis; this is particularly useful in nutritional preparations requiring fast energy mobilization (Duque et al., 2022). In addition, PEF may impact starch retrogradation, a phenomenon in which gelatinized starch recrystallizes over time, leading to undesirable textural changes as well as reduced shelf life in foodstuffs. By disrupting the crystalline structure and altering the amylose-to-amylopectin ratio, PEF processing can decrease the onset of retrogradation, thereby improving starch-based gel and paste stability. This is especially beneficial for frozen foods where maintaining freeze–thaw cycle textures is crucial (Qiu et al., 2021).

In general, PEF-induced starch modification is an energy-saving non-thermal method that improves starch's functional properties without resorting to any chemical additives, thus addressing the rising need for clean-label technology and sustainability in food processing technologies.

10.3.3 Effects of PEF on Starch Structure and Functionality

PEF technology produces high-voltage pulses that change the structure of starch with an electrical frequency; therefore, PEF can be considered an efficient non-thermal method of modifying starch properties in various industries. It is also the first component of destruction to consider because starch granules are still self-assembled at the molecular level after processing (Hong et al., 2016). PEF treatment is based on the mechanism of electroporation and the impact of specific electric fields where electroporation is the first component of starch granule destruction; disorganized partially crystalline granule structures are further destroyed as PEF is applied (Zhu, 2018). Short high-voltage pulses result in faster polarization and movement of starch molecules and in disorganized structures, including internal structural arrangement and the external structure of starch granules. Among all the PEF effects on starch structure, one of the more notable changes is the destabilization and subsequent reorganization of semi-crystalline regions. The semi-crystalline regions are composed of tightly packed amylose and amylopectin, which are entangled and aligned randomly (Li et al., 2019b). PEF-treated starch has increased water absorption, swelling, and solubilization rates, providing functional properties of starch that would not be significantly different in unmodified structures. Starch with PEF modification typically has a greater bond with water and therefore benefits food applications for which consumers prefer and expect enhanced thickening or notable gelling properties (Qiu et al., 2021).

If a starch material has a more significant affinity for water, then the rate of gelatinization and/or hydrating of this starch material will follow suit at lower temperatures than for starch with no treatment (Shrivastava et al., 2024). This property of treated starch permitting faster gelatinization or hydration at lower temperatures could shorten cooking time and create mold in the water. This enhanced property is preferential, depending on the application, for immediate thickening or instant viscosity, especially in instant foods, sauces, and soup products (Maniglia et al., 2021a). Pretreatment with PEF creates starch that has increased solubilization and swelling capabilities that result in a smoother and creamier texture, through potential associated higher-quality severity. The ability to more effectively create a thicker paste that also has improved pasting capabilities is a value-added functionality of the processing speeds resulting from PEF application (Duque et al., 2020). PEF also results in the loss of amylopectin-to-amylose ratios during processing through increased water washing of the starch granules. As a consequence, PEF also enhances the gelling ability of the starch, creating very rigid and high viscosity gels. It also lessens the possibility of starch retrogradation, which is a problem in which gelatinized starch recrystallizes over time, becoming unpleasantly textural through syneresis (water expulsion) and staling of baked foods. For cooked products in pre-made meals—whether baked or frozen—PEF can help extend shelf life, improve texture, and lower the potential for retrogradation. (Grgić et al., 2019).

In conclusion, PEF effects on starch structure and functionality have tremendous benefits for processing efficiency, product quality, and stability; hence, PEF is a potent instrument for use in contemporary food and non-food applications.

10.3.4 Applications and Benefits of PEF in Starch Processing

PEF is an emerging technology to modify starch that has numerous applications and advantages in both food and non-food industries. PEF technology is non-thermal with low energy input application, making it ideally suited as a substitute for chemically modifying starches and cooking processes with impelling heat. The need for clean-label technologies and sustainability increases the application of PEF to modify starch. One notable opportunity for PEF is the functional enhancement of starches used as thickeners, stabilizers, or gelling in food systems. During the thickening process, PEF-modified starch is preferred due to its enhanced water absorption capacity and swelling ability, which promote quick thickening in soups, sauces, and dairy products. Enhanced functional characteristics improve not only the sensory attributes of texture and mouthfeel but also the processing time, leading to increased manufacturing rates (Castro et al., 2023). Furthermore, the application of PEF on starch allows for further enhanced thickening and gelatinization performance, which means that starch is gelatinized at lower temperatures and faster times to develop the desired consistency. This is very important in instant food products where smooth texture and fast reconstitution of starch are important, such as ready-to-eat products, instant noodles, and baby food (Han et al., 2020). PEF treatment potentially reduces starch retrogradation, meaning that previously gelatinized starch granules will recrystallize, changing coagulated texture during frozen and prolonged storage, which is relevant to staling in baked products and syneresis in gels. The reduction of retrogradation with PEF leads to improved product quality over time, with longer shelf life and reduced food wastage (Qiu et al., 2021). The applications of PEF technology also extend to the extraction and transformation of starches for biodegradable packaging materials. Starch-based bioplastics, part of a growing trend towards environmentally friendly substitutes for traditional plastics, take advantage of the improvements in the mechanical and barrier properties of starch films resulting from PEF treatment. PEF-treated starches exhibit higher flexibility and strength while being less affected by moisture than their untreated counterparts, making them more suitable modes of eco-friendly packaging that comply with environmental regulations without sacrificing efficiency (Alabi et al., 2021).

PEF-modified starches are increasingly used beyond the food and packing industries, including in the pharmaceutical industry, because of their improved solubility, binding, and disintegration properties for tablet preparation and personal care products. Because PEF can modify starch properties without using aggressive reagents, the technology is especially appealing for uses that demand high-level cleanliness as well as minimal chemical residues. In summary, the advantages of using PEF in starch transformation include improved functional characteristics, better product quality, reduced energy consumption, and environmental sustainability (Suri & Singh, 2023). The PEF technology enables flexible and efficient change to starch, in turn enabling organizations in several sectors to develop new products and resulting in progress in areas such as food processing, containerization, and pharmacy.

10.4 MICROWAVE-ASSISTED METHODS IN STARCH MODIFICATION

10.4.1 Fundamentals of Microwave Heating

Microwaves are high-frequency electromagnetic waves with frequencies from 300 MHz to 300 GHz and wavelengths between 1 m and 1 mm (Moirangthem et al., 2020). Microwave treatment is based on converting electromagnetic energy into thermal energy by influencing the material's polar molecules. Volumetric heating is the most significant property of microwave heating. Volumetric heating means the direct internal absorption of microwave radiation by materials and transforming that energy into heat. Heat is generated throughout the material, resulting in faster heating rates (Vadivambal & Jayas, 2007). The frequency utilized in household microwave ovens is 2450 MHz, and in industrial microwave ovens, 915 MHz. The main components of microwave heating equipment are a cooking chamber, waveguide, cavity magnetron, capacitor, and transformer. Heat distribution through food is highly dependent on its size, dielectric properties, and the characteristics of the oven. The oven parameters that affect the heating rate and uniformity are oven positioning, oven geometry and size, use of turntables, use of mode stirrers, oven power, feed location, and use of power cycling. Food factors that affect energy absorption are dielectric properties, volume, shape, and aspect ratio (Datta & Rakesh, 2013).

The two heating mechanisms involved in the absorption of microwave energy by food are dipolar rotation and ionic interaction. Microwave power is absorbed by materials with polar molecules such as water. Polar molecules have a positive and a negative charge, and are thus dipole. When a medium is subjected to a microwave oscillating electric field, dipoles in the material try to align their axes with the changing electric fields. As the field oscillates, molecules continuously rotate, producing friction at molecular levels. Hence, when these polar molecules rotate rapidly within themselves, internal friction results in heat generation (Meloni et al., 2023). Microwave energy also causes ions in ion-containing materials (e.g., dissolved salts in water) to flow back and forth as the electric field oscillates. Friction between these ions and nearby molecules causes them to move and produces more heat (Bakshi et al., 2023).

The primary effect of microwaves on a material is dielectric heating. Dielectric characteristics are essential factors to assess the capacity of foods to absorb and convert microwave energy to heat. These characteristics are affected by the microwave frequency, system, and the density, composition, and temperature of the material. A dielectric medium is characterized by numerous parameters, such as the dielectric constant (ε'), dielectric loss factor (ε''), dielectric loss tangent ($\tan \delta\varepsilon$), and equivalent impedance. ε' denotes the ability of a material to store microwave energy. ε'' refers to the capacity of material to dissipate microwave energy into heat. The term $\tan \delta\varepsilon$ ($\varepsilon''/\varepsilon'$) describes the capacity of material to absorb microwave energy and transform it into heat at a given power and temperature. The higher the value of $\tan \delta\varepsilon$, the more microwave energy is efficiently absorbed by a material and the faster the heating rate is (Tao et al., 2020). Penetration depth (D_p) refers to the depth into a material at which the power reduces to 1/e of its original value. Factors affecting D_p are the dielectric constant and loss factor of food. D_p is associated with the dielectric properties and microwave frequency and is used to describe the level to which a material can be penetrated by microwaves at a given frequency. It is an important parameter for evaluating the temperature distribution of food during microwave processing (Venkatesh & Raghavan, 2004).

10.4.2 Mechanisms of Microwave-Induced Starch Modification

In the past few years, microwaves have become of specific interest to researchers as a novel green starch modification technique. Microwaves act as an excellent energy source and allow energy to be transferred rapidly into the food, where the transferred energy is converted into heat with shorter processing times than conventional methods. The microwave field also impacts the properties of the final product (Lewicka et al., 2015). MW heating is a physical modification that has been extensively used in starch modification because of its benefits, such as environmental friendliness, high efficiency, reduced processing time, good penetrability, and lower nutrition loss. Researchers employ microwave heating to modify the physiochemical and functional characteristics of starch, such as crystallization, gelatinization, and digestion, affecting the interaction between starch and other food ingredients to improve the quality of food (Yi et al., 2024).

Microwave heating of starch involves the application of electromagnetic waves with frequencies between 300 GHz and 300 MHz. Polar materials, such as starch, can absorb microwave energy and align themselves with the electric field of the microwave (Fan et al., 2012). The mechanism governing the changes in starch structure during microwave heating has been classified into four stages (Brasoveanu & Nemtanu, 2013). The first stage involves heating the starch, using the dielectric relaxation of water molecules. Subsequently, there is a rapid temperature rise, accompanied by moisture loss from the interior of starch granules. Granule expansion occurs from the center of the granules, which is further caused by high pressure inside the granules due to moisture loss. In the last stage, signs of starch granule degradation are visible. Microwaving appears to be a more convenient technique for starch modification even though the physiochemical changes seen in microwaved starches are comparable to those seen for traditional physical modification methods such as annealing and heat-moisture treatment (Lewandowicz et al., 2000). This is due to the starch sample being heated rapidly, uniformly, and with high microwave energy penetration power. Two parameters that determine microwave heating are the dielectric constant and the dielectric loss factor. The heat produced during microwave treatment of starch has been shown to produce free radicals that can depolymerize large starch into smaller molecules via the cleavage of glycosidic bonds (Kim et al., 2015). The propagation of microwave energy into a material varies with the dielectric and magnetic properties of a medium. But starch reacts only to the electric field of electromagnetic energy and has no magnetic component. This allows for the relative complex permittivity of starch, concerning its absorption, reflection, and transmission of microwave energy, vital electric characteristics in microwave heating (Brasoveanu &

TABLE 10.1
Factors Affecting the Interaction of Starch to Microwave Radiation

Starch Parameters	Microwave Radiation Processing	Reference
Kind of starch	Power level	Brasoveanu and Nemtanu (2013)
Temperature	Frequency	
Moisture content	Type of oven (single/multimode) and its geometry	
Dielectric properties		
Density		
Penetration depth	Exposure time	
Load geometry inside the oven		

Nemtanu, 2013). According to Lewicka et al. (2015), there is a strong heating of the moisture in the material during microwave treatment due to a rapid reorientation of diploes and the dissolution of hydrogen bonds. As a result, structural changes happen in the material. It is crucial to remember that various parameters play a significant role in the interaction of granular starch with microwaves (Table 10.1).

10.4.3 Impact of Microwave Treatment on Starch Properties

The dielectric characteristics are important factors in the microwave heating of starch. The physicochemical properties of starch derived from various botanical sources vary after microwave heating. The physiochemical and functional properties and monomeric nature of starch in aqueous systems in different applications are crucial, particularly for producers of starch-based goods. The literature on the effects of microwave treatment on various starch properties (Table 10.2) is presented here.

10.4.3.1 Swelling Power and Solubility

Swelling of starch involves the interaction between the amorphous and crystalline and portions of starch (Gupta et al., 2024b). Thus, changes in starch solubility and swelling may be due to modifications in the crystalline and amorphous regions (Gupta et al., 2024c). Starch swelling is significantly reduced after microwave heating. This may be due to the rearrangement of crystalline areas within the starch granules, which might become more randomly dispersed within the starch granule (Luo et al., 2006). Zailani et al. (2022) reported better solubility in water (95°C), oil, water binding capacity, and increased amylose content of sago after microwave treatment. The swelling capacity and solubility of mango kernel starches decreased significantly with the microwave treatment (Ramírez-Brewer et al., 2023). The reduction in solubility is due to the microwave treatment altering the hydrogen bonds, resulting in a reduction in the hydrophilicity of the starch.

10.4.3.2 Rheological Properties

The rheological properties of starch are important because a small change in molecular structure and conformation can result in notable changes in rheological behavior (Gupta et al., 2024e). Microwave heating can induce significant changes in starch rheological properties, especially a decrease in viscosity. Lewandowicz et al. (2000) reported a significant reduction in viscosity when corn and wheat starch with 30% moisture were treated using microwave power of 0.5 W/g. Alterations in the starch can result from damage due to the increase in temperature or crystalline structure reorganization brought on by loss of water. Yang et al. (2017) reported a decrease in breakdown, final, trough, and setback viscosities of waxy maize starch after microwave heating.

10.4.3.3 Gelatinization

Gelatinization is one of the most significant functional characteristics of starch, which has specific processing implications. MW treatment can alter the process of gelatinization and influence the rheological properties of the starch. The impact of microwaves on the starch often causes an increase in gelation temperature and a reduction

TABLE 10.2
Studies on the Impact of Microwave Heating on Different Starches

Starch with Moisture Content (mc)	Microwave Parameters	Exposure Time	Impact on Starch	Reference
Waxy maize starch (30% mc)	1600 W	5–20 min	Decrease in starch molecular weight, relative crystallinity, enthalpy, and viscosities; increase in gelatinization temperatures	Yang et al. (2017)
Groundnut starch (30% mc)	700 W	0–60 s	Decrease in relative crystallinity, peak viscosity, cold paste viscosity, hot paste viscosity, pasting temperature, and enthalpy of gelatinization	Oyeyinka et al. (2019)
Barley starch	640, 720, and 800 W	60, 120, and 180 s	Decrease in viscosities; increase in gelatinization temperature, swelling power, and solubility	Ma et al. (2020)
Sago starch (13% mc)	180 W	15 min	Better solubility in water; better oil and water binding capacity Increase in amylose content; changes in granule morphology	Zailani et al. (2022)
Arrowroot starch	400 W	1–4 min	Increase in water and oil holding capacity, swelling power, and solubility	Harni et al. (2023)

of the process enthalpy due to structural modifications of the starch macromolecules. Zhang et al. (2009) observed higher values of gelatinization temperature of canna starch (20%–45% moisture) after microwave treatment, and this could be due to better hydration of amorphous regions. An increase in all parameters (onset, peak, endset peaks, and enthalpy) of thermal properties of corn starch after microwave heating was also reported by Marta et al. (2022).

10.4.3.4 Granule Morphology

Microwave heating altered the morphology of starch granules. The surface of native starch granules has a regular elliptical shape and is smooth (Gupta et al., 2024d). Minor defects and cracks have been observed on the surface of some starch granules after being exposed for 5 s to the microwave at 40 °C. The starch granules show signs of fractures or flaws after heating for 10 s at 55 °C, although the temperature and time do not affect the grain structure integrity. Further microwave radiation for 20 s at 95 °C caused serious deformation cracks and destruction of the majority of the starch granules (Kärkkäinen et al., 2011). Hornung et al. (2018) observed that the granules of Amazon white yam starch (very low moisture) exhibited flaky surfaces and sustained damage after microwave treatment (7000 W/g, 5 min). This might be because of the incredible input of microwave power. Zhong et al. (2019) concluded that microwave treatment did not modify the size and shape of high-amylose maize (30% moisture, wb) starch granules, which was in line with the results of Shevchenko et al. (2013) (potato starch, 15% moisture, wb) and Shah et al. (2016) (Indian horse chestnut starch, 22% moisture, wb). This indicates that the starch granules are not damaged if the moisture content is kept low (no more than 30%) and the heat treatment is properly controlled.

10.4.3.5 Crystallinity

Differences in the degree of starch crystallinity can be detected by X-ray diffration. MW reduces the degree of crystallinity of wheat starch, whereas the crystallinity remains the same in the case of waxy corn starch. The vulnerability of different starch molecules to microwave heating depends on both their crystal structure and their amylose content (Lewandowicz et al., 2000). Ma et al. (2018) found that microwave heating of Laird lentil starch (75% moisture, wb, 3.5 W/g, 30 min) decreased the crystallinity and increased the amorphous and crystalline layer thickness. This is due to the greater energy input into the starch system causing more double helix structural collapse, contributing to the expansion of the amorphous region. It was observed that microwave treatment broke down the molecular chains and decreased polymerization, which affected the crystalline layer thickness. The crystalline structure of normal maize starch (A-type, 30% moisture, wb) remained intact, whereas potato starch (B-type, 30% moisture, wb) changed from B-type to A-type when the same microwave parameters were used. This was partly due to the weaker crystalline pattern of potato starch, with a looser packing mode than normal maize starch (Xu et al., 2019). Furthermore, for any particular starch, various combinations of input power and duration of microwave time, as well as the dielectric properties, can together impact the modifications of starch crystal formations under microwave treatment.

10.4.4 Advantages and Limitations of Microwave-Assisted Methods

Microwave radiation has many benefits, including fast processing time, volumetric heating, selectivity of heating, and environmental friendliness. In addition, it aids in the reduction of operating costs because the energy generated is focused on the material, greatly improving operational efficiency (Vadivambal & Jayas, 2007). Despite the advantages of microwaves, there are limitations. These can include difficulty in optimizing process parameters and the financial costs of specialized equipment. Application of microwaves is still limited by a lack of knowledge of the dielectric properties of heated materials. Microwave heating may result in non-uniform heating and eventually to thermal degradation in the case of materials where the dielectric constant rises with temperature. Achieving the appropriate and uniform temperature during the microwave process can be challenging and demands consistent process monitoring. Factors affecting the behavior of starch in the microwave field are its type, water content, dielectric properties, depth of penetration, and density, plus factors related to the microwave process, such as power, time of exposure, and frequency (Brasoveanu & Nemtanu, 2013). Microwave heating may result in a poor-quality final product if not properly employed. The challenge with microwave treatment is the variation of parameters that impact the behavior of heat transfer, including geometry, thickness, and dielectric characteristics of the food.

10.5 COMPARATIVE ANALYSIS OF PEF AND MICROWAVE TECHNIQUES IN STARCH MODIFICATION

10.5.1 Key Differences and Similarities between PEF and Microwave Techniques

The novel approaches of microwave and PEF techniques, with unique factors, mechanisms, and effects on starch properties, can be used together to modify starch. These techniques have different effects on starch properties even though they share some similarities, such as energy-efficiency and reduced processing time. PEF uses short high-voltage pulses that affect starch by changing its crystallinity and granule morphology (Suri & Singh, 2023). Starches are modified during microwave heating using electromagnetic radiation, which alters the molecular structure and gelatinization qualities (Tao et al., 2020). PEF processing results in reduced gelatinization temperatures and relative crystallinity of starch, promoting extraction rate and digestibility (Achayuthakan et al., 2023). The rheological properties of starch, such as reduced viscosity and increased gelation temperature, are altered by microwave heating (Lewicka et al., 2015). The applications of modified starch in food products are affected by the changes brought upon by each technique. Compared to other physical modification technique, both methods are widely known for their efficiency and short processing times. Due to the lower energy consumption of PEF, it is mostly suitable for industrial applications (Kumari & Sit, 2023). Digestibility of starch can be enhanced using these techniques by modifying the chemical structure and granular morphology. Resistant starch has been shown to be improved by PEF by changing the content of digestible starch (Rostamabadi et al., 2024). Starch digestibility is also impacted by microwave treatment but through different approaches to its structure, potentially influencing how the starch interacts with digestive enzymes. Both techniques avoid the use of chemicals, which is considered safe for the environment.

10.5.2 Influence on Starch Granule Morphology and Crystallinity

Integrating PEF and microwave processing can have a synergistic effect on the morphology and crystallinity of starch granules owing to the complementary mechanism of each process. The structural properties of the starch granule is affected by PEF through electroporation by creating pores and fissures on the surface that increase microwave penetration, swelling, complete disruption, and extraction efficiency (Salee et al., 2023). Microwave processing rapidly heats the starch granules, resulting in swelling, expansion, and subsequent gelatinization. Dual-modified starch granules often exhibit higher surface roughness, cracking, and internal degradation than when each method is applied alone (Suri & Singh, 2023).

PEF treatment often causes moderate decrease in the crystallinity of the granules, since it mainly disturbs the amorphous regions and the surfaces of the granules. This is complemented by microwave treatment, which transforms B-type crystalline starch into A-type crystalline starch. Microwave treatment breaks down the crystallinity to a large extent due to moisture and heat gelatinization of the crystalline regions (Deng et al., 2022). Porosity and surface disruption by PEF treatment renders the crystalline regions more susceptible to microwave heating. All these factors contribute to a rapidly created homogenous, less crystalline structures, which are essential for certain food products where amorphization is required (Maniglia et al., 2021b).

10.5.3 Thermal and Non-Thermal Effects on Starch Modification

Thermal (heat) and non-thermal techniques have an impact on the physiochemical, functional, and structural characteristics of starch. Thermal modification of starch leads to changes in functional and structural properties. This treatment causes reduction of viscosities and alteration of gelatinization behavior, which influence the texture and mouthfeel of food products (Compart et al., 2023). It also improves solubility and water binding capacity owing to gelatinization of starch, and it transforms the crystalline regions of starches into less crystalline areas, thus enhancing digestion and viscosity of the starches.

Non-thermal treatments are an alternative approach; instead of heat, they employ other physical and chemical techniques to change the starch structure. These procedures can facilitate modification processes while maintaining the integrity of heat-sensitive ingredients in food (Mhaske et al., 2024). Changes in starch properties such as reduced crystallinity, enhanced water absorption, partial gelatinization, and granule disruption are observed after the treatment. Non-thermal processes offer more uniform modification with minimized loss of functional properties of starch, while both approaches lead to alterations in attributes such as crystal shape and crystallinity and granular morphology of starch (Wang et al., 2023).

10.5.4 Performance Evaluation in Industrial Applications

To achieve efficient performance, hybrid use of PEF and microwave to modify starch structure, over the application of any one of the techniques, can enhance product quality and application. Microwave processing is improved by being used with PEF, which creates a porous granule structure of granule. Enhanced properties of starch modification where optimal conditions are applied include high water retaining capacity, low crystallinity, and controlled degree of gelatinization. Effective heating and better restructuring of the starch required in maintaining the necessary texture and functionality of food products are obtained by the combined effects of these techniques. Energy effectiency and lower processing times makes this combination better for large-scale industrial use where there precise alteration of the starch is needed (Ali et al., 2024). The comparative processing of starch materials using microwave and PEF offers an understanding of each method's potential and of the combination of the two in relation to their application in food processing.

10.6 SYNERGISTIC EFFECTS OF COMBINED PEF AND MICROWAVE TREATMENTS ON STARCH PROPERTIES

10.6.1 Combination of PEF and Microwave Methods for Enhanced Starch Modification

This new modification approach that combines MW and PEF technologies is assumed to be superior to the application of either technology alone because of the interaction effect observed with their concurrent application. Starch granules have a crystalline structure that can be modified using high-voltage electrical pulse of short duration to make the starch more porous for further treatment (Zhu, 2018). The combination of PEF and MW takes advantage of each individual technique; PEF initiates structural destruction with mild thermal impact, while MW transfers heat effectively, thus increasing gelatinization and solubility (Liu et al., 2019).

The dual application begins with PEF treatment, which serves as a preparatory step by weakening the internal bonds within the starch granules, thereby making them more receptive to subsequent microwave exposure. The PEF-induced electroporation disrupts the granules' semi-crystalline regions, increasing their water binding capacity and creating pathways for deeper microwave penetration. When MW treatment comes after PEF treatment, microwave energy is absorbed penetratingly by already permeabilized granules, which results in faster and more homogeneous gelatinization rates. In other words, sequential treatment increases the starch viscosity and thickening ability while decreasing processing operation time and the amount of energy needed compared to normal thermal processes (Baltacioglu et al., 2024).

One major benefit of integrating PEF and MW technologies is the ability to better control functional properties of starch. This two-stage modification process facilitates fine-tuning of gelatinization temperatures, viscosity, and swelling power, among others, to suit prescribed uses such as in food additives that are viscous or gel-like, films that can decay biologically, and pharmaceutical agents. For instance, in food application areas, the PEF-MW combination can produce starches with improved textures, stabilities, and retrogradation resistance, making them more suitable for instant soup powders or sauces as well as for ready-to-eat meals where quality must remain constant (Org & Ashogbon, 2018).

Also, this double strategy fits into the clean-label movement as less or no chemical additives need to be used during starch alteration, making the product more natural. In addition, the combined PEF-MW technique promotes sustainable processing by cutting down on water and power consumption, thus reducing environmental footprint. This is due to the fact that in non-foods, changes to the amount or construction of starches will improve their mechanical properties, which are important when it comes to flexibility, resistance, and degradation rates in bioplastics and drug delivery systems (Liu et al., 2019).

In general, combining PEF and MW technologies to modify starch is a new, efficient, and sustainable solution that opens up new avenues for using starch in a variety of industries. The integration of these two technologies not only enhances the functionality of starch but also helps in producing superior-quality products that can meet the changing requirements of today's consumers and industries.

10.6.2 Impact on Starch Gelatinization and Retrogradation

The effect of PEF and MW treatments on starch gelatinization and retrogradation is enormous, greatly affecting the structural and functional properties of starch to improve performance under different applications. In gelatinization, a vital process in the modification of starch, granules are swollen, absorb water, and disintegrate structurally through heating leading to loss of crystalline order. This transition is crucial in determining functional properties such as viscosity, texture, and stability (Suri & Singh, 2023). PEF treatment disrupts the semi-crystalline structure of starch granules by way of electroporation, which weakens internal hydrogen bonds resulting in increased penetration of water. This makes gelatinization more likely at lower temperatures over shorter time intervals, therefore enhancing solubility as well as swelling power. As a result, PEF-treated starches gelatinize faster and more effectively than untreated starches, making them suitable for instant food products that require quick thickening or gel formation (Zhu, 2018).

MW treatment also enhances these effects by using electromagnetic waves to provide quick and consistent heating, causing water molecules inside starch granules to vibrate violently. The internal heating hastens the dissolution of crystalline parts and leads to more even gelatinization, thereby resulting in starches with increased viscosity and smoothness. The combination of PEF and MW treatments synergistically increases the gelation properties of starch, producing modified starches that gel at lower energy levels with better functionality. This process is especially important when making foods such as soups, sauces, and some baked products for which uniformity in texture and instantaneous increase in viscosity are vital (Liu et al., 2019).

Degradation of amylose by PEF and destruction of amylopectin chains by MW treatments help prevent retrogradation. While PEF rearranges the structure of starch such that amylose alignment appears impossible, MW treatment attacks the remaining crystalline parts so as to stop the reconstitution of starch molecules (Org & Ashogbon, 2018). As a consequence, retrogradation is reduced, leading to improved quality and shelf life of starchy products such as bread, gels, and foodstuffs meant for refrigerated or frozen storage for which maintenance should last longer than usual. Overall, the combination of PEF and MW leads to modified starches with better functional properties, increased stability, and minor quality loss after processing, so that starch gelatinization and retrogradation are improved. PEF- and MW-modulated starches are widely considered as critical ingredients in food processing and other industries because they meet the requirements to be effective, sustainable, and high-performance components (Schmiele et al., 2018).

10.6.3 Improved Functional Properties: Viscosity, Solubility, and Swelling Power

The process of modifying starches through PEF and MW treatments leads to positive changes in viscosity, solubility, and swelling power. In addition, these non-thermal methods alter the semi-crystalline nature of the starch granules, enhancing their compatibility with water. Further, PEF treatment in starch leads to electroporation of starch granules which in turn disrupts hydrogen bonds that link crystalline regions. This process enhances the porosity of the granules, which can then take up more water and swell and dissolve more than before treatment. This results in modified starches with enhanced solubility and swelling power that are used mostly in the manufacture of instant foods such as soups and sauces that require rapid thickening and gelation (Castro et al., 2023). Research has shown that the effects are further amplified by microwave treatment, which facilitates heating in a rapid and uniform manner, thus speeding up gelatinization. Microwaves make the water molecules oscillate inside the starch so as to create localized heating that tears apart the granule structure from its interior. This results in improved granule swelling patterns characterized by uniformity and thoroughness,greatly increasing their capacity for water retention. The increased ability of the starch to swell rapidly and regularly has positive implications on its viscosity development, enhancing food product texture and mouthfeel (Baltacioglu et al., 2024).

The enhanced solubility of the starches modified by microwave treatment and PEF also allows for better dispersion, making instant powders and drinks requiring rapid hydration an ideal application. Furthermore, these treatments also increased viscosity, allowing manufacturers to obtain target viscosities at lower starch concentrations, thus increasing the efficiency and integrity of the products. On the whole, PEF- and microwave-modified starches have improved functional characteristics that are advantageous in food processing, pharmaceuticals, and other industries creating multifunctional materials to meet current performance demands and enhance sustainability.

10.7 INDUSTRIAL APPLICATIONS AND FUTURE DIRECTIONS OF DUAL MODIFICATION TECHNIQUES

10.7.1 Current Industrial Applications of Dual Modification Techniques

Starches are used in food industries based on various characteristics such as solubility, swelling power, thermal stability, digestibility, and pasting properties (Sudheesh et al., 2020). The popularity of dual modification processing methods has increased in several industries due to the modified starches' improved performance, versatility, and properties. In food industries, dual-modified starches are used as stabilizers, emulsifiers, thickeners, and agglutinants, while in other industries they act as drug carriers for controlled release or adsorbents for heavy metals (Ashogbon & Akintayo, 2014). They are also used in sauces and dressings to give desirable texture and consistency (Kumari & Sit, 2023). The thermal stability of products obtained at high processing temperatures is enhanced through these techniques. Dual modification methods have also been utilized in biodegradable starch-based products owing to heightened demand for green materials. The alterations modify the physical and mechanical properties of starch, making them appropriate for use in packaging material and biofilms. The range of applications of such materials in industry can expanded by modifying starch structure and degree of substitution. Dual-modified starches are becoming popular in the field of 3D printing, where their improved properties facilitate better layering and structural integrity during the printing process (Lagunes-Delgado et al., 2023). This allows for better printability and applications in different designs and shapes. Furthermore, this new approach demonstrates the adaptability of dual-modified starches beyond their conventional use and provides a platform for advancements in technology (Ashogbon, 2021).

10.7.2 Challenges and Opportunities in Scaling Up

In various industries, dual modification techniques, when scaled up, present both hurdles and optimistic prospects. It is vital to understand these factors in order to facilitate the execution of these methods on broader platforms so as to obtain improved quality of goods and operational effectiveness. Integrating numerous modification processes can be quite technologically complex. The optimization of parameters for large-scale applications is challenging because each technique has different conditions, timing, and interaction (Tufvesson et al., 2010). Scaling up dual modifications can pose a financial challenge, especially for small to medium-sized enterprises. For instance, scaling up advanced modifications may require considerable amounts of raw materials, machinery, and even operation costs. Therefore, it can be challenging to maintain consistent quality across huge volumes of products. Variations in final product traits arise from diverse raw material quality or process factors, which necessitates strict quality control measures to maintain standards (Kumari & Sit, 2023).

Despite the significant challenges, dual modification also presents many potential advantages to industries wishing to innovate and enhance their goods. The dual modification processes significantly improves the properties of starches, resulting in a product with better texture, increased nutritional profiles, and improved market competitiveness and consumer appeal. In the current market, there is more demand for innovative food products that can meet the needs of specific consumers. Dual-modified starches with unique properties can meet demands such as enhanced digestibility and sensory qualities. The complexity associated with scaling up dual modification techniques leads to collaborative relationships among industries, research institutions, and technology suppliers. This helps in terms of pooling resources, developing joint ventures, and knowledge sharing, thus promoting innovation and scalability (Prain et al., 2020).

10.7.3 Emerging Trends and Innovations in Starch Modification

The growing need for sustainable, functional, and health-focused components in the food, pharmaceutical, and industrial sectors is driving the new trends and technological advancements in starch modification. New technologies and processes are being discovered in starches, with specific properties like improved solubility, better functionality, and reduced environmental impact to meet increasing demand. Environmental concerns are becoming serious; hence, there is a focus on reducing the environmental footprint of starch modification processes. Many green alternatives are being developed to replace thermal or traditional chemical methods such as enzyme treatment, ultrasound, cold plasma treatment, and supercritical carbon dioxide processing (Chemat et al., 2020). New trends are also emerging in combining multiple modification techniques to achieve starches with unique characteristics. It is possible to enhance process performance and tailor starch properties more accurately by employing hybrid methods. There is an increasing demand for starches that provide health benefits such as blood sugar control or support for digestive health (Bojarczuk et al., 2022). Some starch-based products have been greatly improved through nanotechnology and are now widely used in food packaging as well as in pharmaceutical sectors, thus significantly increasing their application. The use of starch in 3D printing and biopolymers is an innovative field aimed at developing biodegradable and eco-friendly substances (Aziz et al., 2022).

10.7.4 Future Research Directions and Potential Applications

The future of scaling up dual modification techniques is promising because of innovations in technology and sustainable practices. Investing in automation and advanced process control systems can more efficiently overcome the complexities of dual modifications. Automation can enable more precise control over modification processes, resulting in consistent quality, which helps reduce labor costs when dealing with large-scale application (Haleem et al., 2021). Ongoing research and development are critical for optimizing dual modification processes. Focused studies on the interactions between different modification techniques can provide insights into more efficient process designs, ultimately aiding in the scaling efforts. There is an increasing focus on sustainability in food and material manufacturing. Scaling up dual modification techniques in environmentally friendly ways could attract customers and businesses that prioritize sustainability, thus gaining a competitive edge in the market (Lambin et al., 2020). Overcoming the challenges and seizing the opportunities associated with scaling up dual modification techniques is important for fostering innovation across various industries. A strategic approach aimed at technical optimization, market adaptability, and cost management will be key to successful implementation.

10.8 CONCLUSION

Dual modification methods for starch involving PEF and microwave may represent a new and promising route towards improving functional properties of starch in an extensive list of applications. Starch is one of the most versatile and widely applied polysaccharides, but in many instances modification through chemical, physical, or enzymatic modifications is necessary for improved performance in food and non-food sectors. The electrical pulses from PEF impose structural changes on starch granules by changing their molecular arrangement. In general, PEF tends to improve the gelatinization, solubility, and other functional properties of the starch, including its water retention capacity and pasting behavior. Meanwhile, microwave techniques offer rapid uniform heating that assists in the promotion of molecular breakdown without extensive thermal degradation. This is particularly useful for high efficiency and energy saving, with only minimal effects on the nutritional and functional integrity of starch. The combination of PEF and microwave-assisted techniques permits a more controlled and effective starch modification process, providing significant enhancements over the traditional chemical and thermal methods. Where PEF and microwave treatments are applied together, synergy has been reported between the two methods, improving starch digestibility, viscosity, and thermal stability. These modifications make starch more appropriate for use in various food formulations, biodegradable materials, and pharmaceuticals. Dual modification may also have the potential to reduce chemical use, making it more environmentally friendly. These techniques should be fine-tuned, made application-specific, and scaled up for use in industries in future research. In conclusion, dual application of microwave- and PEF-assisted methods offers an innovative, efficient, and sustainable solution for modification of starch, paving the way for broader industrial usage.

REFERENCES

Achayuthakan, P., Wongsagonsup, R., Sriprablom, J., Suphantharika, M., & Intra, P. (2023). Effect of pulsed electric field treatment on the protein, digestibility, and physicochemical properties of starch granules in wheat flour. *Polymers*, *15*(20). https://doi.org/10.3390/polym15204087

Alabi, K. P., Oyebode, O. O., Fadeyibi, A., Obajemihi, O. I., & Olusegun, O. (2021). Effect of pulsed electric field on structural modification and physicochemical properties of starch-based edible films. *LAUTECH Journal of Engineering and Technology*, *15*(1), 99–107.

Ali, M., Liao, L., Zeng, X. A., Manzoor, M. F., & Mazahir, M. (2024). Impact of sustainable emerging pulsed electric field processing on textural properties of food products and their mechanisms: An updated review. *Journal of Agriculture and Food Research*, *15*, 101076. https://doi.org/10.1016/J.JAFR.2024.101076

Amaraweera, S. M., Gunathilake, C., Gunawardene, O. H. P., Fernando, N. M. L., Wanninayaka, D. B., Dassanayake, R. S., & Manipura, A. (2021). Development of starch-based materials using current modification techniques and their applications: A review. *Molecules*, *26*(22), 1–30. https://doi.org/10.3390/molecules26226880

Arshad, R. N., Abdul-Malek, Z., Roobab, U., Munir, M. A., Naderipour, A., Qureshi, M. I., & Aadil, R. M. (2021). Pulsed electric field: A potential alternative towards a sustainable food processing. *Trends in Food Science and Technology*, *111*, 43–54. https://doi.org/10.1016/j.tifs.2021.02.041

Ashogbon, A. O. (2018). Current research addressing physical modification of starch from various botanical sources. *Global Nutrition and Dietetics*, 7, 2.

Ashogbon, A. O. (2021). Dual modification of various starches: Synthesis, properties and applications. *Food Chemistry*, *342*, 128325. https://doi.org/10.1016/j.foodchem.2020.128325

Ashogbon, A. O., & Akintayo, E. T. (2014). Recent trend in the physical and chemical modification of starches from different botanical sources: A review. *Starch - Stärke*, *66*(1–2), 41–57. https://doi.org/10.1002/star.201300106

Aziz, T., Ullah, A., Ali, A., Shabeer, M., Shah, M. N., Haq, F., & Khan, F. U. (2022). Manufactures of bio-degradable and bio-based polymers for bio-materials in the pharmaceutical field. *Journal of Applied Polymer Science*, *139*(29), 1–21. https://doi.org/10.1002/app.52624

Bakr, T. M. A. (2020). Microwave applications in food processing: An overview. *Alexandria Journal of Food Science and Technology*, *17*(2), 11–22. https://doi.org/10.21608/ajfs.2020.150658

Bakshi, N., Jain, S., Raman, A., & Pant, T., Nayik, G. A., Ranjha., Zeng, X. A., Irfan, S., Zahra, S. M. (2023). Microwave: An overview. In *Ultrasound and microwave for food processing: Synergism for preservation and extraction, Elsevier* (pp. 19–59). https://doi.org/10.1016/B978-0-323-95991-9.00014-X

Baltacioglu, C., Yetisen, M., Baltacioglu, H., Karacabey, E., & Buzrul, S. (2024). Pulsed electric fields (PEF) treatment prior to hot-air and microwave drying of yellow- and purple-fleshed potatoes. *Potato Research*, *68*, 1171–1188. https://doi.org/10.1007/s11540-024-09781-6

Bertoft, E. (2017). Understanding starch structure: Recent progress. *Agronomy*, *7*(3). https://doi.org/10.3390/agronomy7030056

Bojarczuk, A., Skąpska, S., Mousavi Khaneghah, A., & Marszałek, K. (2022). Health benefits of resistant starch: A review of the literature. *Journal of Functional Foods*, *93*, 105094. https://doi.org/10.1016/J.JFF.2022.105094

Brasoveanu, M., & Nemtanu, M. R. (2013). Behaviour of starch exposed to microwave radiation treatment. *Starch - Stärke*, *66*(1–2), 3–14. https://doi.org/10.1002/star.201200191

Castro, L. M. G., Alexandre, E. M. C., Saraiva, J. A., & Pintado, M. (2023). Starch extraction and modification by pulsed electric fields. *Food Reviews International*, *39*(4), 2161–2182. https://doi.org/10.1080/87559129.2021.1945620

Chemat, F., Abert Vian, M., Fabiano-Tixier, A. S., Nutrizio, M., Režek Jambrak, A., Munekata, P. E. S., & Cravotto, G. (2020). A review of sustainable and intensified techniques for extraction of food and natural products. *Green Chemistry*, *22*(8), 2325–2353. https://doi.org/10.1039/c9gc03878g

Compart, J., Singh, A., Fettke, J., & Apriyanto, A. (2023). Customizing starch properties: A review of starch modifications and their applications. *Polymers*, *15*(16). https://doi.org/10.3390/polym15163491

Cornejo-Ramírez, Y. I., Martínez-Cruz, O., Del Toro-Sánchez, C. L., Wong-Corral, F. J., Borboa-Flores, J., & Cinco-Moroyoqui, F. J. (2018). The structural characteristics of starches and their functional properties. *CYTA - Journal of Food*, *16*(1), 1003–1017. https://doi.org/10.1080/19476337.2018.1518343

Datta, A. K., & Rakesh, V. (2013). Principles of microwave combination heating. *Comprehensive Reviews in Food Science and Food Safety*, *12*(1), 24–39. https://doi.org/10.1111/j.1541-4337.2012.00211.x

Deng, X., Huang, H., Huang, S., Yang, M., Wu, J., Ci, Z., & Zhang, D. (2022). Insight into the incredible effects of microwave heating: Driving changes in the structure, properties and functions of macromolecular nutrients in novel food. *Frontiers in Nutrition*, *9*. https://doi.org/10.3389/fnut.2022.941527

Duque, S. M. M., Leong, S. Y., Agyei, D., Singh, J., Larsen, N., & Oey, I. (2020). Understanding the impact of pulsed electric fields treatment on the thermal and pasting properties of raw and thermally processed oat flours. *Food Research International*, *129*, 108839. https://doi.org/10.1016/j.foodres.2019.108839

Duque, S. M. M., Leong, S. Y., Agyei, D., Singh, J., Larsen, N., Sutton, K., & Oey, I. (2022). Understanding the mechanism of how pulsed electric fields treatment affects the digestibility and characteristics of starch in oat flour. *Applied Sciences (Switzerland)*, *12*(20). https://doi.org/10.3390/app122010293

Ekezie, C., Glad, F., Sun, D., Han, Z., & Cheng, J. (2017). Microwave-assisted food processing technologies for enhancing product quality and process efficiency: A review of recent developments. *Trends in Food Science & Technology*, 67, 68–69. https://doi.org/10.1016/j.tifs.2017.05.014

Fan, D., Ma, S., Wang, L., Zhao, J., Zhang, H., & Chen, W. (2012). Effect of microwave heating on optical and thermal properties of rice starch. *Starch - Stärke*, *64*(9), 740–744. https://doi.org/10.1002/star.201200029

Grgić, I., Aćkar, D., Barišić, V., Vlainić, M., Knežević, N., & Medverec Knežević, Z. (2019). Nonthermal methods for starch modification: A review. *Journal of Food Processing and Preservation*, *43*(12), 1–13. https://doi.org/10.1111/jfpp.14242

Gupta, R. K., Guha, P., & Srivastav, P. P. (2022). Natural polymers in bio-degradable/edible film: A review on environmental concerns, cold plasma technology and nanotechnology application on food packaging-A recent trends. *Food Chemistry Advances*, *1*, 100135. https://doi.org/10.1016/j.focha.2022.100135

Gupta, R. K., Guha, P., & Srivastav, P. P. (2023). Effect of high voltage dielectric barrier discharge (DBD) atmospheric cold plasma treatment on physicochemical and functional properties of taro (*Colocasia esculenta*) starch. *International Journal of Biological Macromolecules*, *253*, 126772. https://doi.org/10.1016/j.ijbiomac.2023.126772

Gupta, R. K., Guha, P., & Srivastav, P. P. (2024a). Insights into the physicochemical, crystalline, multistructural, thermal and digestibility behavior of taro starch (*Colocasia esculenta*) modified by cold plasma treatment with high voltage dielectric barrier discharge (HVDBD): An alternative technique for chemical modification of starch and environmental sustainability. *Journal of Cleaner Production*, *461*, 142710. https://doi.org/10.1016/j.jclepro.2024.142710

Gupta, R. K., Guha, P., & Srivastav, P. P. (2024b). Dielectric barrier discharge plasma: A green and novel method to change the structure of taro peel starch and improve the physicochemical properties of taro peel starch films. *Plasma Processes and Polymers*, *21*(9), 2400047. https://doi.org/10.1002/ppap.202400047

Gupta, R. K., Guha, P., & Srivastav, P. P. (2024c). Exploring the potential of taro (Colocasia esculenta) starch: Recent developments in modification, health benefits, and food industry applications. *Food Bioengineering*, *3*(3), 365–379. https://doi.org/10.1002/fbe2.12103

Gupta, R. K., Guha, P., & Srivastav, P. P. (2024d). Physical action of nonthermal cold plasma technology for starch modification. *Food Physics*, *1*, 100011. https://doi.org/10.1016/j.foodp.2024.100011

Gupta, R. K., Guha, P., & Srivastav, P. P. (2024e). Effect of cold plasma treatment and plasma-activated water on physicochemical and structural properties of starch: A green and novel approach for environmental sustainability. *Plasma Processes and Polymers*, *21*(4), 2300204. https://doi.org/10.1002/ppap.202300204

Gupta, R. K., Pipliya, S., Patel, J., Srivastav, P. P., Castro-Muñoz, R., Wang, C. K., & Nemţanu, M. R. (2025a). Essential oil-embedded starch based films for active packaging: Insights into migration, stability, and preservation efficiency. *Trends in Food Science & Technology*, 163, 105149. https://doi.org/10.1016/j.tifs.2025.105149

Gupta, R. K., Gnana Moorthy Eswaran, U., Kumar, S., Pipliya, S., Kaur, B., & Srivastav, P. P. (2025b). Synergistic modification of starch polysaccharide via sequential treatments with ultrasonic and cold plasma technology. *Journal of Food Processing and Preservation*, *2025*(1), 9063007. https://doi.org/10.1155/jfpp/9063007

Haleem, A., Javaid, M., Singh, R. P., Rab, S., & Suman, R. (2021). Hyperautomation for the enhancement of automation in industries. *Sensors International*, *2*, 100124. https://doi.org/10.1016/J.SINTL.2021.100124

Han, Z., Han, Y., Wang, J., Liu, Z., Buckow, R., & Cheng, J. (2020). Effects of pulsed electric field treatment on the preparation and physicochemical properties of porous corn starch derived from enzymolysis. *Journal of Food Processing and Preservation*, *44*(3), 1–10. https://doi.org/10.1111/jfpp.14353

Harni, M., Rini, & Suliansyah, I. (2023). The functional properties of starch from arrowroot (*Maranta arundinacea*) tubers using microwave assisted extraction (MAE). *IOP Conference Series: Earth and Environmental Science*, *1182*, 012046. https://doi.org/10.1088/1755-1315/1182/1/012046

Hong, J., Chen, R., Zeng, X. A., & Han, Z. (2016). Effect of pulsed electric fields assisted acetylation on morphological, structural and functional characteristics of potato starch. *Food Chemistry*, *192*, 15–24. https://doi.org/10.1016/j.foodchem.2015.06.058

Hornung, P. S., Barbi, R. C. T., Teixeira, G. L., Ávila, S., da Silva, F. L. A., Lazzarotto, M., & Ribani, R. H. (2018). Brazilian Amazon white yam (*Dioscorea* sp.) starch: Impact on functional properties due to chemical and physical modifications processes. *Journal of Thermal Analysis and Calorimetry*, *134*(3), 2075–2088. https://doi.org/10.1007/s10973-018-7353-1

Kärkkäinen, J., Lappalainen, K., Joensuu, P., & Lajunen, M. (2011). HPLC-ELSD analysis of six starch species heat-dispersed in [BMIM]Cl ionic liquid. *Carbohydrate Polymers*, *84*(1), 509–516. https://doi.org/10.1016/J.CARBPOL.2010.12.011

Kim, H. Y., Park, S. S., & Lim, S. T. (2015). Preparation, characterization and utilization of starch nanoparticles. *Colloids and Surfaces B: Biointerfaces*, *126*, 607–620. https://doi.org/10.1016/J.COLSURFB.2014.11.011

Kumari, B., & Sit, N. (2023). Comprehensive review on single and dual modification of starch: Methods, properties and applications. *International Journal of Biological Macromolecules*, *253*, 126952. https://doi.org/10.1016/J.IJBIOMAC.2023.126952

Lagunes-Delgado, C., Agama-Acevedo, E., & Bello-Pérez, L. A. (2023, November). Dual starch modifications to expand its end-uses: A review. *Starch - Stärke*. https://doi.org/10.1002/star.202300153

Lambin, E. F., Kim, H., Leape, J., & Lee, K. (2020). Scaling up solutions for a sustainability transition. *One Earth*, *3*(1), 89–96. https://doi.org/10.1016/J.ONEEAR.2020.06.010

Leong, S. Y., & Oey, I. (2022). Application of novel thermal technology in foods processing. *Foods*, *11*(1), 1–5. https://doi.org/10.3390/foods11010125

Lewandowicz, G., Jankowski, T., & Fornal, J. (2000). Effect of microwave radiation on physico-chemical properties and structure of cereal starches. *Carbohydrate Polymers*, *42*(2), 193–199. https://doi.org/10.1016/S0144-8617(99)00155-1

Lewicka, K., Siemion, P., & Kurcok, P. (2015). Chemical modifications of starch: Microwave effect. *International Journal of Polymer Science*, *2015*. https://doi.org/10.1155/2015/867697

Li, H., Qi, Y., Zhao, Y., Chi, J., & Cheng, S. (2019a). Starch and its derivatives for paper coatings: A review. *Progress in Organic Coatings*, *135*, 213–227. https://doi.org/10.1016/j.porgcoat.2019.05.015

Li, Q., Wu, Q. Y., Jiang, W., Qian, J. Y., Zhang, L., Wu, M., & Wu, C. S. (2019b). Effect of pulsed electric field on structural properties and digestibility of starches with different crystalline type in solid state. *Carbohydrate Polymers*, *207*, 362–370. https://doi.org/10.1016/j.carbpol.2018.12.001

Liu, C., Grimi, N., Lebovka, N., & Vorobiev, E. (2019). Convective air, microwave, and combined drying of potato pre-treated by pulsed electric fields. *Drying Technology*, *37*(13), 1704–1713. https://doi.org/10.1080/07373937.2018.1536065

Luo, Z., He, X., Fu, X., Luo, F., & Gao, Q. (2006). Effect of microwave radiation on the physicochemical properties of normal maize, waxy maize and amylomaize V starches. *Starch/Staerke*, *58*(9), 468–474. https://doi.org/10.1002/star.200600498

Ma, M., Zhang, Y., Chen, X., Li, H., Sui, Z., & Corke, H. (2020). Microwave irradiation differentially affect the physicochemical properties of waxy and non-waxy hull-less barley starch. *Journal of Cereal Science*, *95*, 103072. https://doi.org/10.1016/J.JCS.2020.103072

Ma, Z., Yin, X., Hu, X., Li, X., Liu, L., & Boye, J. I. (2018). Structural characterization of resistant starch isolated from Laird lentils (*Lens culinaris*) seeds subjected to different processing treatments. *Food Chemistry*, *263*, 163–170. https://doi.org/10.1016/J.FOODCHEM.2018.04.122

Maia, J. L., Cardoso, J. S., Mastrantonio, D. J. S., Bierhals, C. K., Moreira, J. B., Costa, J. A. V., & Morais, M. G. (2020). Microalgae starch: A promising raw material for the bioethanol production. *International Journal of Biological Macromolecules*, *165*, 2739–2749. https://doi.org/10.1016/j.ijbiomac.2020.10.159

Maloney, N., & Harrison, M. (2016). Advanced heating technologies for food processing. In C. E. Leadley (Ed.), *Innovation and future trends in food manufacturing and supply chain technologies* (pp. 203–256). Woodford Publishing. https://doi.org/10.1016/B978-1-78242-447-5.00008-3

Maniglia, B. C., Pataro, G., Ferrari, G., Augusto, P. E. D., Le-Bail, P., & Le-Bail, A. (2021b). Pulsed electric fields (PEF) treatment to enhance starch 3D printing application: Effect on structure, properties, and functionality of wheat and cassava starches. *Innovative Food Science and Emerging Technologies*, *68*, 102602. https://doi.org/10.1016/j.ifset.2021.102602

Mannozzi, C., Fauster, T., Haas, K., Tylewicz, U., Romani, S., Dalla Rosa, M., & Jaeger, H. (2018). Role of thermal and electric field effects during the pre-treatment of fruit and vegetable mash by pulsed electric fields (PEF) and ohmic heating (OH). *Innovative Food Science and Emerging Technologies*, *48*(2017), 131–137. https://doi.org/10.1016/j.ifset.2018.06.004

Marta, H., Cahyana, Y., Bintang, S., Soeherman, G. P., & Djali, M. (2022). Physicochemical and pasting properties of corn starch as affected by hydrothermal modification by various methods. *International Journal of Food Properties*, *25*(1), 792–812. https://doi.org/10.1080/10942912.2022.2064490

Masina, N., Choonara, Y. E., Kumar, P., du Toit, L. C., Govender, M., Indermun, S., & Pillay, V. (2017). A review of the chemical modification techniques of starch. *Carbohydrate Polymers*, *157*, 1226–1236. https://doi.org/10.1016/j.carbpol.2016.09.094

Meloni, E., Iervolino, G., & Palma, V. (2023). Basics of microwave heating and recent advances. *Advances in Microwave-Assisted Heterogeneous Catalysis*, *45*, 1–24. https://doi.org/10.1039/bk9781837670277-00001

Mhaske, P., Farahnaky, A., & Majzoobi, M. (2024). Advancements in pulse starches: Exploring non-thermal modification methods. *Foods*, *13*(16), 1–21. https://doi.org/10.3390/foods13162493

Moirangthem, T. T., Macana, R., & Baik, O. D. (2020). Characterization of 50-ohm radio frequency heating of bulk canola seeds (*Brassica napus*. L) in a tubular applicator with parallel electrodes and post-treatment quality. *Innovative Food Science & Emerging Technologies*, *64*, 102409. https://doi.org/10.1016/J.IFSET.2020.102409

Oyeyinka, S. A., Umaru, E., Olatunde, S. J., & Joseph, J. K. (2019). Effect of short microwave heating time on physicochemical and functional properties of Bambara groundnut starch. *Food Bioscience*, *28*, 36–41. https://doi.org/10.1016/J.FBIO.2019.01.005

Pataro, G., Carullo, D., Bakar Siddique, M. A., Falcone, M., Donsì, F., & Ferrari, G. (2018). Improved extractability of carotenoids from tomato peels as side benefits of PEF treatment of tomato fruit for more energy-efficient steam-assisted peeling. *Journal of Food Engineering*, *233*, 65–73. https://doi.org/10.1016/j.jfoodeng.2018.03.029

Prain, G., Wheatley, C., Odsey, C., Verzola, L., Bertuso, A., Roa, J., & Naziri, D. (2020). Research-development partnerships for scaling complex innovation: Lessons from the Farmer Business School in IFAD-supported loan-grant collaborations in Asia. *Agricultural Systems*, *182*, 102834. https://doi.org/10.1016/J.AGSY.2020.102834

Punia, S. (2020). Barley starch modifications: Physical, chemical and enzymatic - A review. *International Journal of Biological Macromolecules*, *144*, 578–585. https://doi.org/10.1016/j.ijbiomac.2019.12.088

Qiu, S., Abbaspourrad, A., & Padilla-Zakour, O. I. (2021). Changes in the glutinous rice grain and physicochemical properties of its starch upon moderate treatment with pulsed electric field. *Foods*, *10*(2), 1–14. https://doi.org/10.3390/foods10020395

Ramírez-Brewer, D., Quintana, S. E., & García-Zapateiro, L. A. (2023). Effect of microwave treatment on technological, physicochemical, rheological and microstructural properties of mango (*Mangifera indica*) kernel starch variety tommy and sugar. *LWT*, *187*, 115311. https://doi.org/10.1016/J.LWT.2023.115311

Rostamabadi, H., Demirkesen, I., Mert, B., Barua, S., Colussi, R., Frasson, S. F., & Falsafi, S. R. (2024). Starch digestibility: How single, double, and multiple physicochemical modifications change nutritional attributes of starch? *Food Frontiers*, *5*(4), 1410–1444. https://doi.org/10.1002/fft2.396

Salee, N., Naruenartwongsakul, S., Chaiyana, W., Yawootti, A., Hunsakul, K., Tinpovong, B., & Utama-Ang, N. (2023). Comparison of pulse electric field, microwave and ultrasonic pretreatment prior to black rice extraction on antioxidant and sirtuin1 enzyme stimulating activities. *Food Science and Technology (Brazil)*, *43*, 1–10. https://doi.org/10.1590/fst.123022

Schmiele, M., Sampaio, U. M., Gomes, P. T. G., & Silva Clerici, P. M. T. (2018). Physical modifications of starch. In *Starches for food application: Chemical, technological and health properties*. https://doi.org/10.1016/B978-0-12-809440-2.00006-X

Shah, U., Gani, A., Ashwar, B. A., Shah, A., Wani, I. A., & Masoodi, F. A. (2016). Effect of infrared and microwave radiations on properties of Indian horse chestnut starch. *International Journal of Biological Macromolecules*, *84*, 166–173. https://doi.org/10.1016/J.IJBIOMAC.2015.12.020

Shevchenko, T. V., Kondratov, E. A., Ul'rich, E. V., Popov, A. M., Lobanova, V. S., & Tokmakova, I. O. (2013). On the variation of physicochemical properties of starch samples treated by microwave radiation. *Russian Journal of Applied Chemistry*, *86*(1), 127–131. https://doi.org/10.1134/S1070427213010242

Shrivastava, A., Gupta, R. K., & Srivastav, P. P. (2024). Exploring novel frontiers of advancements in purple yam (*Dioscorea alata* L.) starch extraction, modification, characterization, applications in food and other industries. *Measurement: Food*, *15*, 100196. https://doi.org/10.1016/j.meafoo.2024.100196

Soltanzadeh, M., Peighambardoust, S. H., Gullon, P., Hesari, J., Gullón, B., Alirezalu, K., & Lorenzo, J. (2022). Quality aspects and safety of pulsed electric field (PEF) processing on dairy products: A comprehensive review. *Food Reviews International*, *38*(S1), 96–117. https://doi.org/10.1080/87559129.2020.1849273

Sudheesh, C., Sunooj, K. V., Navaf, M., Bhasha, S. A., George, J., Mounir, S., & Sajeevkumar, V. A. (2020). Hydrothermal modifications of nonconventional kithul (*Caryota urens*) starch: Physico-chemical, rheological properties and in vitro digestibility. *Journal of Food Science and Technology*, *57*(8), 2916–2925. https://doi.org/10.1007/s13197-020-04323-7

Suri, S., & Singh, A. (2023). Modification of starch by novel and traditional ways: Influence on the structure and functional properties. *Sustainable Food Technology*, *1*(3), 348–362. https://doi.org/10.1039/d2fb00043a

Tao, Y., Yan, B., Fan, D., Zhang, N., Ma, S., Wang, L., & Zhang, H. (2020). Structural changes of starch subjected to microwave heating: A review from the perspective of dielectric properties. *Trends in Food Science and Technology*, *99*, 593–607. https://doi.org/10.1016/j.tifs.2020.02.020

Tesfaye, T., Gibril, M., Sithole, B., Ramjugernath, D., Chavan, R., Chunilall, V., & Gounden, N. (2018). Valorisation of avocado seeds: Extraction and characterisation of starch for textile applications. *Clean Technologies and Environmental Policy*, *20*(9), 2135–2154. https://doi.org/10.1007/s10098-018-1597-0

Thirumdas, R., Sarangapani, C., & Annapure, U. S. (2015). Cold plasma: A novel non-thermal technology for food processing. *Food Biophysics*, *10*(1), 1–11. https://doi.org/10.1007/s11483-014-9382-z

Tufvesson, P., Fu, W., Jensen, J. S., & Woodley, J. M. (2010). Process considerations for the scale-up and implementation of biocatalysis. *Food and Bioproducts Processing*, *88*(1), 3–11. https://doi.org/10.1016/J.FBP.2010.01.003

Vadivambal, R., & Jayas, D. S. (2007). Changes in quality of microwave-treated agricultural products: A review. *Biosystems Engineering*, *98*(1), 1–16. https://doi.org/10.1016/J.BIOSYSTEMSENG.2007.06.006

Venkatesh, M. S., & Raghavan, G. S. V. (2004). An overview of microwave processing and dielectric properties of agri-food materials. *Biosystems Engineering*, *88*(1), 1–18. https://doi.org/10.1016/J.BIOSYSTEMSENG.2004.01.007

Versteeg, C. (2016). Next generation of innovative food processing technologies: Benefits and challenges. In *Innovative food processing technologies: Extraction, separation, component modification and process intensification*. Elsevier. https://doi.org/10.1016/B978-0-08-100294-0.00018-3

Wang, N., Li, C., Miao, D., Hou, H., Dai, Y., Zhang, Y., & Wang, B. (2023). The effect of non-thermal physical modification on the structure, properties and chemical activity of starch: A review. *International Journal of Biological Macromolecules*, *251*, 126200. https://doi.org/10.1016/J.IJBIOMAC.2023.126200

Wang, Y., & Zhang, G. (2021). The preparation of modified nanostarch and its application in food industry. *Food Research International*, *140*, 110009. https://doi.org/10.1016/j.foodres.2020.110009

Witczak, M., Ziobro, R., Juszczak, L., & Korus, J. (2016). Starch and starch derivatives in gluten-free systems: A review. *Journal of Cereal Science*, *67*(2015), 46–57. https://doi.org/10.1016/j.jcs.2015.07.007

Xu, X., Chen, Y., Luo, Z., & Lu, X. (2019). Different variations in structures of A- and B-type starches subjected to microwave treatment and their relationships with digestibility. *LWT*, *99*, 179–187. https://doi.org/10.1016/J.LWT.2018.09.072

Yang, Q., Qi, L., Luo, Z., Kong, X., Xiao, Z., Wang, P., & Peng, X. (2017). Effect of microwave irradiation on internal molecular structure and physical properties of waxy maize starch. *Food Hydrocolloids*, *69*, 473–482. https://doi.org/10.1016/J.FOODHYD.2017.03.011

Yi, M., Tang, X., Liang, S., He, R., Huang, T., Lin, Q., & Zhang, R. (2024). Effect of microwave alone and microwave-assisted modification on the physicochemical properties of starch and its application in food. *Food Chemistry*, *446*, 138841. https://doi.org/10.1016/J.FOODCHEM.2024.138841

Zailani, M. A., Kamilah, H., Husaini, A., Awang Seruji, A. Z. R., & Sarbini, S. R. (2022). Functional and digestibility properties of sago (*Metroxylon sagu*) starch modified by microwave heat treatment. *Food Hydrocolloids*, *122*(April 2021), 107042. https://doi.org/10.1016/j.foodhyd.2021.107042

Zhang, J., Wang, Z. W., & Shi, X. M. (2009). Effect of microwave heat/moisture treatment on physicochemical properties of *Canna edulis* Ker starch. *Journal of the Science of Food and Agriculture*, *89*(4), 653–664. https://doi.org/10.1002/jsfa.3497

Zhong, Y., Liang, W., Pu, H., Blennow, A., Liu, X., & Guo, D. (2019). Short-time microwave treatment affects the multiscale structure and digestive properties of high-amylose maize starch. *International Journal of Biological Macromolecules*, *137*, 870–877. https://doi.org/10.1016/J.IJBIOMAC.2019.07.025

Zhu, F. (2018). Modifications of starch by electric field based techniques. *Trends in Food Science and Technology*, *75*, 158–169. https://doi.org/10.1016/j.tifs.2018.03.011

11 Effect of Cold Plasma and PEF Treatment on the Functional and Structural Properties of Starch

Priyanka Pandharkar, Sandhya R. Shewale, and Uday S. Annapure
Institute of Chemical Technology, Mumbai-Marathwada Campus, Jalna

ABBREVIATIONS

APPJ	atmospheric pressure plasma jet
CP	cold plasma
DBD	dielectric barrier discharge
DH	dry heating
DS	degree of substitution
DSC	differential scanning calorimetry
FTIR	Fourier transform infrared spectroscopy
G′	storage modulus
G″	loss modulus
HPP	high hydrostatic pressure
OAC	oil absorption capacity
OSA	octenyl succinic anhydride
PAV	peak apparent viscosity
PEF	pulse electric field
RC	relative crystallinity
RDS	rapidly digestible starch
RNS	reactive nitrogen species
ROS	reactive oxygen species
RS	resistant starch
SC	swelling capacity
SDS	slowly digestible starch
TAV	trough apparent viscosity
UHP	ultra-high pressure
WAC	water absorption capacity
WMS	waxy maize starch
WSI	water solubility index

11.1 INTRODUCTION

The world's second-largest biomass is starch, followed by cellulose. In numerous biological and environmental activities, starch is utilized as an energy source and energy store. Starch has been a crucial energy source in human diets but has gained recognition for its nutritional significance as a source of dietary fibre over the last three decades. The common sources of starch in human diet include cereals (e.g., maize, wheat, and red rice), pulses (e.g., mung bean), millets (e.g., kodo, proso, pearl, sorghum, and finger millet), and tubers (e.g., potato, taro, cassava, and tapioca) (Gupta et al., 2023; Shrivastava et al., 2024). Starch in its native form has limitations from an application point of view, such as poor insolubility in water affecting its thickening ability, resistance to enzyme hydrolysis, and quick setting/solidification upon cooling. Therefore, starch often requires modification before or during food processing to give it characteristics suitable for use as a binding, gel-forming agent, stabilizer, thickener, or moisture retention agent (Gupta et al., 2022). To improve starch's functional attributes, it is modified using conventional pre-gelatinization, hydrothermal, chemical and enzymatic methods. However, these techniques have drawbacks such as high temperature requirements, production of waste chemical residues, and longer processing times that cause the degradation of starch by disturbing its chemical structure, thus affecting its functional properties. To overcome the limitation of conventional methods, novel alternative starch modification methods such as pulse electric field (PEF), cold plasma, high hydrostatic pressure processing (HPP), and microwave processing have been explored in the literature (Eliasson et al., 2004). Consumer markets are seeing a growing demand for healthier products, driving the development of modified starches that could fulfil the applicability needs of various food formulations without altering the properties of the final products. According to Bertolin (2010), Whistler and Paschall (1965), Wurzburg (1989), Eliasson (2004), BeMiller and Whistler (2009), and other sources. Starch has been utilized in multiple industries, including packaging, fabrics, medicines, and beauty products(Gupta et al., 2025a).

11.1.1 Overview of Cereal, Millet, Pulse, and Tuber Starches

Starch granules vary based on their origin or source. For instance, tuber starch granules are typically oval-shaped, are copious, and have an eccentric hilum (Jane et al., 1994; Gupta et al., 2024a), while starches with the biggest granules are those found in potatoes. Cereal starch granules (corn, oats, and rice) have round or polygonal shapes. Maize starch has fibrous granules that look like protrusions that mimic buds. The starch granules in legumes and tubers are bigger than those from most cereals. Starch from legume seeds resembles bean pod and contains a central elongation or star pattern. Millet starch granules are mostly polygonal

DOI: 10.1201/9781003564287-11

and spherical. Millet starch particles (e.g., in foxtail, pearl, proso, and finger millet) were found to be in the size range from 2.5 to 24 μm (Marcone et al., 2014). The majority of millet starch granules are polygonal, with a few spherical granules; for example, pearl millet consist of round, oval, and irregularly shaped granules (Sharma et al., 2015). A few starch granules have visible surface holes, which substantially increases the accessible surface area of a microporous structure of starch granules than the boundary surface area.

11.1.2 Importance of Functional and Structural Properties of Starches for Food Application

Starch serves multiple functions in various food formulations; these functions are determined by its structural characteristics, which impact its functional properties. As a result, in selecting a type of starch for incorporation into a food formulation, it is vital to consider the amount of energy or calories increased by starch addition and the starch's effects on the appearance, flavour, and overall quality of the food product. Other important factors to be considered are the specific product type and its composition, manufacturing process, storage approaches, and the product's lifespan. Because of this, to appropriately select starches for given food products, we must have deep knowledge of a product's compositions and its processing procedure. The functional attributes of starch, including gelatinization, viscosity, water absorption capacity (WAC), oil absorption capacity (OAC), and digestibility, influence the texture, stability, and moisture retention of the food products. However, the most essential attribute of starch is gelatinization temperature. Heating starch granules with water causes the granules to swell and gradually rupture; consequently, viscosity increases, which affects the texture and thickness of the products, for instance, in sauces, condiments, desserts, and soups. Viscosity is another crucial functional characteristic that affects the mouthfeel and stability of products made using starch. Viscosity and gel formation are closely linked properties of starch that are important for processed foods, including fillings, gravies, and baked goods. Additionally, important functional properties such as WAC and OAC impact the hydration and fat-binding capacities of starch. These properties are necessary for baked goods, snacks, and sausages even though they alter the food's texture and durability by emulsifying fat or retaining moisture in the food. Digestibility is also a crucial attribute of starches, especially with respect to nutritional value and glycaemic responses during digestion. Rapidly digestible starches severely impact the rise in blood sugar levels, while resistant starches positively affect the gut health; hence the overall health benefits of a starch depend on its form.

The form of the starch is defined based on its structural composition. Thus it can be generalized that the functional properties of the starches are mostly governed by their structural composition. From the structural point of view, the starch granules are composed of amylose and amylopectin. The ratio of amylose to amylopectin plays a significant role in controlling the functional properties of the starches such as gelatinization behaviour, thickness, and clarity of gelatinized solution (Gupta et al., 2024b). Starch containing higher amylopectin produces a softer and more cohesive gel, whereas starch with higher amylose content typically forms harder gels. The arrangement of these structural components within the starch granules affects the granules' surface dimension and size, which in turn considerably affects its functional properties. For instance, the size or dimensions of starch granules in baking flour affect its gelatinization behaviours during baking, the texture of the baked product upon cooling, and the storage life of the baked product. Starch is also utilized as a thickening agent in several food formulations to alter mouthfeel and consistency of products such as gravies, dressings, or soups, custards. Starches find further application as gelling agents in food products because gelatinized starch has potential to solidify and form gel upon cooling. However, the gel flexibility and toughness are largely determined by the ratio of amylose to amylopectin. Many starches are promising for forming and stabilizing foam and emulsion systems, which makes them beneficial in products such as creamy toppings and salad dressings (Gupta et al., 2024c). This characteristic of starches is particularly crucial for products that need a uniform texture and appearance. It can be concluded that starch's structural and functional properties affect food product processing, formed product quality, and stability. Hence, it is essential to understand the structural and functional interaction of the chosen starch for a particular food application.

11.2 WORKING MECHANISMS OF COLD PLASMA AND PULSE ELECTRIC FIELDS

Cold plasma is considered a green technology, and the fourth state of matter is plasma. Plasma is only an ionized gas, nothing more, yet because it defines the laws of gas loss, it cannot be regarded as a gas. In plasma, ionization is a key phase, together with other factors, such as collision types and electron energy distribution. Ionization, excitation, de-excitation, dissociations, and fragmentation are the common gas phase reactions during plasma formation. According to Misra et al. (2016), the feed gas composition is humidity; surrounding phase and amount of power and voltage applied are important variables that affect plasma. Cold plasma can be generated using gases such as atmospheric air, carbon dioxide, and nitrogen or mixtures of these gases; the chemical composition of the produced plasma is dominated by O^-, O_2^-, O^+, N^+, N_2^-, NO+, and CO_2^+ ions (Katsumata and Okazaki, 1967; Gupta et al., 2024d). During practical usage of cold plasma technology, when oxygen and nitrogen gas is combined for the formation of plasma, an intense or energetic electron collision occurs that leads to the starting of a chain of reactions, thus forming nitrogen oxides as well as free radicals in the plasma

(Misra et al., 2016). These free radicals are responsible for the cross-linking of oxygen-containing groups (hydroxyl group) as well as depolymerization at glycosidic linkages within a starch chain. Secondary electrons are also produced when the reactive species of the formed plasma contacts the biopolymer (starch) surface (Gupta et al., 2024e).

In recent decades, the implementation of plasma produced through vacuum or low-pressure processes for the modification of solid materials has gained popularity (Misra et al., 2016). The energy for plasma generation is supplied in various ways, including thermal, electric, magnetic, and radio frequency sources. Among these, an electric field is commonly used for the production of plasma in technical applications. To ionize and break down the gaseous molecules at lower electric power levels, the generation of a vacuum within the plasma chamber is required for direct current (DC); therefore, low-pressure plasma generation requires a costly vacuum apparatus. Due to recent developments in plasma physics engineering, the production of simpler and more affordable atmospheric cold plasma has gained researchers' attention, and the produced plasma can be used on either solid or liquid foods. The practical instrument setup of a pin-to-plate cold plasma system can be seen in Figure 11.1 Instrumental set up of Pin to plate Atmospheric Air Plasma system at ICT Mumbai, India campus; it consists of an array of pins (electrodes) and a bottom plate that acts as the ground. The sample to be treated is placed between them. The pins are usually made of stainless steel and are arranged in a specific pattern to ensure uniform plasma discharge. The device operates at a high voltage, usually in the range of 10–60 kV, and at a frequency of about 50 Hz. This high voltage is necessary to ionize the atmospheric air or a specific gas mixture between the pins and the plate. When the high voltage is applied, it creates an electric field between the pins and the ground plate. This field ionizes the gas, forming a plasma within the chamber that helps in the modification of the sample. Apart from pin-to-plate electrode configuration, dielectric barrier discharge (DBD) or atmospheric pressure plasma jets (APPJs) can also be used based on the specific application and desired changes required in the treated sample. Plasma generation setups include microwave-induced plasma, inductively connected plasma, corona glow discharge, radio frequency, atmospheric glow discharge, and gliding arc discharge (Sarangapani et al., 2015; Thirumdas et al., 2017a). However, among these, due to ready availability in many configurations, simple construction, and ease of adoption, APPJs and DBD are easily accessible as commercial goods that are frequently used in food research.

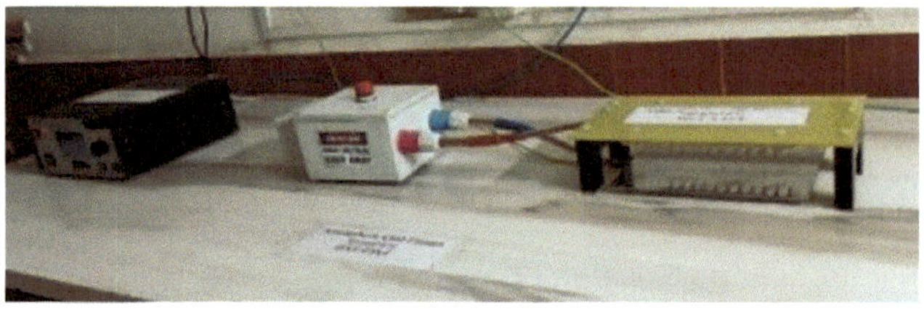

FIGURE 11.1 Pin-to-plate cold plasma instrument setup.

A non-thermal processing method, pulsed electric field technology, is commonly used for biological constituent extraction, enzyme inactivation, pasteurization, and decontamination of food products. In the last decade it has also received a lot of interest from researchers for molecular modification and chemical process enhancement (Han et al., 2020). It provides several benefits, such as being eco-friendly, energy-efficient, cost-efficient, and profitable for a range of industrial applications. A PEF system consists of a treatment chamber with two electrodes between which the sample (usually in liquid form) is placed. The high-voltage pulses (ranging from 1 to 80 kV/cm) are applied for very short durations (microseconds to milliseconds) (Mohamed & Eissa, 2012). These pulses create a strong electric field across the sample, causing electroporation, the development in cell membranes of pores that disrupt cellular functions. The literature suggests that PEF treatment may drastically alter starch properties as it can disrupt the crystalline structure of starch, leading to a decrease in molecular mass and to changes in the surface morphology of starch granules, thus affecting functional properties(Chen et al., 2022; Han et al., 2012b). PEF treatment has possible scalability in industrial applications; the sample flows continuously through the treatment chamber, allowing for efficient processing of large volumes. The schematic representation of the component and working of the continuous PEF system used for starch modification has been given in Figure 11.2. Schematic Representation of Continuous flow of pulse electric field treatment for starch modification; Because the process is non-thermal, it preserves the sensory and nutritional qualities of the food better than traditional thermal methods. PEF technology is being explored for its potential to modify the structure of plant tissues and to detoxify food products. A key working mechanism of this technique is electroporation, which generates a transmembrane potential that can result in either temporary or lasting permeabilization of cell membranes (Zimmermann, 1986).

11.2.1 Extraction of Starches Using PEF

PEF has been used for the extraction of starch from potatoes by Abduh et al. (2019). The potato slurry was exposed to an electric field strength of 0.5 to 1.1 kV/cm and 50 to 150 kJ/kg specific energy. The PEF treatment improved extraction over starch extraction without PEF treatment. The extracted starch samples were characterized to evaluate the changes in the starch compared to the untreated sample. The researchers observed that PEF treatment did not considerably alter the starch's natural characteristics, but slight differences in functional properties of treated and native starch were noticed. The starch extracted using PEF treatment showed reduced digestibility and a more limited range of gelatinization temperatures. This shows that although PEF can improve extraction, it might also change the characteristics of the starch, which is crucial to consider

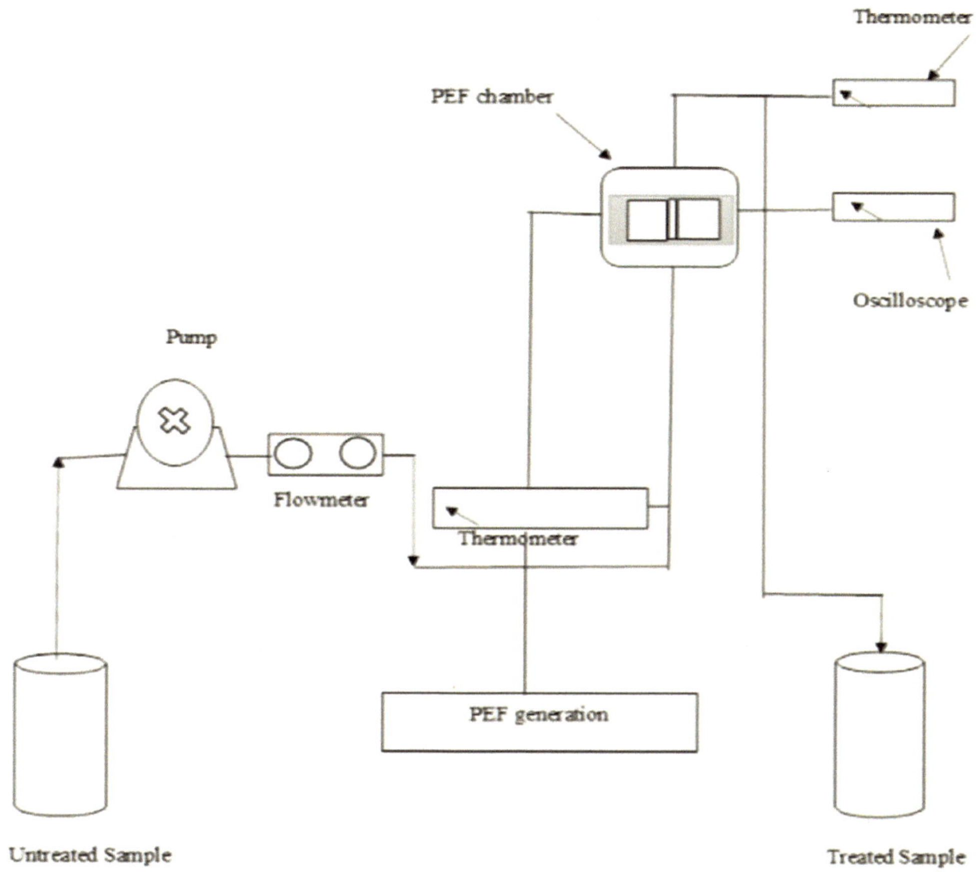

FIGURE 11.2 Continuous pulse electric field system for starch modification.

from an application point of view. The results of the above study showed that PEF technology is a potential method of extracting starch from potato tubers without compromising the quality of the extracted starch. Notably, no studies have been reported so far for the application of cold plasma for the extraction of starch from food matrixes.

11.2.2 Modification of Extracted Starches Using CP and PEF

11.2.2.1 Mechanism of Starch Modification Using CP

Current research studies focus on the application of CP to alter starches. Findings highlight three possible mechanisms for starch modification, namely, (a) cross-linking or grafting, (b) depolymerization, and (c) plasma etching, as schematically represented in Figure 11.3. Schematic representation for starch modification, mechanism using CP, illustrating a) crosslinking of starch chains b) depolymerization of starch chains, and c) etching of starch granules by reactive species; Apart from the above-mentioned three mechanisms, starch modification can also occur due to the addition of functional groups. These processes cause surfaces of biodegradable polymers to change in a variety of ways (Morent et al., 2011). According to Morent et al. (2011), the interactions between the chemically active species generated in gas discharge plasma and biopolymers result in the introduction of polar hydrophilic groups into the biopolymer. Various functional groups are introduced by different plasma treatments during oxygen plasma treatment of starches, including carboxylic acid, peroxides, and hydroxyl compounds. CO_2 gas plasma was shown to introduce hydroxyls, aldehydes, ketones, and esters groups, although the introduction of primary, secondary, and tertiary categories of the amines from the conversion mechanisms of such functional groups (Desmet et al., 2009). However, only free radicals are produced by noble gas plasma, with no functional groups being introduced. It is noteworthy that treatment duration and feed gas type have shown a major impact on the modification mechanism.

11.2.2.1.1 Cross-linking

The cross-linking occurs between the polymeric chain of starch molecules, caused by energetic electrons and free radical formation in the plasma; therefore, it alters the molecular structure of the starch molecule. The mechanism includes breakage between the reducing ends of two polymeric chains (C–OH), then the formation of a new C–O–C linkage between the two polymeric chains as a result of cross-linking and the removal of a water molecule (Zou et al., 2004), as we can see in Figure 11.3a. The generation

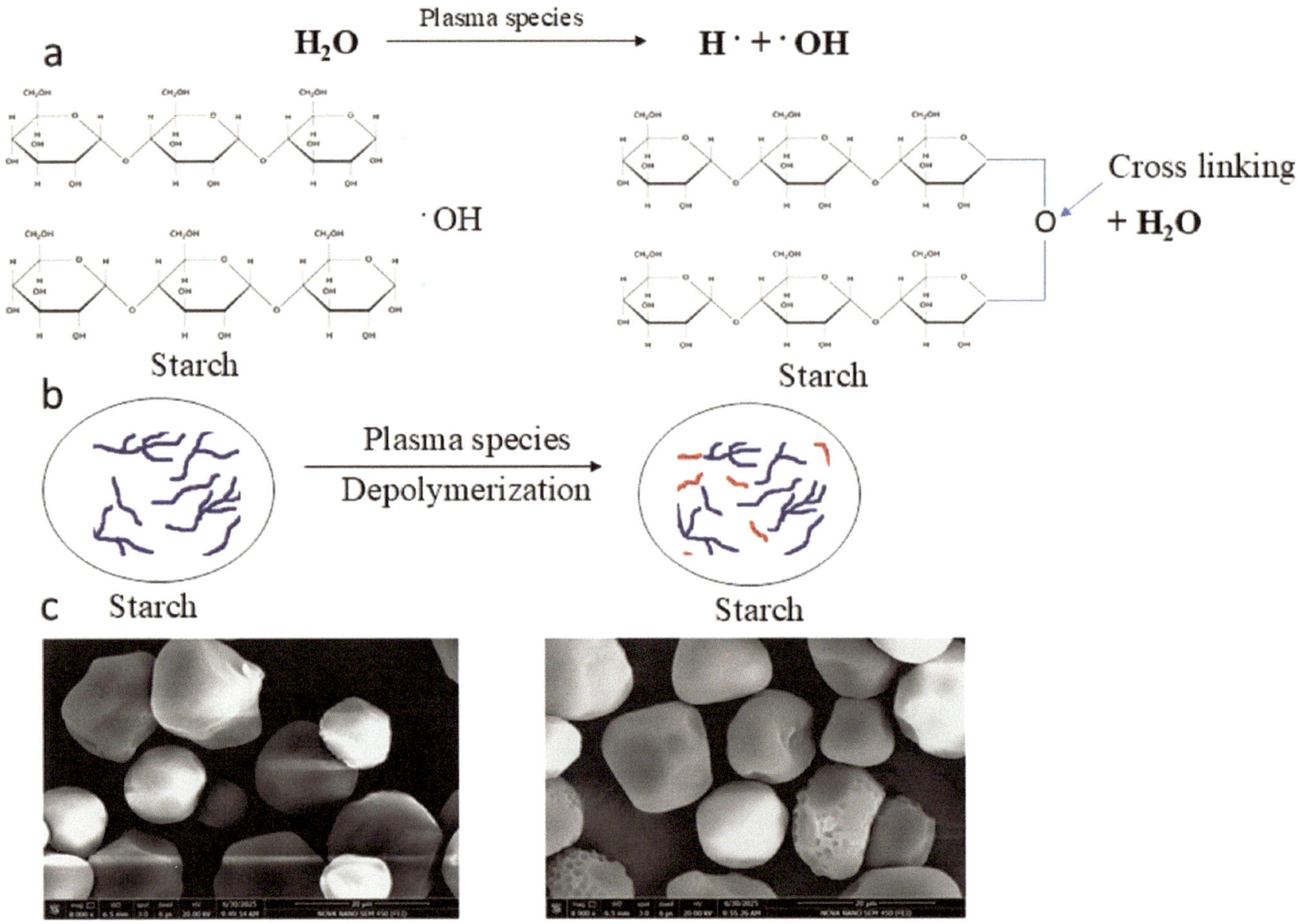

FIGURE 11.3 Mechanisms for starch modification: (a) cross-linking or grafting; (b) depolymerization; (c) plasma etching.

of hydroxyl radicals (*OH) by the decomposition of water molecules in the plasma and the incorporation of oxygen as a process gas generate these free radicals. According to Misra et al. (2016), hydroxyl radicals have an impact on the glycosidic chains of starch that may result in breaking of or change in the molecular structure. This has been further supported by infrared spectroscopy analysis by Deeyai et al. (2013), who reported a rise in the size of the indicative peak (924 cm^{-1}) for the development of new glycosidic linkage upon CP treatment of tapioca starch. Along similar lines, Wongsagonsup et al. (2014) and Bie et al. (2014) observed that cross-linking significantly increases native C–O–C regions inside the starch structure. Hence, it can be stated that increased cross-linking correlates directly with increased glycosidic linkage, particularly at the glucose molecule's C2 position, which is the most reactive site of the three carbons in pyranose. According to Whistler et al. (1978) and Zou et al. (2004), the C2 site is responsible for most cross-linking; it may produce up to 80.6% of the cross-linked structures in treated starch. Therefore, the C2 position is very crucial in promoting cross-linking during plasma treatment. Furthermore, FTIR analysis of plasma-treated samples shows a reduction in the relative strength of OH groups, which indicates the loss of the hydroxyl group during cross-linking (Deeyai et al., 2013; Wongsagonsup et al., 2014). For effective starch modification, treatment parameters should be precisely controlled as prolonged treatment time results in a greater percentage of cross-linking.

11.2.2.1.2 Depolymerization

Ionization is the primary factor involved in the depolymerization process, which alters starch and produces smaller molecules upon the high-energy plasma ions bombardment, that disturb amylose and amylopectin side chains in starch, as schematically represented in Figure 11.2b. Even at moderate energy levels, a small amount of depolymerization can be induced. The end products of starch depolymerization are maltose, maltotriose, and maltotetrose. Depolymerization depending on several factors, such as ion concentration, type of reactive plasma species, and free radicals generated, along with type of treated starch and its structure (Thirumdas et al., 2017a). According to Duan and Kasper (2011), reactive nitrogen species (RNS) and reactive oxygen species (ROS) are the main factors in the depolymerization of polysaccharides during plasma therapy. These are extremely reactive entities that are free radicals due to the presence of unpaired electrons.

A specific type of gas that is utilized to create plasma affects the depolymerization of starches, and Lii et al. (2002a) divided plasma gases into two groups: (1) air and hydrogen plasma, being less active, and (2) ammonia and oxygen plasma, being more active. Moisture content plays important role in the depolymerization process; the breakdown product of water molecules reacts with the secondary electrons to generate stable H^+ and OH^- radicals into hydroxyl (*OH) radicals along with intermediate products such as H_2O^+, H_3O^+, and H_2O_2. Furthermore, when water molecules are ionized by radiation, secondary electrons receive energy and become reactive species. The dissolution of glycosidic bonds at the C1 location of the glucose molecule in a starch macromolecule is facilitated by these reactive species, especially *OH radicals (Misra et al., 2016). According to Zhang et al. (2010), the glow plasma's active species break down the starch crystalline structure's interhelical water molecules, producing a variety of active species such as hydrogen, hydroxyl, and oxygen free radicals. In the end, these active species aggravate the destruction of crystallinity in the treated starch. Furthermore, Misra et al. (2016), along with Lii et al. (2003), shows that oxygen is present during plasma synthesis, which sets off a sequence of events that result in the production of ozone and *OH radicals. In turn, ozone breaks the glycosidic and C2–C3 bonds, which may result in depolymerization.

11.2.2.1.3 Plasma Etching

The surface phenomenon known etching occurs when reactive species produced during plasma generation scratch the starch granule surface, thus reducing its size from macro to nanoscale, as depicted in Figure 11.3c. Etching caused by plasma is the most effective and adaptive technique for altering the surface of the treated material (Gupta et al., 2025b). The process of etching includes physical sputtering or a chemical reaction to effectively reduce the surface material (Whistler et al., 1978). Reactive ion etching and sputter etching systems undergo a higher level of ion bombardment. Surface roughening occurs mainly due to the sputtering effect, in which low-ion energy bombards the surface, removing material and creating a rough texture. Thus, for an effective etching during cold plasma, we need a parallel plate reactor with 0.2–2 Torr pressure operating to maintain the stable plasma. The etching efficiency is influenced by the plasma density and the energy of ions striking the surface of the substrate (Ferreira et al., 2013). According to Wrobel et al. (1988), microwave-excited plasma is the most efficient source for surface etching. However, the etching rate of oxygen plasma is reported to be higher than that of nitrogen plasma, with an increased discharge power level. In the literature, plasma etching has been associated with increased hydrophilicity of the treated sample, attributed to increased surface area and enhanced surface energy. A higher etching rate affects the characteristics of the starch, mainly its hydrophilicity (Morent et al., 2011), because an etched surface allows water to enter starch granules more easily, thus altering the solubility and related functional attributes of the starch.

11.2.2.2 Mechanism of Starch Modification by PEF

Pulsed electric fields were reported to alter the structural and functional characteristics of the starch. The electric field pulses can take various forms during PEF treatments depending on operational conditions, namely, oscillatory, bipolar, square wave, or exponentially decaying pulses (Hong et al., 2016a). PEF treatments' primary impact on starch molecules is due to induced electroporation, which creates the pores in the starch granules that disrupt the granule structure and allow for the penetration of water and other molecules. The applied electric field also cause charged particles to accumulate on the starch granules surface, and these particles create directional movement in the starch solution to produce macroscopic space. The literature suggests that high-intensity electric pulses can alter the structural conformation of starch molecules using the breakage of hydrogen bonds and the reorganization of amylose and amylopectin chains(Hong et al., 2016a). PEF treatment can cause pits and roughness on the surface of starch granules. Studies of PEF-assisted chemical modification of starches have reported that electric potential difference generated between electrodes accelerates the mobility of reaction ions, enhancing the effectiveness of chemical reactions such as acetylation. At higher field strengths, significant deformation of the starch granules can occur. Maintaining a controlled temperature of about 40°C to 45°C is crucial to prevent unintentional denaturation caused due to heat. To summarize, the molecular chains are rearranged due to the applied electric field, which changes the starch granules' ability to interact with water, hence its attributed functional properties such as gelatinization and viscosity.

11.3 PROCESS PARAMETER AFFECTING CP AND PEF EFFICIENCY FOR STARCH EXTRACTION AND MODIFICATION

During cold plasma treatment, process parameters such as feed gas composition, input voltages, exposure time, distance from the electrode, input frequency, and atmospheric temperature affect the extraction and modification of the starch (Misra et al., 2016). Feed gas compositions such as air, oxygen, nitrogen, and argon are responsible for the amount and type of generated reactive species, which significantly affect the changes caused in starch granules. Input voltage denotes how much energy is available for the generation of plasma; high input voltages increase the concentration of reactive species, enhancing processes such as surface etching, cross-linking, and oxidation. Longer exposure times can lead to greater starch modification through enhanced surface roughness, with slight risk for starch degradation. Additionally, careful temperature control is necessary; excessive heat can lead to starch degradation, so maintaining appropriate temperatures is important in preserving the starch's structural integrity. However, in the case of PEF treatment, the process parameters that play an important in influencing starch extraction as well as modification are electric field strength, pulse waveform, number and length of pulses, medium conductivity, temperature, time interval of treatment, electrode spacing, and moisture content of starch suspension (Almeida et al., 2022; Han et al., 2012b). Enhanced electric field strength (measured in kV/cm) effectively disrupts matrix cell walls during PEF-assisted extraction of starch, which promotes the starch granule modification through electroporation. The literature suggests that short, high-frequency pulses tend to produce reversible effects on starch granules, whereas longer pulses lead to lasting structural changes. As pulse form changes, it affects how energy is delivered, thus affecting the respective effects on starch granules and subsequent alterations in starch properties. The efficiency of PEF in modifying starch can be further improved by enhancing medium conductivity, while appropriate temperature control can prevent high swelling or gelatinization of the treated starch. Longer periods of treatment can cause severe changes, but uncontrolled alteration can also increase the chance of starch degradation. Furthermore, reducing the distance between the electrode increases the field strengths and thus the impact on starch modification.

11.4 EFFECTS OF CP AND PEF TREATMENT ON STRUCTURAL PROPERTIES OF CEREAL, MILLET, PULSE, AND TUBER STARCHES

11.4.1 Surface and Granular Morphology

With CP treatment of corn starch at three different frequencies (100, 200, and 300 Hz), research has found that treated starch granules had a smoother surface than untreated starch; treatment at 200 Hz showed significant changes in granules' surface morphology compared to starch treated at 100 and 300 Hz (Shen et al., 2022). However, quinoa and sorghum starch granules preserved their shape upon cold plasma treatment (Gao et al., 2021). Additionally, Shen et al. (2022) observed that the damaged starch surface has significant pores that permit reactive plasma species to pass through the core of the granule, which leads to the depolymerization of starch molecules. The treatment of rice starch samples at 60 W for 10 min resulted in the formation of fissure or cavities in comparison to untreated starch. This change was attributed to etching phenomenon and to the formation of fissures on the surface allowing plasma ions to enter into the granule structure, which causes depolymerization and cross-linking (Thirumdas et al., 2017a; 2017b). Similar fissure formation was observed on surface of plasma-treated cassava, corn, rice japonica, sweet potato, waxy corn, wheat, and high-amylose corn starch (Lii et al., 2002a; 2002b). The treatment of the banana starch at 300 Hz maintained the size of the starch granules, while treatment caused fractures and pores on the outermost part of the starch granules (Yan et al., 2020). A similar result was reported by Ge et al. (2021) and Gao et al. (2019) for cold plasma–treated red adzuki beans, quinoa, sorghum, and kodo millet starch. These surface changes on starch granules affect the structural and functional characteristics of the starch. The structure of the native rice is polyhedral, which was lost for microwave pretreated starch for cold plasma treatment of more than 10 min, with the development of a burnt appearance with a dark colour (Sun et al., 2022). Sequential microwave pretreatment followed by plasma treatment of rice starch produced damaged surface and coarse texture, while the morphology of maize and red adzuki starch remains unaltered. These differences in granule morphology could be attributed to the granular structure and composition of starches that provide greater stability during the CP and microwave treatment (Sun et al., 2022).

PEF treatment of corn starch at 50 kV significantly deformed the granule structure; the majority of starch granules lost their shape, but several fragments gathered together and appeared like gels (Han et al., 2009a). According to Blaszczak et al. (2003), the starch granules' outer parts were significantly less uniform than their internal structure and were made up mostly of a cluster of amylopectin that had a higher molecular weight. The NMR analysis of the PEF-treated maize starch indicated that spiral structure of starch remains unaffected (Han et al., 2012a). SEM examination of a 10% waxy rice starch sample showed that PEF treatment produced deformed starch granules. In a similar study, treated wheat starch granules showed some scratches and damage to the spherical shape, while treated cassava starch showed truncated, circular granules (Maniglia et al., 2021b). For PEF treatment of red rice starch, Almeida et al. (2023) noted that native red rice starch granules had enlarged, smooth oval shapes that became more aggregated and degraded upon PEF treatment due to the electroporation process. As intensity of PEF treatment increases, starch molecules aggregate and develop a gel-like consistency due to increased swelling and water absorption capacity. Exposing potato starch slurry at a temperature below 50°C to various electric field strengths (30, 40, and 50 kV/cm) resulted in no change in granule shape; however, some granules combine to form big particles at a high electric field intensity of 50 kV/cm, with starch granule size increasing from 38 to 114 μm (Han et al., 2009a; 2009b).

11.4.2 Crystalline Structure

Physical modification of starch using CP and PEF has been reported to disturb the amorphous region more than the crystalline region (Luo et al., 2008; Sun et al., 2022). Moreover, these changes in starch degree of crystallinity depend on several variables, such as the amount and size of starch granules, the interaction between double helices within starch molecules, and their orientation (Zhou et al., 2014). Native starch of red adzuki bean showed a C-type crystal structure with a crystallinity of 28.59% and a strong peak for 2θ at 15°, 17.0°, and 23.0° in the pattern of diffraction. However, longer exposure of the same starch to dry heat treatment (130°C for 1, 3, and 9 h) severely changed the crystal structure affecting the degree of crystallinity, which decreased from 28.59 to 26.45%. Similar observations for a reduction in crystallinity were found in maize starch and proso millet starch upon exposure to dry heat treatment in earlier studies (Lei et al., 2020; Sun et al., 2014; Zhou et al., 2019). A slight reduction in the degree of crystallinity (from 25.34% to 25.15%) was reported in red adzuki bean starch upon CP treatment because of depolymerization of the starch (Zhou et al., 2019). A notable reduction in relative crystallinity (RC) in rice starch upon CP treatmen,t from 43.06% to 37.47%, was reported by Thirumdas et al. (2017a; 2017b). On the other hand, wheat and sorghum starch showed an increase in RC upon CP treatment, so these changes in RC of starch samples might be due to differences in the type of starch and CP treatment parameters (Gao et al., 2019). Dual methods such as combined dry heat and CP cause greater loss of crystallinity; preliminary dry heat treatment produces more amorphous starch granules, which improve the efficacy of high-energy particles of CP in causing depolymerization. Bie et al. (2016) found similar results for CP treatments of cassava starch, with more damaged amorphous than crystalline regions. The relative crystallinity of kodo millet starch decreased from 23.06% to 20.51% upon CP treatments, which was prominently visible from the reduced peak's intensity in the diffractogram (Sonkar et al., 2023). CP treatment of corn and tapioca starch also resulted in a notable decrease in crystallinity (Banura et al. in 2018; Ge, Guo, et al., 2022), attributed to depolymerization. The reduction of RC is due mostly to structural dissociation resulting from the interaction of highly explosive plasma species. Also, the hydrogen bonding network within the starch polymeric chain might be disrupted during plasma treatment, which could cause a decrease in crystallinity (Bie et al., 2015). As indicated by Carvalho et al. (2021), ariá starch's crystallinity increased upon DBD plasma treatment, whereby the amorphous portion of the starch was affected by cold plasma, which caused a small crystalline structure to develop, increasing the overall crystallinity of the starch. The increase in RC shows a decrease in amylose chains, which are predominant in crystalline regions. In the case of pearl millet starch, decreasing the CP treatment intensity from 230 V for 30 min to 200 V for 20 min increased crystallinity from 66.2% to 73.2% that of native starch (Carvalho et al., 2021). Similar observation for increase in crystallinity with reduction in treatment intensity was also reported for rice starch (Thirumdas et al., 2017a; 2017b), The treatment of proso millet starch with a combination of ultra-high pressure (UHP; at 200, 400, and 600 MPa) and CP treatment (40 V) for different durations (1, 3, and 5 min) significantly reduced relative crystallinity. The UHP treatment showed more impact on the amorphous and crystalline region of proso millet starch than CP treatment alone. Increased pressure levels pressure destroyed the crystalline structure of starch granules (Zhang et al., 2021).

Li et al. (2019) investigated the effects of PEF on the structural characteristics of three different types of starches, namely, wheat and sorghum (type A), potato (type B), and pea (type C). The most noteworthy changes were reported in potato starch, with observable alterations in the shape of these starch granules. The crystallinity of the corn starch significantly decreased with increasing electric field strength. The relative crystallinity decreased from 27.10%, 24.60%, and 22.20% to 19.91% at electric field strengths 0, 30, 40, and 50 kV/cm, respectively. This decrease occurred due to damage to starch granules affecting crystalline regions in PEF treatment, and similar observations are reported for potato starch (Han et al., 2009a; 2009b). According to Tester et al. (2004), the crystallinity of amylopectin-rich starches arises from the interlinking of amylopectin chains. Similarly, in PEF treatments of maize starch, alteration in amylopectin content have been attributed to a reduction in molecular weight and breakdown of the crystalline structure. Almeida et al. (2023) studied the effect of α-amylase enzyme pretreatment before PEF and simultaneous application of enzyme and PEF on the crystallinity of red rice starch. Only enzyme treatment increased the RC, while simultaneous use of enzymes and PEF produced a notable decrease in the RC of red rice starch. This might be attributed to the enzymes' significant effect on the amorphous area of starch granules, thus bringing the polymeric chain together in a compact form and affecting the crystallinity; however, in combined treatment, the electric pulses disturb the action of the enzyme, decreasing the relative crystallinity. It can be concluded that enzymatic pretreatment before PEF could help control excessive depolymerization during PEF treatment.

11.4.3 Molecular Composition

The effect of CP and PEF on cereal, legume, millet, and tuber starch molecular composition, ratio of amylose to amylopectin, molecular weight, and chain length distribution are highlighted in Table 11.1. Cold plasma treatment generates reactive oxygen and nitrogen species, which break the glycosidic bond present in the starch molecules; this can be responsible for a decrease in molecular weight and chain length distribution. These reactive species oxidize the starch molecules and cause the formation of additional functional groups such as carboxyl, carbonyl, and peroxides, as reported in starches of ariá, buckwheat, quinoa, wheat, and sorghum (Banura et al., 2018; Carvalho et al., 2021; Gao et al., 2019; Kalaivendan et al., 2022). During PEF treatment, owing to electroporation, reduction in molecular weight and chain length distribution in starch granules was reported. The literature suggests that high-intensity electric pulses can alter the structural conformation of starch molecules due to the breakage of hydrogen bonds and the reorganization of amylose and amylopectin chains. This structural change causes a reduction in molecular weight and chain length distribution, thus reflecting the starch FTIR spectra.

TABLE 11.1
FTIR Analysis of the Molecular Composition of Starch: Ratio of Amylose to Amylopectin, Molecular Weight, and Chain Length Distribution

Starch Type	Treatment Type	Operating Parameters	Major Findings	References
Corn	DBD plasma treatment	Voltage: 20 kV Frequency: 100, 200, and 300 Hz Time: 10 and 20 min	• 14% decrease in amylose content. • Increased amylopectin. • Greater conversion of amylose to amylopectin at 200 Hz for 20 min. • Modification in molecular structure at 200 Hz. • Small modifications in both ordered and amorphous areas. • Small changes in crystallinity affecting change in swelling and gelatinization of starch. • FTIR spectrum showed impurities of protein and phenolics at band at 1337, 762, and 703 cm^{-1}.	Goiana & Fernandes (2023)
Rice	Bell jar type plasma	Voltage: 40 and 60 W Time: 5 and 10 min	• Decreased amylose content from 29.93% to 22.83% with decreasing trend of plasma power and increased treatment duration. • FTIR showed additional functional groups such as peroxides, carbonyl groups, and carboxyl groups with acidic character.	Sun et al. (2022); Thirumdas et al. (2017a; 2017b)
Pearl millet	Pin-to-plate plasma	Voltage: 170, 210, and 230 V Time: 10, 20, and 30 min	• Most effective treatment reported at 230 V for 20 and 30 min decreased amylose from 31.32 to 17.99 g/100. • Some changes in crystalline and ordered structure bands at 1047 and 1022 cm^{-1}, thus observed cross-linking at lower voltage and depolymerization at higher voltage.	Kaur & Annapure (2023)
Ariá	DBD plasma	Voltage: 7, 10, 14, and 20 kV Time: 15 min	• Decreased amylose content with increased input voltage, indicating differences in their absorption peaks related to glycosidic bonds. The peak at 924 cm^{-1}, glycosidic bond, showed a decline at lower voltage but then an increase at 14 kV.	Carvalho et al. (2021)
Arrow root	CP	Voltage: 190, 210, and 230 V Time: 5, 10, and 15 min	• Increased molecular weight and chain length in later CP treatment due to cross-linking. • CP treatment did not change the complete backbone of starch; with chemical structure maintained, only position shifted. Amine peak observed at 1251 cm^{-1}, produced by the interaction with nitrogen species of plasma.	Devi et al. (2023)
Red rice	PEF	α-Amylase about 9.0 U mg cm^{-1} Electric field strength: 30 kV cm^{-1} (PEF30)	• Increased amylose content, attributed to greater enzyme accessibility of structural modification. • FTIR band showed increase in the double helix degree at PEF30 and AA0 with decreased degree of order.	Almeida et al. (2023)
Cassava	PEF	Acetylated starch assisted with PEF Electric field strength: 1.5 and 2.5 kV/cm	• Reduced molecular weight, thus change in ratio of amylose and amylopectin. • FTIR indicated an acetyl group in treated starch at 1730 cm^{-1}; intensity of ester peak increased, with peak at 1733 cm^{-1}, as they rising the degree of substitution because of the low DS value of the ester.	Lu et al. (2013); Shogren et al. (2003)
Wheat and cassava	PEF	Electric field strength: 15 and 25 kV/cm Energy: 25 and 50 kJ/kg	• Reduces molecular weight due to depolymerization changed the distribution of chain length, influenced viscosity and gel formation, and affected stability in applications such as 3D printing. • Changed degree of short-range order possibly because it induced a decrease in the part of crystalline structures.	Maniglia et al. (2021a)

Notes: CP, cold plasma; DBD, dielectric barrier discharge; DS, degree of substitution; FTIR, Fourier transform infrared spectroscopy; PEF, pulse electric field.

11.5 EFFECT OF CP AND PEF TREATMENT ON FUNCTIONAL PROPERTIES OF CEREAL, MILLETS, AND TUBER STARCHES

The modification of the starch improves its usefulness due to its effect on functional properties. When the starch granules come into contact with water, water first enters into the amorphous regions of starch granule, thus enabling it to expand slightly. The associations that occur between amylopectin and amylose chains within starch granules significantly affect the functional properties of starch. Important structural components, including the percentage of amylose, the arrangement of amylopectin's chain length, and the degree of crystalline structure inside the granule, are crucial in gelatinization and retrogradation of starch. Starch granules do not dissolve in water at room temperature but hydrate rapidly at high temperatures. Hydrated starch undergoes swelling of starch granules that disturb their molecular organization and crystal structure, and molecular disorganization further increases with heating (Moita et al., 2008). The structural characteristics of the starches produced from different natural sources or crops differ, leading to variations in swelling power (Singh et al., 2010). High temperatures and humidity also disturb the hydrogen bonding within the polymeric amylose and amylopectin chain organization inside the starch granules. The starch granules' capacity to retain moisture and expand depends upon molecular packing inside the starch granules, such as the amount of amylose, lateral sequences of amylopectin, and the presence of cracks and channels (Uthumporn et al., 2010).

11.5.1 Hydration Properties

11.5.1.1 Water Absorption Capacity

The water absorption capacity of corn starch has been found to slightly increase compared to the control sample when treated with DBD plasma treatment at 200 and 300 Hz frequencies, attributed to improved hydrophilicity due to surface etching (Goina et al., 2023). Similar observations of increased WAC were found for DBD-modified banana starch and bell jar–modified rice starch (Gao et al., 2022). Banura et al. (2018) attributed the rise in WAC to the depolymerization effect during CP. Plasma etching can also lead to the higher-treated starch WAC comparison to untreated starch (Thirumdas et al., 2017a) because an etched surface enhances the entry of heat and water into the starch granules. Sonkar et al. (2023) also found a significant rise in the WAC of kodo millet starch, attributed to a change in particle structure due to the creation of microchannels, and similar results were found in rice starch (Hera et al., 2013). A noticeable decrease in WAC has been reported for higher-intensity cold plasma treatment, especially at high voltage. Increased energy can induce stronger interactions between starch molecules, such as cross-linking between amylose and amylopectin chains, which may reduce the availability of hydrophilic sites. A gradual decrease in WAC was observed in pearl millet starch when CP treatment time increased (Jaddu et al., 2022; Kaur & Annapure, 2023; Thirumdas et al., 2017a; 2017b). On the other hand, arrowroot starch showed increased WAC with an increase in CP treatment voltage and time. This result was correlated to the presence of more hydrophilic groups (hydroxyl, carbonyl, and carboxyl groups) as a result of plasma-induced starch oxidation (Nieves et al., 2020). It can be concluded that the type of starch treated and the treatment parameters significantly influence the efficacy of CP treatment for altering the hydration properties.

11.5.1.2 Oil Absorption Capacity

Upon cold plasma treatment, prior surface modifications in starch granules such as etching and pores could affect the oil absorption capacity of starches. The OAC of a kodo millet starch sample increased from 1.01 to 1.34 g/g after plasma treatment due to reduced particle size and increased surface area of starch granules (Sonkar et al., 2023). Similarly, CP treatment of arrowroot starch also resulted in a significant increase in OAC compared to the control, especially with CP treatment at 230 V for 15 min. This was mainly due to the breakdown of starch granules, which increased the overall surface area that was easily accessible for oil binding. Similar observations have been reported for rice and mango seed kernel starches upon CP treatment (Kalaivendan et al., 2022; Thirumdas et al., 2017a; 2017b).

11.5.1.3 Swelling Capacity

A slight enhancement in the swelling capacity of plasma-treated starch has been reported upon CP treatment. Research suggests that amylose content in starch has a major impact on swelling properties: the higher the amylose content, the lower the ability to swell, and vice versa (Banura et al., 2018; Singh Sandhu & Singh, 2005; Thirumdas et al., 2017b). CP treatment of tapioca, red adzuki bean, and corn starches by Banura et al. (2018), Ge et al., 2021), and Thirumdas et al. (2017b), respectively, resulted in an increase in swelling power. The generation of plasma reactive species and the depolymerization of amylose chains during the plasma treatment may increase swelling capacity, and different observations could be noted for different treatment parameters. The effect of CP on starch's crystallinity positively impacts the swelling power of the treated sample, as reported for rice starch by Pal et al. (2016) and Sarangapani et al. (2016). Arrowroot starch showed the highest swelling capacity (46.82% at 70°C) compared to the control after treatment with plasma at 230 V for 15 min. These changes imitate the increased hydrophilicity of the starch granules upon plasma treatment, which allows them to absorb more water and swell more effectively (Deeyai et al., 2013).

11.5.1.4 Water Solubility Index

Starch solubility is measured in terms of its water solubility index (WSI). It has been reported that starch solubility can be improved by surface ruptures caused by the bombardment of higher-energy particles and plasma species during CP. Either an increase or a decrease in the WSI could be caused by depolymerization, cross-linking, and etching of starch molecules during CP treatment. According to Sun et al. (2022), dielectric barrier discharge air and microwave-generated plasma treatments of rice starch improved solubility. In contrast, rice starch WSI was significantly decreased during CP treatment, attributed to differences in treatment parameters or types of plasma (Thirumdas et al., 2017a; 2017b). Potato starch exposed to plasma treatment at 200 and 300 Hz frequencies exhibited an 8.5% reduction in solubility (Shen et al., 2021) due to molecular breakdown and oxidative damage of starch granules by plasma. The change in the ratio of amylose to amylopectin, attributed to a change in concentration of amylose content due to depolymerization during CP treatment, is most likely responsible for the WSI reduction. Similar reduction in WSI have been noted for banada and corn starches upon CP treatment (Goiana et al., 2021; Marenco et al., 2018). A notable rise in starch solubility was reported for combined treatment, such as microwave pretreatment followed by plasma treatment for rice starch, which increased solubility (Sun et al., 2022; Thirumdas et al., 2017a; 2017b). Studies reported so far for PEF-assisted modification of starch have not examined the hydration properties of treated starch samples.

11.5.2 Rheological Properties

As reported so far, exposing starch to ionized gas during CP treatment causes modifications in surface characteristics, molecular interactions, and granule structures, which may significantly influence the rheological properties of starch. Due to the inherent differences in molecular structure and granule composition between starches from different sources, the effect of cold plasma treatment on different starches may be distinct. As a result, CP-treated starch flow behaviour and potential applications may vary across industries. According to Devi et al. (2023), pin-to-plate atmospheric pressure CP treatment of arrowroot starch caused surface etching and rough texture, thus enhancing the surface friction between particles and lowering the ability to bind efficiently, causing a reduction in bulk density (14.7%) and tap density (10.25%). An enhancement of the Hausner ratio and compressibility index (Carr's index) indicated poor to very poor flow behaviour. Elevated exposure times (5, 10, and 15 min) and voltage (190, 210, and 230 V) enhanced the reduction in density and deterioration of flow ability. The findings of this investigation were consistent with the results of Kaur and Annapure (2023) for CP treatment of pearl millet. That study reported a notable increase in Hausner ratio and Carr's index upon CP treatment due to a decrease in bulk density. Research by Wongsagonsup et al. (2014) on the application of jet atmospheric argon plasma to two types of tapioca starch slurries, granular (G) and cooked (C), indicates that modification in molecular and structural integrity affected rheological properties. Starch samples were exposed to plasma treatment at input powers of 50 W (G50 and C50) and 100 W (G100 and C100) for an interval of 5 min. The G50 and C50 samples demonstrated lower paste clarity, indicating significant changes in the structural integrity of tapioca starch. On the other hand, the G50 sample showed enhanced thermal stability and led to the development of stronger gels compared to its untreated substitutes, suggesting a favourable outcome from 50 W plasma treatment on gel strength. In contrast, the G100 and C100 samples showed a decline in these appealing traits. These studies highlight a trend in which higher-energy input leads to compromised rheological properties due to induced alterations in molecular interactions and structural changes at varying plasma intensities. Thirumdas et al. (2017a; 2017b) found that all studied plasma treatment parameters significantly reduce the storage modulus (G′) of rice starch with the exception of the 60 W, 5 min treated sample, which showed improved G′ and loss modulus (G″) compared to native samples. This enhancement could be attributed by the formation of a cross-linking network leading to a more rigid gel structure and, thus, higher moduli. The study also found that cooling of these starches may consistently increase the G′ and G″, suggesting better stronger gel network structuring over time. Variations in these moduli can be associated with the amylose content of the rice starch, which directly influences retrogradation rates and the viscoelastic properties of starches.

11.5.3 Pasting Properties

The point at which when the viscosity of the starch rapidly increases through the continuous heating is called pasting temperature. The pasting characteristics measured using a Rapid Visco Analyzer (RVA) are pasting temperature (PT), peak viscosity (PV), trough viscosity (TV), and breakdown viscosity (BV). A wide range of factors influence starch pasting properties, including swelling index, solubility, crystalline characteristics, interaction between water molecules and leaching amylose and non-gelatinized particles, and ratio of amylose to amylopectin. An increase in CP treatment intensity from 40 W (10 min) to 60 W (10 min) resulted in a decrease in the peak temperature of the rice starch from 78.02°C to 77.11°C. A similar decline in peak temperature with increasing in CP treatment intensity was reported for mung bean, pearl millet, and potato starches (Kaur & Annapure, 2023; Shen et al., 2021), attributed to a decrease in amylose content. Hydroxypropyl starch modification decreased PT, attributed to improved hydrophilicity and easier water penetration within the granule. It is widely recognized that materials become more hydrophilic upon treatment with air plasma (Rekha et al., 2002). CP treatment at 40 W for 5 min increased the peak viscosity of rice starch to 4542 cP, compared to untreated starch at 3113 cP. Pal et al. (2016) reported increased PV because of the cross-linking of starch molecules due to plasma oxidation. Chung, Lee, Hong, Lee, and Lim (2015) observed that granules expanded more, which enhances PV and reduces PT; the higher PV values may occur due to the weakening or breaking down of bonds by reactive plasma species, which allows the amylose molecules to seep out more easily. Similar observations have been reported for starch that underwent ultrasound modification (Park & Han, 2016). TV and FV were found to increase at 40 W for 5 min CP treatment. The cooling process causes a rise in ultimate viscosity due to amylose, which starts to retrograde and re-form links between chain molecules. Higher viscosity is the consequence of water molecules being more easily incorporated as starch modification breaks hydrogen bonding. According to Lee et al. (2015), cross-linking during CP increases the pasting viscosity of japonica and indica rice starches. Similarly, Chaiwat et al. (2016) found that cassava starch's viscosity increased upon CP treatment; starch type and CP input parameters may be causes of differences in observed viscosity. Sonkar et al. (2023) used a combination of CP and UHP treatment on proso millet starch; they observed that dual treatment improved the starch's resistance to temperature and shearing due to enhanced cross-linking and thus improved gelatinization stability. Liang et al. (2023) reported that the maximum PV of the maize starch granules was reached at a surface gelatinization rate of 10%, along with significant swelling and similar results found in pearl millet starch (Kaur & Annapure, 2023). Falade et al. (2014) noticed a strong correlation between PV and starch degradation; while a higher percentage of B chains within granules might be cause of the subsequent decrease in PV, shorter internal B chains tend to disrupt the parallel stacking of the double helix, which leads to starch granules that are less stable during swelling.

The same result has been found in red adzuki bean starch. In addition, similar decreasing tendencies were observed with longer treatment intervals for the setback (SB), TV, and FV. The significant decrease in TV indicates that the starch granules' ability to swell was hampered due to structural alterations by the treatments. In general, findings show that starch pasting characteristics are affected by dry heat and CP treatments, leading the treated starch to act differently from untreated starch in food applications (Ge et al., 2021).

In PEF treatment of the tapioca starch, as the electric field strength increased, peak, breakdown, and final viscosities decreased (Han et al., 2009a). The highest decrease in viscosity, from 250 to 49 BU, was observed at 50 kV cm^{-1}. The reaction mechanism of starch to shear thinning has been studied using BV. Additionally, granule swelling ability just before breaking down physically was observed at maximum peaks at a specific concentration. Low-swelling starches showed a significant decrease in viscosity before obtaining their optimum level, when they were more susceptible to separating down after cooking (Massaux et al., 2008). As the electric field intensity rises, the breakdown and peak viscosity of the treated starch decrease (Han et al., 2012a; 2012b). Corn starch was shown to be more susceptible to gelatinization because PEF treatment disrupted the molecular structure of the starch. Therefore, this technique offers the potential for modifying the structure of starch and meeting the needs of food processing industries (Maniglia et al., 2021a). Similarly, wheat and cassava starch showed lowered final viscosity and trough apparent viscosity (TAV) but increased pasting temperature and peak apparent viscosity (PAV); hence, one must expect distinct behaviours from these starch sources through gelatinization and retrogradation processes. According to Almeida et al. (2023), combined PEF and α-enzyme treatment obtained the best pasting behaviour, with maximum stability and enhanced viscosity. This increases the starch's capacity to retain water during heating, which provided the best pasting profiles compared to the control.

11.5.4 Thermal Properties

Thermal properties of starch, including gelatinization and enthalpy, are important properties of functional behaviour during food processing. Gelatinization temperature is that at which starch granules absorb water and swell, losing their crystalline structure. Enthalpy is the energy required for the starch granule disruption. Changes in starch structure, particularly after techniques such as cold plasma or pulsed electric field, could result in modifications to these thermal properties. To improve texture, consistency in behaviour, and quality in food processing, it is essential that we understand these modifications in order to accurately forecast how starch will behave during heat processing. Thermal properties of starch, which include gelatinization and enthalpy, are usually measured using differential scanning calorimetry (DSC). The result of CP and PEF on thermal properties of treated starches are summarized in Table 11.2.

Most starches studied showed a reduction in gelatinization temperature, including the onset, peak, and conclusion temperatures, after plasma treatment. Certain treatments, such as low-voltage plasma applied to wheat starch, resulted in a slight rise in gelatinization temperatures, due to structural changes caused by reactive species in plasma. These species disrupt the starch crystalline structure, thus reducing the energy needed for gelatinization due to granule structure breakdown. Literature reviews suggest that the efficacy of CP treatments for alternating the thermal properties of treated starches is closely associated with the CP intensity voltage, frequency, and exposure time (Thirumdas et al., 2017a; 2017b). On the other hand, a partial reformation of the granule structure may result from a specific interaction with starch components such as amylose, which might explain the slight rise observed in certain treatments. PEF treatment was also reported to decrease gelatinization temperature. PEF causes the starch granules' crystalline structure to be broken due to electroporation, which decreases the temperature at which gelatinization occurs and decreases the starch's crystallinity. The amylose-to-amylopectin ratio can also be changed by PEF, usually with a decrease in thermal stability. Gelatinization enthalpy is further decreased by PEF-induced hydrolysis, which changes chain length distribution and produces smaller molecular weight fragments. Increased water absorption and the formation of amylose–lipid interactions are additional factors contributing to the decrease in thermal characteristics (Shogren et al., 2003).

11.5.5 Digestibility

Starch's nutritional composition and health benefits depend mainly on how quickly it can be digested. Starch can be differentiated into three main types, namely, resistant starch (RS), slowly digestible starch (SDS), and rapidly digestible starch (RDS). SDS releases glucose slowly, which is advantageous for longer-lasting energy, while RDS breaks quickly into glucose, producing a sudden rise in blood sugar. RS acts like dietary fibre: it can withstand digestion in the small intestine but helps in promoting gut health.

TABLE 11.2
Thermal Properties of Starch Using Differential Scanning Calorimetry (DSC)

Starch Type	Treatment	Operating Parameters	Thermal Properties				References
			T_o	T_p	T_c	ΔH	
Wheat	DBD plasma	Voltage: 40, 60, and 90 V Time: 2, 6, and 10 min	–	Increased from 62.39°C to 62.94°C at 40 V, 10 min.	–	Increased from 9.82 to 11.76 J/g at 40V for 10 min.	Shen et al. (2022)
Rice	Bell jar type plasma	Voltage: 40 and 60 V Time: 5 and 10 min	Decreased from 63.1°C to 59.9°C at 60W for 10 min due to plasma species greater damage of the starch.	Decreased from 73.5°C to72.6°C at 60W for 5 min.	Decreased from 79.6°C to 77.9°C.	Decreased from 14.9 to 13.06 J/g at 60W for 10 Min.	Zhang et al. (2015) Nemtanu and Bras¸oveanu (2010)
Ariá	Atmospheric air plasma	Voltage: 7, 10, 14, and 20 kV Average pressure: 101 kpa Frequency: 200 Hz Electrode distance: 15 mm	Decreased from 85.6°C to 80.5°C at 14 kV.	Decreased from 87.4°C to 82.4°C at 14 kV.	Decreased from 92.1°C to 86.7°C to at 14 kV and increased at 20 kV, but this was not acceptable.	Decreased from 22.7 to 20.1 J/g to, and retrogradation ratio was more significantly reduced at 14 kV.	Carvalho et al. (2021)
Pearl millet	Atmospheric pin-to-plate plasma	Voltage: 170, 200, and 230 V Time: 10, 20, and 30 min Frequency: 55.51 Hz.	Reduced when the treatment parameter increased, at 200 V for 10 min (30.3°C) and at 230 V for 30 min (29.5°C).	Significant reduction from 77.3°C to 40.5°C with increasing intensity of plasma.	Decreased from 128.6°C to 53.7°C at 200 V for 30 min.	Significantly reduction of enthalpy from 327.5 to 6.68 J/g.	Kaur & Annapure (2023)
Red adzuki bean	Plasma treatment	DH treatment: 3 h in combination with CP (40 V) Duration: 1, 5, and 10 min	Increased from 56.13°C to 58.33°C with increasing intensity of plasma.	Significantly increased from 59.63°C to 61.82°C.	Increased from 65.47°C to 67.16°C.	Low energy required for gelatinize the starch, 12.62 to 12.13 J/g.	Ge et al. (2021)
Cassava	PEF-assisted acylation	Voltage: 450 to 750 V Electric field strength: 1.5–2.5 kV/cm Acetic anhydride content: 6%, 8%, and 10%	Decreased from 65.8°C to 61.5°C with DS of 0.131, possible weak internal hydrogen bonding.	Decreased from 71.9°C to 67.1°C with DS of 0.099.	Significantly decreased from 82.5°C to 73.9°C with DS of 0.131.	Decreased from 18.1 to 10.1 J/g with DS of 0.131.	Hong et al. (2016a; 2016b)
Corn	PEF	Electric field strength: 30, 40, and 50 kV m^{-1} Flow rate: 60 mL/min Sample temperature: 50°C	Decreased at 30, 40, and 50 kV cm^{-1} to 68.60°C, 67.46°C, and 67.34°C, respectively.	Slightly decreased from 73.38°C, 72.21°C, and 71.51°C at 30, 40, and 50 kV cm^{-1}.	Decreased at 30, 40, and 50 kV cm^{-1}, to 78.49°C, 77.68°C, and 77.40°C, respectively.	Decreased from 1.61 to 1.22 and 0.95 J/g. As the strength of the electric field increased	Han et al. (2009a)
Maize	PEF	Electric field strength: 30, 40, and 50 kV/cm Flow rate: 60 mL/cm Time: 424, 848, and 1272 μs.	Decreased from 67.15°C to 64.42°C at 50 kV/cm, 1272 μs.	Decreased from 72.23°C to 68.86°C at 50 kV/cm, 1272 μs.	–	Decreased from 4.51 to 2.74 J/g at electric field strength 50 kV/cm, 1272 μs. Decreased with increasing electric field strength treatment duration.	Han et al. (2012a; 2012b)

(Continued)

TABLE 11.2 *(Continued)*
Thermal Properties of Starch Using Differential Scanning Calorimetry (DSC)

Starch Type	Treatment	Operating Parameters	Thermal Properties				References
			T_o	T_p	T_c	ΔH	
Potato	PEF	Electric field strength: 30, 40, and 50kV cm^{-1} Frequency: 1008 Hz Sample flow rate: 60 mL/min Time: 806 µs	Decreased from 61.20°C to 60.35°C.	Decreased from 65.15°C to 64.61°C.	Decreased from 73.17°C to 72.22°C at 30 kV/cm.	Decreased significantly 11.61 to 10.04 and 9.52 J/g at 40 and 50 kV.	Han et al. (2009a; 2009b)
Potato and tapioca	PEF-assisted with octenyl succinylate	Electric field intensity: 0, 2, 3, 4, 5, and 6 kV/cm Duration: 42.5 ms Flow rate: 1 mL/s	Decreased from 63.1°C to 55°C at 6 kV/cm for 42.5 ms.	Decreased from 64.3°C to 61.1°C at 6 kV/cm.	Decreased from 71.8°C to 59.4°C more than peak temperature.	The low energy required at 6 kV/cm such as 1.5J /g.	Chen et al. (2021), Han et al. (2012a; 2012b)
Red rice	PEF-assisted α-amylase	Electric field strength: 30 kV/cm Frequency: 600 Hz Pulses: 6 µs Time: 90 s	Reduced from 63.58°C to 60.59°C.	Decreased from 69.54°C to 65.50°C.	Decreased from 76.57°C to 70.16°C.	The low energy was found AA0 that was 7.18 J/g.	Almeida et al. (2023)
Wheat and cassava starch	PEF	Electric field strength: 15 and 25 kV/cm Energy: 25 and 50 kJ/kg	Decreased to 47.73°C in wheat and 55.23°C in cassava, as compared to native starch.	The peak temperature of cassava is greater than wheat starch by 4°C.	Decreased slightly to 73.91°C in wheat and 80.85°C in cassava.	Cassava starch (11.74 J/g) required higher energy than wheat (9.15 J/g).	Maniglia et al. (2021b)

Notes: ΔH, enthalpy; DBD, dielectric barrier discharge; DS, degree of substitution; PEF, pulse electric field; T_o, onset temperature; T_p, peak temperature; T_c, conclusion temperature.

Researchers simulate the human digestive process to understand the behaviour of starch inside the human digestive tract through in vitro studies to quantify the amount and rate of starch breakdown. Plasma treatment is reported to significantly increase RDS content and reduce SDS content (Zou et al., 2020). The amorphous regions of SDS can easily targeted by plasma reactive species, which weaken the helicoidal structure. These species break down the starch chains into RDS by depolymerizing them. However, there were no notable changes reported in the RS crystalline region after exposure to CP. This is because CP treatment leads to only minor molecular rearrangements in RS types 1 and 2 and to integrated structures characteristic of type 4 RS. This restructuring at the molecular level could be due to the induced formation of new covalent bonds without considerably breaking down the crystalline regions of RS, which could also explain the observed increase in starch crystallinity after treatment.

When red adzuki bean starch were treated with dry heating (DH) and CP, these treatments altered the starch digestibility due to observable changes in molecular configuration. The DH treatment caused marginal decrease in RDS content from 62.82% to 61.40%, with slight increase in SDS content from 26.45% to 28.00% and RS from 37.18% to 38.60% compared with native starch. The reorganization of the amylopectin and amylose chains is the primary cause of the observed change. The occurrence of amorphous regions in DH-treated starch inhibits the amylase's ability to attach to its active sites, leading to either partial or slow enzymatic hydrolysis. The starch treated with combined DH and CP exhibited a greater rate of decreased enzyme digestion, thus decreasing the values obtained for RDS by 2% to 5%. On the other hand, longer exposure to CP treatment at studied CP intensity resulted in an increase of SDS and SDS+RS ratios by 2% to 3% compared to native starch. During CP treatment, starch molecules undergo depolymerization, which lowers their molecular weight (Mw) and degree of polymerization. The smaller fragments generated by CP treatment form double helices, and their sensitivity to enzymatic hydrolysis decreases because they inhibit amylose–amylose interactions (Sharma et al., 2019). In addition, the structure is considerably rearranged by CP, which disorders and disrupts the amylopectin particles. As a portion of RDS and RS becomes SDS, this conversion is accelerated by short chain lengths as well as changed molecular structures that hinder enzyme access. The alteration of starch digestibility and specific dietary modifications that provide health benefits, such as better glucose regulation, is made possible by these structural changes in molecules.

11.6 EFFECT OF PHYSIO-CHEMICAL PRETREATMENTS ON STRUCTURAL AND FUNCTIONAL PROPERTIES OF STARCHES EXPOSED TO CP AND PEF TREATMENTS

Starch modification is continually being developed to create innovative starches with improved functionality and value-added properties. To widen the application of starches in the food and non-food industries, native starch granules are often altered to enhance their properties through different physical, chemical, and enzymatic strategies. But each of these treatments has its pros and cons, leading researchers to explore the use of pretreatments to improve the efficiency of starch modification and thus achieve the target values for functional attributes. Pretreatments with physical methods such as ultrasound and chemical methods such as esterification and acetylation have been explored to increase the efficacy of CP and PEF (Chen et al., 2022; Hong et al., 2016a; 2016b; Kaur & Gill, 2019).

11.6.1 Physical Pretreatment by Ultrasound

Ultrasound is a quick, easy, and effective physical method that uses sound waves at a frequency over 20 kHz to cause change in the target material. The primary effect of ultrasonic waves on starch granules is disruption of amorphous region while maintaining granule integrity (Kaur & Gill, 2019). During ultrasound treatment, the most crucial factors are frequency, time, power, and moisture content in starches, which can influence the efficacy of treatment in terms of altering starch functional properties. Shen et al. (2021) treated 50% mung bean starch slurry with ultrasound at 300 W for 10, 30, and 50 min before CP treatment at 40 V; results indicated that molecular weight and chain length decreased more significantly with combined treatment than with CP treatment alone. This change was attributed to the fast formation and collapse of bubbles in the starch slurry due to cavitation phenomena during ultrasound, which produced cracks and grooves on the starch granules' surface. These ease the penetration of plasma active species into the interior of the starch granules, resulting in etching and depolymerization and thus decreasing the molecular weight of the starch. This process changes the internal molecular organization in the starch granule, which results in increasing functional properties, especially swelling power and solubility. An increase in gelatinization temperatures and ΔH was observed with increased exposure to combined ultrasound and plasma treatments due to the increased cross-linking of amylopectin side chains induced by plasma active species, which could stabilize the starch crystallites.

11.6.2 Chemical Pretreatment Methods

11.6.2.1 Calcium Chloride–Assisted Cross-linking

Cross-linking has also been explored to modify starch. The most frequently used cross-linking agents are sodium tripolyphosphate (STPP), sodium trimetaphosphate (STMP), epichlorohydrin (ECH), calcium chloride ($CaCl_2$), and phosphoryl chloride ($POCl_3$). The cross-linked starch is found to be water-insoluble and brittle with limited swelling power and enhanced thermal, chemical, and acid-resistance properties (Sharma et al., 2020). The hydroxyl groups of the glucose molecule of starch polymeric chains are involved in covalent bonding during the cross-linking process. Liang et al. (2023) treated waxy maize starch with $CaCl_2$ followed by heating for 10, 20, and 30 min to achieve 10%, 30%, and 50% surface gelatinization. The dried $CaCl_2$-treated pre-gelatinized starch was further subjected to CP treatment for modification. Reduction in crystallinity, molecular weight, and amylose content of starch after the CP treatment was observed due to disruption of the ordered structure. Consequently, these changes in structural properties enhance solubility and decrease swelling capacity, thermal and pasting behaviour, and digestibility. The disruption of the starch shell structure enhances the degree of starch modification and enhances enzymatic sensitivity by facilitating the entry of water molecules into the starch lamellar structure, thus increasing the digestibility of the starch granules (Gao et al., 2021). Therefore, cross-linking pretreatment followed by CP treatment generated a waxy maize starch with improved functional properties, increasing the possibility for use in a range of food applications.

11.6.2.2 Acetylation

During acetylation, starch is mixed with acetic anhydride and vinyl acetate with water as well as alkaline catalysts such as sodium hydroxide (NaOH). The total amylose content rises with respect to the degree of substitution; the level of substitution increases the sugar's hydroxyl group while preserving its acetyl group. Increasing the amylose content is beneficial for human health because amylose breaks down more slowly than glucose, thus decreasing glycaemic load and insulin impact. Recently, the U.S. Food and Drug Administration approved a 0.2 substitute level of synthesized starch for usage in the pharmaceutical and food sectors (Suri & Singh, 2023). Acetylated starches at a low degree of substitution provide desirable characteristics, including low gelatinization temperature, desired thickening, low stability of temperature (reduced syneresis), more effective film-forming characteristics, and a decreased susceptibility to retrogradation (Chi et al., 2008; Lawal, 2004; Zięba et al., 2011). The insertion of acetic groups disturbs the naturally organized starch structure and reduces the electrostatic interaction among the amylopectin chains. Acetylated starch is composed of dual-functioning hydrophobic and hydrophilic groups, so it may act as an effective emulsifier (Bao et al., 2003).

Potato starch acetylation using acetic anhydride concentration of 6% to 10% with the help of PEF at electric field strengths from 1.5 to 2.5 kV/cm and durations of 27 to 51 milliseconds (ms). At 10% acetic anhydride concentration with a PEF at 2 kV/cm for 51 ms, a greater degree of substitutions (0.131) was observed (Hong et al., 2016a; 2016b). PEF-assisted starch acetylation significantly increased surface roughness as observed through atomic force microscopy. Hong et al. (2016a; 2016b) revealed that with increased treatment duration from 15 to 60 min at an electric field intensity of about 3 kV/cm, the DS significantly increased from 0.054 to 0.130, suggesting that the acetylation process was substantially accelerated by PEF treatment. Morphological examination showed more prominent surface variations in high-DS starches, including swelling and tougher regions, which could be indications of structural modifications made by PEF-assisted acetylation. FTIR showed that substituted acetyl groups had a unique band at 1730 cm^{-1}, verifying the molecular changes. The modified starch had decreased breakdown and setback viscosity, which may increase stability and improve cooking characteristics, with a reduced rate of retrogradation (39%).

11.6.2.3 Esterification

Extracted starches, or native starches, are poor emulsifiers due to their somewhat large particle dimensions and their hydrophilicity; thus, several modifications are essential for improving their emulsifying efficiency. Research has been completed on the esterification of starch utilizing octenyl succinic anhydride (OSA). Zhang, Mei, Chen, and Chen (2017) showed that starches modified with OSA have the potential to stabilize emulsions. The substitution of OSA imparts hydrophobicity to the starch granules, thus enhancing emulsifying capacity and stability. The degree of substitution, or quantity of hydrophobic groups, determines OSA-modified starch emulsion capacity (Li et al., 2018). Lin, Zeng, Yu, and Sun (2012) studied the PEF-assisted esterification of waxy maize starch (WMS) at 1000 Hz pulse frequency and 80 μs pulse duration for 50 min. Enhanced esterification sites are produced when PEF treatment changes the granular surfaces of the crystallized WMS structure, thus achieving a higher degree of substitution (46.9%) than without PEF (3%), increasing the hydrophobic nature of the starch.

Chen et al. (2021) studied the octenyl succinylate modification of potato starch using PEF at electric field strengths of 2 to 6 kV/cm, flow rate of 1 mL/s for starch suspension in the PEF chamber, 42.5 Ms and 50 min processing time. The results indicated that treating starch using PEF significantly alters pasting properties; the pasting temperature decreased from 15.1°C to 7.6°C to under the studied conditions. The swelling power, solubility, freeze–thaw stability, and retrogradation of the treated starch were lower than those of the untreated starch. This was attributed to the introduction of abundant bulky acetyl groups increasing DS values and hydrophobicity and consequently decreasing the starch granules' capacity to absorb and hold water. The positive effects of PEF treatment may influence the efficacy of chemical reaction used for starch modification, widening the applicability of the modified starches.

11.7 ADVANTAGES OF CP AND PEF TREATMENTS OVER CONVENTIONAL THERMAL TREATMENTS FOR MODIFICATION OF STARCHES

Conventional commonly used thermal methods for starch modification are heat-moisture treatment and dry heat treatment. However, starch modification using CP and PEF treatments shows significant benefits (Maniglia et al., 2021a) as the extremely high temperatures used for conventional thermal processing usually decrease the functional characteristics of starch. CP is reported to better retain starch structure and to improve functional properties during modification of starch as it performs at low temperatures, thus avoiding heat-associated degradation. As CP does not generate hazardous by-products, it is a "greener" approach to modifying starch than conventional techniques. In contrast, PEF treatment modifies starch characteristics faster, using high-intensity electric pulses, which minimizes the chance of degradation and reduces energy requirement. These techniques are excellent options for industrial applications of starch modification as various types of starches could be treated with these method, and as process variables allow for the perfect modification of starch characteristics.

11.8 LIMITATIONS OF CP AND PEF FOR STARCH TREATMENT AND COMMERCIALIZATION STRATEGIES

The food industry has not implemented CP technology in large-scale operations, despite the technology's potential for multiple applications, as several important features are not well understood. International research initiatives are currently in progress to delve further into the safety features of CP technology before its official commercial release (Awad et al., 2012). The treatment capacity of industrial PEF equipment has limitations and the equipment itself is expensive. The commercialization of CP and PEF technologies has been restricted by the lack of innovation and the absence of reliable, affordable industrial equipment. Although significant differences in processing costs could be mapped to quality advantages that may support the equipment investment costs, additional equipment costs could be lowered by increasing the equipment capacity for successful implementation of these techniques (Toepfl & Heinz, 2007). Numerous commercialization strategies can be implemented to overcome these limitations, such as the following: (1) Developing trust between stakeholders in the industry through effective applications can be accomplished through pilot investigations that demonstrate the positive effects of CP and PEF treatment. (2) Concise cost-effectiveness examinations that highlight long-term advantages and money saving, improved product quality, and energy conservation may promote implementation. (3) Continuous spending on research and development are essential for improving treatment conditions and developing equipment that can be scaled up, including university collaboration to encourage innovation. (4) Training programs focused on stakeholders in the industry will enhance understanding and accelerate the transition to these leading-edge methods of processing. (5) Finally, adopting an active role when communicating with regulatory organizations may be helpful in addressing compliance-related challenges and accelerating the marketplace acceptance approvals.

11.9 APPLICATION OF CP- AND PEF-TREATED STARCHES

Starch is a significant biopolymer used in the food sector (e.g., as binding agent or stabilizer) and in the paper, textile, pharmaceutical, and other industries (Abduh et al., 2019). Cold plasma–treated starches are extensively used in food formulation, particularly in the preparation of delicious dairy-based foods such as milk desserts (Bahmanpour et al., 2024). The addition of treated starch to milk desserts has a significant effect on sensory qualities such as taste and consistency; uniform texture and smooth, transparent surface; and overall acceptance. Improved gel strength and viscosity of treated starch leads to their application as excellent stabilizer and thickeners for preserving integrity and avoiding separation of liquids in food preparations. Increased resistant starch content provides additional health benefits. Modified starches also find application in food formulations owing to their improved stability and shelf life. PEF-treated cassava and wheat starch have been used for 3D printing application owing to their lower viscosity increasing the printability of different shapes such as hearts, stars, and cylinders (Maniglia et al., 2021b). PEF treatment not only enhances printability but also reduces the syneresis and produces smoother, harder, and less sticky printed products. The PEF and CP treatment of cereals, millets, tubers, and pulses starch has been shown to enhance their properties, thus improving their application in bakery products for improving texture as well as moisture retention in cakes and pastries. These starches could be used to replace gluten in gluten-free product formulations, to provide greater viscosity to sauces and dressings, or to increase resistant starch content levels for better digestion. Modified starches could be explored in food coatings to inhibit oxygen and moisture transfer rates and thus preserve the crunchy texture of snacks, prolonging product storage.

11.10 FUTURE DIRECTIONS

In the future, research on the modification of starch much continue to focus on enhancing starch characteristics with the goal of meeting the growing consumer demand for improved nutritional benefits. Research should also look at the functional uses of other modified starches in many different kinds of food products, 3D printing, and health applications. Research must also look into the synergistic benefits of combining CP and PEF treatments, as well as the effectiveness of this practice within food systems. More research is important to understand the rheological behaviour of modified starches in food applications, in vivo starch digestibility, and hydration characteristics.

11.11 SUMMARY

This chapter explores the effects of CP and PEF treatments on the functional and structural properties of various starches including cereals (maize, red rice, wheat), pulses (mung bean), millets (kodo, proso, pearl, sorghum, finger millet), and tubers (potato, cassava, tapioca). A particular focus is on the effects of CP and PEF treatment parameters, such as voltage, time, and field intensity, on structural changes such as starch granular shape, amylose-to-amylopectin ratio, crystallinity, molecular weight, and chain length distribution. These structural modifications enhance functional properties such as solubility, viscosity, absorption capacity, and gelatinization. Furthermore, the chapter includes pretreatment methods such as cross-linking, ultrasound, acetylation, and esterification that can also enhance the properties of starches, contributing to functional improvements and significant changes in the relative crystallinity of starch. This chapter also highlights the energy efficiency of these treatments, noting that lower energy inputs can achieve desired modifications in starch properties. CP and PEF treatments, as innovative starch processing techniques, can be possible advantages in modified starch for food applications in response to customer demand for healthy food products.

11.12 CONCLUSION

The chapter provides deeper insight into the underlining mechanism of starch modification during CP and PEF treatments, with detailed description of alterations at starch structural and thereby in the functional properties of starch. The process parameters of CP and PEF treatment, such as treatment voltage, exposure time, and field intensity, significantly affect the structural properties of starch. Modifications in molecular structure, crystallinity, amylose-to-amylopectin ratio, granule size, and surface morphology directly impact functional properties, including gelatinization, water and oil absorption capacity, viscosity, and solubility. The structural changes caused by the CP and PEF treatments are crucial in producing starch that meets increasing customer demand for functional and healthy food products and for exploring the function of modified starches in food preparation and potential health benefits.

ACKNOWLEDGEMENTS

The Institute of Chemical Technology, Mumbai-Marathwada Campus, Jalna is acknowledged by the authors for its support. They also appreciate the guidance of colleagues and mentors and the encouragement of family and friends.

REFERENCES

Abduh, S. B. M., Leong, S. Y., Agyei, D., & Oey, I. (2019). Understanding the properties of starch in potatoes (*Solanum tuberosum* var. Agria) after being treated with pulsed electric field processing. *Foods*, *8*(5), 159. https://doi.org/10.3390/foods8050159

Almeida, R. L. J., Santos, N. C., Muniz, C. E. S., da Silva Eduardo, R., de Almeida Silva, R., Ribeiro, C. A. C., da Costa, G. A., de Figueiredo, M. J., Galdino, P. O., & dos Santos, E. S. (2023). Red rice starch modification: Combination of the non-thermal method with a pulsed electric field (PEF) and enzymatic method using α-amylase. *International Journal of Biological Macromolecules*, *253*, 127030.https://doi.org/10.1016/j.ijbiomac.2023.127030

Annor, G. A., Marcone, M., Bertoft, E., & Seetharaman, K. (2014). Physical and molecular characterization of millet starches. *Cereal Chemistry*, *91*(3), 286–292. https://doi.org/10.1094/CCHEM-08-13-0155-R

Awad, T. S., Moharram, H. A., Shaltout, O. E., Asker, D. Y. M. M., & Youssef, M. M. (2012). Applications of ultrasound in analysis, processing and quality control of food: A review. *Food Research International*, *48*(2), 410–427. https://doi.org/10.1016/j.foodres.2012.05.004

Bahmanpour, H., Asefi, N., Alizadeh, A., & Pirsa, S. (2024). Assessment of the impact of cold plasma technology on physicochemical properties of corn starch flour and the associated modified corn starch incorporated into milk dessert. *Heliyon*, *10*(17), e37399. https://doi.org/10.1016/j.heliyon.2024.e37399

Bao, J., Xing, J., Phillips, D. L., & Corke, H. (2003). Physical properties of octenyl succinic anhydride modified rice, wheat, and potato starches. *Journal of Agricultural and Food Chemistry*, *51*(8), 2283–2287. https://doi.org/10.1021/jf020371u

Banura, S., Thirumdas, R., Kaur, A., Deshmukh, R. R., & Annapure, U. S. (2018). Modification of starch using low pressure radio frequency air plasma. *LWT*, *89*, 719–724. https://doi.org/10.1016/j.lwt.2017.11.056

Bertolini, A. C., Bello-Pérez, L. A., Méndez-Montealvo, G., Almeida, C. A., & Lajolo, F. (2010). Rheological and functional properties of flours from banana pulp and peel. *Starch-Stärke*, *62*(6), 277–284. https://doi.org/10.1002/star.200900216

Błaszczak, W., Valverde, S., Fornal, J., Amarowicz, R., Lewandowicz, G., & Borkowski, K. (2003). Changes in the microstructure of wheat, corn and potato starch granules during extraction of non-starch compounds with sodium dodecyl sulfate and mercaptoethanol. *Carbohydrate Polymers*, *53*(1), 63–73. https://doi.org/10.1016/S0144-8617(03)00004-3

Bie, P., Li, X., Xie, F., Chen, L., Zhang, B., & Li, L. (2016). Supramolecular structure and thermal behavior of cassava starch treated by oxygen and helium glow-plasmas. *Innovative Food Science & Emerging Technologies*, *34*, 336–343. https://doi.org/10.1016/j.ifset.2016.03.005

Carvalho, A. P. M. G., Barros, D. R., da Silva, L. S., Sanches, E. A., Pinto, C. da C., de Souza, S. M., Clerici, M. T. P. S., Rodrigues, S., Fernandes, F. A. N., & Campelo, P. H. (2021). Dielectric barrier atmospheric cold plasma applied to the modification of Ariá (*Goeppertia allouia*) starch: Effect of plasma generation voltage. *International Journal of Biological Macromolecules*, *182*, 1618–1627. https://doi.org/10.1016/j.ijbiomac.2021.05.165

Chen, B. R., Wang, Z. M., Lin, J. W., Wen, Q. H., Xu, F. Y., Li, J., Wang, R., & Zeng, X. A. (2022). Improving emulsification performance of waxy maize starch by esterification combined with pulsed electric field. *Food Hydrocolloids*, *129*, 107655. https://doi.org/10.1016/j.foodhyd.2022.107655

Chen, B. R., Wen, Q. H., Zeng, X. A., Abdul, R., Roobab, U., & Xu, F. Y. (2021). Pulsed electric field assisted modification of octenyl succinylated potato starch and its influence on pasting properties. *Carbohydrate Polymers*, *254*, 117294. https://doi.org/10.1016/j.carbpol.2020.117294

Chen, J. J., Lii, C. Y., & Lu, S. H. I. N. (2003). Physicochemical and morphological analyses on damaged rice starches. *Journal of Food and Drug Analysis*, *11*(4), 10. https://doi.org/10.38212/2224-6614.2684

Chi, H., Xu, K., Wu, X., Chen, Q., Xue, D., Song, C., ... & Wang, P. (2008). Effect of acetylation on the properties of corn starch. *Food Chemistry*, *106*(3), 923–928. https://doi.org/10.1016/j.foodchem.2007.07.002

De La Hera, E., Gomez, M., & Rosell, C. M. (2013). Particle size distribution of rice flour affecting the starch enzymatic hydrolysis and hydration properties. *Carbohydrate Polymers*, *98*(1), 421–427. https://doi.org/10.1016/j.carbpol.2013.06.002

Deeyai, P., Suphantharika, M., Wongsagonsup, R., & Dangtip, S. (2013). Characterization of modified tapioca starch in atmospheric argon plasma under diverse humidity by FTIR spectroscopy. *Chinese Physics Letters*, *30*(1), 018103.https://doi.org/10.1088/0256-307X/30/1/018103

Desmet, T., Morent, R., De Geyter, N., Leys, C., Schacht, E., & Dubruel, P. (2009). Nonthermal plasma technology as a versatile strategy for polymeric biomaterials surface modification: A review. *Biomacromolecules*, *10*(9), 2351–2378. https://doi.org/10.1021/bm900186s

Devi, E., Kalaivendan, R. G. T., Eazhumalai, G., & Annapure, U. S. (2023). Impact of atmospheric pressure pin-to-plate cold plasma on the functionality of arrowroot starch. *Journal of Agriculture and Food Research*, 12, 100567. https://doi.org/10.1016/j.jafr.2023.100567

Eliasson, A. C. (Ed.). (2004). *Starch in food: Structure, function and applications.* CRC press.

Falade, K. O., Semon, M., Fadairo, O. S., Oladunjoye, A. O., & Orou, K. K. (2014). Functional and physico-chemical properties of flours and starches of African rice cultivars. *Food Hydrocolloids*, *39*, 41–50. https://doi.org/10.1016/j.foodhyd.2013.11.002

Ferreira, C. M. (2013). Departamento de Fisica and Centro de Electrodinamica da Universidade Tecnica de Lisboa (INIC), Instituto Superior Tecnico, Lisbon, Portugal. *Radiative Processes in Discharge Plasmas*, *149*, 431.

Gao, S., Liu, H., Sun, L., Liu, N., Wang, J., Huang, Y., ... & Wang, M. (2019). The effects of dielectric barrier discharge plasma on physicochemical and digestion properties of starch. *International Journal of Biological Macromolecules*, *138*, 819–830. https://doi.org/10.1016/j.ijbiomac.2019.07.147

Gao, L., Wu, Y., Wan, C., Wang, P., Yang, P., Gao, X., ... & Gao, J. (2022). Structural and physicochemical properties of pea starch affected by germination treatment. *Food Hydrocolloids*, *124*, 107303. https://doi.org/10.1016/j.foodhyd.2021.107303

Ge, X., Shen, H., Su, C., Zhang, B., Zhang, Q., Jiang, H., & Li, W. (2021). The improving effects of cold plasma on multiscale structure, physicochemical and digestive properties of dry heated red adzuki bean starch. *Food Chemistry*, *349*, 129159. https://doi.org/10.1016/j.foodchem.2021.129159

Ge, X., Guo, Y., Zhao, J., Zhao, J., Shen, H., & Yan, W. (2022). Dielectric barrier discharge cold plasma combined with cross-linking: An innovative way to modify the multiscale structure and physicochemical properties of corn starch. *International Journal of Biological Macromolecules*, *215*, 465–476. https://doi.org/10.1016/j.ijbiomac.2022.06.060

Godina-Rodríguez, Garay-Martínez, J. R., J. E., Maldonado-Jáquez, J. A., Lucio-Ruíz, F., Joaquín-Cancino, S., Bautista-Martínez, Y., & Granados-Rivera, L. D. (2024). Nutritive value of maize and soybean silages at different ratio in a subtropical climate condition. *Chilean Journal of Agricultural Research*, *84*(4), 540–547.

Goiana, M. L., de Brito, E. S., Alves Filho, E. G., de Castro Miguel, E., Fernandes, F. A. N., de Azeredo, H. M. C., & de Freitas Rosa, M. (2021). Corn starch-based films treated by dielectric barrier discharge plasma. *International Journal of Biological Macromolecules*, *183*, 2009–2016. https://doi.org/10.1016/j.ijbiomac.2021.05.210

Goiana, M. L., & Fernandes, F. A. N. (2023). Influence of dielectric barrier discharge plasma treatment on corn starch properties. *Processes*, *11*(7), 1966. https://doi.org/10.3390/pr11071966

Gupta, R. K., Guha, P., & Srivastav, P. P. (2022). Natural polymers in bio-degradable/edible film: A review on environmental concerns, cold plasma technology and nanotechnology application on food packaging-A recent trends. *Food Chemistry Advances*, *1*, 100135. https://doi.org/10.1016/j.focha.2022.100135

Gupta, R. K., Guha, P., & Srivastav, P. P. (2023). Effect of high voltage dielectric barrier discharge (DBD) atmospheric cold plasma treatment on physicochemical and functional properties of taro (*Colocasia esculenta*) starch. *International Journal of Biological Macromolecules*, *253*, 126772. https://doi.org/10.1016/j.ijbiomac.2023.126772

Gupta, R. K., Guha, P., & Srivastav, P. P. (2024a). Insights into the physicochemical, crystalline, multistructural, thermal and digestibility behavior of taro starch (*Colocasia esculenta*) modified by cold plasma treatment with high voltage dielectric barrier discharge (HVDBD): An alternative technique for chemical modification of starch and environmental sustainability. *Journal of Cleaner Production*, *461*, 142710. https://doi.org/10.1016/j.jclepro.2024.142710

Gupta, R. K., Guha, P., & Srivastav, P. P. (2024b). Dielectric barrier discharge plasma: A green and novel method to change the structure of taro peel starch and improve the physicochemical properties of taro peel starch films. *Plasma Processes and Polymers*, *21*(9), 2400047. https://doi.org/10.1002/ppap.202400047

Gupta, R. K., Guha, P., & Srivastav, P. P. (2024c). Exploring the potential of taro (Colocasia esculenta) starch: Recent developments in modification, health benefits, and food industry applications. *Food Bioengineering, 3*(3), 365–379. https://doi.org/10.1002/fbe2.12103

Gupta, R. K., Guha, P., & Srivastav, P. P. (2024d). Physical action of nonthermal cold plasma technology for starch modification. *Food Physics*, *1*, 100011. https://doi.org/10.1016/j.foodp.2024.100011

Gupta, R. K., Guha, P., & Srivastav, P. P. (2024e). Effect of cold plasma treatment and plasma-activated water on physicochemical and structural properties of starch: A green and novel approach for environmental sustainability. *Plasma Processes and Polymers*, *21*(4), 2300204. https://doi.org/10.1002/ppap.202300204

Gupta, R. K., Pipliya, S., Patel, J., Srivastav, P. P., Castro-Muñoz, R., Wang, C. K., & Nemţanu, M. R. (2025a). Essential oil-embedded starch based films for active packaging: Insights into migration, stability, and preservation efficiency. *Trends in Food Science & Technology*, 163, 105149. https://doi.org/10.1016/j.tifs.2025.105149

Gupta, R. K., Gnana Moorthy Eswaran, U., Kumar, S., Pipliya, S., Kaur, B., & Srivastav, P. P. (2025b). Synergistic modification of starch polysaccharide via sequential treatments with ultrasonic and cold plasma technology. *Journal of Food Processing and Preservation*, *2025*(1), 9063007. https://doi.org/10.1155/jfpp/9063007

Han, X., Zhang, M., Zhang, R., Huang, L., Jia, X., Huang, F., & Liu, L. (2020). Physicochemical interactions between rice starch and different polyphenols and structural characterization of their complexes. *LWT*, *125*, 109227. https://doi.org/10.1016/j.lwt.2020.109227

Han, Z., Yu, Q., Zeng, X. A., Luo, D. H., Yu, S. J., Zhang, B. S., & Chen, X. D. (2012a). Studies on the microstructure and thermal properties of pulsed electric fields (PEF)-treated maize starch. *International Journal of Food Engineering*, *8*(1). https://doi.org/10.1515/1556-3758.2375

Han, Z., Zeng, X. A., Fu, N., Yu, S. J., Chen, X. D., & Kennedy, J. F. (2012b). Effects of pulsed electric field treatments on some properties of tapioca starch. *Carbohydrate Polymers*, *89*(4), 1012–1017. https://doi.org/10.1016/j.carbpol.2012.02.053

Han, Z., Zeng, X., Zhang, B., & Yu, S. (2009a). Effects of pulsed electric fields (PEF) treatment on the properties of corn starch. *Journal of Food Engineering*, *93*(3), 318–323. https://doi.org/10.1016/j.jfoodeng.2009.01.040

Han, Z., Zeng, X. A., Yu, S. J., Zhang, B. S., & Chen, X. D (2009b). Effects of pulsed electric fields (PEF) treatment on physicochemical properties of potato starch. *Innovative Food Science and Emerging Technologies*, *10*(4), 481–485. https://doi.org/10.1016/j.ifset.2009.07.003

Hong, J., Chen, R., Zeng, X. A., & Han, Z. (2016a). Effect of pulsed electric fields assisted acetylation on morphological, structural and functional characteristics of potato starch. *Food Chemistry*, *192*, 15–24. https://doi.org/10.1016/j.foodchem.2015.06.058

Hong, J., Zeng, X. A., Buckow, R., Han, Z., & Wang, M. (2016b). Nanostructure, morphology and functionality of cassava starch after pulsed electric fields assisted acetylation. *Food Hydrocolloids*, *54*, 139–150. https://doi.org/10.1016/j.foodhyd.2015.09.025

Jane, J.-L., Kasemsuwan, T., Leas, S., Ia, A., Zobel, H. Il, & Robyt, D., J. F. (1994). Anthology of starch granule morphology by scanning electron microscopy. *Starch - Stärke*, *46*(4), 121–129. https://doi.org/10.1002/star.19940460402

Jaddu, S., Pradhan, R. C., & Dwivedi, M. (2022). Effect of multipin atmospheric cold plasma discharge on functional properties of little millet (*Panicum miliare*) flour. *Innovative Food Science & Emerging Technologies*, *77*, 102957. https://doi.org/10.1016/j.ifset.2022.102957

Kalaivendan, R. G. T., Mishra, A., Eazhumalai, G., & Annapure, U. S. (2022). Effect of atmospheric pressure non-thermal pin to plate plasma on the functional, rheological, thermal, and morphological properties of mango seed kernel starch. *International Journal of Biological Macromolecules*, *196*, 63–71. https://doi.org/10.1016/j.ijbiomac.2021.12.013

Katsumata, I., & Okazaki, M. (1967). Ion sensitive probe-a new diagnostic method for plasma in magnetic fields. *Japanese Journal of Applied Physics, 6*(1), 123. https://doi.org/10.1143/JJAP.6.123

Kaur, H., & Gill, B. S. (2019). Effect of high-intensity ultrasound treatment on nutritional, rheological and structural properties of starches obtained from different cereals. *International Journal of Biological Macromolecules*, *126*, 367–375. https://doi.org/10.1016/j.ijbiomac.2018.12.149

Kaur, P., & Annapure, U. S. (2023). Effects of pin-to-plate atmospheric cold plasma for modification of pearl millet (*Pennisetum glaucum*) starch. *Food Research International*, *169*, 112930. https://doi.org/10.1016/j.foodres.2023.112930

Lee, S. J., Hong, J. Y., Lee, E. J., Chung, H. J., & Lim, S. T. (2015). Impact of single and dual modifications on physicochemical properties of japonica and indica rice starches. *Carbohydrate Polymers*, *122*, 77–83. https://doi.org/10.1016/j.carbpol.2015.01.009

Lee, Y., & Chang, Y. H. (2015). Effects of galactomannan addition on rheological, pasting and physical properties of water chestnut starch. *Journal of Texture Studies*, *46*(1), 58–66. https://doi.org/10.1111/jtxs.12113

Lei, X., Wang, S., Li, Y., Han, H., Zhang, X., Mao, X., & Ren, Y. (2024). The multi-scale structure changes of γ-ray irradiated potato starch to mitigate pasting/digestion properties. *Food Research International*, *178*, 113931. https://doi.org/10.1016/j.foodres.2024.113931

Li, J., Ye, F., Lei, L., Zhou, Y., & Zhao, G. (2018). Joint effects of granule size and degree of substitution on octenylsuccinated sweet potato starch granules as Pickering emulsion stabilizers. *Journal of Agricultural and Food Chemistry*, *66*(17), 4541–4550. https://doi.org/10.1021/acs.jafc.7b05507

Liang, W., Zhang, Q., Duan, H., Zhou, S., Zhou, Y., Li, W., & Yan, W. (2023). Understanding $CaCl_2$ induces surface gelatinization to promote cold plasma modified maize starch: Structure-effect relations. *Carbohydrate Polymers*, *320*, 121200. https://doi.org/10.1016/j.carbpol.2023.121200

Lii, C. Y., Liao, C. D., Stobinski, L., & Tomasik, P. (2002a). Behaviour of granular starches in low-pressure glow plasma. *Carbohydrate Polymers, 49*(4), 499–507. https://doi.org/10.1016/S0144-8617(01)00365-4

Lii, C. Y., Liao, C. D., Stobinski, L., & Tomasik, P. (2002b). Exposure of granular starches to low-pressure glow ethylene plasma. *European Polymer Journal*, *38*(8), 1601–1606. https://doi.org/10.1016/S0014-3057(02)00022-8

Lin, Z. R., Zeng, X. A., Yu, S. J., & Sun, D. W. (2012). Enhancement of ethanol–acetic acid esterification under room temperature and non-catalytic condition via pulsed electric field application. *Food and Bioprocess Technology, 5*(7), 2637–2645. https://doi.org/10.1002/marc.200600751

Lu, X., Wang, Y., Zhen, X., Che, Y., Yu, H., Ge, Y., ... & Yao, Y. (2025). Editing of the soluble starch synthase gene MeSSIII-1 enhanced the amylose and resistant starch contents in cassava. *Carbohydrate Polymers*, *348*, 122903. https://doi.org/10.1016/j.foodhyd.2015.09.025

Luo, S. (2008). Using measurement-induced disturbance to characterize correlations as classical or quantum. *Physical Review A—Atomic, Molecular, and Optical Physics*, *77*(2), 022301. https://doi.org/10.1103/PhysRevA.77.022301

Maniglia, B. C., Castanha, N., Le-Bail, P., Le-Bail, A., & Augusto, P. E. D. (2021a). Starch modification through environmentally friendly alternatives: A review. *Critical Reviews in Food Science and Nutrition*, *61*(15), 2482–2505. https://doi.org/10.1080/10408398.2020.1778633

Maniglia, B. C., Pataro, G., Ferrari, G., Augusto, P. E. D., Le-Bail, P., & Le-Bail, A. (2021b). Pulsed electric fields (PEF) treatment to enhance starch 3D printing application: Effect on structure, properties, and functionality of wheat and cassava starches. *Innovative Food Science and Emerging Technologies*, *68*, 102602. https://doi.org/10.1016/j.ifset.2021.102602

Marenco Orozco, G. A. (2021). Produção de filmes de amido extraído de banana prata verde (*Musa sapientum*) modificados com plasma DBD em múltiplas etapas. http://www.repositorio.ufc.br/handle/riufc/60001

Massaux, C., Sindic, M., Lenartz, J., Sinnaeve, G., Bodson, B., Falisse, A., ... & Deroanne, C. (2008). Variations in physicochemical and functional properties of starches extracted from European soft wheat (*Triticum aestivum* L.): The importance to preserve the varietal identity. *Carbohydrate Polymers*, *71*(1), 32–41. https://doi.org/10.1016/j.carbpol.2007.05.012

Misra, N. N., Pankaj, S. K., Segat, A., & Ishikawa, K. (2016). Cold plasma interactions with enzymes in foods and model systems. *Trends in Food Science and Technology*, *55*, 39–47. https://doi.org/10.1016/j.tifs.2016.07.001

Mohamed, M. E., & Eissa, A. H. A. (2012). Pulsed electric fields for food processing technology. In *Structure and function of food engineering*. IntechOpen. https://doi.org/10.5772/48678

Moita, A. S., & Moreira, A. L. N. (2008, September). Boiling morphology and heat removal of impinging coolant droplets. In *Proceedings of 22nd European conference on liquid atomization and spray systems, ILASS2008, Como Lake, Italy*. https://doi.org/10.1109/TVLSI.2007.915434

Morent, R., De Geyter, N., Desmet, T., Dubruel, P., & Leys, C. (2011a). Plasma surface modification of biodegradable polymers: A review. *Plasma Processes and Polymers*, *8*(3), 171–190. https://doi.org/10.1002/ppap.201000153

Nemţanu, M. R., & Braşoveanu, M. (2010). Functional properties of some non-conventional treated starches. *Biopolymers (Edited by M. Eknashar)*, 319. https://doi.org/10.1016/S0969-806X(00)00346-7

Pal, P., Kaur, P., Singh, N., Kaur, A., Misra, N. N., Tiwari, B. K., ... & Virdi, A. S. (2016). Effect of nonthermal plasma on physico-chemical, amino acid composition, pasting and protein characteristics of short and long grain rice flour. *Food Research International*, *81*, 50–57. https://doi.org/10.1016/j.foodres.2015.12.019

Park, D. J., & Han, J. A. (2016). Quality controlling of brown rice by ultrasound treatment and its effect on isolated starch. *Carbohydrate Polymers*, *137*, 30–38. https://doi.org/10.1016/j.carbpol.2015.10.045

Sandhu, K. S., Singh, N., & Malhi, N. S. (2005). Physicochemical and thermal properties of starches separated from corn produced from crosses of two germ pools. *Food Chemistry*, *89*(4), 541–548. https://doi.org/10.1016/j.foodchem.2004.03.007

Sarangapani, C., Devi, Y., Thirundas, R., Annapure, U. S., & Deshmukh, R. R. (2015). Effect of low-pressure plasma on physico-chemical properties of parboiled rice. *LWT-food Science and Technology*, *63*(1), 452–460. https://doi.org/10.1016/j.lwt.2015.03.026

Sarangapani, C., Thirumdas, R., Devi, Y., Trimukhe, A., Deshmukh, R. R., & Annapure, U. S. (2016). Effect of low-pressure plasma on physico–chemical and functional properties of parboiled rice flour. *LWT-Food Science and Technology*, *69*, 482–489. https://doi.org/10.1016/j.lwt.2016.02.003

Sharma, M., Yadav, D. N., Singh, A. K., & Tomar, S. K. (2015). Rheological and functional properties of heat moisture treated pearl millet starch. *Journal of Food Science and Technology*, *52*(10), 6502–6510. https://doi.org/10.1007/s13197-015-1735-1

Sharma, A., Sharma, T., Sharma, T., Sharma, S., & Kanwar, S. S. (2019). Role of microbial hydrolases in bioremediation. *Microbes and Enzymes in Soil Health and Bioremediation*, 149–164. http://www.imtech.res.in/raghava/oxdbase

Sharma, V., Kaur, M., Sandhu, K. S., & Godara, S. K. (2020). Effect of cross-linking on physico-chemical, thermal, pasting, in vitro digestibility and film forming properties of Faba bean (*Vicia faba* L.) starch. *International Journal of Biological Macromolecules*, *159*, 243–249. https://doi.org/10.1016/j.ijbiomac.2020.05.014

Shen, H., Yu, J., Bai, J., Ge, X., Liang, W., Ospankulova, G., ... & Li, W. (2022). Electron beam irradiation regulates the structure and functionality of ball-milled corn starch: The related mechanism. *Carbohydrate Polymers*, *297*, 120016. https://doi.org/10.1016/j.carbpol.2022.120016

Shen, H., Guo, Y., Zhao, J., Zhao, J., Ge, X., Zhang, Q., & Yan, W. (2021). The multi-scale structure and physicochemical properties of mung bean starch modified by ultrasound combined with plasma treatment. *International Journal of Biological Macromolecules*, *191*, 821–831. https://doi.org/10.1016/j.ijbiomac.2021.09.157

Shogren, R. L., Doane, W. M., Garlotta, D., Lawton, J. W., & Willett, J. L. (2003). Biodegradation of starch/polylactic acid/poly(hydroxyester-ether) composite bars in soil. *Polymer Degradation and Stability*, *79*(3), 405–411. https://doi.org/10.1016/S0141-3910(02)00356-7

Shrivastava, A., Gupta, R. K., & Srivastav, P. P. (2024). Exploring novel frontiers of advancements in purple yam (*Dioscorea alata* L.) starch extraction, modification, characterization, applications in food and other industries. *Measurement: Food*, *15*, 100196. https://doi.org/10.1016/j.meafoo.2024.100196

Sifuentes-Nieves, I., Flores-Silva, P. C., Gallardo-Vega, C., Hernández-Hernández, E., Neira-Velázquez, G., Mendez-Montealvo, G., & Velazquez, G. (2020). Films made from plasma-modified corn starch: Chemical, mechanical and barrier properties. *Carbohydrate polymers*, *237*, 116103. https://doi.org/10.1016/j.carbpol.2020.116103

Sonkar, S., Jaddu, S., Pradhan, R. C., Dwivedi, M., Seth, D., Goksen, G., Sarangi, P. K., & Lorenzo, J. M. (2023). Effect of atmospheric cold plasma (pin type) on hydration and structure properties of kodo-millet starch. *LWT*, *182*. https://doi.org/10.1016/j.lwt.2023.114889

Sun, X., Saleh, A. S. M., Sun, Z., Ge, X., Shen, H., Zhang, Q., Yu, X., Yuan, L., & Li, W. (2022). Modification of multi-scale structure, physicochemical properties, and digestibility of rice starch via microwave and cold plasma treatments. *LWT*, *153*. https://doi.org/10.1016/j.lwt.2021.112483

Suri, S., & Singh, A. (2023). Modification of starch by novel and traditional ways: Influence on the structure and functional properties. *Sustainable Food Technology*, *1*(3), 348–362. https://doi.org/10.1039/d2fb00043a

Singh, J., Dartois, A., & Kaur, L. (2010). Starch digestibility in food matrix: A review. *Trends in Food Science & Technology*, *21*(4), 168–180. https://doi.org/10.1016/j.tifs.2009.12.001

Tan, X., Zhang, B., Chen, L., Li, X., Li, L., & Xie, F. (2015). Effect of planetary ball-milling on multi-scale structures and pasting properties of waxy and high-amylose cornstarches. *Innovative Food Science & Emerging Technologies*, *30*, 198–207. https://doi.org/10.1016/j.ifset.2015.03.013

Tester, R. F., Karkalas, J., & Qi, X. (2004). Starch—composition, fine structure and architecture. *Journal of Cereal Science*, *39*(2), 151–165. https://doi.org/10.1016/j.jcs.2003.12.001

Thirumdas, R., Kadam, D., & Annapure, U. S. (2017a). Cold plasma: An alternative technology for the starch modification. *Food Biophysics*, *12*, 129–139. https://doi.org/10.1007/s11483-017-9468-5

Thirumdas, R., Trimukhe, A., Deshmukh, R. R., & Annapure, U. S. (2017b). Functional and rheological properties of cold plasma treated rice starch. *Carbohydrate Polymers*, *157*, 1723–1731. https://doi.org/10.1016/j.carbpol.2016.11.050

Toepfl, S., Heinz, V., & Knorr, D. (2007). High intensity pulsed electric fields applied for food preservation. *Chemical Engineering and Processing: Process Intensification*, *46*(6), 537–546. https://doi.org/10.1016/j.cep.2006.07.011

Uthumporn, U., Zaidul, I. S., & Karim, A. A. (2010). Hydrolysis of granular starch at sub-gelatinization temperature using a mixture of amylolytic enzymes. *Food and Bioproducts Processing*, *88*(1), 47–54. https://doi.org/10.1016/j.fbp.2009.10.001

Whistler, R. L., BeMiller, J. N., & Paschall, E. F. (1965). Fractionation of starch. *Starch: Chemistry and Technology*, *1*, 331–347. https://doi.org/10.1016/B978-0-12-746270-7.50014-8

Whistler, R. L., & Daniel, J. R. (1984). Molecular structure of starch. In *Starch: chemistry and technology* (pp. 153–182). Academic press. https://doi.org/10.1016/B978-0-12-746270-7.50012-4

Wongsagonsup, R., Deeyai, P., Chaiwat, W., Horrungsiwat, S., Leejariensuk, K., Suphantharika, M., ... & Dangtip, S. (2014). Modification of tapioca starch by non-chemical route using jet atmospheric argon plasma. *Carbohydrate Polymers*, *102*, 790–798. https://doi.org/10.1016/j.carbpol.2013.10.089

Wrobel, V. (1988). Joint spectra and joint numerical ranges for pairwise commuting operators in Banach spaces. *Glasgow Mathematical Journal*, *30*(2), 145–153. https://doi.org/10.1016/0924-2716(91)90017-P

Wurzburg, O. B. (2006). Modified starches. *Food Polysaccharides and Their Applications*, 87–118.

Yan, Y., Feng, L., Shi, M., Cui, C., & Liu, Y. (2020). Effect of plasma-activated water on the structure and in vitro digestibility of waxy and normal maize starches during heat-moisture treatment. *Food Chemistry*, *306*, 125589. https://doi.org/10.1016/j.foodchem.2019.125589

Zhang, B., Mei, J. Q., Chen, B., & Chen, H. Q. (2017). Digestibility, physicochemical and structural properties of octenyl succinic anhydride-modified cassava starches with different degree of substitution. *Food Chemistry*, *229*, 136–141. https://doi.org/10.1016/j.foodchem.2017.02.061

Zhang, Y., Zhu, K., He, S., Tan, L., & Kong, X. (2016). Characterizations of high purity starches isolated from five different jackfruit cultivars. *Food Hydrocolloids*, *52*, 785–794. https://doi.org/10.1016/j.foodhyd.2015.07.037

Zhang, Y., Zhao, X., Bao, X., Xiao, J., & Liu, H. (2021). Effects of pectin and heat-moisture treatment on structural characteristics and physicochemical properties of corn starch. *Food Hydrocolloids*, *117*, 106664. https://doi.org/10.1016/j.foodhyd.2021.106664

Zhou, D., Ma, Z., Yin, X., Hu, X., & Boye, J. I. (2019). Structural characteristics and physicochemical properties of field pea starch modified by physical, enzymatic, and acid treatments. *Food Hydrocolloids*, *93*, 386–394. https://doi.org/10.1016/j.foodhyd.2019.02.048

Zięba, T., Szumny, A., & Kapelko, M. (2011). Properties of retrograded and acetylated starch preparations: Part 1. Structure, susceptibility to amylase, and pasting characteristics. *LWT-Food Science and Technology*, *44*(5), 1314–1320. https://doi.org/10.1016/j.lwt.2010.12.018

Zimmermann, J. L., & Rutherford, A. W. (1986). Electron paramagnetic resonance properties of the S2 state of the oxygen-evolving complex of photosystem II. *Biochemistry*, *25*(16), 4609–4615. https://doi.org/10.1021/bi00364a023

Zou, J. J., He, H., Cui, L., & Du, H. Y. (2004). Highly efficient Pt/TiO2 photocatalyst for hydrogen generation prepared by a cold plasma method. *International Journal of Hydrogen Energy*, *32*(12), 1762–1770. https://doi.org/10.1016/j.ijhydene.2006.11.030

Zou, J., Xu, M., Tang, W., Wen, L., & Yang, B. (2020). Modification of structural, physicochemical and digestive properties of normal maize starch by thermal treatment. *Food Chemistry*, *309*, 125733. https://doi.org/10.1016/j.foodchem.2019.125733

12 Starch Nanoparticles

Extraction Technology, Modification, Characterization, and Application in Food Industry

Paushali Mukherjee
Department of Chemical Engineering, Indian Institute of Technology, Guwahati, Guwahati, India

Kamal Narayan Baruah
Department of Biotechnology, Sharda School of Bioscience and Technology, Sharda University, Greater Noida, Uttar Pradesh, India

12.1 INTRODUCTION

Starch is a biodegradable, environmentally friendly polymer synthesized by plants for energy sources. Starchy substances have long served as a primary food source for humans and their ancestors (Le & Angellier-Coussy, 2014). Plants store starch in the form of granules with various forms and with particle sizes ranging from 2 to 50 µm in tubers and roots, fruits, grains, and seeds. Plants such as maize, potato, wheat, and rice are abundant in starch, a polysaccharide that is composed of amylose and amylopectin. Starch is the primary energy source for plants (Alves et al., 2021; Gupta et al., 2022). As a method of storing energy from photosynthesis, plant amyloplasts typically produce starch, one of the most prevalent biopolymers. Starch's versatile nature, biocompatibility, and ability to form nanoparticles make it an ideal candidate for the development of functional materials (Xie et al., 2024). In general, the polymers of anhydrous glucose units, which are referred to as starch molecules, are accumulated in separate and autonomous granules (Kim et al., 2015). Investigations by Gallant et al. (1997) and Tang et al. (2006) found that the starch granules comprised several semicrystalline blocks with nanoscale dimension. Although starch is somewhat common in the food industry, this macromolecule is not soluble in water at normal temperature. Moreover, many starches are affected by relative humidity, temperature, and storage duration (retrogradation process), which influences their weak-bodied and unpleasant gels when cooked (Campelo et al., 2020). The nano-blocks could be broken from the starch by using acids and/or enzymes for a mild hydrolysis (García et al., 2011; Namazi & Dadkhah, 2010). The food industry has been working to enhance the physicochemical properties of starches for many years because native starches lack sufficient qualities to ensure the quality of finished food items. Transforming these macromolecules from the micro to the nano scale is a current alternative for enhancing physicochemical properties. It also introduces new functionalities into the starches. These days, the scientific world is paying more attention to starch nanoparticles (SNPs) because of their many uses and adaptability.

Starch can be utilized to synthesize bionanoparticles, referred to as SNPs. SNP typically have particle sizes of tens to hundreds of nanometers and a spherical form, in contrast to native starches (Chavan et al., 2021). As a biodegradable, renewable, and non-toxic material, starch has been widely used in food applications (Kumari et al., 2019). The prospective applications of starch nanoparticles in the food, pharmaceutical, biomedical, and packaging industries have garnered significant interest in recent years (Lin et al., 2011). Derived from natural starch, these nanoparticles offer a green and decomposable alternative to synthetic materials.

Starch nanoparticles have gained prominence in food technology owing to their distinctive functional attributes and natural provenance. The emergence of nanotechnology has facilitated the advancement of SNPs, creating new opportunities for improving food quality, stability, and nutritional value (Marta et al., 2023). The high surface-to-volume ratio of these nanoscale particles enables the encapsulation of bioactive compounds, flavors, and nutrients, resulting in a controlled release and enhanced bioavailability. This enhanced interaction with other dietary components is facilitated by the nanoscale size (Marta et al., 2023). The capacity to encapsulate bioactive chemicals, increased mechanical strength, increased solubility, and a wide surface area are among the distinctive physicochemical features of SNPs. Owing to their nanoscale size and

DOI: 10.1201/9781003564287-12

high surface-to-volume ratio, these particles can encapsulate bioactive substances, tastes, and minerals, allowing for regulated release and higher bioavailability. They also interact better with other food components (Marta et al., 2023). These characteristics have led to research on the synthesis, modification, and application of SNPs. SNPs have demonstrated significant potential in the advancement of food technology by modifying the functional properties of food products, such as by encapsulating constituents for controlled release, modifying texture, and modifying to serve as a fat substitute in low-calorie foods (Wang et al., 2022). Moreover, they are potential nutrient delivery vehicles because of their non-toxicity and compatibility with a range of food matrices, which enhances the stability and bioavailability of encapsulated bioactives (Gorantla et al., 2021). Given their multifunctional potential and sustainable origin, SNPs present a burgeoning area of research with numerous applications across diverse fields. Furthermore, they enhance the physical characteristics of starch-based films, rendering SNPs particularly well suited for the creation of sustainable food packaging solutions that minimize deterioration and extend the shelf life of food. The retention of volatile compounds is further improved by their capacity to form inclusion complexes with lipophilic molecules, thereby enhancing the sensory attributes of food products. Due to their multidimensional capability and everlasting origin, SNPs constitute a rapidly advancing study domain towards various applications across a wide range of commercial sectors (Qin et al., 2020). Besides, SNPs are environmentally benign, are non-toxic, and exhibit distinctive characteristics, including a substantial surface area–to–mass ratio, reduced diffusion constraints, enhanced absorption capacity, superior biological penetration rate, and greater solubility compared to natural starch granules (García-gurrola et al., 2019).

Several physical and chemical techniques are available for the synthesis of SNPs (Wang et al., 2018). Kim et al. (2014) classified the various manufacturing methods into "top-down" (produced from bulk materials) and "bottom-up" (created through nanoscale matter manipulation). In the "top-down" approach, nanoparticles are generated by refining the structure and size through the disintegration of larger particles. In a "bottom-up" approach, nanoparticles are synthesized through the controlled aggregation of atoms or molecules, guided by thermodynamic methods such as self-assembly.

Several technologies, including nanoprecipitation, high-pressure homogenization, and enzymatic treatment, have been developed to yield SNPs with varied properties (Morán et al., 2021). Recently, Campelo et al. (2020) developed a comprehensive classification system that differentiates the preparation methods into two categories: traditional (acid hydrolysis) and unusual methods. These unconventional approaches include enzyme debranching, milling, sonication, thermosonication, nanoprecipitation, and ultrasound–non-thermic plasma combination. Physical approaches involve subjecting starch to high pressure, temperature, radiation, or shear forces in order to disrupt glycosidic bonds, which ultimately results in a reduction in particle size (Sun, 2018). Hydroxonium ions attack glycosidic bonds in starch and hydrolyze the linkage (Singh & Ali, 2000), which reduces the size of starch particles.

The spectrum of use or performance of as-prepared SNPs can be expanded with further treatments (Qin et al., 2020). Many investigations have reported on grafting, physical alterations, and chemical modifications of starch nanoparticles (Namazi & Dadkhah, 2010). Specifically, modification techniques that decrease the hydrophilic nature of SNPs have been the object of special interest (Cazotti et al., 2020). Graft modification of SNPs has been utilized to enhance their hydrophobicity, improving interfacial adsorption in Pickering emulsions. Conversely, core-shell SNPs synthesized using emulsion polymerization serve as effective fillers to augment the durability of fragile materials (Wang et al., 2018). SNPs continue to be essential in food technology because of their versatility and the increasing demand for natural, health-promoting foods. They provide novel approaches to food processing, preservation, and delivery methods. While SNP manufacturing routes have been widely covered in a number of works, the methods for modifying SNPs and the uses of changed SNPs in both conventional and non-traditional settings are not as well covered (Torres & De-la-Torre, 2022).

12.2 EXTRACTION PROCEDURES

Starch nanoparticles can be extracted or synthesized through various methods, including acid hydrolysis, enzymatic hydrolysis, ultrasonication, and various top-down and bottom-up approaches that are explained in detail in the following subsections (Figure 12.1).

12.2.1 Acid Hydrolysis

Acid hydrolysis is a simple procedure and can be easily regulated by modifying variables such as acid content, temperature, and hydrolysis duration. These parameters substantially affect the dimensions, shape, and yield of the SNPs. The resulted nanoparticles are often highly crystalline and exhibit distinctive features, including enhanced dispersibility and thermal stability, rendering them advantageous for various applications, such as in food, medicines, and biocomposites (Bras & Dufresne, 2010). The acid hydrolysis procedure is typically conducted using mineral acids such as H_2SO_4, HNO_3, or H_3PO_4 at temperatures below the gelatinization threshold for many days with constant agitation (Gatto et al., 2016; Lewicka et al., 2015). Acid concentration, temperature, and reaction time play crucial roles in determining the size, shape, and properties of the resulting SNPs. For example, high acid concentrations and extended hydrolysis times can produce smaller, more uniform particles with enhanced thermal stability and crystallinity (Zhou et al., 2020). Moderate acid hydrolysis, which is typically performed with hydrochloric or sulfuric

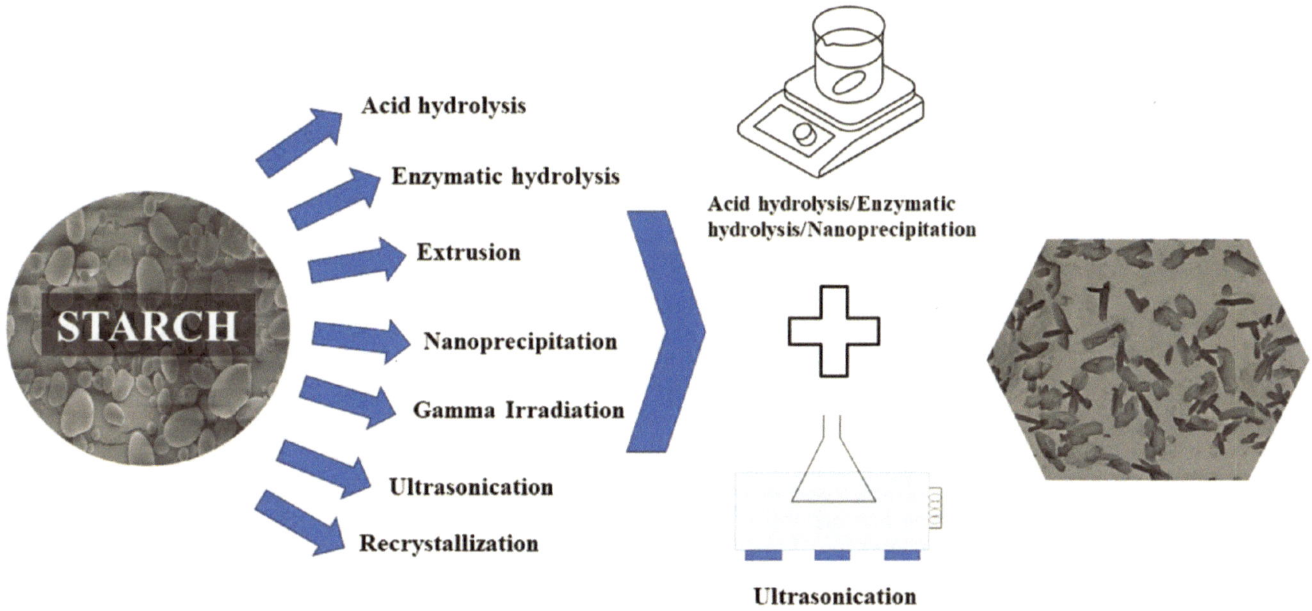

FIGURE 12.1 Different methods for the extraction of starch nanoparticles.

acid, can be used to isolate crystalline moieties from starch granules, as the crystalline regions are more resistant to acid hydrolysis than the amorphous regions. The starch nanoparticles in the reaction solution have a high crystallinity due to the potential for moderate acid hydrolysis to selectively erode the amorphous regions. The hydrolysis of starch granules is typically monitored in two ways: the recovery of insoluble starch residues and the content of soluble sugar in solution (phenol–sulfuric acid). The majority of the carbohydrates demonstrate a two-stage hydrolysis pattern, with a rapid initial hydrolysis and a slower subsequent hydrolysis. The hydrolysis of the amorphous components within starch granules is hypothesized to be the initial stage of acid hydrolysis, while the erosion of the crystalline regions is attributed to the sluggish stage. Nevertheless, the recovery yield is comparatively low due to the fact that this selective hydrolysis employing acids typically necessitates an extended process.

One of the disadvantage of this approach, though, is the possibility that acid residues might linger in the finished good and call for careful washing and neutralizing procedures. Furthermore, if not closely regulated, acid hydrolysis could cause starch to break down and result in a loss of its useful characteristics (Jayakody & Hoover, 2002). Notwithstanding these constraints, acid hydrolysis continues to be a favored method owing to its straightforwardness and capacity to generate SNPs with advantageous properties for food applications.

12.2.2 Enzymatic Hydrolysis

For SNP extraction, enzymatic hydrolysis provides a less harmful and more ecologically responsible option. By using enzymes such as α-amylase, β-amylase, and glucoamylase, starch molecules' glycosidic linkages are selectively broken, resulting in the formation of smaller nanoparticles (Chorfa et al., 2022). The utilization of enzymes facilitates more regulated hydrolysis, preserving the original structure and functionality of starch, which enhances its potential applications in food systems (Falkowska & Molga, 2014). In order to produce the desired nanoparticle size and characteristics, it is necessary to optimize parameters such as enzyme concentration, temperature, pH, and reaction time (Falkowska & Molga, 2014). The specificity of enzymes is one of the primary benefits of enzymatic hydrolysis as it allows for the production of SNPs with high purity and minimal by-products. In addition, the diversity of SNPs available for food formulations is increased by the suitability of enzymatic methods for processing a variety of carbohydrate sources, including those from non-conventional crops (Chorfa et al., 2022). Nonetheless, enzymatic hydrolysis may be prolonged and more costly than acid hydrolysis owing to the expense of enzymes and the requirement for meticulous process regulation.

12.2.2.1 Combination of Enzymatic and Acid Hydrolysis

The production of SNPs has utilized a combination of enzymatic and acid hydrolysis. The SNP preparation might be completed in a shorter duration by employing the combination approach. A two-hour pretreatment of starch with glucoamylase can significantly reduce the duration of acid hydrolysis. This preparation established channels for the acids to permeate the granules, rendering the amorphous portions more susceptible to hydrolysis. At an equivalent degree of hydrolysis (~70%), the SNPs derived from pretreatment starch were produced at a greater yield but exhibited a bigger size (~145 nm) in comparison to those obtained from untreated starch (typically ranging from 50 to 100 nm). The typical duration for acid hydrolysis, which was five days, was achieved in about 45 hours, yielding around 15%.

12.2.3 Mechanical Processes

Mechanical processes, such as high-pressure homogenization, ultrasonication, and milling, are physical methods used to disrupt starch granules and produce nanoparticles. High-pressure homogenization involves forcing starch suspensions through a narrow gap at high pressure, resulting in shear forces that break down the granules into nanoparticles (Ding & Kan, 2017). The process of ultrasonication involves the use of high-frequency sound waves to create cavitation, which in turn generates powerful mechanical forces that can fragment starch granules into nanoscale particles (Ahmad et al., 2020). Milling processes, such ball milling, reduce the size of starch particles by subjecting them to mechanical stress. By eliminating the need for chemical additions, these mechanical procedures yield SNPs devoid of any lingering solvents or chemicals, which is a major advantage (Mat Nor et al., 2013). Nevertheless, they frequently necessitate substantial energy input and may induce alterations in the structural and functional characteristics of starch, including partial gelatinization or diminished crystallinity, thereby affecting the efficacy of SNPs in food applications. The integration of mechanical methods with techniques like acid or enzymatic hydrolysis can improve extraction efficiency and customize the characteristics of SNPs for particular applications in the food industry.

12.2.4 Physical Treatments

The process of high-pressure homogenization using a microfluidizer relies on the control of a constant liquid flow via microfabricated channels. External pressure sources, external mechanical pumps, integrated mechanical micropumps, or electrokinetic mechanisms are used to implement the liquid flow's actuator. When an electronic-hydraulic intensifier pump is used in a system, the product stream accelerates to high speeds, which causes the product stream to shear excessively. This reduction in size was ascribed to the mechanical shear forces breaking the hydrogen bonds inside the big particles. Reactive extrusion-based nanoparticle production has also been studied. Reactive extrusion is characterized as a simultaneous reaction occurring during the extrusion manufacturing of polymers. The premixed starch and plasticizer were introduced into a twin-screw extruder, and reversible cross-linkers, such as glyoxal, were incorporated. The SNPs may be formed via cross-linking during reactive extrusion and are regarded as "regenerated" starch nanoparticles.

12.2.4.1 Extrusion

In the process of extrusion, starch is subjected to a significant amount of heat, mechanical shear forces, and pressures that can reach up to 103 pounds per square inch. Under high shear forces and temperatures, the hydrogen bonds between starch chains and responsible for maintaining the integrity of starch granules can be disrupted. As a consequence, major structural changes may take place, such as gelatinization, melting, and fragmentation, which may result in the partial or entire disintegration of the crystalline structure. Applying shear forces throughout extrusion decreases starch's molecular mass by causing chain scission, breaking the long chains into shorter fragments. This depolymerization modifies starch's functional and physicochemical properties, containing solubility, water absorption, and viscosity. Gelatinization, a key transformation taking place all through extrusion, comprises starch granule enlargement, amylose leaking, and damage of birefringence, converting starch structure from a methodical crystal-like arrangement to a tangled amorphous state. This irreversible progression considerably controls the extruded product's digestibility and textural properties.

Melting starch granules promotes fundamental disintegration, enhancing the mobility of starch molecules and facilitating their interaction with other components in the extruder. The high pressure and temperatures in the extrusion process may also result in dextrinization, in which starch molecules are fragmented into smaller glucose chains or oligosaccharides. These changed starch molecules demonstrated altered functional properties, making them suitable for unambiguous foodstuffs, feed, and industrial applications. The fast chilling displays the extrusion of the amorphous starch and locks in the structural alterations initiated for the giving-out period. Thus, extrusion converts native starch into a highly functional element with superior solubility, altered viscosity, and upgraded digestibility, personalized to meet miscellaneous product requirements.

12.2.4.2 Gamma Irradiation

Gamma irradiation has been proposed as a quick and convenient modification approach for fragmenting big molecules by cleaving glycosidic bonds. Gamma irradiation can generate free radicals capable of hydrolyzing chemical bonds, resulting in smaller starch pieces known as dextrin. According to Habib et al. (2023), SNPs with sizes of approximately 20 and 30 nm were generated by applying a dosage of 20 kGy (at an irradiation rate of 14 kGy/h) to cassava and waxy maize starches, respectively. Thermal characterization using simultaneous instrumental analysis found that the irradiated SNPs were more vulnerable to thermal degradation than their parent native starches. This suggests that SNPs have many hydroxyl groups on their surface where thermal breakdown begins, similar to the thermal characteristics of SNPs generated via acid hydrolysis. However, the nanoparticles produced by gamma irradiation are comparable to SNPs obtained from other physical treatments. The enriched reactivity of irradiated starch nanoparticles makes them suitable for applications demanding amplified interaction with other compounds, such as emulsification or encapsulation. Furthermore, the method permits accurate particle size and molecular weight regulated by amending the irradiation dose. Despite the resemblances, gamma-irradiated SNPs may reveal exclusive functional properties due to the specific mechanisms of bond cleavage and free radical formation. Gamma irradiation suggests a multi-purpose and proficient technique for fabricating improved starches with tailored functional characteristics.

12.2.4.3 Acid Hydrolysis and Ultrasonic Treatment

For acid hydrolysis alone, the SNP preparation time takes several days. However, the acid hydrolysis for two days has been shown to be sufficient to remove most of the amorphous regions in starch (50%–70%). The relative crystallinity of waxy maize starch has been reported to range from 30% to 50%. Subsequently, ultrasonic treatment (60% vibration amplitude, 3 min) was applied to the redispersed suspension of the large microparticles of starch hydrolyzates, which had been recovered by a mild centrifugation (500 rpm, 10 min). Using this simple process, the microparticles could be completely transformed to nanoparticles. However, ultrasonication often disrupts the crystalline structure of starch, and recovery yield of the crystalline SNPs becomes low (less than 30%). Kim et al. recently reported a procedure for SNP preparation without disrupting the crystallinity by using acid hydrolysis and ultrasonication. In their study, a low temperature (4°C) was applied to facilitate the association of the starch chains, the crystallinity disruption, and the formation of new crystalline structures in the SNPs by chain association. The starch hydrolyzates, obtained after six days of acid hydrolysis, were more resistant to the subsequent ultrasonication than those obtained after two or four days, regardless of the hydrolysis temperature. Compared to the ordinary process of acid hydrolysis alone, combining acid hydrolysis and ultrasonication increased the recovery yield of SNPs (15% vs. 78%). In addition, the low-temperature acid hydrolysis in the combined process was an effective process for the mass production of crystalline SNPs.

12.2.4.4 Nanoprecipitation

The nanoprecipitation process entails the incremental introduction of a dilute polymer solution into a solvent, resulting in the nanoscale precipitation of the polymer. This approach fundamentally relies on the interfacial deposition of polymers after the displacement of a semipolar solvent, which is miscible with water, from a lipophilic solution. This method offers numerous benefits: it eliminates the need for substantial quantities of hazardous solvents and external energy inputs while achieving submicron particle sizes with narrow size distributions. A dispersion of waxy maize starch acetate in acetone was combined with distilled water to create nanospheres, and acetone was subsequently evaporated from the aqueous suspension. The dimensions of the nanospheres are contingent upon the polymer content in the acetone. The average diameter rises from 249 to 720 nm when the starch concentration in acetone escalates from 1 to 20 mg/mL. A new process to obtain SNPs based on complex formation between amylomaize starch and *n*-butanol and subsequent enzymatic hydrolysis. First, a dilute starch solution was placed in the upper compartment of a glass filtration apparatus (Millipore, Billerica, MA, USA) and allowed to gravimetrically pass through a membrane filter (polytetrafluoroethylene [PTFE], 10 μm pore size, 47 mm diameter; Millipore) into the bottom compartment, which was filled with *n*-butanol. The filtration apparatus was kept at 70°C in a convection oven for six days. The starch–butanol complex was isolated as the precipitate and subjected to α-amylolysis (porcine pancreatic alpha-amylase, EC 3.2.1.1, activity 1122 units/mg). Because the amylose–butanol complex formed in the butanol layer contained a large portion of amorphous matrices, selective removal of these was required to isolate the nanoparticles. The subsequent enzymatic treatment made the SNPs spherical or oval in shape and 10–20 nm diameter. However, there was a significant loss by hydrolysis (85%–90%) of the starch, and thus the overall yield of the nanoparticles was very low.

12.2.4.5 Recrystallization

Cooked starch underwent treatment with pullulanase, after which the solution was preserved at 4°C to facilitate starch recrystallization. The recrystallized starch can be extracted via centrifugation as starch nanoparticles measure between 20 and 100 nm. The research demonstrated that debranching using pullulanase enhanced the recrystallization of starch chains. This method is significantly faster than traditional acid hydrolysis and achieves a greater yield of roughly 55%. This approach also benefits from not incorporating any chemical reagents during preparation. It avoids using harsh chemical reagents, and assembly is an eco-friendly and harmless process for fabricating SNPs. The SNPs arranged through this methodology show amended functional properties such as higher thermal stability, better dispersibility in aqueous systems, and potential as delivery systems for bioactive compounds. This technique holds great potential for applications in food, pharmaceutical, and other industrial sectors due to its easiness, efficiency, and sustainability.

12.2.4.6 Emulsion–Cross-linking

The synthesis of SNPs via the emulsion–cross-linking approach has been reported lately. The process entails the dispersion of an aqueous phase containing hydrophilic natural substances, such as starch and cross-linkers, inside an oil phase facilitated by emulsifiers. The emulsion can produce tiny particles via a cross-linking reaction. Nonetheless, the particles generated from this emulsion-cross-linking method were rather large on a micro scale. Atyabi et al. (2006) and Hamdi & Ponchel, (1999) reported starch microspheres with an average diameter of 19 μm, varying from 2.5 to 25 μm, produced via process emulsion–cross-linking.

12.2.4.7 Nanoemulsion

To diminish the dimensions of these particles, the emulsion must have nanoscale droplets, commonly termed mini-emulsion, submicron emulsion, or nanoemulsion.

The emulsion droplets retain their shape and size within the dispersed phase. Sharma et al. (2021) endeavored to diminish particle size by employing an ionic liquid-in-oil (IL/O) microemulsion system rather than the conventional water-in-oil (W/O) emulsion method. By replacing the aqueous phase with 1-octyl-3-methylimidazolium acetate, starch nanoparticles were produced, with an average diameter of 96.9 nm. A study by Zhou et al. (2014) proposed the potential generation of SNPs within a water-in-ionic-liquid (W/IL) microemulsion system. The technique was economical and environmentally sustainable, as it did not emit harmful organic reagents such as cyclohexane as the oil phase. Mini-emulsion can be generated using specialized apparatuses that deliver adequate mechanical energy, such as rotor-stator systems, sonicators, high-pressure homogenizers, and membrane technologies. The utilization of mechanical devices enhances emulsification effectiveness and decreases costs by requiring a reduced quantity of surfactant. Shi et al. (2011) utilized a high-pressure homogenizer at pressures between 10 and 60 MPa to generate a mini-emulsion, whereby the nanoparticles exhibited favorable sphericity, favorable sphericity, and somewhat uniform size distribution. The particle size fluctuated based on process parameters such as surfactant concentration, water-to-oil ratio, starch concentration, and homogenization pressure and cycles. Increasing energy application may result in a larger emulsion size due to the re-coalescence of droplets. The size of these microspheres is influenced by aspects such as the type and concentration of emulsifier, stirring speed, and the nature of cross-linkers used during the process. Higher stirring speeds and lower oil-to-water ratios can potentially condense particle size, but completing nanoparticles remains challenging with this method. Although the particles are larger than conventional SNPs, the emulsion–cross-linking methodology offers excellent control over particle uniformity and shape. Moreover, the starch microspheres fashioned via this procedure are robust, are chemically constant, and exhibit superior encapsulation efficiency, making them suitable for specific applications such as drug delivery and controlled-release systems. Further adjustments, such as altering the emulsifier or leading additional cross-linking agents, could advance the likelihood of this method for nanoparticle synthesis. Despite its limits, the emulsion–cross-linking process holds promise for tailoring starch-based materials to meet diverse industrial and biomedical needs.

12.3 MODIFICATION TECHNIQUES

To improve the functionality, stability, and applicability range of starch nanoparticles in a variety of various food products, modification techniques are implemented. These techniques can substantially impact the texture, viscosity, and overall quality of food formulations by modifying the physicochemical properties of SNPs. Chemical, physical, and enzymatic processes are the primary methods of modification, each of which provides distinct benefits and the properties of SNPs to suit specific dietary applications (Otache et al., 2021).

12.3.1 Chemical Modification

Chemical modifications are the process of modifying starch's molecular structure by reacting with various chemical agents. Acetylation, esterification, oxidation, and cross-linking are the most frequently employed chemical modification techniques (Otache et al., 2021; Shrivastava et al., 2024). The esterification is the process by which an organic acid reacts with an alcohol to produce an ester. An ester group can replace at least one of the hydroxyl groups (C-2, C–3, and/or C-6) in starch. An esterification reaction with acetic acid or a derivative, such as acetic anhydride, is known as acetylation and involves replacing hydroxyl groups with acetyl groups. Starch acetylation with acetic anhydride, typically performed in the presence of an alkaline catalyst (e.g., NaOH), is a straightforward method for functionalizing SNPs and modifying their characteristics (Saartrat et al., 2005). Acetylation, for example, incorporates acetyl groups into starch molecules, diminishing intermolecular interactions and enhancing the dispersibility and hydrophobicity of SNPs (Oliveira et al., 2017). Oxidation is a widely utilized chemical modification of starch. The oxidation of primary or secondary hydroxyl groups in starch leads to the production of aldehyde or carboxyl groups. Oxidation, typically conducted with chemicals such as sodium hypochlorite, incorporates carbonyl and carboxyl groups into the starch structure, improving its capacity to create stable gels and emulsions in food systems (Garrido et al., 2014) (Koo et al., 2010). Mostly, organic or inorganic peroxides handle this procedure. Among starch oxidizers, hydrogen peroxide is the most often used; sodium hypochlorite and sodium periodate are also relatively common. Smaller particle sizes of oxidized SNPs than of natural ones were evident. Shorter glucose chains resulting from starch oxidation—which induces partial depolymerization—could be responsible for this size decrease. Agents such as phosphorous oxychloride or epichlorohydrin cross-link to improve starch's molecular stiffness, producing SNPs with more resilience to heat, acid, and shear (Zhang et al., 2011). Although chemical modifications can considerably enhance the functional properties of SNPs, the safety and residual levels of modifying agents must be carefully assessed due to regulatory concerns regarding the use of specific chemical agents in food products.

12.3.2 Physical Modification

Physical modification techniques are eco-friendly options that exclude the use of chemical reagents. Methods include heat-moisture treatment, annealing, and high-pressure processing modify the physical characteristics of SNPs by influencing their crystalline structure and molecular configuration (Das & Sit, 2021). Heat-moisture treatment is heating starch at low moisture content to cause starch molecules to reorganize and changes in gelatinization characteristics, improving the thermal stability and resistance to enzymatic degradation of the SNPs (Hoover & Vasanthan, 1994). The pasting qualities and crystallinity of SNPs are enhanced through annealing, which is heating starch in excess water at a temperature lower than its gelatinization threshold (Campelo et al., 2020). High-pressure processing, employing techniques such as high-pressure homogenization or ultrasonication, physically disaggregates starch granules into smaller nanoparticles, modifying their surface properties and enhancing their interaction with other dietary constituents. Physical alterations provide a clean-label method for customizing SNPs' characteristics, rendering them especially attractive for incorporation in natural and organic food product (Marta et al., 2023).

Cold plasma (CP) is a non-thermal method employed for disinfection, particularly for microbe inactivation and the improvement of surface hydrophobicity (Han et al., 2014;). Sample processing utilizing CP does not necessitate hazardous chemicals and is considered an eco-friendly method (Chang et al., 2019; Gupta et al., 2024a). CP processing of SNPs has revealed alterations in their structure and functionality according to the processing gas and duration (Chang et al., 2019). CP-treated SNPs subjected to oxygen processing gas underwent oxidation, evidenced by the emergence of additional FTIR spectrum peaks at 1720 cm^{-1} (carbonyl/carboxyl groups). Negative charges escalated from –9.1 to –21.6 mV (CP treated for 21 minutes in oxygen gas) due to oxidation.

Ultrasound (sound waves exceeding frequencies of 15–20 kHz) is a physical modification technique that can result in depolymerization and alterations in the surface structure of starch granules (Sujka & Jamroz, 2013; Tomasik & Zaranyika, 1995; Gupta et al., 2025b). Like CP, ultrasonic modification is considered an eco-friendly procedure creating less need for supplementary chemicals (Sujka, 2017). It has benefits such as condensed treating time, lower energy consumption, and boosted regulator over the modification process. Ultrasonication can also improve starch's solubility, swelling power, and enzymatic digestibility, making it a versatile tool for tailoring starch properties for specific applications. Furthermore, due to upgraded dispersion and interaction with other constituents, ultrasonic behavior can increase the functional properties of starch-based materials, such as film-forming ability and adhesive strength.

The degree of modification induced by ultrasound depends on parameters such as frequency, power, duration, and the properties of the treated starch. For example, higher ultrasound power and longer treatment times result in greater depolymerization and structural changes. Moreover, ultrasound can be combined with other techniques, such as heat or enzymatic treatments, to achieve synergistic effects and further enhance the modification of starch. Due to its eco-friendly nature and effectiveness, ultrasonic modification has gained attention in various industries, including food, pharmaceuticals, and biodegradable packaging.

Grafting is a technique that integrates covalently bound monomers onto a polymer backbone. Grafting is classified into two methods: "grafting onto," in which polymer chains are attached to the surface via reactive groups, and "grafting from," in which new chains are synthesized from the surface using initiators. This technique has been examined utilizing SNP as grafting surfaces with various copolymer kinds. Although starch retained its crystalline structure, the grafted polylactic acid (PLA) surface exhibited predominant amorphous patches. The thermal stability of PLA-grafted SNPs was enhanced.

12.3.3 Enzymatic Modification

Enzymatic modifications produce SNPs with distinct functional characteristics by hydrolyzing or rearranging the molecular structure of starch using particular enzymes, such as pullulanase, phosphorylases, and amylases (S. O. B. Lee et al., 2020). For instance, α-amylase can partially hydrolyze starch to produce SNPs with changed viscosity and gel-forming characteristics. These can be useful for thickening and stabilizing sauces and dressings, among other food applications (Yang et al., 2023; Gupta et al., 2024c). Enzymatic phosphorylation incorporates phosphate groups into starch molecules, enhancing their solubility, swelling capacity, and freeze–thaw durability. Enzymatic alterations are preferred in the food sector due to their specificity, their gentle operational conditions, and the elimination of potentially hazardous substances. This approach facilitates the generation of SNPs devoid of undesirable by-products, in accordance with the clean-label trend in food formulation.

12.4 CHARACTERIZATION METHODS

Characterization of starch nanoparticles is crucial for understanding their morphology, structure, thermal properties, and functional behavior in various food applications. Several analytical techniques are used to assess the size, shape, crystallinity, chemical composition, and thermal stability of SNPs. This section discusses the primary characterization methods: scanning electron microscopy (SEM), transmission electron microscopy (TEM), X-ray diffraction (XRD), Fourier transform infrared spectroscopy (FTIR), differential scanning calorimetry (DSC), and thermal gravimetric analysis (TGA).

12.4.1 Scanning Electron Microscopy (SEM)

Scanning electron microscopy is widely used to study SNPs' surface morphology and particle size. SEM provides detailed, high-resolution images by scanning the sample surface with a focused beam of electrons. This technique reveals information about nanoparticles' surface texture, shape, and size distribution, which is essential for optimizing their functionality in food products (Cui et al., 2020; Gupta et al., 2024b). The images obtained from SEM show the uniformity and degree of aggregation of SNPs. For instance, SEM analysis can highlight the effect of modification techniques, such as acid hydrolysis, on the formation of smaller, more spherical particles (Liu et al., 2016). However, sample preparation techniques, such as coating with a conductive layer, must be performed to avoid charging effects in non-conductive samples like SNPs.

12.4.2 Transmission Electron Microscopy (TEM)

Transmission electron microscopy provides more detailed insights into SNPs' internal structure, shape, and size at the nanometer scale. TEM operates by transmitting a beam of electrons through a thin sample, producing images with higher resolution than SEM. This technique is particularly useful for characterizing the crystalline and amorphous regions within SNPs (Rao & Biswas, 2009). TEM can reveal the dispersion state of nanoparticles in a medium, which is crucial for understanding nanoparticles' interactions in food systems. For example, TEM has been used to and visualize the spherical nature and nanoscale size of starch particles prepared via acid hydrolysis or enzymatic treatment (Mourdikoudis et al., 2018). Due to its high resolution, TEM is also effective for identifying core-shell structures in modified starch particles.

12.4.3 X-ray Diffraction (XRD)

X-ray diffraction is a critical technique for determining the crystalline structure and degree of crystallinity of SNPs. SNPs typically consist of both crystalline and amorphous regions, and XRD helps identify specific crystal forms, such as the A-type, B-type, or C-type polymorphs in starch granules (Kumari et al., 2024; Gupta et al., 2024d),. XRD works by directing X-rays onto the SNP sample and measuring the diffracted rays' intensity and angle, which reflects the crystal lattice's arrangement. The peaks in an XRD spectrum indicate the presence of specific crystalline structures, while the peak intensity provides information about the degree of crystallinity. In SNPs, acid hydrolysis often increases crystallinity by removing amorphous regions, as revealed by sharper XRD peaks (Kim et al., 2012). This method is essential for correlating the crystalline nature of SNPs with their functional properties, such as water absorption and swelling power.

12.4.4 Fourier Transform Infrared Spectroscopy (FTIR)

Fourier transform infrared spectroscopy is used to identify the chemical composition and functional groups present in SNPs. FTIR spectra provide characteristic absorption bands corresponding to various molecular vibrations, including O–H stretching, C–H bending, and C–O stretching, which help determine any chemical modifications in SNPs (Faghihzadeh et al., 2016). For example, FTIR analysis can confirm the presence of ester or acetyl groups in chemically modified SNPs, indicating successful modification processes such as acetylation or phosphorylation (Phukaoluan et al., 2021). Additionally, changes in the intensity or shift of specific bands in the FTIR spectra can reveal the formation of hydrogen bonds or the interaction of SNPs with other food components. This information is valuable for understanding how SNPs behave in complex food matrices.

12.4.5 Differential Scanning Calorimetry (DSC)

Differential scanning calorimetry measures the thermal properties of SNPs, including gelatinization temperature, melting point, and enthalpy changes. This technique provides insights into the stability of SNPs when subjected to thermal processing, which is critical for food applications (Menczel & Prime, 2009). During DSC analysis, a sample is heated at a controlled rate, and the heat flow associated with physical and chemical changes is recorded. The gelatinization properties of SNPs, such as the onset, peak, and conclusion temperatures, can be determined from the DSC thermogram. These properties are affected by factors such as particle size, crystallinity, and amylose-to-amylopectin ratio (Gupta et al., 2024e). For instance, SNPs produced by enzymatic hydrolysis often exhibit lower gelatinization temperatures than native starch, as observed in DSC analysis (Abraham et al., 2018). Understanding these thermal characteristics is vital for designing SNP-based food products that undergo various heat treatments during processing.

12.4.6 Thermal Gravimetric Analysis (TGA)

Thermal gravimetric analysis is used to assess the thermal stability and degradation behavior of SNPs. In TGA, the sample is heated at a constant rate in an inert or oxidative atmosphere, and the mass loss is monitored as a function of temperature. This method provides valuable information on the moisture content, decomposition temperature, and thermal stability of SNPs (Loganathan, 2017). SNPs typically exhibit two-step weight loss: an initial weight loss due to moisture evaporation, followed by thermal degradation of the starch polymer. The onset and peak temperatures of degradation indicate the SNPs' thermal stability, which can be influenced by modification techniques. For example, chemically modified SNPs, such as acetylated starch, often display higher thermal stability than native starch, as evidenced by TGA analysis (Shapi'i et al., 2022). TGA results are crucial for optimizing processing conditions and predicting the behavior of SNPs in food products subjected to high-temperature processing.

12.5 APPLICATIONS IN THE FOOD INDUSTRY

Starch nanoparticles with their enhanced physicochemical properties have found a multitude of applications in the food process engineering sector. The utility of starch nanoparticles is largely focused on its application as a wall material for encapsulation, as a nutrient enhancer, for improving

TABLE 12.1
Application of Starch Nanoparticles in Nanoencapsulation, Edible Packaging/Coating, and Biosensors

Nanoencapsulation

Starch Source	Material Loaded	Method	Particle Size (nm)	Encapsulation Efficiency	Functional Benefits	Reference
Horse chest nut	Resveratrol	Ultrasonication	419	81.46%	Higher anti-obesity and anti-diabetic activities	Guo et al., 2022
Corn starch	β-carotene	Nanoprecipitation and freeze drying	<120 nm	–	Controlled release in the small intestine	Lee et al., 2022
Potato starch, pea starch	Quercetin	Nanoprecipitation and freeze drying	–	49%, potato starch 44%, pea starch	Better controlled release of 57%–59% at 175 mins	Farrag et al., 2018
Potato starch	Vitamin D3	Ultrasonication and freeze drying	99 nm	94.8%	Enhanced bioavailability	Hasanvand et al., 2018
Potato starch	EGCG	Inclusion complexation	20–120 nm	–	Enhanced bioactivity	Baruah et al., 2024

Edible Films and Food Packaging

Starch Source	Type of Application	Method	% SNPs	Mechanical Properties	Reference
Pea starch	Edible coating nanocomposite film	Self-assembly	6% (w/w) in starch films	Enhanced tensile strength by 70%	Jiang et al., 2016
Mango seed waste	Edible packaging of kiwis and tomatoes	Composite reinforcement	5% (w/w)	Increased bursting strength, reduced vapor transmission rate	Singh et al., 2024
Cassava starch	Edible packaging of Huangguan pears	Cross-linking nanocomposite reinforcement	6% (w/w)	Enhanced mechanical and water barrier properties	Ghosh et al., 2021

Biosensors

Starch Source	Type of application	Method	Detection Material	Lowest lImit of Detection	Reference
Rice flour	Colorimetric sensor	Immobilization on starch cryogels	Iron	0.48 mg/L	Phatthanawiwat et al., 2022
Soluble potato starch	Chromium speciation	Starch-stabilized Ag nanoparticles	Chromium	48 μg/L	Sapyen et al., 2022

texture and shelf life, for the development of food packaging materials, and in biosensors. Table 12.1 shows the various applications of SNPs in the food processing industry. The following sections delve deeper into these applications of SNPs.

12.5.1 Application as Wall Material

Encapsulation is a process by which functional bioactives that suffer from poor bioaccessibility are protected in the harsh digestion environments (Cai et al., 2018; Neilson et al., 2007). Figure 12.2 illustrates the multifaceted applications of starch nanoparticles as a wall material applied through various methods. Nanoencapsulation offers enhanced absorption in the small intestine due to its reduced size along with enhanced bioaccessibility (Zhu & Huang, 2019). Starch nanoparticles offer diverse applications in the field of encapsulation as a wall material for different nutraceuticals ranging from hydrophilic to hydrophobic compounds (Fathi et al., 2014). Various methods such as inclusion complex, nanoprecipitation, emulsion, and ultrasonic treatment combined with various drying techniques such as spray drying, freeze drying, and vacuum drying have been explored for the application of starch nanoparticles as carapace for the nanoencapsulation of bioactive compounds (Contreras-Jiménez et al., 2019; Fathi et al., 2014; Morán et al., 2021).

12.5.1.1 Inclusion Complex–Based Encapsulation

Considerable research has been carried out in the nanoencapsulation of functional bioactives or nutraceuticals using the inclusion complex (IC) method. In a related prior art, epigallocatechin gallate (EGCG) was encapsulated using potato starch nanoparticles using inclusion complexation. The process yielded nanosized structures of 20–120 nm with enhanced thermal and crystalline properties (Baruah et al., 2024). In another study, the researchers explored a scalable method for preparation of inclusion complexes of amylose nanoparticles with oleic acid. The authors used steam jet cooking and prepared ICs of amylose and oleic acid. Nanoencapsulated complex structures smaller than 375 nm were synthesized. Amylose formed V-type helical structures, which formed stable complexes with oleic acid and formed nanostructures (Fanta et al., 2015). In some of these processes, the researchers prepared starch nanoparticles separately and then encapsulated the bioactives, while many other adopted methods to synthesize starch nanoparticles during the encapsulation process itself, such as by using nanoprecipitation methods.

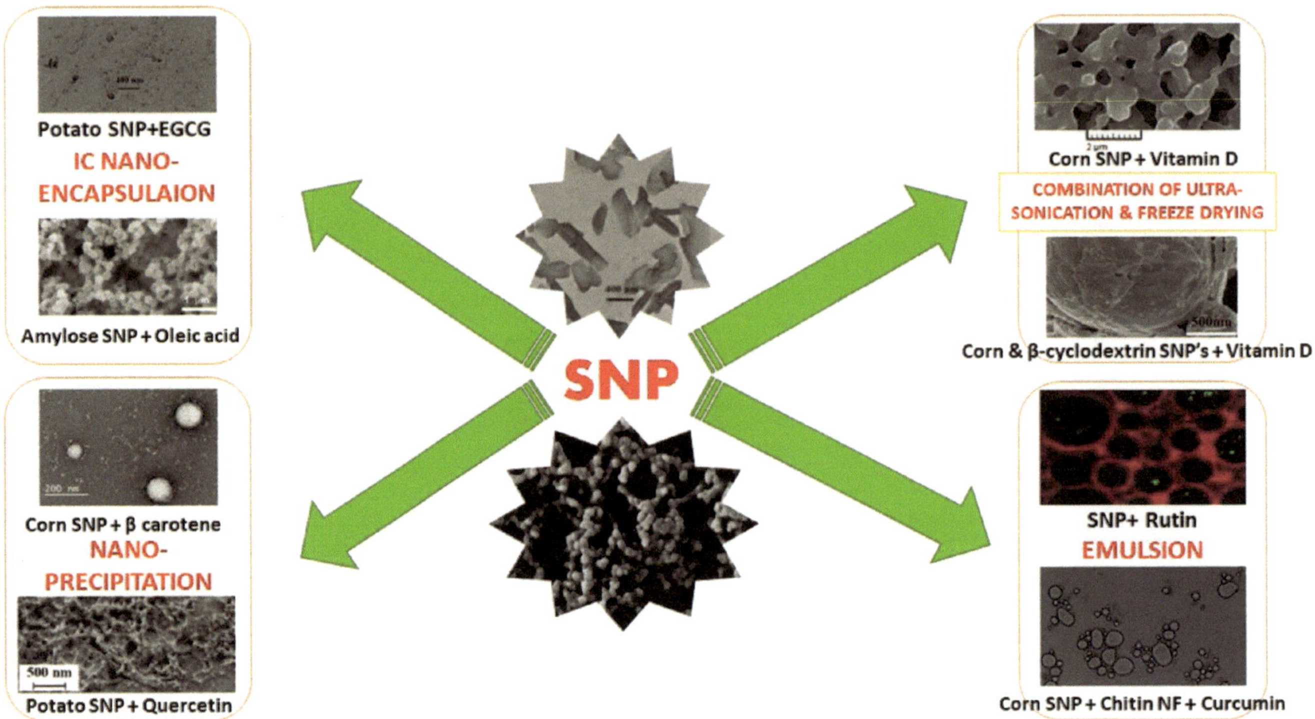

FIGURE 12.2 Various types of nanoencapsulation systems using starch nanoparticles classified based on their method of encapsulation. EGCG, epigallocatechin gallate; NF, nanofibers; SNP, starch nanoparticle.

12.5.1.2 Nanoprecipitation-Based Encapsulation

In another relevant prior art, starch nanoparticles prepared from corn starch was used to encapsulate β-carotene using nanoprecipitation. Starch nanoparticles from corn starch were prepared using a combination of acid and heat treatment coupled with freeze drying. Starch nanoparticles varied from 10 to 100 mg and β-carotene from 10–200 μg. It was observed that at 50 mg concentration of corn SNP, approximately 80% of the β-carotene was encapsulated (Lee et al., 2022). Another important bioactive curcumin was encapsulated using starch nanoparticles prepared from sago starch and banana starch nanoparticles using nanoprecipitation. The nanoencapsulated bioactives using starch nanoparticles always produced nanostructures with lower particle sizes, which enhances their physicochemical properties as well as their bioabsorption and bioavailability (Morán et al., 2021). In sago starch, nanoparticle-based encapsulation of curcumin obtains nanostructure with as low as 87 nm average particle size with high encapsulation efficiency, at 78%. In a similar work, the researchers used native and acetylated banana starch nanoparticles to encapsulate curcumin. The resulting nanostructures exhibited particle sizes smaller than 250 nm average along with an acceptable polydispersity index. Encapsulation efficiency was found to be better with acetylated banana starch as it allowed increased formation of H bonds that facilitated nanostructure formation (Acevedo-Guevara et al., 2018; Chin et al., 2014; Miskeen et al., 2021).

12.5.1.3 Emulsion-Based Encapsulation

Starch nanoparticles were also utilized in the form of emulsions to encapsulate hydrophobic bioactives. In a relevant prior art, SNPs prepared from maize and quinoa were utilized as a stabilizing component for the emulsion-based encapsulation of ferulic acid. In this case, the nanoparticles were prepared in advance using an ultrasonic-assisted antisolvent precipitation method and then modified using octenyl succinic acid and nonenyl succinic anhydride. Significant encapsulation of 99% was achieved with the modified starch nanoparticles (Remanan & Zhu, 2023). The same group of researchers replicated these results for the nanoencapsulation of rutin in emulsions stabilized using the same sources of starch nanoparticles (Remanan & Zhu, 2023). In another related study, the authors examined the efficacy of encapsulating curcumin immersed in soyabean oil with a mix of waxy corn starch nanoparticles and nanofibers (NFs) prepared from chitin. The SNP and chitin NF enhanced the stability of the emulsion, which enjoyed 60% retention of the curcumins, and even after 16 days of storage and UV exposure retained over 50% of the curcumins (Lee et al., 2021).

12.5.1.4 Combination Methods with Ultrasonication

Apart from the conventional methods, recent studies have explored a combination of ultrasonication, nanoprecipitation, and freeze drying methods for the encapsulation of different bioactives. In one such study, vitamin D3 was encapsulated using corn starch nanoparticles by sonication-assisted nanoprecipitation. The nanoencapsulation resulted in nanostructures having less than 99 nm average particle size and 94% encapsulation efficiency. A probe-type sonicator was used for the sonication process, which assisted in the formation of H bonds and increased encapsulation efficiency (Hasanvand et al., 2018). Similarly, encapsulation of EGCG was achieved using a mix of corn starch and β-cyclodextrin nanoparticles. The principles behind the encapsulation process adopted a self-assembly mechanism that required the inclusion complex method coupled with ultrasonication and freeze drying. The prepared encapsulated EGCG exhibited 87% encapsulation rate and was found to be stable under acidic and basic pH conditions (Li et al., 2024).

12.5.2 Starch Nanoparticles in Edible Films and Food Packaging

Recently, much research has applied starch nanoparticles or nanocrystals for the preparation of biodegradable edible coatings. Research has revealed that carbohydrate-based nanoparticles have excellent properties that propagate their application in edible coatings. In one interesting application of starch nanoparticles, potato starch nanoparticles were applied to prepare an biodegradable organic edible nanocomposite film coating made from pea starch. Potato SNPs were incorporated within a range of 3%–12%, with 6% SNP incorporation yielding the best results with a significant enhancement of 70% in tensile strength along with enhanced thermal stability (Jiang et al., 2016). In another study, the authors prepared a sago starch edible film that was then strengthened using carboxymethyl cellulose nanoparticles. These resulted in enhanced mechanical as well as physical properties of the edible film (Tabari, 2018).

In a similar study, researchers used starch nanocrystals to enhance the mechanical and physical properties of edible coatings and further studied its efficacy in terms of enhancement of the shelf life of Huangguan pears. Cassava starch was cross-linked for the preparation of the edible coating, and the results showed that incorporation of 6% starch nanocrystals exhibited the best properties in terms of vapor permeability, morphology and texture, color indices, and tensile/compression strengths (Dai et al., 2020). Starch nanoparticles can also be prepared from waste materials such as mango seeds, as observed in research that further explored its utility in the preparation of edible coatings. Researchers prepared starch nanoparticles using acid hydrolysis method and utilized at concentrations of 1%–10% of the native starch concentration. The prepared edible film with 5% substitution showed significant enhancement in its mechanical properties with decreased vapor permeability and enhanced bursting strength. When applied for the storage of tomatoes and kiwifruits, shelf life was enhanced by 43.5% and 50%, respectively, compared to control (Singh et al., 2024).

12.5.3 Application in Biosensors

Outside its traditional usage in food industry, starch nanoparticles have recently been utilized as a key component in the development of biosensors for various applications. These include detection of minerals such as iron and detection of toxic heavy metals such as Cr(VI), Pb_2^+, $Cu2^+$, and Hg_2^+ (Alves et al., 2021). Starch from rice flour has a low gelatinization temperature, and researchers have taken the advantage of this property to produce starch cryogels. These were used to immobilize catechin and curcumin nanoparticles, and the combined structure was used as a green colorimetric sensor to detect iron. The cryogels showed high accuracy ($R^2 > 0.99$) and sensitivity at 0.48 mg/L (Phatthanawiwat et al., 2022). Another study utilized soluble starch for the synthesis of a colorimetric sensor using Ag nanoparticles to detect Cr(VI). Soluble starch has a highly ordered structure due to its repetitive glucose units, and it functioned as a base to stabilize the Ag nanoparticles. Presence of Cr(VI) was detected through a color alteration from yellow to no color, which occurred due to the redox reaction between Cr and Ag nanoparticles. The method was combined with ICP-OES for the detailed speciation and showed high sensitivity up to 48 μg/L (Sapyen et al., 2022). Similar research reported the application of starch nanocrystals in combination with reduced graphene oxide for the preparation of a composite film for a multifunctional sensor of bending or stretching, relative humidity, and temperature. The starch nanocrystals were prepared using acid hydrolysis and sourced from waxy maize starch (Li et al., 2022). In a related study, the authors prepared a reversible multifunctional sensor using carboxymethyl starch, silver nanoparticles, and polyvinyl alcohol. The sensor was used to detect and evaluate complex human behavior and patterns of handwriting. With high sensitivity (0.5%) and high antibacterial activity and 3R circulation, the sensors were found to have excellent utility (Li et al., 2023).

12.6 ADVANTAGES OF STARCH NANOPARTICLES

Starch nanoparticles are versatile in application, starting from their use as a wall material in food packaging for edible film preparation to their application in biosensors. Current research trends focus on the application of both conventional and non-conventional sources of starch. Conventional sources such as potato, taro, corn, maize, sago, and quinoa are easily available (Gupta et al., 2025a), and much research has been done to synthesize starch nanoparticles from them. Different sources have different advantages; for instance, high-amylose starch sources such as maize, peas, and potatoes have been utilized for nanoencapsulation using self-assembly mechanism. Amylose-based nanoparticles form a V-type helical structure that helps in the entrapment and encapsulation of bioactives such as tea catechin and curcumin by inclusion complexation method (Fanta et al., 2015; Wang et al., 2021).

Compared to natural starch, starch nanoparticles have superior thermal, crystallographic, mechanical, and chemical properties, facilitating their varied uses in the food industry. Thermal analyses indicate that native starch has a melting start temperature of 245°C, but starch nanoparticles produced using sonication-assisted acid hydrolysis exhibit a melting point of 319°C (Baruah et al., 2023, 2024). Comparable improvements were noted for crystallinity, rendering an ideal material for the encapsulation of bioactives and the fabrication of edible films. Recently, non-traditional sources of starch have been investigated for their potential as reinforcing materials in the manufacture of edible films. Starch polymer provides a sustainable, biodegradable solution for food packaging, which has long been dominated by non-biodegradable plastics. Starch was isolated from mango seeds and transformed into nanoparticles using acid hydrolysis for use in the formulation of edible coatings.

12.7 CHALLENGES AND FUTURE DIRECTIONS

Despite the advantages of starch nanoparticles, modern researchers face specific obstacles regarding their manufacture and utilization. Starch nanoparticles are challenging to replicate with respect to shape and crystalline structure, resulting in variance in their thermal, physical, and chemical properties (Jambrak et al., 2010; Rajisha et al., 2014). Starch is obtained from several traditional and unconventional sources, exhibiting variable rheological, morphological, and crystalline characteristics. For example, potato starch possesses a B-type crystalline structure, whereas corn exhibits an A-type crystalline structure. Maize alters its crystalline structure in accordance with variations in the amylose-to-amylopectin ratio. The synthesis of starch nanoparticles from several sources results in nanoparticles exhibiting differences in size, polydispersity index, structure, morphology, amylose-to-amylopectin ratio, crystallinity, and thermal characteristics. Certain researchers have synthesized starch nanoparticles by amalgamating various starch sources, resulting in increased complexities and ambiguities concerning the regulation of the quality of the produced starch nanoparticles (Torres & De-la-Torre, 2022). Future researchers must remain vigilant about these variables and conduct comprehensive analyses on the production of nanoparticles, enabling more effective control over their resultant qualities.

Secondly, traditional research has concentrated on the creation of starch nanoparticles using acid hydrolysis via sulfuric and hydrochloric acids, nanoprecipitation techniques, and thermal cooking methods. Among these options, processes such as acid hydrolysis are detrimental to the environment due to the production of concentrated acidic waste. Consequently, future researchers should focus on innovative methods for the regulated and efficient synthesis of starch nanoparticles. Recently, research on combination approaches for the manufacture of starch nanoparticles has increased. These approaches encompass a combination of sonication, nanoprecipitation, steam cooking, microwave processing, and hydrothermal treatments. The amalgamation of these technologies may facilitate the effective synthesis of starch nanoparticles (Baruah et al., 2023; Lee et al., 2022). Thirdly, concerns persist about the safety and consumer acceptance of various starch nanoparticles. Starch nanoparticles utilized for encapsulating bioactive substances are often synthesized via sulfuric acid–mediated hydrolysis or nanoprecipitation using several chemicals, which may leave residual traces in the nanoparticles. The safety implications of utilizing starch nanoparticles as wall material for bioactive chemicals may be a concern.

12.8 SUMMARY

Starch nanoparticles are a valuable, easily available, multifunctional biomaterial that can be utilized in various aspects of food processing including not limited to encapsulation, edible packaging, and biosensors. There has been extensive research done in the extraction and synthesis of starch nanoparticles and their modification. SNPs as a biomaterial are very versatile and can be modified according to the requirements of the specific application. In this respect, the major findings of the book chapter can be summarized in the following points:

1. Starch nanoparticle extraction is conventionally carried out through acid hydrolysis whereby starch from different sources os exposed to 3.16 M sulfuric acid concentration at 15% w/v for seven to nine days.
2. Acid hydrolysis has been combined with ultrasonication, enzymatic hydrolysis, and freeze drying to extract starch nanoparticles faster and with better storage stability. A few studies also attempted increasing the sulfuric acid concentration, which led to the formation of SNPs with enhanced thermal and crystalline attributes.
3. The versatility of SNPs can be attributed to the ease with which they can be modified by chemical, physical, and enzymatic treatments. Chemical methods such as acetylation are applied to SNPs for the encapsulation of lipophilic compounds, while enzymatic and acid hydrolysis can enhance the concentration of resistant starch in SNPs.
4. SNPs are used to encapsulate a variety of bioactive compounds such as EGCG, quercetin, vitamin D, and β-carotene using various encapsulation methods such as inclusion complexation, nanoprecipitation, ultrasonication, and freeze drying.
5. SNPs are used as reinforcement material in edible films and packaging. Edible coatings prepared using SNPs have significantly enhanced mechanical and thermal attributes.
6. Starch nanoparticles can be used in making biosensors to detect iron and chromium speciation. Starch nanoparticles can be converted into cryogels and used as a base for the preparation of biosensors.

12.9 CONCLUSION

This book chapter provides an in-depth analysis into the multidimensional application of starch nanoparticles or nanocrystals in the field of food processing. Current research trends in starch nanoparticle synthesis indicates a shift towards sustainable and combined methods that can lead to better control of the nanoparticle quality. Starch nanoparticles can easily be modified and used for various applications, and researchers are looking for new ways to utilize them. The most important hurdle in the preparation of starch nanoparticles is replicability. It is important for future researchers to focus on replicable and consistent quality of starch nanoparticles. Only then will it be possible to take the application of starch nanoparticles from lab scale to industrial scale. Another important aspect to look out for is the green and sustainable synthesis of starch nanoparticles. Adopting greener methods that use less hazardous chemicals and leave no metal or non-metal residues in the starch nanoparticles could lead the way to application in functional food product formulations.

REFERENCES

Abraham, J., Arif P., M., Kumar, M. P. A., George, S. C., & Thomas, S. (2018). Thermoanalytical techniques of nanomaterials. In S. M. Bhagyaraj, O. S. Oluwafemi, N. Kalarikkal, & S. Thomas (Eds.), *Characterization of nanomaterials: Advances and key technologies* (pp. 213–236). Elsevier Ltd. https://doi.org/10.1016/B978-0-08-101973-3.00008-0

Acevedo-Guevara, L., Nieto-Suaza, L., Sanchez, L. T., Pinzon, M. I., & Villa, C. C. (2018). Development of native and modified banana starch nanoparticles as vehicles for curcumin. *International Journal of Biological Macromolecules*, *111*, 498–504. https://doi.org/10.1016/j.ijbiomac.2018.01.063

Ahmad, M., Gani, A., Hassan, I., Huang, Q., & Shabbir, H. (2020). Production and characterization of starch nanoparticles by mild alkali hydrolysis and ultra-sonication process. *Scientific Reports*, 10(1), 1–11. https://doi.org/10.1038/s41598-020-60380-0

Alves, M. J. S., Chacon, W. D. C., Gagliardi, T. R., Agudelo Henao, A. C., Monteiro, A. R., & Ayala Valencia, G. (2021). Food applications of starch nanomaterials: A review. *Starch - Stärke*, *73*(11–12), 2100046. https://doi.org/10.1002/star.202100046

Baruah, K. N., Nagaoka, S., Banno, A., Singha, S., & Uppaluri, R. V. S. (2024). Nano-encapsulation of epigallocatechin gallate using starch nanoparticles: Characterization and insights on in vitro micellar cholesterol solubility. *Journal of Food Science*, *89*(9), 5701–5711. https://doi.org/10.1111/1750-3841.17260

Baruah, K. N., Singha, S., & Uppaluri, R. V. S. (2023). Preparation of potato starch nanoparticles using acid hydrolysis and ultrasonic post-treatment. *ACS Food Science and Technology*, *3*(4), 626–634. https://doi.org/10.1021/acsfoodscitech.2c00369

Bras, J., & Dufresne, A. (2010). Starch nanoparticles : A review. *Biomacromolecules*, *11*(5), 1139–1153.

Cai, Z. Y., Li, X. M., Liang, J. P., Xiang, L. P., Wang, K. R., Shi, Y. L., Yang, R., Shi, M., Ye, J. H., Lu, J. L., Zheng, X. Q., & Liang, Y. R. (2018). Bioavailability of tea catechins and its improvement. *Molecules*, *23*(9), 10–13. https://doi.org/10.3390/molecules23092346

Campelo, P. H., Sant'Ana, A. S., & Pedrosa Silva Clerici, M. T. (2020). Starch nanoparticles: Production methods, structure, and properties for food applications. *Current Opinion in Food Science*, *33*, 136–140. https://doi.org/10.1016/j.cofs.2020.04.007

Cazotti, J. C., Fritz, A. T., Smeets, N. M. B., Dubé, M. A., Garcia-valdez, O., & Cunningham, M. F. (2020). Graft modification of starch nanoparticles with pH-responsive polymers via nitroxide-mediated polymerization. *Journal of Polymer Science*, *58*, 2211–2220. https://doi.org/10.1002/pol.20200337

Chang, R., Ji, N., Li, M., Qiu, L., Sun, C., Bian, X., & Qiu, H. (2019). Sonochemistry green preparation and characterization of starch nanoparticles using a vacuum cold plasma process combined with ultrasonication treatment. *Ultrasonics Sonochemistry*, *58*, 104660. https://doi.org/10.1016/j.ultsonch.2019.104660

Chavan, P., Sinhmar, A., Nehra, M., Thory, R., Kumar, A., Allwyn, A., & Nain, V. (2021). Impact on various properties of native starch after synthesis of starch nanoparticles: A review. *Food Chemistry*, *364*, 130416. https://doi.org/10.1016/j.foodchem.2021.130416

Chin, S. F., Mohd Yazid, S. N. A., & Pang, S. C. (2014). Preparation and characterization of starch nanoparticles for controlled release of curcumin. *International Journal of Polymer Science*, *2014*. https://doi.org/10.1155/2014/340121

Chorfa, N., Nlandu, H., Belkacemi, K., & Hamoudi, S. (2022). Physical and enzymatic hydrolysis modifications of potato starch granules. *Polymers*, *14*(10), 2027. https://doi.org/10.3390/polym14102027

Contreras-Jiménez, B., Torres-Vargas, O. L., & Rodríguez-García, M. E. (2019). Physicochemical characterization of quinoa (*Chenopodium quinoa*) flour and isolated starch. *Food Chemistry*, *298*, 124982. https://doi.org/10.1016/j.foodchem.2019.124982

Cui, H., et al. (2020). Impact of nanoparticle shape, size, and properties of the sustainable nanocomposites. In *Sustainable polymer composites and nanocomposites* (pp. 313–336). Springer.

Dai, L., Zhang, J., & Cheng, F. (2020). Cross-linked starch-based edible coating reinforced by starch nanocrystals and its preservation effect on graded Huangguan pears. *Food Chemistry*, *311*, 125891. https://doi.org/10.1016/j.foodchem.2019.125891

Das, A., & Sit, N. (2021). Modification of taro starch and starch nanoparticles by various physical methods and their characterization. *Starch - Stärke*, *73*(5–6), 2000227. https://doi.org/10.1002/star.202000227

Ding, Y., & Kan, J. (2017). Optimization and characterization of high pressure homogenization produced chemically modified starch nanoparticles. *Journal of Food Science and Technology*, *54*(13), 4501–4509. https://doi.org/10.1007/s13197-017-2934-8

Faghihzadeh, F., Anaya, N. M., & Schifman, L. A. (2016). Fourier transform infrared spectroscopy to assess molecular-level changes in microorganisms exposed to nanoparticles. *Nanotechnology for Environmental Engineering*, *1*(1), 1–16. https://doi.org/10.1007/s41204-016-0001-8

Falkowska, M., & Molga, E. J. (2014). Nanosilver: A catalyst in enzymatic hydrolysis of starch. *Polish Journal of Chemical Technology*, (3), 111–113.

Fanta, G. F., Kenar, J. A., & Felker, F. C. (2015). Nanoparticle formation from amylose-fatty acid inclusion complexes prepared by steam jet cooking. *Industrial Crops and Products*, *74*, 36–44. https://doi.org/10.1016/j.indcrop.2015.04.046

Fathi, M., Martín, Á., & McClements, D. J. (2014). Nanoencapsulation of food ingredients using carbohydrate based delivery systems. *Trends in Food Science and Technology*, *39*(1), 18–39. https://doi.org/10.1016/j.tifs.2014.06.007

Gallant, D. J., Bouchet, B., & Baldwin, P. M. (1997). Microscopy of starch: Evidence of a new level of granule organization. *Carbohydrate Polymers*, *32*(3–4), 177–191.

García, N. L., Ribba, L., Dufresne, A., Aranguren, M., & Goyanes, S. (2011). Effect of glycerol on the morphology of nanocomposites made from thermoplastic starch and starch nanocrystals. *Carbohydrate Polymers*, *84*, 203–210. https://doi.org/10.1016/j.carbpol.2010.11.024

García-Gurrola, A., Rincón, S., Escobar-Puentes, A. A., Zepeda, A., Pérez-Robles, J. F., & Martínez-Bustos, F. (2019). Synthesis and succinylation of starch nanoparticles by means of a single step using sonochemical energy. *Ultrasonics Sonochemistry*, *56*, 458–465. https://doi.org/10.1016/j.ultsonch.2019.04.035

Gatto, F., Troncoso, O. P., Brunetti, V., Malvindi, M. A., Pompa, P. P., Torres, F. G., & Bardi, G. (2016). Human monocyte response to Andean-native starch nanoparticles. *Starch - Stärke*, *68*(9–10), 1016–1023. https://doi.org/10.1002/star.201600105

Garrido, L. H., Schnitzler, E., Boff Zortéa, M. E., Rocha, T. S., & Demiate, I. M. (2014). Physicochemical properties of cassava starch oxidized by sodium hypochlorite. *Journal of Food Science and Technology*, *51*(12), 2640–2647. (https://link.springer.com/article/10.1007/s13197-012-0794-9)

Gorantla, S., Wadhwa, G., Jain, S., Sankar, S., Nuwal, K., Mahmood, A., Dubey, S. K., Taliyan, R., Kesharwani, P., & Singhvi, G. (2021). Recent advances in nanocarriers for nutrient delivery. *Drug Delivery and Translational Research*, *12*, 2359–2384.

Gupta, R. K., Guha, P., & Srivastav, P. P. (2022). Natural polymers in bio-degradable/edible film: A review on environmental concerns, cold plasma technology and nanotechnology application on food packaging-A recent trends. *Food Chemistry Advances*, *1*, 100135. https://doi.org/10.1016/j.focha.2022.100135

Gupta, R. K., Guha, P., & Srivastav, P. P. (2023). Effect of high voltage dielectric barrier discharge (DBD) atmospheric cold plasma treatment on physicochemical and functional properties of taro (*Colocasia esculenta*) starch. *International Journal of Biological Macromolecules*, *253*, 126772. https://doi.org/10.1016/j.ijbiomac.2023.126772

Gupta, R. K., Guha, P., & Srivastav, P. P. (2024a). Insights into the physicochemical, crystalline, multistructural, thermal and digestibility behavior of taro starch (*Colocasia esculenta*) modified by cold plasma treatment with high voltage dielectric barrier discharge (HVDBD): An alternative technique for chemical modification of starch and environmental sustainability. *Journal of Cleaner Production*, *461*, 142710. https://doi.org/10.1016/j.jclepro.2024.142710

Gupta, R. K., Guha, P., & Srivastav, P. P. (2024b). Dielectric barrier discharge plasma: A green and novel method to change the structure of taro peel starch and improve the physicochemical properties of taro peel starch films. *Plasma Processes and Polymers*, *21*(9), 2400047. https://doi.org/10.1002/ppap.202400047

Gupta, R. K., Guha, P., & Srivastav, P. P. (2024c). Exploring the potential of taro (*Colocasia esculenta*) starch: Recent developments in modification, health benefits, and food industry applications. *Food Bioengineering, 3*(3), 365–379. https://doi.org/10.1002/fbe2.12103

Gupta, R. K., Guha, P., & Srivastav, P. P. (2024d). Physical action of nonthermal cold plasma technology for starch modification. *Food Physics*, *1*, 100011. https://doi.org/10.1016/j.foodp.2024.100011

Gupta, R. K., Guha, P., & Srivastav, P. P. (2024e). Effect of cold plasma treatment and plasma-activated water on physicochemical and structural properties of starch: A green and novel approach for environmental sustainability. *Plasma Processes and Polymers*, *21*(4), 2300204. https://doi.org/10.1002/ppap.202300204

Gupta, R. K., Pipliya, S., Patel, J., Srivastav, P. P., Castro-Muñoz, R., Wang, C. K., & Nemţanu, M. R. (2025a). Essential oil-embedded starch based films for active packaging: Insights into migration, stability, and preservation efficiency. *Trends in Food Science & Technology*, 163, 105149. https://doi.org/10.1016/j.tifs.2025.105149

Gupta, R. K., Gnana Moorthy Eswaran, U., Kumar, S., Pipliya, S., Kaur, B., & Srivastav, P. P. (2025b). Synergistic modification of starch polysaccharide via sequential treatments with ultrasonic and cold plasma technology. *Journal of Food Processing and Preservation*, *2025*(1), 9063007. https://doi.org/10.1155/jfpp/9063007

Habib, M., Jan, K., Qureshi, I., Rani, S., & Bashir, K. (2023). Gamma irradiation of starch. In *Starch: Advances in modifications, technologies and applications* (pp. 385–407). Cham: Springer International Publishing.

Han, C., Zhao, R., & Liu, X. (2014). DDB2 suppresses tumorigenicity by limiting the cancer stem cell population in ovarian cancer. *Molecular Cancer Research*, *12*(5), 784–794. https://doi.org/10.1158/1541-7786.MCR-13-0638

Hasanvand, E., Fathi, M., & Bassiri, A. (2018). Production and characterization of vitamin D3 loaded starch nanoparticles: Effect of amylose to amylopectin ratio and sonication parameters. *Journal of Food Science and Technology*, *55*(4), 1314–1324.

Hoover, R., & Vasanthan, T. (1994). The effect of heat-moisture treatment on the structure and physicochemical properties of cereal, legume, and tuber starches. *Food Research International*, *27*(5), 445–456.

Jambrak, A. R., Herceg, Z., Šubarić, D., Babić, J., Brnčić, M., Brnčić, S. R., Bosiljkov, T., Čvek, D., Tripalo, B., & Gelo, J. (2010). Ultrasound effect on physical properties of corn starch. *Carbohydrate Polymers*, *79*(1), 91–100. https://doi.org/10.1016/j.carbpol.2009.07.051

Jayakody, L., & Hoover, R. (2002). The effect of lintnerization on cereal starch granules. *Food Research International*, *35*, 665–680.

Jiang, S., Liu, C., Wang, X., Xiong, L., & Sun, Q. (2016). Physicochemical properties of starch nanocomposite films enhanced by self-assembled potato starch nanoparticles. *LWT*, *69*, 251–257. https://doi.org/10.1016/j.lwt.2016.01.053

Kim, H.-Y., Park, S. S., & Lim, S.-T. (2015). Biointerfaces Preparation, characterization and utilization of starch nanoparticles. *Colloids and Surfaces B: Biointerfaces*, *126*, 607–620. https://doi.org/10.1016/j.colsurfb.2014.11.011

Kim, H.-Y., Lee, J. H., Kim, J. Y., Lim, W.-J., & Lim, S.-T. (2012). Characterization of nanoparticles prepared by acid hydrolysis of various starches. *Starch - Stärke*, *64*(5), 367–373. https://doi.org/10.1002/star.201100105

Kumari, S., Yadav, B. S., & Yadav, R. B. (2019). Synthesis and modification approaches for starch nanoparticles for their emerging food industrial applications: A review. *Food Research International*, *128*, 108765. https://doi.org/10.1016/j.foodres.2019.108765

Kumari, P., Luthra, A., Sethi, P., & Banyal, S. (2024). X-ray diffraction analysis for starch. In *Standardized procedures and protocols for starch* (pp. 197–220). New York, NY: Springer US.

Le, D., & Angellier-Coussy, H. (2014). Reactive & functional polymers preparation and application of starch nanoparticles for nanocomposites: A review. *Reactive and Functional*

Polymers, *85*, 97–120. https://doi.org/10.1016/j.reactfunctpolym.2014.09.020

Lee, D. H., Kwon, K. S., Jeong, D., Kim, I. H., Nam, H. S., & Kim, J.-Y. (2022). Structural and physicochemical properties of composites between starch nanoparticles and β-carotene prepared via nanoprecipitation. *International Journal of Biological Macromolecules*, *214*, 100–110. https://doi.org/10.1016/j.ijbiomac.2022.06.062

Lee, S. O. B., Choi, D. S. H., Lee, B., & Kim, B. (2020). Starch nanoparticles prepared by enzymatic hydrolysis and self-assembly of short-chain glucans. *Food Science and Biotechnology*, *29*(5), 585–598. https://doi.org/10.1007/s10068-020-00768-w

Lee, Y. S., Tarté, R., & Acevedo, N. C. (2021). Curcumin encapsulation in Pickering emulsions co-stabilized by starch nanoparticles and chitin nanofibers. *RSC Advances*, *11*(27), 16275–16284. https://doi.org/10.1039/d1ra01622a

Lewicka, K., Siemion, P. B., & Kurcok, P. (2015). Chemical modifications of starch : Microwave effect.

Li, H., Song, J., Liu, C., Wang, X., Liu, Y., Han, M., Liang, J., & Gao, Z. (2024). Corn starch/β-cyclodextrin composite nanoparticles for encapsulation of tea polyphenol and development of oral targeted delivery systems with pH-responsive properties. *Food Hydrocolloids*, *151*. https://doi.org/10.1016/j.foodhyd.2024.109823

Li, J., Deng, Y., Fu, H., Zhang, Y., Zhang, Y., Fu, L., Xu, C., & Lin, B. (2023). Multifunctional starch-based sensor with non-covalent network to achieve "3R" circulation. *Small*, *19*(23), 1–14. https://doi.org/10.1002/smll.202208116

Li, S., Zhou, W., Hu, Y., Huang, C., Gao, Q., & Chen, Y. (2022). Preparation of graphene-starch composite film and its application in sensor materials. *International Journal of Biological Macromolecules*, *207*(9), 365–373. https://doi.org/10.1016/j.ijbiomac.2022.03.028

Lin, N., Huang, J., Chang, P. R., Feng, L., & Yu, J. (2011). Effect of polysaccharide nanocrystals on structure, properties, and drug release kinetics of alginate-based microspheres. *Colloids and Surfaces B: Biointerfaces*, *85*(2), 270–279. https://doi.org/10.1016/j.colsurfb.2011.02.039

Liu, C., Qin, Y., Li, X., Sun, Q., Xiong, L., & Liu, Z. (2016). Preparation and characterization of starch nanoparticles via self-assembly at moderate temperature. *International Journal of Biological Macromolecules*, *84*, 354–360. https://doi.org/10.1016/j.ijbiomac.2015.12.040

Loganathan, S. (2017). Thermogravimetric analysis for characterization of nanomaterials. In S. Thomas, R. THomas, A. K. Zachariah, & R. K. Mishra (Eds.), *Thermal and rheological measurement techniques for nanomaterials characterization* (pp. 67–108). Elsevier. https://doi.org/10.1016/B978-0-323-46139-9.00004-9

Marta, H., Rizki, D. I., Mardawati, E., Djali, M., Mohammad, M., & Cahyana, Y. (2023). Starch nanoparticles: Preparation, properties and applications. *Polymers*, *15*(5), 1167.

Mat Nor, N. A., et al. (2013). Starch nanoparticles: Their preparation and applications. In *Plant biotechnology: Recent advancements and developments* (pp. 213–232). Springer. https://doi.org/10.1007/978-981-10-4732-9_11

Menczel, J. D., & Prime, R. B. (2009). Polymers thermal analysis of fundamentals and applications.

Miskeen, S., An, Y. S., & Kim, J. Y. (2021). Application of starch nanoparticles as host materials for encapsulation of curcumin: Effect of citric acid modification. *International Journal of Biological Macromolecules*, *183*, 1–11. https://doi.org/10.1016/j.ijbiomac.2021.04.133

Morán, D., Gutiérrez, G., Blanco-López, M. C., Marefati, A., Rayner, M., & Matos, M. (2021). Synthesis of starch nanoparticles and their applications for bioactive compound encapsulation. *Applied Sciences (Switzerland)*, *11*(10). https://doi.org/10.3390/app11104547

Mourdikoudis, S., Pallares, R. M., & Thanh, N. T. K. (2018). Characterization techniques for nanoparticles: Comparison and complementarity upon studying nanoparticle properties. *Nanoscale*, *10*(27). https://doi.org/10.1039/C8NR02278J

Namazi, H., & Dadkhah, A. (2010). Convenient method for preparation of hydrophobically modified starch nanocrystals with using fatty acids. *Carbohydrate Polymers*, *79*(3), 731–737. https://doi.org/10.1016/j.carbpol.2009.09.033

Neilson, A. P., Hopf, A. S., Cooper, B. R., Pereira, M. A., Bomser, J. A., & Ferruzzi, M. G. (2007). Catechin degradation with concurrent formation of homo- and heterocatechin dimers during in vitro digestion. *Journal of Agricultural and Food Chemistry*, *55*(22), 8941–8949. https://doi.org/10.1021/jf071645m

Oliveira, N. R., De, Fornaciari, B., & Carvalho, G. M. (2017). Acetylated starch-based nanoparticles: Synthesis, characterization, and studies of interaction with antioxidants. *Starch - Stärke*, 70(3–4), 1700170. https://doi.org/10.1002/star.201700170

Phatthanawiwat, K., Boonkanon, C., Wongniramaikul, W., & Choodum, A. (2022). Catechin and curcumin nanoparticle-immobilized starch cryogels as green colorimetric sensors for on-site detection of iron. *Sustainable Chemistry and Pharmacy*, *29*, 100782. https://doi.org/10.1016/j.scp.2022.100782

Phukaoluan, A., Khantachawana, A., Dechkunakorn, S., Anuwongnukroh, N., Tunthawiroon, P., & Srirussamee, K. (2021). Influence of mouthwash rinsing on the mechanical properties of polymeric ligature ties used for dental applications. *Polymers*, *13*(14), 2236.

Qin, Y., Xue, L., Hu, Y., Qiu, C., Jin, Z., & Xu, X. (2020). Green fabrication and characterization of debranched starch nanoparticles via ultrasonication combined with recrystallization. *Ultrasonics Sonochemistry*, *66*, 105074. https://doi.org/10.1016/j.ultsonch.2020.105074

Rajisha, K. R., Maria, H. J., Pothan, L. A., Ahmad, Z., & Thomas, S. (2014). Preparation and characterization of potato starch nanocrystal reinforced natural rubber nanocomposites. *International Journal of Biological Macromolecules*, *67*, 147–153. https://doi.org/10.1016/j.ijbiomac.2014.03.013

Rao, C. N. R., & Biswas, K. (2009). Characterization of nanomaterials by physical methods. *Annual Review of Analytical Chemistry*, *2*, 435–462. https://doi.org/10.1146/annurev-anchem-060908-155236

Remanan, M. K., & Zhu, F. (2023). Encapsulation of rutin in Pickering emulsions stabilized using octenyl succinic anhydride (OSA) modified quinoa, maize, and potato starch nanoparticles. *Food Chemistry*, *405*, 134790. https://doi.org/10.1016/j.foodchem.2022.134790

Saartrat, S., Puttanlek, C., Rungsardthong, V., & Uttapap, D. (2005). Paste and gel properties of low-substituted acetylated canna starches. *Carbohydrate Polymers*, *61*, 211–221. https://doi.org/10.1016/j.carbpol.2005.05.024

Sapyen, W., Toonchue, S., Praphairaksit, N., & Imyim, A. (2022). Selective colorimetric detection of Cr(VI) using starch-stabilized silver nanoparticles and application for chromium speciation. *Spectrochimica Acta - Part A: Molecular and Biomolecular Spectroscopy*, *274*, 121094. https://doi.org/10.1016/j.saa.2022.121094

Shapi'i, R. A., Othman, S. H., Basha, R. K., & Naim, M. N. (2022). Mechanical, thermal, and barrier properties of starch films incorporated with chitosan nanoparticles. *Nanotechnology Reviews*, *11*, 1464–1477. https://doi.org/10.1515/ntrev-2022-0094

Shrivastava, A., Gupta, R. K., & Srivastav, P. P. (2024). Exploring novel frontiers of advancements in purple yam (*Dioscorea alata* L.) starch extraction, modification, characterization, applications in food and other industries. *Measurement: Food*, *15*, 100196. https://doi.org/10.1016/j.meafoo.2024.100196

Singh, G. P., Aayush, K., Chavan, P., Chiu, I., Yan, S., Verma, R., Kuca, K., Kumar, D., & Yang, T. (2024). Valorization of agro-industrial wastes: Converting mango seed into starch nanoparticles-based edible coating for fresh produce preservation. *Innovative Food Science and Emerging Technologies*, *95*. https://doi.org/10.1016/j.ifset.2024.103722

Singh, V., & Ali, S. Z. (2000). Acid degradation of starch: The effect of acid and starch type. *Carbohydrate Polymers*(2), *41*, 191–195.

Sujka, M. (2017). Ultrasonic modification of starch: Impact on granules porosity. *Ultrasonics Sonochemistry*, *37*, 424–429. https://doi.org/10.1016/j.ultsonch.2017.02.001

Sujka, M., & Jamroz, J. (2013). Food hydrocolloids ultrasound-treated starch: SEM and TEM imaging, and functional behaviour. *Food Hydrocolloids*, *31*(2), 413–419. https://doi.org/10.1016/j.foodhyd.2012.11.027

Sun, Q. (2018). Starch nanoparticles. In M. Sjöö & L. Nilsson (Eds.), *Starch in food: Structure, function and applications* (2nd ed., pp. 691–745). Elsevier. https://doi.org/10.1016/B978-0-08-100868-3.00018-4

Tabari, M. (2018). Characterization of a new biodegradable edible film based on sago starch loaded with carboxymethyl cellulose nanoparticles. *Nanomedicine Research Journal*, *3*(1), 25–30. https://doi.org/10.22034/NMRJ.2018.01.004

Tang, H., Mitsunaga, T., & Kawamura, Y. (2006). Molecular arrangement in blocklets and starch granule architecture. *Carbohydrate Polymers*, *63*, 555–560. https://doi.org/10.1016/j.carbpol.2005.10.016

Tomasik, P., & Zaranyika, M. F. (1995). Nonconventional methods of modification of starch. *Advances in Carbohydrate Chemistry and Biochemistry*, *51*, 243–318. https://doi.org/10.1016/S0065-2318(08)60195-X

Torres, F. G., & De-la-Torre, G. E. (2022). Synthesis, characteristics, and applications of modified starch nanoparticles: A review. *International Journal of Biological Macromolecules*, *194*, 289–305. https://doi.org/10.1016/j.ijbiomac.2021.11.187

Xie, Y., Wang, Y., & Torres-León, C. (2024). Starch-based ingredients and functional foods. In *Starch Nanomaterials and Food Applications* (pp. 165–182). Springer.

Wang, H., Feng, T., Zhuang, H., Xu, Z., Ye, R., & Sun, M. (2018). A review on patents of starch nanoparticles: Preparation, applications, and development. *Recent Patents on Food, Nutrition & Agriculture*, *9*(1), 23–30. https://doi.org/10.2174/2212798410666180321101446

Wang, Y., et al. (2022). Starch-based ingredients and functional foods. In *Starch nanomaterials and food applications* (pp. 165–182). Springer. https://doi.org/10.1007/978-3-031-60086-9_8

Wang, Y., Zhang, Y., Guan, L., Wang, S., Zhang, J., Tan, L., Kong, L., & Zhang, H. (2021). Lipophilization and amylose inclusion complexation enhance the stability and release of catechin. *Carbohydrate Polymers*, *269*, 118251. https://doi.org/10.1016/j.carbpol.2021.118251

Wang, Y., Xie, Y., & Torres-León, C. (2024). Starch-based ingredients and functional foods. In *Starch nanomaterials and food applications* (pp. 165–182). Springer. https://doi.org/10.1007/978-3-031-60086-9_8

Zhang, L., Liu, P., Wang, Y., & Gao, W. (2011). Study on physicochemical properties of dialdehyde yam starch with different aldehyde group contents. *Thermochimica Acta*, *512*(1–2), 196–201. https://doi.org/10.1016/j.tca.2010.10.006

Zhou, L., Fang, D., Wang, M., Li, M., Li, Y., Ji, N., Dai, L., Lu, H., Xiong, L., & Sun, Q. (2020). Preparation and characterization of waxy maize starch nanocrystals with a high yield via dry-heated oxalic acid hydrolysis. *Food Chemistry*, *318*, 126479. https://doi.org/10.1016/j.foodchem.2020.126479

Zhu, J., & Huang, Q. (2019). Nanoencapsulation of functional food ingredients. In L.-T. Lim & M. Rogers (Eds.), *Advances in food and nutrition research* (Vol. 88, pp. 129–165). Elsevier. https://doi.org/10.1016/bs.afnr.2019.03.005

13 Practical Application of Modified Starch in Food and Packaging Systems

Vigneshwaran V
Department of Bio-Technology, Anna University, Chennai, India

Devanampriyan Rajan
Department of Food Technology, Kongu Engineering College, Erode, Tamil Nadu, India

Chitra Devi Venkatachalam
Department of Food Technology, Kongu Engineering College, Erode, Tamil Nadu, India

Sankari Rajan
Department of Food Technology, Kongu Engineering College, Erode, Tamil Nadu, India

Kalaivani Sundaravadivelu
Department of Food Technology, Kongu Engineering College, Erode, Tamil Nadu, India

13.1 INTRODUCTION

Starch is the second most plentiful renewable bioenergy resource, after cellulose. Starch, a crucial natural polymer, has a complicated structure, a wide range of uses, and a substantial global output. Globally, more than 56 million tons of starch are generated, predominantly from potatoes, rice, and maize, and this adds to the roughly \$1.12 trillion world trade of starch and starch derived products (Adigwe et al., 2022). Plants store starch in their grains or root tubers. For food starch, major sources are sweet potatoes, wheat, cassava, corn, and potatoes (Gupta et al., 2022). Around the world, sorghum, barley, rice, millet, yam, and similar crops are used as minor sources (Rashwan et al., 2024). Amylose and amylopectin combine to make up starch. The amount of amylose in starch ranges between 20% and 70% (Rashwan et al., 2024). Modified starch physiochemical characteristics are largely influenced by the semicrystalline granules within its hierarchical structure. (Damager et al., 2010; Wang et al., 2014; Gupta et al., 2023). In recent years, the use of starch in manufacturing has expanded, mostly in reformulated goods to improve technical qualities or to reduce and/or eliminate certain components. Starch can be used in variety of products such as as a bread preservative; a thickener and stabilizer for fruit cakes, soups, deserts, and sauces; a binder in the paper industry, because it increases mechanical resistance; and a textile sizing agent (Guillén, 2021). Starch is widely utilized as a primary commodity; it acts as a backbone of the food manufacturing process, providing ingredients ahd animal feed; and it is used in a vast range of non-food goods (e.g., bioplastics, medicines, bioethanol). The market for modified starch in plant-based meat alternatives has a growth rate of close to 30% (Damager et al., 2010; Rashwan et al., 2024).

Despite the fact that starches have special qualities (such as gel strength, pasting ability, varying temperatures for gelatinization, and varying water absorption rates), their production cannot keep up with the growing demand for budget-friendly, functional properties, which calls for more study into the biosynthesis and modification of starch. To overcome starch drawbacks, starch modification was developed to enhance starch's functional properties and thereby widened industrial applications (Rashwan et al., 2024; Wang et al., 2014). The alternate to native starches is modified starches because of their flexibility in various industrial processes, such as by providing increased resistance to retrogradation, maintaining sensory properties, better retaining process conditions such as freezing and thawing, and stability, thereby maintaining the shelf life of products much better than native starch (Guillén, 2021). "Starch modification" refers to a process of change in starch by several operational, environmental, and processing factors. These modifications may cause improvement or decrement in starch properties. Various reports have noted that physical, mechanical, and chemical modification may increase starch functional properties, while some processing and environmental modifications decrease starch functional properties during further processing or during storage (Nawaz et al., 2020; Gupta et al., 2025a). Nevertheless, the presence of conventional chemical residues or traces in the

DOI: 10.1201/9781003564287-13

chemically modified starches restricts the products' use in food and medicine. Therefore, environmentally friendly techniques are favoured over chemical modifications. The changed starches, both enzymatically and physically modified, are referred to as "clean-label ingredients." Among the texturizers used in food, nearly 70% are a modified starch (Rashwan et al., 2024). Modified starches have wide application in the food sector, including as fat replacers, thickeners, stabilizers, biofilm, and edible films, and are also used for improved starch functions in non-food industries (Shrivastava et al., 2024). Overall, the extensive usage of modified starches in food processing and packaging represents a significant advancement in materials science, overcoming the limitations of native starches in order to satisfy the evolving demands of global markets. This chapter will provide an overview of the utilization of modified starch in food manufacturing and packaging.

13.2 GLOBAL COMPANIES AND MARKET ANALYSIS FOR SPECIALTY MODIFIED STARCHES

The global market for specialty modified starches is driven by increasing demand in various industries such as foods around the globe (Figure 13.1). Import volume in 2022 was 19,500 tons, and in 2023, close to 23,000 tons. The significant increase of 3500 tons was due to high consumer demand for processed foods. Modified starch was a key ingredient because of its multiple versatile functionalities in ready-to-eat food products. Global companies have responded to this demand by further investing in research to create multiple modifications which can meet specific needs of different applications (Vilpoux & Silveira Junior, 2023). The major consumption of modified starch in food processing industries as per domestic imports in 2022 were from sources such as waxy corn, tapioca, potato, corn-based resistant starch, dextrin (food grade), pregelatinized starches, oxidized starch (E 1404), and acetylated starch (E 1420), followed by acetylated distarch adipate (E 1422) and hydroxypropyl starch (E 1440), hydroxypropyl distarch phosphate (E 1442) and starch sodium octenyl succinate (E 1450). The global market size of modified starches was closely estimated at nearly USD 11.8 billion in 2021, with a compound annual growth of 5.3% from 2022 to 2027. Global suppliers of modified starches were Ingredion, Cargill, Roquette, Tate & Lyle, Avebe, Grain Processing Corporation (GPC), and SMS Corporation (Adewale et al., 2022).

The leading importers of modified starches were the United States (2669 MT), Thailand (1908 MT), Netherlands (188 MT), and Brazil (172 MT) others contribute close to 70 MT. Higher import volumes result from several factors, including convenience foods. These countries possess strong industrial demand, particularly in food and beverages, pharmaceuticals, and non-food applications such as textiles, clothing, and paper products, where modified starches play a critical role. Their advanced research and development efforts drive the need for high-quality ingredients such as modified starches. Additionally, stringent quality standards, strategic trade agreements, and growing consumer demand for processed foods further boost import volumes (Adigwe et al., 2022). The diversified industrial base and global supply chains in these countries also contribute to their significant reliance on modified starch imports from India, at close to about 5000 MT in business.

13.3 EFFECTS OF MODIFICATION ON STARCH PROPERTIES

Starches are sensitive to several factors, including high temperature, mechanical stress, pressure, ultrasound treatment, light, and very low or high pH. Heating starch with water causes gelatinization of starch, and microwave treatment

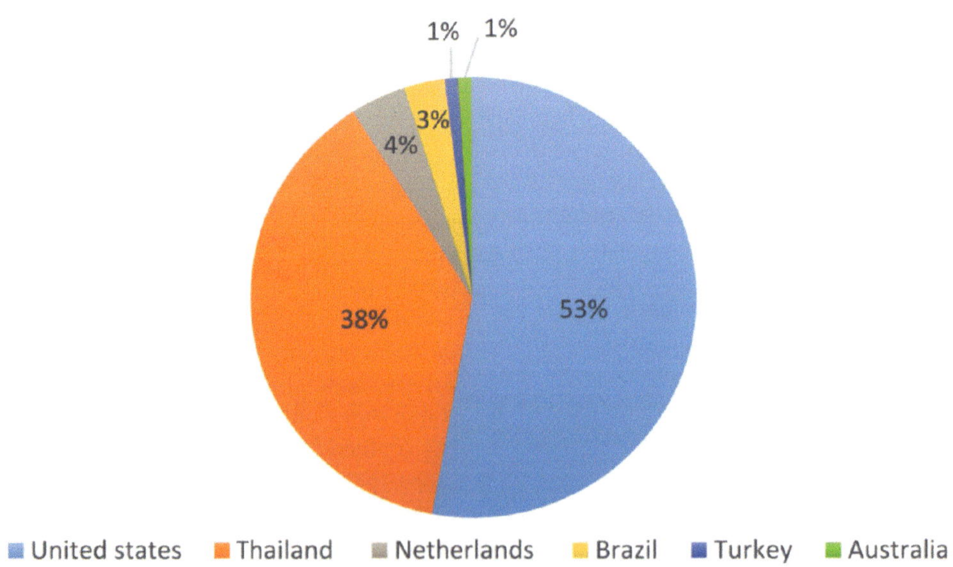

FIGURE 13.1 Import data for modified starch by country, 2022.

changes the its crystalline structure. Gelatinization of starch results due to change in starch structure from the amorphous to crystalline form (Nawaz et al., 2020; Gupta et al., 2024a). Factors that affect gelatinization include temperature and starch properties (origin, variety, concentration of sugars, salt, lipids, protein, moisture content) (Chakraborty et al., 2022). The final viscosity (equilibrium viscosity after cooling) depends on the starch type, shear rate, and amylose and amylopectin rearrangement (Singh et al., 2020). Freezing after gelatinization causes recrystallization and enhances toughness of corn starch. Interaction with non-aqueous phases such as water and oil causes starch granules to swell, which is initially reversible to some extent but eventually irreversible. Chemicals such as hydroxy derivatives or carboxy derivatives of hydrocarbon compounds, oxidizing agents, phosphate salts, some carboxylic acids, various acids and base, cross-linking agents, synthetic polymers, and some cationic and anionic molecules modify properties and structure of starches and enzymatic degradation also alter starch characteristics, affecting its use in a variety of applications. Enzymatic and acid hydrolysis results in degradation of intact starch components, which leads to change in the structural properties by altering functional properties of starch (Nawaz et al., 2020).

The following re effects or reactivity changes in starch caused by modification. Physical and chemical alterations alter the molecular and granule structure, resulting in changes in reactivity and functioning behaviour. Physical alterations such as mechanical processes, acid-catalyzed hydrolytic degradation, and thermal radiolytic modification of starch increase the reactivity of starch by exposing reactive functional groups via breaking amylose and amylopectin chains (Gupta et al., 2024b). Microwave modification of starch changes functional properties and affects crystalline structure due to birefringence loss and deformation, respectively. Chemical modifications such as acetylation, oxidation, phosphorylation, cationization, carboxymethylation, and copolymerization add new functional groups to starch, causing a change in reactivity with non-aqueous phase such as water, oil, and enzymes (Nawaz et al., 2020; Gupta et al., 2024c). In particular, in various chemically modified food-grade potato and tapioca starches, oxidation led to decreased viscosity and to thixotropic pastes with a rigid texture. Acetylation generated very slight modifications, resulting in the stability of rheological characteristics through thermal processing. Cross-linking can increases molecular weight, which improves the thickening capacities of the starch preparation; however, the type of cross-linking agents employed for the modification has no influence on texture or rheological qualities (Lewandowicz et al., 2022). Some recent methods of starch modification of various starches and their effect on starch properties are presented in Table 13.1.

In terms of combined products made with modified starches, such as biodegradable plastic made from acetylated tapioca, the effect adding starch was an improvement in the water resistant property without affecting the tensile strength properties, biodegradability and elongation, or stretchability of the packaging material (Islamiyah et al., 2022). In pizza crust, pregelatinized starch showed the same rheological properties of the dough even at higher water content, and textural properties showed it can withstand strain with less stress (Matsumoto et al., 2022).

TABLE 13.1
Some Recent Modification Methods and Their Effects on Properties of Various Starches

Starch Type	Modification Method(s)	Effect on Starch Properties	Reference
Elephant yam (EFS)	Heat-moisture treatment (HMT) methods, hot air oven (HA), autoclave, and microwave	All HMTs decreased solubility and swelling but exhibited shear thinning effect. HA increased EFS amylose content, elasticity, structural rigidity, and rheological property over other HMTs and native EFS.	Barua et al. (2024)
Pigeon pea (PPS)	Chemical modifications (CMs) by cross-linking (CL), hydroxypropylation (HP), and oxidation	All CMs decreased swelling power (except HP), paste clarity, viscosities (peak, hot paste, cold paste, setback), and retrogradation but increased viscoelastic (elasticity) behaviour and did not change starch structure. CL PPS showed high thermal stability and overall improved functional property.	Guleria and Yadav (2022)
Corn	Heat-moisture treated, autoclave heated, microwave heated, and osmotic pressured (OP)	All of these modifications proceeded without full gelatinization and maintained integrity. OP minimized retrogradation but maximized thermal stability; increased water gel strength, water absorption capacity and swelling volume.	Marta et al. (2022)
Yam (lesser)	Heat-moisture treated (HMT), annealed (ANN)	HMT changed X-ray diffraction pattern, but no change in structure and morphology for ANN. HMT and ANN increased amylose but reduced overall starch content.	Ayuningtyas et al. (2022)
Wheat	*Ligustrum robustum* (Rxob.) Blume extract	Decreased gelatinization enthalpy, crystallinity, loss and storage moduli. Changed gelatinization temperature. Improved loss tangent, roughness and hole size.	Chen et al. (2022)
Rice (Japonica)	Octenyl succinic anhydride (OSA)	Reduced crystallinity and the pasting temperature. Increased swelling properties and resistant starch content.	Zhang et al. (2021)

13.4 FOOD ADDITIVE NUMBERING OF MODIFIED STARCHES WITH EXISTING MARKET BRANDS

Modified starches are commonly utilized for various purposes, including thickening, stabilizing, and emulsifying, for their versatile functionalities. Modified starches are created by physically, enzymatically, or chemically altering the structure of native starches (Gupta et al., 2024d). The purpose of these modifications is to enhance properties such as solubility, texture, stability, and ability to withstand processing conditions such as heat and shear. Legal compliances and regulatory policies have been adopted for modified starch labelling on food products. In India, the regulation of food additives uses the International Numbering System (INS). Chemically treated starches fall under the modified starch section of INS numbers. Table 13.2 lists a variety of modifications of starches and their applications. The table data were compiled from a market analysis by Mintel (2024), and the commercial brands were sources from Innova market insights (Mintel, 2024).

TABLE 13.2
E-numbers of Different Starch Modification with the Application

INS Number	Starch Modified	Process	Food Applications	Existing Brands in Market
E 1400	Dextrins, roasted starch white and yellow	Polymer of dextrose prepared by partial degradation of starch	Foam stabilizer in beer, artificial sweetener base from corn, tapioca	Bisk Farm Tiffun Slice
E 1401	Acid-treated starch	Partial degradation of starch or starch granules with inorganic acids	Sauces, starch-based gummies	Wrigley's Orbit Chupa Chups
E 1402	Alkaline-based starch	Treatment of starch with alkaline base sodium or potassium hydroxide	Meat-based products, sour breads	Paldo Hwa Ramyun
E 1403	Bleached starch	Partial degradation of starch and oxidation followed by treatment with bleaching agents such as sulphur dioxide	Baby foods, jelly-type sweets, wine gums, batter mixes (ready to cook)	Tata Simply Better plant-based nuggets
E 1404	Oxidized starch	Partial degradation of starch and oxidation followed by treatment with bleaching agents	Coffee and coffee substitutes, infants' food, cooked fish, fermented milks	Godrej Tyson Yummiez Veg Range
E 1405	Enzyme-treated starch	Treatment of starch granules with enzymes (maltodextrin, cyclodextrins)	Fat replacement applications, low-calorie sugar-free applications, drug formulation, pharmaceuticals	Perfetti Van Malle Center Fresh
E 1410	Monostarch phosphate	Treatment of starch granules with phosphoric acid by hydrolysis	Low-calorie margarines, cottage cheese, puddings, dessert sauces, soups, salad dressings, pie fillings	Godrej Tyson Yummiez
E 1411	Distarch glycerol	Treatment of starch granules with glycerol, which is partially degraded and combined with glycerol	soup, sauce and ketchup mixes, pizza crusts, pellets, meat sausages, low-fat foods	Behrooz frozen pizza dough
E 1412	Distarch phosphate esterified with sodium trimetaphosphate	Treatment of starch granules with trimetaphosphate followed by treatment with phosphoroxychloride	Dairy and dairy subproducts (cheese, ripened cheese), dried pasta and other noodles, coffee, tea, hot cereals, grain beverages	Frish potato products
E 1413	Phosphated distarch phosphate	Treatment of starch with a phosporylating agent and phosphoric acid (increased resistant to acid, increased stability)	Baby foods, frozen dough, pizza, frozen foods, refrigerated sweets, soy products, infant formulas	Mooz Formaggio tofu garlic sauce
E 1414	Acetylated distarch phosphate	Treatment of starch granules with a phosporylating agent and acetic acid (increased stability and better solubility at low temperatures)	Canned meat, other canned foods, soup mixes, cereal-based foods, soy-based products, noodles	Unibic Royal vanilla cake
E 1420	Starch acetate esterified with acetic anhydride	Treatment of starch granules with acetic acid anhydride (increased stability at high temperatures and low pH)	Mayonnaise, yogurts, fermented milk products, acidified milk products, cooked ketchup, sauces	Dr. Oetker Kuppies Choco Orange Supreme Brownie Bar
E 1421	Starch acetate esterified with vinyl acetate	Treatment of starch with vinyl acetate (increased stability at high temperatures and low pH)	Low-fat cheese, crispy snacks, muffins, dairy desserts, mayonnaise	Knorr Korean Meal Pot ramen noodles
E 1422	Acetylated distarch adipate	Treatment of starch granules with acetic acid anhydride and adipic acid anhydride (resistance to stirring and high temperatures)	Relishes, Pickles, fruit pie fillings, baby foods, seasonings, sauce and dips, ready-to-eat products	Kissan fresh tomato ketchup Milken yogurts

(Continued)

TABLE 13.2 *(Continued)*
E-numbers of Different Starch Modification with the Application

INS Number	Starch Modified	Process	Food Applications	Existing Brands in Market
E 1423	Acetylated distarch glycerol	Treatment of starch with acetic acid anhydride and glycerol (high temperature after cooling)	Tablet disintegrants, jelly-type sweets, wine gums, iced lollies, yogurts, egg white premix, fruit-flavoured fillings	WhyFryy popped snacks
E 1440	Hydroxypropyl starch	Treatment of starch with propylene oxide	Cereal-based drinks, extruded products, cassava roots	Sprinng Gluten Free Multiseed Sandwich Bread
E 1441	Hydroxypropyl distarch glycerine	Treatment of starch granules with propylene oxide epichlorhydrine and glycerol (better glossiness)	Batter mixes, tablet coating, confectionery premixes, water jellies	Frylo golgappa mixes
E 1442	Hydroxypropyl distarch phosphate	Treatment of starch with propylene oxide and phosphoric acid (increased stability and better glossiness)	Frozen products, refrigerated doughs, frozen pizza	Del Monte roasted garlic mayonnaise
E 1450	Starch sodium octenyl succinate	Treatment of starch granules with octenyl succinate	Oil and water emulsions (mayonnaise, salad dressings, drink whiteners, dry drink bases, flavour emulsions)	Veeba eggless veg mayonnaise
E 1451	Acetylated oxidized starch	Treatment of starch granules with acetic acid anhydride and oxygen	Disintegrants, coatings, compressed tablets, binders, stabilizers in tablet blends	Kari Kari Salt Pepper Roasted Rice Cracker Peanut Mix
E 1452	Starch aluminium octenyl succinate	Treatment with aluminium salt of the reaction product with succinic acid and octanol	Personal care items such as cosmetics, nanocapsules, Pickering emulsion	Less explored in food applications

Sources: Gałkowska et al (2023), Mintel (2024).

13.5 FOOD INDUSTRIAL APPLICATIONS OF MODIFIED STARCHES

Modified starches are widely used due to their enhanced properties and versatile functions compared to native starches. Examples of their use include as thickeners and stabilizers in soups, sauces, and gravies. In dairy applications, they are used as for stabilization of yogurts, puddings, and ice creams; for bakery goods, to retain freshness and improve volume. In meat industries, they are widely used as tenderizers and for marinating, which can enhance water holding capacity and moisture retention in meats and sausages. Among their versatile functions, modified starches can withstand high temperature conditions in baking, cooked sauces and mayonnaise, and high shear homogenizer application, whereas native starches can perform these functions. Modified starches offer customized or tailored functionalities which elevate the stability and shelf life of various food products, making these starches indispensable in modern food processing and formulation (Gupta et al., 2024e). The following sections discuss the major functions of modified starches in various foods and their applications in the food industry.

13.5.1 Baked Goods

Cake glazes which remain soft for an extended period of time appeal to bakers and consumers alike. Generally, modified starch improves the shelf life of the glaze and is suitable for cinnamon rolls, pastries, and similar applications. Major versatile functions of modified starches build solids without a significant increase in viscosity, allowing for easy application and quick drying, and also preventing moisture migration to baked goods and the atmosphere through the glaze coating. The film-forming properties of hydroxypropyl-modified starch (E-1440) provides shine or a barrier coating. The strength and integrity of film coatings made with E 1440 are superior to gum acacia. The addition of maltodextrins shortens the drying and polishing time while favourably changing the attributes of the coating. Modified starches can provide a glossy appearance, and refrigeration-stable hydroxypropyl distarch phosphate (E 1442) has been widely used for pregelatinized starches for this application (Dar & Embuscado, 2014). Sugar frostings are much consumed with baked goods. Cake decorators need a frosting that flows smoothly from the decorating tip without tearing or leaving a ragged appearance. Frosting with this modified starch fulfils that requirement and also provides a slightly hard exterior and a soft interior (Mason, 2009). This is a simple process with a no-cook formulation that will have wide consumer appeal. Because of increasing consumer concern about obesity and type II diabetes, there is greater demand for baked goods with reduced sugar content. This recent development in frosting can be applied to a reduced-sugar baked good to improve eating quality and flavour profile. The reduction of sugars using the maltodextrin as dried glucose syrup builds solids and provides a creamy mouthfeel in the frosting. The pregelatinized modified food starch (E 1442) improves moisture retention and whipping properties of the frosting.

Bakery products such as breads, cakes, and muffins and frozen products such as cakes, pizza, doughs, and other baked goods can utilize pregelatinized cold-water-swelling starches (Bashirian & Abdolmaleki, 2020). Hydroxypropyl distarch phosphate (E 1442) provides good freeze–thaw stability without leaching water (Abbas et al., 2010). Pregelatinized starches can be used to reduce the levels of hydrocolloids such as xanthan gum (E 415), guar gum (E 414), and carrageenan (E 407) in centre-filled baked goods. Pregelatinized modified food starch has been designed for mildly acidic to neutral pH bakery applications such as sourdough breads and fermented doughs (Miller et al., 2008). Gluten-free baked goods are in increasing demand by consumers with celiac disease and other medical conditions, creating a need for quality. The main challenge is finding a wheat flour substitute. A combination of dent corn starch, which has a highly absorbent starch that has excellent oil loading characteristic, with pregelatinized modified food starches E 1442 and E 1440 works well with sorghum flour to replace wheat flour (Roman et al., 2020). Whole-grain and high-fibre crackers are popular, but their textural qualities tend to suffer with the increased proportion of fibre in the formulations. Products such as whole-grain biscuits, cookies, and crackers are healthy snacks for adults and children. Instant-grade hydroxypropyl starch (E 1440) has been added to improve crispness and crunchiness in the cracker. Its film-forming properties help improve cracker strength. In production, these products can lead to the improvement in machinability of the dough in the finished product (Rapaille and Vanhemelrijck 1992). Incorporation of hydroxypropyl starch in the whole-grain cracker at 3% to 4% without no additional starch content and a positive control of unmodified dent corn starch resulted in improved crispness and crunchiness (Abdel-Aal & Rabalski, 2008; Walsh et al., 2022).

13.5.2 Confectionery Applications

Jelly candies using modified starch are popular and can be an intermediate for later enrobing or panning. Jellies enjoy a tolerance to warm storage that many confections lack, making them a logical choice for warm climatic conditions. Modified starch moulding allows an impressive amount of detail to be incorporated into the piece, providing a variety of characters and shapes. The possibilities of colour, flavour, and piece characteristics are virtually unlimited. Acid-treated starches (E 1401), or thin boiling starches, provide a chewy texture and excellent flavour release which are designed for optimal cook viscosity for easy depositing while providing optimal gel strength (Kibar et al., 2024). The major functionality of these starches are the medium conversion and good textures achieved at lower solids, and high viscosity at medium solids where a firmer set is required to create a chewy texture. In confectioneries such as high cooked jellies, they can provide lower hot viscosity with good gel strength which has the versatile properties of high conversion and low viscosity. Higher solids usage is possible with moderate firmness and a strong gel.

Hard-boiled candies require coatings and gum layers for the appearance of sheen. Benefits range from purely shine coats to scuff coats to barrier layers for the inhibition of the migration of moisture and emulsion flavour oils used in the formulation. Combining coating-grade starches with maltodextrin demonstrated the ability to form films with excellent oil and oxygen barrier properties to produce transparent, strong, flexible films (Kaya, 2019). Choosing maltodextrin to accompany the hydroxypropyl starch (E 1440) allows for versatile performance. A maltodextrin with higher dextrose equivalence (DE) added to modified starches will increase tackiness and adhesive properties when particulate is to be applied. A lower DE combined with the modified starch E 1440 forms stronger but more opaque films. In some cases, this solution can be used for centre-filled hard-boiled candies (Dar & Embuscado, 2014). Centre fruit fillings such as in candies, jellies, and fruit bars require a stabilizer system which can withstand the heat and shear of processing, pumping, depositing, and baking. A high-solids filling for cookies or bars needs to have a very high viscosity, short texture, and low water activity. The centre-filled baked good or confectionery filling must remain stable over long storage periods (up to 12 months) with no separation of water prior to use by the bakery or in the package after baking (Agudelo et al., 2014). Hydroxypropyl distarch phosphate (E 1442) helps to provide the heat, shear, and acid stability needed during processing and storage of the high-solids filling. Also, it helps to maintain viscosity and textural properties of the filling during depositing of fillings and baking, and throughout the shelf life of the baked good.

13.5.3 Snacks and Extruded Foods

Snack and cereal manufacturers are continually looking for ways to improve the adherence and distribution of seasonings and particulates on snack or cereal pieces using an even coating of adhesive solution which improves the adherence of seasonings and resists cracking and peeling better than current dextrin-based coatings (E 1400). Starch should be modified to have a translucent coating on snack or cereal, to resist peeling and cracking compared to dextrin-based coatings, and for low viscosity to provide easy application to the snack or cereal in a tumbler pan. In some cases, coatings of snack products such as honey-coated snacks and cereals are popular with consumers. Among the versatile functions of coatings of snack products are building solids without the stickiness speeding the drying of the coating, and helping prevent crystallization of sugars in the snack coating. Deep-fried chips (tortilla type) are a popular snack in which starches modified with hydroxypropyl (E 1440) are added to increase crispness. This combined with low-DE maltodextrin helps bind the dough during mixing and controls browning during the frying process (Arunachalam et al., 2025). Instant-grade modified starch E 1440 enables snack manufacturers to lower the fat added to snacks, carry an oil-based flavour, and increase the level of seasoning which can be added to the snack; it possesses versatile functionalities such as hydrating easily without heating and compatibility with an oil-based system (Sajilata & Singhal, 2005).

13.5.4 Savouries and Seasonings Applications

Sauces and gravies include a wide range of products which draw from an even larger inventory of ingredients and processing methods. The desired viscosity, pH of the formula, serving temperature, and preparation and handling conditions all dictate the requirements for the modified food starch in the system. Corn-based food starches, which encompass products from native or unmodified corn starch to the most functional modified food starch, can contribute significant viscosity to a gravy or sauce formulation, especially unmodified corn starch. The limitation of the unmodified starch is the lack of viscosity stability with prolonged or elevated temperatures, high-acid formulations, or freeze–thaw handling conditions. The stabilized type of pregelatinized modified food starches (E 1442) consists of a series of starches from lightly modified to very heat stable. These starches have been designed to cover the wide range of processing and handling parameters encountered in the manufacture of sauces and gravies (Arocas et al., 2009). A wide variety of processing and handling equipment is utilized for mixing, cooking, cooling, pumping, and filling or depositing sauces and gravies. Whether using open kettle batch cooking, using a scrape surface heat exchanger, or homogenizing a high-fat cream sauce, the functional property of these stabilized starches includes good viscosity, instant hydration temperature, acid tolerance, and temperature stability over freeze–thaw cycles (Iftikhar & Dutta, 2020).

Ready-to-prepare instant soup mix done for moringa leaf by Shaikh et al. (2020) has a bowl of steamy soup satisfies and nourishes lifestyle. In limited preparation time, this smooth and creamy chicken rice soup using pregelatinized modified starches (E 1442) is ready to serve. In tomato soups, 30% total solid content combining with agglomerated maltodextrin could increase the solid content in soups with fine dispersion during mixing. Ethnic foods and sauces have become popular to satisfy the tastes of populations worldwide. Salad dressings are consumed with fruits and vegetables for a healthy diet (Huang, 1995). Pregelatinized modified starch E 1442 combined with unmodified starch aids a cold-water-swelling or hot-water-dispersible agglomerated instant starch combining with quick dispersible (QD) maltodextrin aids in the dispersion. The overall Brabender gel temperature of this starch varies from 52°C to 58°C on functional property.

Seasoning dips were designed as a dairy spread for fresh vegetables, snacks, and chips or as a dipping sauce for deep-fried chicken or meat nuggets. Dips are suitable for retail or food service applications. Pregelatinized food starch is designed to hydrate rapidly in cold water or when mixed with sour cream (Egharevba, 2020). Mayonnaise is one of the major usages of modified starch in the savouries category. Manufacturers have shown interest in using lower-fat mayonnaise, with 6% to 8% fat (max up to 25% fat), in the formulation. Non-vegan mayonnaise uses egg yolk as the protein source in the formulation. India has a largely vegetarian population and cost concerns in the market; eggless mayonnaise was utilizing whey protein concentrate (WPC), which has a minimum of 80% protein and is widely consumed in the Indian market. Experts in this industry forecast about 50% growth in the mayonnaise segment in India, especially for low-fat mayonnaise (Teklehaimanot, 2011). Manufacturers using regular WPC (non-lecithinated) would prefer cold processing. A manufacturer using a high-gelling WPC would prefer hot processing that can largely withstand the protein folding and unfolding at higher temperatures. In some cases, manufacturers use both cold and hot processes in the manufacturing, which they can conduct in two batches, for example, heating of ingredients and then bringing them down to room temperature followed by a cooling process which occurs mainly to form a good emulsion. For mayonnaise, most manufacturers will use two starches: one, E 1422 (acetylated distarch adipate), for thickening and stabilizing, and the second, E 1450 (starch sodium octenyl succinate), for emulsifying functions (Phillips & Williams, 2011).

Mayonnaise manufacturers have largely used waxy corn starch of about 60%–75% of amylopectin contents. Due to its functionality, this provides much higher viscosity than dent corn and it is more resistant to heat and shear rough and type starches can withstand higher processing conditions. It helps in reducing the usage of hydrocolloids such as xanthan gum (E 415). Because market prices fluctuate for starches, manufacturers consider low-cost starches such as potato, tapioca, and unmodified corn starches with stable pricing because most market product sales are for street foods and fast-food restaurants (Thaiwong & Thaiudom, 2021).

13.5.5 Flavour Emulsions

Flavour emulsions are thick, viscous flavours in concentrated form which can be used in beverages, including soft drink or carbonated beverage formulations. Enzyme-treated modified starches (E 1405) are used in flavour emulsion applications. Generally, these have highly absorbent starch and excellent oil loading characteristics. Due to their intact granules, they can increase surface area that binds with solid particles and are able to create emulsion-type liquids. They can possess functionalities ideal for plating and carrying aqueous and non-aqueous ingredients and help to maintain free-flowing powder characteristics. Honey emulsions are one such application where E 1405 has been used at up to 33% in the formulation. Honey can be incorporated, and the final product is a free-flowing powder. Beyond 33%, the emulsion becomes very sticky and the powder is not free flowing (Umesh Hebbar et al., 2008). A particular specialized absorbent starch was developed for plating and dry-mix applications and other difficult-to-handle ingredients (Popil, 2010).

13.5.6 Coated Peanut Application

Crispy-coated peanut snack items are popular throughout the Far East and in many other parts of the world. Soybeans, dried peas, and peanuts are commonly coated with a cereal-based coating to produce this unique type of snack food. Fried coated peanuts have more consumers in Asian countries as "cracker nuts." There is also a final step of applying a seasoning to the surface and making the snack more similar to American savoury snack foods. This coating is applied without the use of additional oil, avoiding increasing the overall fat content without loss in palatability. The coating agent should produce a clear, continuous film on the peanuts prior to adding the seasoning mix. Hydroxypropyl starch (E 1440) is largely used to provide a continuous film to peanuts and the seasonings, and it improves the crispness of the shell. It can improve adhesion between product and shell, decreases drying time of the coated crisp product with an even coating on peanut and good sheen (Pranoto et al., 2021). Gum coatings are typically used to prevent migration of off-flavours in waxes and solvents from contaminating the coated layer underneath. The film-forming properties of E 1440–modified food starch provide this shine or barrier coating. The strength and integrity of film coatings made with hydroxypropyl starch are superior those from gum acacia. The addition of maltodextrins shortens the drying and polishing time while favourably changing the attributes of the coating (Durch, 2020).

13.5.7 Other Food Products

Retort-processed meat products are packed in many forms and containers. Some of the terms used in the industry are "potted," "tinned," or "canned" in aluminium cans in a shelf stable condition. (Durch, 2020). Stabilized modified food starch E 1442 has a unique low hydration temperature requirement as compared to many cook-up starches (needs heat to get activated) and is not affected by the mechanical input of heat and shear in the process of grinding, blending, chopping, emulsifying, pumping, or stuffing until after the gelation temperature is reached and the starch granules are hydrating. E 1442 reaches its activation temperature when the meat product reaches 125°F. Some processors have realized the advantage of a dual starch system with two thermal processes. In a simple formulation, E 1442 and E 1404 are used in combination at a 1:1 ratio. This dual starch system performs very well in many retort applications, ensuring both enhanced yields and texture for economic and consumer acceptability (Rather et al., 2024).

The plant-based meat sector needs modified starches for their texture binding properties. Sodium hydroxypropyl starch phosphate–modified food starch (E 1450) helps maintain texture via moisture management from processing through the frozen distribution of these products (Venkatachalam et al., 2024). Extruded products such as pastas and other noodles are known to be high in carbohydrates and calories. Some pastas have a process modification with a cross-linked phosphorylated type-4 resistant starch (RS4) produced from corn which is cross-linked. Phosphorylated RS4 starches have known health benefits such as lowering postprandial insulin levels and increasing dietary fibre intake. A pasta featuring the modified food starch was found to contain 85% dietary fibre with 25% fewer calories than traditional pasta. It had major functionalities such as lower water holding capacity than bran-based fibre (Sharma et al., 2008).

13.6 APPLICATION OF MODIFIED STARCH IN NUTRACEUTICALS AND PHARMACEUTICALS

Modified starches have emerged as versatile and indispensable ingredients in the nutraceutical and pharmaceutical industries due to their enhanced functional properties and wide-ranging applications. Corn starch has been modified to produce clear, flexible films. These films have been utilized in nutraceutical, food, agricultural, and cosmetic applications, including in aqueous film coating, oral thin films, and soft gel capsules; for texture modification; and in adhesives. Modification of starches can perform versatile functions such as improving their stability, solubility, and bioavailability, making them suitable for a diverse array of health-promoting products. In nutraceuticals, modified starches serve as effective carriers for bioactive compounds, ensuring controlled release and improved absorption of nutrients and dietary supplements. Meanwhile, in pharmaceuticals, they play crucial roles as excipients, binders, and disintegrants in tablet formulations, enhancing the delivery and efficacy of active pharmaceutical ingredients. The unique properties of modified starches include resistance to digestion, prebiotic effects, and the ability to form protective matrices. Coatings can also mask flavours and odours and can be used to regulate the release of the active ingredient. The following sections discuss the major functions of modified starches in various nutraceuticals and pharmaceutical sectors and their applications.

13.6.1 Nutraceutical Supplements

Whey protein supplements are the most common nutraceutical supplements, consumed widely by athletes and others who work out heavily. They can be categorized as nutritious protein drinks and are popular as meal replacements and as a recovery drink after exercising. Flavours are customized according to consumer preferences. Whey supplements can contain protein in the form of concentrates or isolates and low-DE maltodextrins for improving the dispersibility of the drink in water or milk. Hydroxypropyl distarch phosphate–modified food starch (E-1442) helps improve the mouthfeel of the drink and replaces some of the guar gum (E 414) or xanthan gum (E 415) in the formulation to lower the cost of the drink. The usage of maltodextrin in combination with the modified starches is quite common to reduce sweetness, reduce bulk density for volume filling, help control viscosity of the whey-based beverages, provide a high level of digestible carbohydrates, and help controls hygroscopicity (Comunian et al., 2020). Due to lifestyle, consumers turn to beverages to replace a meal with a tasty and nutritious substitute. Formulations with skim milk can provide up to 25% of RDI for most vitamins and minerals.

Meal replacement drinks using plant-based sources such as pea proteins and soy proteins are considered as alternative protein sources and are popular as meal replacements. A pea protein beverage was developed containing maltodextrin and modified starches (E 1442) to provide a readily digestible carbohydrate and to improve dispersibility in the beverage (Altuna et al., 2018). Pregelatinized food-grade modified starches can be used for their film-forming properties, which are desired for coatings, barriers, binding, adhesives, and oral thin films. They can be used to replace or extend other polymers, such as hypromellose, hydroxypropyl cellulose, hydrocolloid gums, and gelatine. Instant-grade modified starches disperse and hydrate more quickly and create less foam than other typically used polymers. In coating formulations, modified starches can be utilized for primary or secondary film-forming polymers to coat tablets, capsules, granules, and so on. As a primary film former dosages were typically used at a 5% to 25% level in the coating solution. As a secondary film former, pregelatinized starch has been used at levels of 2% to 10%. Similarly, in oral thin films where the film delivers an active ingredient or flavour, higher concentrations of polymer are used to minimize drying time (Garcia et al., 2020).

Plasticizers can be used to provide the coatings and films with more elasticity. Propylene glycol, glycerine, and sodium citrate or combinations of these are most commonly used. Plasticizer-to-polymer ratio usually range from 10% to 20%. Sweeteners, flow agents, colourants, flavours, preservatives, and other ingredients, including active ingredients, can also be used with E 1442 to give coatings and films additional properties. Some combinations of plasticizers and these other ingredients may leave the coating slightly tacky (Limpongsa & Jaipakdee, 2020). If desired, detackifiers such as polysorbate 80, magnesium stearate, corn starch, or talc may be used to reduce the tackiness. A combination of maltodextrin and corn syrup solids can be added to the hydroxypropyl starch solution to increase the sheen or tackiness or to speed the drying time of the coating. Due to the binding properties of this pregelatinized modified starch, it can also be used in granulations, including for fluid bed, continuous, and conventional granulation (Bangar et al., 2022).

13.6.2 Pharmaceutical Applications

In multivitamin tablets and capsule preparations, functional starches such as E 1420, E 1442, E 1440, and E 1422 can be used as excipients for thickening and binding applications. Specifically, esterified starch acetate (E 1420) and acetylated distarch adipate (E 1422) can be used in tablet and capsule formulations. In the pharmaceutical industry, commonly used modified starches include E 1404 (oxidized starch), E 1420 (acetylated distarch adipate), and E 1450 (cross-linked sodium starch glycolate). For instance, acetylated starch (often derived from E 1420) can be used as a pharmaceutical polymer. Acetylated starch enhances stability, resistance to retrogradation, and customization for controlled or sustained release in pharmaceutical formulations. In some cases, modified starch can be utilized as an excipient and directly as drug delivery vehicles. Additionally, chemical modifications such as esterification and etherification enhance hydrophobicity and solubility. Moreover, starch derivatives such as carboxymethyl starch, cross-linked carboxymethyl starch, and hydroxypropyl starch enhance solubility, swelling, oil absorption, and film formation. Modified starch has been used for targeting specific sites in the body for treating peptic ulcers and in in vivo studies for coatings and encapsulations. Pharmaceuticals are a minor segments for modified starch usage, with very limited applications (Garcia et al., 2020).

13.6.3 Oral Suspensions and Syrups

In applications for oral suspensions and syrups, several modified starches are commonly utilized, including E 1404 (oxidized starch), E 1410 (monostarch phosphate), E 1412 (distarch phosphate), and E 1450 (starch sodium octenyl succinate). These modified starches are favoured for liquid formulations. Their acceptable daily intake (ADI), indicating a safety margin for consumption, is "not specified." Typical exposure assessments indicate that the combined intake of E 1404 and E 1451 can be up to 3053 mg/kg of body weight per day (Ochubiojo & Rodrigues, 2012). The use and safety of these additives in oral suspensions and pharma syrups must comply with standards from *United States Pharmacopoeia* (*USP*) monographs. They can be typically used at concentrations ranging from 1% to 10% to be considered physiologically safe. However, partially hydrolyzed starches can be used for oral suspensions due to their low cost and various functionalities (Chiu & Solarek, 2009).

13.6.4 Encapsulation and Spray Drying

Modified food starch derived from waxy maize is particularly well suited for encapsulating flavours, clouding agents, vitamins, and spices. It serves as an excellent alternative to the expensive gums and proteins typically used for these purposes. Due to its binding properties, the modified starch E-1440 can also be used in granulations, including high shear, top spray, and rotor granulations, prior to spray drying. A distinctive advantage of starches compared to other encapsulating agents such as gum arabic is their ability to create highly stable oil-in-water emulsions. The fine particle size of these emulsions minimizes losses during spray drying. Encapsulated starches increase the viscosity, allowing for the preparation of higher-solids emulsions, which can be spray dried. However their use is limited to certain concentration. This characteristic also leads to lower energy costs. Waxy maize starch compatibility with water-insoluble substances enables the creation of spray-dried powders with a higher concentration of active ingredients (Santana et al., 2014).

The enzyme-treated starch E 1405 is extensively used for flavour encapsulation of spray-dried actives and extract powders which are incorporated into a wide range of finished nutraceutical products. For cloud encapsulation, E 1405 is highly effective in producing clouds with excellent opacifying properties and stability. It facilitates the preparation of clouds with higher fat content, such as medium-chain triglyceride (MCT) oils, for which optimal opacity is desired, making them ideal for use in pharmaceutical applications (Nimbkar et al., 2022). Additionally, enzyme starches are utilized to encapsulate dry mixes in their original form (Moreton, 2010). Modified food starch derived from waxy maize is known for its exceptional emulsion-stabilizing and encapsulating properties as a cold-water-soluble modified starch. It is particularly effective as a substitute for caseinate salts in coffee encapsulation, bioactive compounds, vitamins, antibiotics, extracts, and active pharmaceutical ingredients, ensuring that emulsions remain stable without separating or increasing in micelle size over time. Spray-dried emulsions made with waxy maize starches reconstitute instantly and without clumping, which guarantees non-genetically modified organism (non-GMO) status (Zhang, Zheng et al., 2022).

13.6.5 Soft Gel Capsules

In soft gel applications, native starch does not produce a usable film for encapsulation. Iota carrageenan and hydroxypropylated acid–modified corn starch can interact synergistically to provide films of sufficient strength and elasticity. In the production of soft gel capsules, E 1404 (oxidized starch), E 1410 (monostarch phosphate), a combination of E 1412 (distarch phosphate) and E 1413 (phosphated distarch phosphate), E 1414 (acetylated distarch phosphate), and E 1450 (starch sodium octenyl succinate) are frequently utilized. For instance, hydroxypropyl-modified tapioca starch is used at 18%–25% by weight, while iota carrageenan (E 407), often included as a gelling agent, is present at 2%–4%. These concentrations highlight the versatility and safety of modified starches in soft gel capsule applications, for example, enhancing properties such as film performance and dissolution rate (Labelle et al., 2020). Moreover, hydroxypropyl tapioca starch, cross-linked potato starch, and hydroxypropyl corn starch can be used at concentrations of 18–25 wt% in starch-based soft capsules. Also, refined iota melts at a lower temperature (40°C–50°C) than modified starch; the correct sealing time and temperature depends on the hydration temperatures of the starch (below about 90°C) and its film-forming capabilities (Reich, 2004). The base composition for preparing empty capsules with modified starch matrices includes modified starch with acetyl (0–2.5 wt%) and hydroxypropyl starch E 1442 (0.5–6.5 wt%) for soft gel capsules. Carrageenan provides good gel strength to the capsule at 2.7%–3.3%, and iota carrageenan (E 407), at 0.45%–0.55%, for rotary die process (Misic et al., 2012). In soft gel capsules, hydroxylpropyl starch (E 1440) provides not only film-forming properties but also adhesion for good seal formation. Because it is an intact granule, a high concentration of polymer is heated to hydrate the starch granules for a flowable gel texture.

13.7 APPLICATION OF MODIFIED STARCH IN PACKAGING

Modified starches, derived from natural sources such as corn, potato, rice, and tapioca, are chemically or physically altered to enhance packaging performance and sustainability. They improve barrier properties, making packaging more resistant to moisture, oxygen, and grease, which extends food shelf life (Gupta et al., 2025a). Additionally, modified starches enhance the flexibility and strength of packaging materials and can be combined with other biopolymers to create high-performance composites, addressing environmental concerns and ensuring effective product protection. Strong, flexible films have been successfully created in many studies using a 15% to 20% solution of hydroxypropyl starch in water. To ensure film clarity, the starch must be fully hydrated; it is recommended to heat to 160°F–180°F (71°C–82°C) and hold for 20 minutes. Films can be made by casting the solution onto plastic or glass using a drawdown gauge. Typical film thickness is 0.5 to 1.0 mm. Films can be dried at room temperature overnight or by using a cool air oven (García-Guzmán et al., 2022).

13.7.1 Edible Packaging Film

Modified starches enhance the properties of biodegradable packaging materials, improving characteristics such as water resistance and mechanical strength. Typical concentrations of modified starch in packaging applications may vary depending on plasticizers, like glycerol concentration, with some studies indicating effective use at levels such as 1 to 2 grams for biodegradable films. Overall, modified starches represent a sustainable alternative to traditional plastics, aligning with environmental goals for packaging. A study utilized glycerol, sorbitol, or a glycerol–sorbitol combination as plasticizer in concentrations ranging from 0% to 45% in edible fruit and vegetable coating films. In general, packaging and edible coatings are prepared from aqueous solutions of various coating agents such as polysaccharides (chitosan, hemicellulose, alginate, cellulose, carrageenan, pectin, modified starch, etc.), proteins (whey, casein, soy protein composites, etc.), and lipids (carnauba, beeswax, fatty acid esters). Along with these agents, certain active ingredients such as antioxidants, antimicrobials, oxygen scavengers, and enzymes are added to enhance safety, nutritional, and sensory attributes as they are consumed along with foods (Sharma & Rao, 2015). This active agent can be added directly into the edible coating solution or coated onto the product surface for better functionality (Bastarrachea et al., 2015). Edible-coated fresh-cut mango with a combination of alginate and ascorbic acid shows better preservation than uncoated samples (Robles-Sánchez et al., 2013). The tomato ripening process was delayed by coating it with 10% gum arabic. Edible coating acts as a barrier, which can reduce the rate of ethylene production and respiration of tissues. Modified starch applications for packaging are polysorbate (Tween) derivatives. They are focused on nanofillers such as nanoclays, polysaccharides, and carbonaceous nanofillers to enhance starch nanocomposites. Modified starch packagings are maltodextrins (up to 70% w/v) and destrins. Typical concentrations range from 10%–20% w/v for maltodextrins.

13.7.2 Active Packaging Film

Fruits and vegetables, being highly perishable, are susceptible to spoilage caused by biochemical processes, enzymatic breakdown, and microbe proliferation on their surfaces. Edible coatings with active compounds such as antimicrobial or antifungal agents are employed as a method for protecting fruits and vegetables from deleterious effects (Ghidelli & Pérez-Gago, 2018). Starch-based biodegradable non-perforated active packaging film loaded with an inorganic ethylene scavenger (zeolite) reduced the decline in the colour (hue) of broccoli during storage. An active film produced from starch and gluten, along with turmeric and cinnamon essential oils, reduced the weight loss of carrots by 44% during storage (Rivera Leiva et al., 2022). This was attributed to increased hydrophobicity of film to retain water. In other research, modified starch active packaging film loaded with an inorganic ethylene scavenger (zeolite) resulted in a high retention of vitamin C in broccoli for 7 days storage at 12°C (Marzano-Barreda et al., 2021). Banana carbon fibre, with a high surface area and microporous nature, was more effective than activated charcoal in preserving the chlorophyll level. When banana carbon fibre obtained from pyrolysis of banana fibre was used, it reduced chlorophyl loss from 25% to 1% in broccoli and from 13% to 5% in guava (Kalyani et al., 2024). Similar to the effect on lettuce, other green leafy vegetables such as spinach, when packed in thyme essential oil–loaded starch nanocomposite film, exhibited a significant lowering of *E. coli* and *Salmonella typhimurium*, whereas the control sample had 4.5 log cfu/g (Issa et al., 2017). Non-climacteric fruits such as papaya will undergo biochemical and colour change after harvest; the papaya becomes highly prone to anthracnose disease, which is caused by *Colletotrichum gloeosporioides*. Hydroxypropyl methyl cellulose film incorporated with silver nanoparticles (0.25%) was found to be effective in inhibition of *C. gloeosporioides* and extended the shelf life of papaya. Tapioca starch films combined with oregano, Vietnamese mint, and curry leaf essential oils can be an effective barrier against oxidation and prevents the spoilage of fruits and vegetables in active packaging.

13.7.3 Biodegradable Packaging Material

Modified starch–based biodegradable packaging materials have significant potential applications in the food and beverage industry. Generally, these materials derived from renewable resources offer biodegradability, which helps reduce plastic waste and environmental pollution and can act as eco-friendly solutions. Some research has indicated that starch-based films can be enhanced through the addition of other biopolymers and bioactive components, improving their mechanical strength, thus extending shelf life of the packaged foods. For instance, the addition of cross-linking agents such as citric acid and acetic acid acts antimicrobially for food packaging applications. The ability to incorporate antimicrobial agents into these films can improve food safety and preservation, enhancing the security of food products until it reaches consumers (Saeed et al., 2024).

Research by Rupérez et al. (2024) focused on analyzing oligoesters migrating from starch-based food packaging, with potential applications for fruit and vegetable packaging due to their biopolymer films complying with European legislation and standards. Modified cassava starch–based films with cross-linking show enhanced water resistance, mechanical properties, and antioxidant activity with a controlled bio-deterioration (Khadsai et al., 2024). Soluble and insoluble portions carbohydrate polymers such as hemicellulose, cellulose, and lignin have been isolated and used for biodegradable packaging. Incorporating lignin offers enhanced tensile strength and water vapour barrier prop, antioxidant, UV barrier, and antibacterial properties (Wang et al., 2024). Materials such as polyhydroxyalkanoates (PHAs) used in food packaging application, when blended with modified starches, offer improved thermal stability, elasticity, and reduced permeability (Paul et al., 2024). Moreover, the natural forms of insoluble fibres from fruits such as mango, banana stem, and date seed extracts possess various functional benefits in packaging and can replace petroleum-based plastics (Murtaza et al., 2024). Biodegradable packaging materials with modified starch E 1405, reinforced with treated banana fibres, show enhanced mechanical strength. Some advanced techniques such as nanogold-imprinted starch bionanocomposites offer enhanced shelf life and antibacterial properties (Picolotto et al. 2024).

13.8 SHELF LIFE OF PRODUCTS IN APPLICATION ANALYSIS

In the ever-changing area of food preservation, the use of modified starches offers a potential advantage in the form of films or coatings for product. Two alternative biodegradable, edible, and intelligent guava coatings enhanced shelf life to 15 days after harvest at room temperature. A coating composed of thermally modified and gelatinized starch, gelatine by-product, and glycerol showed outstanding physicochemical qualities. Another coating was created by adding peppermint oil into the earlier coating mixture at increasing concentrations (0.5%–1.5%), resulting in decreased microbial growth and ripening but no effect on fruit weight loss or firmness. These two coatings improved the shelf life of guava for two main reasons: reducing weight loss by limiting respiration and keeping acidity levels low by inhibiting ethylene synthesis (Moreira et al., 2021). For the storage of strawberries, biodegradable films made from native and modified tapioca starch (oxidized and acetylated) along with gelatine and sorbitol plasticizer resulted in eight days of storage with an increment of five days of shelf life. This was due to the low opacity of the film and minimal water loss in the fruit during storage; mass loss was drastically reduced from 53.44% in the control (no film) to 28.17%–32.96% for the sorbitol film, with no change in colour and an increase from the initial vitamin C content (except for oxidized tapioca–sorbitol film) after eight days of storage (Franco et al., 2017).

In cucumber, the combinations of composite films (modified corn starch and gelatine) in ratios of 1:1, 1:4, and 4:1, the 4:1 film demonstrated superior film qualities by extends shelf life up to 16 days. The 4:1 composite film resulted in minimal change in weight, texture, colour, or pH compared to the uncoated cucumber. The suggested cause for the increase in shelf life was due to the semi-permeable layer, minimal moisture permeability, slower gas migration during respiration, and a low metabolic rate. (Kumar et al., 2021). For sweet cherry, a composite film made with modified starch (octenyl succinyl cassava starch), chitosan, zinc oxide, and ε-polylysine reduced cherry deterioration, stem drying, and weight loss rate (to 40%) on day 8. Preservation was due to the addition of ε-polylysine; improved thermal stability was due to the nanoparticle; and the other properties such as reduced water loss and oxygen and water vapour barriers were attributed to the modified starch film. Overall, they improved the colour by preventing oxidation and dehydration and with a non-disturbing ripening process (Zhang, Zhu, et al., 2022). An edible coating for grapes made with modified starch (heat treatment of wheat starch with octenyl succinic anhydride) has shown ability to retain phenols, total carotenoids, firmness, and radical scavenging activity and to prolong shelf life by microbial protection for 13 days. These effects were mainly a result of the strong barrier to gas and water loss provided by the modified starch (Punia et al., 2019).

In a study, a film created using a mixture of acetylated cassava starch, maltitol, and glycerol for butter cake showed that maltol applied at 1%–5% and 10% inhibited fungal growth by 2.7 times and mould development by 6 times (up to 19 days of storage), respectively, and provided better preservative effect as a packaging material. This preservative effect was mainly due to release of maltol from the modified starch (acetlyated cassava starch) film (Promhuad et al., 2022).

13.9 CONSUMER ACCEPTABILITY OF MODIFIED STARCHES IN FOOD APPLICATIONS

The consumer acceptability of starch largely results from major functions provided by starch, such as improved functional properties, without affecting sensory properties. Different type of starch modifications have been shown to improve the consumer acceptability in specific starches. Among physical methods of modification, heat-moisture treated (HMT) starch had been shown to improve the consumer acceptability in various cases; for example, HMT rice starch, at 1.0%, improved the texture and stability of non-fat yogurt and increased customer satisfaction (Rashwan et al., 2024). When jack bean starch was completely replaced with HMT jack bean starch in biscuits, the sensory properties did not differ significantly from control samples (Ariyantoro et al., 2023). Yam starch modified by two hydrothermal methods, heat-moisture treatment and annealing, improved starch functional property for prebiotic and diabetic-friendly products by slowing digestibility (Ayuningtyas et al., 2022).

Chemically modified starches have also been shown to have health benefits for the consumer, apart from acting as a functional replacer (fat substitute) and fulfilling sensory requirements. When starch is modified with polyphenols, it promotes polyphenol delivery, improves consumer health, and their application to diverse consumer-friendly (Ngo et al. 2022). The use of modified starch at 1% in ice cream as a fat replacer achieved sensory acceptability compared to full-fat control ice cream and had no effect on microbiological stability (Chen et al., 2020). Octenyl succinic anhydride (OSA)-modified starch obtained from potato, mung bean, and pearl millet was used as a fat replacer at 75%–100%, 30%, and 1%–2% in mayonnaise, cake, and ice cream respectively. Even OSA-modified starch made with gelatine could be used for encapsulating as a low-calorie provider without affecting sensory properties of the product (Anggriawan et al., 2020; Chen et al., 2020).

As a part of enzymatic modification, consumer-acceptable and specifically reactive starches are made. Two different cassava starches, starch modified by amyloglucosidase and pregelatinized starch, can successfully substitute for fat in crackers. Amyloglucosidase-modified starch and control crackers are chosen for their colour and texture, with the highest sensory acceptance (Otero-Guzmán et al., 2023). The enzymes pullulanase and transglucosidase decrease starch digestibility while increasing amylose concentration, making the starch less susceptible to breakdown for regulated glucose release, appealing to health-conscious customers (Rashwan et al., 2024). Apart from single modification of starches, dual enzymatic modification of oat starch by α-amylase and cellulase improved sensory evaluation of starch (He et al., 2023). In another case, banana starch modified by ultrasound and resveratrol treatment had a positive effect on starch digestion (Sun et al., 2022).

13.10 TOXICITY, SAFETY, AND LEGAL COMPLIANCE OF MODIFIED STARCHES

Regulatory frameworks for modified starches in food products are established primarily through international standards and regional legislation in the United States, the European Union (EU), and China. In the EU, regulation and its usage is governed by strict regulations that include maximum limits for specific substances, as outlined in Codex STAN 192–1995, developed by the Food and Agricultural Organization (FAO) and the World Health Organization (WHO) (Vilp, 2023). Starch used for pharmaceutical applications in India must compile with *British Pharmacopeia* (*BP*) and *USP* regulations. In 2009, the use of modified food starches was allowed in Appendix B by the Food Safety and Standards Authority of India (FSSAI) up to a maximum concentration of nearly 0.5% by weight. In 2016, the FSSAI changed the legislation on modified starches in processed foods to be Good Manufacturing Practices (GMP). Less restrictive regulation on the use of modified starches has provided new opportunities for companies. The EU distinguishes intentionally added substances such as pesticides, veterinary drugs, biocides, and hormones in meat products, ensuring that modified starches meet safety and quality standards (Schebesta & Purnhagen, 2024). Overall, the regulatory landscape is designed to ensure that modified starches are safe for consumption and compliant with legal standards across jurisdictions and demographic locations. Modified food starch products have to meet the requirements guidelines in 21 CFR 172.892 with a label declaration including "food starch modified" or "modified food starch." All labels use E-numbers as per European legislation.

Regulatory frameworks in the United States, Canada, and the EU oversee genome editing, synthetic biology, and gene silencing methods in food products in which the source is a genetically modified organism (GMO) or non-GMO would be efficient and evolving approaches. For GMOs, proper certifications and documentations and legal frameworks, objectives, risk assessment, traceability, and labelling requirements can ensure safety and compliance of modified starches in food products. However, GMO-based sources have been banned in countries like Europe, the United States, and India due to the health risks associated with the products (Schebesta & Purnhagen, 2024). The international regulatory framework for GMOs in food aid includes the World Trade Organization (WTO), human rights law, and Codex Alimentarius Commission (CAC) food safety standards (Vezzani, 2018). All the testing methods for nutritional information and standard specifications were derived from a laboratory information management system (LIMS). Testing methods can be adopted from the FDA's *Bacteriological Analytical Manual* (*BAM*), from modified Corn Refiners Association (CRA; formerly the Corn Industries Research Foundation) methods, and from the Association of Official Analytical Chemists (AOAC); and for pharmaceutical compliance, they can be adopted from the *USP* and various sources listed by the Cereals & Grains Association (formerly the American Association of Cereal Chemists) and the CRA. Allergen declarations do not contain egg products, dairy products, lactose, peanuts, oils and extracts, wheat gluten, seafood, antibiotics, colours, animal fats, monosodium glutamate (MSG), tartrazine, and cottonseed flour. If SO_2 is used, a sensitizer level in ppm should be mentioned on the certificate of analysis (COA) (Schmiele et al., 2019).

Non-conventional starch modification techniques such as plasma treatment offer eco-friendly and chemical-free options, aligning with safety and legal compliance for the modified starches. The material safety data sheet (MSDS) for the modified starch product contains hazard information, composition, and other regulatory information related to the U.S. Occupational Safety and Health Administration (OSHA) should be mentioned. Modified starches are not absorbed whole by the human body; instead, they are hydrolyzed and fermented by enzymes and microbiota, respectively. In silico studies have shown that modified starches are not genotoxic. No risk relevant to humans was shown when modified starch was incorporated into rat feed at 31 g/kg body weight, and in humans the modified starches were well accepted up to 25 g/kg of body weight even in one dose (Mortensen et al., 2017). In neonatal farm piglets, concentrations of 20 mg were well tolerated and had no adverse effect on health or piglet growth (Mahadevan et al., 2014). When safety aspects of the chemically and enzymatically modified starches were compared for minced meat, the chemically modified starch oxyamyl starch (OPV-1) showed greater genotoxicity in both fractions (acetone and water-soluble). Using enzyme-modified starches such as those modified with α-amylase (α-8) and β-amylase (β-8) at 5% concentration resulted in decreased genotoxicity and higher stability, making them safer choices (Nikitina et al., 2015).

13.11 SWOT ANALYSIS

The strengths, weaknesses, opportunities, and threats of modified starches are reviewed here. Modified starches present significant strengths and opportunities for food applications due to their enhanced functional properties. Major key strengths include improved emulsification in mayonnaise application, water-binding capacity, and resistance to retrogradation in breads. For instance, modified starches can exhibit superior performance compared to native starches, making them ideal for moisture retention in food formulations. Cross-linking further enhances the stability of starch granules, allowing for better performance under thermal and mechanical stress. Opportunities lie in the development of innovative applications in different food products to reduce the fat absorption or act as fat replacements/substitutes. In the area of encapsulation of bioactive ingredients, which align with growing consumer demand for sustainable and health-promoting food products in targeted nutraceutical applications for infants and pregnant women, integrated chemical and enzymatic modifications can lead to good pumpability and flowability and to high viscosity by improving resistance to acid, syneresis, and enzymatic hydrolysis (Vitolo, 2020). Additionally, the exploration of non-traditional starches from food waste offers potential for creating low-digestible starches that can aid in addressing obesity and related health issues. For example, modified starches from *Oxalis tuberosa* offer improved functional properties in multiple food applications in combination with gums, mucilage, polyphenols, and salts. This synergy provides improved texture, stability, and nutritional benefits, presenting significant opportunities for innovation. Tailoring starch properties for specific food needs provides major functional benefits specific to the application. Waxy maize starch has the highest benefits among starches due to high processing conditions.

New emerging sources from which modified starches can be derived include banana starches, which when modified offer enhanced functionalities such as increased resistant starch content, improved gel strength, and lower glycaemic index for specialized diets, making them valuable for diverse food applications. Modified starches offer advantages such as low toxicity, rapid breakdown, and limited environmental impact. They have diverse sources and biodegradability. Modified starchy flours derived from pulses, minor cereals, Andean grains, tubers, and roots from various sources offer diverse nutritional attributes and micronutrients for healthier food products, presenting opportunities for novel foodstuffs and improved diets. Emerging trends focusing on polyphenol-modified starches offer reduced caloric impact and potential for novel functional foods. Moreover, modified starches efficiently adsorb microelements such as copper, iron, and zinc, making them suitable carriers for food fortification. With respect to processing, some non-thermal modification methods are considered to be eco-friendly; however, yields will be lower than for other commercially available methods.

In the packaging sector, modified starches exhibit various functions in fruits and vegetables. Active packaging delays ripening and supports international transport. The inhibition or delay of ripening and oxidative browning and preserved freshness of fruits increase their acceptance by consumers (Gul et al., 2018). Active packaging can be tool to supplement or fortify probiotics and nutrients. Moreover, active packaging prevents oxidation of nutrients such as vitamin C and preserves the phytochemicals in fruits. Active packaging has been shown to inhibit numerous pathogens, such as *Listeria monocytogens* and *Clostridium botulinum* (Ramos et al., 2013). Therefore, active packaging can ensure the safety of fresh produce. It is challenging to recycle active packaging from the supply chain, separate the active material layer from the laminate, and reactivate the active constituents for fresh use. Some active packaging materials (i.e., anthocyanins, ascorbic acid) have superior antioxidant/antimicrobial properties but poor thermal or light stability. Moreover, the nanomaterials increasingly studied as active packaging require extensive toxicity studies before clearance for use. The potential threat of active packaging must be addressed to improve the safety of fruits for consumption (Figure 13.2).

13.12 FUTURE PROSPECTS FOR MODIFIED STARCHES IN VARIOUS APPLICATIONS

With the growing demands for starches and modified starch products, there a wide range of future prospects or opportunities to work on or improve the modified starches obtained through different technologies. Beginning with starch production sources, research has produced limited data on starch derived from fruit seeds and waste materials. With this future research, we could reduce costs and waste and produce very useful starch products (Syuhada et al., 2018). There need to be more variety of starches produced from various natural sources such as turmeric and annatto (Maniglia et al., 2022; Silveira & Tapia-Blácido, 2018), and in the future these could be modified. One such starch is cocoyam starch, in which annealing and citric acid treatment are key parameters (He et al., 2023).

In terms of modification methods, physical modification by superheated steam, a hydrothermal method of starch modification, demonstrated improved starch gel stability and reduced swelling, suggesting its future importance in terms of efficiency and low loss (He et al., 2023). Future research should investigate the effects of thermal and non-thermal treatments on these complexes. Comprehensive in vitro and in vivo research is required to completely comprehend and optimize these starch–polyphenol interactions for commercial usage (Ngo et al., 2022). Similarly, the mechanisms of OSA modification of starch and the factors influencing them need to be explored (Anggriawan et al., 2020). Starch chemically modified with citric acid has been shown to have potential, but there are still limitations on commercial uses. With respect to application, unlike starch, citric acid–modified starch has a positive outlook, with enhanced resistance to heat and moisture, increased mechanical strength, and diverse degradation properties which could be utilized (M. Zhang et al., 2023). Gamma-irradiated and annealed sago starch, a novel dual modification, showed improved starch properties but suggested the need for in-depth investigation of the changes in starch structure and its digestibility (Lee et al., 2022). Enzymatic modification of starch by pullulanase and transglucosidase increased resistance to breakdown of starch and its qualities for regulated glucose release and industrial applications. CRISPR-Cas9 genetic changes improve starch properties such as high amylose content and resistance to retrogradation (Rashwan et al., 2024). Aspects of enzymatic modification yet to be explored include how enzymes affect lamellar structure, gelatinized and native starches, and different starches (Zhong et al., 2022).

Further research areas for starch modification extend to emerging green, sustainable, and eco-friendly technologies. These methods involve techniques such as regioselective

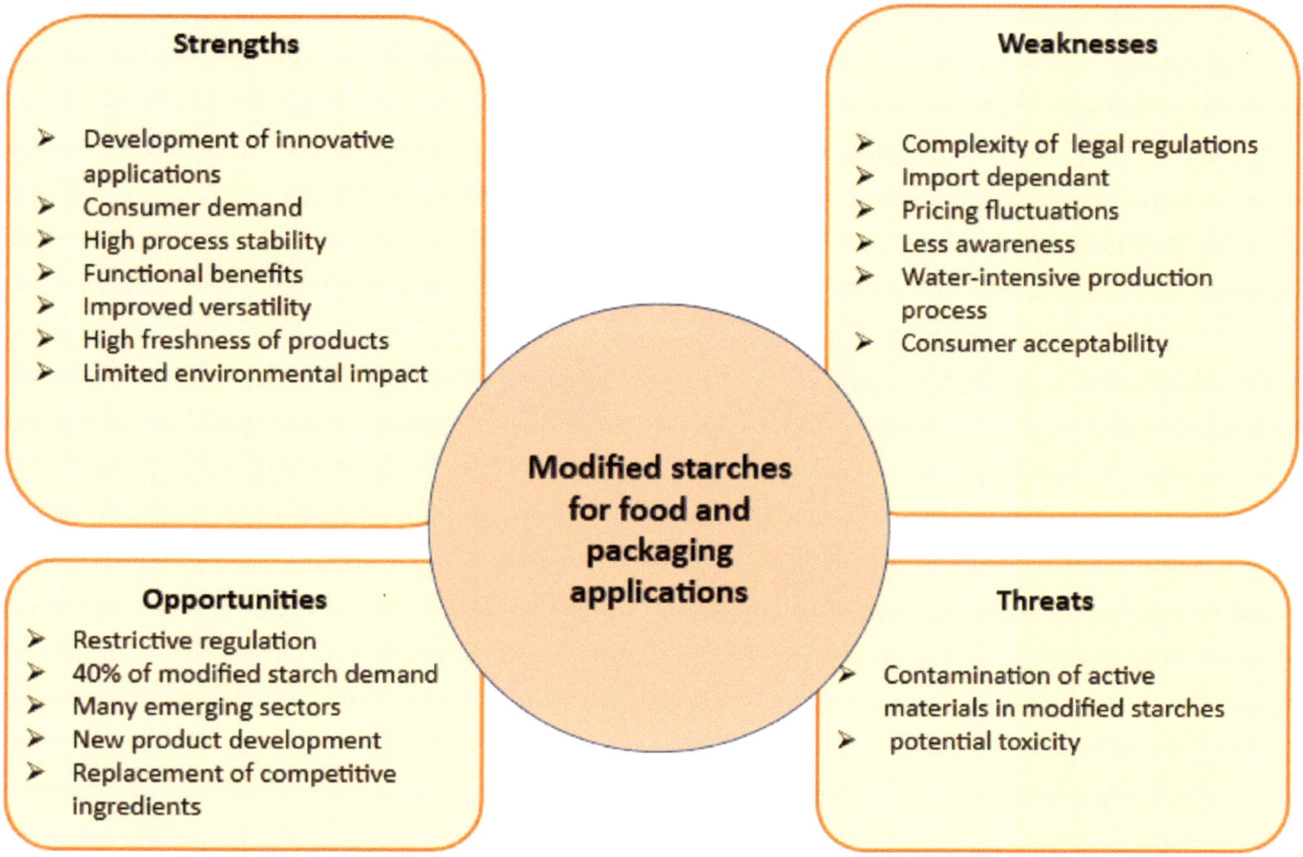

FIGURE 13.2 SWOT analysis of modified starches in food and packaging applications.

functionalization and controlled grafting for specific starch derivative production. The three methods of regioselective functionalization are protection deprotection, site-selective replacement, and catalyzed functionalization. Important methods of grafting are atom transfer radical polymerization (ATRP), copper-mediated Living Radical Polymerization (LRP), reversible addition–fragmentation chain transfer (RAFT), and chemo-enzymatic approaches. There is scope to merge these methods to create sustainable and high-value starch-based materials (Fan & Picchioni, 2020). Radiation techniques are also green methods for natural polymer modification, for example, electron beams and gamma rays. Molecular weight (analyzed by chromatographic analysis and thermal characteristics (measured by differential scanning calorimetry) are expected in the near future to enable modified starches to enter new areas of application (Brașoveanu & Nemțanu, 2018). Ultrasonic modification with customizable settings can change the starch structure from double to single helix, but additional study is needed to maintain its effects (He et al., 2023). Apart from the modification of starches by methods such as ultrasonication, hydrostatic pressure, and microwaves, detailed investigation into emerging techniques such as high-pressure homogenization and cold plasma modification of starch is lacking and needed (Guillén, 2021; Marta et al., 2022). These challenges could be overcome in the future and, along with technological improvements, will further extend the applications of modified starch.

13.13 SUMMARY

- Demand for modified starches had been created for companies globally by trends in various applications and quality standards, strategic trade agreements, and growing consumer demand for processed foods.
- Unlike other food additives, modified starches have been widely used in food industry, nutraceuticals, pharmaceuticals, and packaging applications along with various other functional uses.
- The consumer acceptance of modified starch-based products has been widespread due to the safety regulations of the modified starch being highly stringent. The legal safety standards of the product give consumers confidence in these products.
- Active packaging using modified starches reduces respiration rate, ripening, loss of firmness, and weight loss and preserves nutrients of fresh produce. It has been shown to effectively inhibit both spoilage and pathogenic microbes.
- Active packaging with modified starches has great scope to be integrated with intelligent and

modified atmospheric packaging to improve the safety of food products.

- Future prospects for and innovations to modified starches across various industries have been widely used and accepted among consumers and food processors.

13.14 CONCLUSION

Modified starches have been used as an alternative to native starches. The production of modified starches is expected to grow greatly in the upcoming years, and many products are now commercially available with specific E-numbers. Although modifications can be made by a variety physical, chemical, enzymatic, and novel methods and even by dual modification methods, the chosen method of modification depends solely on factors such as the desired structural and functional properties and the sustainability of starch production. The effect of modifications on starches can be on structural or functional properties or on both. Thus, structural and functional properties in a variety of food and non-food products can be improved to withstand a variety of processing conditions and to give the desired final output. Modified starch has been shown to protect the product through enhanced film properties. The toxicity and consumer acceptability of these modified starches have been shown to be acceptable and even better, respectively, in several food products. The future scenario for modified starches shows a vast area for exploring modified starches, from changes in production sources to various dual modifications combinations, towards modified starches with better performance from the existing methods. Overall, modified starch will be an upcoming in-demand product, effectively used in various processes with regulatory guidelines and acceptability, to improve a variety of products.

REFERENCES

Abbas, K. A., Khalil, S. K., & Hussin, A. S. M. (2010). Modified starches and their usages in selected food products: A review study. *Journal of Agricultural Science*, *2*(2), 90.

Abdel-Aal, E. S. M., & Rabalski, I. (2008). Effect of baking on nutritional properties of starch in organic spelt whole grain products. *Food Chemistry*, *111*(1), 150–156.

Adewale, P., Yancheshmeh, M. S., & Lam, E. (2022). Starch modification for non-food, industrial applications: Market intelligence and critical review. *Carbohydrate Polymers*, *291*, 119590.

Adigwe, O. P., Egharevba, H. O., & Emeje, M. O. (2022). In M. O. O. Emeje & A. M. Carmona-Ribeiro (Eds.), Starch: A veritable natural polymer for economic revolution. In *Starch: Evolution and recent advances*. IntechOpen.

Agudelo, A., Varela, P., Sanz, T., & Fiszman, S. (2014). Formulating fruit fillings. Freezing and baking stability of a tapioca starch–pectin mixture model. *Food Hydrocolloids*, *40*, 203–213.

Altuna, L., Herrera, M. L., & Foresti, M. L. (2018). Synthesis and characterization of octenyl succinic anhydride modified starches for food applications: A review of recent literature. *Food Hydrocolloids*, *80*, 97–110.

Anggriawan, R., Maksum, A., & Nurhayati, N. (2020). Production and application of octenyl succinic-modified starch as fat replacer: A review of established and recent research. Preprints.org; 2020. https://doi.org/10.20944/preprints202005.0283.v1. PPR:PPR164115

Ariyantoro, A. R., Ishartani, D., Yulviatun, A., & Attaqiyyah, F. (2023). Analysis of acceptability, microstructure, and resistant starch content of biscuits substitution of jack beans (*Canavalia ensiformis*) with starch modification of annealing-HMT. *IOP conference series: Earth and environmental science*, *1200*(1), 012003.

Arocas, A., Sanz, T., & Fiszman, S. M. (2009). Influence of corn starch type in the rheological properties of a white sauce after heating and freezing. *Food Hydrocolloids*, *23*(3), 901–907.

Arunachalam, S., Rajan, D., Giridharaprasad, S., & Subramani, D. (2024). Optimization of texture and quality of deep-fried prawn chips using response surface methodology. *Journal of Culinary Science & Technology*, *23*(3), 512–530.

Ayuningtyas, L. P., Benita, A. M., & Triastuti, D. (2022). Crystalline and digestive characteristics of heat moisture treated and annealed lesser yam (*Dioscorea esculenta*) starch. *Journal Pangan dan Agroindustrir, 10*(1), 1–8.

Bangar, S. P., Ashogbon, A. O., Singh, A., Chaudhary, V., & Whiteside, W. S. (2022). Enzymatic modification of starch: A green approach for starch applications. *Carbohydrate Polymers*, *287*, 119265.

Barua, S., Luciano, G., Ahmed, J., Srivastav, P. P., & Vilgis, T. A. (2024). Physics of starch system: Rheological and mechanical properties of hydrothermally modified elephant foot yam starch. *Food Biophysics*, *19*(1), 71–84.

Bashirian, A., & Abdolmaleki, F. (2020). Use of modified hydroxypropyl starch diphosphate starch (E1442) as a fat substitute in low calorie cakes. *Journal of Innovation in Food Science and Technology*, *12*(1), 97–108.

Bastarrachea, L. J., Wong, D. E., Roman, M. J., Lin, Z., & Goddard, J. M. (2015). Active packaging coatings. *Coatings*, *5*(4), 771–791.

Braşoveanu, M., & Nemţanu, M. R. (2018). Aspects on starches modified by ionizing radiation processing. *Applications of Modified Starches*, 49–68.

Chakraborty, I., N, P., Mal, S. S., Paul, U. C., Rahman, M. H., & Mazumder, N. (2022). An insight into the gelatinization properties influencing the modified starches used in food industry: A review. *Food and Bioprocess Technology*, *15*(6), 1195–1223.

Chen, N., Gao, H. X., He, Q., & Zeng, W. C. (2022). Wheat starch modified with *Ligustrum robustum* (Rxob.) Blume extract and its action mechanism. *Foods*, *11*(20), 3187.

Chen, Y., She, Y., Zhang, R., Wang, J., Zhang, X., & Gou, X. (2020). Use of starch-based fat replacers in foods as a strategy to reduce dietary intake of fat and risk of metabolic diseases. *Food Science & Nutrition*, *8*(1), 16–22.

Chiu, C. W., & Solarek, D. (2009). Modification of starches. In J. BeMiller & R. Whistler (Eds.), *Starch* (pp. 629–655). Academic Press.

Comunian, T. A., Roschel, G. G., da Silva Anthero, A. G., de Castro, I. A., & Hubinger, M. D. (2020). Influence of heated, unheated whey protein isolate and its combination with modified starch on improvement of encapsulated pomegranate seed oil oxidative stability. *Food Chemistry*, *326*, 126995.

Damager, I., Engelsen, S. B., Blennow, A., Lindberg Møller, B., & Motawia, M. S. (2010). First principles insight into the α-glucan structures of starch: Their synthesis, conformation, and hydration. *Chemical Reviews*, *110*(4), 2049–2080.

Dar, Y. L., & Embuscado, M. (2014). Starches as food texturizing systems. *Functionalizing carbohydrates for food applications: Texturizing and bioactive/flavor delivery systems* (pp. 41–79). DEStech Publications.

Durch, N. (2020). *Characterization of starches for retort applications*. [Doctoral dissertation, University of Minnesota Twin Cities]. https://conservancy.umn.edu/items/df97ba2b-a10c-4d5d-b84a-babf2ab96577

Egharevba, H. O. (2020). Chemical properties of starch and its application in the food. In M. Emeje (Ed.),*Chemical properties of starch*. IntechOpen.https://www.intechopen.com/chapters/68437

Fan, Y., & Picchioni, F. (2020). Modification of starch: A review on the application of "green" solvents and controlled functionalization. *Carbohydrate Polymers*, *241*, 116350.

Franco, M. J., Martin, A. A., Bonfim, L. F. Jr, Caetano, J., Linde, G. A., & Dragunski, D. C. (2017). Effect of plasticizer and modified starch on biodegradable films for strawberry protection. *Journal of Food Processing and Preservation*, *41*(4), e13063.

Garcia, M. A. V. T., Garcia, C. F., & Faraco, A. A. G. (2020). Pharmaceutical and biomedical applications of native and modified starch: A review. *Starch - Stärke*, *72*(7-8), 1900270.

Gałkowska, D., Kapuśniak, K., & Juszczak, L. (2023). Chemically modified starches as food additives. *Molecules*, *28*(22), 7543.

García-Guzmán, L., Cabrera-Barjas, G., Soria-Hernández, C. G., Castaño, J., Guadarrama-Lezama, A. Y., & Rodríguez Llamazares, S. (2022). Progress in starch-based materials for food packaging applications. *Polysaccharides*, *3*(1), 136–177.

Ghidelli, C., & Pérez-Gago, M. B. (2018). Recent advances in modified atmosphere packaging and edible coatings to maintain quality of fresh-cut fruits and vegetables. *Critical Reviews in Food Science and Nutrition*, *58*(4), 662–679.

Guillén, J. S. (2021). Modification of potato starch by emerging technologies. *Tropical Agricultural Research*, *32*(4).

Gul, K., Mir, N. A., Singh, P., & Wani, A. A. (2018). Postharvest biology and technology of apple. In S. A. Mir, M. A. Shah, & M. M. Mir (Eds.), *Postharvest biology and technology of temperate fruits* (pp. 223–243). Springer.

Guleria, P., & Yadav, B. S. (2022). Effect of chemical treatments on the functional, morphological and rheological properties of starch isolated from pigeon pea (*Cajanus cajan*). *Current Research in Food Science*, *5*, 1750–1759.

Gupta, R. K., Guha, P., & Srivastav, P. P. (2022). Natural polymers in bio-degradable/edible film: A review on environmental concerns, cold plasma technology and nanotechnology application on food packaging-A recent trends. *Food Chemistry Advances*, *1*, 100135. https://doi.org/10.1016/j.focha.2022.100135

Gupta, R. K., Guha, P., & Srivastav, P. P. (2023). Effect of high voltage dielectric barrier discharge (DBD) atmospheric cold plasma treatment on physicochemical and functional properties of taro (*Colocasia esculenta*) starch. *International Journal of Biological Macromolecules*, *253*, 126772. https://doi.org/10.1016/j.ijbiomac.2023.126772

Gupta, R. K., Guha, P., & Srivastav, P. P. (2024a). Insights into the physicochemical, crystalline, multistructural, thermal and digestibility behavior of taro starch (*Colocasia esculenta*) modified by cold plasma treatment with high voltage dielectric barrier discharge (HVDBD): An alternative technique for chemical modification of starch and environmental sustainability. *Journal of Cleaner Production*, *461*, 142710. https://doi.org/10.1016/j.jclepro.2024.142710

Gupta, R. K., Guha, P., & Srivastav, P. P. (2024b). Dielectric barrier discharge plasma: A green and novel method to change the structure of taro peel starch and improve the physicochemical properties of taro peel starch films. *Plasma Processes and Polymers*, *21*(9), 2400047. https://doi.org/10.1002/ppap.202400047

Gupta, R. K., Guha, P., & Srivastav, P. P. (2024c). Exploring the potential of taro (*Colocasia esculenta*) starch: Recent developments in modification, health benefits, and food industry applications. *Food Bioengineering, 3*(3), 365–379. https://doi.org/10.1002/fbe2.12103

Gupta, R. K., Guha, P., & Srivastav, P. P. (2024d). Physical action of nonthermal cold plasma technology for starch modification. *Food Physics*, *1*, 100011. https://doi.org/10.1016/j.foodp.2024.100011

Gupta, R. K., Guha, P., & Srivastav, P. P. (2024e). Effect of cold plasma treatment and plasma-activated water on physicochemical and structural properties of starch: A green and novel approach for environmental sustainability. *Plasma Processes and Polymers*, *21*(4), 2300204. https://doi.org/10.1002/ppap.202300204

Gupta, R. K., Pipliya, S., Patel, J., Srivastav, P. P., Castro-Muñoz, R., Wang, C. K., & Nemţanu, M. R. (2025a). Essential oil-embedded starch based films for active packaging: Insights into migration, stability, and preservation efficiency. *Trends in Food Science & Technology*, *163*, 105149. https://doi.org/10.1016/j.tifs.2025.105149

Gupta, R. K., Gnana Moorthy Eswaran, U., Kumar, S., Pipliya, S., Kaur, B., & Srivastav, P. P. (2025b). Synergistic modification of starch polysaccharide via sequential treatments with ultrasonic and cold plasma technology. *Journal of Food Processing and Preservation*, *2025*(1), 9063007. https://doi.org/10.1155/jfpp/9063007

He, R., Li, S., Zhao, G., Zhai, L., Qin, P., & Yang, L. (2023). Starch modification with molecular transformation, physicochemical characteristics, and industrial usability: A state-of-the-art review. *Polymers*, *15*(13), 2935.

Huang, D. P. (1995). New perspectives on starch and starch derivatives for snack applications. *Cereal Foods World*, *40*(8), 528–531.

Iftikhar, S. A., & Dutta, H. (2020). Use of raw and physically modified rice starches as fat replacer in whipping cream. *Current Research in Nutrition and Food Science Journal*, *8*(1), 122–130.

Islamiyah, H. S., Alamsjah, M. A., & Pujiastuti, D. Y. (2022). Application of modified starch in the carragenan-based biodegradable packaging from *Eucheuma cottonii* on biodegradablility and mechanical properties. *IOP Conference Series: Earth and Environmental Science*, *1036*(1), 012072.

Issa, A., Ibrahim, S. A., & Tahergorabi, R. (2017). Impact of sweet potato starch-based nanocomposite films activated with thyme essential oil on the shelf-life of baby spinach leaves. *Foods*, *6*(6), 43.

Kalyani, P., Das, P. P., & Khandelwal, M. (2024). Utilization of natural fiber–derived active agents for shelf-life extension of broccoli (*Brassica oleracea* l.) and guava (*Psidium guajava*). *Biomass Conversion and Biorefinery*, *14*(19), 24753–24764.

Kaya, D. (2019). *Determination of the best drying conditions for gelatin-based candies* [Master's thesis, Middle East Technical University (Turkey)].

Khadsai, S., Janmanee, R., Sam-Ang, P., Nuanchawee, Y., Rakitikul, W., Mankhong, W., & Ninjiaranai, P. (2024). Influence of crosslinking concentration on the properties of biodegradable modified cassava starch-based films for packaging applications. *Polymers*, *16*(12), 1647.

Kibar, E. A. A., Aslan, Ö, Bakan, A., & Özer, H. (2024). Determination of physicochemical and functional properties of cross-linked, hydroxypropylated, and dual modified corn starch and investigation of its use as gelling agent in soft confectionery. *Starch - Stärke*, *76*(7-8), 2300036.

Kumar, R., Ghoshal, G., & Goyal, M. (2021). Biodegradable composite films/coatings of modified corn starch/gelatin for shelf-life improvement of cucumber. *Journal of Food Science and Technology*, *58*, 1227–1237.

Labelle, M. A., Ispas-Szabo, P., & Mateescu, M. A. (2020). Structure-functions relationship of modified starches for pharmaceutical and biomedical applications. *Starch - Stärke*, *72*(7-8), 2000002.

Lee, J. S., Akanda, J. H., Fong, S. L., Siew, C. K., & Ho, A. L. (2022). Effects of annealing on the properties of gamma-irradiated sago starch. *Molecules*, *27*(15), 4838.

Lewandowicz, J., Le Thanh-Blicharz, J., & Szwengiel, A. (2022). The effect of chemical modification on the rheological properties and structure of food grade modified starches. *Processes*, *10*(5), 938.

Limpongsa, E., & Jaipakdee, N. (2020). Physical modification of Thai rice starch and its application as orodispersible film former. *Carbohydrate Polymers*, *239*, 116206.

Mahadevan, B., Thorsrud, B. A., Brorby, G. P., & Ferguson, H. E. (2014). A 3-week dietary safety study of octenyl succinic anhydride (OSA)-modified starch in neonatal farm piglets. *Food and Chemical Toxicology*, *72*, 83–89.

Maniglia, B. C., Silveira, T. M. G., & Tapia-Blácido, D. R. (2022). Starch isolation from turmeric dye extraction residue and its application in active film production. *International Journal of Biological Macromolecules*, *202*, 508–519.

Marta, H., Cahyana, Y., Bintang, S., Soeherman, G. P., & Djali, M. (2022). Physicochemical and pasting properties of corn starch as affected by hydrothermal modification by various methods. *International Journal of Food Properties*, *25*(1), 792–812.

Marta, H., Cahyana, Y., Djali, M., & Pramafisi, G. (2022). The properties, modification, and application of banana starch. *Polymers*, *14*(15), 3092.

Marzano-Barreda, L. A., Yamashita, F., & Bilck, A. P. (2021). Effect of biodegradable active packaging with zeolites on fresh broccoli florets. *Journal of Food Science and Technology*, *58*, 197–204.

Mason, W. R. (2009). Starch use in foods. In J. BeMiller & R. Whistler (Eds.), *Starch* (pp. 745–795). Academic Press.

Matsumoto, A., Nakai, K., & Kawai, K. (2022). Effects of water and gelatinized starch on the viscoelasticity of pizza dough and the texture of pizza crust. *Journal of Applied Glycoscience*, *69*(1), 1–7.

Miller, R. A., Maningat, C. C., & Hoseney, R. C. (2008). Modified wheat starches increase bread yield. *Cereal Chemistry*, *85*(6), 713–715.

Mintel. (2024). Search by modified starch market brands: Mintel India. https://www.gnpd.com/sinatra/analysis/chart_results/search/FyS6DqJvk0/?analysis_id=216f0be4-7bcd-4179-9c4a-b28281ccb780&current_tab=216f0be4-7bcd-4179-9c4a-b28281ccb780/. Accessed October 10, 2024.

Misic, Z., Muffler, K., Sydow, G., & Kuentz, M. (2012). Novel starch-based PVA thermoplastic capsules for hydrophilic lipid-based formulations. *Journal of Pharmaceutical Sciences*, *101*(12), 4516–4528.

Moreira, E. S., Silva, N. M. C., Brandão, M. R. S., Santos, H. C., & Ferreira, T. A. P. C. (2021). Effect of modified starch and gelatin by-product based edible coating on the postharvest quality and shelf life of guava fruits. *Food Science and Technology*, *42*, e26221.

Moreton, C. (2010). Trends in pharmaceutical excipients: An update. *American Pharmaceutical Review*, *13*, 22–27.

EFSA Panel on Food Additives and Nutrient Sources added to Food (ANS), Mortensen, A., Aguilar, F., Crebelli, R., Di Domenico, A., Dusemund, B., … & Woutersen, R. A. (2017). Re-evaluation of oxidised starch (E 1404), monostarch phosphate (E 1410), distarch phosphate (E 1412), phosphated distarch phosphate (E 1413), acetylated distarch phosphate (E 1414), acetylated starch (E 1420), acetylated distarch adipate (E 1422), hydroxypropyl starch (E 1440), hydroxypropyl distarch phosphate (E 1442), starch sodium octenyl succinate (E 1450), acetylated oxidised starch (E 1451) and starch aluminium octenyl succinate (E 1452) as food additives. *EFSA Journal*, *15*(10), e04911.

Murtaza, S., Shah, F. U. H., Shahbaz, M., Murtaza, M. A., Farooq, U., & Ma, Y. (2024). Starch-nanostructured-based active packaging for food applications. In G. A. Nayik & A. H. Dar (Eds.), *Starch based nanomaterials for food packaging* (pp. 103–160). Academic Press.

Nawaz, H., Waheed, R., Nawaz, M., & Shahwar, D. (2020). Physical and chemical modifications in starch structure and reactivity. In M. O. O. Emeje & A. M. Carmona-Ribeiro (Eds.), *Chemical properties of starch*. IntechOpen. https://www.intechopen.com/chapters/68720

Ngo, T. V., Kusumawardani, S., Kunyanee, K., & Luangsakul, N. (2022). Polyphenol-modified starches and their applications in the food industry: Recent updates and future directions. *Foods*, *11*, 3384.

Nikitina, E. V., Gabdukaeva, L. Z., & Egkova, M. S. (2015). The effect of enzyme modified starches on the quality and biosafety of meat products. *Вестник Казанского технологического университета/Journal of Agriculture and Environment*, *18*(16), 265–269. https://doi.org/10.23649/JAE.2023.40.12

Nimbkar, S., Leena, M. M., Moses, J. A., & Anandharamakrishnan, C. (2022). Medium chain triglycerides (MCT): State-of-the-art on chemistry, synthesis, health benefits and applications in food industry. *Comprehensive Reviews in Food Science and Food Safety*, *21*(2), 843–867.

Ochubiojo, E. M., & Rodrigues, A. (2012). Starch: From food to medicine. *Scientific, Health and Social Aspects of the Food Industry*, *18*, 355–380.

Otero-Guzmán, N., Cadena-Chamorro, E., Ciro-Velásquez, H. J., Mejía-Villota, A., & Rodriguez-Sandoval, E. (2023). Application of modified cassava starch as a fat substitute in cracker production. *Journal of Food Processing and Preservation*, *2023*(1), 6888180.

Paul, V., Rai, S., Vikranta, U., Nandan, A., Agarwal, A., & Tripathi, A. D. (2024). Starch-PHA blend-based biopolymers with potential food applications. *Starch - Stärke*, *76*(11-12), 2300131.

Phillips, G. O., & Williams, P. A. (Eds.). (2011). *Handbook of food proteins*. Elsevier.

Picolotto, A., Rossa Beltrami, L. V., Dallé, D., Ornaghi Júnior, H. L., Zattera, A. J., Catto, A. L., & Borsoi, C. (2024). Development and evaluation of biodegradable starch-based films containing cellulose nanocrystals/titanium dioxide nanoparticles as an alternative for food packaging. *Journal of Thermoplastic Composite Materials*, *37*(6), 1991–2012.

Popil, R. E. (2010). High stiffness surface coating optimization through starch encapsulation of platy kaolin. *BioResources*, *5*(4), 2738–2750.

Pranoto, Y., Paramita, B. L., Cahyanto, M. N., & Benjakul, S. (2021). Properties of ozone-oxidized tapioca starch and its use in coating of fried peanuts. *Molecules*, *26*(20), 6281.

Promhuad, K., Bumbudsanpharoke, N., Wadaugsorn, K., Sonchaeng, U., & Harnkarnsujarit, N. (2022). Maltol-incorporated acetylated cassava starch films for shelf-life-extension packaging of bakery products. *Polymers, 14*(24), 5342.

Punia, S., Sandhu, K. S., Dhull, S. B., & Kaur, M. (2019). Dynamic, shear and pasting behaviour of native and octenyl succinic anhydride (OSA) modified wheat starch and their utilization in preparation of edible films. *International Journal of Biological Macromolecules, 133*, 110–116.

Ramos, B., Miller, F. A., Brandão, T. R. S., Teixeira, P., & Silva, C. L. M. (2013). Fresh fruits and vegetables: An overview on applied methodologies to improve its quality and safety. *Innovative Food Science & Emerging Technologies, 20*, 1–15.

Rapaille, A., & Vanhemelrijck, J. (1992). Modified starches. In *Thickening and gelling agents for food* (pp. 171–201). Boston, MA: Springer US.

Rashwan, A. K., Younis, H. A., Abdelshafy, A. M., Osman, A. I., Eletmany, M. R., Hafouda, M. A., & Chen, W. (2024). Plant starch extraction, modification, and green applications: A review. *Environmental Chemistry Letters, 22*(5), 2483–2530.

Rather, J. A., Akhter, N., Rather, S. A., Masoodi, F. A., & Dar, B. N. (2024). Effect of xanthan gum treatment on the shelf-life enhancement of retorted meatballs (Goshtaba): A traditional meat product of India. *Measurement: Food, 13*, 100127.

Reich, G. (2004). Formulation and physical properties of soft capsules. *Pharmaceutical capsules* (pp. 201–212). Pharmaceutical Press.

Rivera Leiva, A. F., Hernández-Fernández, J., & Ortega Toro, R. (2022). Active films based on starch and wheat gluten (*Triticum vulgare*) for shelf-life extension of carrots. *Polymers, 14*(23), 5077.

Robles-Sánchez, R. M., Rojas-Graü, M. A., Odriozola-Serrano, I., González-Aguilar, G., & Martin-Belloso, O. (2013). Influence of alginate-based edible coating as carrier of antibrowning agents on bioactive compounds and antioxidant activity in fresh-cut kent mangoes. *LWT, 50*(1), 240–246.

Roman, L., Reguilon, M. P., Martinez, M. M., & Gomez, M. (2020). The effects of starch cross-linking, stabilization and pre-gelatinization at reducing gluten-free bread staling. *LWT, 132*, 109908.

Rupérez, D., Rivière, M., Lebreton, J., Aznar, M., Silva, F., Tessier, A., & Nerín, C. (2024). Synthesis and quantification of oligoesters migrating from starch-based food packaging materials. *Journal of Hazardous Materials, 476*, 135202.

Saeed, F., Afzaal, M., Khan, M. A., Niaz, B., Rasheed, A., Raza, M. A., & Arooj, H. (2024). Cereal starch-based nanoparticles in food packaging applications. In G. A. Nayik & A. H. Dar (Eds.), *Starch Based Nanomaterials for Food Packaging* (pp. 227–246). Academic Press.

Sajilata, M. G., & Singhal, R. S. (2005). Specialty starches for snack foods. *Carbohydrate Polymers, 59*(2), 131–151.

Santana, A. A., Oliveira, R. A. D., Kurozawa, L. E., & Park, K. J. (2014). Microencapsulation of pequi pulp by spray drying: Use of modified starches as encapsulating agent. *Engenharia Agrícola, 34*, 980–991.

Schebesta, H., & Purnhagen, K. (2024). *EU food law.* Oxford University Press.

Schmiele, M., Sampaio, U. M., Gomes, P. T. G., & Clerici, M. T. P. S. (2019). Physical modifications of starch. In M. T. P. S. Clerici & M. Schmiele (Eds.), *Starches for food application* (pp. 223–269). Academic Press.

Shaikh, M., Ali, T. M., & Hasnain, A. (2020). Comparative study on the application of chemically modified corn and pearl millet starches in white sauce. *Journal of Food Processing and Preservation, 44*(4), e14393.

Sharma, A., Yadav, B. S., & Ritika (2008). Resistant starch: Physiological roles and food applications. *Food Reviews International, 24*(2), 193–234.

Sharma, S., & Rao, T. R. (2015). Xanthan gum based edible coating enriched with cinnamic acid prevents browning and extends the shelf-life of fresh-cut pears. *LWT-Food Science and Technology, 62*(1), 791–800.

Shrivastava, A., Gupta, R. K., & Srivastav, P. P. (2024). Exploring novel frontiers of advancements in purple yam (*Dioscorea alata* L.) starch extraction, modification, characterization, applications in food and other industries. *Measurement: Food, 15*, 100196. https://doi.org/10.1016/j.meafoo.2024.100196

Silveira, T. M. G., & Tapia-Blacido, D. R. (2018). Is isolating starch from the residue of annatto pigment extraction feasible? *Food Hydrocolloids, 77*, 117–125.

Sun, Y., Yang, Y., Zheng, L., Zheng, X., Xiao, D., Wang, S., ... & Sheng, Z. (2022). Physicochemical, structural, and digestive properties of banana starch modified by ultrasound and resveratrol treatments. *Foods, 11*(22), 3741.

Syuhada, N., Yazid, M., Abdullah, N., Muhammad, N., & Matias-peralta, M. (2018). Application of starch and starch-based products in food industry. *Journal of Science and Technology,10*(2), 144–174.

Teklehaimanot, W. H. (2011). *Rheology and microscopy of low-calorie mayonnaise-type emulsions made with stearic acid-modified maize and teff starches.* [Master's thesis, University of Pretoria]. https://repository.up.ac.za/server/api/core/bitstreams/53545fc6-ebbe-4e5b-a560-404f2a442e02/content

Thaiwong, N., & Thaiudom, S. (2021). Stability of oil-in-water emulsion influenced by the interaction of modified tapioca starch and milk protein. *International Journal of Dairy Technology, 74*(2), 307–315.

Umesh Hebbar, H., Rastogi, N. K., & Subramanian, R. (2008). Properties of dried and intermediate moisture honey products: A review. *International Journal of Food Properties, 11*(4), 804–819.

Venkatachalam, C. D., Rajan, D., & Arunachalam, S. (2024). Optimization of legume based meat alternative using freeze structuring and evaluation of its quality indices. *Suranaree Journal of Science & Technology, 31*(1).

Vezzani, S. (2018). The international regulatory framework for the use of GMOs and products thereof as food aid. *European Journal of Risk Regulation, 9*(1), 120–136.

Vilpoux, O. F. (2023). International legislations for the use of modified starches in food and standards for native starches. In *Starchy crops morphology, extraction, properties and applications* (pp. 131–163). Academic Press.

Vilpoux, O. F., & Silveira Junior, J. F. S. (2023). Global production and use of starch. In Marney Pascoli Cereda and Olivier François Vilpoux (Eds.), *Starchy crops morphology, extraction, properties and applications* (pp. 43–66). Academic Press.

Vitolo, M. (2020). Enzymatic modification of starch. *World Journal of Pharmacy and Pharmaceutical Sciences, 9*(4), 1341–1358.

Walsh, S. K., Lucey, A., Walter, J., Zannini, E., & Arendt, E. K. (2022). Resistant starch: An accessible fiber ingredient acceptable to the Western palate. *Comprehensive Reviews in Food Science and Food Safety, 21*(3), 2930–2955.

Wang, K., Henry, R. J., & Gilbert, R. G. (2014). Causal relations among starch biosynthesis, structure, and properties. *Springer Science Reviews, 2*, 15–33.

Wang, S., Hao, Y., He, Q., & Gao, Q. (2024). Biodegradable starch-polyvinyl alcohol composite films by the incorporation of lignin for packaging applications. *Journal of Thermoplastic Composite Materials*, *37*(11), 3413–3435.

Zhang, B., Zheng, L., Liang, S., Lu, Y., Zheng, J., Zhang, G., & Jiang, H. (2022). Encapsulation of capsaicin in whey protein and OSA-modified starch using spray-drying: Physicochemical properties and its stability. *Foods*, *11*(4), 612.

Zhang, J., Zhu, L., Li, K. M., Ye, J., Xiao, X., Xue, M., & Chen, Y. H. (2022). Preparation of bio-based modified starch film and analysis of preservation mechanism for sweet cherry. *Food Chemistry: X*, *16*, 100490.

Zhang, M., Jia, H., Wang, B., Ma, C., He, F., Fan, Q., & Liu, W. (2023). A prospective review on the research progress of citric acid modified starch. *Foods*, *12*(3), 458.

Zhang, W., Cheng, B., Li, J., Shu, Z., Wang, P., & Zeng, X. (2021). Structure and properties of octenyl succinic anhydride-modified high-amylose japonica rice starches. *Polymers*, *13*(8), 1325.

Zhong, Y., Xu, J., Liu, X., Ding, L., Svensson, B., Herburger, K., & Blennow, A. (2022). Recent advances in enzyme biotechnology on modifying gelatinized and granular starch. *Trends in Food Science & Technology*, *123*, 343–354.

14 Practical Application of Modified Starch as Edible Ink in 3D/4D Printing in Food Industry

G Sai Preetham
R&D Division, Vidya Insta Coffee, Chikamagaluru, Karnataka, India

14.1 INTRODUCTION

New innovations continue to emerge in food engineering, bringing more competencies to the manufacturing of foods to increase customer choices and meet constantly growing consumer desire to enjoy foods in different ways. Foods are printed layer by layer using optimized rheological parameters to turn virtual human ideas into real structures of particular foods to consume for pleasure. Applications of 3D printing (3DP) and 4D printing (4DP) techniques have emerged through research. The cost of production, convenience, and nutrition, along with the most palatable flavours, are attracting consumers with dynamic food lifestyles (Singhal et al., 2020). Additive manufacturing (AM) techniques begin with predesigned computer aided design (CAD) of virtual shapes and use food sources as ink and additives to extrude the material from a printer (Liu et al., 2017). These techniques can also be called "food layer manufacturing" (Nachal et al., 2019). The various process principles involved in printing are discussed in this chapter. This process of 3D manufacturing is the layering of food along three spatial axes (x, y, z). Another dimension which of great importance to this technique is time, which produces dynamic structure using 3DP to create properties that are time dependent (Fahmy et al., 2023). This process of AM is called "4D printing of foods", or "time-responsive food printing". Based on extensive research, starch is modified to meet the requirements of specific applications by altering its functionality. Modifying starch to improve printability and structural integrity for 3DP/4DP of foods is encouraging for providing customization of texture and mouthfeel, and its compatibility with other ingredients can improve nutrition. As it is important for the starch to retain its fluid nature for printability (plastic behaviour) and strength after printing (pressure release), starch requires modification of its gelation behaviour and crystallization of starch components (Gulzar et al., 2023). This chapter offers comprehensive starch modification techniques for improved printability as edible ink in AM.

14.2 FUNDAMENTALS OF 3D/4D FOOD PRINTING

The approach for printing edible food follows multiple concepts to produce the final virtual-reality food product. Figure 14.1 illustrates how different levels of printing along different dimensions produce diverse food items with varied functionality. To better understand these printing properties, the concepts and methods of 3D and 4D printing will be explored in the following subsections.

14.2.1 Basic Principles of 3D Printing

The methodology of printing the edible foods has various principles to achieve the final food product. Extrusion-based printing techniques can be used for materials that have the ability to form gel at varied concentrations and that show thermoplastic behaviour. Droplets of a controlled-accumulation, liquid-like slurry material supply are used in inkjet printing. Binder jetting (BJ), also known as "liquid binding", involves selectively pouring a liquid binder to deposit powdered material layer by layer (Anandharamakrishnan et al., 2022). Selective laser sintering (SLS) involves the employment of a laser to fuse powder particles (Jonkers, van Dijk, et al., 2022). Fused deposition modelling (FDM) is the most prevalent AM method for polymers. This process involves extruding a filament of thermoplastic polymer through a nozzle and depositing it layer by layer to generate the final product. Multi-printing systems employ a coaxial-based system with two or more nozzles to print various materials on a single object (Tang et al., 2024).

14.2.2 Evolution of 4D Printing

3D food printing (3DFP) generally develops food structures in static conditions and has difficulty achieving dynamic changes over time. To address these constraints, researchers have turned their focus to 4D food printing (4DFP). Unlike traditional shape morphing of foods, 4DFP uses a printing method to accurately manipulate the shape, structure, and content of the food. Hybrid printing techniques

DOI: 10.1201/9781003564287-14

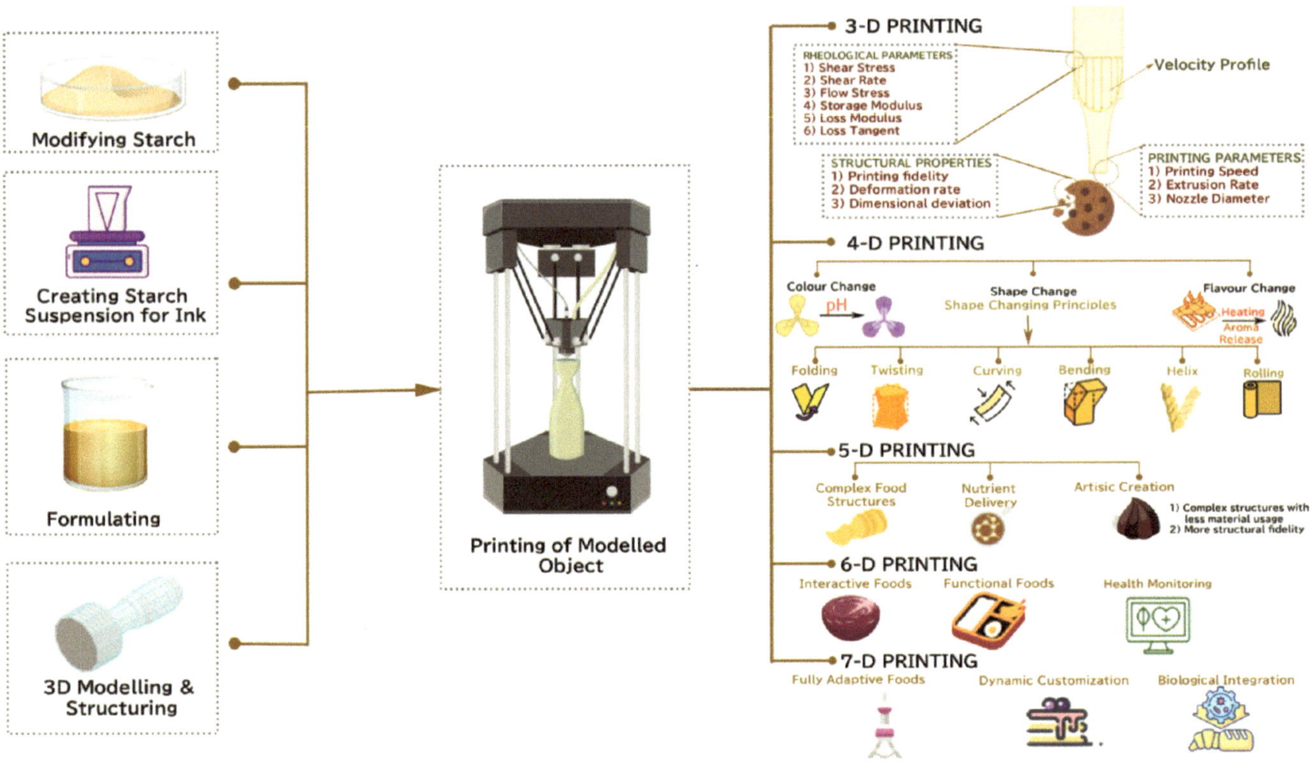

FIGURE 14.1 Approach for printing edible food follows multiple concepts to produce the final products.

involve 4DP techniques to obtain the programmable texture by using specialized formulations and smart materials in a printing ink that changes based on time. In response to changes in external factors such as temperature, pH, pressure, 3DP food properties such as colour, shape, and flavour are changed (Oral et al., 2021). Because of its self-transformation ability, the end product brings distinctive characters that improve the visual experience throughout the eating experience. For example, the edges of pasta petals may curl up and form a flower-like shape during boiling; the final dish can be served with the pasta arranged to resemble a blooming flower, adding a beautiful and artistic touch to a meal. The benefit of this technique is that it can construct suspended structures, which are not achievable with 3DP technology (Teng et al., 2021). This advantage helps advance this technology to improve culinary experiences, enabling reductions in waste and degradation (Singh et al., 2024).

14.2.3 Key Technologies and Techniques

14.2.3.1 Extrusion-Based Technologies

The pressurized material is printed layer on layer and heated to above its melting temperature so that it can act as a binder or adhesive when a new layer is printed. The system supports multi-material printing with the help of single or multiple nozzles (Sun et al., 2018). Pressure over the aperture of nozzle can be applied either using a motor-driven mechanism or with a pneumatic application. The ink should be colloidal with a highly viscous semisolid structure (Anandharamakrishnan et al., 2022); it should not lose texture and should not flow over the nozzle aperture without application of external pressure. Process variables for this technology are extrusion rate, nozzle speed, feed pressure, and temperature. Rheological variables are viscosity, consistency index (K), flow behaviour index (n), and shear force of ink. Product variables are degree of gelatinization, gel-forming ability, degree of emulsification, and plasticity of the ink. The configurations vary according to the properties of inks used for printing. Pneumatic systems can be employed for low-viscosity liquid foods, and syringe-based systems for high-viscosity inks (Anandharamakrishnan et al., 2022). Screw extrusion systems are used for printing starch gels from corn, potato, and wheat (Zheng et al., 2019). Extrusion is further categorized into hot and cold extrusion. Hot extrusion occurs by application of higher temperatures, whereas cold extrusion is done at room temperature. Products formulated with food ingredients such as chocolates, natural starches, and modified starches use hot extrusion, whereas ready-to-eat products use cold extrusion.

14.2.3.2 Inkjet Printing

In inject printing, the target material is forced through the smallest opening, or orifice, of the nozzle. This type of technology has two main configurations: drop-on-drop demand (DOD) (subclassified as drop-on-drop configuration and drop-on-powder configuration) and continuous inkjet printing (CIJP). In CIJP, ink splits into a stream of droplets due to capillary action when forced through orifice (Guo et al., 2017). Once the drops are out of orifice, their position can be controlled using electric field plates; ink used in this process should therefore be conducive to electricity. The droplets can be in the range of 50–80 μm and are created using a high-pressure pump for continuous flow of ink (Godoi et al., 2019). This process typically produces drop velocities of more than 10 m s^{-1} (Sun, Zhou, et al., 2015). When powdered flavours are to be printed on the substrate, particles can be infused together with ink droplets and expedited from the nozzle. DOD inkjet printing generates drops as needed by heat or piezoelectric actuation, making it more often employed. The normal falling velocity is between 5 and 8 m s^{-1}. DOD printing cause ink to dry at the nozzle during downtime, leading to particle deposition and blockage (Hsiao et al., 2014). In addition to providing for food artistry and high-precision printing, electrostatic inkjet printing enhances internal structure to produce texture in food. Additionally, it can print chocolate without the need for additives because it can print highly viscous materials (Vadodaria & Mills, 2020).

14.2.3.3 Binder Jetting

The principle behind binder jetting is the binding of powdered materials with a printing ink that act as a binder. The process consists of two major phases that are repeated using a layout from a CAD file until the product is completed. In the first phase, the recoater applies the target material layer by layer, distributing the powdered material to the construction area with great precision. In the second phase, the printhead applies the liquid binder and allows the binder material to come into contact with each grain of the moulding (Varvara et al., 2021). During post-processing, the printing structures are heated by radiation to provide structural integrity and strength (Sun, Peng, et al., 2015). The typical food materials used in this printing are starches (corn and wheat starch), cellulose (ethyl and methyl cellulose), sugars, dextrins, gums (xanthan and guar), brown rice gel, brown rice flour (Huang et al., 2020), and pea protein isolate (Hussain et al., 2021), which can be aqueous or alcohol-based. The binder variables should be optimized during printing to avoid bleeding of excess binder and to construct the structure with good mechanical strength. The process variables include nozzle type and diameter, pressure of jetting binder, layer shape and thickness, printhead types, and printing rate. Product variables include fluid properties of the binder such as viscosity, wettability, flowability (Godoi et al., 2016); droplet size, shape, spacing, and velocity (Shirazi et al., 2015); binder concentration, surface tension. and contact angle; and target material particle size distribution.

14.2.3.4 Selective Layer Sintering

One of the most widely used additive manufacturing techniques is selective laser sintering (SLS). It is a solvent-free, reasonably fast method. Metal components and tissue scaffolding are the most typical uses of this technique (Farooq & Ranjan, 2023). It involves utilizing a laser as a power source to fuse or sinter small layers of powdered materials over a printing bed platform (Farahani et al., 2018). This method makes it possible to precisely regulate the mechanical qualities of items and create their form. Therefore, SLS provides special opportunities for food product texture adjustment (Jonkers, van Dommelen, et al., 2022). The two phases of this method are layering of target material, using either a blade or a roller, and using a laser as a source that is scanned using a previously created CAD file to fuse or stinter the particles to create a first overlayer as a 2D cross-section. Then new powdered material is layered on the first layer to create a new layer along the *z*-axis. This procedure is repeated until a fully developed structure is formed. SLS provides better spatial resolution at the small scale, which increases the freedom to locally modify mechanical characteristics. Safety regulations are still being developed for foods produced using SLS systems as laser at high intensities may induce chemical reactions in foods that may make the final product inedible. The process variables include the laser preheat temperature, laser power and velocity, scanning direction of layers, energy density, hatch space (distance between two sintered lines) (Gueche et al., 2021), chamber temperature (Barakh Ali et al., 2019), hot air temperature, and velocity (Zhang et al., 2019). The product variables include melting temperature and particle size distribution. Rheological variables include fracture, yield stress of the ink, and stiffness.

14.2.3.5 Fused Deposition Modelling

Multiple methods are based in thermal extrusion-based technology to melt a filament and then allow it to harden. These include fused deposition modelling and fused filament fabrication, which both use a process of melting solid thermoplastic filaments and depositing them layer by layer until the desired structure is formed when the filaments solidify (Mustapha & Metwalli, 2021). The FDM method is now commonly utilized in 3D food printers, where slurry, such as liquid materials or paste, is continually emerging from the moving nozzle and layered while cooling (Lee, 2021). Some studies have quantitatively linked stable printed goods and sufficient extrusion performance to the physical characteristics of food compositions. Schutyser et al. (2018) used an FDM printer that contains a dispenser set on an *xyz* table to extrude material for the creation of solid objects along the third dimension, printing sodium caseinate with different additives and modifications. As this method is extrusion-based, it is relatively cheap, and polishing of surfaces post-processing is required to give a shiny finish. The different configurations of this type of method include auger-based, in which a screw-based auger system drives the ink through the nozzle using to the force exerted by moving volume; syringe-based, using air to push a plunger, exerted after filling the ink (continuous production of printing is not possible in this configuration, but it can handle high-viscous inks such as dough); and pressure-based, which utilized air pressure to push the ink through the nozzle (this method is able to handle low-viscosity foods). The process parameters for the printer include build orientation, air gap, temperature (extrusion and platform), raster angle, speed (extrusion and printing head), and flow rate (Schutyser et al., 2018). The product parameters include glass transition temperature (T_g), viscosity, flowability and shear stress, degree of gelation, moisture content, shape retention, phase angle (δ), and yield stress (τ_0) (In et al., 2021). The rheological parameters include fracture strength, structure stability, hardness, adhesion between layers, impact strength, tensile, and compression nature of product.

14.2.3.6 Multi-Material Printing

Multi-material 3D printing is largely based on extrusion printing, using multiple printheads to print different ingredients or multiple heads sharing nozzle design to print different ingredients simultaneously. Coaxial printing involves coordinating multiple materials to print a stable structure and a double extruder system in which two nozzles are mounted on a single printhead to print simultaneously in a shared nozzle system. The shared nozzle design was introduced by Skylar-Scott et al. (2019). More extruders are necessary to provide regulated material deposition and the production of multi-textured and multi-flavour goods. A multi-printhead printer may be used to create very appealing multi-material structures with more geometric complexity and an appealing look (Liu et al., 2018). In general, separate CAD files create .stl formats and merge them together to feed into the printers; a second option is to assign separate roles to be performed by separate extruders. As different inks have different fluid properties, it's difficult to program and print multiple materials with differing extrusion parameters, and difficult to control the material distribution. To overcome these challenges, Lee et al. (2024) introduced a multi-channelled nozzle with a rapid ink-changing system, providing a continuous ink supply of different materials. The authors also reported that the differences in the rheological parameters of two different inks can cause backflow of a higher-yield stress ink into the channel of a lower-yield one, and this can be overcome by different nozzle channel sizes. In the case of starch, Yang et al. (2019) used Polyflow to simulate the effect of lemon juice gel formulations on different sources of starch (wheat, potato, corn, and sweet potato) for process parameters that included pressure field distribution, velocity, and shear rate. Unusual taste combinations can also be delivered using this system when printing ingredients such as doughs, gels, and pastes (Waseem et al., 2024).

14.2.3.7 Hybrid Printing Technologies

In the context of 3DP/4DP, hybrid printing techniques are defined as the integration of various printing processes and technologies into a single system, allowing the construction of complex food structures that would be impractical to create with a single printing approach. Combining extrusion and inkjet printing techniques, for instance, a food item's foundation shape can be formed via extrusion, then layers of distinct flavours or colours can be accurately printed on top using inkjet printing to provide a multisensory experience. Combining BJ and SLS makes it possible to produce food products with precisely regulated textures and intricate interior structures. The production of complex shapes is made possible by the binder jetting step, while SLS offers the structural stability and crispness that some food products may need. A hybrid printer might employ SLS for the exterior layer and extrusion for the inside to create a combination of a crunchy outer shell and a soft, creamy interior. This multifaceted technique provides for greater flexibility in food design by combining the benefits of several printing processes to improve the functionality, aesthetics, and nutritional content of printed foods. Guo et al. (2022) investigated the shape transformation in starch products printed through 4DP and employing microwaves (200 W) for shape transformation and simultaneous colour and aroma change. He et al. (2020) investigated a purple potato puree for shape transformation, where the important factors such as salt content and microwave power increased the dehydration and deformation angle in printed samples, and increased butter content decreased the deformation angle. This feature creates new opportunities for dynamic culinary experiences in which consumers may engage with food in novel ways, such as watching it change shape on the plate.

14.2.4 Comparison with Traditional Food Manufacturing

14.2.4.1 Efficiency and Precision

3DFP and 4DFP are more efficient and precise than traditional food manufacturing procedures. Traditional methods sometimes require many processes, such as mixing, moulding, and cooking, which can be time-consuming and inconsistent. In contrast, 3D/4D printing allows for the accurate layer-by-layer deposition of food ingredients, allowing detailed shapes and complicated structures to be created with minimum operator involvement. 3DP/4DP automates the manufacturing process, eliminating the requirement for human labour and lowering the risk of mistakes. The capacity to regulate the printing process at the tiny level guarantees that each food item is manufactured to exact specifications, resulting in consistent shape, size, and texture.

14.2.4.2 Customization and Personalization

. 3D printing has revolutionized manufacturing by enabling highly customized and personalized product development to meet specific consumer needs. Unlike conventional mass production, it allows for the design and fabrication of unique items with tailored shapes, sizes, and functionalities, making it especially valuable in fields such as healthcare, food, packaging, and consumer goods. In food systems, 3D printing can personalize nutrition by designing meals with precise compositions based on individual dietary requirements. Similarly, in packaging and biomedical applications, it facilitates the production of bespoke structures that enhance functionality and user experience. This flexibility not only supports on-demand production and reduced waste but also opens avenues for innovation and consumer-centric design.

14.2.4.3 Waste Reduction and Sustainability

Starch-based inks are precisely deposited during the printing process, ensuring that only the required amount of material is used where starch and other ingredients could be wasted due to excess material, offcuts, or inefficient shaping processes. Modified starch can be derived from various agricultural by-products, such as potato peels, corn husks, and other starchy residues. Incorporating these starches into 3DFP/4DFP allows for the upcycling of what would in traditional manufacturing be waste, which not only reduces overall waste but also adds value to the by-products, contributing to a more circular economy within the food industry. Traditional manufacturing often involves producing large quantities of starch-based foods, which may require significant packaging to preserve freshness during transportation and storage. By contrast, on-demand printing enables the production of fresh, ready-to-consumer products, thereby reducing the environmental impact associated with packaging and food waste from unsold products.

14.3 PREPARATION OF MODIFIED STARCH FOR EDIBLE INK

14.3.1 Selection of Starches

Selection of starch sources is very important for the efficiency of printing and final product characteristics (Table 14.1). Efficiency is largely linked to the degree of gelatinization, as the crystalline and amorphous structure of granules has a high impact on rheological properties such as storage modulus (G′) and yield stress (τ_y) (Gupta et al. 2024e). Starch which has a more self-supporting structure is better for printability, which can be attributed to better shear recovery rate during gelatinization. Cheng et al. (2023) showed that leached amylose content of corn starch increased the formation of new crystalline structures by cross-linking and new crystal formation increased G′. The same was observed in the rice starch, where increased gelatinization decreased crystallinity, resulting in lower printability (Zheng, Tang, et al., 2022).

14.3.2 Modification Process

14.3.2.1 Physical Modification

Physical alterations of starch are those caused by physical treatments that do not result in any chemical changes to the starch other than restricted glycosidic bond cleavages (BeMiller, 2018). Schmiele et al. (2019) reported changes including in solubility, swelling power, polymorphism, freeze and shear stability, temperature of gelatinization, and paste viscosity. Typically, physically modified starches are manufactured by heating native starches by spray drying, extrusion, or drum drying. The crucial component, residence time, determines the degree of gelatinization in all three instances (Zia-ud-Din et al., 2017)(Zia-ud-Din et al., 2017; Gupta et al., 2023; Shrivastava et al., 2024). Drum drying normally consists of passing a starch slurry over steam-heated rollers, where the thermal characteristics rapidly gelatinize and dry the starch slurry and annihilated into powder (BeMiller & Huber, 2015). The most adaptable method is extrusion, in which the slurry goes through a screw die assembly system via various temperature zones, causing starch to rapidly expand owing to internal pressure differences and to gelatinize due to temperature, lowering the cost of manufacturing (Zia-ud-Din et al., 2017).

Hydrothermal modifications include annealing (ANN), and heat-moisture treatment (HMT). AN involves pretreatment of starch with a solvent and approaching glass-transition temperature (T_g), at which the susceptibility of amorphous regions increases due to initial wetting, resulting in reduction of imperfect starch crystals; recrystallization occurs, making a more perfect crystalline nature. HMT involves heating the starch above T_g at temperatures of about 80°C–140°C and with low moisture content (10%–35%). Starch with tuber and root origins are more susceptible to HMT, where major changes occur in type B starches that result in type A crystals due to the difference in water molecules present in type A and type B starches. Cold water soluble (CWS) treament involves the treating of starches at atmospheric pressure with water and polyhydride alcohol at temperatures ranging from 145°C to 155°C for minutes to hours (Rajagopalan & Seib, 1992), or at elevated pressures

TABLE 14.1
Application of Different Modified Starches in 3D/4D Printing of Foods

Type of Starch	Method of Modification	Improved Printing Parameters	Reference
Quinoa and brown rice starch	Dry heating	Good shear thinning properties than control with ease of extrusion	Huang et al., 2024
Wheat starch	Dielectric plasma treatment	Better printing accuracy and surface texture	Ma, Ma, et al., 2024
Pea starch	Cross-linking	Good oil holding capacity with better printability	W. Liu et al., 2024
Potato starch	Spray drying	Structural stability during printing and cooking	P. Liu et al., 2024
Maize starch	Addition of sodium alginate	Higher G′ value	Cui et al., 2024
Rice starch	Effect of addition of peanut protein	Improved viscosity, firmness of paste for printing	Kadival et al., 2024
Quinoa starch	Addition of xanthum gum	Improved microstructure, texture properties with enhanced printing accuracy and stability	Cheng et al., 2024
Wheat starch	Annealing	Decrease in printing efficiency and rougher edges	Zheng et al., 2023

up to 35–52 atm (Majzoobi & Farahnaky, 2021), to obtain a cold solubility of up to 95% (Sarifudin et al., 2021).

Novel techniques include cold plasma generation, pulsed electric fields (PEFs), ultrasound, ozone, and irradiation. Ozone has an advantage of being processable in aqueous and gaseous phases where changes that occur due to the different forces caused by the carbonyl and carboxyl groups lead to depolymerization, decreased apparent viscosity, and modification in molecular size; this process improved strength of hydrogels by 250%–300%, thereby improving 3D printing efficiency (Maniglia et al., 2019). Granular morphology, lamellar and crystalline structures, and double helices experience variable degrees of modification/disorganization during HHP; however, these changes are distinct from thermal modification by an improvement at the same gelatinization degree.(Ma et al., 2022). The PEF approach involves exposing a sample held between two electrodes to electric fields by applying brief high-voltage pulses (μs to ms). Zeng et al. (2016) observed surface roughness and pits in PEF-treated waxy rice starch, and the same was observed in esterified potato starch (Hong et al., 2018). Major characteristics include surface damage; decreased crystallinity, viscosity, and gelatinization temperature; and structure breakage. Under ultrasound, the starch–water system undergoes cavitation, generating intense shear stress, high temperatures, and free radicals, which alter the structure and characteristics of starch. The amount of change relies on the frequency and strength of ultrasound, the length and settings of ultrasonication, the system's temperature and moisture content, and the starch type. Cold plasma, or non-thermal plasma, is a partially ionized gas created by applying sufficient energy to a neutral gas at low or atmospheric pressure. Cold plasma causes surface etching, irregular shapes, fissures and cavities, reduced thermal properties, and crystallinity with increased peak viscosity.

14.3.2.2 Chemical Modification

14.3.2.2.1 Esterification

Acetylation is to the process of adding an acetyl group to a starch hydroxyl group by removing hydrogen, resulting in an ester attachment as a functional group. Potential sources include the addition of acetic anhydride, octenyl succinic anhydride, and other acetyl groups for various sources of starches with NaOH as a catalyst (Acevedo-Guevara et al. 2018; Ojukwu et al. 2023; Zhao et al. 2018; Gupta et al., 2024c). Major changes include decreased crystallinity but enhanced solubility and swelling power based on the percentage of substitution and the ratio of amylose to amylopectin. Incorporation of acid also improves the rate of degree of substitution of starch (chain length 6–8), showing increased solubility and swelling power. So based on the degree of substitution (DS), this method can be classified as low DS (0.01–0.2), medium DS (0.1–0.3), and high DS (2–3). Esterification methods include attachment of various hydrophilic functional groups such as carboxymethylation, hydroxypropylation, and hydroxyethylation.

14.3.2.2.2 Etherification

In 1850 Alexander Williamson's invention of a nucleophilic substitution reaction between an alkoxide and an alkyl halide etherifies the hydroxyl groups of starch (Kim & Jung, 2022). The etherification of starch's hydroxyl group can yield considerably diverse results depending on the reaction solvent employed to assess starch particle size, substitution rate, and deformation force. As stated in above section, this reaction also requires a catalyst such as NaOH, and its concentration can decide the degree of etherification in some cases. Starch ethers are easier to synthesis into DS substitutions in water than starch esters because of their compatibility with the alkylating reagent and water solubility (J. Xu et al., 2020).

14.3.2.2.3 Cross-Linking

Cross-linking involves addition to native starches of agents that create intermolecular links of esters and ethers between the starch molecules and hydroxyl groups, yielding a stabilized granular structure and controlled swelling. It is presumed that cross-linking strengthens the hydrogen bonds in the starch granules, which serve as a connection between the molecules (Tang et al., 2018; Gupta et al., 2024a). Factors affecting the degree of cross-linking include agent concentration, source of starch, pH, temperature of reaction, and scale of substitution. Agents include epichlorohydrin, sodium trimetaphosphate, phosphoryl chloride, and sodium tripolyphosphate (Koo et al., 2010). After the cross-linking reaction, the molecules are joined by various bonds: ionic bonds, covalent bonds, and other intermolecular forces (Siroha et al., 2021; Gupta et al., 2024d).

14.3.2.2.4 Acid Modification

Acid modification involves treating native starch with acids to hydrolyze the amorphous regions, enhancing the double helices and thereby increasing the crystallinity of starch. Acid-hydrolyzed starches are typically made by subjecting starch granules to varying durations of treatment with diluted aqueous or alcoholic solutions of mineral acid (sulphuric or hydrochloric acid) below the temperature of gelatinization (Wang & Copeland, 2015). Acid modification increases relative crystallinity in the initial stages, especially in the central region. Decreased swelling power is also observed during acid hydrolysis of starches, with a trend of initial increase then decrease over reaction time being common in potato and cereal starches. It is possible that the lower swelling capacity of acid-hydrolyzed starch, which delays the formation of viscosity, is the cause of the rise in pasting temperature.

14.3.2.3 Enzymatic Modification

Enzymatic reactions with gelatinized starch can modify molecular mass, branch chain length distribution, and amylose-to-amylopectin ratio (Punia Bangar et al., 2022). The disadvantage of acid modification is recovery of acid from the modified starch, and there is no control of the formation of side products generated due to the reaction, but in the case of enzyme treatment, reaction occurs at low activation energy and there is no formation of side products with ease of enzyme retrieval. Mechanisms of enzymatic modifications include branching and debranching, transfer, and hydrolytic cleavage. Glycosyltransferases improve the high branch modification of the distribution and ratio of glycosidic linkages (Wu et al., 2024). The change mediated by glucan branched enzymes resulted in a decrease in average chain length and relative crystallinity, as well as a time-dependent rise in the ratios of α-1,6 links to total glycosidic linkages, which improved the characteristics of slowly digestible starch and resistant starch (Li et al., 2018).

14.3.2.4 Mechanical Processing

Mechanical processing involves size reduction measures for native starches by several principles such as impact force, abrasion, shear, and compressive forces. The changes that occur on the starch include surface roughness, cracks, and increased granule size. In addition, compared to native starch, mechanically modified starch exhibits reduced enthalpy and temperature and greater rates of water solubility, water absorption, and enzymatic availability (Wasserman et al., 2022). Degraded crystallinity with reduced viscosity and shear thinning of cassava starch was observed during mechanical activation (Huang et al., 2007).

14.3.3 Effect of Additives on Starch Modification

Recently, heightened attention has been given to modifying the physicochemical characteristics of starch through various processing techniques and the incorporation of additives such as sugars, hydrocolloids, proteins, and lipids. Different starch and non-starch components also react, either those present naturally in the ink or those intermittently added to improve the ink properties for ease of printing, to bring a change in rheological characteristic of ink or to get a desired characteristic in the final product for end use. These additives are utilized in synergy, enhancing storage stability and delaying starch digestion by establishing physical obstructions and steric hindrance. The effect of utilization of these additives has been a focus of interest for many scientists, and many new fascinating results have been shown for end application use. Additives from natural sources have received keen interest as clean-label ingredients for finished consumer products.

14.3.3.1 Effect of Polysaccharides

The use of polysaccharide-modified starch has implications for the pasting behaviour, rheological properties, texture, and in vitro digestibility of starch gels. Key factors affecting these properties include the type of starch used and the specific polysaccharides incorporated. Incorporation of soluble soyabean polysaccharide reduced peak viscosity and formed instable gel structures in kudzu and lotus starches (Liu et al., 2019). The impact of konjac glucomannan on corn starch was studied by Ma et al. (2019), who reported that both storage module (G′) and loss modulus (G″) were improved. The impact of cordyceps polysaccharides on wheat starch reduced digestibility (Kong et al., 2020). In similar way, Huangjing polysaccharides on sweet potato starch reduced crystallinity and hardness (Zhang et al., 2024). The reduced breakdown of peak viscosity and hardness for potato starch was increased with its ratio when it was treated with fig seed polysaccharide (Liu et al., 2021).

14.3.3.2 Effect of Salt Ions

The influence of salt on modified starch plays a crucial role in altering structural, functional, and physicochemical properties, impacting starch behaviour in various food and industrial applications. The addition of salt can affect the fluid phase and gelatinization process of starch granules. Ca^{+2} after sonication enhanced pasting viscosity, which was reduced in the presence of Na^{+} ions (especially after sonication) due to increased relative crystallinity in starch granules in the case of tapioca starch (Abedi & Pourmohammadi, 2020). Unlike Na^{+} ions, Ca^{+2} ions significantly increase viscosity, as Na^{+} ions reduce yield stress by up to 60.4% whereas Ca^{+2} ions increase it by up to 100.8% (Abedi et al., 2024). In comparison to the starch–water system, the addition of calcium chloride raises T_g levels, and Ca^{+2} ions were effective in controlling gelatinization temperature in the case of potato starch (Chuang et al., 2015).

14.3.3.3 Effect of Lipid Content

A substantial increase in the glass transition temperature of starch gel can be observed upon the addition of fatty acid. Two considerations explained this: the first was the amylose–lipid complexes' physical cross-linking action in the starch–water system; the second was the possible impact of the simple fatty acids' amphiphilic nature on the water distribution in the gel structure (Arik Kibar et al., 2014). Fatty acid addition significantly increased the T_g'' of starch gel in corn starch. Increasing the percentage of added fatty acids (0.5%–1.5%) decreased the peak viscosity due to the restraining of starch granules from hydration and further swelling, but the final viscosity showed a consequent increase on the addition of both stearic and linoleic acids. Starch–lipid interaction forms, with amylose creating a V-shaped crystalline structure by establishing an inclusion complex with amylose's helical structures.

14.3.3.4 Effect of Polyphenols and Protein Content

Protein also has a similar nature of starch when hydrated and has similar influencing factors, such as pH, temperature, heating curve, additives, and salts. Proteins of dairy origin have been a major ingredient of interest for much research due to ease of availability and lower costs. Ji, Xu, Liu, et al. (2022) reported the addition of casein protein in cassava starch causing the relaxation time to decrease and the free water content to rise. This gave the gel system a higher elastic modulus and, as a result, improved the accuracy of 3DP foods. The G′ was lowest and tan δ values higher in the case of protein-rich faba bean flour, which was attributed due to the excess swelling and structure instability. When oleogel at 20% was added to the starch–protein system (potato starch:whey protein:water = 3:1:5), the greatest combination of parameters for the best printing outcome were 20% for 30°C printing temperature, 40 mm/s printing speed, and a 1.2 mm nozzle diameter (Shi et al., 2021). The interactions of corn starch with salmon protein isolate were investigated, and it was reported that different concentrations raised the temperature of protein denaturation, which affected gel shear recovery. With a 0.91 mm nozzle diameter and 30 mm/s printing speed, 15% corn starch and 3%–4% salmon protein isolate produced the best dimensional and textural stability (Carvajal-Mena et al., 2023). Whereas polysaccharides such as catechins and procyanidins also improve extrudability in rice starch, where increasing concentration decreased the G′ value, layer deposition was hindered, perhaps due to a reduction in gel hardness and stiffness, necessitating a balance between the durability of the layers and their extrudability (Zeng et al., 2021). Catechins also improved the resistance to digestion during 3D printing of rice starch (Zheng, Liu, et al., 2022). The addition of wheat protein to wheat starch in combination with caffeic acid enhanced the structural recovery rate of yield stress (τ_y), flow stress (τ_f), and consistency factor (K), leading to smooth extrusion during 3DP.

14.4 PROPERTIES AND TYPES OF MODIFIED STARCH

14.4.1 Chemical Structure and Composition

14.4.1.1 Amylose and Amylopectin

When starch undergoes gelation in starch–water interaction systems, the changes that occur in amylose and amylopectin drives many properties of starch for application (Gupta et al., 2024b). During thermal degradation of corn starch, the leaching of amylose from starch granules was detected, resulting in a filamentous network, which is responsible for the rise in perceived viscosity (Rincón-Londoño et al., 2016). Amylose reduces grain swelling, averting morphological changes (Kong et al., 2015) due to the amylose's double helicoidal shape preventing water from entering. Amylose was more susceptible toANN treatment than amylopectin due to higher molecular ordering in case of rice starch (Guo et al., 2020), which was confirmed by FTIR. Starch gels (cassava) with a greater proportion of short amylopectin chains ($13 < x \leq 24$) and a peak polymerization degree of amylose chains ($5000 < x \leq 20{,}000$), were shown to improve the rheological characteristics of starch gels, facilitating smoother extrusion in 3D printing. A smaller overall molecular size of whole starch ($\bar{R}h$) and a higher ratio of amylose chains ($100 < x \leq 1000$) enhanced water distribution within the gels, leading to greater precision in 3D-printed outputs (Ji, Xu, Li, et al., 2022). Ma et al. (2023) also reported that the short chain ratio in amylopectin increased following radio frequency treatment. The introduction of amylose in combination with native starch showed lower flow stress (τ_f), indicating better printability, but upon increasing the ratio of amylose starch to native starch, the extrusion lines slanted, which reduced the self-supporting capability of the structure (Xian et al., 2024; Gupta et al., 2025a).

14.4.1.2 Starch Concentration

Starch concentration is an important factor in deciding the necessary viscosity for extrusion after gelation. At a starch content of 3.6%, ozonated starches have been shown to have lower viscosity than native ones, with cassava starch showing the lowest viscosity after 15 and 30 minutes of ozonation. After 15 minutes of ozonation, cassava starch with a concentration of 10.7% gelatinized similarly to native starch at 85 and 95 minutes (Maniglia et al., 2019). Potato starch samples with a concentration of 15%–25% at 70°C showed outstanding printability, with the degradation of the original granule, crystalline, and lamellar structures and the development of a homogeneous, compact gel network (Liu et al., 2020). The G′ and G″ considerably increased with the increase in potato starch content from 10 g/100g to 20.224 g/100 g when characterized using small-amplitude oscillatory frequency sweep mode, indicating stronger network architectures and more solid gel behaviour (Yang, Zhang, Bhandari, et al., 2018). Potato, rice, and maize starch paste with concentrations of 15% to 25%, if heated to 70°C–85°C, had acceptable 3D printability. It can be concluded that a suitable starch suspension can be affixed at results of tan $\delta < 1$, showing elasticity dominant behaviour with higher yield stress values to avoid broken threads during printing (Liu et al., 2020).

14.4.1.3 Effect of Non-Starch Constitutes

During 3D printing, different non-starch constitutes are added as additives such as gums (guar gum, xanthan gum, κ-carrageenan, sodium alginate, agar, gum arabic, locust bean gum, gellan gum), proteins (pea protein, soya protein, wheat protein isolates, egg white protein), salts (NaCl, $CaCl_2$, magnesium stearate, citric acid, transglutaminase), lipids (cheese, butter, egg yolk, olive oil), and carbohydrates (chocolate, mango powders, sucrose) to bring about desired characteristics for printing. Adding gums increases apparent viscosity, G′, and G″. Objects printed with blends containing κ-carrageenan have shown maximum integrity to the desired geometry and strong loading-bearing capability, preventing collapse over time due to the appropriate G′ value, while gum arabic decreased apparent viscosity in the case of orange concentrate wheat starch samples (Azam et al., 2018). The addition of konjac and xanthan gum for 3DP of sturgeon fish improved shear thinning and viscoelastic behaviour through hydrogen bonding (Yu et al., 2023). Adding guar gum to starch can improve its viscosity, increase stability during heating and stirring, and minimize starch regeneration (Chen et al., 2019). Incorporating gums into mashed potato enhanced its resistance to deformation, resulting in a lower maximum creep %strain except in the case of xanthum gum (Liu, Zhang, & Bhandari, 2018). But up to 0.3% xanthan gum showed that printing the foods for dysphagia diets at lower concentrations improved G′ and yield stress, forming uniform microstructure (Liu et al., 2023). The heat resistance of a 3DP cookie sample was increased by adding 0.5 g/100 g of xanthan gum to the raw material (Kim et al., 2019). Rheology studies revealed that adding peach gum polysaccharide at a rate of 6% might significantly increase the apparent viscosity and thermal stability of the product, improving 3D printability and supportability; the result exceeded printing accuracy by 90% (Zhou et al., 2023). Overall, inclusion of gums improves mechanical properties with increased apparent viscosity and helps to ease extrusion for high-expansion base inks of vegetable origin (carrot, broccoli, and spinach) (Kim et al., 2019), which would help in printing foods of different shapes for people with dysphagia. Different protein origins interact with the starch-based gels to further improve the properties of ink to create the optimized texture for efficient 3DP. The viscosity of soya protein isolate (SPI) doughs increased with greater concentrations due to increased entanglements in the samples. Increasing the concentration of SPI in the food dough formulation might improve the self-supporting ability and structure preservation of printed structures (Carranza et al., 2023). The inclusion of gelatine (2% concentration) and higher printing temperatures increased the creation of strong linkages between soy protein particles, resulting in a dense 3D structure in SPI–based pastes with improvement in G', τ_y, and τ_f (J. Chen et al., 2022). Whey protein isolate reduced the apparent viscosity of milk protein concentrate and provided the best extrudability (Y. Liu et al., 2018). Pea protein isolate at 60% intensified the interactions between protein and polysaccharide, resulting in a more strong and stable gel network, and improved the rheological properties and textural characteristics of the mango pulp, showing gel structure at 0.84 mm nozzle diameter and 35 mm/s printing speed (Ren et al., 2023). Approximately 1.3 g of NaCl per 100 g of dry material (wheat starch and egg white powder) is a strong enough concentration to improve the impression of salt in layered 3D-printed items (Fahmy et al., 2021). Denaturation of the myosin tail in surimi-based gel was caused by the addition of NaCl; viscosity was lowered by increasing the salt content, which made it easier for the gel to extrude from the nozzle (Wang et al., 2018). High liquid oil concentration affects paste viscosity, making 3DFP challenging. Improper oil-to-water ratios can lead to phase separation in multi-component systems (Liu & Ciftci, 2021). The combination of rice starch and glycerol monostearate increased hydrophobic contacts between starch and lipids, generating a V-type starch–lipid complex with a compact and organized structure, thereby improving enzymatic resistance to thermal shear in hot extrusion-3DP (Liu et al., 2022).

14.4.2 Starch Esters

Starch esters can be modified to possess particular flow behaviour, which is important for the extrusion-based method prevalent in 3D food printing. Microwave-assisted potato starch modification exhibited 52.8% less swelling power compared to native starch and showed the highest power at a lower temperature of 50°C. (Zhao et al., 2018). Buckwheat starch with hydroxy methyl pectin was microwave-treated with Ca^{+2} ions, and the different degrees of esterification in the pectin carboxylate groups were confirmed from FTIR results. The interaction between Ca^{2+} and high methoxy pectin reduced the strength of hydrogen bonds, lowering the viscosity of gels, and this property can be an advantage in 3DFP (Guo et al., 2021b). Starch nanocellulose esterification with citric acid resulted in increased tensile yield strength and Young's modulus (Ma et al., 2008) while reducing elongation (Ghanbarzadeh et al., 2011) to create perfect parameters for printing. Kudzu starch–gelatine coacervates modified by octenyl succinic acid esterification improved the gelatinization temperature (Zhao et al., 2019), and similar outcomes were observed in potato starch–whey protein fibrils (Chen et al., 2020). Starch treatment after esterification exhibits good thermoplasticity and hydrophobicity, and this improvement of thermoplasticity represents the self-supporting characteristic for the printing structure (Wu et al., 2023). Combining alginates with varying guluronic acid–to–mannuronic acid ratios with pectin with varying esterification degrees might lead to the discovery of a novel printable material for the construction of food structures (Godoi et al., 2016).

14.4.3 Starch Ethers

Introduction of ether groups can enhance starch's stability, water retention, and viscosity, which are critical factors in food processing and 3DP. The main advantage of etherification is the reduction of inter- and intramolecular hydrogen bonding (Fonseca et al., 2015), which helps in maintaining the starch stability (Schmitz et al., 2006) and facilitating easy gelatinization of starch (Haroon et al., 2016), showing the advantage of introducing the water-and-oil-based emulsion. Starch etherification has the capability to produce clear paste exhibiting enhanced freeze–thaw stability and resisting extreme conditions for shear and temperature (Sinhmar et al., 2023), creating a path into a new era of applications in 3DP. However, due to use of chemical additives to modify starch, not many applications have been researched compared to physical and other green modification technologies.

14.4.4 Cross-Linked Starches

Cross-linked starches are molecules that have been chemically modified with the purpose of forming intermolecular bonding amongst the starch molecules (Gupta et al., 2022). This greatly enhances their stability and textural functionality under varying conditions of processing, making these types of starches uniquely suited for the production of structured and stable printed food products. Rice starch gel cross-linked with other starches at 12% concentration at 4.475% amylose content exhibited improved printing accuracy (Xiao et al., 2024). Oxidized starch cross-linking with chitosan using the Schiff base method underwent rearrangement throughout the printing environment, demonstrating a greater degree of cross-linking and resilient high structural recovery (> 95%) (Zhu et al., 2024). Dry spraying transglutaminase for surface cross-linking to print surimi gels improved the texture, creating hydrophobic fine and dense interactions, suitable water distribution, and enhanced printable height efficiency, by 97.50% (Zhu et al., 2023). Microwave-assisted cross-linking of tapioca starch with glutaraldehyde as a cross-linking agent, printed using BJ 3DP, exhibited high mechanical strength up to 32 MPa after microwaving for 70 seconds (Cheng & Liao, 2022). Maize starch photo-cross-linked with methacrylic anhydride with a UV light source exhibited improved cross-link density, which resulted in lower swelling capacity (Noè et al., 2020). Cassava starch also exhibited improved peak viscosities over native starch when sodium trimetaphosphate and sodium tripolyphosphate were added as cross-linking agents at low degrees (1%) (Sriprablom et al., 2023); when used for 3DP, this can produce improved flow rates at higher temperatures for easy extrusion. Thermal stability and adsorption capacities were also increased after cross-linking of highly porous potato starch (Han et al., 2022). Post-processing during microwave frying with less oil absorption can produced fried foods with appealing structures if it can be used in 3DP.

14.5 CHARACTERIZATION OF MODIFIED STARCH EDIBLE INK

14.5.1 Rheological Properties for Extrusion

14.5.1.1 Viscosity

Generally, at a particular concentrations with increased temperature, gelatinization occurs and viscosity increases, making it difficult to print a structure (Liu et al., 2020). Printing temperature and gelatinization temperature thus also affect the modified starch used as ink. Peak apparent viscosity is the maximum viscosity attained during the heating stage. Final apparent viscosity is the apparent viscosity achieved at the end of the run and setback is the difference of first two viscosities (Maniglia et al., 2020a). For 3DP, the paste has to be viscous enough to pass through tiny nozzles and attach layer by layer. This demands composition with greater extrudability and form retention (Lille et al., 2018). Increasing overall mass leads to greater viscosity, which supports the printed shape (Liu & Ciftci, 2021).

14.5.1.2 Shear Stress and Shear Rate

As the majority of the starches follow non-Newtonian behaviour after gelation, the relation between the shear stress and shear rate can be described by the power law equation (Equation 14.1) and the Herschel–Bulkley model (Equation 14.2).

$$\tau = K \cdot \gamma^{\mathrm{n}} \tag{14.1}$$

$$\tau = \tau_{\mathrm{y}} + K_{\gamma}{}^{\mathrm{n}} \tag{14.2}$$

where τ is the shear stress (Pa), K is the consistency index (Pa s^{-1}), n is the flow behaviour index (dimensionless), γ is the shear rate (s^{-1}), and τ_y is the yield stress. A printed paste should have a yield stress of 200–300 Pa but can go up to 1000–2000 Pa (Zhong et al., 2017). The ink for printing should possess shear thinning behaviour; this is the major criterion. The shear stress generated during the extruded printing process has the ability to disrupt micro-domains and change the network topology inside the starch gel (Xian et al., 2024). The viscosity of the substance must be low enough at high shear rates to enable flow through a tiny nozzle. The pace of structural recovery after shear withdrawal also influences form stability after printing, and it has been shown that a high yield stress is necessary to obtain good shape stability after printing (Lille et al., 2018).

14.5.1.3 Yield Stress (τ_y)

Materials with a high yield stress might resist extrusion until a certain pressure is applied. The material's capacity to maintain its form in the face of gravity and of the pressures caused by material layers put on top of it is correlated with the yield stress. A greater yield stress value, which indicates a material's mechanical strength, is essential for preserving printed forms and supporting layers that are deposited later on during the deposition process (Agunbiade et al., 2022). Addition of rice starch and carrageenan increased the yield stress of surimi gel (Li et al., 2022). Temperature and potato content increases cause yield shear stress to decrease, with apparent viscosity calculated based on Equation 14.3 at a constant shear rate of 50 s^{-1} (Martínez-Monzó et al., 2019).

$$\eta_{ap} = \left(\frac{\tau_{\mathrm{y}}}{\gamma}\right) + \mathrm{k}\gamma^{\mathrm{n}-1} \tag{14.3}$$

where η_{ap} is apparent viscosity, τ_y is yield stress, K is the consistency index, γ is shear rate, and n is the flow behaviour index.

Semi-skimmed milk protein printed at 60% concentration showed better printability (Lille et al., 2018), suggesting that high yield stress was necessary for optimum form stability after printing by cold extrusion. In contrast, high temperatures enhance the breakdown of intermolecular hydrogen bonds in the starch granule, resulting in a starch paste with low yield stress. Similar results were observed in case of potato starch (Martínez-Monzó et al., 2019).

14.5.1.4 Flow Index (n) and Consistency Index (K)

The flow behaviour exponent (n) and consistency behaviour index (K) play a crucial role in defining a material's compatibility with syringe extrusion and desired extrusion rate (Martínez-Monzó et al., 2019). To describe the relation between the viscosity profiles and the consistency index, Álvarez-Castillo et al. (2021) developed Equation 14.4 to understand the shear thinning nature of materials.

$$\eta = \frac{\eta_0}{1 + (K \cdot \gamma)^{\mathrm{m}}} = \frac{\eta_0}{1 + (\mathrm{k} \cdot \gamma)^{\mathrm{m}}} \tag{14.4}$$

where K = consistency index, m = dimensionless shear thinning index, γ is shear rate, and η_0 represents zero shear rate limiting viscosity.

14.5.2 Shear Rheology

Shear rheology focuses on the behaviour of materials when subjected to shear forces.

14.5.2.1 Elastic Modulus (G′) and Loss Modulus (G″)

The mathematical representations for G′ and G″ can be given from Equations 14.5 and 14.6, respectively, from frequency data using the power law relationship.

$$G' = K'\omega^{n'} \quad (14.5)$$

$$G'' = K''\omega^{n''} \quad (14.6)$$

where G′ represents the storage (or elastic) modulus, and G″ represents the loss modulus, K' and K'' represent the modulus at 1 Hz frequency rate, and n' and n'' represents modulus dependant on frequencies.

An increase in red cabbage content was found to increase the G′ value of the prepared dough (Carranza et al., 2023), and inks with G′ greater than G″ showed solid-like behaviour and the potential to form gel-like structures to facilitate printing (Pérez et al., 2019). Starch with reduced swelling capacity has a larger G′ and smaller loss tangent, allowing for better 3D model retention (Chen et al., 2019). But the addition of salts such as NaCl decreased viscosity due to denaturation and rearrangements causing weak gel structure in soya protein isolate (Phuhongsung et al., 2020b). Untreated cheese has a larger elastic modulus than melted cheese, but the loss tangent is smaller due to protein network breakdown during extrusion and melting (Bareen et al., 2021). Cereal-based materials with higher flour content and sucrose had higher viscosity, storage modulus, and mechanical strength. This increased shape stability after deposition of the 3D model but decreased flowability through the nozzle (Yang, Zhang, Prakash, et al., 2018).

14.5.2.2 Loss Tangent (tan δ)

The loss tangent (tan δ) offers information on how a material reacts during cyclic loading, which is useful for forecasting performance during the printing process and in the finished product. The relation between the loss tangent and the storage and loss moduli is described in Equation 14.7.

$$\tan\delta = \frac{G''}{G'} \quad (14.7)$$

A high loss tangent (tan $\delta > 1$) means that the material is more viscous than elastic, indicating it will flow more easily and may not retain its shape well after deposition. A low loss tangent (tan $\delta < 1$) indicates that the material is more elastic than viscous, meaning it will hold its shape better and has greater resistance to deformation. For proper printing of ink, a material should possess lower tan δ values to exhibit solid-like behaviour with less fluidity (Liu & Ciftci, 2021). For obtaining a stable 3D printed structure, tan $\delta \leq 0.268$ and G′ $\geq$ 300 Pa are preferable (Outrequin et al., 2023).

14.5.3 Stability and Shelf Life

Starches generally undergo retrogradation, decreasing palatability and shelf life, so modifications to native starches are required to restrict retrogradation. Ultrasonication of modified starch showed better structural stability in wheat–papaya starches (K. Xu et al., 2020). Microwave pretreatment along with the use of saline of strawberry–soya protein isolate ink exhibited dimensional stability (Fan et al., 2020). As 3D printing involves both heating and cooling at the final product phase, there is less microbial stability due to attack by different microorganisms (Tomašević et al., 2021). Alongside regular food preservation methods, various novel technologies are being used as a post-processing step to improve shelf life. "Hurdle" technologies are the innovative technologies with the ability to not only increase overall quality but also positively impact textural and sensory aspects by utilizing lower operating conditions (Tomašević et al., 2021). However, future research is required on variability of properties during ink printing and especially during 4D printing.

14.5.4 Safety and Regulatory Considerations

The majority of 3D food applications use extrusion technology, including printable materials such as chocolate and cereal-derived products, for which the native material shows some disadvantages in parameters that can be improved by modifications for an improved final product (Zamboulis et al., 2022). Due to the complexity of printing structures, it is especially challenging for regulatory bodies to define specific regulations considering safety and quality of the product reaching the end consumer. Product safety in the food industry is a critical problem that must be addressed by regulation enacted by the U.S. Food and Drug Administration or the UK Food Standards Agency (Hai Alami et al., 2024). The generation of off-flavours, especially such as by lipids during interaction with modified starch during printing with and without post-processing, is also important and should be carefully assessed. Acrylamide formation is also an important factor to consider during the baking and frying application, even though 3DP samples have exhibited its reduction (Derossi et al., 2023). Modifying the microbial flora of foods may cause unforeseen food safety hazards, since customers prioritize nutrition over safety. 3DFP, whether industrial or at home, requires strict regulation to mitigate potential health concerns and compensate for lack of familiarity with the technology (Portanguen et al., 2019).

14.6 PRINTING AND PROCESS VARIABLES

14.6.1 Viscosity and Shear Rheology

Modified starch, viscosity, and shear rheology are crucial for smooth extrusion, structural integrity, and exact layer deposition. Modified starch often shows shear thinning behaviour, meaning that its viscosity drops under shear stress, allowing it to flow through the nozzle more easily. This feature is critical for achieving a balance of flowability during printing and stiffness following deposition. The viscoelastic characteristics of starch, which are controlled by additives such as plasticizers and cross-linkers, may be modified to improve printability by increasing elasticity and decreasing deformation under stress. Temperature and shear rate have a major effect on rheology, necessitating precision process control. Furthermore, gelation dynamics during cooling might aid in solidifying printed structures, making modified starch a flexible material for applications ranging from food construction to biodegradable goods.

14.6.2 Impact on Flow and Printability

Specific gelation mechanisms (cross-linking), thermal characteristics (melting point and glass transition temperature), bulk density, wettability, and flowability of the materials can all have an impact on the food printing process. As discussed in the previous sections, the flowability of ink depends on the behaviour of ink in the viscoelastic region, based on the loss tangent. The effect of the loss tangent on printability has been discussed above, but the visual appearance, tactile perception (stickiness, roughness, hardness), initial bite, chewing, swallowing (flow characteristics, roughness or smoothness), and aftertaste on the mouth are all included in the overall assessment of food design (Sun et al., 2018). Highly modified starches, for example, may reduce viscosity and facilitate material extrusion via a nozzle. But very low viscosity might cause problems such as sagging or spreading after deposition, which can lessen printing accuracy. Starch's shear thinning property helps with 3DP by facilitating smooth material flow through the nozzle and a speedy solidification process that preserves the created structure. A smaller nozzle might require starches with lower viscosity to ensure smooth flow, while faster printing speeds might necessitate starches that set quickly to maintain structure.

14.6.2.1 Adjustments for Different Food Inks

In order to achieve consistent deposition without nozzle obstruction, it may be necessary to adjust the nozzle's diameter and extrusion speed for some food inks that are made from more viscous liquids like gels or paste. Due to their low viscosity, liquid-based food inks must be printed more slowly or with smaller nozzles in order to be printed effectively without smearing or leaking. Maniglia et al. (2019) employed a 0.8 mm diameter nozzle with an 18 mm nozzle height, 20 mm/s speed, and 30 mm/s extrusion rate at 20°C for cassava starch modified by ozone. Lu et al. (2024) specfied the printing speed (20 mm/s) and layer height (0.84 mm) for starch modified by radio frequency air cold plasma. Ma, Zhang, et al. (2024) set print settings at 15 mm/s for the nozzle speed, 30 mm for layer height, and 60% for infill density for wheat starch modified by plasma-activated water.

14.6.2.2 Printing Speed, Extrusion Rate, and Nozzle Diameter

During printing, pressure is utilized to propel the starch material through the nozzle, and the desired shape is achieved when a layer is constantly deposited. The printing flow and speed are determined by the compressive pressure. Because at low pressure there isn't adequate compressive force to drive the sample ink pushed out of the nozzle, the pattern is uneven. At high pressure, there is severe sample deformation, curling and stacking, and poor resolution; rearrangement of disorder starch due to notably high compressive pressure may be negative, but this would further reduce the structure fidelity. Equation 14.8 relates the desired printing speed to the extrusion rate (Yang et al., 2019).

$$V_d = \frac{\pi}{4} v_n D_n^2 = \frac{\pi}{4} v_n h_c^2 \quad (14.8)$$

where V_d is the volume extruded per second (mm^3/s), v_n is nozzle movement speed (mm/s), D_n is the nozzle diameter (mm), and h_c is the nozzle height (mm). The extruding rate, on the other hand, specifies how much sample is extruded per unit time. Printing speed and extruding rate impact 3D printing by varying the amount of produced paste per unit length and time (Liu & Ciftci, 2021). The extrusion rate in terms of layer height (*h*) was also calculated from Equation 14.9 by Derossi et al. (2020).

$$\text{Extrusion rate} = \mathrm{D_n} \times \mathrm{h} \times \mathrm{V_n} \quad (14.9)$$

where extrusion rate is in mm^3/s, v_n is nozzle movement speed (mm/s), D_n is the nozzle diameter (mm), and h_c is the nozzle height (mm).

The nozzle, a key component of 3D printing equipment, significantly impacts the quality of starch ink-printed edifice (Rong et al., 2023). Nozzle height, which describes the span between the tip and the deposited top layer, exhibits a significant influence on the geometry and shape of extrudates, and Equation 14.10 predicts the critical height for the nozzle (h_c).

$$h_c = \frac{v_d}{v_n D_n} \quad (14.10)$$

where v_n is nozzle movement speed (mm/s), v_d is slurry extrusion speed per second.

D_n is the nozzle diameter (mm), and h_c is the nozzle height (mm).

The nozzle speed controls the movement rate of the nozzle when printing contours, infills, bridges, and support material.

Printing contours, infills, bridges, and support materials, The nozzle speed is a parameter that controls how quickly the nozzle moves and this can be calculated by Equation 14.11, given by Guo et al. (2019).

$$v_n = \frac{4Q}{\pi D_n^2} \tag{14.11}$$

where v_n is nozzle movement speed (mm/s), D_n is the nozzle diameter (mm), and Q is filament extrusion rate.____.

The optimal printing speed has been determined to be between 600 and 800 mm/min, regardless of the material supply (Anukiruthika et al., 2020). The shear rate varied with nozzle diameter, with the smallest diameter (0.8 mm) resulting in the maximum shear rate (952.75 s^{-1}) (Oyinloye & Yoon, 2021). In 3DP, particularly with starch-based materials, the phrase "non-printing movement" refers to the motions made by the printer's extruder or printhead when the material is not actively being deposited. Retraction is a technique in which the printer draws the material back slightly into the nozzle before performing a non-printing movement. This helps to prevent oozing and stringing by lowering the pressure in the nozzle and stopping material flow when the printhead passes over the print without depositing the material.

14.6.3 Temperature and Humidity Control

14.6.3.1 Extrusion Temperature

Extrusion 3D printing causes starch fragmentation, which relies on 3DP factors such as temperature, nozzle diameter, moisture content, and source (Shahbazi et al., 2023), along with degree of modification. Printing the modified starch below gelatinization temperature helps in keeping the starch structure intact and shows less strength due to limited free starch chains, whereas temperature above gelatinization enhanced collision and swelling, increases viscosity and strength of the gel. During extrusion 3DP, starch fragmentation is relatively inevitable, depending on 3DP factors such as temperature, nozzle diameter, moisture content, and the starch supply. At increasing temperatures, the mobility of starch chains increases, making chain interactions less dominant, resulting in decreased strength (Shahbazi et al., 2022).

14.6.3.2 Build Plate Temperature

The effect of build plate temperature on 3DP of modified starch has undergone very limited research concerning the adherence of the printed layers to the build plate. If the construction plate becomes too cool, the starch may not adhere properly, resulting in curling or moving, which can damage the entire print. The temperature of the construction plate also influences the rate at which the starch substance cools after extrusion. The effect of layer adhesion can be calculated from Equation (14.12).

$$\sigma = \tau \times \cos\theta \tag{14.12}$$

where σ is the normal stress, τ is the shear stress, and θ is the contact angle between the layers. The best build plate temperature for printing starch-based materials can typically be between 50°C and 70°C, depending on the type of starch and its modification.

14.6.3.3 Cooling Mechanisms

The cooling process affects how well each layer adheres, the final texture of the printed product, and the prevention of faults such as warping and cracking. There has been no research on how the cooling method affects the structural fidelity of modified starches; however, some of the principles and mechanisms are presented in this part. Proper cooling ensures that each layer of starch cools down sufficiently before the next layer is applied. Rapid cooling could cause the material to contract too quickly, resulting in warping or fracture. Slow cooling, on the other hand, allows for progressive solidification, which helps the printed object retain its shape and structure. Active cooling is the employment of fans or cooling devices to move air over the printed layers and hasten cooling. This is especially advantageous for printing with materials such as modified starches that would otherwise remain soft for too long. Passive cooling relies on natural air cooling without any additional mechanical aid.

14.6.4 Printing Speed and Resolution

14.6.4.1 Effect on Build Time

The build time in 3D printing is the entire amount of time necessary to finish a print, and it may have a considerable impact on the quality, structure, and overall performance of modified starch prints. Longer build periods may alter how well modified starch layers stick to one another. During a lengthy print, the starch substance may thicken or gel more as it cools or dries, making it more difficult to extrude uniformly and producing errors in layer deposition. Insufficient adhesive force between the deposited material and the build plate might increase the stipulated delay (Fahmy et al., 2020). For example, raising the printing velocity reduced the overall printing time of the 3D structure and the cooling time of the deposited chocolate (Rando & Ramaioli, 2021).

14.6.4.2 Impact on Structural Integrity

Formulation issues during extrusion can impact both the process and the stability of the extruded product. Here, product stability is briefly discussed. The 3D printing process is divided into two sections that involve balancing optimal material qualities (Wilms et al., 2021): (1) product flow during extrusion, also known as "extrudability", and (2) product stability following extrusion, also known as "buildability". Although the effect of yield stress on structural integrity is discussed above, it possesses a dynamic time-dependent behaviour. When it comes to 3DP, the material's dynamic yield stress is present during layer deposition, and as time passes, the structure gradually returns to the static yield stress. Printing fidelity was calculated Guo et al. (2021b) as given in Equation 14.13, and the percent of dimensional deviation (DD) after printing was calculated by Dick et al. (2021) as given in Equation 14.14.

$$\text{printing fidelity} = \left(1 - \frac{\text{Printed area-Designed area}}{\text{Designed area}}\right) \times 100\% \quad (14.13)$$

$$\text{DD} = \frac{(\text{D}_m - \text{D}_t) + (\text{H}_m - \text{H}_t)}{(\text{D}_m + \text{H}_m)} \times 100 \quad (14.14)$$

where D_m is measured diameter, D_t is target diamter, H_m is measured height, and H_t is target height.

The deformation rate (DR%) after printing the object can be calculated by the formulae given by X. Chen et al. (2022a) mentioned in Equation 14.15.

$$\text{DR\%} = \frac{2\Delta L + \Delta T}{2L + T} \times 100 \quad (14.15)$$

where L is length of printed object, T is thickness of printed object, ΔL is length difference before post-printing treatment, and ΔT is thickness difference before post-printing treatment.

14.6.4.3 Optimization for Different Food Materials

Different food materials possess unique properties that affect how they can be printed. Parameters employed for starch and modified starch have been discussed above; this section gives a comprehensive overview of printing resolution. Smaller nozzles give better resolution, although they might clog easily with certain food items. The printed threads must be mechanically strong enough to sustain stacked layers without buckling or sagging. Additionally, the threads must be stable after deposit to guarantee adequate resolution (Chen et al., 2019).

14.6.5 Layer-by-Layer Deposition

14.6.5.1 Layer Thickness and Mouthfeel

Thinner layers form a structure with finer resolution in items that require softness (e.g., puddings or gels), which improves smoothness during mastication. Thinner layers can contribute to a faster melt-in-your-mouth feel for materials with a high moisture content or that are quickly dissolvable (such as modified starch gels). This is perfect for printing textures that resemble desserts, such as puddings or cream cheese, using thicker layers (0.6–1.2 mm) of higher viscosity starch formulations. This is ideal for items such as snack bars, crackers, and baked goods that require a solid structure. Proper adhesion can help to create a consistent, cohesive structure, resulting in a more pleasing mouthfeel. Thicker layers may result in a more layered, stratified feel, creating the sense of biting through several levels. This can be beneficial in foods such as pastries and stacked snacks.

14.6.5.2 Precision in Layer Deposition

SLS printing precision and accuracy rely on both material qualities and processing parameters (e.g., laser types, power, and spot diameter) (Shirazi et al., 2015). To achieve precise and accurate printing, factors such as ink–substrate compatibility, viscosity, rheological characteristics, temperature, and printing rate play a crucial role. Because of their distinct rheological characteristics, modified starches necessitate precise control over the printing process in order to guarantee uniform layer formation. Retraction settings (i.e., slightly pushing the material back between motions) can help to prevent material seeping and guarantee smooth layer deposition begins and stops. This is especially crucial when working with materials that exhibit a high degree of flow following extrusion. Modern slicing software allows for real-time flow rate adjustments based on printing circumstances and object shape. This can help to ensure continuous layer deposition even in complicated designs.

14.6.5.3 Achieving Fine Details and Balancing Speed

To improve process productivity, printing speed can be increased and larger nozzles or laser diameters can be employed. However, 3DFP suffers from reduced accuracy and resolution, making these solutions unattractive (Liu et al., 2017). Some 3D printers come equipped with dual nozzles (Sun et al., 2018) of different sizes and engage multiple materials. A small nozzle can be used for fine details, while a larger nozzle handles broader, less detailed sections, allowing for faster overall printing without sacrificing precision in critical areas. Fine details are necessary for generating further clarify patterns, improving product aesthetics, and managing texture, yet speed is critical for efficiency, particularly in large-scale production. Printers with better resolution can precisely position the printhead, allowing for the deposition of individual layers.

14.7 APPLICATIONS IN 3D FOOD PRINTING

14.7.1 Edible Structures and Designs

The first challenge in 3DFP is to develop digital 3D models and generate the codes that regulate the printer's operation. Apps for novices (e.g., Tinkercad) and more sophisticated applications (e.g., Rhino 3D, Blender, Fusion 360) may be used to create nearly any structure (Derossi et al., 2024). For example, infill pattens improved chewing time from 26 to 45 seconds in the case of cookies (Lin et al., 2020). When ice cream ink made with hydroxypropyl distarch phosphate had greater hydrogen bonding support and a stable protein structure, it had the best accuracy (94.6%) (Xu et al., 2024). A 12% potato starch gel coaxially printed with surimi was successful in imitating the crab meat cylindrically stable structure (Kim et al., 2021). Different stable structures of pyramid, anchor, tetrahedron, and gecko shapes were printed at 24 mm^3/s with 30 mm/s velocity and a 1.0 mm nozzle diameter using lemon juice and potato starch at different concentrations (Yang et al., 2019). Dynamic printed structures composing potato starch and red cabbage had colour changes induced by varying pH solutions when printed hexagonal designs (Ghazal et al., 2021). Modified wheat starch by dry heat treatment resulted in stable cylindrical structures and reproducibility during printing (Maniglia et al., 2020a).

14.7.2 Customization and Personalization

3D food printers can use modified starch to create food items tailored to individual dietary needs, such as low-carb, gluten-free, and high-fibre foods. To stabilize the structures with less infill density and complex structures, novel ingredients with enhanced properties are necessary, and modified starch can play a role in this case. Customization of chocolates has been developed due to the popularity of chocolate in confectionary, and many different shapes have been printed from chocolate, varying the infill (Mantihal et al., 2019). But consumer preferences for textural attributes vary from person to person. Other customizations that can be performed for individual consumer preferences include shape changes (He, Zhang, & Devahastin, 2020), colour changes during post-processing (Derossi et al., 2024), texture changes based on porosity (Derossi et al., 2021), changes in chewiness based on infill pattern modification (Zhu et al., 2021), perception of aroma release during eating (Guo et al., 2021a), sweetness and sourness (Chow et al., 2021), and saltiness (Fahmy et al., 2021).

14.7.3 Case Studies and Practical Examples

14.7.3.1 Functional Foods

The innovative technology of functional foods for 3D printing using modified starch has great prospects in foods when the health aspects of functional ingredients are combined with the personalization possibilities of 3D printing. During 3DP, modified starch can be used to encapsulate and shield probiotics bioactive bacteria. 3DP foods can be fortified with specific vitamins, minerals, or bioactive compounds (e.g., omega-3 fatty acids, antioxidants) using modified starch as a carrier. Biomaterials such as starch can also be used to incorporate certain vitamins, minerals, or bioactive agents into 3D printed foods. Modified starch can be combined with protein powders, fibres, and other functional ingredients to create personalized high-protein, high-fibre snacks. Modified starch can help in structuring these bars while incorporating functional ingredients such as caffeine, B-vitamins, or electrolytes. 3D-printed foods can be fortified with nootropics (substances that enhance cognitive function) such as caffeine, L-theanine, or ginseng, using modified starch to stabilize and deliver these compounds effectively.

14.7.3.2 Starch-Based Films

Although starch is not inherently thermoplastic, it can display thermoplastic properties when plasticizers such as water and polyols are introduced, either by applying heat and shear stresses or without them (García-Guzmán et al., 2022). 3DP films can be made with adjusted permeability to gases and moisture by altering the starch, which makes them appropriate for packing items such as fresh vegetables or meats that need particular atmospheric conditions. Edible films that may be used to cover sweets, fruits, or vegetables can be made using modified starch (Gupta et al., 2025b). Edible films provide a regulated method of delivering useful substances by encapsulating tastes, colours, or vitamins. Further research is needed to develop smart films that respond to environmental changes (e.g., temperature, pH) using 4DP technologies.

14.7.3.3 Personalized Foods

Modified starch can be used as a foundation to make meals that fulfil certain dietary requirements, such as for gluten-free, low-carb, or high-protein diets. For example, someone with celiac illness may have gluten-free starch mixed using 3DP into bread or pasta. Modified starch can be used as a flavour transporter, allowing for the development of individualized flavour profiles. Customized meals with precise textures and nutrient profiles may be made for older patients who have difficulty chewing or swallowing, ensuring they get the nutrients they need in a safe way. Athletes may have energy bars 3D printed with modified starch and personalized with the exact carbohydrate, protein, and fat ratios required for their individual training plan or competition day. 4DP gels or beverages that gradually release electrolytes and recovery nutrients, suited to an athlete's exact requirements during or after physical exercise.

14.8 APPLICATIONS IN 4D FOOD PRINTING

14.8.1 Concept and Potential of 4D Printing

Skylar Tibbits of the Massachusetts Institute of Technology (MIT) first put forth the concept of four-dimensional (4D) printing in 2013 (Aldawood, 2023). Presently, the utilization of 4DP technology in the food sector entails the amalgamation of dissimilar food components (recombined food) according to a structural template and meticulous formula (Teng et al., 2021). Under specific stimuli, such as heat, water, pH, light, and magnetic field, the properties of printed items (such as colour, shape, nutrition, texture, and taste) might manifest predictable alterations (X.Chen et al., 2022). A key difference between 3DP and 4DP is in the integration of intelligent materials and intelligent design. 3DP works by superimposing layers on top of each other, so these complicated forms need a support layer below them to hold up the top structure. Indeed, 4DP has the capability to fabricate suspended structures that are beyond the capabilities of 3DP, such as blossoming flowers (Teng et al., 2021). Food made with 4DP provides creative ideas for creating interactive meals and increases interaction between food components and diners. The colour, taste, or nutrition of the 4DP product can be incorporated at consumption by the user, rather than being retained during storage.

14.8.2 4D Food Printing Conditions

The fundamental elements of 4D food printing include printing software, inks, 3D printers, structural design, and stimulants such as heat, water, pH, and light (Phuhongsung et al., 2022). These components allow for the precise and foreseeable development of the printed items along time.

14.8.2.1 Software for Printing

A key goal of 4D food manufacturing research is to develop a 3D representation that integrates various details such as patterns, material, and stimulation signals (Ghazal et al., 2019). The optimal 3D model, containing several pieces of information, is divided into slices, and the software for slicing directs the printers to produce slices (Wang et al., 2024). In essence, the standard of 4DP is determined by simulation, printing, modelling, and slicing software. Comprehensive 4D modelling systems can exactly replicate the printing process; increase physical attributes such as viscosity, pressure, and homogeneity; and limit the possibility of printing errors (Kokane et al., 2024). Currently, two types of simulation software are used: (1) simulation using software for finite element analysis libraries (including Polyflow, Abaqus, and COMSOL) and (2) simulation through programming languages (such as C^{++} and JavaScript) (Ghazal et al., 2023). The equivalent to a written blueprint is modelling software. Modelling tools such as AutoCAD, SolidWorks, CATIA,, Blender, and Rhinoceros are suitable for utilization in 4D food printing (Singh et al., 2024).

14.8.2.2 Printers

3D food printing primarily utilizes stereolithography, direct ink writing, diffused deposition modelling (FDM), select laser melting, digital light processing, and inkjet printing (Singh et al., 2024). The FDM technique has emerged as the predominant method for 4D food printing owing to its user-friendly operation, extensive affordability, and applicability (Shi et al., 2024). Differentiated based on the arrangement of displacement interfaces, FDM is divided into Cartesian, polar, SCARA, and delta configurations, with Cartesian being the most commonly utilized. The main idea behind 4DFP is that making parts out of more than one material makes them more flexible and useful. Printers that use more than one material can make hierarchical structures by using multiple materials together in one print job (He et al., 2021). In the case of microwave-assisted printers, the dielectric characteristics of food materials are altered by varying the composition of the materials (e.g., by altering the amount of salt or sugar) or microwave power. This contributes to a variation in the rate at which heat is absorbed, which in turn alters the properties of the printed object (Jiang et al., 2022).

14.8.2.3 Printing Inks

The food ingredients utilized by printers in 4DFP are called "printing ink". Optimal rheological characteristics for mechanical strength and seamless extrusion to maintain texture and mass during printing are combined to enable food printing ink to be printable (Cheng et al., 2022). The flowability, post-printing durability, and extrudability of structure may be affected by printability of food ink. Prior to printing, the ink made for food applications must exhibit both liquid and solid-like properties (Patoliya et al., 2023). The functional and physical characteristics (such as density and water absorption ability) of the printed objects are influenced by the particle size. Food ink includes various food components such as starch, chocolate, food hydrocolloids, fruits and vegetables, and soybean are utilized (Wang et al., 2024). Pigments in printing ink induce 4D colour alterations in foods at varying pH levels (Kang et al., 2023). Among the stimulus-responsive materials employed in 4D printing, curcumin reveals a red hue at pH more than 7 and a yellow hue at pH lower than or equal to 7. The complex of arabic- oil and gelatin gum has a thermal response and has been used to print buckwheat dough with yellow flesh peach (Guo et al., 2021a).

14.8.3 Stimulations in 4D Printing

The requirement for inducing modifications in the properties, attributes, and configuration of 4D printed materials is stimulation. Stimuli can manifest through several mechanisms, including physical, chemical, or biological actions (Guo et al., 2022). Physical stimuli include humidity, ambient temperature, electromagnetic radiation, magnetic fields, and electrical potential. Chemical stimuli include ionic strength, pH, and the chemicals themselves. Biological stimuli are enzymes and glucose.

14.8.3.1 Light

The application of light as a stimulant in 4DP enables rapid response and non-intrusive stimulation. It is possible to illuminate a particular area of a sample in order to induce localized alterations in the printed item. Printed structures in shape-memory polymers can be illuminated by light and subsequently produce thermal energy (Pingale et al., 2023).

14.8.3.2 Temperature

Thermal exposure can cause alterations or deterioration of the characteristics of polymers. Two parameters taken into account when subjected to thermal treatments are the melting temperature and the glass transition temperature of the ink composition. A non-invasive technique that raises the temperature of 4DP things is microwave heating. This alteration affected by temperature is reversible. Its benefits include great energy efficiency, quick heating, ease of use, hygienic conditions, and some benefits in post-printing procedures. It significantly affects 4D object shape-changing. Dewatering from microwave heating bends or shrinks printed materials (Narupai et al., 2021).

14.8.3.3 pH

pH fluctuations in 4DP structures and surrounding mediums affect functional qualities and structural and are extensively explored. Many plant-based printed food materials have had their colour variations caused by pH changes extensively examined. The primary pigments in fruits and vegetables are anthocyanin, carotenoids, and chlorophyll, and they are good for human health. The pH-sensitive pigment chlorophyll, which is responsible for the green hue, exhibits significant colour deterioration at lower pH levels. Similarly, a decrease in carotenoid concentration has been observed at lower pH levels. At lower pH, anthocyanin displayed a red hue, but at higher pH, it developed a yellow hue (Chen, Zhang, Mujumdar, et al., 2021).

14.8.4 Time-Dependent Shape and Property Changes

14.8.4.1 Nutrition Change

Homan et al. (2016) created a type of ready-to-eat snack known as "edible growth food". This involved incorporating fungal yeast or spores and bacterial cells with seeds from edible plants into a highly digestible nutrient matrix such as fruits, vegetables, agar, or nuts with light stimulation and FDM technology. Alongside alterations in nutrients, the palatability of flavours also intensifies as a dish ages. Currently, edible water colloids such as sodium alginate, gelatin, and agar are the predominant matrix materials (Shen et al., 2023). Fibre from fungi, fruits and vegetables (e.g., mango, ophiocordyceps sinensis, celery) can be included into the gel system to strengthen its self-supporting ability, resulting in a more favourable impact and shear thinning behaviour. However, larger fibre particles may clog the nozzle; the fibres thus need to be removed, dried, and ground into nanofibres. The hydrogel's nanofibres are then uniformly distributed using ultrasonic technology.

14.8.4.2 Texture Change

Within the context of customer acceptance, texture of food is as important as visual appearance and taste. 3D additive manufacturing is the process of creating food by aligning the food ink layer by layer. A 4D shape of mashed potato with different inner components was fabricated, and the gumminess and hardness of printed items were significantly influenced by the concentrations of infill and the perimeters (Liu, Bhandari, et al., 2018). In the same vein, a 4D deformation of the double-layer framework made of paper and pumpkin was also observed (Chen, Zhang, Guo, et al., 2021). This work focused mainly on the directionality of the plane layer and the anisotropy of the structure in order to accomplish the objective of 4D deformation. The researchers observed that the material hardness exhibited exponential growth with time, demonstrating a negative association with time. Microwave-induced stimulation was also shown to greatly enhance the hardness, cohesion, springiness, and gumminess of a 4DP lotus root powder gel. Microwaves cause food to lose moisture, which in turn makes the food tougher and more compact during the 4DP process. When cohesiveness increases, the internal force becomes stronger, and when springiness increases, a greater amount of force is required to break down the food (Fan et al., 2020).

14.8.4.3 Shape Change

A few examples of deformation behaviours seen in 4DFP can be seen in Figure 14.2, including surface bending, linear and nonlinear expansion and contraction, distortion, and

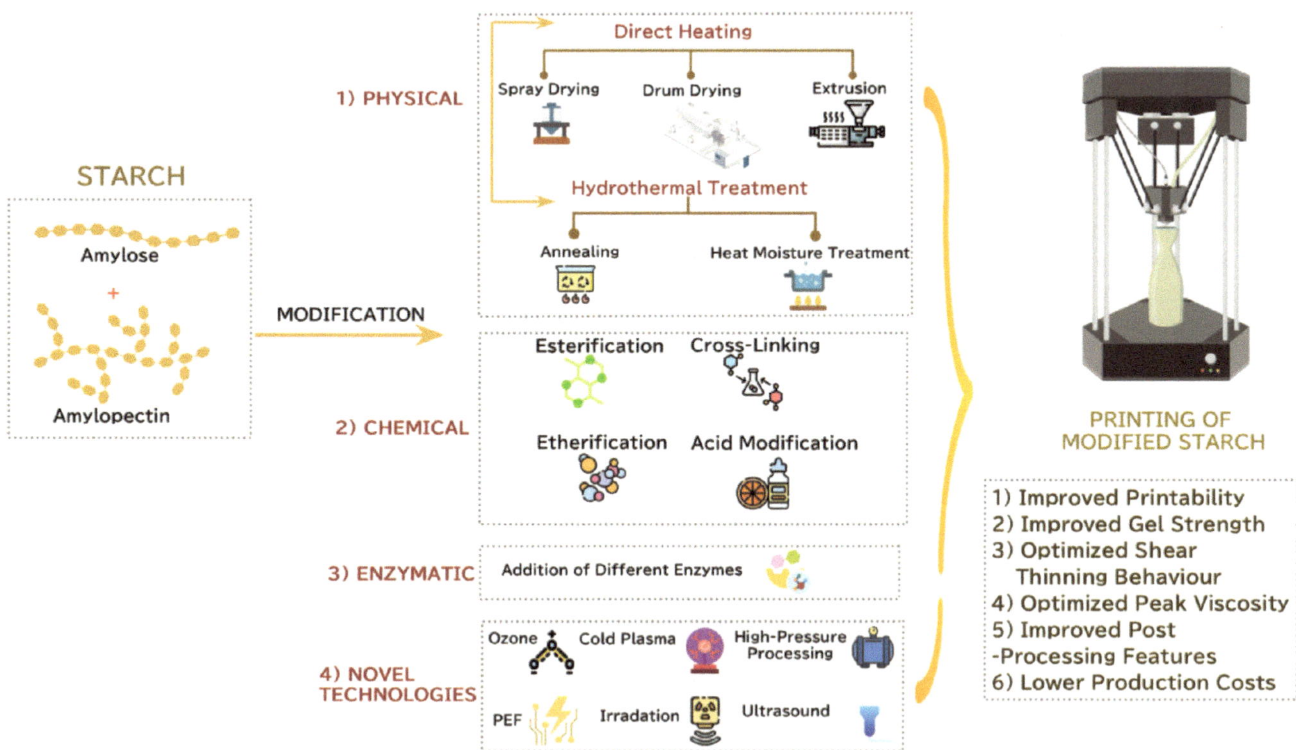

FIGURE 14.2 Different methods of starch modification and their applicability in printing. PEF, pulsed electric field.

bending (Chen et al., 2023). When it comes to 4DFP, one of the causes of distortion is water stimulation (J. Chen et al., 2022). Both stiff and expandable hydrophilic active materials make up the printing framework. The material's form is altered as a result of the forces generated by its expansion. The complicated deformation behaviour can be achieved by arranging the expandable and rigid materials correctly. Tao et al. (2019) employed two distinct types of flour (oat fibre and plain flour) to construct a dual-layer framework, aiming to induce a transformation in the physical form of the flour. However, the dough yielded poor outcomes because of its inadequate gluten network. Remarkably, a groove structure was devised that effectively altered the form of the dough. He, Zhang, and Devahastin (2020) investigated the deformation of purple potato gel with varying salt or syrup concentrations under variable microwave power conditions. Elevated hardeners on the product surface result in reduced bending angle and delayed deformation with increased dehydration rate. The inadequate water evaporation can result in the formation of voids and cracks. Moreover, the configuration of the internal structure (including the angle of filling and rate) of the printed samples seems to be another significant determinant in the extent of deformation caused by dehydration.

14.8.4.4 Colour Change

The transformation of colour of the recombined meal is produced mostly by the existence of plant pigments in diverse chemical states according to different pH levels. Ghazal et al. (2019) employed a dual-nozzle printer to create a recombined structure consisting of two layers: one having a tailored lemon–potato starch at the bottom, and the other carrying anthocyanin–potato starch at the top. The hue of this recombined system was seen to undergo separate changes over time, attributable to molecular diffusion. Indeed, the reconstituted food presented itself as a readily consumable snack. The printed output derived from this study was seen to necessitate a considerable duration of colour transfer, exceeding 30 minutes. Further investigation was carried out on the chromatic alteration of 3DP objects utilizing starch gels (He, Zhang, & Fang, 2020). The researchers formed potato gel systems that were weakly acidic (pH 6.5), strongly acidic (pH 2.5), and alkaline (pH 7.8) by combining mashed potatoes with citric acid, sodium alginate, and sodium bicarbonate, respectively. The materials were fabricated in the shape of a cone, stratified into four layers. The bottom three portions comprised a purple potato gel encased in potato gel with different pH values, while the top layer was composed of pure mashed potatoes. Over time, the anthocyanin from the interior purple potato penetrated the exterior mashed potato and manifested a discolouration response. In other research, a technique was developed to selectively print 4D jelly with dynamic colour changes using blueberry anthocyanin (Khanal et al., 2010). Anthocyanin-containing potato starch–pectin compounds served as the colour-loading layer, while potato starch–pectin compounds containing lemon juice functioned as the alkaline layer. The acid layer consisted of potato starch, pectin, and sodium bicarbonate.

14.8.4.5 Flavour Change

The use of printing technology is essential to achieving flavour shifts in 4D food. Food components may be carefully arranged thanks to the use of printing, which enhances the chemical interaction of flavour, and the distinctive structure design can uniformizes flavour changes or lessens the likelihood of undesirable changes such as growth. Right now, there are two primary methods to make 4D food taste different. To start, printers divide the various food ingredients in a ratio that produce tasty flavour. Following that, the necessary flavour change happens when pH or microwaves are used as stimulus. Phuhongsung et al. (2020b) sprayed it with a variety of pH solutions, including beetroot powder, pumpkin powder, distilled water, and soy protein isolates, to alter the taste of the reconstituted food. At pH 8 and 10, the electronic nose revealed the desired alteration in volatile components, primarily terpene esters and aromatic chemicals. Guo et al. (2021b) synthesized microcapsules using gelatine and gum arabic as complex coacervates, together with pepper red pigment, cinnamaldehyde, and maize oil as core components. A combination of peach flesh, buckwheat flour, and microcapsules was printed in cylindrical specimens. The samples were subjected to a 200 W microwave to induce a progressive release of red pigment and cinnamaldehyde, resulting in simultaneous changes in aroma and colour.

14.9 FUTURE PROSPECTS

14.9.1 Market Trends

A food company recently launched the first 3D-food-printing restaurant, which has gained significant popularity among consumers. The utilization of 3DP in pancake production in Holland indicates a growing trend in the adoptability of 3D food printers (Sandhu & Singh, 2022). 4DP was projected to be 8.38 times higher in 2025 than in 2019, reaching a total of \$64.2 million (Y. Chen et al., 2022). Incorporating 4DFP technology enables innovative business concepts to align with prevailing market trends through time savings, reduction of inventory space, capital needs, and enhancement of company effectiveness. The primary limitations of 4DP have been the high cost of printed materials and printers. A conference will address investment prospects and legal concerns specific to printed food, aiming to establish external safeguards for the introduction of printed food into the food industry.

14.9.2 Future Challenges and Opportunities

Currently, 4DFP methods depend on well-established 3DP equipment and technology. The establishment of a material–process–performance function correlation model is necessary to fulfil the demands of food denaturation, distortion, and flexible activity. Moreover, the advent of advanced technology and novel materials will undoubtedly provide fresh obstacles to the computational modelling of 4DP. Development of novel finite element models and techniques is necessary to achieve precise modelling. The development by researchers of intelligent food products should be integrated with the realtimeindustrial food sector and product commercialization. To make 4DFP possible, new stimulation factors such as electromagnetic fields, osmotic dehydration, light, and combinations of these should be investigated along with the printing structures and characteristics of food ingredients. Crucially, it is necessary to attain the simultaneous appearance of several characteristics, including shape, colour, and flavour. Investigation should be conducted using online monitoring technology for important metrics, including stimulation of each layer and dimensional correctness during the 4D printing procedure.

14.9.3 Future Opportunities for Printing: 5D/6D/7D

3D printing has already revolutionized food manufacturing, particularly in areas such as personalized nutrition, elaborate food designs, and functional food products. Looking beyond 3D, the future of printing using modified starches (and other biomaterials) may include the development of 5D, 6D, and perhaps 7D printing technologies. These extra "dimensions" would extend beyond spatial control, possibly combining time, functionality, and dynamic changes in printed materials to enable new applications and capabilities in food printing. The 5DP process may be utilized to create exceedingly complex-shaped confectionery goods and curved snack products. 6DP may also be used in the food industry for stronger items that respond better to environmental stimuli (Nida et al., 2022). In the context of 6DP, extra factors such as interactivity (items that alter or respond to environmental or sensory stimuli) or multi-functional materials may be included. Modified starches might be combined with sensors that detect certain biological signals (such as saliva or sweat pH) and adjust characteristics to meet nutritional requirements. 7DP may improve production efficiency, lower prices, and create more sophisticated goods than traditional manufacturing methods (Srivastava et al., 2023). 7DP might result in meals that adapt in real time to a consumer's health measurements, personal preferences, or ambient factors.

14.9.4 Sustainability and Environmental Impact

3D printing provides exact control over the shape, size, and quantity of food generated. This implies that only the necessary amount of food is printed, eliminating waste and surplus resources. Starch-based products originating from crops such as maize, potatoes, and tapioca are both renewable and biodegradable. Starch modification for 3DP applications (such as hydrolysis, esterification, or extrusion) can be energy intensive. To increase the sustainability of modified starch production, more energy-efficient processing technologies must be developed. 3DP requires less energy than traditional manufacturing, but it is still necessary to power printers with renewable energy sources to reduce their carbon footprint. Sustainability requires minimizing failed prints and allowing waste materials to be recycled or disposed of. As 3DFP becomes more widely used in restaurants, bakeries, and food hubs, there is an opportunity to power printers with renewable energy. Food printers powered by solar or wind energy have the potential to significantly lower the overall carbon footprint. 3D food printers might be used in remote places where traditional food production and delivery are challenging, decreasing dependency on global supply systems and maintaining food availability during interruptions.

14.10 CONCLUSIONS

Summary of Key Points

- The innovative application of modified starch as an edible ink in 3D and 4D food printing represents substantial advances in accuracy, personalization, and sustainability in the food sector.
- Modified starch has good viscosity control, structural integrity, and flexibility to different printing conditions, making it perfect for developing complex, individualized food items.
- The ability to eliminate food waste, optimize production, and employ sustainable resources will make modified starch an important component in the future of food manufacturing.
- This chapter emphasizes the revolutionary impact of 3D and 4D food printing in enabling unique culinary experiences and more sustainable approaches to food production.

ABBREVIATIONS

G′	storage modulus
G″	loss modulus
K	consistency index
m	dimensionless shear thinning index
n	flow behaviour index
T_g	glass transition temperature
γ	shear rate
δ	phase angle
τ_0	yield stress
τ_y	yield stress
τ_f	flow stress
3DP	3D printing
3DFP	3D food printing
4DP	4D printing
4DFP	4D food printing
AM	additive manufacturing
ANN	annealing
BJ	binder jetting

CAD computer aided design
CIJP continuous inkjet printing
CWS cold water soluble
DOD drop-on-drop demand
DS degree of substitution
FDM fused deposition modelling
HE hot extrusion
HMT heat-moisture treatment
PEF pulsed electric field
SLS selective laser sintering
SPI soya protein isolate

REFERENCES

Abedi, E., Altemimi, A. B., Roohi, R., Hashemi, S. M. B., & Conte, F. L. (2024). Understanding starch gelatinization and rheology modeling of tapioca starch–NaCl/$CaCl_2$ blends: Thermodynamic properties and gelatinization reaction kinetics during pre- and post-ultrasonication. *International Journal of Biological Macromolecules*, *272*, 132865. https://doi.org/10.1016/j.ijbiomac.2024.132865

Abedi, E., & Pourmohammadi, K. (2020). Aggregation behaviors of sonicated tapioca starch with various strengths of Hofmeister salts under pre- and post-ultrasonic treatment. *Food Hydrocolloids*, *105*, 105826. https://doi.org/10.1016/j.foodhyd.2020.105826

Acevedo-Guevara, L., Nieto-Suaza, L., Sanchez, L. T., Pinzon, M. I., & Villa, C. C. (2018). Development of native and modified banana starch nanoparticles as vehicles for curcumin. *International Journal of Biological Macromolecules*, *111*, 498–504. https://doi.org/10.1016/j.ijbiomac.2018.01.063

Agunbiade, A. O., Song, L., Agunbiade, O. J., Ofoedu, C. E., Chacha, J. S., Duguma, H. T., Hossaini, S. M., Rasaq, W. A., Shorstkii, I., Osuji, C. M., Owuamanam, C. I., Okpala, C. O. R., Korzeniowska, M., & Guine, R. P. F. (2022). Potentials of 3D extrusion-based printing in resolving food processing challenges: A perspective review. *Journal of Food Process Engineering*, *45*(4). https://doi.org/10.1111/jfpe.13996

Aldawood, F. K. (2023). A comprehensive review of 4D printing: State of the arts, opportunities, and challenges. *Actuators*, *12*(3). https://doi.org/10.3390/act12030101

Álvarez-Castillo, E., Oliveira, S., Bengoechea, C., Sousa, I., Raymundo, A., & Guerrero, A. (2021). A rheological approach to 3D printing of plasma protein based doughs. *Journal of Food Engineering*, *288*, 110255. https://doi.org/10.1016/j.jfoodeng.2020.110255

Anandharamakrishnan, C., Moses, J. A., & Anukiruthika, T. (2022). *3D printing of foods*. Wiley. https://doi.org/10.1002/9781119669838

Anukiruthika, T., Moses, J. A., & Anandharamakrishnan, C. (2020). 3D printing of egg yolk and white with rice flour blends. *Journal of Food Engineering*, *265*, 109691. https://doi.org/10.1016/j.jfoodeng.2019.109691

Arik Kibar, E. A., Gönenç, I., & Us, F. (2014). Effects of fatty acid addition on the physicochemical properties of corn starch. *International Journal of Food Properties*, *17*(1), 204–218. https://doi.org/10.1080/10942912.2011.619289

Azam, R. S. M., Zhang, M., Bhandari, B., & Yang, C. (2018). Effect of different gums on features of 3D printed object based on vitamin-D enriched orange concentrate. *Food Biophysics*, *13*(3), 250–262. https://doi.org/10.1007/s11483-018-9531-x

Barakh Ali, S. F., Mohamed, E. M., Ozkan, T., Kuttolamadom, M. A., Khan, M. A., Asadi, A., & Rahman, Z. (2019). Understanding the effects of formulation and process variables on the printlets quality manufactured by selective laser sintering 3D printing. *International Journal of Pharmaceutics*, *570*, 118651. https://doi.org/10.1016/j.ijpharm.2019.118651

Bareen, M. A., Joshi, S., Sahu, J. K., Prakash, S., & Bhandari, B. (2021). Assessment of 3D printability of heat acid coagulated milk semi-solids "soft cheese" by correlating rheological, microstructural, and textural properties. *Journal of Food Engineering*, *300*, 110506. https://doi.org/10.1016/j.jfoodeng.2021.110506

BeMiller, J. N. (2018). Physical modification of starch. In *Starch in food* (pp. 223–253). Elsevier. https://doi.org/10.1016/B978-0-08-100868-3.00005-6

BeMiller, J. N., & Huber, K. C. (2015). Physical modification of food starch functionalities. *Annual Review of Food Science and Technology*, *6*(1), 19–69. https://doi.org/10.1146/annurev-food-022814-015552

Carranza, T., Guerrero, P., Caba, K., & Etxabide, A. (2023). Texture-modified soy protein foods: 3D printing design and red cabbage effect. *Food Hydrocolloids*, *145*, 109141. https://doi.org/10.1016/j.foodhyd.2023.109141

Carvajal-Mena, N., Tabilo-Munizaga, G., Pérez-Won, M., Herrera-Lavados, C., Lemus-Mondaca, R., & Moreno-Osorio, L. (2023). Evaluation of physicochemical properties of starch-protein gels: Printability and postprocessing. *LWT*, *182*, 114797. https://doi.org/10.1016/j.lwt.2023.114797

Chen, C., Zhang, M., Guo, C., & Chen, H. (2021). 4D printing of lotus root powder gel: Color change induced by microwave. *Innovative Food Science and Emerging Technologies*, *68*, 102605. https://doi.org/10.1016/j.ifset.2021.102605

Chen, C., Zhang, M., Mujumdar, A. S., & Phuhongsung, P. (2021). Investigation of 4D printing of lotus root-compound pigment gel: Effect of pH on rapid colour change. *Food Research International*, *148*, 110630. https://doi.org/10.1016/j.foodres.2021.110630

Chen, D., Fang, F., Federici, E., Campanella, O., & Jones, O. G. (2020). Rheology, microstructure and phase behavior of potato starch-protein fibril mixed gel. *Carbohydrate Polymers*, *239*, 116247. https://doi.org/10.1016/j.carbpol.2020.116247

Chen, H., Xie, F., Chen, L., & Zheng, B. (2019). Effect of rheological properties of potato, rice and corn starches on their hot-extrusion 3D printing behaviors. *Journal of Food Engineering*, *244*, 150–158. https://doi.org/10.1016/j.jfoodeng.2018.09.011

Chen, J., Sun, H., Mu, T., Blecker, C., Richel, A., Richard, G., Jacquet, N., Haubruge, E., & Goffin, D. (2022). Effect of temperature on rheological, structural, and textural properties of soy protein isolate pastes for 3D food printing. *Journal of Food Engineering*, *323*, 110917. https://doi.org/10.1016/j.jfoodeng.2021.110917

Chen, X., Zhang, M., & Tang, T. (2023). Microwave-induced rapid shape change of 4D printed vegetable-based food. *Foods*, *12*(11), 1–15. https://doi.org/10.3390/foods12112158

Chen, X., Zhang, M., Teng, X., & Mujumdar, A. S. (2022a). Internal structure design for improved shape fidelity and crispness of 3D printed pumpkin-based snacks after freeze-drying. *Food Research International*, *157*, 111220. https://doi.org/10.1016/j.foodres.2022.111220

Chen, X., Zhang, M., Teng, X., & Mujumdar, A. S. (2022b). Recent progress in modeling 3D/4D printing of foods. *Food Engineering Reviews*, *14*(1), 120–133. https://doi.org/10.1007/s12393-021-09297-6

Chen, Y., Zhang, M., Sun, Y., & Phuhongsung, P. (2022). Improving 3D/4D printing characteristics of natural food gels by novel additives: A review. *Food Hydrocolloids*, *123*, 107160. https://doi.org/10.1016/j.foodhyd.2021.107160

Cheng, T.-Y., & Liao, Y.-C. (2022). Rapid hardening process for starch-based powder bed 3D printing. *Journal of Manufacturing Processes*, *75*, 259–267. https://doi.org/10.1016/j.jmapro.2021.12.053

Cheng, Y., Fu, Y., Ma, L., Yap, P. L., Losic, D., Wang, H., & Zhang, Y. (2022). Rheology of edible food inks from 2D/3D/4D printing, and its role in future 5D/6D printing. *Food Hydrocolloids*, *132*(2). https://doi.org/10.1016/j.foodhyd.2022.107855

Cheng, Y., Liang, K., Chen, Y., Gao, W., Kang, X., Li, T., & Cui, B. (2023). Effect of molecular structure changes during starch gelatinization on its rheological and 3D printing properties. *Food Hydrocolloids*, *137*, 108364. https://doi.org/10.1016/j.foodhyd.2022.108364

Cheng, Y., Wang, B., Lv, W., Zhong, Y., & Li, G. (2024). Effect of xanthan gum on physicochemical properties and 3D printability of emulsion-filled starch gels. *Food Hydrocolloids*, *149*, 109613. https://doi.org/10.1016/j.foodhyd.2023.109613

Chow, C. Y., Thybo, C. D., Sager, V. F., Riantiningtyas, R. R., Bredie, W. L. P., & Ahrné, L. (2021). Printability, stability and sensory properties of protein-enriched 3D-printed lemon mousse for personalised in-between meals. *Food Hydrocolloids*, *120*, 106943. https://doi.org/10.1016/j.foodhyd.2021.106943

Chuang, L., Panyoyai, N., Shanks, R., & Kasapis, S. (2015). Effect of sodium chloride on the glass transition of condensed starch systems. *Food Chemistry*, *184*, 65–71. https://doi.org/10.1016/j.foodchem.2015.03.031

Cui, Y., Yang, F., Wang, C., Blennow, A., Li, C., & Liu, X. (2024). 3D printing windows and rheological properties for normal maize starch/sodium alginate composite gels. *Food Hydrocolloids*, *146*, 109178. https://doi.org/10.1016/j.foodhyd.2023.109178

Derossi, A., Caporizzi, R., Paolillo, M., & Severini, C. (2021). Programmable texture properties of cereal-based snack mediated by 3D printing technology. *Journal of Food Engineering*, *289*, 110160. https://doi.org/10.1016/j.jfoodeng.2020.110160

Derossi, A., Corradini, M. G., Caporizzi, R., Oral, M. O., & Severini, C. (2023). Accelerating the process development of innovative food products by prototyping through 3D printing technology. *Food Bioscience*, *52*, 102417. https://doi.org/10.1016/j.fbio.2023.102417

Derossi, A., Paolillo, M., Caporizzi, R., & Severini, C. (2020). Extending the 3D food printing tests at high speed. Material deposition and effect of non-printing movements on the final quality of printed structures. *Journal of Food Engineering*, *275*, 109865. https://doi.org/10.1016/j.jfoodeng.2019.109865

Derossi, A., Spence, C., Corradini, M. G., Jekle, M., Fahmy, A. R., Caporizzi, R., Devahastin, S., Moses, J. A., Le-Bail, A., Zhou, W., Zhang, M., Bhandari, B., & Severini, C. (2024). Personalized, digitally designed 3D printed food towards the reshaping of food manufacturing and consumption. *Npj Science of Food*, *8*(1), 54. https://doi.org/10.1038/s41538-024-00296-5

Dick, A., Bhandari, B., & Prakash, S. (2021). Printability and textural assessment of modified-texture cooked beef pastes for dysphagia patients. *Future Foods*, *3*, 100006. https://doi.org/10.1016/j.fufo.2020.100006

Fahmy, A. R., Amann, L. S., Dunkel, A., Frank, O., Dawid, C., Hofmann, T., Becker, T., & Jekle, M. (2021). Sensory design in food 3D printing – Structuring, texture modulation, taste localization, and thermal stabilization. *Innovative Food Science & Emerging Technologies*, *72*, 102743. https://doi.org/10.1016/j.ifset.2021.102743

Fahmy, A. R., Becker, T., & Jekle, M. (2020). 3D printing and additive manufacturing of cereal-based materials: Quality analysis of starch-based systems using a camera-based morphological approach. *Innovative Food Science & Emerging Technologies*, *63*, 102384. https://doi.org/10.1016/j.ifset.2020.102384

Fahmy, A. R., Derossi, A., & Jekle, M. (2023). Four-dimensional (4D) printing of dynamic foods: Definitions, considerations, and current scientific status. *Foods*, *12*(18), 3410. https://doi.org/10.3390/foods12183410

Fan, H., Zhang, M., Liu, Z., & Ye, Y. (2020). Effect of microwave-salt synergetic pre-treatment on the 3D printing performance of SPI-strawberry ink system. *LWT*, *122*, 109004. https://doi.org/10.1016/j.lwt.2019.109004

Farahani, R. D., Therriault, D., Dubé, M., Bodkhe, S., & Mahdavi, M. (2018). Additive manufacturing of multifunctional nanocomposites and composites. In E. E. Gdoustos (Eds), *Comprehensive composite materials II* (Vol. 6, pp. 380–407). Elsevier. https://doi.org/10.1016/B978-0-12-803581-8.10019-0

Farooq, S., & Ranjan, N. (2023). A review on binder jetting and selective laser sintering: A novel assessment of the processes for 3D insect food printing materials. In D. Singh, R. Kumar, S. Singh, & S. Ramniwas (Eds.), *3D printing of sustainable insect materials* (pp. 173–190). Springer International Publishing. https://doi.org/10.1007/978-3-031-25994-4_11

Fonseca, L. M., Gonçalves, J. R., El Halal, S. L. M., Pinto, V. Z., Dias, A. R. G., Jacques, A. C., & Zavareze, E. R. (2015). Oxidation of potato starch with different sodium hypochlorite concentrations and its effect on biodegradable films. *LWT*, *60*(2), 714–720. https://doi.org/10.1016/j.lwt.2014.10.052

García-Guzmán, L., Cabrera-Barjas, G., Soria-Hernández, C. G., Castaño, J., Guadarrama-Lezama, A. Y., & Rodríguez Llamazares, S. (2022). Progress in starch-based materials for food packaging applications. *Polysaccharides*, *3*(1), 136–177. https://doi.org/10.3390/polysaccharides3010007

Ghanbarzadeh, B., Almasi, H., & Entezami, A. A. (2011). Improving the barrier and mechanical properties of corn starch-based edible films: Effect of citric acid and carboxymethyl cellulose. *Industrial Crops and Products*, *33*(1), 229–235. https://doi.org/10.1016/j.indcrop.2010.10.016

Ghazal, A. F., Zhang, M., Bhandari, B., & Chen, H. (2021). Investigation on spontaneous 4D changes in color and flavor of healthy 3D printed food materials over time in response to external or internal pH stimulus. *Food Research International*, *142*, 110215. https://doi.org/10.1016/j.foodres.2021.110215

Ghazal, A. F., Zhang, M., & Liu, Z. (2019). Spontaneous color change of 3D printed healthy food product over time after printing as a novel application for 4D food printing. *Food and Bioprocess Technology*, *12*(10), 1627–1645. https://doi.org/10.1007/s11947-019-02327-6

Ghazal, A. F., Zhang, M., Mujumdar, A. S., & Ghamry, M. (2023). Progress in 4D/5D/6D printing of foods: Applications and R&D opportunities. *Critical Reviews in Food Science and Nutrition*, *63*(25), 7399–7422. https://doi.org/10.1080/10408398.2022.2045896

Godoi, F. C., Bhandari, B. R., Prakash, S., & Zhang, M. (2019). An introduction to the principles of 3D food printing. In *Fundamentals of 3D food printing and applications* (pp. 1–18). Elsevier. https://doi.org/10.1016/B978-0-12-814564-7.00001-8

Godoi, F. C., Prakash, S., & Bhandari, B. R. (2016). 3d printing technologies applied for food design: Status and prospects. *Journal of Food Engineering*, *179*, 44–54. https://doi.org/10.1016/j.jfoodeng.2016.01.025

Gueche, Y. A., Sanchez-Ballester, N. M., Cailleaux, S., Bataille, B., & Soulairol, I. (2021). Selective laser sintering (SLS), a new chapter in the production of solid oral forms (SOFs) by 3D printing. *Pharmaceutics*, *13*(8), 1212. https://doi.org/10.3390/pharmaceutics13081212

Gulzar, S., Narciso, J. O., Elez-Martínez, P., Martín-Belloso, O., & Soliva-Fortuny, R. (2023). Recent developments in the application of novel technologies for the modification of starch in light of 3D food printing. *Current Opinion in Food Science*, *52*, 101067. https://doi.org/10.1016/j.cofs.2023.101067

Guo, B., Wang, Y., Pang, M., Wu, J., Hu, X., Huang, Z., Wang, H., Xu, S., Luo, S., & Liu, C. (2020). Annealing treatment of amylose and amylopectin extracted from rice starch. *International Journal of Biological Macromolecules*, *164*, 3496–3500. https://doi.org/10.1016/j.ijbiomac.2020.08.245

Guo, C., Zhang, M., & Bhandari, B. (2019). Model building and slicing in food 3D printing processes: A review. *Comprehensive Reviews in Food Science and Food Safety*, *18*(4), 1052–1069. https://doi.org/10.1111/1541-4337.12443

Guo, C., Zhang, M., Bhandari, B., & Devahastin, S. (2022). Investigation on simultaneous change of deformation, color and aroma of 4D printed starch-based pastes from fruit and vegetable as induced by microwave. *Food Research International*, *157*, 111214. https://doi.org/10.1016/j.foodres.2022.111214

Guo, C., Zhang, M., & Devahastin, S. (2021a). Color/aroma changes of 3D-printed buckwheat dough with yellow flesh peach as triggered by microwave heating of gelatin-gum arabic complex coacervates. *Food Hydrocolloids*, *112*, 106358. https://doi.org/10.1016/j.foodhyd.2020.106358

Guo, C., Zhang, M., & Devahastin, S. (2021b). Improvement of 3D printability of buckwheat starch-pectin system via synergistic Ca_2^+-microwave pretreatment. *Food Hydrocolloids*, *113*, 106483. https://doi.org/10.1016/j.foodhyd.2020.106483

Guo, Y., Patanwala, H. S., Bognet, B., & Ma, A. W. K. (2017). Inkjet and inkjet-based 3D printing: Connecting fluid properties and printing performance. *Rapid Prototyping Journal*, *23*(3), 562–576. https://doi.org/10.1108/RPJ-05-2016-0076

Gupta, R. K., Guha, P., & Srivastav, P. P. (2022). Natural polymers in bio-degradable/edible film: A review on environmental concerns, cold plasma technology and nanotechnology application on food packaging-A recent trends. *Food Chemistry Advances*, *1*, 100135. https://doi.org/10.1016/j.focha.2022.100135

Gupta, R. K., Guha, P., & Srivastav, P. P. (2023). Effect of high voltage dielectric barrier discharge (DBD) atmospheric cold plasma treatment on physicochemical and functional properties of taro (*Colocasia esculenta*) starch. *International Journal of Biological Macromolecules*, *253*, 126772. https://doi.org/10.1016/j.ijbiomac.2023.126772

Gupta, R. K., Guha, P., & Srivastav, P. P. (2024a). Insights into the physicochemical, crystalline, multistructural, thermal and digestibility behavior of taro starch (*Colocasia esculenta*) modified by cold plasma treatment with high voltage dielectric barrier discharge (HVDBD): An alternative technique for chemical modification of starch and environmental sustainability. *Journal of Cleaner Production*, *461*, 142710. https://doi.org/10.1016/j.jclepro.2024.142710

Gupta, R. K., Guha, P., & Srivastav, P. P. (2024b). Dielectric barrier discharge plasma: A green and novel method to change the structure of taro peel starch and improve the physicochemical properties of taro peel starch films. *Plasma Processes and Polymers*, *21*(9), 2400047. https://doi.org/10.1002/ppap.202400047

Gupta, R. K., Guha, P., & Srivastav, P. P. (2024c). Exploring the potential of taro (*Colocasia esculenta*) starch: Recent developments in modification, health benefits, and food industry applications. *Food Bioengineering*, *3*(3), 365–379. https://doi.org/10.1002/fbe2.12103

Gupta, R. K., Guha, P., & Srivastav, P. P. (2024d). Physical action of nonthermal cold plasma technology for starch modification. *Food Physics*, *1*, 100011. https://doi.org/10.1016/j.foodp.2024.100011

Gupta, R. K., Guha, P., & Srivastav, P. P. (2024e). Effect of cold plasma treatment and plasma-activated water on physicochemical and structural properties of starch: A green and novel approach for environmental sustainability. *Plasma Processes and Polymers*, *21*(4), 2300204. https://doi.org/10.1002/ppap.202300204

Gupta, R. K., Gnana Moorthy Eswaran, U., Kumar, S., Pipliya, S., Kaur, B., & Srivastav, P. P. (2025a). Synergistic modification of starch polysaccharide via sequential treatments with ultrasonic and cold plasma technology. *Journal of Food Processing and Preservation*, *2025*(1), 9063007. https://doi.org/10.1155/jfpp/9063007

Gupta, R. K., Pipliya, S., Patel, J., Srivastav, P. P., Castro-Muñoz, R., Wang, C. K., & Nemţanu, M. R. (2025b). Essential oil-embedded starch based films for active packaging: Insights into migration, stability, and preservation efficiency. *Trends in Food Science & Technology*, 163, 105149. https://doi.org/10.1016/j.tifs.2025.105149

Hai Alami, A., Ghani Olabi, A., Khuri, S., Aljaghoub, H., Alasad, S., Ramadan, M., & Ali Abdelkareem, M. (2024). 3D printing in the food industry: Recent progress and role in achieving sustainable development goals. *Ain Shams Engineering Journal*, *15*(2), 102386. https://doi.org/10.1016/j.asej.2023.102386

Han, X., Wen, H., Luo, Y., Yang, J., Xiao, W., & Xie, J. (2022). Effects of chitosan modification, cross-linking, and oxidation on the structure, thermal stability, and adsorption properties of porous maize starch. *Food Hydrocolloids*, *124*, 107288. https://doi.org/10.1016/j.foodhyd.2021.107288

Haroon, M., Wang, L., Yu, H., Abbasi, N. M., Zain-ul-Abdin, Z.-A., Saleem, M., Khan, R. U., Ullah, R. S., Chen, Q., & Wu, J. (2016). Chemical modification of starch and its application as an adsorbent material. *RSC Advances*, *6*(82), 78264–78285. https://doi.org/10.1039/C6RA16795K

He, C., Zhang, M., & Devahastin, S. (2020). Investigation on spontaneous shape change of 4D printed starch-based purees from purple sweet potatoes as induced by microwave dehydration. *ACS Applied Materials & Interfaces*, *12*(34), 37896–37905. https://doi.org/10.1021/acsami.0c10899

He, C., Zhang, M., & Devahastin, S. (2021). Microwave-induced deformation behaviors of 4D printed starch-based food products as affected by edible salt and butter content. *Innovative Food Science and Emerging Technologies*, *70*, 102699. https://doi.org/10.1016/j.ifset.2021.102699

He, C., Zhang, M., & Fang, Z. (2020). 3D printing of food: Pretreatment and post-treatment of materials. *Critical Reviews in Food Science and Nutrition*, *60*(14), 2379–2392. https://doi.org/10.1080/10408398.2019.1641065

Homan, K. A., Kolesky, D. B., Skylar-Scott, M. A., Herrmann, J., Obuobi, H., Moisan, A., & Lewis, J. A. (2016). Bioprinting of 3D convoluted renal proximal tubules on perfusable chips. *Scientific Reports*, *6*(1), 13. https://doi.org/10.1038/srep34845

Hong, J., Zeng, X.-A., Han, Z., & Brennan, C. S. (2018). Effect of pulsed electric fields treatment on the nanostructure of esterified potato starch and their potential glycemic digestibility. *Innovative Food Science & Emerging Technologies*, *45*, 438–446. https://doi.org/10.1016/j.ifset.2017.11.009

Hsiao, W.-K., Martin, G. D., & Hutchings, I. M. (2014). Printing stable liquid tracks on a surface with finite receding contact angle. *Langmuir*, *30*(41), 12447–12455. https://doi.org/10.1021/la502490p

Huang, J., Zhang, M., Mujumdar, A. S., & Li, C. (2024). Modulation of starch structure, swallowability and digestibility of 3D-printed diabetic-friendly food for the elderly by dry heating. *International Journal of Biological Macromolecules*, *264*, 130629. https://doi.org/10.1016/j.ijbiomac.2024.130629

Huang, M., Zhang, M., & Guo, C. (2020). 3D printability of brown rice gel modified by some food hydrocolloids. *Journal of Food Processing and Preservation*, *44*(7). https://doi.org/10.1111/jfpp.14502

Huang, Z., Lu, J., Li, X., & Tong, Z. (2007). Effect of mechanical activation on physico-chemical properties and structure of cassava starch. *Carbohydrate Polymers*, *68*(1), 128–135. https://doi.org/10.1016/j.carbpol.2006.07.017

Hussain, S., Arora, V. K., & Malakar, S. (2021). Formulation of protein-enriched 3D printable food matrix and evaluation of textural, rheological characteristics, and printing stability. *Journal of Food Processing and Preservation*, *45*(2). https://doi.org/10.1111/jfpp.15182

In, J., Jeong, H., Song, S., & Min, S. C. (2021). Determination of material requirements for 3D gel food printing using a fused deposition modeling 3D printer. *Foods*, *10*(10), 2272. https://doi.org/10.3390/foods10102272

Ji, S., Xu, T., Li, Y., Li, H., Zhong, Y., & Lu, B. (2022). Effect of starch molecular structure on precision and texture properties of 3D printed products. *Food Hydrocolloids*, *125*, 107387. https://doi.org/10.1016/j.foodhyd.2021.107387

Ji, S., Xu, T., Liu, Y., Li, H., Luo, J., Zou, Y., Zhong, Y., Li, Y., & Lu, B. (2022). Investigation of the mechanism of casein protein to enhance 3D printing accuracy of cassava starch gel. *Carbohydrate Polymers*, *295*, 119827. https://doi.org/10.1016/j.carbpol.2022.119827

Jiang, Q., Binks, B. P., & Meng, Z. (2022). Double scaffold networks regulate edible Pickering emulsion gel for designing thermally actuated 4D printing. *Food Hydrocolloids*, *133*, 107969. https://doi.org/10.1016/j.foodhyd.2022.107969

Jonkers, N., van Dijk, W. J., Vonk, N. H., van Dommelen, J. A. W., & Geers, M. G. D. (2022). Anisotropic mechanical properties of selective laser sintered starch-based food. *Journal of Food Engineering*, *318*, 110890. https://doi.org/10.1016/j.jfoodeng.2021.110890

Jonkers, N., van Dommelen, J. A. W., & Geers, M. G. D. (2022). Selective laser sintered food: A unit cell approach to design mechanical properties. *Journal of Food Engineering*, *335*, 111183. https://doi.org/10.1016/j.jfoodeng.2022.111183

Kadival, A., Mitra, J., Machavaram, R., & Kaushal, M. (2024). 3D printing of rice starch incorporated peanut protein isolate paste: Rheological characterization and simulation of flow properties. *Innovative Food Science & Emerging Technologies*, *94*, 103669. https://doi.org/10.1016/j.ifset.2024.103669

Kang, Z., Wang, Z., Zhang, W., & Fang, Z. (2023). 4D printing of edible insect snacks: Color changes with external pH stimulation. *Food Bioscience*, *55*, 102971. https://doi.org/10.1016/j.fbio.2023.102971

Khanal, R. C., Howard, L. R., & Prior, R. L. (2010). Effect of heating on the stability of grape and blueberry pomace procyanidins and total anthocyanins. *Food Research International*, *43*(5), 1464–1469. https://doi.org/10.1016/j.foodres.2010.04.018

Kim, H. W., Lee, I. J., Park, S. M., Lee, J. H., Nguyen, M.-H., & Park, H. J. (2019). Effect of hydrocolloid addition on dimensional stability in post-processing of 3D printable cookie dough. *LWT*, *101*, 69–75. https://doi.org/10.1016/j.lwt.2018.11.019

Kim, H. W., Lee, J. H., Park, S. M., Lee, M. H., Lee, I. W., Doh, H. S., & Park, H. J. (2018). Effect of hydrocolloids on rheological properties and printability of vegetable inks for 3D food printing. *Journal of Food Science*, *83*(12), 2923–2932. https://doi.org/10.1111/1750-3841.14391

Kim, S. M., Kim, H. W., & Park, H. J. (2021). Preparation and characterization of surimi-based imitation crab meat using coaxial extrusion three-dimensional food printing. *Innovative Food Science & Emerging Technologies*, *71*, 102711. https://doi.org/10.1016/j.ifset.2021.102711

Kim, Y., & Jung, C. (2022). Reaction mechanisms applied to starch modification for biodegradable plastics: Etherification and esterification. *International Journal of Polymer Science*, *2022*, 1–10. https://doi.org/10.1155/2022/2941406

Kokane, S. B., Arora, V. K., & Thangalakshmi, S. (2024). Potential of bioactive ingredients for healthy 4D food printing: Applications and challenges. *Food Reviews International*, *40*(10), 1–24. https://doi.org/10.1080/87559129.2024.2374808

Kong, X., Zhu, P., Sui, Z., & Bao, J. (2015). Physicochemical properties of starches from diverse rice cultivars varying in apparent amylose content and gelatinisation temperature combinations. *Food Chemistry*, *172*, 433–440. https://doi.org/10.1016/j.foodchem.2014.09.085

Kong, X.-R., Zhu, Z.-Y., Zhang, X.-J., & Zhu, Y.-M. (2020). Effects of cordyceps polysaccharides on pasting properties and in vitro starch digestibility of wheat starch. *Food Hydrocolloids*, *102*, 105604. https://doi.org/10.1016/j.foodhyd.2019.105604

Koo, S. H., Lee, K. Y., & Lee, H. G. (2010). Effect of cross-linking on the physicochemical and physiological properties of corn starch. *Food Hydrocolloids*, *24*(6–7), 619–625. https://doi.org/10.1016/j.foodhyd.2010.02.009

Lee, C. P., Ng, M. J. Y., Chian, N. M. Y., & Hashimoto, M. (2024). Multi-material direct ink writing 3D food printing using multi-channel nozzle. *Future Foods*, *10*, 100376. https://doi.org/10.1016/j.fufo.2024.100376

Lee, J. (2021). A 3D food printing process for the new normal era: A review. *Processes*, *9*(9), 1495. https://doi.org/10.3390/pr9091495

Li, G., Zhan, J., Hu, Z., Huang, J., Luo, X., Chen, J., Yuan, C., Takaki, K., & Hu, Y. (2022). 3D printing properties and printability definition of *Pennahiaargentata* surimi and rice starch. *Food Bioscience*, *48*, 101748. https://doi.org/10.1016/j.fbio.2022.101748

Li, Y., Ren, J., Liu, J., Sun, L., Wang, Y., Liu, B., Li, C., & Li, Z. (2018). Modification by α-D-glucan branching enzyme lowers the in vitro digestibility of starch from different sources. *International Journal of Biological Macromolecules*, *107*, 1758–1764. https://doi.org/10.1016/j.ijbiomac.2017.10.049

Lille, M., Nurmela, A., Nordlund, E., Metsä-Kortelainen, S., & Sozer, N. (2018). Applicability of protein and fiber-rich food materials in extrusion-based 3D printing. *Journal of Food Engineering*, *220*, 20–27. https://doi.org/10.1016/j.jfoodeng.2017.04.034

Lin, Y.-J., Punpongsanon, P., Wen, X., Iwai, D., Sato, K., Obrist, M., & Mueller, S. (2020). FoodFab: Creating food perception illusions using food 3D printing. *Proceedings of the 2020 CHI Conference on Human Factors in Computing Systems*, 1–13. https://doi.org/10.1145/3313831.3376421

Liu, C., Zhang, H., Chen, R., Chen, J., Liu, X., Luo, S., & Chen, T. (2021). Effects of creeping fig seed polysaccharide on pasting, rheological, textural properties and in vitro digestibility of potato starch. *Food Hydrocolloids*, *118*, 106810. https://doi.org/10.1016/j.foodhyd.2021.106810

Liu, D., Li, Z., Fan, Z., Zhang, X., & Zhong, G. (2019). Effect of soybean soluble polysaccharide on the pasting, gels, and rheological properties of kudzu and lotus starches. *Food Hydrocolloids*, *89*, 443–452. https://doi.org/10.1016/j.foodhyd.2018.11.003

Liu, L., & Ciftci, O. N. (2021). Effects of high oil compositions and printing parameters on food paste properties and printability in a 3D printing food processing model. *Journal of Food Engineering*, *288*, 110135. https://doi.org/10.1016/j.jfoodeng.2020.110135

Liu, P., Zhou, Z., Zhou, W., Liu, C., Wang, N., & Zhang, F. (2024). Spray-drying assisted 3D printing to manufacture beef products. *Starch - Stärke*, *76*(9–10). https://doi.org/10.1002/star.202300286

Liu, W., He, K., McClements, D. J., Jin, Z., & Chen, L. (2024). Pea starch-based oleogels based on capillary water cross-linking: Physicochemical properties and 3D printing performance. *Food Hydrocolloids*, *154*, 110161. https://doi.org/10.1016/j.foodhyd.2024.110161

Liu, Y., Liu, D., Wei, G., Ma, Y., Bhandari, B., & Zhou, P. (2018). 3D printed milk protein food simulant: Improving the printing performance of milk protein concentration by incorporating whey protein isolate. *Innovative Food Science & Emerging Technologies*, *49*, 116–126. https://doi.org/10.1016/j.ifset.2018.07.018

Liu, Z., Bhandari, B., Prakash, S., & Zhang, M. (2018). Creation of internal structure of mashed potato construct by 3D printing and its textural properties. *Food Research International*, *111*, 534–543. https://doi.org/10.1016/j.foodres.2018.05.075

Liu, Z., Chen, H., Zheng, B., Xie, F., & Chen, L. (2020). Understanding the structure and rheological properties of potato starch induced by hot-extrusion 3D printing. *Food Hydrocolloids*, *105*, 105812. https://doi.org/10.1016/j.foodhyd.2020.105812

Liu, Z., Chen, L., & Zheng, B. (2022). Control of starch–lipid interactions on starch digestibility during hot-extrusion 3D printing for starchy foods. *Food & Function*, *13*(9), 5317–5326. https://doi.org/10.1039/D2FO00558A

Liu, Z., Chen, X., Dai, Q., Xu, D., Hu, L., Li, H., Hati, S., Chitrakar, B., Yao, L., & Mo, H. (2023). Pea protein-xanthan gum interaction driving the development of 3D printed dysphagia diet. *Food Hydrocolloids*, *139*, 108497. https://doi.org/10.1016/j.foodhyd.2023.108497

Liu, Z., Zhang, M., Bhandari, B., & Wang, Y. (2017). 3D printing: Printing precision and application in food sector. *Trends in Food Science & Technology*, *69*, 83–94. https://doi.org/10.1016/j.tifs.2017.08.018

Liu, Z., Zhang, M., & Bhandari, B. (2018). Effect of gums on the rheological, microstructural and extrusion printing characteristics of mashed potatoes. *International Journal of Biological Macromolecules*, *117*, 1179–1187. https://doi.org/10.1016/j.ijbiomac.2018.06.048

Liu, Z., Zhang, M., & Yang, C. (2018). Dual extrusion 3D printing of mashed potatoes/strawberry juice gel. *LWT*, *96*, 589–596. https://doi.org/10.1016/j.lwt.2018.06.014

Lu, K., Du, Z., Chen, J., Kong, Y., & Huang, Q. (2024). Radio frequency air cold plasma pretreatment to enhance 3D printing performance of tapioca starch by altering its rheological behavior and water migration. *Food Hydrocolloids*, *153*, 110032. https://doi.org/10.1016/j.foodhyd.2024.110032

Ma, H., Liu, M., Liang, Y., Zheng, X., Sun, L., Dang, W., Li, J., Li, L., & Liu, C. (2022). Research progress on properties of pre-gelatinized starch and its application in wheat flour products. *Grain & Oil Science and Technology*, *5*(2), 87–97. https://doi.org/10.1016/j.gaost.2022.01.001

Ma, S., Ma, T., Tsuchikawa, S., Inagaki, T., Wang, H., & Jiang, H. (2024). Effect of dielectric barrier discharge (DBD) plasma treatment on physicochemical and 3D printing properties of wheat starch. *International Journal of Biological Macromolecules*, *269*, 132159. https://doi.org/10.1016/j.ijbiomac.2024.132159

Ma, S., Zhang, M., Wang, X., Yang, Y., He, L., Deng, J., & Jiang, H. (2024). Effect of plasma-activated water on the quality of wheat starch gel-forming 3D printed samples. *International Journal of Biological Macromolecules*, *274*, 133552. https://doi.org/10.1016/j.ijbiomac.2024.133552

Ma, S., Zhang, Q., Lin, Q., Pan, L., Yu, X., & Jiang, H. (2023). Performance of 3D-printed samples based on starch treated by radio frequency energy. *Innovative Food Science & Emerging Technologies*, *85*, 103337. https://doi.org/10.1016/j.ifset.2023.103337

Ma, S., Zhu, P., & Wang, M. (2019). Effects of konjac glucomannan on pasting and rheological properties of corn starch. *Food Hydrocolloids*, *89*, 234–240. https://doi.org/10.1016/j.foodhyd.2018.10.045

Ma, X., Jian, R., Chang, P. R., & Yu, J. (2008). Fabrication and characterization of citric acid-modified starch nanoparticles/plasticized-starch composites. *Biomacromolecules*, *9*(11), 3314–3320. https://doi.org/10.1021/bm800987c

Majzoobi, M., & Farahnaky, A. (2021). Granular cold-water swelling starch; Properties, preparation and applications, a review. *Food Hydrocolloids*, *111*, 106393. https://doi.org/10.1016/j.foodhyd.2020.106393

Maniglia, B. C., Lima, D. C., da Matta Júnior, M., Oge, A., Le-Bail, P., Augusto, P. E. D., & Le-Bail, A. (2020a). Dry heating treatment: A potential tool to improve the wheat starch properties for 3D food printing application. *Food Research International*, *137*, 109731. https://doi.org/10.1016/j.foodres.2020.109731

Maniglia, B. C., Lima, D. C., da Matta Júnior, M., Le-Bail, P., Le-Bail, A., & Augusto, P. E. D. (2019). Hydrogels based on ozonated cassava starch: Effect of ozone processing and gelatinization conditions on enhancing 3D-printing applications. *International Journal of Biological Macromolecules*, *138*, 1087–1097. https://doi.org/10.1016/j.ijbiomac.2019.07.124

Mantihal, S., Prakash, S., & Bhandari, B. (2019). Textural modification of 3D printed dark chocolate by varying internal infill structure. *Food Research International*, *121*, 648–657. https://doi.org/10.1016/j.foodres.2018.12.034

Martínez-Monzó, J., Cárdenas, J., & García-Segovia, P. (2019). Effect of temperature on 3D printing of commercial potato puree. *Food Biophysics*, *14*(3), 225–234. https://doi.org/10.1007/s11483-019-09576-0

Mustapha, K. B., & Metwalli, K. M. (2021). A review of fused deposition modelling for 3D printing of smart polymeric materials and composites. *European Polymer Journal*, *156*, 110591. https://doi.org/10.1016/j.eurpolymj.2021.110591

Nachal, N., Moses, J. A., Karthik, P., & Anandharamakrishnan, C. (2019). Applications of 3D printing in food processing. *Food Engineering Reviews*, *11*(3), 123–141. https://doi.org/10.1007/s12393-019-09199-8

Narupai, B., Smith, P. T., & Nelson, A. (2021). 4D printing of multi-stimuli responsive protein-based hydrogels for autonomous shape transformations. *Advanced Functional Materials*, *31*(23), 1–9. https://doi.org/10.1002/adfm.202011012

Nida, S., Moses, J. A., & Anandharamakrishnan, C. (2022). Emerging applications of 5D and 6D printing in the food industry. *Journal of Agriculture and Food Research*, *10*, 100392. https://doi.org/10.1016/j.jafr.2022.100392

Noè, C., Tonda-Turo, C., Chiappone, A., Sangermano, M., & Hakkarainen, M. (2020). Light processable starch hydrogels. *Polymers*, *12*(6), 1359. https://doi.org/10.3390/polym12061359

Ojukwu, M., Tan, H. L., Murad, M., Nafchi, A. M., & Easa, A. M. (2023). Improvement of cooking and textural properties of rice flour-soy protein isolate noodles stabilised with microbial transglutaminase and glucono-δ-lactone and dried using superheated steam. *Food Science and Technology International*, *29*(8), 799–808. https://doi.org/10.1177/10820132221121169

Oral, M. O., Derossi, A., Caporizzi, R., & Severini, C. (2021). Analyzing the most promising innovations in food printing. Programmable food texture and 4D foods. *Future Foods*, *4*, 100093. https://doi.org/10.1016/j.fufo.2021.100093

Outrequin, T. C. R., Gamonpilas, C., Siriwatwechakul, W., & Sreearunothai, P. (2023). Extrusion-based 3D printing of food biopolymers: A highlight on the important rheological parameters to reach printability. *Journal of Food Engineering*, *342*, 111371. https://doi.org/10.1016/j.jfoodeng.2022.111371

Oyinloye, T., & Yoon, W. (2021). Application of computational fluid dynamics (CFD) in the deposition process and printability assessment of 3D printing using rice paste. *Processes*, *10*(1), 68. https://doi.org/10.3390/pr10010068

Patoliya, J., Makwana, M., Moradiya, P., & Vala, K. (2023). Shape deformation/transformation in 4D printed food: A review. *12*(6), 25–32.

Pérez, B., Nykvist, H., Brøgger, A. F., Larsen, M. B., & Falkeborg, M. F. (2019). Impact of macronutrients printability and 3D-printer parameters on 3D-food printing: A review. *Food Chemistry*, *287*, 249–257. https://doi.org/10.1016/j.foodchem.2019.02.090

Phuhongsung, P., Zhang, M., & Devahastin, S. (2020a). Influence of surface pH on color, texture and flavor of 3D printed composite mixture of soy protein isolate, pumpkin, and beetroot. *Food and Bioprocess Technology*, *13*(9), 1600–1610. https://doi.org/10.1007/s11947-020-02497-8

Phuhongsung, P., Zhang, M., & Devahastin, S. (2020b). Investigation on 3D printing ability of soybean protein isolate gels and correlations with their rheological and textural properties via LF-NMR spectroscopic characteristics. *LWT*, *122*, 109019. https://doi.org/10.1016/j.lwt.2020.109019

Phuhongsung, P., Zhang, M., Devahastin, S., & Mujumdar, A. S. (2022). Defects in 3D/4D food printing and their possible solutions: A comprehensive review. *Comprehensive Reviews in Food Science and Food Safety*, *21*(4), 3455–3479. https://doi.org/10.1111/1541-4337.12984

Pingale, P., Dawre, S., Dhapte-Pawar, V., Dhas, N., & Rajput, A. (2023). Advances in 4D printing: From stimulation to simulation. *Drug Delivery and Translational Research*, *13*(1), 164–188. https://doi.org/10.1007/s13346-022-01200-y

Portanguen, S., Tournayre, P., Sicard, J., Astruc, T., & Mirade, P.-S. (2019). Toward the design of functional foods and biobased products by 3D printing: A review. *Trends in Food Science & Technology*, *86*, 188–198. https://doi.org/10.1016/j.tifs.2019.02.023

Punia Bangar, S., Ashogbon, A. O., Singh, A., Chaudhary, V., & Whiteside, W. S. (2022). Enzymatic modification of starch: A green approach for starch applications. *Carbohydrate Polymers*, *287*, 119265. https://doi.org/10.1016/j.carbpol.2022.119265

Rajagopalan, S., & Seib, P. A. (1992). Granular cold-water-soluble starches prepared at atmospheric pressure. *Journal of Cereal Science*, *16*(1), 13–28. https://doi.org/10.1016/S0733-5210(09)80076-3

Rando, P., & Ramaioli, M. (2021). Food 3D printing: Effect of heat transfer on print stability of chocolate. *Journal of Food Engineering*, *294*, 110415. https://doi.org/10.1016/j.jfoodeng.2020.110415

Ren, S., Tang, T., Bi, X., Liu, X., Xu, P., & Che, Z. (2023). Effects of pea protein isolate on 3D printing performance, nutritional and sensory properties of mango pulp. *Food Bioscience*, *55*, 102994. https://doi.org/10.1016/j.fbio.2023.102994

Rincón-Londoño, N., Millan-Malo, B., & Rodríguez-García, M. E. (2016). Analysis of thermal pasting profile in corn starch rich in amylose and amylopectin: Physicochemical transformations, part II. *International Journal of Biological Macromolecules*, *89*, 43–53. https://doi.org/10.1016/j.ijbiomac.2016.04.057

Rong, L., Chen, X., Shen, M., Yang, J., Qi, X., Li, Y., & Xie, J. (2023). The application of 3D printing technology on starch-based product: A review. *Trends in Food Science & Technology*, *134*, 149–161. https://doi.org/10.1016/j.tifs.2023.02.015

Sandhu, K., & Singh, S. (2022). Food printing: 3D printing in food industry. In *Food printing: 3D printing in food industry*. https://doi.org/10.1007/978-981-16-8121-9

Sarifudin, A., Sholichah, E., Setiaboma, W., & Tongta, S. (2021). Cold water swelling starch: Methods to prepare and recent applications. *IOP Conference Series: Materials Science and Engineering*, *1011*(1), 012025. https://doi.org/10.1088/1757-899X/1011/1/012025

Schmiele, M., Sampaio, U. M., Gomes, P. T. G., & Clerici, M. T. P. S. (2019). Physical modifications of starch. In *Starches for food application* (pp. 223–269). Elsevier. https://doi.org/10.1016/B978-0-12-809440-2.00006-X

Schmitz, C. S., De Simas, K. N., Santos, K., João, J. J., De Mello Castanho Amboni, R. D., & Amante, E. R. (2006). Cassava starch functional properties by etherification–hydroxypropylation. *International Journal of Food Science & Technology*, *41*(6), 681–687. https://doi.org/10.1111/j.1365-2621.2005.01136.x

Schutyser, M. A. I., Houlder, S., de Wit, M., Buijsse, C. A. P., & Alting, A. C. (2018). Fused deposition modelling of sodium caseinate dispersions. *Journal of Food Engineering*, *220*, 49–55. https://doi.org/10.1016/j.jfoodeng.2017.02.004

Shahbazi, M., Jäger, H., & Ettelaie, R. (2022). Kinetic evaluation of the starch molecular behavior under extrusion-based or laser powder bed fusion 3D printing systems: A systematic structural and biological comparison. *Additive Manufacturing*, *57*, 102934. https://doi.org/10.1016/j.addma.2022.102934

Shahbazi, M., Jäger, H., Ettelaie, R., & Ulbrich, M. (2023). Insights into the supramolecular structure and degradation mechanisms of starch from different botanical sources as affected by extrusion-based 3D printing. *Biomacromolecules*, *24*(1), 69–85. https://doi.org/10.1021/acs.biomac.2c00881

Shen, C., Chen, W., Li, C., Chen, X., Cui, H., & Lin, L. (2023). 4D printing system stimulated by curcumin/whey protein isolate nanoparticles: A comparative study of sensitive color change and post-processing. *Journal of Food Engineering*, *342*, 111357. https://doi.org/10.1016/j.jfoodeng.2022.111357

Shi, H., Zhang, M., & Mujumdar, A. S. (2024). 3D/4D printed super reconstructed foods: Characteristics, research progress, and prospects. *Comprehensive Reviews in Food Science and Food Safety*, *23*(2), 1–32. https://doi.org/10.1111/1541-4337.13310

Shi, Y., Zhang, M., & Bhandari, B. (2021). Effect of addition of beeswax based oleogel on 3D printing of potato starch-protein system. *Food Structure*, *27*, 100176. https://doi.org/10.1016/j.foostr.2021.100176

Shirazi, S. F. S., Gharehkhani, S., Mehrali, M., Yarmand, H., Metselaar, H. S. C., Adib Kadri, N., & Osman, N. A. A. (2015). A review on powder-based additive manufacturing for tissue engineering: Selective laser sintering and inkjet 3D printing. *Science and Technology of Advanced Materials*, *16*(3), 033502. https://doi.org/10.1088/1468-6996/16/3/033502

Shrivastava, A., Gupta, R. K., & Srivastav, P. P. (2024). Exploring novel frontiers of advancements in purple yam (*Dioscorea alata* L.) starch extraction, modification, characterization, applications in food and other industries. *Measurement: Food*, *15*, 100196. https://doi.org/10.1016/j.meafoo.2024.100196

Singh, P., Pandey, V. K., Tripathi, A., Singh, R., Net, T. S. S., Ramniwas, S., & Pandiselvam, R. (2024). 4D food printing technology: Structural changes to culinary art and beyond. *Journal of Food Process Engineering*, *47*(1), 1–15. https://doi.org/10.1111/jfpe.14535

Singhal, S., Rasane, P., Kaur, S., Garba, U., Bankar, A., Singh, J., & Gupta, N. (2020). 3D food printing: Paving way towards novel foods. *Anais Da Academia Brasileira De Ciências*, *92*(3). https://doi.org/10.1590/0001-3765202020180737

Sinhmar, A., Pathera, A. K., Sharma, S., Nehra, M., Thory, R., & Nain, V. (2023). Impact of various modification methods on physicochemical and functional properties of starch: A review. *Starch - Stärke*, *75*(1–2). https://doi.org/10.1002/star.202200117

Siroha, A. K., Bangar, S. P., Sandhu, K. S., Trif, M., Kumar, M., & Guleria, P. (2021). Effect of cross-linking modification on structural and film-forming characteristics of pearl millet (*Pennisetum glaucum* L.) starch. *Coatings*, *11*(10), 1163. https://doi.org/10.3390/coatings11101163

Skylar-Scott, M. A., Mueller, J., Visser, C. W., & Lewis, J. A. (2019). Voxelated soft matter via multimaterial multinozzle 3D printing. *Nature*, *575*(7782), 330–335. https://doi.org/10.1038/s41586-019-1736-8

Sriprablom, J., Tatikunakorn, P., Lerdpriyanun, P., Suphantharika, M., & Wongsagonsup, R. (2023). Effect of single and dual modifications with cross-linking and octenylsuccinylation on physicochemical, in-vitro digestibility, and emulsifying properties of cassava starch. *Food Research International*, *163*, 112304. https://doi.org/10.1016/j.foodres.2022.112304

Srivastava, S., Pandey, V. K., Singh, R., & Dar, A. H. (2023). Recent insights on advancements and substantial transformations in food printing technology from 3 to 7D. *Food Science and Biotechnology*, *32*(13), 1783–1804. https://doi.org/10.1007/s10068-023-01352-8

Sun, J., Peng, Z., Zhou, W., Fuh, J. Y. H., Hong, G. S., & Chiu, A. (2015). A review on 3D printing for customized food fabrication. *Procedia Manufacturing*, *1*, 308–319. https://doi.org/10.1016/j.promfg.2015.09.057

Sun, J., Zhou, W., Huang, D., Fuh, J. Y. H., & Hong, G. S. (2015). An overview of 3D printing technologies for food fabrication. *Food and Bioprocess Technology*, *8*(8), 1605–1615. https://doi.org/10.1007/s11947-015-1528-6

Sun, J., Zhou, W., Yan, L., Huang, D., & Lin, L. (2018). Extrusion-based food printing for digitalized food design and nutrition control. *Journal of Food Engineering*, *220*, 1–11. https://doi.org/10.1016/j.jfoodeng.2017.02.028

Tang, H., Qu, Y., Li, Y., & Dong, S. (2018). Surface modification mechanism of cross-linking and acetylation, and their influence on characteristics of high amylose corn starch. *Journal of Food Science*, *83*(6), 1533–1541. https://doi.org/10.1111/1750-3841.14161

Tang, T., Zhang, M., Bhandari, B., & Li, C. (2024). Personalized transformation of 3D printing for traditional multi-material food with stuffing: A review. *Food Bioscience*, *59*, 104112. https://doi.org/10.1016/j.fbio.2024.104112

Tao, Y., Do, Y., Yang, H., Lee, Y. C., Wang, G., Mondoa, C., Cui, J., Wang, W., & Yao, L. (2019). Morphlour: Personalized flour-based morphing food induced by dehydration or hydration method. *UIST 2019: Proceedings of the 32nd Annual ACM Symposium on User Interface Software and Technology*, 329–340. https://doi.org/10.1145/3332165.3347949

Teng, X., Zhang, M., & Mujumdar, A. S. (2021). 4D printing: Recent advances and proposals in the food sector. *Trends in Food Science & Technology*, *110*, 349–363. https://doi.org/10.1016/j.tifs.2021.01.076

Tomašević, I., Putnik, P., Valjak, F., Pavlić, B., Šojić, B., Bebek Markovinović, A., & Bursać Kovačević, D. (2021). 3D printing as novel tool for fruit-based functional food production. *Current Opinion in Food Science*, *41*, 138–145. https://doi.org/10.1016/j.cofs.2021.03.015

Vadodaria, S., & Mills, T. (2020). Jetting-based 3D printing of edible materials. *Food Hydrocolloids*, *106*, 105857. https://doi.org/10.1016/j.foodhyd.2020.105857

Varvara, R.-A., Szabo, K., & Vodnar, D. C. (2021). 3D food printing: Principles of obtaining digitally-designed nourishment. *Nutrients*, *13*(10), 3617. https://doi.org/10.3390/nu13103617

Wang, L., Zhang, M., Bhandari, B., & Yang, C. (2018). Investigation on fish surimi gel as promising food material for 3D printing. *Journal of Food Engineering*, *220*, 101–108. https://doi.org/10.1016/j.jfoodeng.2017.02.029

Wang, N., Li, R., Wang, X., & Yang, X. (2024). 4D food printing: Key factors and optimization strategies. *Trends in Food Science and Technology*, *145*, 104380. https://doi.org/10.1016/j.tifs.2024.104380

Wang, S., & Copeland, L. (2015). Effect of acid hydrolysis on starch structure and functionality: A review. *Critical Reviews in Food Science and Nutrition*, *55*(8), 1081–1097. https://doi.org/10.1080/10408398.2012.684551

Waseem, M., Tahir, A. U., & Majeed, Y. (2024). Printing the future of food: The physics perspective on 3D food printing. *Food Physics*, *1*, 100003. https://doi.org/10.1016/j.foodp.2023.100003

Wasserman, L. A., Papakhin, A. A., Krivandin, A. V., Filatova, A. G., Borodina, Z. M., & Plashchina, I. G. (2022). Changes in the structure, thermodynamic, and functional properties of maize starch during mechanical processing. *Russian Journal of Physical Chemistry B*, *16*(1), 141–147. https://doi.org/10.1134/S1990793122010328

Wilms, P., Daffner, K., Kern, C., Gras, S. L., Schutyser, M. A. I., & Kohlus, R. (2021). Formulation engineering of food systems for 3D-printing applications: A review. *Food Research International*, *148*, 110585. https://doi.org/10.1016/j.foodres.2021.110585

Wu, H., Sang, S., Weng, P., Pan, D., Wu, Z., Yang, J., Liu, L., Farag, M. A., Xiao, J., & Liu, L. (2023). Structural, rheological, and gelling characteristics of starch-based materials in context to 3D food printing applications in precision nutrition. *Comprehensive Reviews in Food Science and Food Safety*, *22*(6), 4217–4241. https://doi.org/10.1111/1541-4337.13217

Wu, K., Li, C., Li, Z., Li, Z., Gu, Z., Ban, X., Hong, Y., Cheng, L., & Kong, H. (2024). Highly-branched modification of starch: An enzymatic approach to regulating its properties. *Food Hydrocolloids*, *147*, 109433. https://doi.org/10.1016/j.foodhyd.2023.109433

Xian, D., Wu, L., Lin, K., Liu, P., Wu, S., Yuan, Y., & Xie, F. (2024). Augmenting corn starch gel printability for architectural 3D modeling for customized food. *Food Hydrocolloids*, *156*, 110294. https://doi.org/10.1016/j.foodhyd.2024.110294

Xiao, X., Yang, L., Xu, Z., Huang, P., Shu, C., Song, S., Zhang, Y., & Pei, H. (2024). Research on rice starch gel preparation and crosslink network structure-rheological property based on direct-writing 3D printing. *Heliyon*, *10*(2). https://doi.org/10.1016/j.heliyon.2024.e24057

Xu, J., Andrews, T. D., & Shi, Y. (2020). Recent advances in the preparation and characterization of intermediately to highly esterified and etherified starches: A review. *Starch - Stärke*, *72*(3–4). https://doi.org/10.1002/star.201900238

Xu, K., Zhang, M., & Bhandari, B. (2020). Effect of novel ultrasonic- microwave combined pretreatment on the quality of 3D printed wheat starch-papaya system. *Food Biophysics*, *15*(2), 249–260. https://doi.org/10.1007/s11483-019-09615-w

Xu, M., Xu, Y., Ji, S., Amrouche, A. T., Li, Y., Zhou, Z., Shen, J., Li, K., & Lu, B. (2024). Investigation of the 3D printability of modified starch-based inks and their formation mechanism: Application in ice cream. *Food Hydrocolloids*, *154*, 110038. https://doi.org/10.1016/j.foodhyd.2024.110038

Yang, F., Guo, C., Zhang, M., Bhandari, B., & Liu, Y. (2019). Improving 3D printing process of lemon juice gel based on fluid flow numerical simulation. *LWT*, *102*, 89–99. https://doi.org/10.1016/j.lwt.2018.12.031

Yang, F., Zhang, M., Bhandari, B., & Liu, Y. (2018). Investigation on lemon juice gel as food material for 3D printing and optimization of printing parameters. *LWT*, *87*, 67–76. https://doi.org/10.1016/j.lwt.2017.08.054

Yang, F., Zhang, M., Prakash, S., & Liu, Y. (2018). Physical properties of 3D printed baking dough as affected by different compositions. *Innovative Food Science & Emerging Technologies*, *49*, 202–210. https://doi.org/10.1016/j.ifset.2018.01.001

Yu, X., Wang, Y., Zhao, W., Li, S., Pan, J., Prakash, S., & Dong, X. (2023). Hydrophilic colloids (konjac gum/xanthan gum) in 3D printing of transitional food from fish paste. *Food Hydrocolloids*, *137*, 108333. https://doi.org/10.1016/j.foodhyd.2022.108333

Zamboulis, A., Michailidou, G., Koumentakou, I., & Bikiaris, D. N. (2022). Polysaccharide 3D printing for drug delivery applications. *Pharmaceutics*, *14*(1), 145. https://doi.org/10.3390/pharmaceutics14010145

Zeng, F., Gao, Q., Han, Z., Zeng, X., & Yu, S. (2016). Structural properties and digestibility of pulsed electric field treated waxy rice starch. *Food Chemistry*, *194*, 1313–1319. https://doi.org/10.1016/j.foodchem.2015.08.104

Zeng, X., Li, T., Zhu, J., Chen, L., & Zheng, B. (2021). Printability improvement of rice starch gel via catechin and procyanidin in hot extrusion 3D printing. *Food Hydrocolloids*, *121*, 106997. https://doi.org/10.1016/j.foodhyd.2021.106997

Zhang, H., Fang, M., Yu, Y., Liu, Q., Hu, X., Zhang, L., Hu, H., Zhang, C., Huang, F., Tian, F., Zhang, N., & Dai, X. (2019). *Application prospect of 3D printing technology in the food intelligent manufacturing* (pp. 974–984). https://doi.org/10.1007/978-3-030-00214-5_120

Zhang, M., Golding, J. B., Pristijono, P., Yu, Y., Wang, P., Chen, G., Li, Y., Si, J., & Yang, H. (2024). Effects of Huangjing polysaccharides on the properties of sweet potato starch. *LWT*, *204*, 116474. https://doi.org/10.1016/j.lwt.2024.116474

Zhao, K., Li, B., Xu, M., Jing, L., Gou, M., Yu, Z., Zheng, J., & Li, W. (2018). Microwave pretreated esterification improved the substitution degree, structural and physicochemical properties of potato starch esters. *LWT*, *90*, 116–123. https://doi.org/10.1016/j.lwt.2017.12.021

Zhao, Y., Khalid, N., Shu, G., Neves, M. A., Kobayashi, I., & Nakajima, M. (2019). Complex coacervates from gelatin and octenyl succinic anhydride modified kudzu starch: Insights of formulation and characterization. *Food Hydrocolloids*, *86*, 70–77. https://doi.org/10.1016/j.foodhyd.2018.01.040

Zheng, B., Liu, Z., Chen, L., Qiu, Z., & Li, T. (2022). Effect of starch-catechin interaction on regulation of starch digestibility during hot-extrusion 3D printing: Structural analysis and simulation study. *Food Chemistry*, *393*, 133394. https://doi.org/10.1016/j.foodchem.2022.133394

Zheng, B., Tang, Y., Xie, F., & Chen, L. (2022). Effect of pre-printing gelatinization degree on the structure and digestibility of hot-extrusion 3D-printed starch. *Food Hydrocolloids*, *124*, 107210. https://doi.org/10.1016/j.foodhyd.2021.107210

Zheng, L., Yu, Y., Tong, Z., Zou, Q., Han, S., & Jiang, H. (2019). The characteristics of starch gels molded by 3D printing. *Journal of Food Processing and Preservation*, *43*(7). https://doi.org/10.1111/jfpp.13993

Zheng, L., Zhang, Q., Yu, X., Luo, X., & Jiang, H. (2023). Effect of annealing and heat-moisture pretreatment on the quality of 3D-printed wheat starch gels. *Innovative Food Science & Emerging Technologies*, *84*, 103274. https://doi.org/10.1016/j.ifset.2023.103274

Zhong, J., Zhou, G.-X., He, P.-G., Yang, Z.-H., & Jia, D.-C. (2017). 3D printing strong and conductive geo-polymer nanocomposite structures modified by graphene oxide. *Carbon*, *117*, 421–426. https://doi.org/10.1016/j.carbon.2017.02.102

Zhou, L., Meng, F.-B., Li, Y.-C., Shi, X.-D., Yang, Y.-W., & Wang, M. (2023). Effect of peach gum polysaccharide on the rheological and 3D printing properties of gelatin-based functional gummy candy. *International Journal of Biological Macromolecules*, *253*, 127186. https://doi.org/10.1016/j.ijbiomac.2023.127186

Zhu, J., Cheng, Y., Ouyang, Z., Yang, Y., Ma, L., Wang, H., & Zhang, Y. (2023). 3D printing surimi enhanced by surface crosslinking based on dry-spraying transglutaminase, and its application in dysphagia diets. *Food Hydrocolloids*, *140*, 108600. https://doi.org/10.1016/j.foodhyd.2023.108600

Zhu, J., Xie, F., Qiu, Z., & Chen, L. (2024). Effect of active carbonyl-carboxyl ratio on dynamic Schiff base crosslinking and its modulation of high-performance oxidized starch-chitosan hydrogel by hot extrusion 3D printing. *Carbohydrate Polymers*, *343*, 122438. https://doi.org/10.1016/j.carbpol.2024.122438

Zhu, S., Ruiz de Azua, I. V., Feijen, S., van der Goot, A. J., Schutyser, M., & Stieger, M. (2021). How macroscopic structure of 3D printed protein bars filled with chocolate influences instrumental and sensory texture. *LWT*, *151*, 112155. https://doi.org/10.1016/j.lwt.2021.112155

Zia-ud-Din, Xiong, H., & Fei, P. (2017). Physical and chemical modification of starches: A review. *Critical Reviews in Food Science and Nutrition*, *57*(12), 2691–2705. https://doi.org/10.1080/10408398.2015.1087379

15 Functional and Nutritional Aspects of Modified Starches

Swathika E. and Dr. Sundaramoorthy Haripriya
Department of Food Science and Technology, Pondicherry University, Pondicherry, India

15.1 INTRODUCTION

Starch, a naturally abundant polysaccharide, plays a significant role in various industries, particularly in food processing. Its applications include gelling, thickening and stabilising foods as well as serving as a cost-effective substitute for more expensive ingredients (Gupta et al., 2022; Shrivastava et al., 2024). Notably, the integration of starch into food products often occurs with minimal impact on the overall taste and aroma, making it a versatile additive in culinary formulations. Its unique properties enable food manufacturers to enhance product texture and consistency while maintaining sensory attributes. Figure 15.1 shows the detailed structure of a starch granule.

Native starch is limited in its industrial applications due to its susceptibility to retrogradation, primarily attributed to its high amylose content. Additionally, it faces challenges related to structural stability when exposed to pH levels, shear conditions and extreme temperatures (Manzoor et al., 2020). This highlights the need for modifications or alternatives that can enhance the performance of starch in various industrial contexts. Starch can be modified using different methods to overcome such limitations and achieve the desired alterations (Fonseca et al., 2021). Modification methods comprise physical, chemical and enzymatic processes, as portrayed in Figure 15.2. All these modifications help enhance the overall physicochemical and functional qualities of the starch, making it more flexible and thus widening its food industry applications (Saboonchi et al., 2021). Each modification method has different functionalities and limitations. For example, although conventional chemical modification methods exist, they might be detrimental to the environment and not consumer-safe due to unfavourable chemical residues within the products. Manufacturers therefore prefer green (physical and enzymatic) modification methods as they are highly efficient in tailoring the

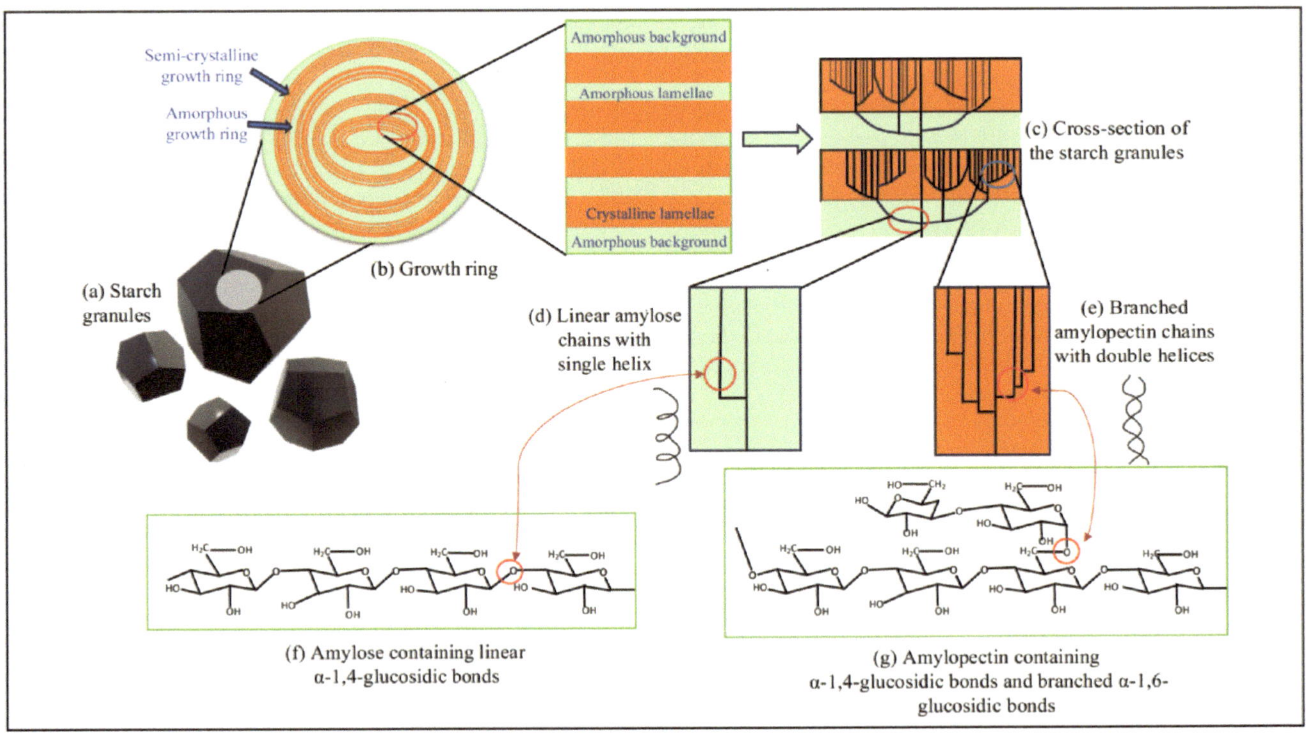

FIGURE 15.1 Schematic diagram of a starch granule.

DOI: 10.1201/9781003564287-15

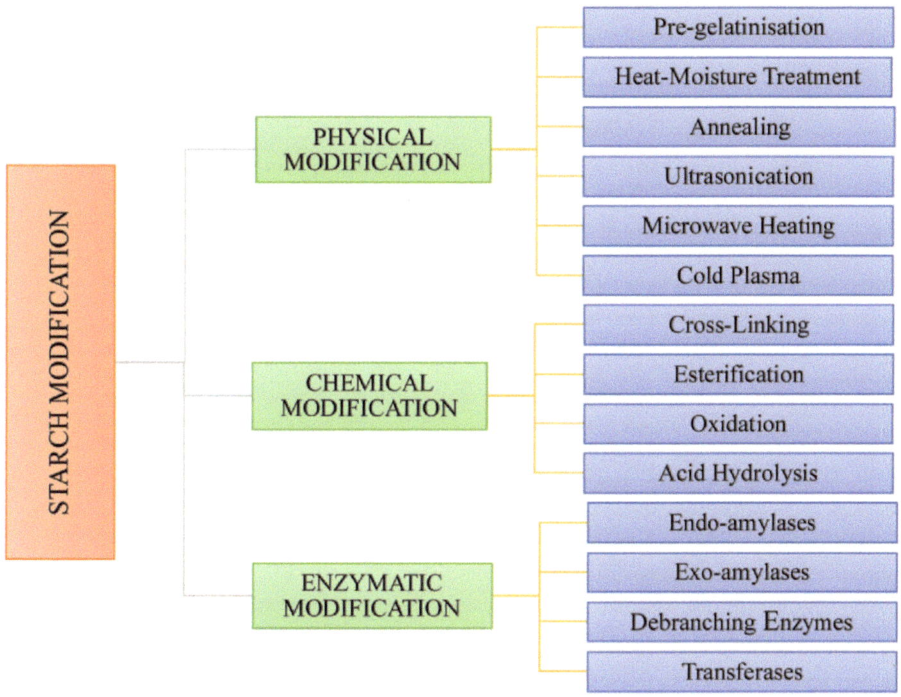

FIGURE 15.2 Various starch modification methods.

intended physicochemical properties and manipulating them without artificial reactants (Punia Bangar et al., 2022; Rostamabadi et al., 2024; Gupta et al., 2023). Different modification procedures are presented in Table 15.1

Physical modification, including thermal and non-thermal treatments, primarily utilises thermal or mechanical force, radiation, or electric fields on the starch granules, imparting no change to the D-glucopyranosyl units of starch polymers. It largely impacts the starch's structure granule size, thereby improving functionality, textural and pasting properties and sometimes even digestibility (BeMiller & Huber, 2015; Wang et al., 2020; Gupta et al., 2024a). Chemical modification, such as esterification, etherification, cross-linking and oxidation, involves introducing or altering a functional group in the starch polymer, leading to optimisation of the physicochemical properties of the polymer without having an impact on its size and distribution, facilitating the enhancement of the starch's desirability for food processing. Chemical modification also aids in stabilising the inter- and intramolecular bonds in various positions and locations (Saboonchi et al., 2021; Zia-ud-Din et al., 2017). Enzymatic modification utilises enzymes to depolymerise the starch into oligosaccharides or smaller units by cutting the glycosidic linkages or transferring the starch molecules by transferring such glycosidic linkages and residues through the creation of new bonds (Hii et al., 2012). By using enzymatic modification, one can achieve the desired behaviours of starch while improving its digestibility nutritionally (Wang & Copeland, 2015). Enzymatic modification is primarily used in the food industry since it is an environmentally friendly method with no artificial chemical ingredients (Choton et al., 2024).

15.2 FUNCTIONAL PROPERTIES OF MODIFIED STARCHES

"Functional properties" denotes the primary aspects that determine the behaviour, quality and acceptability of a material during cooking and processing, as well as their influence on prepared food products in terms of structure, texture, appearance and taste (Godswill et al., 2019; Gupta et al., 2024d). Starch possesses several functional characteristics, such as gelatinisation, swelling, solubility, pasting and retrogradation properties, but has restricted applications when subjected to extreme conditions such as shear, heat and freezing (Choton et al., 2024; Maniglia et al., 2021). Fortunately, modification methods can alter the functional shortcomings into desired attributes to increase their suitability in the food industries (Maniglia et al., 2021). Table 15.2 presents the impact of various modifications on the functional properties of various starches.

15.2.1 Swelling and Solubility

"Swelling" refers to the water-holding capacity of the starch granules, that is, their ability to absorb adequate water to swell and increase the viscosity of the starch suspension, whereas "solubility" indicates the amount of starch that leaches out into the supernatant during swelling (Jia et al., 2023; Verma & Srivastav, 2022; Gupta et al., 2024e). The swelling property estimates the degree of crystallinity and the amount of amylose leaching. High swelling power indicates lower crystallinity and high amylose leaching out, increasing solubility (Kumoro et al., 2019). Pre-gelatinisation increases the swelling power and solubility

TABLE 15.1
Different Modifications of Starches

Modification	Method/Procedure	Reference
	Physical Modification: Hydrothermal Treatments	
Pre-gelatinisation	Disruption of the starch granules by cooking, gelatinising and recovering them as a powder using various drying methods; results in less molecular reassociation	Saboonchi et al., 2021
Heat-moisture treatment	Modification of starch by heating above the glass transition temperature (80–140 °C) combined with low moisture levels of 10% to 35% for periods from 15 minutes to 16 hours	Schmiele et al., 2019
Annealing	Modification of starch by heating and holding in intermediate (~40%, w/w) or excessive (~70%, w/w) water levels at a temperature higher than glass transition and lower than gelatinisation over some time	Fonseca et al., 2021
	Physical Modification: Non-thermal Treatments	
Ultrasonication	Modification using sound waves, transversing the medium to generate a vibration that will displace the particles and create cavitations	Schmiele et al., 2019
Microwave	Application of electromagnetic waves with 300 GHz–300 MHz frequency to heat and modify the starch	Oyeyinka et al., 2021
Cold plasma	Production of cold plasma by applying energy in various forms, to ionise and disintegrate the neutral gas to form new electrons, ions and radicals that bombard and alter the starch	Okyere et al., 2022
	Chemical Modification	
Cross-linking	Molecular reaction between the reactive hydroxyl radical in the starch through the use of multi-functional reagents	Compart et al., 2023
Esterification	Substitution reaction by combining an organic acid (RCOOH) with alcohol (ROH) groups present in the starch to form ester (RCOOR) and water	Compart et al., 2023
Oxidation	Depolymerisation of starch by oxidising the hydroxyl groups into carbonyl or carboxyl groups by cleaving the α-1,4-glycosidic bonds	Chan et al., 2011
Acid hydrolysis	Hydrolysis by protic acids (provides protons) such as H_2SO_4, HNO_3, HCl and H_3PO_4 at temperatures from ambient to below the gelatinisation temperature for a particular duration	Biduski et al., 2017; Pratiwi et al., 2018
	Enzymatic Modification	
Endo-amylase	Hydrolysis of starch by cleaving the α-1,4-glycosidic bonds located in the interior (endo-) part of the amylose and amylopectin chains	Hii et al., 2012
Exo-amylase	Cleavage of the non-reducing ends of the starch chains. Cleaves the α-1,4 glycosidic bonds as well as α-1,6-glycosidic bonds	Hashim, 2019
Debranching enzymes	Hydrolysis of α-1,6-glycosidic bonds in glycogen, amylopectin and various other macromolecules yielding glucans and dextrins	Bensaad et al., 2022
Transferases	Hydrolysis of the α-1,4-glycosidic bonds of a donor molecule and transfer of the cleaved residue to the glycosidic acceptor by creating a new glycosidic bond	Compart et al., 2023

of the starch by partially gelatinising the starch molecules, disintegrating the crystalline structure and breaking down the starch chains, forming smaller starch molecules that are readily soluble in water (Obadi et al., 2020). These results have been demonstrated on various starch sources, including rice starches (He et al., 2020; Liu et al., 2017), maize and potato starches (Kankate et al., 2020), sweet potato starches (Ulfa et al., 2020, 2021) and buckwheat starch (Obadi et al., 2020). In contrast, heat-moisture treatment (HMT) and annealing decreased the swelling and solubility (Zeng et al., 2015). The decrease in swelling and solubility of starches subjected to HMT is attributed to a variety of factors, including (1) alteration of crystalline rearrangement; (2) prevention of amylose leaching by facilitating the interaction of amylose with lipids and linear regions of amylopectin-forming complexes; (3) formation of ordered double helices, denser complexes and strong intermolecular bonds; and (4) further reduction in the swelling property as treatment time and temperature increase (Kumoro et al., 2019). Annealing leads to decreased swelling and solubility for the following reasons: (1) increased perfection of crystallites owing to boosted interactions between amylopectin chains, (2) decreased amylose leaching reinforced by increased amylose interactions within the starch granules, (3) reduction in hydration of amorphous region caused by the interactions between amylose–amylose chains and amylose–amylopectin chains, and (4) increased intra-granular binding forces (Tahira & Abid, 2016; Zeng et al., 2015).

Ultrasound causes the augmentation of the swelling index and solubility for the following reasons: (1) reactive species causing physical damage to starch granules, exposing hydrophilic groups to water, thereby increasing the water absorption and swelling; (2) the enlargement of cavities on the surface of the granules, enhancing easier

TABLE 15.2
Implications of Various Modification Methods on Starch's Functional Properties

Modification	Starch Source	Implications	Reference
		Physical Modification: Hydrothermal Treatments	
Pre-gelatinisation	Sweet potato	Low viscosity High swelling, solubility and WAC	Ulfa et al., 2020
Heat-moisture treatment	Pearl millet	Lower swelling, solubility, pasting viscosities, and gelatinisation enthalpy Higher pasting temperature, stability ratio, gelatinisation temperatures and OAC	Sharma et al., 2015
Annealing	Indigenous white sorghum	Decreased swelling, solubility, WAC and pasting viscosities Improved pasting temperature and paste clarity	Tahira & Abid, 2016
		Physical Modification: Non-thermal Treatments	
Ultrasonication	Potato	Higher swelling, WAC, crystallinity, gelatinisation parameters, setback Decreased light transmittance, gelatinisation enthalpy	Cao & Gao, 2020
Microwave	Indigenous rice varieties	Decreased swelling, solubility, relative crystallinity and syneresis Increased WAC and OAC	Anjali et al., 2024
Cold plasma	Rice	Higher swelling, solubility, WAC, OAC and syneresis. Decreased relative crystallinity, gelatinisation enthalpy	Thirumdas et al., 2017
		Chemical Modification	
Cross-linking	Maize	Decreased solubility and gelatinisation enthalpy Increased gelatinisation temperatures, syneresis and paste clarity	Liu et al., 2015
Esterification	Potato and rice	Reduced swelling, solubility, relative crystallinity, gelatinisation enthalpy Higher gelatinisation temperatures	Martins et al., 2018
Oxidation	Sorghum	Higher swelling, solubility, gelatinisation temperatures, peak viscosity and breakdown Lower pasting temperature	Biduski et al., 2017
Acid hydrolysis	Maize	Decreased swelling, WAC, OAC, pasting temperature, pasting viscosities and paste clarity Increased solubility and setback	Lawal et al., 2005
		Enzymatic Modification	
Transglucosidase (TGAN) in combination with maltogenic α-amylase (MABS) and β-amylase (BA) (MABS→BA→TGAN)	Sweet potato	Increased solubility Decreased relative crystallinity and gelatinisation temperatures and enthalpy	Guo, Tao, et al., 2019
1,4-α-glucan branching enzyme	Waxy corn starch	Higher onset temperature Decreased enthalpy Reduced pasting viscosities, breakdown and setback	Ren et al., 2017
Glucoamylase	Potato starch	Increased solubility Decreased gelatinisation enthalpy and range	Wang et al., 2017
Pullulanase	Black chickpea	Higher water solubility, WAC, OAC, syneresis and pasting temperature Reduced swelling and pasting viscosities, including breakdown and setback Decreased gelatinisation temperature and enthalpy	Demirkesen-Bicak et al., 2018

Notes: OAC, oil absorption capacity; WAC, water absorption capacity.

ingression of water and amylose dissolution; (3) disruption of ordered molecular structure including crystalline regions; and (4) reduction in the molecular weight by depolymerisation, allowing water percolation into the overall structure (Kaur & Gill, 2019; Wang et al., 2017; H. Wang et al., 2020; Gupta et al., 2025a). The increasing of swelling power and solubility relies on the starch source and the ultrasonication parameters (Zhu, 2015). It was observed that dual-frequency ultrasonication (25 kHz and 80 kHz) amplifies swelling power and solubility more than single-frequency modification (Zheng et al., 2013). Another study, by Kaur and Gill (2019), proved that increased treatment time significantly increases the swelling and solubility values.

Microwave heating leads to decreased swelling index and solubility. This modification reduces swelling as a consequence of the starch granules' disintegration and degradation of the amylopectin chains, restricting the water-holding capacity (Das & Sit, 2021). It also facilitates the increase of intramolecular and intermolecular hydrogen bonds, causing the free hydroxyl radicals present in the starch to bind with inherent water molecules. As a result, there is a limitation in the absorption of water from the outside, making it harder for the starch granules to swell. The molecular rearrangement led to decreased solubility (Anjali et al., 2024). In contrast, Zheng et al. (2020) reported increased solubility in microwaved proso millet starch. Cold plasma modification results in the decomposition or depolymerisation of the starch and disrupts its crystallinity. This creates loose surface structures of starch granules, encouraging easier penetration of water into the granules and increasing the swelling power (Sun et al., 2022; Yan et al., 2020). Granular corrosion and degradation produce low-molecular-weight fragments that seep out and dissolve readily in the water (Guo et al., 2022; Shokrollahi et al., 2023). Meanwhile, Yan et al. (2020) observed a reduction in the swelling power due to the disintegration of amylopectin chains, weakening the three-dimensional structure of the granules during the swelling process.

In cross-linking, the introduction of cross-linking reagents reinforces additional covalent bonds that aid in retaining the structural integrity and forming a higher density of cross-links in the starch structure, thereby maintaining the intact structure and restricting water ingression. Thus, a prominent reduction in the swelling factor and solubility is noticed (Bhosale & Singhal, 2007; Heo et al., 2017; Kaur et al., 2006). In esterification, the swelling index is influenced by the type of esterification agent utilised. For instance, using succinic anhydride increases the swelling power induced by its hydrophilic nature, and octenyl succinic anhydride increases by weakening the intramolecular bonds through the introduction of bulky groups (Bhosale & Singhal, 2007; Bushra et al., 2013). Employing citric acid as an esterifying agent reduces swelling power and solubility by triggering the esterification in reactive hydroxyl groups present in amylose and amylopectin, leading to enhanced water retention (Martins et al., 2018; Mei et al., 2015). The solubility of the esterified starch decreases owing to the cross-linking and lower exposure to reactive hydroxyl groups (Martins et al., 2018). Oxidation decreases swelling power by destroying the granular structure and integrity of the starch molecules (Fonseca et al., 2015; Sandhu et al., 2021; Sukhija et al., 2016). The increase in solubility of oxidised starches is the consequence of oxidising agents penetrating deep into the granules and depolymerising the starch chains, impairing the granular structure and facilitating easy percolation of water (Lawal et al., 2005). Acid hydrolysis disintegrates the starch molecules and breaks the α-1,4 and β-1,6 bonds in amylose and amylopectin. This makes it harder for the granules to retain water, thereby causing a reduction in swelling power (Martins et al., 2018). Another reason for decreased swelling is increased crystallinity, as acid hydrolysis occurs mainly in the amorphous regions, caused by reduced accessibility for the acids into the intact crystalline region (Lawal et al., 2005). The depolymerisation and increment in the smaller fragments of starch molecules increases the solubility of acid-treated starches (Abdorreza et al., 2012; Lawal et al., 2005).

Enzymatic modification alters swelling and solubility based on the enzymes used for such changes. Amylases such as α-amylase and β-amylase promote solubility as they tend to depolymerise the starch and decrease their chain lengths, causing damage to the starch's robustness. The increased branched points due to starch fragmentation prevent crystallisation (Guo, Li, et al., 2019; Guo, Tao, et al., 2019). Pullulanase-treated enzymes present decreased swelling and increased solubility compared to native starches. These changes can be attributed to pullulanase debranching the starch into shorter linear chains and the hydrophilicity of the debranched starch to retain more water-forming hydrogels (Ge et al., 2021; Li et al., 2019). Keeratiburana et al. (2020) studied the impact of maltogenic α-amylase and transglucosidases. Both enzymes decreased the swelling of the starches. The swelling index decreased due to the introduction of the pores on the granular surface, weakening the overall granular structure. Another reason could be the strengthening of double helices due to molecular rearrangement. The amyloglucosidases decreased the solubility, while the α-amylase increased it. The transglucosidases continuously generate new branching points, thereby restricting the recrystallisation and realignment of the amylose and amylopectin chains, producing more small branched starch molecules that are efficiently soluble in water (Guo, Deng, et al., 2019; Guo et al., 2021).

15.2.2 Relative Crystallinity

X-ray diffraction is used to study the crystalline structure and behaviour of the starch (Verma & Srivastav, 2022). Pregelatinisation tends to disintegrate the crystalline structures present in the starch, reducing the chance of recrystallisation and increasing the amorphous regions, contributing to lower relative crystallinity (He et al., 2020; Kankate et al., 2020; Liu et al., 2017). HMT reduces the relative crystallinity due to the destruction and disorganisation of double helices in the treated starches. The rupture of amylopectin double helices leads to the amorphisation of the starch granules, thus lowering the crystallinity (De Oliveira et al., 2018; Kong et al., 2015; Zeng et al., 2015). In specific cases, HMT can induce higher crystallinity as the smaller crystalline regions may reassociate into an ordered, novel crystalline matrix (Liu et al., 2016; Marboh et al., 2022). The boosting of interactions between the amylose–amylose and amylose–amylopectin chains and the enhanced double helices' rearrangement in the annealed starches prompt better crystalline organisation. Therefore, annealed starches have higher relative crystallinity than their respective native starches (Marboh et al., 2022; Rocha et al., 2012; Zeng et al., 2015).

Regarding ultrasonication, treated samples have increased or decreased crystallinity based on the modification parameters (Kaur & Gill, 2019). It had been noted that ultrasonication increases the relative crystallinity in potato starch (Cao & Gao, 2020) and wheat starch (Karwasra et al., 2020) due to the recombination of smaller crystals that were already disintegrated by the treatment's cavitation effects. However, occasionally, the destruction of the amorphous region arising from intense mechanical pressure and cavitation can lead to double helices' reorientation, thereby ultimately reducing the relative crystallinity of the treated starches (Rahaman et al., 2021; H. Wang et al., 2020). Microwave treatment decreases crystallinity through the destabilisation of starch's layered arrangement and reorganisation of the double helices towards amorphic form (Kumar et al., 2020; M. Wang et al., 2020). Cold plasma–treated starches also exhibit a notable reduction in relative crystallinity. The decrease is attributed to the following factors: (1) decomposition or depolymerisation of the starch caused by the interaction between the reactive species of the plasma and starch molecules, (2) prevention of recrystallisation and (3) subsequent loose rearrangement of the damaged starch chains (Guo et al., 2022; Trimukhe et al., 2017; Yan et al., 2020).

Cross-linking increases the compactness of the starch molecules, thus resulting in better relative crystallinity (Sandhu et al., 2021; Shi et al., 2013). Esterification reduces the relative crystallinity for these possible reasons: (1) esterification primarily occurs in the amorphous zone, leading to the reorientation of such zones, and (2) disruption of the crystalline structure caused by deep penetration of the esterification agents into the cavities of the starch granules (Martins et al., 2018; Mei et al., 2015). Oxidised starches showed an increase of relative crystallinity when treated with the lower concentrations of the oxidising agents, but with higher concentrations, the reduction in the crystallinity is evident. The oxidising agents initially attack the amorphous zones, but with increasing concentrations, the degradation of the amylopectin happens, indicating a decrease in the relative crystallinity (Vanier et al., 2017; Zhou et al., 2016). Acid hydrolysis increases the relative crystallinity for various reasons, including (1) preferential hydrolysis in the amorphous regions of starch by cleaving the amylose chains, (2) facilitating the reordering of such cleaved chains into compact crystallites and (3) the retrogradation of free hydrolysed amylose into double helices, enabling the formation of acid-resistant crystallites (Wang & Copeland, 2015).

The impact of enzymatic modification on the relative crystallinity depends on the type of enzyme used and other modification conditions. Starches treated with maltogenic α-amylase and β-amylase treated have been shown to present reduced crystallinity as they produced smaller fragments by hydrolysis of the starch chains and prevented crystallite aggregation, whereas transglucosidase increases the crystallinity by inducing the chain rearrangements (Guo, Tao, et al., 2019). Pullulanase-treated starches had decreased crystallinity index as a result of the debranching of starch chains, and with increased treatment time, the crystallinity also increased (P. Li et al., 2017).

15.2.3 Gelatinisation and Thermal Properties

"Gelatinisation" refers to heating the starch granules in water, contributing to the reduction or elimination of semi-crystalline structures, thereby breaking down the granules and forming a viscous solution. When heated in the water, the amylopectin double helix in the starch granules disassociates and swells due to water absorption, and amylose leaches into the water (Ratnayake & Jackson, 2008; Verma & Srivastav, 2022). Gelatinisation is an endothermic process and has various parameters, including gelatinisation temperature, transition temperatures (onset, peak, gelatinisation and final temperatures) and gelatinisation enthalpy (amount of energy required for gelatinisation), which differential scanning calorimetry measures (Bushra et al., 2013; Schafranski et al., 2021). The gelatinisation temperature range depicts the double helices' heterogeneity in the amylopectin (Li & Hu, 2021).

Pre-gelatinised starch prepared by improved extrusion cooking technology has been shown to have higher gelatinisation than native starch with increasing moisture content. The increased moisture content facilitates increased water absorption and swelling of the granules, leading to rapid gelatinisation (Liu et al., 2017). Any mechanical or thermal agitation used to produce the pre-gelatinised starch weakens the molecular hydrogen bonds and disrupts the crystalline structures of the granules with enhancing amorphisation, allowing easier water penetration. This ultimately results in a higher degree of gelatinisation (Santos et al., 2019; Zhang et al., 2013). HMT influences the gelatinisation parameters and depends on the treatment's moisture and heat intensities. HMT promotes the strengthening of intramolecular bonds of the starch molecules and of amylose–amylose and amylose–lipid interactions, resulting in better structural improvement and requiring higher temperatures for gelatinisation of the granules. Elevated gelatinisation temperatures, such as onset (T_o), peak (T_p) and conclusion (T_c) gelatinisation temperatures, indicate that the treated starch has better thermal stability (Wang et al., 2021). There is a decline in the gelatinisation enthalpy (ΔH) as a result of exposing the starch to high temperatures during HMT, triggering the rupture of double helices and collapse in the crystalline and amorphous regions (Adawiyah et al., 2017; Fonseca et al., 2021; Wang et al., 2021). The gelatinisation temperature range (T_c–T_o) usually drops in HMT-treated starch, indicating the melting of crystallites with diverse stabilities and their increased homogeneity (Molavi et al., 2018; Zeng et al., 2015).

Similar results have been presented in the case of annealing. The increase in the gelatinisation temperatures of the annealed starch arises from the strengthening of crystalline networks, attributed to the aggregation of amylopectin via amylose rearrangements and the formation of heterogenous crystalline, while the decrease in ΔH is a result of double helices' disassociation and the melting of imperfect amylopectin crystals (Fonseca et al., 2021; Hu et al., 2020; Song et al., 2014). Also, the stabilisation of the crystallites in annealed sweet potato starch depends on the duration of the annealing treatment. Longer annealing time (five days) presented better outcomes in thermal stability with a significant rise in T_o, T_p and T_c, to 70.20, 73.07 and 76.97 °C from

66.64, 69.12 and 74.65 °C, respectively, of native starch (Hu et al., 2020). Ultrasound typically increases the gelatinisation temperatures with a decrease in the gelatinisation temperature range and ΔH. The improved T_o and T_c imply that ultrasonication destroys weak crystallites (Cao & Gao, 2020). Thus, a higher temperature is necessary to melt the stronger crystallites of the treated starches (Hu et al., 2014). The decrease in ΔH is the result of ultrasonication disrupting the double helices present in both amorphous and crystallite regions of the starch granules (Hu et al., 2014; Luo et al., 2008). The ultrasonicated starches have reduced T_c–T_o owing to the melting of crystallites with different stabilities and homogenisation of the crystallites (Hu et al., 2014; H. Wang et al., 2020). Concerning microwave treatment, there is an increase in the onset temperature, implying that the melting/destruction of weak crystallites present in the starch granules is becoming more stable. At the same time, a decrease is noticed in T_p, T_c and enthalpy on account of the intensified starch molecules' interaction and the degradation of double helices and perfect crystallites (Anjali et al., 2024; Kumar et al., 2020; M. Wang et al., 2020). In some cases, all gelatinisation temperatures increase due to the rearrangement of starch molecules, thereby strengthening the amylose–amylose and amylose–amylopectin chains (Luo et al., 2006; Oyeyinka et al., 2019). The thermal stability also depends on the amylopectin chains' length and molecular structure. A paper by Yang et al. (2017) found an elevation in T_o, T_p and T_c owing to the rising of longer amylopectin chains. The longer amylopectin chains (DP > 36) produce longer double helices that need higher temperatures for breakdown, whereas shorter chains (DP 6–12) impair the crystalline lamellar structure.

Cold plasma bombards the reactive species and causes plasma etching, destroying the starch's structure, crystallinity and depolymerisation, facilitating easy water penetration into the granules. This ultimately reduces the gelatinisation parameters. With increased treatment times and plasma power, the gelatinisation temperatures decreased further due to improved surface energy and hydrophilicity of the starch granules prompted by the active species (Thirumdas et al., 2017). Thirumdas et al. (2017) treated starch with lower plasma power (40 W for 5 minutes); the resulting sample had higher T_p than the untreated sample. The active species stimulate cross-linking at lower plasma power levels, while higher power levels activate depolymerisation. Similar results were produced by Zhang et al. (2015); when samples were treated with glow plasma, they showed decreased gelatinisation temperature and enthalpy.

In cross-linking, thermal properties are highly dependant on the type and concentration of the reagent. When starch was treated with various concentrations of epichlorohydrin (EPI) and phosphoryl chloride ($POCl_3$), the highest concentrations produced the highest T_o, T_p, T_c and enthalpy. In 1 gKg^{-1} $POCl^3$-treated starches, there was a notable decrease in the thermal parameters, indicating fragility; however, 2 gKg^{-1} $POCl^3$-treated starches showed greater values than native starch. This increase is associated with the forming of covalent bonds by phosphate groups that hold the starch granules together, diminishing their mobility and thus needing higher temperatures to gelatinise (Kaur et al., 2006). Shi et al. (2013) and Liu et al. (2015) reported similar observations regarding the increase in T_o, T_p and T_c, although some papers have reported decreased enthalpy compared to the untreated samples. This decrease in enthalpy could be attributed to substituting functional groups primarily in the amorphous region, resulting in promoted swelling and disruption of the crystalline phase, hence the lower temperature requirement for melting.

Esterification introduces hydrophobic bulky groups into the starch polymer, thus inducing a structural reorganisation that weakens the intramolecular and intermolecular hydrogen bonds. As a result, they require less energy and readily undergo gelatinisation in lower temperatures, leading to decreased T_o, T_p, T_c and enthalpy (Bushra et al., 2013; Mei et al., 2015). A study by Mei et al. (2015) implicates that citric acid esterification disrupts the crystalline region and increases the amorphous regions of the starch, leading to reduced gelatinisation temperatures and enthalpy. Octenyl succinic anhydride–treated cassava starch presented similar results: with the increased degree of substitutions, the T_o, T_p, T_c and enthalpy decreased prominently (Zhang et al., 2017).

The starch source, reagent type and processing conditions have a strong influence on the gelatinisation properties of oxidised starch. Sangseethong et al. (2009) noted that the gelatinisation temperatures reduce as the alkalinity of the starch increases. This could be due to the introduction of carboxyl groups that are negatively charged, which makes the starch hydrophilic, thereby increasing the swelling and decreasing the gelatinisation temperature. Another reason for reduced gelatinisation temperatures for oxidised starch could be the mechanical damage and degradation of the starch's structure caused by oxidation (Adebowale et al., 2006; Sandhu et al., 2021). In the case of the ozone treatment, the gelatinisation percentage increased from 3.1% to 11% as the treatment duration increased from 0 to 60 seconds. The extended periods of ozone oxidation cause depolymerisation of the amylopectin chains and swell rapidly from the hilium to the periphery region of the granules (Çatal & İbanoğlu, 2014). Generally, oxidation causes partial gelatinisation in the crystal lamella and weakens the starch granules, demanding less gelatinisation energy than native starches (Vanier et al., 2017). Despite this, no significant change in the enthalpy was noted in the oxidised corn, sago and tapioca starches from their respective untreated starches (Chan et al., 2011).

Acid hydrolysis introduces carbonyl and carboxyl groups that interact with the hydroxyls, subsequently promoting the surge of hydrogen bonds. This requires a higher temperature to disrupt such bonds (Biduski et al., 2017). The increase in T_c is because of the elimination of the amorphous regions and the melting of the remaining crystalline structures at greater temperatures. The widened gelatinisation temperature range indicates crystalline heterogeneity of acid-treated starch (Hernández-Jaimes et al., 2014). A similar trend of increased gelatinisation temperature range (T_c–T_o) was noticed in waxy maize, maize and sago starches after subjecting them to acid hydrolysis, revealing the heterogeneity in the amylopectin's double helices (Li & Hu, 2021). Additionally, a decrease in enthalpy was noticed, highlighting the acid's impact on the disruption of structural integrity and loss of ordered double helices of amylopectin (Biduski et al., 2017).

The impact of enzymatic modification is that the enzymes disrupt the α 1,4 linkages and reduce the longer chains of double helices to shorter chains, affecting crystallinity that directly impairs the thermal resistance (Guo, Tao, et al., 2019; Wang et al., 2017). This was observed in pullulanase-treated starch, for which the T_c–T_o range was significantly higher than the native starch. Pullulanase distorted the double helices by debranching, decreasing the gelatinisation enthalpy, T_c and expanding the T_c–T_o range (Zhou et al., 2019). Guo et al. (2019) studied the effects of sequential enzymatic treatment with maltogenic amylase → β-amylase → transglucosidase enzymes on sweet potato starch. These enzyme-treated starches possessed lesser double helices and weaker starch chain interactions as the enzymes deteriorated the perfect crystallites, ultimately reducing the gelatinisation temperature and enthalpy.

15.2.4 Pasting Properties

Pasting properties are essential to estimate the behaviour of the starches while they undergo cooking and cooling processes (Verma & Srivastav, 2022). Pasting temperature denotes thermostability, as it is an indicator of starch's structural resistance to heat-induced swelling and rupture of the granules (Liu et al., 2016; H. Wang et al., 2020). Peak viscosity represents the equilibrium between the swelling and the onset of rupture in the starch granules (Liu et al., 2017). Higher starch damage yields a higher peak viscosity value (Akubor & Igba, 2019). The setback value indicates the ability of the starch to retrograde (Sharma et al., 2015). The disruption of the starch molecules in pre-gelatinisation modification reduced viscosities compared to their respective untreated starches (Kankate et al., 2020). When starches arre pre-gelatinised with mechanical activation (ball milling), it has been shown to lead to intensive destruction of the starch granules, allowing easier water percolation and decreased pasting temperature. The increased ball milling time reduced viscosities further as it promoted (1) breakage of intermolecular and intramolecular hydrogen bonds of starch chains, (2) rupture of the granules and (3) disintegration of the compact crystalline structure (Y. Zhang et al., 2013). The pre-gelatinised starches' source and drying methods significantly influence the pasting parameters. Modified rice starches have higher peak and final viscosities facilitated by their smaller granular size and increased swelling power, while modified corn and wheat starches exhibit lower viscosities than their corresponding starches. Spray drying of starches yields higher cold paste viscosity than drum drying (He et al., 2020; S. Lee et al., 2017).

HMT increases the pasting temperature, peak time and stability ratio while decreasing the peak, setback and breakdown viscosities. The formation of cross-links, rearrangement of amorphous regions into a more crystalline region and decrease in the swelling power demand longer times and higher pasting temperatures. The strengthening of cross-links and lower amylose leaching decrease swelling, setback and peak viscosities (Molavi et al., 2018; Sharma et al., 2015). These changes in the modified starches were amplified with increasing moisture content from 15% to 35% (Y. Li et al., 2017). Likewise, annealing also increases the pasting temperature, peak time and stability owing to restricted structural disintegration and paste formation as annealing promotes higher crystalline stability. The reorganisation of the starch chains within the granules and restriction of the swelling capacity results in decreased peak viscosity and cold and hot paste viscosities. The decrease in breakdown and setback viscosities in the annealed starches indicates the shear stability. Annealing encourages amylose–amylose and amylose–amylopectin interactions, facilitating resistance to breakdown. Annealed starches have lower setbacks, which have been attributed to reduced solubility because of declination in amylose leaching (Tahira & Abid, 2016; Yadav et al., 2013).

Ultrasonication improves the pasting temperature and viscosities while decreasing the setback. The sonicated starches swell at higher temperatures, indicating higher pasting temperatures. The destruction of crystallite regions leading to micro-crystallisation and reorientation of starch chains aids in forming a rigid gel network, increasing the peak viscosity. Ultrasonication increases the interactions between different starch chains, prompting decreased retrogradation and setback. The increase in the breakdown implies less stability in the treated starches (Cao & Gao, 2020; Krishnakumar & Sajeev, 2018). H. Wang et al. (2020) investigated different treatment times of ultrasonication on sweet potato starch, and the results contrasted earlier results. The pasting temperature and the viscosities decreased for the first 15 minutes, except for setback (which increased). However, as the treatment time was extended, there were no significant changes except for the setback, which further increased with increased treatment time. The decrease in the pasting temperature denoted the weaker resistance to the pores, cracks and pits induced by the ultrasonication.

Microwaves decrease viscosities while increasing the pasting temperature. The decrease in the viscosities results from the disruption of the starch granules, thereby reducing the structural integrity and the ability to retain water molecules. A higher pasting temperature implies a decreased capacity for taking in water (Oyeyinka et al., 2019). Cold plasma decreases viscosities while increasing the pasting temperature. The decrease in the peak viscosity results from starch reacting with non-starch compounds and the formation of amylose–lipid complexes preventing granular swelling. The plasma reactive species act on the starch, weakening or breaking down the starch chains, thus causing an overall reduction in the viscosities. The decrease in the breakdown implies increased heating and shear stability (Sun et al., 2022; Wu et al., 2019). The treatment time, intensity and type of starch also influence the impact of cold plasma on the pasting properties (Banura et al., 2018; Okyere et al., 2019; Yan et al., 2020). When tapioca and corn starches were treated with different power levels (40 and 60 W) at two different time intervals (10 and 20 minutes), the viscosity increased for the starches treated at 60 W and 20 minutes. This increase was ascribed to the depolymerisation of starch into a simple sugar that can hold water and increase in viscosity (Banura et al., 2018).

Cross-linking led to increased granular integrity due to the binding of the starch molecules. This resulted in increased pasting viscosities and a reduction in the breakdown, denoting heat and shear stability (Mehboob et al., 2020). Sandhu et al. (2021) reported that increased epichlorohydrin concentrations decreased pasting viscosities and breakdown while pasting temperature increased. This was due to additional covalent bonds between the starch chains, strengthening the swollen granules and reducing their breakage. Concerning esterification, the pasting viscosities increased with a decrease in the pasting temperature, specifically in starches treated with octynyl succinic anhydride (OSA). Incorporating OSA groups into the starch facilitated structural reorganisation caused by repulsion and steric hindrance, aiding in easier water penetration and swelling of the granules. The bulky functional groups also increased the molecular weight of the starch, thereby increasing the viscosities (Bhosale & Singhal, 2007; J. Guo et al., 2020). Oxidised starches have decreased viscosities and pasting temperatures. Oxidation leads to structural weakness and disintegration, increasing starch sensitivity to heat and shearing, resulting in lower viscosities. Introducing the carboxyl and carbonyl groups leads to less reassociation of chains during cooling, decreasing the setback and final viscosities (Adebowale et al., 2005; Sukhija et al., 2016). Biduski et al. (2017) note that an elevation in peak viscosity contributed by the repulsion between the carboxyl, carbonyl and hydroxyl groups resulted in more swelling. With respect to acid hydrolysis, Lim et al. (2023) found a decrease in all pasting parameters, including the pasting temperatures, due to depolymerisation and degradation of the starches, producing weak starch pastes. The hydrolysed shorter amylose chains were less likely to reassociate, decreasing the pastes' setback. Lawal et al. (2005) reported similar results except for the increment in the setback, suggesting that the acid-thinned starches underwent recrystallisation.

Enzymatic modification influences the pasting properties of starches based on the type of enzymes used for the modification. When starches were treated with 1,4-α-glucan branching enzyme for 4 and 8 hours, the pasting properties decreased except for the pasting temperature. The decreased breakdown and setback viscosities indicated the treated starch's thermostability and resistance to retrogradation. The pullulanase modification decreased the pasting parameter, except for the pasting temperatures, compared to their native starches (Demirkesen-Bicak et al., 2018). The decrease is attributed to the loss of starch chains during enzymatic hydrolysis, leading to reduced swelling and viscosities in the pullulanase-treated starches (P. Li et al., 2017). Keeratiburana et al. (2020) observed that maltogenic α-amylase and amyloglucosidase altered pasting properties. The increased pasting temperatures in enzyme-treated starches are linked with their decreased swelling index. The peak, final viscosities, setback and breakdown were reduced compared to the native starches. The enzyme treatments led to the disintegration of the starch's surface and fragmentation of the granules, decreasing the peak viscosities. The disability of retrogradation due to lower amylose content in the starch and the shear stability of the disrupted molecules indicated lower setback and breakdown, respectively.

15.2.5 Water Absorption, Holding and Binding Capacities

Water absorption capacity (WAC) is to the amount of water absorbed by the starch granules to acquire desirable consistency to create a good-quality food product (Godswill et al., 2019). Pre-gelatinisation increases WAC for the following reasons: (1) disruption of the molecular integrity, (2) deterioration of crystalline regions due to breakage of inter- and intramolecular bonds, and (3) increased exposure of hydroxyl groups to react with water molecules (Kankate et al., 2020; Liu et al., 2017). HMT decreases treated starches' WAC due to the increased crystalline area (Ogundiran & Afolabi, 2018; Subroto et al., 2018). However, there are cases in which increased WAC has been observed in starches as a result of increased hydrophilic tendency and expansion of amorphous regions as a consequence of the disruption of the hydrogen bonds (Liu et al., 2016; Olayinka et al., 2008). Annealing had both negative and positive effects on WAC. The increase in WAC of annealed starches has been attributed to the expansion of amorphous regions (Adebowale et al., n.d.; Liu et al., 2016). In contrast, a decrease has been attributed to interactions between amylose–amylose and amylose–amylopectin chains, thus reducing the water-binding sites in the starch granules (Ogundiran & Afolabi, 2018; Tahira & Abid, 2016).

Both ultrasonication and microwave treatments increase WAC. These treatments degrade the starch into smaller units such as dextrin, glucose and maltose, which have a higher affinity to water compared to starch (M. Wang et al., 2020). Ultrasonication destroys the inner components of the starch and places the low-molecular-mass amylose on the outer layer, allowing easier water penetration inside the granules (Cao & Gao, 2020). Microwave treatment strains the inter-crystallites and increases the amorphous region. It also causes partial gelatinisation of the starch, promoting the augmentation of the hydrophilic groups on the outer edge of the glucose units, increasing water absorption (Anjali et al., 2024). Cold plasma increases WAC due to the following factors: (1) debranching of the amylose and amylopectin chains of the starch, converting them into simple sugars such as glucose and maltose (have increased water affinity); (2) increased surface area and water-binding capacity arising from the plasma etching; and (3) granular corrosion and the formation of microchannels that allow ready percolation of water into the granules (Banura et al., 2018; Sonkar et al., 2023).

Cross-linking decreases WAC at lower reagent concentrations, but with increased concentrations, WAC also increases (Núñez-Bretón et al., 2019). The introduction of hydrophilic functional groups such as acetic anhydride and succinic anhydride for esterification increases WAC by improving the absorption and binding of water (Adebowale et al., 2006; Bushra et al., 2013). Esterification with octenyl succinic anhydride reduces WAC because its hydrophobic constituent replaces the starch's hydroxyl groups (Marta et al., 2023). Oxidation improves WAC for the same reason as esterification: the introduction of hydrophilic functional groups (Adebowale et al., 2006; Bushra et al., 2013). Another reason for the increase of WAC in oxidised starches is that the functional groups weaken the hydrogen bonds and allow easier water permeation into the granules (Sukhija et al., 2016). Acid hydrolysis hydrolyses the glycosidic linkages, thereby increasing the hydrophilic hydroxyl groups that hold on to water and also impairing the associative forces between the polymers, thus increasing WAC (Falade & Ayetigbo, 2015; Lim et al., 2023). But at times, a decrease in WAC had been noted in acid-thinned starches as increased crystallinity restricts water absorption (Lawal, 2004). WAC decreased for cassava starches treated with a combination of α-amylase and glucoamylase, while the water-holding and water-binding capacities increased slightly (Cornejo et al., 2022). When hydrolysed with α-amylase or glucoamylase, the starches had increased water-holding capacity, whereas treating them with β-amylase showed a significant decrease after eight hours of treatment (Jung et al., 2017). The water-holding capacity was increased in pullulanase-treated starch (Demirkesen-Bicak et al., 2018).

15.2.6 Oil Absorption Capacity

Oil absorption capacity (OAC) is the amount of oil absorbed by a material while undergoing processing and serves a vital role in enhancing the mouthfeel of any food while retaining its flavours (Godswill et al., 2019; Okekunle et al., 2020). Pre-gelatinisation and annealing increase OAC due to the increased density of the lipophilic/hydrophobic molecules on the starch surface after the modification (Adebowale et al., n.d.; Akubor & Igba, 2019; Arinola, 2019). HMT causes surface roughness, cavities and granular expansion that lead to higher lipophilic tendencies, thereby increasing OAC (Dewi et al., 2022; Iqbal et al., 2024; Sharma et al., 2015). However, in buckwheat, maize and acha starch, a reduction in the OAC value has been observed (Alimi & Workneh, 2018; Liu et al., 2015; Sindhu et al., 2019). Ultrasonication improves OAC by aiding in the physical entrapment of oil caused by capillary action and the exposure of hydrophobic amylose–lipid complexes on the starch surface (Majeed et al., 2017; M. Wang et al., 2020). Microwave treatment also increases OAC, for the same reasons as ultrasonication (Anjali et al., 2024; M. Wang et al., 2020). However, at times, microwave-modified starch possesses too-large pores, which reduces capillary actions and decreases the oil-binding capacity (Zailani et al., 2021). Cold plasma tends to increase the surface area by striking high-energy electrons on the granules' surface, facilitating more oil interaction with the granules (Sonkar et al., 2023). Cross-linking increases OAC with an increase in the epichlorohydrin concentration. Incorporating the functional groups from epichlorohydrin creates changes in the micro-structure of the starch (Núñez-Bretón et al., 2019). Xie et al. (2019) reported that the enhanced porosity of the starches cross-linked using sodium trimetaphosphate led to increased OAC. The formation of strong chemical bonds due to cross-linking with sodium trimetaphosphate aids in strengthening lipid retention inside the starch molecules (Núñez-Bretón et al., 2024). Esterification with octenyl succinic anhydride positively impacted OAC by facilitating lipophilic absorption in the starches (Núñez-Bretón et al., 2024). In contrast, starches treated with succinic anhydride decreased the oil absorption as the succinic constituents are hydrophilic (Bushra et al., 2013). Oxidation elevated OAC due to the functional group integration with more binding capacities. However, acid thinning decreased OAC as it increased the crystallinity and restricted the absorption of oil (Lawal, 2004; Lawal et al., 2005).

OAC decreased for cassava starches subjected to dual enzymatic treatment with α-amylase and glucoamylase. The enzymatic attack on the cassava starches largely caused exo-hydrolysis, underexposing the starch's lipophilic helices, reducing the OAC (Cornejo et al., 2022). The oil-holding capacity of the corn starch increased when treated with α-amylase, glucoamylase or transglucosidase. The OAC amplifies with increased treatment time (Jung et al., 2017). Likewise, pullulanase-treated starch has increased OAC compared to its native starch (Demirkesen-Bicak et al., 2018). The dual enzymatic modification of the starch using α-amylase and glucoamylase or glycosyltransferase and glucoamylase drastically increased the oil absorption. A porous starch prepared with all three enzymes showed the highest OAC, indicating that the combined action of glycosyltransferase and α-amylase yielded a larger amorphous area with abundant macro-sized pores. On the other hand, glycosyltransferase formed more branching chains that serve as binding sites in the crystalline region of the starch, thus increasing the OAC (L. Guo et al., 2020).

15.2.7 Light Transmittance/Paste Clarity

Light transmittance estimates the behaviour of the starch, especially the granular rigidity, when the light passes through it, highlighting the clarity of the starch paste. It is essential to know the paste clarity of starch to determine its application in food products, including jellies and fruit pastes (Verma & Srivastav, 2022). Pre-gelatinised starches have higher paste clarity, which has been attributed to the degradation of the macromolecules during the thermal modification (Li et al., 2014). When pre-gelatinisation is combined with dry heating, it induces the production of amylose–lipid complexes that decrease amylose leaching while improving the transparency of the paste (Lee et al., 2017). The increased flexibility of the starch chains in the amorphous regions and the destruction of amylopectin chains that produce smaller chain starch fragments in HMT starches decrease light transmittance (Dewi et al., 2022; Y. Li et al., 2017). Annealing results in lower apparent amylose due to the formation of amylose–amylopectin and amylose–lipid complexes while reducing amylose leaching. This increases the paste clarity along with increased disintegration and fragility of the annealed starches (Almeida et al., 2020; Yadav et al., 2013).

The impact of ultrasonication on paste clarity depends on the processing conditions. Ultrasonication decreases light transmittance due to higher amplitude inducing higher disintegration, water absorption and swelling, thereby restricting light from passing through the paste (Krishnakumar & Sajeev, 2018; Sit et al., 2014). Ultrasonication also improves paste clarity due to decreased retrogradation and breaking of the covalent bonds in amylopectin (Majeed et al., 2017; Zheng et al., 2013). Microwave heating increased the opacity of the treated starches as they were subjected to reassociation of amylose and reduced disintegration (Das & Sit, 2021) Zailani et al., 2021). In cold plasma treatment, the following factors lead to increased paste clarity in treated starches: (1) depolymerisation of starch arising in smaller starch granules that have increased solubility and (2) increased hydroxyl groups having a higher affinity for water molecules as a result of depolymerisation (Sonkar et al., 2023; Wu et al., 2019).

Cross-linking decreases light transmittance because of strengthened granular integrity and reduced degradation of the dense treated starches, reflecting or refracting the light instead of transmitting it (Kaur et al., 2006; Liu et al., 2014). Esterified starches with OSA improves light transmittance as octenylsuccination aids in better water penetration and swelling while reducing the residual amount of the starch granules, allowing more light to pass through the treated starch paste (Bhosale & Singhal, 2007; J. Guo et al., 2020). Acetylation improves the hydrophilicity of the starch, enhancing water penetration and increasing light transmittance (Mehboob et al., 2020). Oxidation replaces the hydroxyl groups with carbonyl and carboxyl groups, causing repulsion between the adjacent starch molecules. This reduces interchain association and increases the light transmittance percentage (Sandhu et al., 2008). Additionally, the disruption of the starch granules caused by excessive swelling permits light to pass through the granules effectively (Liu et al., 2014). Acid hydrolysis increases light transmittance by reducing the retrogradation of the modified starch and improving the interactive bond formation of amylopectin, allowing light to flow through quickly (Singh Sandhu et al., 2007; Thys et al., 2013). In some cases, however, acid hydrolysis has been shown to decrease paste clarity (Lawal et al., 2005; Sobini et al., 2022).

A single enzymatic modification using transglucosidase and combined treatment of α-amylase, β-amylase and transglucosidase increased paste clarity drastically. The modified starches are hydrolysed, giving rise to smaller fragmented starches with increased short chains and branching points. This continuous introduction of branch points (formed by transglucosidase) in the amylose and amylopectin prevents starch recrystallisation and realignment of glucan chains, respectively (Guo, Li, et al., 2019). Pullulanase decreased the paste clarity due to increased retrogradation, forming smaller and stronger crystallites by the interactions between the amylose chains (Lee et al., 2023; Van Hung et al., 2012).

15.2.8 Syneresis/Freeze–Thaw Stability

While storing the starch gels at lower temperatures, the starch excludes water as a result of the increased molecular association between the chains. This process is called syneresis (Kaur et al., 2006). In short, syneresis indicates the amount of water the starch gel releases during the freeze–thaw process (Sobini et al., 2022). Starch gels' low syneresis value implies slower retrogradation due to stronger interactions between dispersed amylose/amylopectin and water molecules (Vanier et al., 2017). Pre-gelatinised starches show improved freeze–thaw stability as they exhibit a lower syneresis percentage than native starches (Sobini et al., 2022). Hedayati et al. (2016) studied the impact of sugars as sweeteners on pre-gelatinised starches, and the results revealed higher freeze–thaw stability, indicating that sugars bind the hydroxyl groups present in the starch, thus retarding ice crystallisation. HMT positively impacts freeze–thaw stability by improving the crystalline structure. The amylose–amylopectin crystalline structure binds the water in the gel matrix, making it harder for the water to release, decreasing the syneresis percentage (Subroto et al., 2018). A decrease in syneresis has also been reported for chestnut starch (Yadav et al., 2013) and taro starch (Das & Sit, 2021). Annealing enhances the freeze–thaw stability (Dutta & Sit, 2022; Nwaogazie et al., 2022). However, Tahira et al. (2016) observed an increase in the syneresis percentage owing to structural rearrangement and reduced water-binding capacity.

Ultrasonicated starches have a lower syneresis percentage than untreated starches, and this could be linked to various factors, such as excessive breakdown of granules, extensive reordering of the chains and increased exposure of hydroxyl groups to water molecules, thereby increasing the retention of water during the freeze–thaw cycles (Luo et al., 2008; Sit et al., 2014). The decreased syneresis percentage in microwave-treated starches results from the production of smaller starch fragments that possess enhanced water absorption. Moreover, the increased interactions between the amylose–amylose and amylose–amylopectin result in lower separated amylose chains. This prevents more water from oozing while thawing (Anjali et al., 2024; Luo et al.,

2006). Cold plasma typically increases the syneresis percentage due to depolymerisation and cleaving of the glycosidic bonds. It also facilitates starch oxidation, forming weaker gels (Gao et al., 2019; Thirumdas et al., 2017). Increased cross-linking and retrogradation in plasma-treated starches also increases syneresis (Shokrollahi et al., 2023).

Cross-linked starch using sodium trimetaphosphate showed increased freeze–thaw stability as the positively charged hydrophilic phosphate groups repelled each other and swelled, thus increasing the size of the granules. The frozen paste formed a stable three-dimensional structure (Liu et al., 2015; Gupta et al., 2025b). Starches cross-linked with phosphoryl chloride exhibited increased syneresis. The increase was attributed to the enhanced molecular reassociation between the starch chains and increased retrogradation during refrigeration (Hazarika & Sit, 2016; Mirmoghtadaie et al., 2009). Esterification increases freeze–thaw stability. Succinylation leads to increased water-holding capacity due to incorporating hydrophilic succinyl groups in the starch, reducing retrogradation and syneresis (Bushra et al., 2013). OSA-modified starches have shown improved resistance to syneresis. The reduction in retrogradation as a result of the steric effect imposed by bulky OSA groups prevents the rearrangement of the starch chains (Bhosale & Singhal, 2007). The increase in the degree of substitution of OSA groups further amplifies the decrease in the syneresis (J. Guo et al., 2020). Oxidised starches show increased freeze–thaw stability due to increased carboxyl and carbonyl content, which have stronger hydrophilicity, thereby improving water retention (Liu et al., 2015; Rao et al., 2022). Acid hydrolysis of starch leads to enhanced syneresis. The degradation of the starch by acids brings about higher starch fragmentations, encouraging molecular reassociation, retrogradation and high water liberation (Abdorreza et al., 2012; Thys et al., 2013).

Enzymatic modification influences freeze–thaw stability based on the enzymes used, and α-amylase has been shown to drastically increase freeze–thaw stability (Almeida et al., 2023). It reduces the interchain bonds in the modified starches, causing a decrease in syneresis (Balasubramanian et al., 2014). In contrast, pullulanase-treated starches have shown decreased stability. Pullulanase produces long unbranched amyloses that aggregate due to the substantial intra- and intermolecular bonds, leading to decreased freeze–thaw ability (Lee et al., 2010).

15.3 STARCH NANOPARTICLES AND THEIR FOOD INDUSTRIAL APPLICATIONS

Several studies have ventured into various modification methods to enhance the physicochemical attributes of the starch to attain remarkable end products, and one of these modifications is the reduction of the starch into nanoscale (less than 1000 nm) products, yielding starch nanoparticles. The breakdown of the starch granules into nanoparticles can be achieved by physical, chemical, enzymatic and other mechanical methods. Altering the particle size of the native starch particles down to their respective nanoparticles results in beneficial functional, pasting and rheological properties, making the nanoparticles suitable as fillers and binding agents. They find typical roles largely in active packaging, stabilising food emulsions, creating biocomposite films and encapsulating bioactive compounds (Chavan et al., 2021; Rizki et al., 2023). Table 15.3 presents various studies on food industry applications of starch nanoparticles.

Incorporating SNPs and starch nanocrystals (crystalline residues of the starch granules) into polymeric films enhances the mechanical and thermal properties with enhanced barrier performance. These improvements are attributed to their higher surface area–to–volume ratio facilitating easier dispersion, increased interaction with the starch molecules and compatibility with the starch matrices (Chavan et al., 2024; Velásquez-Castillo et al., 2023). SNPs can also be added in biocomposite films as nanofillers as they increase the elasticity modulus, elongation at break and tensile strength while reducing water vapour permeability (Rizki et al., 2023). The application of the SNPs in edible or composite packaging films is preferred in the food industries as the production of nanomaterials from the starch imposes a lesser threat to the environment while being cost-effective and widely available (Gujral et al., 2021).

The fabrication of Pickering emulsions using SNPs as stabilisers has been gaining popularity among researchers as they act as strong barriers on the surface of the oil droplets and inhibit demulsification resulting from oil droplet aggregation (Jiang et al., 2023). They enhance the physical stability of the emulsion while reducing its oil oxidation when combined with any bioactive compounds. This makes the emulsions a potential food-grade Pickering emulsifier (Wang & Zhou, 2022).

Encapsulation is the process of enclosing a sensitive core material within a protective wall material to secure the core component from any physical, chemical or biological degradation. The incorporation of SNPs in encapsulated systems serves to produce better carriers of the bioactive compounds (Campelo et al., 2020). This can be attributed to biocompatibility, large surface area, low cost and precise entrapment of bioactive materials (Rizki et al., 2023). SNPs are non-toxic to cells, which makes them ideal for food and pharmaceutical applications (Hu et al., 2020).

15.4 NUTRITIONAL ASPECTS OF MODIFIED STARCH

Dietary interventions have increasingly become a focal point of research and practice, emphasising their significance in promoting host health and mitigating disease risk. Such interventions aim to modify dietary patterns to enhance nutritional intake, thereby influencing multiple physiological and biochemical processes. By fostering a better understanding of the relationship between nutrition and health, these interventions seek to provide evidence-based

TABLE 15.3
Recent Studies on Applications of Starch Nanoparticles (SNPs) in Food Industries

Applications	SNP Source	Highlights	Reference
Nanocomposite films	Potato starch nanoparticles	Increased burst strength with increasing concentrations of SNPs Increased film thickness, biodegradability Decreased water vapour transmission rate	Gujral et al., 2021
	Quinoa starch nanocrystals	Increased tensile strength, film opacity Decreased water vapour permeability	Velásquez-Castillo et al., 2023
	Waxy corn starch nanoparticles	Increased tensile strength with increasing concentration of NSPs Decreased water vapour permeability and oxygen transmission rate	Lin et al., 2020
	Potato starch nanoparticles	Increased UV blocking, tensile and barrier attributes	Xie et al., 2023
Bioactive edible film	Quercetin-encapsulated carboxymethyl lotus root starch nanoparticles	Increased mechanical strength, thermal stability, barrier property and antioxidant activity	Y.-F. Zeng et al., 2024
	Potato starch nanoparticles and tea polyphenols	Increased elongation at break, water resistance, tensile strength and antioxidant activity	Yang et al., 2023
Edible films	Pearl millet starch nanoparticles	Increased bursting strength, biodegradability Decreased water vapour transmission rate, solubility, moisture content	Chavan et al., 2024
	Mango seed starch nanoparticles	Increased bursting strength Decreased water vapour transmission rate, moisture content	Singh et al., 2024
Emulsifiers	Waxy maize starch (tea polyphenols and camellia seeds oil)	Increased physical and oxidative stability of the emulsion	Wang & Zhou, 2022
	Resistant starch nanoparticles from lotus stem (flax seed oil and ferulic acid)	Increased emulsion stability, nutraceutical delivery	Noor et al., 2022
	Quinoa starch nanoparticles (corn oil)	Increased stability against pH of 3–9 and ionic strength of 0–200 mM Increased oxidative stability with increasing concentration	Jiang et al., 2023
	Rice starch nanoparticles (sunflower oil and curcumin)	Increased physical stability Curcumin slowly released at low pH and rapidly released at higher pH	Kamwilaisak et al., 2022
Encapsulation (carriers)	Water chestnut, horse chestnut and lotus stem nanoparticles (loaded with resveratrol)	Horse chestnut nanoparticles showed highest efficacy Increased in anti-obesity and anti-diabetic activities over free resveratrol after digestion	Ahmad & Gani, 2021
	Debranched waxy maize starch and β-cyclodextrin nanoparticles (loaded with tangeretin)	Increased tangeretin stability in human stomach conditions ionic strength, temperature (37 °C) and acidic level	Qin et al., 2020
	quinoa and maize starch nanoparticles (loaded with rutin)	Increased bioavailability and in vitro antioxidant activities in quinoa SNPs compared to maize SNPs	Remanan & Zhu, 2021
	Waxy maize starch nanoparticles (loaded with curcumin)	Increased curcumin photostability, solubility and sustained release compared to free curcumin	Wang et al., 2022
	Corn starch nanoparticles (loaded with curcumin)	Increased UV, pH and thermal stability of curcumin	Miskeen et al., 2021

strategies for disease prevention and health optimisation across diverse populations (X. Zeng et al., 2024).

15.4.1 Glucose Metabolism

Postprandial glycemia, which refers to the blood sugar levels that occur after eating, has been identified as a significant factor associated with several chronic diseases. Research suggests that elevated post-meal blood sugar levels may contribute to the development of type 2 diabetes, increase the risk of cardiovascular disease, and are linked to various forms of cancer. Understanding the implications of postprandial glycemia is crucial for managing these health conditions effectively (Zhang & Bao, 2023).

The glycemic index (GI) is a critical metric quantifying the postprandial rise in blood glucose levels in response to carbohydrate consumption. This index categorises foods based on their effect on blood sugar, providing insights into how quickly and significantly they can elevate glucose levels after ingestion. Understanding the GI is essential for managing blood sugar levels, particularly for individuals with diabetes, as it aids in making informed dietary choices that can influence overall health and metabolic response (Zhang & Bao, 2023).

Insulin resistance is a significant disorder associated with obesity, characterised by elevated insulin levels that fail to regulate normal blood glucose levels effectively. This condition impairs glucose utilisation in target tissues, leading to metabolic dysfunction. As a result, insulin-resistant individuals may experience difficulty maintaining glycemic control, which can have further implications for their overall health. Understanding the mechanisms underlying insulin resistance is crucial for developing effective strategies for prevention and treatment in the context of obesity-related health issues. High-GI meals are characterised by their ability to rapidly deliver glucose into the bloodstream, leading to quick spikes in blood sugar levels. In contrast, low-GI foods release glucose more gradually, which is beneficial for maintaining stable blood glucose control and improving insulin sensitivity. Many researchers have underscored the significant health benefits of modified starches, particularly resistant starch (RS), in promoting better glycemic control and enhancing insulin sensitivity. A range of studies strongly supports this evidence in the literature (Z. Zhang & Bao, 2023).

Additionally, RS is noteworthy because it is not digested or absorbed in the small intestine, contributing to various health benefits, including improved glycemic response and enhanced satiety. Understanding the distinction between high- and low-GI foods and the role of RS is essential for developing dietary strategies to manage blood sugar levels and promote overall metabolic health.

The encapsulation of RS with *Ganoderma lucidum* spores (EGLS) exhibited significant hypoglycemic properties in a study conducted on rats with type 2 diabetes mellitus (T2DM). Over a 28-day experimental duration, the EGLS-treated group maintained consistently lower blood glucose levels than the control group. Remarkably, by the end of the third week, a reduction of 21.9% in blood glucose levels was observed in the EGLS-treated rats. This finding suggests the potential therapeutic benefits of EGLS in managing hyperglycemia associated with T2DM. Further research could elucidate these effects' mechanisms and explore clinical applications in human populations. The hypoglycemic effects of EGLS may be linked to a reduction in insulin secretion alongside an improvement in glucose homeostasis. This improvement could be achieved through the inhibition of glycogen synthesis and the suppression of hepatic glucose production. Glucose metabolism consists of glycogen synthesis and gluconeogenesis (Z. Zhang & Bao, 2023).

Research has revealed that individuals with T2DM exhibit notable dysbiosis in their gut microbiota. Specifically, T2DM patients have been found to harbour elevated levels of pathogenic bacteria, such as *Clostridium symbiosum*, *Clostridium hathewayi* and *Escherichia coli*. In contrast, healthy control subjects demonstrate a higher presence of beneficial butyrate-producing bacteria in their gut (Zhou et al., 2022). Given the crucial role of short-chain fatty acids (SCFAs), particularly butyrate, in maintaining gut health and homeostasis, incorporating RS into the diet is suggested as a preventive measure against T2DM. This dietary adjustment may help modulate gut microbiota composition and enhance metabolic health (Jiali et al., 2024).

The consumption of oat RS has been observed to significantly influence gut microbiota composition. Specifically, it increases the abundance of the genera *Clostridium* and *Butyricicoccus*, while concurrently decreasing the populations of *Bacteroides*, *Lactobacillus*, *Oscillospira* and *Ruminococcus*. This shift in microbial profiles may affect gut health and metabolic functions, highlighting the potential prebiotic effects of oat-resistant starch. Proinflammatory cytokines, secreted primarily by adipose tissue, regulate glucose metabolism. These cytokines can influence insulin sensitivity and metabolic processes, potentially leading to conditions such as insulin resistance and T2DM. Understanding the interplay between adipose tissue, proinflammatory cytokines and glucose metabolism is essential for developing therapeutic strategies to mitigate metabolic disorders (Zhang & Bao, 2023).

15.4.2 Effect of Phenolic Acids on Digestibility

The impact of different phenolic acids on starch digestibility has been a focal point of recent studies. Research indicates that starch–phenolic acid complexes exhibit lower digestibility than native starch and their combinations. Notably, after the gelatinisation process of these complexes, a small fraction of phenolic acids remains associated with the starch molecules. This observation suggests that bound phenolic acids could significantly hinder the hydrolysis of starch. This insight into the interaction between phenolic acids and starch could have implications for nutritional and food science, particularly in developing functional food products (Jiali et al., 2024).

The impact of gallic acid on starch modification and the potential effects of alkyl gallates on starch-digesting enzyme activity are significant areas of study. Research indicates that gallic acid can alter the properties of starch, which, combined with alkyl gallates, may affect the digestibility of starches. The presence of alkyl chains in these gallates has been shown to enhance the formation of starch inclusion complexes, subsequently leading to an increase in resistant starch concentration. This complex interaction poses challenges to starch digestion (Jiali et al., 2024).

The study examined the modification of tapioca starch using three different phenolic acids: ferulic acid, sinapic acid and *p*-coumaric acid. The results indicated that the presence of these acids in starch esters contributed to reduced digestibility. This phenomenon can be attributed to the steric hindrance and increased hydrophobicity associated with the phenolic acid moiety. These modifications could potentially enhance the functional properties of tapioca starch, leading to novel applications in food systems and beyond (Jiali et al., 2024).

15.4.3 Lipid Metabolism

The serum lipid composition includes various lipid profiles such as triglycerides (TG), total cholesterol (TC), high-density lipoprotein cholesterol (HDL-c) and low-density lipoprotein cholesterol (LDL-c).The dysregulation of lipid metabolism is often associated with decreased levels of

HDL-c, highlighting the complex interplay between obesity and lipid profiles in the human body. This relationship is crucial for understanding the pathophysiological mechanisms that contribute to cardiovascular diseases and metabolic dysfunctions linked to obesity. Research has shown that low HDL-c levels can exacerbate the adverse effects of obesity, leading to heightened risks of atherosclerosis and other metabolic disorders. Further exploration of this connection is essential for developing targeted interventions aimed at improving lipid profiles in individuals struggling with obesity. Obesity is significantly linked to the dysregulation of lipid metabolism, often resulting in increased serum triglyceride levels. Understanding of these interconnections is crucial for effectively addressing metabolic disorders and enhancing cardiovascular health (Zhang & Bao, 2023).

The intervention of chlorogenic acid combined with extruded rice starch (CGA-ES) considerably reduces the projected glycemic index (pGI) via enhancing resistant starch, alleviating lipid metabolism problems caused by a high-fat diet in rats (X. Zeng et al., 2024). *Pueraria lobata* resistant starch (PLRS) synthesised by altering *Pueraria lobata* starch with pullulanase mitigated lipid metabolic diseases by greatly increasing the prevalence of *Akkermansia*, *Lactobacillus*, *Blautia* and *Dubosiella* while substantially decreasing *Staphylococcus* in mice with a high-fat diet. Molecular processes have shown that PLRS may diminish fat accumulation and stabilise gut microbiota by modulating the fatty acid metabolic signalling pathway and the peroxisome proliferator–activated receptor signalling (PPAR) pathway (Lai et al., 2024).

A double-blind, randomised controlled experiment examined the impact of type 3 resistant starch derived from *Canna edulis* (Ce-RS3) on cholesterol levels in moderate hyperlipidemia. Human trials were conducted with either Ce-RS3 or a placebo (native starch from *Canna edulis*) for a duration of 12 weeks (20 g/day). Targeted quantitative metabolomics (TQM) was used to identify alterations in metabolites. Compared to the placebo, Ce-RS3 markedly reduced TC, LDL-c and non-high-density lipoprotein cholesterol, while enhancing glutathione peroxidase levels (Miao et al., 2024). Rosado et al. demonstrated that when incorporating 15% green banana–RS into a high-fat diet, AMP-activated protein kinase (AMPK), HMG-CoA reductase and fatty acid synthas (FAS) participate in lipid synthesis, while ABCG5 and ABCG8 function as sterol transporters in the liver, responsible for inhibiting the deposition of dietary sterols. Green banana RS mitigated hepatic steatosis, TG and cholesterol accumulation. Additionally, it enhanced the expression levels of AMPK phosphorylation, ABCG5 and ABCG8, while reducing those of HMGCoA reductase and FAS. Results indicate that green banana RS decreased the risk of liver lipogenesis by down-regulating protein levels associated with lipid synthesis while enhancing the expression of sterol transport proteins linked to lipid excretion. The examination of lipid metabolism in obese mice subjected to a high-fat diet revealed that RS supplementation resulted in a reduction of TG, TC and LDL-c levels while simultaneously increasing HDL-c to levels comparable to the control group after six weeks. This suggests that RS intervention may effectively mitigate hyperlipidemia in obese mice. Acox1 is the rate-limiting enzyme in lipid breakdown, whereas Insig1 and Insig2 are crucial for the control of intracellular cholesterol levels. RS encapsulated with *Ganoderma lucidum* spores in T2DM rats increased Insig1, InsIG2 and Acox1 (Zhang & Bao, 2023). Modified starches have the potential to influence lipid oxidation and support cholesterol homeostasis, possibly linked to the enhanced composition of the fecal microbial community.

15.4.4 Gut Microbiota Modulation

Gut microbiota composition and diversity significantly influence chronic diseases such as cardiovascular disease, diabetes, obesity and inflammatory bowel diseases. Research indicates that the microbial ecosystem in the gut influences metabolic processes and immune responses, which may contribute to the pathogenesis of these conditions. Grasping the connection between gut microbiota and chronic diseases is essential for crafting innovative therapeutic approaches and effective preventive solutions (Zhang & Bao, 2023).

The utilisation of modified starch in dietary formulations exhibits a higher efficacy in exhibiting prebiotic properties than other starch variants. Its functionality is particularly significant in the upper gastrointestinal tract, where it facilitates the transport of probiotic organisms. This process enhances the host's immune response but also plays a crucial role in inhibiting the proliferation of potentially pathogenic microorganisms, thereby contributing to overall gut health and immune function (Ağagündüz et al., 2023).

The role of starches, particularly resistant starches, as well as other undigested carbohydrates and endogenous secretions, is crucial in large intestinal digestion. These components are fermented by the host microbiome within the large intestine, a process highlighted by Cione et al. (2021). The fermentation process in the gut microbiota is crucial for producing short-chain fatty acids such as acetate, propionate and butyrate. These SCFAs serve essential functions in maintaining the physiological health of internal organs and contribute to overall wellness. They have been shown to influence various metabolic pathways, support immune function and enhance gut barrier integrity, thereby playing a significant role in both gastrointestinal and systemic health. The modulation of SCFA levels through diet and microbiome composition highlights the importance of understanding the dynamics of fermentation in promoting health and preventing disease. Various types of starch serve as essential dietary energy sources and are vital for their interactions with the intestinal microbiome throughout the digestive tract. This relationship underscores the importance of nutritional choices on gut health and metabolic processes (Ağagündüz et al., 2023).

Modified starch exhibits limited gastrointestinal side effects, including minimal occurrences of bloating, while also causing negligible changes to the physicochemical and organoleptic properties of the final food product. This characteristic makes modified starch a suitable ingredient in food formulations, maintaining the desired texture and sensory attributes without significantly impacting consumer digestion or perception (Ağagündüz et al., 2023).

Resistant starch from native pea starch (type 2 RS, or RS2) primarily decreased the Firmicutes-to-Bacteroidetes ratio while promoting the growth of *Bifidobacterium*, a known acetate producer. In contrast, RS derived from acid-hydrolysed and pullulanase-debranched pea starches (type 3 RS, or RS3) significantly impacted the proliferation of Faecalibacterium, a key butyrate producer. This distinction underscores the diverse impacts that various forms of resistant starch can have on the composition of gut microbiota as well as on the production of SCFAs. Understanding these effects is crucial for elucidating the relationship between diet, gut health and overall metabolic function (Z. Zhang & Bao, 2023).

Utilising specific modified starches has been linked to positive outcomes in body composition and lipid profiles. This association is primarily attributed to their ability to promote the production of SCFAs within the colon. SCFAs play a crucial role in various metabolic processes, influencing both gut health and overall lipid metabolism. As a result, incorporating these modified starches into dietary regimens may offer a strategic approach to improve health markers related to obesity and cardiovascular risk factors. The mechanisms through which SCFAs exert their effects are multifaceted, highlighting the importance of understanding the dietary components that contribute to metabolic health. Further research may provide valuable insights into dietary strategies for enhancing gut health and reducing metabolic disorders. This increase in SCFA production plays a significant role in metabolic health, contributing positively regulation of lipid metabolism and overall body composition. Most studies have examined the in vitro prebiotic effects of resistant starch on gut microbiota, with limited evidence found in vivo (Ağagündüz et al., 2023).

The findings of a study conducted by Le Thanh-Blicharz et al. (2014) on animals indicate that acetylation and cross-linking serve as prebiotics by promoting the production of SCFAs during the digestion process in the cecum. This suggests that these processes play a significant role in enhancing gut health and metabolic functions associated with SCFA production. Moreover, there is a clear need for thoroughly designed studies that can simultaneously evaluate the impact of starch as a functional and adaptable food component on the gut microbial community of the host, along with its metabolic functions and bacterial metabolites in individuals with metabolic syndrome (Ağagündüz et al., 2023).

Research involving rodent models has demonstrated that starches esterified with specific SCFAs, achieved through propionic, acetic, or butyric anhydride treatment, exhibit increased resistance to digestion in the small intestine. Furthermore, these SCFAs play a significant role in modulating the gut microbiota in the large intestine (Ağagündüz et al., 2023).

Qin et al. investigated the prebiotic effects of RS3, demonstrating its role in modulating gut microbiota metabolism. Their study identified significant enrichment of specific bacterial genera, including *Megamonas*, *Bifidobacterium*, *Dialister*, *Romboutsia* and *Collinsella*, in the RS5 group during in vitro fecal fermentation. These genera are known for their capacity to produce butyric acid, which is beneficial for maintaining gut health. These findings highlight the potential of specific dietary interventions in influencing gut microbiota and their subsequent impact on metabolic health (Zhang & Bao, 2023).

15.5 INNOVATIVE TRENDS AND FUTURE PROSPECTS

The market for modified starch was estimated to be worth USD 10.8 billion in 2023 and is expected to increase at a compound annual growth rate of 4.3% from 2024 to 2030, reaching USD 17.9 billion (Verified Market Research, n.d.). The increasing use of modified starches in contemporary food science and the food industry signifies a crucial shift in food production, packaging and diverse industrial uses. Modified starches, obtained from conventional starches by physical, enzymatic or chemical methods, have seen significant change, providing a range of functions and uses that address various consumer requirements. The primary catalyst for extensively using modified starches is their improved functional diversity. Modified starches, traditionally acknowledged for their thickening and gelling qualities, have surpassed conventional uses. Through technological breakthroughs, scientists have effectively modified starch characteristics to enhance heat stability, freeze–thaw stability and shear force resistance. Technological innovations have facilitated new modifications, allowing starches to function as stabilisers, fat replacers, texture enhancers and encapsulation agents, transforming various industries, including food, pharmaceuticals and textiles. The attributes have significant applications in producing convenience foods, frozen meals and diverse culinary uses, demonstrating superior performance under stimulating conditions. In response to the need for natural and clean-label goods, modified starch technology has developed cleaner, more natural modification procedures. The quest for sustainability has sparked the investigation into biodegradable and eco-friendly materials. This transition corresponds with consumer inclinations towards familiar, minimally processed components, leading to the creation of modified starches via non-chemical methods, including physical or enzymatic alterations. In recent years, modified starches have emerged as a potentially helpful choice. These starches provide biodegradability and renewability, which helps to reduce the ecological imprint that diverse businesses leave behind. These refined methods enhance the natural properties of starches while adhering to the rigorous standards of clean-label trends. The advancements in this field include the creation of starch-based bioplastics, packaging materials and biodegradable films, which play an enhancing role in the global sustainability agenda. The future of modified starch technology is expected to emphasise precision modifications, allowing for customising starches to fulfill particular industrial needs. The convergence of biotechnology and materials science presents significant potential for developing starch-based nanomaterials, facilitating innovative applications in medicine and pharmaceutical fields.

15.6 SUMMARY

- Ultrasound increases swelling index and solubility due to the interaction of reactive species, formation of ordered double helices and reduction in molecular weight.
- Cold plasma modification disrupts the crystallinity, creating loose surface structures and increasing the swelling power.
- Acid hydrolysis disintegrates starch molecules and breaks the α-1,4 and β-1,6 bonds in amylose and amylopectin, making it harder for granules to retain water.
- Amylases such as α-amylase and β-amylase promote solubility by depolymerism and decreasing chain lengths.
- Pullulanase-treated enzymes decrease swelling and increase solubility due to the debranching of starch into shorter linear chains.
- Maltogenic α-amylase and transglucosidase decrease the swelling of starches due to the introduction of pores on the granular surface and strengthening of double helices.
- Pre-gelatinisation disintegrates crystalline structures, reducing recrystallisation and increasing amorphous regions, contributing to lower relative crystallinity.
- Heat-moisture treatment (HMT) reduces relative crystallinity due to the destruction and disorganisation of double helices in treated starches.
- Microwave treatment decreased viscosities while increasing pasting temperature, reducing structural integrity and water retention capacity.
- Increased epichlorohydrin concentrations decreased pasting viscosities and breakdown while the pasting temperature increased.
- Esterification increased pasting viscosities with a decrease in the pasting temperature, specifically in octenyl succinic anhydride–treated starches.
- Pre-gelatinisation increases water absorption capacity (WAC) due to disruption of molecular integrity, deterioration of crystalline regions and increased exposure of hydroxyl groups to react with water molecules.
- Annealing has both negative and positive effects on WAC.
- Ultrasonication destroys the inner components of the starch, allowing easier water penetration inside the granules.
- Pre-gelatinisation and annealing increase oil absorption capacity (OAC) due to increased density of lipophilic/hydrophobic molecules on the starch surface.
- HMT causes surface roughness, cavities and granular expansion, leading to higher lipophilic tendencies, increasing OAC.
- Microwave-modified starch possesses too large pores, which reduce capillary actions and decrease the oil binding capacity.
- Cold plasma increases the surface area by striking high-energy electrons on the granules' surface, facilitating more oil interaction with the granules.
- Dual enzymatic treatment with α-amylase and glucoamylase reduces OAC.
- Corn starch's oil-holding capacity increases when treated with α-amylase, glucoamylase or transglucosidase.
- Modified starches, particularly resistant starch (RS), promote better glycemic control and enhance insulin sensitivity.
- RS is not digested or absorbed in the small intestine, contributing to improved glycemic response and enhanced satiety.
- *Ganoderma lucidum* spore encapsulation of RS showed significant hypoglycemic properties in rats with type 2 diabetes mellitus (T2DM).
- T2DM patients have elevated levels of pathogenic bacteria in their gut microbiota, suggesting the potential therapeutic benefits of incorporating RS into the diet.
- Consumption of oat RS significantly influences gut microbiota composition, highlighting potential prebiotic effects.
- Starch–phenolic acid complexes exhibit lower digestibility than native starch and their combinations.
- Gallic acid can alter the properties of starch, potentially affecting the digestibility of starches.
- The presence of phenolic acids in tapioca starch esters contributes to reduced digestibility.
- These modifications could enhance the functional properties of tapioca starch, potentially leading to novel applications in food systems.
- Chlorogenic acid combined with extruded rice starch reduces the projected glycemic index and alleviates lipid metabolism problems caused by a high-fat diet.
- *Pueraria lobata* RS can mitigate lipid metabolic diseases by increasing the prevalence of certain bacteria while decreasing *Staphylococcus* in mice on a high-fat diet.
- A double-blind, randomised controlled experiment examined the impact of type 3 RS derived from *Canna edulis* on cholesterol levels in moderate hyperlipidemia.

- Green banana–RS mitigated hepatic steatosis, triglyceride and cholesterol accumulation.
- Modified starches have the potential to influence lipid oxidation and support cholesterol homeostasis, possibly linked to the enhanced composition of the fecal microbial community.
- Modified starch in dietary formulations exhibits higher efficacy in exhibiting prebiotic properties.
- Modified starches have been linked to positive outcomes in body composition and lipid profiles due to their ability to promote the production of short-chain fatty acids within the colon.

15.7 CONCLUSION

Native starches limit their application due to their insolubility in cold water, low gelatinisation temperature, high propensity for retrogradation, susceptibility to syneresis and reduced viscosity. Furthermore, the increased reduced tensile strength, hydrophilicity and solubility of native starches inhibit their use as food packaging materials. Advancements in modified starch technology have surpassed traditional limits, advancing starches into diverse applications across multiple sectors. Physical modifications influence starch digestibility by altering the structural arrangement of amylose and amylopectin molecules. Altering starches by single or multiple changes may enhance their physical, morphological, optical and barrier properties, affecting their nutritional characteristics and digestibility.

Additionally, the absorption and metabolism of modified starch are influenced by numerous physiological factors, including enzyme inhibitors, viscosity and gut microbiota, along with its interaction with dietary matrices and its transit through the gastrointestinal tract. The specific type of modification also plays a role in determining the susceptibility of modified starch to amylolysis.

Modified starch is a widely utilised ingredient in the food industry due to its unique properties. This substance is non-digestible, meaning it passes through the digestive system without being broken down. As it travels through the gut, modified starch plays a significant role in shaping the intestinal microbial community, fostering a balanced ecosystem of beneficial bacteria. Additionally, its consumption is linked to an increase in short-chain fatty acids, which are essential for gut health and exhibit prebiotic characteristics. These traits make modified starch an important component in promoting overall digestive well-being.

Consequently, in any research examining the alterations in starch characteristics throughout its modification, this essential quality should be accurately assessed using dependable in vitro and in vivo experiments. As we further explore the possibilities of modified starches, their many uses and role in sustainability will undoubtedly influence the future of food industries, addressing both consumer demands and environmental awareness. The evolution of modified starches, influenced by technological advancements and market demands, indicates a future with endless possibilities.

REFERENCES

Abdorreza, M. N., Robal, M., Cheng, L. H., Tajul, A. Y., & Karim, A. A. (2012). Physicochemical, thermal, and rheological properties of acid-hydrolyzed sago (*Metroxylon sagu*) starch. *LWT*, *46*(1), 135–141. https://doi.org/10.1016/j.lwt.2011.10.015

Adawiyah, D. R., Akuzawa, S., Sasaki, T., & Kohyama, K. (2017). A comparison of the effects of heat moisture treatment (HMT) on rheological properties and amylopectin structure in sago (*Metroxylon sago*) and arenga (*Arenga pinnata*) starches. *Journal of Food Science and Technology*, *54*(11), 3404–3410. https://doi.org/10.1007/s13197-017-2787-1

Adebowale, K. O., Afolabi, T. A., & Olu-Owolabi, B. I. (2006). Functional, physicochemical and retrogradation properties of sword bean (*Canavalia gladiata*) acetylated and oxidized starches. *Carbohydrate Polymers*, *65*(1), 93–101. https://doi.org/10.1016/j.carbpol.2005.12.032

Adebowale, K. O., Olu-Owolabi, B. I., Olawumi, E. K., & Lawal, O. S. (2005). Functional properties of native, physically and chemically modified breadfruit (*Artocarpus artilis*) starch. *Industrial Crops and Products*, *21*(3), 343–351. https://doi.org/10.1016/j.indcrop.2004.05.002

Adebowale, K. O., Olu-Owolabi, B. I., & Olayinka, O. O. (2005). Effect of heat moisture treatment and annealing on physicochemical properties of red sorghum starch. *African Journal of Biotechnology*, *4*(9), 928–933.

Ağagündüz, D., Özata-Uyar, G., Kocaadam-Bozkurt, B., Özturan-Şirin, A., Capasso, R., Al-Assaf, S., & Özoğul, F. (2023). A comprehensive review on food hydrocolloids as gut modulators in the food matrix and nutrition: The hydrocolloid-gut-health axis. *Food Hydrocolloids*, *145*, 109068. https://doi.org/10.1016/j.foodhyd.2023.109068

Ahmad, M., & Gani, A. (2021). Ultrasonicated resveratrol loaded starch nanocapsules: Characterization, bioactivity and release behaviour under in-vitro digestion. *Carbohydrate Polymers*, *251*, 117111. https://doi.org/10.1016/j.carbpol.2020.117111

Akubor, P. I., & Igba, T. (2019). Effect of pre-gelatinization and annealing on the chemical composition, functional and pasting properties of starch prepared from unripe banana fruits. *South Asian Journal of Food Technology and Environment*, *05*(01), 807–816. https://doi.org/10.46370/sajfte.2019.v05i01.08

Alimi, B. A., & Workneh, T. S. (2018). Structural and physicochemical properties of heat moisture treated and citric acid modified acha and iburu starches. *Food Hydrocolloids*, *81*, 449–455. https://doi.org/10.1016/j.foodhyd.2018.03.027

Almeida, M. C. B. D. M., Costa, S. D. S., Cavalcanti, M. T., & Almeida, E. L. (2020). Characterization of *prata* banana (*Musa* AAB-*Prata*) starch: Native and modified by annealing. *Starch - Stärke*, *72*(3–4), 1900137. https://doi.org/10.1002/star.201900137

Almeida, R. L. J., Santos, N. C., Muniz, C. E. S., Da Silva Eduardo, R., De Almeida Silva, R., Ribeiro, C. A. C., Da Costa, G. A., De Figueiredo, M. J., Galdino, P. O., & Dos Santos, E. S. (2023). Red rice starch modification: Combination of the non-thermal method with a pulsed electric field (PEF) and enzymatic method using α-amylase. *International Journal of Biological Macromolecules*, *253*, 127030. https://doi.org/10.1016/j.ijbiomac.2023.127030

Anjali, K. U., Kamatchi, A. R., Haripriya, S., & Nasha, P. (2024). An insight into the impact of microwave treatment on the morphology, functionality, and resistant starch content from conventional rice landraces. *Biomass Conversion and Biorefinery*. https://doi.org/10.1007/s13399-024-05552-y

Arinola, S. O. (2019). Physicochemical properties of pregelatinized and microwave radiated white and red cocoyam (*Colocasia esculenta*) starches. *Croatian Journal of Food Science and Technology*, *11*(2), 251–258. https://doi.org/10.17508/CJFST.2019.11.2.15

Balasubramanian, S., Sharma, R., Kaur, J., & Bhardwaj, N. (2014). Characterization of modified pearl millet (*Pennisetum typhoides*) starch. *Journal of Food Science and Technology*, *51*(2), 294–300. https://doi.org/10.1007/s13197-011-0490-1

Banura, S., Thirumdas, R., Kaur, A., Deshmukh, R. R., & Annapure, U. S. (2018). Modification of starch using low pressure radio frequency air plasma. *LWT*, *89*, 719–724. https://doi.org/10.1016/j.lwt.2017.11.056

BeMiller, J. N., & Huber, K. C. (2015). Physical modification of food starch functionalities. *Annual Review of Food Science and Technology*, *6*(1), 19–69. https://doi.org/10.1146/annurev-food-022814-015552

Bensaad, D. E., Saleh, M., Ismail, K., & Lee, Y. (2022). Recent advances in physical, enzymatic, and genetic modifications of starches. *Jordan Journal of Agricultural Sciences*, *18*(3), 245–258. https://doi.org/10.35516/jjas.v18i3.474

Bhosale, R., & Singhal, R. (2007). Effect of octenylsuccinylation on physicochemical and functional properties of waxy maize and amaranth starches. *Carbohydrate Polymers*, *68*(3), 447–456. https://doi.org/10.1016/j.carbpol.2006.11.011

Biduski, B., Silva, F. T. D., Silva, W. M. D., Halal, S. L. D. M. E., Pinto, V. Z., Dias, A. R. G., & Zavareze, E. D. R. (2017). Impact of acid and oxidative modifications, single or dual, of sorghum starch on biodegradable films. *Food Chemistry*, *214*, 53–60. https://doi.org/10.1016/j.foodchem.2016.07.039

Bushra, M., Yun, X. X., Pan, S. Y., Hydamaka, A., Hua, M. W., & Feng, W. L. (2013). Effect of oxidation and esterification on functional properties of mung bean (*Vigna radiata* (L.) Wilczek) starch. *European Food Research and Technology*, *236*(1), 119–128. https://doi.org/10.1007/s00217-012-1857-x

Campelo, P. H., Sant'Ana, A. S., & Pedrosa Silva Clerici, M. T. (2020). Starch nanoparticles: Production methods, structure, and properties for food applications. *Current Opinion in Food Science*, *33*, 136–140. https://doi.org/10.1016/j.cofs.2020.04.007

Cao, M., & Gao, Q. (2020). Effect of dual modification with ultrasonic and electric field on potato starch. *International Journal of Biological Macromolecules*, *150*, 637–643. https://doi.org/10.1016/j.ijbiomac.2020.02.008

Çatal, H., & İbanoğlu, Ş. (2014). Effect of aqueous ozonation on the pasting, flow and gelatinization properties of wheat starch. *LWT - Food Science and Technology*, *59*(1), 577–582. https://doi.org/10.1016/j.lwt.2014.04.025

Chan, H., Leh, C. P., Bhat, R., Senan, C., Williams, P. A., & Karim, A. A. (2011). Molecular structure, rheological and thermal characteristics of ozone-oxidized starch. *Food Chemistry*, *126*(3), 1019–1024. https://doi.org/10.1016/j.foodchem.2010.11.113

Chavan, P., Singh, G. P., Aayush, K., Sharma, S., Chauhan, P. K., Thory, R., Shandilya, M., & Kumar, D. (2024). Development and characterization of pearl millet (AHB 1200) starch nanoparticle-based edible films: A paradigm shift in food packaging. *Journal of Food Process Engineering*, *47*(3), e14575. https://doi.org/10.1111/jfpe.14575

Chavan, P., Sinhmar, A., Nehra, M., Thory, R., Pathera, A. K., Sundarraj, A. A., & Nain, V. (2021). Impact on various properties of native starch after synthesis of starch nanoparticles: A review. *Food Chemistry*, *364*, 130416. https://doi.org/10.1016/j.foodchem.2021.130416

Choton, S., Bandral, J. D., Singh, J., Bhat, A., Sood, M., Gupta, N., Reshi, M., & Kaur, D. (2024). Enzymatic modification of starch: A review. *Saudi Journal of Medical and Pharmaceutical Sciences*, *10*(01), 1–8.

Cione, E., Fazio, A., Curcio, R., Tucci, P., Lauria, G., Cappello, A. R., & Dolce, V. (2021). Resistant starches and non-communicable disease: A focus on mediterranean diet. *Foods*, *10*(9), 2062. https://doi.org/10.3390/foods10092062

Compart, J., Singh, A., Fettke, J., & Apriyanto, A. (2023). Customizing starch properties: A review of starch modifications and their applications. *Polymers*, *15*(16), 3491. https://doi.org/10.3390/polym15163491

Cornejo, F., Maldonado-Alvarado, P., Palacios-Ponce, S., Hugo, D., & Rosell, C. M. (2022). Impact of cassava starch varieties on the physiochemical change during enzymatic hydrolysis. *Molecules*, *27*(18), 6098. https://doi.org/10.3390/molecules27186098

Das, A., & Sit, N. (2021). Modification of taro starch and starch nanoparticles by various physical methods and their characterization. *Starch - Stärke*, *73*(5–6), 2000227. https://doi.org/10.1002/star.202000227

Demirkesen-Bicak, H., Tacer-Caba, Z., & Nilufer-Erdil, D. (2018). Pullulanase treatments to increase resistant starch content of black chickpea (*Cicer arietinum* L.) starch and the effects on starch properties. *International Journal of Biological Macromolecules*, *111*, 505–513. https://doi.org/10.1016/j.ijbiomac.2018.01.026

De Oliveira, C. S., Bet, C. D., Bisinella, R. Z. B., Waiga, L. H., Colman, T. A. D., & Schnitzler, E. (2018). Heat-moisture treatment (HMT) on blends from potato starch (PS) and sweet potato starch (SPS). *Journal of Thermal Analysis and Calorimetry*, *133*(3), 1491–1498. https://doi.org/10.1007/s10973-018-7196-9

Dewi, A. M. P., Santoso, U., Pranoto, Y., & Marseno, D. W. (2022). Dual modification of sago starch via heat-moisture treatment and octenyl succinylation to improve starch hydrophobicity. *Polymers*, *14*(6), 1086. https://doi.org/10.3390/polym14061086

Dutta, D., & Sit, N. (2022). Comparison of properties of films prepared from potato starch modified by annealing and heat–moisture treatment. *Starch - Stärke*, *74*(11–12), 2200110. https://doi.org/10.1002/star.202200110

Falade, K. O., & Ayetigbo, O. E. (2015). Effects of annealing, acid hydrolysis and citric acid modifications on physical and functional properties of starches from four yam (*Dioscorea* spp.) cultivars. *Food Hydrocolloids*, *43*, 529–539. https://doi.org/10.1016/j.foodhyd.2014.07.008

Fonseca, L. M., Gonçalves, J. R., El Halal, S. L. M., Pinto, V. Z., Dias, A. R. G., Jacques, A. C., & Zavareze, E. D. R. (2015). Oxidation of potato starch with different sodium hypochlorite concentrations and its effect on biodegradable films. *LWT*, *60*(2), 714–720. https://doi.org/10.1016/j.lwt.2014.10.052

Fonseca, L. M., Halal, S. L. M. E., Dias, A. R. G., & Zavareze, E. D. R. (2021). Physical modification of starch by heat-moisture treatment and annealing and their applications: A review. *Carbohydrate Polymers*, *274*, 118665. https://doi.org/10.1016/j.carbpol.2021.118665

Gao, S., Liu, H., Sun, L., Liu, N., Wang, J., Huang, Y., Wang, F., Cao, J., Fan, R., Zhang, X., & Wang, M. (2019). The effects of dielectric barrier discharge plasma on physicochemical and digestion properties of starch. *International Journal of Biological Macromolecules*, *138*, 819–830. https://doi.org/10.1016/j.ijbiomac.2019.07.147

Ge, X., Shen, H., Su, C., Zhang, B., Zhang, Q., Jiang, H., Yuan, L., Yu, X., & Li, W. (2021). Pullulanase modification of granular sweet potato starch: Assistant effect of dielectric barrier discharge plasma on multi-scale structure, physicochemical properties. *Carbohydrate Polymers*, *272*, 118481. https://doi.org/10.1016/j.carbpol.2021.118481

Godswill, C., Somtochukwu, V., & Kate, C. (2019). The functional properties of foods and flour. *International Journal of Advanced Academic Research*, *5*(11), 139–160.

Gujral, H., Sinhmar, A., Nehra, M., Nain, V., Thory, R., Pathera, A. K., & Chavan, P. (2021). Synthesis, characterization, and utilization of potato starch nanoparticles as a filler in nanocomposite films. *International Journal of Biological Macromolecules*, *186*, 155–162. https://doi.org/10.1016/j.ijbiomac.2021.07.005

Guo, J., Tang, W., Quek, S. Y., Liu, Z., Lu, S., & Tu, K. (2020). Evaluation of structural and physicochemical properties of octenyl succinic anhydride modified sweet potato starch with different degrees of substitution. *Journal of Food Science*, *85*(3), 666–672. https://doi.org/10.1111/1750-3841.15031

Guo, L., Deng, Y., Lu, L., Zou, F., & Cui, B. (2019). Synergistic effects of branching enzyme and transglucosidase on the modification of potato starch granules. *International Journal of Biological Macromolecules*, *130*, 499–507. https://doi.org/10.1016/j.ijbiomac.2019.02.160

Guo, L., Li, H., Lu, L., Zou, F., Tao, H., & Cui, B. (2019). The role of sequential enzyme treatments on structural and physicochemical properties of cassava starch granules. *Starch - Stärke*, *71*(7–8), 1800258. https://doi.org/10.1002/star.201800258

Guo, L., Li, J., Li, H., Zhu, Y., & Cui, B. (2020). The structure property and adsorption capacity of new enzyme-treated potato and sweet potato starches. *International Journal of Biological Macromolecules*, *144*, 863–873. https://doi.org/10.1016/j.ijbiomac.2019.09.164

Guo, L., Li, J., Yuan, Y., Gui, Y., Zou, F., Lu, L., & Cui, B. (2021). Structural and functional modification of kudzu starch using α-amylase and transglucosidase. *International Journal of Biological Macromolecules*, *169*, 67–74. https://doi.org/10.1016/j.ijbiomac.2020.12.099

Guo, L., Tao, H., Cui, B., & Janaswamy, S. (2019). The effects of sequential enzyme modifications on structural and physicochemical properties of sweet potato starch granules. *Food Chemistry*, *277*, 504–514. https://doi.org/10.1016/j.foodchem.2018.11.014

Gupta, R. K., Guha, P., & Srivastav, P. P. (2022). Natural polymers in bio-degradable/edible film: A review on environmental concerns, cold plasma technology and nanotechnology application on food packaging-A recent trends. *Food Chemistry Advances*, *1*, 100135. https://doi.org/10.1016/j.focha.2022.100135

Gupta, R. K., Guha, P., & Srivastav, P. P. (2023). Effect of high voltage dielectric barrier discharge (DBD) atmospheric cold plasma treatment on physicochemical and functional properties of taro (*Colocasia esculenta*) starch. *International Journal of Biological Macromolecules*, *253*, 126772. https://doi.org/10.1016/j.ijbiomac.2023.126772

Gupta, R. K., Guha, P., & Srivastav, P. P. (2024a). Insights into the physicochemical, crystalline, multistructural, thermal and digestibility behavior of taro starch (*Colocasia esculenta*) modified by cold plasma treatment with high voltage dielectric barrier discharge (HVDBD): An alternative technique for chemical modification of starch and environmental sustainability. *Journal of Cleaner Production*, *461*, 142710. https://doi.org/10.1016/j.jclepro.2024.142710

Gupta, R. K., Guha, P., & Srivastav, P. P. (2024b). Dielectric barrier discharge plasma: A green and novel method to change the structure of taro peel starch and improve the physicochemical properties of taro peel starch films. *Plasma Processes and Polymers*, *21*(9), 2400047. https://doi.org/10.1002/ppap.202400047

Gupta, R. K., Guha, P., & Srivastav, P. P. (2024c). Exploring the potential of taro (*Colocasia esculenta*) starch: Recent developments in modification, health benefits, and food industry applications. *Food Bioengineering, 3*(3), 365–379. https://doi.org/10.1002/fbe2.12103

Gupta, R. K., Guha, P., & Srivastav, P. P. (2024d). Physical action of nonthermal cold plasma technology for starch modification. *Food Physics*, *1*, 100011. https://doi.org/10.1016/j.foodp.2024.100011

Gupta, R. K., Guha, P., & Srivastav, P. P. (2024e). Effect of cold plasma treatment and plasma-activated water on physicochemical and structural properties of starch: A green and novel approach for environmental sustainability. *Plasma Processes and Polymers*, *21*(4), 2300204. https://doi.org/10.1002/ppap.202300204

Gupta, R. K., Gnana Moorthy Eswaran, U., Kumar, S., Pipliya, S., Kaur, B., & Srivastav, P. P. (2025a). Synergistic modification of starch polysaccharide via sequential treatments with ultrasonic and cold plasma technology. *Journal of Food Processing and Preservation*, *2025*(1), 9063007. https://doi.org/10.1155/jfpp/9063007

Gupta, R. K., Pipliya, S., Patel, J., Srivastav, P. P., Castro-Muñoz, R., Wang, C. K., & Nemţanu, M. R. (2025b). Essential oil-embedded starch based films for active packaging: Insights into migration, stability, and preservation efficiency. *Trends in Food Science & Technology*, *163*, 105149. https://doi.org/10.1016/j.tifs.2025.105149

Guo, Z., Gou, Q., Yang, L., Yu, Q., & Han, L. (2022). Dielectric barrier discharge plasma: A green method to change structure of potato starch and improve physicochemical properties of potato starch films. *Food Chemistry*, *370*, 130992. https://doi.org/10.1016/j.foodchem.2021.130992

Hashim, S. O. (2019). Starch-modifying enzymes. In G. Mamo & B. Mattiasson (Eds), *Alkaliphiles in biotechnology* (Vol. 172, pp. 221–244). Springer International Publishing. https://doi.org/10.1007/10_2019_91

Hazarika, B. J., & Sit, N. (2016). Effect of dual modification with hydroxypropylation and cross-linking on physicochemical properties of taro starch. *Carbohydrate Polymers*, *140*, 269–278. https://doi.org/10.1016/j.carbpol.2015.12.055

He, X., Xia, W., Chen, R., Dai, T., Luo, S., Chen, J., & Liu, C. (2020). A new pre-gelatinized starch preparing by gelatinization and spray drying of rice starch with hydrocolloids. *Carbohydrate Polymers*, *229*, 115485. https://doi.org/10.1016/j.carbpol.2019.115485

Hedayati, S., Shahidi, F., Koocheki, A., Farahnaky, A., & Majzoobi, M. (2016). Comparing the effects of sucrose and glucose on functional properties of pregelatinized maize starch. *International Journal of Biological Macromolecules*, *88*, 499–504. https://doi.org/10.1016/j.ijbiomac.2016.04.026

Heo, H., Lee, Y.-K., & Chang, Y. H. (2017). Effect of cross-linking on physicochemical and in vitro digestibility properties of potato starch. *Emirates Journal of Food and Agriculture*, 463. https://doi.org/10.9755/ejfa.2017-01-237

Hernández-Jaimes, C., Utrilla-Coello, R. G., Carrillo-Navas, H., García-Márquez, E., Meraz, M., Bello-Pérez, L. A., Vernon-Carter, E. J., & Alvarez-Ramirez, J. (2014). Corn starch acid hydrolysis at the onset gelatinization temperature: Morphology, crystallinity, viscoelasticity, and thermal properties. *Starch - Stärke*, *66*(7–8), 636–644. https://doi.org/10.1002/star.201300215

Hii, S. L., Tan, J. S., Ling, T. C., & Ariff, A. B. (2012). Pullulanase: Role in starch hydrolysis and potential industrial applications. *Enzyme Research*, *2012*, e921362. https://doi.org/10.1155/2012/921362

Hu, A., Li, L., Zheng, J., Lu, J., Meng, X., Liu, Y., & Rehman, R. (2014). Different-frequency ultrasonic effects on properties and structure of corn starch. *Journal of the Science of Food and Agriculture*, *94*(14), 2929–2934. https://doi.org/10.1002/jsfa.6636

Hu, A., Wang, X., Li, L., Xu, T., & Zheng, J. (2020). Effects of annealing time on structure and properties of sweet potato starch. *Cereal Chemistry*, *97*(3), 573–580. https://doi.org/10.1002/cche.10272

Hu, Y., Qin, Y., Qiu, C., Xu, X., Jin, Z., & Wang, J. (2020). Ultrasound-assisted self-assembly of β-cyclodextrin/debranched starch nanoparticles as promising carriers of tangeretin. *Food Hydrocolloids*, *108*, 106021. https://doi.org/10.1016/j.foodhyd.2020.106021

Iqbal, U., Saini, P., & Ahmed, M. (2024). Effect of heat moisture treatment, annealing, and gelatinization-retrogradation modifications on physico-chemical, functional, and structural properties of starch from elephant foot yam (*Amorphophallus paeoniifolius*). *Starch - Stärke*, *76*(9–10), 2300220. https://doi.org/10.1002/star.202300220

Jia, R., Cui, C., Gao, L., Qin, Y., Ji, N., Dai, L., Wang, Y., Xiong, L., Shi, R., & Sun, Q. (2023). A review of starch swelling behavior: Its mechanism, determination methods, influencing factors, and influence on food quality. *Carbohydrate Polymers*, *321*, 121260. https://doi.org/10.1016/j.carbpol.2023.121260

Jiali, L., Wu, Z., Liu, L., Yang, J., Wang, L., Li, Z., & Liu, L. (2024). The research advance of resistant starch: Structural characteristics, modification method, immunomodulatory function, and its delivery systems application. *Critical Reviews in Food Science and Nutrition*, *64*(29), 10885–10902. https://doi.org/10.1080/10408398.2023.2230287

Jiang, F., Zhu, Y., Hu, W.-X., Li, M., Liu, Y., Feng, J., Lv, X., Yu, X., & Du, S. (2023). Characterization of quinoa starch nanoparticles as a stabilizer for oil in water Pickering emulsion. *Food Chemistry*, *427*, 136697. https://doi.org/10.1016/j.foodchem.2023.136697

Jung, Y., Lee, B.-H., & Yoo, S.-H. (2017). Physical structure and absorption properties of tailor-made porous starch granules produced by selected amylolytic enzymes. *PLoS One*, *12*(7), e0181372. https://doi.org/10.1371/journal.pone.0181372

Kamwilaisak, K., Rittiwut, K., Jutakridsada, P., Iamamorphanth, W., Pimsawat, N., Knijnenburg, J. T. N., & Theerakulpisut, S. (2022). Rheology, stability, antioxidant properties, and curcumin release of oil-in-water Pickering emulsions stabilized by rice starch nanoparticles. *International Journal of Biological Macromolecules*, *214*, 370–380. https://doi.org/10.1016/j.ijbiomac.2022.06.032

Kankate, D., Panpalia, S. G., Kumar, K. J., & Kennedy, J. F. (2020). Studies to predict the effect of pregelatinization on excipient property of maize and potato starch blends. *International Journal of Biological Macromolecules*, *164*, 1206–1214. https://doi.org/10.1016/j.ijbiomac.2020.07.170

Karwasra, B. L., Kaur, M., & Gill, B. S. (2020). Impact of ultrasonication on functional and structural properties of Indian wheat (*Triticum aestivum* L.) cultivar starches. *International Journal of Biological Macromolecules*, *164*, 1858–1866. https://doi.org/10.1016/j.ijbiomac.2020.08.013

Kaur, H., & Gill, B. S. (2019). Effect of high-intensity ultrasound treatment on nutritional, rheological and structural properties of starches obtained from different cereals. *International Journal of Biological Macromolecules*, *126*, 367–375. https://doi.org/10.1016/j.ijbiomac.2018.12.149

Kaur, L., Singh, J., & Singh, N. (2006). Effect of cross-linking on some properties of potato (*Solanum tuberosum* L.) starches. *Journal of the Science of Food and Agriculture*, *86*(12), 1945–1954. https://doi.org/10.1002/jsfa.2568

Keeratiburana, T., Hansen, A. R., Soontaranon, S., Blennow, A., & Tongta, S. (2020). Porous high amylose rice starch modified by amyloglucosidase and maltogenic α-amylase. *Carbohydrate Polymers*, *230*, 115611. https://doi.org/10.1016/j.carbpol.2019.115611

Kong, X., Qiu, D., Ye, X., Bao, J., Sui, Z., Fan, J., & Xiang, W. (2015). Physicochemical and crystalline properties of heat–moisture-treated rice starch: Combined effects of moisture and duration of heating. *Journal of the Science of Food and Agriculture*, *95*(14), 2874–2879. https://doi.org/10.1002/jsfa.7028

Krishnakumar, T., & Sajeev, M. S. (2018). Effect of ultrasound treatment on physicochemical and functional properties of cassava starch. *International Journal of Current Microbiology and Applied Sciences*, *7*(10), 3122–3135. https://doi.org/10.20546/ijcmas.2018.710.362

Kumar, Y., Singh, L., Sharanagat, V. S., Patel, A., & Kumar, K. (2020). Effect of microwave treatment (low power and varying time) on potato starch: Microstructure, thermo-functional, pasting and rheological properties. *International Journal of Biological Macromolecules*, *155*, 27–35. https://doi.org/10.1016/j.ijbiomac.2020.03.174

Kumoro, A. C., Retnowati, D. S., Ratnawati, R., & Widiyanti, M. (2019). Effect of temperature and reaction time on the swelling power and solubility of gadung (*Dioscorea hispida* Dennst) tuber starch during heat moisture treatment process. *Journal of Physics: Conference Series*, *1295*, 012062. https://doi.org/10.1088/1742-6596/1295/1/012062

Lai, M., Li, N., Li, Q., Ge, X., Huang, Z., Chen, F., Liu, B., & Zeng, F. (2024). Pueraria lobata resistant starch regulating lipid metabolism and gut microbiota in high-fat diet mice. *Starch - Stärke*, *76*(9–10), 2300123. https://doi.org/10.1002/star.202300123

Lawal, O. (2004). Composition, physicochemical properties and retrogradation characteristics of native, oxidised, acetylated and acid-thinned new cocoyam (*Xanthosoma sagittifolium*) starch. *Food Chemistry*, *87*(2), 205–218. https://doi.org/10.1016/j.foodchem.2003.11.013

Lawal, O. S., Adebowale, K. O., Ogunsanwo, B. M., Barba, L. L., & Ilo, N. S. (2005). Oxidized and acid thinned starch derivatives of hybrid maize: Functional characteristics, wide-angle X-ray diffractometry and thermal properties. *International Journal of Biological Macromolecules*, *35*(1–2), 71–79. https://doi.org/10.1016/j.ijbiomac.2004.12.004

Lee, K. Y., Lee, S., & Lee, H. G. (2010). Effect of the degree of enzymatic hydrolysis on the physicochemical properties and in vitro digestibility of rice starch. *Food Science and Biotechnology*, *19*(5), 1333–1340. https://doi.org/10.1007/s10068-010-0190-z

Lee, S., Yang, Y., Chung, H., & Lim, S. (2017). Effect of dry heating on physicochemical properties of pregelatinized rice starch. *Cereal Chemistry*, *94*(6), 928–933. https://doi.org/10.1094/CCHEM-03-17-0045-R

Lee, T., Lee, Y., Shin, J., & Chang, Y. H. (2023). Physicochemical and prebiotic properties of waxy rice flour modified by pullulanase. *Food Biotechnology*, *37*(2), 89–105. https://doi.org/10.1080/08905436.2023.2200835

Le Thanh-Blicharz, J., Anioła, J., Kowalczewski, P., Przygoński, K., Zaborowska, Z., & Lewandowicz, G. (2014). Type IV resistant starch increases cecum short chain fatty acids level in rats. *Acta Biochimica Polonica*, *61*(1), 109–114.

Li, C., & Hu, Y. (2021). Effects of acid hydrolysis on the evolution of starch fine molecular structures and gelatinization properties. *Food Chemistry*, *353*, 129449. https://doi.org/10.1016/j.foodchem.2021.129449

Li, M.-N., Zhang, B., Xie, Y., & Chen, H.-Q. (2019). Effects of debranching and repeated heat-moisture treatments on structure, physicochemical properties and in vitro digestibility of wheat starch. *Food Chemistry*, *294*, 440–447. https://doi.org/10.1016/j.foodchem.2019.05.040

Li, P., He, X., Dhital, S., Zhang, B., & Huang, Q. (2017). Structural and physicochemical properties of granular starches after treatment with debranching enzyme. *Carbohydrate Polymers*, *169*, 351–356. https://doi.org/10.1016/j.carbpol.2017.04.036

Li, W., Cao, F., Fan, J., Ouyang, S., Luo, Q., Zheng, J., & Zhang, G. (2014). Physically modified common buckwheat starch and their physicochemical and structural properties. *Food Hydrocolloids*, *40*, 237–244. https://doi.org/10.1016/j.foodhyd.2014.03.012

Li, Y., Liu, S., Liu, X., Tang, X., & Zhang, J. (2017). The impact of heat-moisture treatment on physicochemical properties and retrogradation behavior of sweet potato starch. *International Journal of Food Engineering*, *13*(5), 20170001. https://doi.org/10.1515/ijfe-2017-0001

Lim, X. X., Zulkurnain, M., Yussof, N. S., & Utra, U. (2023). Effects of dry heating, acetylation, and acid pre-treatments on modification of potato starch with octenyl succinic anhydride (OSA). *E-Polymers*, *23*(1), 20228090. https://doi.org/10.1515/epoly-2022-8090

Lin, Q., Ji, N., Li, M., Dai, L., Xu, X., Xiong, L., & Sun, Q. (2020). Fabrication of debranched starch nanoparticles via reverse emulsification for improvement of functional properties of corn starch films. *Food Hydrocolloids*, *104*, 105760. https://doi.org/10.1016/j.foodhyd.2020.105760

Liu, H., Lv, M., Peng, Q., Shan, F., & Wang, M. (2015). Physicochemical and textural properties of tartary buckwheat starch after heat–moisture treatment at different moisture levels. *Starch - Stärke*, *67*(3–4), 276–284. https://doi.org/10.1002/star.201400143

Liu, H., Lv, M., Wang, L., Li, Y., Fan, H., & Wang, M. (2016). Comparative study: How annealing and heat-moisture treatment affect the digestibility, textural, and physicochemical properties of maize starch. *Starch - Stärke*, *68*(11–12), 1158–1168. https://doi.org/10.1002/star.201500268

Liu, J., Wang, B., Lin, L., Zhang, J., Liu, W., Xie, J., & Ding, Y. (2014). Functional, physicochemical properties and structure of cross-linked oxidized maize starch. *Food Hydrocolloids*, *36*, 45–52. https://doi.org/10.1016/j.foodhyd.2013.08.013

Liu, Y., Chen, J., Luo, S., Li, C., Ye, J., Liu, C., & Gilbert, R. G. (2017). Physicochemical and structural properties of pregelatinized starch prepared by improved extrusion cooking technology. *Carbohydrate Polymers*, *175*, 265–272. https://doi.org/10.1016/j.carbpol.2017.07.084

Luo, Z., Fu, X., He, X., Luo, F., Gao, Q., & Yu, S. (2008). Effect of ultrasonic treatment on the physicochemical properties of maize starches differing in amylose content. *Starch - Stärke*, *60*(11), 646–653. https://doi.org/10.1002/star.200800014

Luo, Z., He, X., Fu, X., Luo, F., & Gao, Q. (2006). Effect of microwave radiation on the physicochemical properties of normal maize, waxy maize and amylomaize V starches. *Starch - Stärke*, *58*(9), 468–474. https://doi.org/10.1002/star.200600498

Majeed, T., Wani, I. A., & Hussain, P. R. (2017). Effect of dual modification of sonication and γ-irradiation on physicochemical and functional properties of lentil (*Lens culinaris* L.) starch. *International Journal of Biological Macromolecules*, *101*, 358–365. https://doi.org/10.1016/j.ijbiomac.2017.03.110

Maniglia, B. C., Castanha, N., Le-Bail, P., Le-Bail, A., & Augusto, P. E. D. (2021). Starch modification through environmentally friendly alternatives: A review. *Critical Reviews in Food Science and Nutrition*, *61*(15), 2482–2505. https://doi.org/10.1080/10408398.2020.1778633

Manzoor, M., Singh, J., Bandral, J. D., Gani, A., & Shams, R. (2020). Food hydrocolloids: Functional, nutraceutical and novel applications for delivery of bioactive compounds. *International Journal of Biological Macromolecules*, *165*, 554–567. https://doi.org/10.1016/j.ijbiomac.2020.09.182

Marboh, V., Gayary, M. A., Gautam, G., & Mahanta, C. L. (2022). Comparative study of heat-moisture treatment and annealing on morphology, crystallinity, pasting, and thermal properties of sohphlang (*Flemingia vestita*) starch. *Starch - Stärke*, *74*(7–8), 2100294. https://doi.org/10.1002/star.202100294

Marta, H., Rismawati, A., Soeherman, G. P., Cahyana, Y., Djali, M., Yuliana, T., & Sondari, D. (2023). The effect of dual-modification by heat-moisture treatment and octenylsuccinylation on physicochemical and pasting properties of arrowroot starch. *Polymers*, *15*(15), 3215. https://doi.org/10.3390/polym15153215

Marta, H., Rizki, D. I., Mardawati, E., Djali, M., Mohammad, M., & Cahyana, Y. (2023). Starch nanoparticles: Preparation, properties and applications. *Polymers*, *15*(5), 1167. https://doi.org/10.3390/polym15051167

Martins, P. C., Gutkoski, L. C., & Martins, V. G. (2018). Impact of acid hydrolysis and esterification process in rice and potato starch properties. *International Journal of Biological Macromolecules*, *120*, 959–965. https://doi.org/10.1016/j.ijbiomac.2018.08.170

Mehboob, S., Ali, T. M., Sheikh, M., & Hasnain, A. (2020). Effects of cross linking and/or acetylation on sorghum starch and film characteristics. *International Journal of Biological Macromolecules*, *155*, 786–794. https://doi.org/10.1016/j.ijbiomac.2020.03.144

Mei, J.-Q., Zhou, D.-N., Jin, Z.-Y., Xu, X.-M., & Chen, H.-Q. (2015). Effects of citric acid esterification on digestibility, structural and physicochemical properties of cassava starch. *Food Chemistry*, *187*, 378–384. https://doi.org/10.1016/j.foodchem.2015.04.076

Miao, T., Zhang, X., Zhang, C., Wu, J., Zhu, Y., Xiao, M., Zhang, N., Zhong, Y., Liu, Y., Lin, Y., Wu, Y., Li, W., Song, C., Liu, Y., & Wang, X. (2024). Type 3 resistant starch from *Canna edulis* reduce lipid levels in patients with mild hyperlipidemia through altering gut microbiome: A double-blind randomized controlled trial. *Pharmacological Research*, *205*, 107232. https://doi.org/10.1016/j.phrs.2024.107232

Mirmoghtadaie, L., Kadivar, M., & Shahedi, M. (2009). Effects of cross-linking and acetylation on oat starch properties. *Food Chemistry*, *116*(3), 709–713. https://doi.org/10.1016/j.foodchem.2009.03.019

Miskeen, S., An, Y. S., & Kim, J.-Y. (2021). Application of starch nanoparticles as host materials for encapsulation of curcumin: Effect of citric acid modification. *International Journal of Biological Macromolecules*, *183*, 1–11. https://doi.org/10.1016/j.ijbiomac.2021.04.133

Molavi, H., Razavi, S. M. A., & Farhoosh, R. (2018). Impact of hydrothermal modifications on the physicochemical, morphology, crystallinity, pasting and thermal properties of acorn starch. *Food Chemistry*, *245*, 385–393. https://doi.org/10.1016/j.foodchem.2017.10.117

Noor, N., Gani, A., Jhan, F., Ashraf Shah, M., & Ul Ashraf, Z. (2022). Ferulic acid loaded pickering emulsions stabilized by resistant starch nanoparticles using ultrasonication: Characterization, in vitro release and nutraceutical potential. *Ultrasonics Sonochemistry*, *84*, 105967. https://doi.org/10.1016/j.ultsonch.2022.105967

Núñez-Bretón, L. C., Cruz-Rodríguez, L. C., Tzompole-Colohua, M. L., Jiménez-Guzmán, J., Perea-Flores, M. D. J., Rosas-Flores, W., & González-Jiménez, F. E. (2019). Physicochemical, functional and structural characterization of Mexican *Oxalis tuberosa* starch modified by cross-linking. *Journal of Food Measurement and Characterization*, *13*(4), 2862–2870. https://doi.org/10.1007/s11694-019-00207-3

Núñez-Bretón, L. C., Torres-González, C. E., Del Ángel-Zumaya, J. A., Peredo-Lovillo, A., Rivera-Villanueva, J. M., Perea-Flores, M. D. J., Guzmán-Gerónimo, R. I., Manero, O., & González-Jiménez, F. E. (2024). Functionalization of starches from Mexican *Oxalis tuberosa* using dual chemical modification. *Food Hydrocolloids*, *149*, 109500. https://doi.org/10.1016/j.foodhyd.2023.109500

Nwaogazie, F. O., Akinwande, B. A., & Oyeyinka, S. A. (2022). Physicochemical properties of bambara groundnut (*Vigna subterranea*) starch annealed at different temperatures. *Journal of Food Processing and Preservation*, *46*(12). https://doi.org/10.1111/jfpp.17183

Obadi, M., Chen, Y., Qi, Y., Liu, S., & Xu, B. (2020). Effects of different pre-gelatinized starch on the processing quality of high value-added Tartary buckwheat noodles. *Journal of Food Measurement and Characterization*, *14*(6), 3462–3472. https://doi.org/10.1007/s11694-020-00572-4

Ogundiran, O. O., & Afolabi, T. A. (2018). Effects of annealing and heat moisture treatments on the physicochemical properties of white sorghum (*Sorghum bicolor*) starch. *Journal of Chemical Society of Nigeria*, *43*(3), Article 3. https://www.journals.chemsociety.org.ng/index.php/jcsn/article/view/177

Okekunle, M. O., Adebowale, K. O., Olu-Owolabi, B. I., & Lamprecht, A. (2020). Physicochemical, morphological and thermal properties of oxidized starches from lima bean (*Phaseolus lunatus*). *Scientific African*, *8*, e00432. https://doi.org/10.1016/j.sciaf.2020.e00432

Okyere, A. Y., Bertoft, E., & Annor, G. A. (2019). Modification of cereal and tuber waxy starches with radio frequency cold plasma and its effects on waxy starch properties. *Carbohydrate Polymers*, *223*, 115075. https://doi.org/10.1016/j.carbpol.2019.115075

Okyere, A. Y., Rajendran, S., & Annor, G. A. (2022). Cold plasma technologies: Their effect on starch properties and industrial scale-up for starch modification. *Current Research in Food Science*, *5*, 451–463. https://doi.org/10.1016/j.crfs.2022.02.007

Olayinka, O., Adebowale, K., & Olu-Owolabi, B. (2008). Effect of heat-moisture treatment on physicochemical properties of white sorghum starch. *Food Hydrocolloids*, *22*(2), 225–230. https://doi.org/10.1016/j.foodhyd.2006.11.004

Oyeyinka, S. A., Akintayo, O. A., Adebo, O. A., Kayitesi, E., & Njobeh, P. B. (2021). A review on the physicochemical properties of starches modified by microwave alone and in combination with other methods. *International Journal of Biological Macromolecules*, *176*, 87–95. https://doi.org/10.1016/j.ijbiomac.2021.02.066

Oyeyinka, S. A., Umaru, E., Olatunde, S. J., & Joseph, J. K. (2019). Effect of short microwave heating time on physicochemical and functional properties of bambara groundnut starch. *Food Bioscience*, *28*, 36–41. https://doi.org/10.1016/j.fbio.2019.01.005

Pandiselvam, R., Manikantan, M. R., Divya, V., Ashokkumar, C., Kaavya, R., Kothakota, A., & Ramesh, S. V. (2019). Ozone: An advanced oxidation technology for starch modification. *Ozone: Science & Engineering*, *41*(6), 491–507. https://doi.org/10.1080/01919512.2019.1577128

Pratiwi, M., Faridah, D. N., & Lioe, H. N. (2018). Structural changes to starch after acid hydrolysis, debranching, autoclaving-cooling cycles, and heat moisture treatment (HMT): A review. *Starch - Stärke*, *70*(1–2), 1700028. https://doi.org/10.1002/star.201700028

Punia Bangar, S., Ashogbon, A. O., Singh, A., Chaudhary, V., & Whiteside, W. S. (2022). Enzymatic modification of starch: A green approach for starch applications. *Carbohydrate Polymers*, *287*, 119265. https://doi.org/10.1016/j.carbpol.2022.119265

Rahaman, A., Kumari, A., Zeng, X.-A., Adil Farooq, M., Siddique, R., Khalifa, I., Siddeeg, A., Ali, M., & Faisal Manzoor, M. (2021). Ultrasound based modification and structural-functional analysis of corn and cassava starch. *Ultrasonics Sonochemistry*, *80*, 105795. https://doi.org/10.1016/j.ultsonch.2021.105795

Rao, H., Sindhu, R., & Panwar, S. (2022). Morphology and functionality of dry heat-treated and oxidized quinoa starches. *Journal of Food Processing and Preservation*, *46*(7). https://doi.org/10.1111/jfpp.16672

Ratnayake, W. S., & Jackson, D. S. (2008). Starch Gelatinization. In *Advances in food and nutrition research, Academic Press*, (Vol. 55, pp. 221–268). Elsevier. https://doi.org/10.1016/S1043-4526(08)00405-1

Remanan, M. K., & Zhu, F. (2021). Encapsulation of rutin using quinoa and maize starch nanoparticles. *Food Chemistry*, *353*, 128534. https://doi.org/10.1016/j.foodchem.2020.128534

Ren, J., Li, Y., Li, C., Gu, Z., Cheng, L., Hong, Y., & Li, Z. (2017). Pasting and thermal properties of waxy corn starch modified by 1,4-α-glucan branching enzyme. *International Journal of Biological Macromolecules*, *97*, 679–687. https://doi.org/10.1016/j.ijbiomac.2017.01.087

Rocha, T. S., Felizardo, S. G., Jane, J., & Franco, C. M. L. (2012). Effect of annealing on the semicrystalline structure of normal and waxy corn starches. *Food Hydrocolloids*, *29*(1), 93–99. https://doi.org/10.1016/j.foodhyd.2012.02.003

Rostamabadi, H., Yildirim-Yalcin, M., Demirkesen, I., Toker, O. S., Colussi, R., do Nascimento, L., Şahin, S., & Falsafi, S. R. (2024). Improving physicochemical and nutritional attributes of rice starch through green modification techniques. *Food Chemistry*, *458*, 140212. https://doi.org/10.1016/j.foodchem.2024.140212

Saboonchi, S., Mehran, A., Bahramizadeh, P., & Massoud, R. (2021). Applications of modified starch in food. 3rd International Congress on Engineering Technology and Innovation, 1–8. https://www.researchgate.net/publication/351194178

Sandhu, K. S., Kaur, M., Singh, N., & Lim, S.-T. (2008). A comparison of native and oxidized normal and waxy corn starches: Physicochemical, thermal, morphological and pasting properties. *LWT*, *41*(6), 1000–1010. https://doi.org/10.1016/j.lwt.2007.07.012

Sandhu, K. S., Siroha, A. K., Punia, S., Sangwan, L., Nehra, M., & Purewal, S. S. (2021). Effect of degree of cross linking on physicochemical, rheological and morphological properties of sorghum starch. *Carbohydrate Polymer Technologies and Applications*, *2*, 100073. https://doi.org/10.1016/j.carpta.2021.100073

Sangseethong, K., Lertphanich, S., & Sriroth, K. (2009). Physicochemical properties of oxidized cassava starch prepared under various alkalinity levels. *Starch - Stärke*, *61*(2), 92–100. https://doi.org/10.1002/star.200800048

Santos, T. P. R. D., Franco, C. M. L., Mischan, M. M., & Leonel, M. (2019). Improvement in spray-drying technology for preparation of pregelatinized cassava starch. *Food Science and Technology*, *39*(4), 939–946. https://doi.org/10.1590/fst.10418

Schafranski, K., Ito, V. C., & Lacerda, L. G. (2021). Impacts and potential applications: A review of the modification of starches by heat-moisture treatment (HMT). *Food Hydrocolloids*, *117*, 106690. https://doi.org/10.1016/j.foodhyd.2021.106690

Schmiele, M., Sampaio, U. M., Gomes, P. T. G., & Clerici, M. T. P. S. (2019). Physical modifications of starch. In *Starches for food application* (pp. 223–269). Elsevier. https://doi.org/10.1016/B978-0-12-809440-2.00006-X

Sharma, M., Yadav, D. N., Singh, A. K., & Tomar, S. K. (2015). Rheological and functional properties of heat moisture treated pearl millet starch. *Journal of Food Science and Technology*, *52*(10), 6502–6510. https://doi.org/10.1007/s13197-015-1735-1

Shi, M., Gu, F., Wu, J., Yu, S., & Gao, Q. (2013). Preparation, physicochemical properties, and in vitro digestibility of cross-linked resistant starch from pea starch. *Starch - Stärke*, *65*(11–12), 947–953. https://doi.org/10.1002/star.201300008

Shokrollahi, F., Shahidi, F., Varidi, M. J., Koocheki, A., & Sohbatzadeh, F. (2023). The effect of non-thermal plasma on improvement of functional properties of sorghum starch. *Iranian Food Science and Technology Research Journal*, *19*(2), 217–229. https://doi.org/10.22067/ifstrj.2021.70973.1062

Shrivastava, A., Gupta, R. K., & Srivastav, P. P. (2024). Exploring novel frontiers of advancements in purple yam (*Dioscorea alata* L.) starch extraction, modification, characterization, applications in food and other industries. *Measurement: Food*, *15*, 100196. https://doi.org/10.1016/j.meafoo.2024.100196

Sindhu, R., Devi, A., & Khatkar, B. S. (2019). Physicochemical, thermal and structural properties of heat moisture treated common buckwheat starches. *Journal of Food Science and Technology*, *56*(5), 2480–2489. https://doi.org/10.1007/s13197-019-03725-6

Singh, G. P., Aayush, K., Chavan, P., Chiu, I., Yan, S., Verma, R., Kuca, K., Kumar, D., & Yang, T. (2024). Valorization of agro-industrial wastes: Converting mango seed into starch nanoparticles-based edible coating for fresh produce preservation. *Innovative Food Science & Emerging Technologies*, *95*, 103722. https://doi.org/10.1016/j.ifset.2024.103722

Singh Sandhu, K., Singh, N., & Lim, S.-T. (2007). A comparison of native and acid thinned normal and waxy corn starches: Physicochemical, thermal, morphological and pasting properties. *LWT*, *40*(9), 1527–1536. https://doi.org/10.1016/j.lwt.2006.12.012

Sit, N., Misra, S., & Deka, S. C. (2014). Yield and functional properties of taro starch as affected by ultrasound. *Food and Bioprocess Technology*, *7*(7), 1950–1958. https://doi.org/10.1007/s11947-013-1192-7

Sobini, N., Darsiga, S., Kananke, T. C., & Srivijeindran, S. (2022). Characterization of modified palmyrah tuber starch by pre-gelatinization, acid and dextrinization processes and its applicability. *Food Chemistry Advances*, *1*, 100143. https://doi.org/10.1016/j.focha.2022.100143

Song, H. Y., Lee, S. Y., Choi, S. J., Kim, K. M., Kim, J. S., Han, G. J., & Moon, T. W. (2014). Digestibility and physicochemical properties of granular sweet potato starch as affected by annealing. *Food Science and Biotechnology*, *23*(1), 23–31. https://doi.org/10.1007/s10068-014-0004-9

Sonkar, S., Jaddu, S., Pradhan, R. C., Dwivedi, M., Seth, D., Goksen, G., Sarangi, P. K., & Lorenzo, J. M. (2023). Effect of atmospheric cold plasma (pin type) on hydration and structure properties of kodo-millet starch. *LWT*, *182*, 114889. https://doi.org/10.1016/j.lwt.2023.114889

Subroto, E., Indiarto, R., Marta, H., & Shalihah, S. (2018). Effect of heat-moisture treatment on functional and pasting properties of potato (*Solanum tuberosum* L. var. Granola) starch. *Food Research*, *3*(5), 469–476. https://doi.org/10.26656/fr.2017.3(5).110

Sukhija, S., Singh, S., & Riar, C. S. (2016). Effect of oxidation, cross-linking and dual modification on physicochemical, crystallinity, morphological, pasting and thermal characteristics of elephant foot yam (*Amorphophallus paeoniifolius*) starch. *Food Hydrocolloids*, *55*, 56–64. https://doi.org/10.1016/j.foodhyd.2015.11.003

Sun, X., Saleh, A. S. M., Sun, Z., Ge, X., Shen, H., Zhang, Q., Yu, X., Yuan, L., & Li, W. (2022). Modification of multi-scale structure, physicochemical properties, and digestibility of rice starch via microwave and cold plasma treatments. *LWT*, *153*, 112483. https://doi.org/10.1016/j.lwt.2021.112483

Tahira, M. A., & Abid, H. (2016). Effect of annealing on morphological and functional properties of Pakistani white sorghum (*Sorghum bicolor*) starch. *Journal of Basic & Applied Sciences*, *12*, 358–364. https://doi.org/10.6000/1927-5129.2016.12.55

Taslikh, M., Abbasi, H., Mortazavian, A. M., Ghasemi, J. B., Naeimabadi, A., & Nayebzadeh, K. (2022). Effect of cold plasma treatment, cross-linking, and dual modification on corn starch. *Starch - Stärke*, *74*(5–6), 2200008. https://doi.org/10.1002/star.202200008

Thirumdas, R., Kadam, D., & Annapure, U. S. (2017). Cold plasma: An alternative technology for the starch modification. *Food Biophysics*, *12*(1), 129–139. https://doi.org/10.1007/s11483-017-9468-5

Thirumdas, R., Trimukhe, A., Deshmukh, R. R., & Annapure, U. S. (2017). Functional and rheological properties of cold plasma treated rice starch. *Carbohydrate Polymers*, *157*, 1723–1731. https://doi.org/10.1016/j.carbpol.2016.11.050

Thys, R. C. S., Aires, A. G., Marczak, L. D. F., & Noreña, C. P. Z. (2013). The effect of acid hydrolysis on the technological functional properties of pinhão (*Araucaria brasiliensis*) starch. *Ciência e Tecnologia De Alimentos*, *33*, 89–94. https://doi.org/10.1590/S0101-20612013000500014

Ulfa, G. M., Putri, W. D. R., Fibrianto, K., Prihatiningtyas, R., & Widjanarko, S. B. (2020). The influence of temperature in swelling power, solubility, and water binding capacity of pregelatinised sweet potato starch. *IOP Conference Series: Earth and Environmental Science*, *475*(1), 012036. https://doi.org/10.1088/1755-1315/475/1/012036

Ulfa, G. M., Putri, W. D. R., Fibrianto, K., & Widjanarko, S. B. (2021). Optimization studies on pre-gelatinized sweet potato starch influenced by temperature and time. *Food Research*, *5*(S2), 25–30. https://doi.org/10.26656/fr.2017.5(S2).017

Van Hung, P., Lan Phi, N. T., & Vy Vy, T. T. (2012). Effect of debranching and storage condition on crystallinity and functional properties of cassava and potato starches. *Starch - Stärke*, *64*(12), 964–971. https://doi.org/10.1002/star.201200039

Vanier, N. L., El Halal, S. L. M., Dias, A. R. G., & Da Rosa Zavareze, E. (2017). Molecular structure, functionality and applications of oxidized starches: A review. *Food Chemistry*, *221*, 1546–1559. https://doi.org/10.1016/j.foodchem.2016.10.138

Velásquez-Castillo, L. E., Leite, M. A., Tisnado, V. J. A., Ditchfield, C., Sobral, P. J. D. A., & Moraes, I. C. F. (2023). Cassava starch films containing quinoa starch nanocrystals: Physical and surface properties. *Foods*, *12*(3), 576. https://doi.org/10.3390/foods12030576

Verified Market Research. (n.d.). *Modified starch market size, share, trends, scope & forecast*. Verified Market Research.

Retrieved 29 October 2024 from https://www.verifiedmarketresearch.com/product/modified-starch-market/

Verma, D. K., & Srivastav, P. P. (2022). Isolation, modification, and characterization of rice starch with emphasis on functional properties and industrial application: A review. *Critical Reviews in Food Science and Nutrition*, *62*(24), 6577–6604. https://doi.org/10.1080/10408398.2021.1903383

Wang, D., Ma, X., Yan, L., Chantapakul, T., Wang, W., Ding, T., Ye, X., & Liu, D. (2017). Ultrasound assisted enzymatic hydrolysis of starch catalyzed by glucoamylase: Investigation on starch properties and degradation kinetics. *Carbohydrate Polymers*, *175*, 47–54. https://doi.org/10.1016/j.carbpol.2017.06.093

Wang, H., Xu, K., Ma, Y., Liang, Y., Zhang, H., & Chen, L. (2020). Impact of ultrasonication on the aggregation structure and physicochemical characteristics of sweet potato starch. *Ultrasonics Sonochemistry*, *63*, 104868. https://doi.org/10.1016/j.ultsonch.2019.104868

Wang, M., Wu, Y., Liu, Y., & Ouyang, J. (2020). Effect of ultrasonic and microwave dual-treatment on the physicochemical properties of chestnut starch. *Polymers*, *12*(8), 1718. https://doi.org/10.3390/polym12081718

Wang, Q., Li, L., & Zheng, X. (2021). Recent advances in heat-moisture modified cereal starch: Structure, functionality and its applications in starchy food systems. *Food Chemistry*, *344*, 128700. https://doi.org/10.1016/j.foodchem.2020.128700

Wang, R., Qin, X., Du, Y., Shan, Z., Shi, C., Huang, K., Wang, J., & Zhi, K. (2022). Dual-modified starch nanoparticles containing aromatic systems with highly efficient encapsulation of curcumin and their antibacterial applications. *Food Research International*, *162*, 111926. https://doi.org/10.1016/j.foodres.2022.111926

Wang, R., & Zhou, J. (2022). Waxy maize starch nanoparticles incorporated tea polyphenols to stabilize Pickering emulsion and inhibit oil oxidation. *Carbohydrate Polymers*, *296*, 119991. https://doi.org/10.1016/j.carbpol.2022.119991

Wang, S. (Ed.). (2020). *Starch structure, functionality and application in foods*. Springer Singapore. https://doi.org/10.1007/978-981-15-0622-2

Wang, S., & Copeland, L. (2015). Effect of acid hydrolysis on starch structure and functionality: A review. *Critical Reviews in Food Science and Nutrition*, *55*(8), 1081–1097. https://doi.org/10.1080/10408398.2012.684551

Wu, T.-Y., Chang, C.-R., Chang, T.-J., Chang, Y.-J., Liew, Y., & Chau, C.-F. (2019). Changes in physicochemical properties of corn starch upon modifications by atmospheric pressure plasma jet. *Food Chemistry*, *283*, 46–51. https://doi.org/10.1016/j.foodchem.2019.01.043

Xie, Y., Xu, H., Xu, S., Ge, S., Chang, X., Xu, Y., Luo, Z., Shan, Y., & Ding, S. (2023). How to effectively and greenly prepare multi-scale structural starch nanoparticles for strengthening gelatin film (ultrasound-Fenton system). *International Journal of Biological Macromolecules*, *247*, 125848. https://doi.org/10.1016/j.ijbiomac.2023.125848

Xie, Y., Zhang, B., Li, M.-N., & Chen, H.-Q. (2019). Effects of cross-linking with sodium trimetaphosphate on structural and adsorptive properties of porous wheat starches. *Food Chemistry*, *289*, 187–194. https://doi.org/10.1016/j.foodchem.2019.03.023

Yadav, B. S., Guleria, P., & Yadav, R. B. (2013). Hydrothermal modification of Indian water chestnut starch: Influence of heat-moisture treatment and annealing on the physicochemical, gelatinization and pasting characteristics. *LWT*, *53*(1), 211–217. https://doi.org/10.1016/j.lwt.2013.02.007

Yan, S., Chen, G., Hou, Y., & Chen, Y. (2020). Improved solubility of banana starch by dielectric barrier discharge plasma treatment. *International Journal of Food Science & Technology*, *55*(2), 641–648. https://doi.org/10.1111/ijfs.14318

Yang, Q., Hu, X., Bao, Q., Zhao, Y., Zhang, S., Li, S., & Li, T. (2023). Preparation of starch nanoparticles by jet cavitation and enzymatic hydrolysis and their application in starch films. *LWT*, *185*, 115138. https://doi.org/10.1016/j.lwt.2023.115138

Yang, Q., Qi, L., Luo, Z., Kong, X., Xiao, Z., Wang, P., & Peng, X. (2017). Effect of microwave irradiation on internal molecular structure and physical properties of waxy maize starch. *Food Hydrocolloids*, *69*, 473–482. https://doi.org/10.1016/j.foodhyd.2017.03.011

Zailani, M. A., Kamilah, H., Husaini, A., & Sarbini, S. R. (2021). Physicochemical properties of microwave-heated sago (*Metroxylon sagu*) starch. *CyTA - Journal of Food*, *19*(1), 596–605. https://doi.org/10.1080/19476337.2021.1934550

Zeng, F., Ma, F., Kong, F., Gao, Q., & Yu, S. (2015). Physicochemical properties and digestibility of hydrothermally treated waxy rice starch. *Food Chemistry*, *172*, 92–98. https://doi.org/10.1016/j.foodchem.2014.09.020

Zeng, X., Chen, L., & Zheng, B. (2024). Extrusion and chlorogenic acid treatment increase the ordered structure and resistant starch levels in rice starch with amelioration of gut lipid metabolism in obese rats. *Food & Function*, *15*(10), 5224–5237. https://doi.org/10.1039/D3FO05416K

Zeng, Y.-F., Chen, Y.-Y., Deng, Y.-Y., Zheng, C., Hong, C.-Z., Li, Q.-M., Yang, X.-F., Pan, L.-H., Luo, J.-P., Li, X.-Y., & Zha, X.-Q. (2024). Preparation and characterization of lotus root starch based bioactive edible film containing quercetin-encapsulated nanoparticle and its effect on grape preservation. *Carbohydrate Polymers*, *323*, 121389. https://doi.org/10.1016/j.carbpol.2023.121389

Zhang, B., Chen, L., Li, X., Li, L., & Zhang, H. (2015). Understanding the multi-scale structure and functional properties of starch modulated by glow-plasma: A structure-functionality relationship. *Food Hydrocolloids*, *50*, 228–236. https://doi.org/10.1016/j.foodhyd.2015.05.002

Zhang, B., Mei, J.-Q., Chen, B., & Chen, H.-Q. (2017). Digestibility, physicochemical and structural properties of octenyl succinic anhydride-modified cassava starches with different degree of substitution. *Food Chemistry*, *229*, 136–141. https://doi.org/10.1016/j.foodchem.2017.02.061

Zhang, Y., Huang, Z., Yang, C., Huang, A., Hu, H., Gong, Z., Sun, G., & Huang, K. (2013). Material properties of partially pregelatinized cassava starch prepared by mechanical activation. *Starch - Stärke*, *65*(5–6), 461–468. https://doi.org/10.1002/star.201200174

Zhang, Z., & Bao, J. (2023). Recent advances in modification approaches, health benefits, and food applications of resistant starch. *Starch - Stärke*, *75*(9–10), 2100141. https://doi.org/10.1002/star.202100141

Zheng, J., Li, Q., Hu, A., Yang, L., Lu, J., Zhang, X., & Lin, Q. (2013). Dual-frequency ultrasound effect on structure and properties of sweet potato starch. *Starch - Stärke*, *65*(7–8), 621–627. https://doi.org/10.1002/star.201200197

Zheng, M., Xiao, Y., Yang, S., Liu, H., Liu, M., Yaqoob, S., Xu, X., & Liu, J. (2020). Effects of heat–moisture, autoclaving, and microwave treatments on physicochemical properties of proso millet starch. *Food Science & Nutrition*, *8*(2), 735–743. https://doi.org/10.1002/fsn3.1295

Zhou, D., Ma, Z., Yin, X., Hu, X., & Boye, J. I. (2019). Structural characteristics and physicochemical properties of field pea starch modified by physical, enzymatic, and acid treatments. *Food Hydrocolloids*, *93*, 386–394. https://doi.org/10.1016/j.foodhyd.2019.02.048

Zhou, F., Liu, Q., Zhang, H., Chen, Q., & Kong, B. (2016). Potato starch oxidation induced by sodium hypochlorite and its effect on functional properties and digestibility. *International Journal of Biological Macromolecules*, *84*, 410–417. https://doi.org/10.1016/j.ijbiomac.2015.12.050

Zhou, Z., Sun, B., Yu, D., & Zhu, C. (2022). Gut microbiota: An important player in type 2 diabetes mellitus. *Frontiers in Cellular and Infection Microbiology*, *12*, 834485. https://doi.org/10.3389/fcimb.2022.834485

Zhu, F. (2015). Impact of ultrasound on structure, physicochemical properties, modifications, and applications of starch. *Trends in Food Science & Technology*, *43*(1), 1–17. https://doi.org/10.1016/j.tifs.2014.12.008

Zia-ud-Din, Xiong, H., & Fei, P. (2017). Physical and chemical modification of starches: A review. *Critical Reviews in Food Science and Nutrition*, *57*(12), 2691–2705. https://doi.org/10.1080/10408398.2015.1087379

16 Regulatory Considerations and Safety of Modified Starch

Sumanta Bhattacharya
Asian International University, Imphal, Ghari, Manipur

16.1 INTRODUCTION

Starches are structurally modified with the aid of physical, enzymatic, or chemical treatments so as to obtain better functional characteristics required for a variety of uses. The primary starch sources are maize, potato, wheat, and tapioca, which are rich in supply and have wide application (Gupta et al., 2022). Some of the properties that can be controlled, altered, or enhanced relative to native starches through starch modification are stability, texture, thickening properties, and shelf life. Chemical modification is done by changing the molecular structure by using agents such as acids or alkalis that change properties such as viscosity, solubility, and freeze–thaw stability (Shrivastava et al., 2024). Examples of common chemical changes are acetylation, cross-linking, and oxidation. While chemical modification has its benefits, enzymatic modification is also carried out according to specific guidelines that utilize enzymes such as amylase to partially hydrolyze starch molecules, which is especially suitable for food industries and other industries needing tailored textures and solubility patterns. Physical modification processes include heat treatment, pre-gelatinization, cold plasma and extrusion, in which the structure of starch molecules is reorganized but the chemical components are not changed. These methods assist in increasing the stability of the starch under other processing stresses, such as heat or change in pH (Bashir & Aggarwal, 2019; Gupta et al., 2023; Gupta et al., 2024b).

Different types of modification have different advantages for applications. Chemical alteration of starch enables the presentation of an exact character in experience control and pharmaceuticals and enables texture management in sauces and dressings. Enzymatic methods are closer to natural approaches at altering the structures and are typically applied in food and drink products where during processes such as brewing or baking, specific rates at which starches break down are required. These changes are particularly applied to instant foods and convenience products for which cooking has to be done quickly. This makes modified starches versatile and ensures that there is a suitable one for every industry need, hence their popularity across many industries (Compart et al., 2023).

In the food industry, modified starches are used to perform specific powerful functions stabilizing the textural quality of products. As such, they are widely applied as thickeners of soups, sauces, and dairy products for the specific viscosity and the mouthfeel that consumers like. In baked products, modified grain starches contribute to the preservation of bread freshness and may add to shelf stability by minimizing staling, particularly in frozen foods (Gupta et al., 2025a). Stabilizers prepared from modified starch are widely used in dairy products, salad dressings, and sauces, acting as anti-phase agents to prevent separation and keep liquids emulsified (Gupta et al., 2024c). Modified starchy products are also essential in confectioneries as they are used to regulate both chewiness of candy, such as gummies, and to adjust elasticity. In addition to thickening and fianchetting, modified starches are used as adhesives in meat products and snacks and to create homogeneity of texture and flavor (Novasol Ingredients, 2021). Recent applications of various nanomaterials in the food industry are listed in Table 16.1 (Biswas et al., 2022).

In the pharmaceutical industry, modified starches act as binders and diluents in the preparation of tablets. The changes in geometry of starch properties improve the

TABLE 16.1
Application of Different Nanomaterials in Food Industry

Nanomaterial	Applications in the Food Industry
Nanoencapsulation of nutrients like vitamins, minerals, antioxidants, and flavors	Improved stability of processed food products and controlled delivery
Nanosilver	Antimicrobial protection for food products and prevention of contamination
Nanoscale self-assembled structures of lipids, and proteins	Formulation of low-fat food products
Nanoencapsulation	Increase bioavailability of nutrient in body
Oxygen scavenging nonpackaging	Prevention of putrefaction of food items
Nanofilters	Minimization of bacterial contamination
Food nanosensors	Identification of contaminants and allergens
Nanosensor	Detection of poisonous items in packaged food products

DOI: 10.1201/9781003564287-16

mechanical communication and weaken the bonding in tablets for controlled drug release (Gupta et al., 2024d). Starches used include pre-gelatinized starch, which is a physically modified starch commonly used as a tablet disintegrating agent to facilitate the breakdown of tablets in the body. For capsule fillings, modified starches provide stability and enhance bulk to support product standards (Gupta et al., 2024e). In addition to usage in the pharmaceuticals and food industries, modified starches are used in the paper industry to enhance the strength of paper and smoothness of its outer surface. In textiles, modified starches as sizing agents provide increase fabric rigidity and protection from wear during processing. The adhesive industry uses modified starches as natural adhesives or green sources of adhesives due to their biodegradable nature. Modification of starches also provides added advantages to packaging, especially biodegradable packaging (Gupta et al., 2025b), as the product using modified starch does not depend solely on synthetic polymers. The flexibility and functionality of modified starches enable them to meet specific requirements in these industries to offer high-quality, consistent manufacturing systems (Manek et al., 2012).

16.2 REGULATORY FRAMEWORK BY REGION

In the United States, modified starches are governed mainly by the Food and Drug Administration (FDA) as products that need to be generally recognized as safe (GRAS). This status means that professionals concur, either scientifically or historically or both, that a given compound is safe to use for a specific purpose. While GRAS substances are not entirely analogous to food additives, including modified starches, and are exempted from premarket approval processes, using them in food products means that their safety has to be backed by sound scientific evidence. The list of GRAS products covered by the FDA contains many types of modified starches, listing concentrations and permitted uses. The FDA also regulates labeling, including when companies need to include modified starches in ingredient declarations. This labeling often includes the source of the starch, for example, corn or potato, and the particular modification made if it has an influence on potential allergens or additives so that consumers can feed themselves appropriately. The FDA also limits the use of certain chemically modified starches, mostly for specified uses, including for sensitive groups such as babies or people with sensitive stomachs. For modified starches that are to be used in products in the pharmaceutical industry, the FDA has a standard that these modified starches should meet *United States Pharmacopoeia* requirements related to purity, quality, and safety.

In the European Union (EU), modifications to starches fall under the European Food and Safety Authority (EFSA), which has strict safety benchmarks for additives to food products. Where modified starches for the EU market meet the criteria set by EFSA, the additive is given an E-number, a unique code assigned to each approved additive product. E-numbers assist consumers and regulatory authorities to easily distinguish between modified starches that have been approved and the uses that they will be used for. The EFSA evaluates food supplements using such attributes as toxicity, digestibility, and allergenicity before allowing their use on the market. This involves assessment of toxicological data and estimates of food consumption and exposure in order to gain an insight into whether each starch modification is safe for use in EU countries. Examples of approved modified starches in the EU are those that have been treated with acids or enzymes, oxidized starches, and cross-linked starches, which are all labeled using E-numbers; for instance, E 1404 is oxidized starch, while E 1422 is acetylated distarch adipate. Another aspect of the legislation is labeling, according to which modified starches included in foods must be declared in the list of ingredients by their E-number and source if the modification in the structure of starch is important for the consumer, for example, due to allergy. Furthermore, the EFSA continuously revises the approved list due to new findings on products, advanced technologies, and emerging disease trends. This regulatory approach demonstrates the EU's continued concern with keeping high food hygiene standards and passing this information to consumers (European Food and Safety Authority, 2024).

Other parts of the world, especially the East, Australia, and other developing countries, have a variety of mechanisms of keeping key players in line with different levels of standard checks and consumer protection. In Japan, the Ministry of Health, Labor, and Welfare closely monitors modified starches, especially genetically modified organism (GMO)-derived food additives. Safety assessment is mandatory and specific, and only limited changes in starch are permitted in foods, for which enhanced labeling is a priority, especially of GMOs. Australia and New Zealand share a single code authority, Food Standards Australia New Zealand (FSANZ), which has a list of permitted modified starches like the EFSA's, with special E-numbers and norms of utilization in foods. While introducing new provisions to ensure consumer safety and to clarify the issue of labeling food ingredients, FSANZ generally follows the EU legislation. The protective laws regarding modified starches can differ greatly one from another in the regions of the developing nations, some of them use the intergovernmental *Codex Alimentarius*, while others lack sufficient legal requirements because they simply cannot afford them due to lack of resources. This brings into focus the common understanding of the need for aligned regulation around the world since divergent regulatory approaches necessarily influence both the development of commerce and the protection of consumers. Unfortunately, most developing nations seek the guidance of such international regulatory agencies as the World Health Organization or the Food and Agricultural Organization on matters to do with food safety and which modifications are acceptable. These variations show the general differences in policies relating to modification of starch between countries and put into focus the need for set general universal safety measures to protect consumers around the world (Ueda et al., 2024).

16.3 SAFETY ASSESSMENTS AND TOXICOLOGICAL STUDIES

Most available safety reviews of modified starches have been based on assessing the level of toxicity, allergenicity, and effects on the digestive system and metabolism. Experimental investigations on modified starches with respect to acute, subchronic, and chronic toxicity using in vitro and in vivo models have been carried out with the aid of dose response and safety limits as well as acceptable daily intake (ADI). The existing acute toxicity investigations test the direct effects of short-term intake of high concentrations of modified starches, and most indicate low toxicity of the substances tested, that is, they do not cause death even when consumed in large quantities. For example, acute tests in rodents of cross-linked starches or starch acetates indicate that the modified starches are not toxic; no unpleasant effects were observed during the testing of levels of consumption that are considerably higher than those normally experienced by humans (Mortensen et al., 2017).

Like acute toxicity, tests of chronic toxicity are important since they target effects that may arise from consumption of modified starches over a long period of time and whether continued consumption of the products might have negative impacts on organ function and integrity, cancer, or metabolic function. Overall, such tests indicate that with minor modifications, starches do not deposit in tissues since most are broken down in the digestive system. Animal and clinical investigations of chronic consumption of modified starches show that they are free from carcinogenic, genotoxic, or teratogenic impacts if consumed in permitted rations. For instance, in lifetime studies of the effects of modified starches such as octenyl succinate–starch fed to rats, there was no noteworthy difference in tumor risk, organ size, and death in comparison with control groups. The outcomes of these studies reinforce the perception of the safety of modified starches for human consumption in food products that have concentrations of modified starches acceptable in regular usage; toxicogenomic profiling of rats fed a diet containing 30% of modified waxy starches for three weeks showed that no depressed organ weights, histopathological lesions, or biomarkers of systemic toxicity were elicited.(Erhirhie et al., 2018).

Some conclusions from these studies reflect the digestibility, allergenicity, and metabolic features of modified starches. Recent modifications to starches do not change the basic characteristic of natural starches in that they molecularly degrade in the small intestine into glucose. Cross-linking or enzyme treatments make the starch slightly less digestible, implying slower presence of glucose, which can be beneficial for people suffering from diabetes. The literature reveals that modified starches, in general, are inert structures that tend to withstand the vagaries of the gastrointestinal tract without causing inflammation or other complications. However, people deficient in the digestive enzyme amylase may experience slight gut irritation with high consumption since modified starches may not be so easily digested (Singh et al., 2018).

Modified starches are safe to use in food products, but care must be taken to avoid use of modified starch where the raw material comes from an allergenic product such as wheat and corn. Research shows that allergenicity is not increased by the process of modification of starch. However, labeling rules tend to indicate the source of the main input for the starch, such as wheat, to sensitive consumers. In general metabolic analysis, it has been observed that as with natural starches, modified starches are absorbed and consumed as glucose to act as energy. Among other types of modified starches, those with slower digestion rates are likely to offer benefits such as enhanced satiety and better glycemic regulation that are always beneficial for people with diabetes or other metabolic syndromes. Examination of the Practical digestion of these modified starches finds no negative effect on blood insulin or lipids, further affirming their safety for human consumption (Gałkowska et al., 2023).

It has been found that these two categories of starches are harmless for humans; at the same time, differences in the digestibility of these products and their potential health-boosting properties are noticeable. Natural starches, including corn, potatoes, tapioca, wheat, and beans, are quickly reduced into glucose for easy digestion in the small intestine. This process provides fast energy, though often followed by rapid fluctuations of blood glucose levels, especially for products containing starches with a high glycemic index (GI). Depending on the treatment applied to the starches, they can display retarded digestion and reduced GI, which can be of benefit for managing blood sugar levels after ingesting the modified starches. For instance, physically processed (e.g., starches that have been heat-moisture treated) and chemically modified starches remain mostly undigested in the small intestine and are fermented only in the large intestine. This can be positive for gut health due to the use of beneficial bacteria along with being a good way to feel full without adding calories (da Rosa Zavareze & Dias, 2011).

At the same time, modified starches may have differences in digestibility from natural starches. Chemical starches such as cross-linked and acetylated starches are produced for heat and acid stability during processing and thus are considered prehydrolytically active or partially digested. Although they can be beneficial in certain circumstances, such as controlling GI, they may from time to time lead to a some stomach discomfort in people whose digestive system is sensitive. Several scientific investigations have indicated that the food chemical risks presented by modified starches are similar to those posed by natural starches, in that they are not dangerous at levels regularly consumed in daily diets. Further, like natural starches, modified starches can be subject to toxicological safety assessment, assessment of allergic properties, and digestibility, which helps to maintain a high level of quality. Overall, modified starches

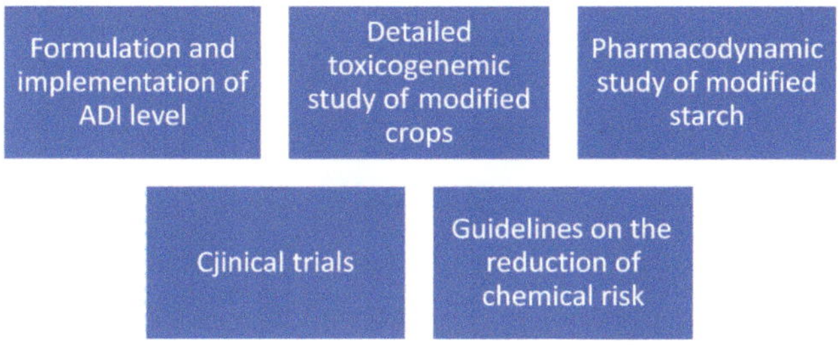

FIGURE 16.1 Safety assessment and toxicological studies of the use of modified crops. ADI, acceptable daily intake.

are not only functionally enhanced but also considered safe for various food and other consumer product uses as they do not have substantially different metabolism or safety profiles from natural starches (Gałkowska et al., 2023). Components of safety assessment and toxicological study of modified crops are depicted in Figure 16.1.

16.4 HEALTH IMPLICATIONS

Digestibility and metabolic properties of most modified starches are slightly different from those of native starches because of the changes in structure achieved through different modifications. Native starches, including corn, wheat, and potato starches, are largely composed of two polymers, amylose and amylopectin, which are hydrolyzed into glucose in the digestive system by enzymes, including amylase. This breakdown starts in the mouth and continues in the small intestine, in which glucose molecules are taken for use as source of energy. Native starches are highly digestible and can be metabolized with relative ease and efficiency; a number of them are quickly hydrolyzed to glucose with subsequent increase in blood glucose levels, especially high-GI starches (Rong et al., 2024). Some modified starches, however, show differences in digestibility according to the method of modification. For example, cross-linked or acetylated starches may behave differently from native starches; that is, they may undergo digestion slowly and reduce the glucose release rate. Carbohydrates take longer to be broken down and absorbed in the blood, which might prove useful for those with certain diabetes conditions. Digestibility is higher in starches modified enzymatically, such as by using amylase; however, enzymatic modification is adjustable, so starches can be made to have only partial resistance (Magallanes-Cruz et al., 2017).

Some structural modification processes, such as heat-moisture treatment, create resistant starches that largely progress intact through the small intestine and then ferment in the colon. These modified starches have a similarity to dietary fiber and are fermented by gut bacteria to produce short-chain fatty acids (SCFAs), including butyrate, which has beneficial impact on colonic health. Modified starches can therefore mimic soluble fibers in providing feelings of fullness, ease of bowel movement, and blood glucose regulation, which is associated with metabolic health. Chemically modified starches may vary in starch digestibility depending on the type and degree of modification. For example, starches that have been modified by the addition of phosphate groups (phosphorylated starches) swell, absorb water, and are generally only partially digestible; they are useful as fillers and as ingredients in low-calorie foods (Birt et al., 2013). Some of these changes in starch structures will predispose the modified starch to respond differently in the digestive system from the native starch. Most modified starches fully soluble in water eventually depolymerize to glucose, yet the modification slows the glucose release, reduces the post-meal glycemic peak, and makes the feeling of satiation more robust, which renders that molecule a fundamental element for various eating regimens, including but not limited to special diets (Alcázar-Alay & Meireles, 2015).

Recent studies to the effects of modified starches have looked at their effects on the gut. The indigestible components of modified starches are similar to dietary fibers in their effects on the microbiota. Resistant starches, or retrograded starches, are modified and serve as prebiotic starches because they do not undergo digestion in the small intestine but rather enter the large intestine colonic stomach Churchwell (Gupta et al., 2024a). Once in the colon, they are resistant to the digestion by human enzymes but readily attacked by gut bacteria, which turn the starches into SCFAs such as butyrate, acetate, and propionate. These are involved in maintaining colonic health; butyrate can serve as a source of energy to colonic cells and is anti-inflammatory. Scientific research has revealed that concentrations of butyrate enhance the integrity of the gastrointestinal tract and thus may be protective against such diseases as inflammatory bowel disease (IBD) and colorectal cancer (Bojarczuk et al., 2015). Research has also analyzed effects of modified starches on gut microbiota and its composition. Resistant starches can enhance the number of specific bacterial strains that positively affect the large intestine, including *Bifidobacteria* and *Lactobacillus*. These bacteria ferment resistant starches, reduce gut pH, and thus may reduce the opportunity for pathogenic bacteria to grow and increase risk of inflammation in the gut. A recent study has shown evidence that some modified starches possess prebiotic properties and positively influence the growth of beneficial bacteria and the synthesis of SCFAs, which might lead to an amelioration in gut functions, bowel movement, and general digestive tract function (Dobranowski & Stintzi, 2022).

In addition, some new products of starch modification and hydrolysis can promote the growth of beneficial microbiota and improve nutrient transport. In secured gut-associated disease or for people with inflammatory bowel syndrome (IBS), modified starches have been found to reduce symptoms by increasing the colonic concentration of SCFAs, which are useful for soothing symptoms such as bloating, discomfort, and abnormal bowel habits. Nevertheless, modified starches are not created equal; categories of chemically altered starches might have the prebiotic properties described above or might have none and cause slight gastrointestinal distress in certain consumers. In general, modified starches, especially those with the properties similar to dietary fibers, are beneficial to the gut microbiota and its associated health, such as improved digestive function, suggesting benefits for the metabolic and immune systems (González-Bermúdez et al., 2018).

Allergenicity and sensitivity to modified starches are low, although people with pre-existing allergy to a starch or grain would respond to certain modified starches. Wheat or corn, which are native sources of starch, should be avoided by allergic patients, and modifications do not remove this allergenicity. For instance, people with wheat allergies have their allergy triggered by modified wheat starch despite the processing. Thus, labeling requirements typically specify the source of the starch so that people can avoid products that trigger allergies (Goodman & Tetteh, 2011). Overall, starch modification, which can be chemical, physical, or enzymatic, does not increase allergenicity, unlike proteins. Many sensitivity issues linked to modified starches stem from the base raw materials, not from the modifications. For instance, cross-linked starches produced from non-allergenic sources such as tapioca or potato are not expected to elicit an allergic outbreak in most people. However, modified starches may include a small amount of proteins from the original products, so people with extreme allergies may be at risk (Gałkowska et al., 2023).

Apart from allergic reactions, in some cases modified starches, especially those that are resistant to digestion, cause mild gastrointestinal upset in people with starch sensitivity or deficient in certain enzymes involved in digestion. When starches are rendered more resistant, for instance, in cross-linked starches, they may not be digested in the small intestine and thus cause bloating or gas in the colon. In most individuals, these effects are not dramatic and are masked when continued consumption of the product creates gut tolerance. Some reactions to chemically modified starches have been reported, for instance, to acetylated starches, but reactions to chemically modified starches are very rare and have been observed only in people with very sensitive digestion systems. Allergenicity is a function of the source of the starch rather than the process of modification, and most modified starches are approved for use (Birt et al., 2013). Implications of modified starches in the health sector are presented in Figure 16.2.

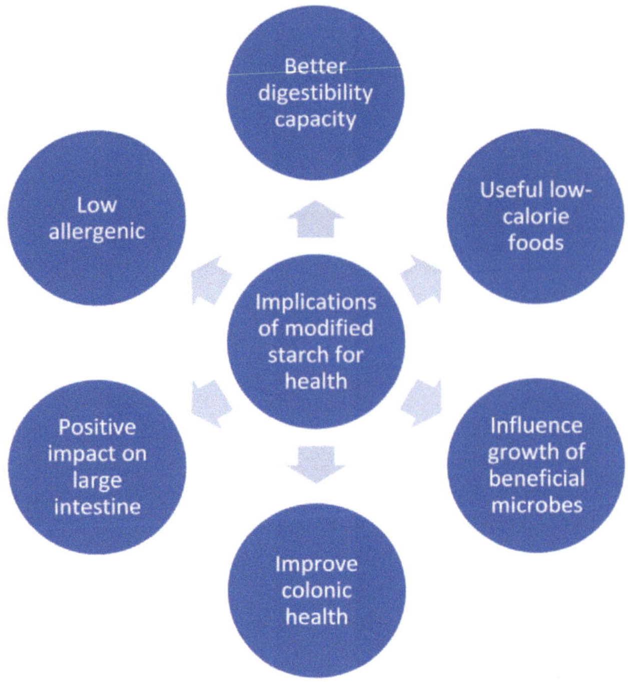

FIGURE 16.2 Implication of modified starch for health.

16.5 LABELING AND CONSUMER AWARENESS

The labeling rules that dictate the way modified starches are referred to as additives are standardized globally to provide stakeholders with correct information that will enable consumers to make good decisions about eating these products. In most regions, modified starches used as additives are required to be labeled, and this labeling requires information on the kind of modification made on the starch and, where possible, the original source of the starch. Such labeling is crucial where the modification is likely to have adverse impact on some people, including where the source of the starch, such as wheat or corn, may cause allergy. Regulations vary from country to country; nevertheless, most require indication that a starch is modified and what exactly it thickens/stabilizes, and some countries use an internationally recognized sequence number (for example, E-numbers from the EU) to define the type of the modification (Gałkowska et al., 2023).

In the United States, the FDA mandates labeling of modified starch by its functional classification, such as modified food starch or stabilizer in a product. Although the FDA does require that the kind of alteration made be disclosed if it has a significant impact on either safety or nutritional profile of the product, it can be vague about what kind of modification it requires to be stated. But if a starch is derived from a potential allergen source, such as wheat, then it must have the allergen label according to the U.S. Food Allergen Labeling and Consumer Protection Act. This extra labeling measure benefits consumers with food allergies since they can avoid products that will cause negative repercussions. The EFSA in the European Union expects modified starchy foods to be labeled by their E-number of the type of extraction used. For instance, E 1404 stands for oxidized starch, and E 1422, for acetylated distarch adipate. The labeling regulations also require that the ingredient list indicate the source if the starch has come from a specific material that is an allergen. This level of detail and transparency makes it easier for the consumers to consider and distinguish the type of modified starch in foods. Similar procedures of identifying any source material containing an allergen and naming modified starches according to their intended function are practiced in other countries, including Japan, Australia, and New Zealand (EFSA, 2024).

Clear labeling for modified structures in food and pharmaceuticals is critical for vulnerable groups such as infant, elderly people, and clients with poor health who need to know what products they are consuming. Infants have peculiar dietary needs and may be more vulnerable to the deleterious effects of some modified starches, especially those with high RD kits. As a result, food products targeted to the infants, for instance, baby formula, that contain modified starches raise issues of labeling and regulatory scrutiny. Labeling helps caregivers make the right choice of products to buy based on the particular nutrition requirements and intolerance level of children (Díaz et al., 2020). Another population that benefits from clear labeling is the elderly, who may have changed digestion rates or dietary limitations because of chronic disease. Many elderly people require modified starches in food supplements or low-energy foods to ensure nutrient intake without energy density. The type and source of modified starches need to be properly indicated so that these consumers do not suffer from digestive problems or allergic reactions. Proper labeling also assists healthcare providers to recommend good products for use by the elderly, especially for nutritional management of patients in eldercare facilities (Qin et al., 2024).

In addition, consumers whose diets are limited for reasons such as celiac disease, gluten intolerance, or allergen-related disorders accurately labels. For instance, to avoid gluten, an individual with celiac disease must distinguish modified starches that come from wheat. Due to transparent labeling, such consumers are able to make sound decisions and identify safe products, strengthening the bonds between consumer and manufacturer. As a result of emerging consumer awareness concerning allergens and diets, companies have improved their labeling strategies; in part, this has led to the use of labels that identify product freedom from certain allergens or gluten (Makovicky et al., 2020). Labeling is also done for ethics reasons since it assists in the creation of trust with consumers while meeting current worldwide trends in consumer demand for more extensive information, including ingredient lists, due to concerns for nutritional value and environmental sustainability. Where labeling is mandatory for GMOs, affected modified starches will also indicate this, since there are people who have ethical and health beliefs about genetically modified (GM) foods. This is most useful to consumers who want a diet without GMOs and to producers who want to meet this market demand by using and declaring non-GMO modified starches (Priya & Alur, 2023). In general, labeling of modified starches is important to ensure the safety of consumers and manufacturers and should clearly indicate the use of the modification on starches in foods. In addition, manufacturers can improve consumer trust for products aimed at vulnerable groups, as well as follow the worldwide tendency toward clearer, more transparent information on food packaging. The key reasons behind the labeling of starch and increased consumer awareness are provided in Figure 16.3.

FIGURE 16.3 Labeling and increased consumer awareness.

16.6 INNOVATION AND EMERGING TRENDS

Improved technology has enabled new modified starches that address functional, environmental, and health-related requirements in the food, pharmaceutical, and manufacturing industries. The main advancement in this area is that bioengineered and sustainable raw materials are used to develop new functional additives based on modified starch. Products from vegetable origin such as corn, potato, and wheat are still popular, while bioengineering opens new approaches for direct change of starch in the plant, which may decrease the demands on processing. Molecular breeding technologies enable researchers to change amylose content at the source plant level, creating high levels of amylose or resistant starch directly in crops. For instance, high-amylose corn can be developed to generate raw starch that is extremely resistant to amylase and thus useful in low-glycemic or high-fiber foods (Yudhistira et al., 2024). Another development in modified starches is the increasing concern for the sustainability and efficient use of raw materials. Recently, increased attention has been paid to the modification of starches obtained from underutilized crops, for example, cassava, sorghum, and quinoa, that have significantly less impact on the environment compared to, for example, wheat or corn. These crops are normally distinguished by using less water and fertilizers and can be cultivated in a variety of environmental conditions, which is good news for food production. Also, novel extraction and modification techniques are expected to be established to decrease the starch production footprint. For example, enzymatic treatments are replacing chemical modifications because they fundamentally do not involve the use of chemicals and thus energy consumption during processing (Rashwan et al., 2024).

In the recent past, the demand for environmentally degradable and renewable materials shaped the enhancement of modified starches as a replacement for petroleum products. The use of bio-based plastics and packaging materials from modified starch is being worked on, especially to develop packaging that can biodegrade. These materials are based on modified starches that originate from renewable sources and are biodegradable. Because they can be composted, they fit into a circular economy. Bioplastics derived from some starch alterations, including acetylation and esterification, are known to possess packaging features of flexibility, durability, and water resistance, like conventional plastics. Modified starches also have application as binders in construction industries and in adhesive preparations; they enhance sustainable developments in all sectors (Jiang et al., 2020).

The other category of innovation includes modified starches developed to meet nutritional and functional requirements in nutrition, including prebiotic modified starches. New research is being conducted to create specific types of starch that will have lower digestibility, be resistant to digestion, or gradually release glucose into the bloodstream, especially for patients with diabetes, among other diseases. These starches, which have an energy-supplying characteristic akin to that of dietary fiber, help in the management of body weight and metabolism. New developments in this area are appropriate since consumers today are looking for functional foods and beverages more than ever before and since bioengineered modified starches can be considered healthy food processing (Farooq & Yu, 2024). However, with new trends in starch modification, socio-regulatory and safety issues are outstanding obstacles. Starch modification through genetic engineering is another key regulatory concern because the public is increasingly wary of genetically modified foods. National and international bodies such as the FDA and EFSA are responsible for evaluating GM starches for safety, allergenicity, and nutritional value. Nonetheless, GM starches need to be tested for their effects on human beings and the overall environment since the production of genetically modified foods demands more and costlier research than other modified foods. For instance, scientific and safety assessments of GM starches include those on the ability to induce allergic reactions, changes in the nutritional quality of the food, and other impacts on human health. Such assessments are useful in keeping modified starches safe for consumption across populations, at the cost of stretching the time to market new starches and adding to product development costs (Bawa & Anilakumar, 2012).

Other regulatory concerns are presented in connection to labeling requisites and GMOs. In particular, the European Union requires notification of consumers if a product contains GM-derived items. The labeling of GM starches as GMOs can affect acceptance since different societies have different perception of GMOs. In areas where consumers have low acceptance of GM products, this translates to low market prospects even though bioengineered starches have many advantages. As a result, manufacturers are forced into a conflict between creativity and tradition, between the need to appeal to the consumer and the necessity of meeting legal requirements when introducing and developing modified starches that are up to par with the expectations of the general public (Buiatti et al., 2012). Another area regulated for sustainability and safety is the effect that bioengineered and chemically modified starches pose on environment. Chemical modifications may often comprise the use of acids or alkalis or other reagents, which yield unwanted products that are negative for the environment if they are discharged. Pressure groups continue to lobby regulatory authorities for green processing methodology that are less damaging to the environment; enzymatic and physical modifications are among the sought-after methods presently. Nevertheless, these standards can be difficult for companies to achieve because not only do firms have to develop new ways to modify starch, they must also develop new processes to produce them (Nasrabadi et al., 2021).

Other regulatory and safety issues concern the determination of the health impacts of the new modified starches. Most modified starches are on the GRAS list in the United States or have an E-number in the EU; novel starches, however, are comprehensively reviewed. Regulatory agencies are interested in the chronic effects of exposure and in effects on the digestive system; hence, data are needed on how such starches affect gut health, nutrient bioavailability, and metabolism. Research needs to explain whether bioengineered starchy products are digested in a way different from natural or otherwise modified starchy products, which will matter to consumers with sensitive gastrointestinal tracts or pre-existing primary diseases (Gałkowska et al., 2023). Thus, for both bioengineered and sustainably modified starches that could form the basis of a plethora of products within different industries, it is important to properly comprehend the regulatory realities to guarantee both safety and acceptability of the end products. Key constraints when marketing these modified starches include regional labeling laws, safety testing of new properties, and sustainable processing needs. Such measures help to regulate relations in the industry and achieve equilibrium between innovative developments that will bring profits to companies, the safety of consumers and the environment, and the provision of clear and comprehensible information. Innovations and emerging trends in the field of modified starch are shown in Figure 16.4.

16.7 ENVIRONMENTAL AND ETHICAL CONSIDERATIONS

During the production of modified starches, there are several steps that may affect the environment, both in relation to the acquisition of raw materials and in relation to the modification of the starch. The most common traditional starch sources include corn, potato, wheat, and tapioca, which are cultivated using intensive methods that demand large amounts of water, fertilizers, and pesticides. Such

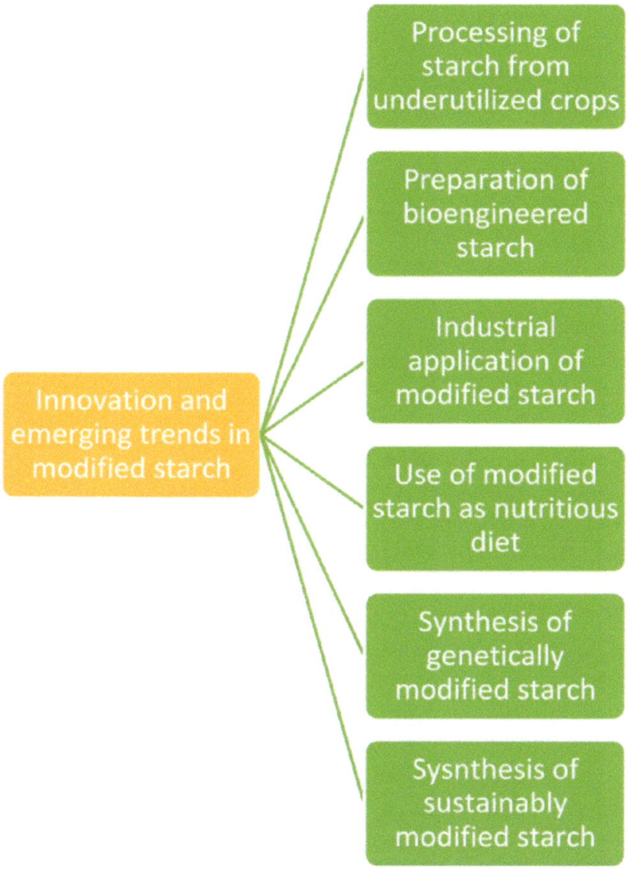

FIGURE 16.4 Innovations and emerging trends in modified starch.

inputs make a higher contribution to production of greenhouse gases, deterioration of soils, and water pollution, which are sustainability issues. Action to address these effects relies on the acquisition of starches from less invasive crops or using practices with small impact on the earth. For instance, cassava and sorghum are increasingly seen as first-generation feedstock because they need less water and fewer chemicals than corn or potatoes. Moreover, principal starch sources grown in crop rotation or in polycultures provide benefits to the soil, promote reduced erosion, and improve biodiversity, making starch production processes more sustainable.

The process of modifying products and services can also affect the environment depending to the procedure that is used. Chemical modifications, for instance, require the use of agents such as acids, alkalis, or cross-linking chemicals on the starch, causing a change in the physical characteristic of the starch; they may produce risky byproducts. Handling of wastes, especially chemicals, is a crucial matter because their improper discharge into the ground and water sources may cause significant harm to the environment and to human health. In a bid to protect the environment and living organisms, manufacturers increasingly are opting for less toxic modification methods, such as enzymatic or physical processes, which do not require a large amount of toxic chemicals and energy. Organic processes employ enzymes, which are biodegradable and leave minimal residues; mechanical techniques exclude chemicals in favor of heat and moisture. All these green ideas assist in the achievement of sustainability objectives since they reduce carbon emission and waste production in starch production. Also, advances in the use of modified starches as materials in bioplastics help in the campaign to replace petroleum plastics with biodegradable plastics in an effort to improve the quality of wastes and the environment (Fan & Picchioni, 2020).

Public concerns about the ethics of producing and researching modified starches and their relationship to ecological consequences of GM variants have been rising steadily. Genetically modified starches, modified in the plant genome for histopathological starch, pose many positive questions about biodiversity, ecological cruciality, and sustainability. Some of the ethical concern is based on the effects of GM crops on related plants or nearby crops and other plants where the effects could be the non-intentional transfer of modified genes. This gene transfer, when influencing natural plant populations within ecosystems, may diminish genetic variation and alter robustness to adapt to current and future environmental change. In addition, GM crops are largely grown because farmers engage in monoculture agriculture, which reduces soil fertility and leaves cash crops vulnerable to disease, among other problems. These practices differ from sustainable agriculture principles, in which crop diversification and soil conservation are emphasized (Karalis et al., 2020).

Another issue is the patent rights of companies that develop GM starch sources, as most large firms in biotechnology own the seeds with the necessary genes. This can confine farmers' access to seeds, augment reliance on standards crops, and affect local economies, particularly in the developing world, where smallholder farming is more prevalent. Opponents state that these processes may result in emergence of the dependency on expensive GMOs and correlated products such as particular fertilizers or pesticides developed to be used with these plants. These ethical issues raise questions of fairness, autonomy, and food sovereignty for consumers and of negative socioeconomic effects where economies are based on agriculture (Buiatti et al., 2012).

Consumer boycotts are another issue in many countries due to ethical issues related to GMO products' effects on human health and the environment. It has been noted that from a human health perspective, GM starch is safe, but the long-term ecological and biological consequences remain ambiguous with respect to effects on non-target organisms, including insects and microbes in the soil. Ethical research standards entail disclosure of results, result validation, and monitoring of the surroundings for effects. Many regulatory agencies and biotech businesses now need to perform environmental assessments and engage in public discussions about these ethical concerns. However, some companies focus on the production of "second-generation" GM crops that are environmentally sustainable, for instance, crops developed to be pest resistant, limiting the use of

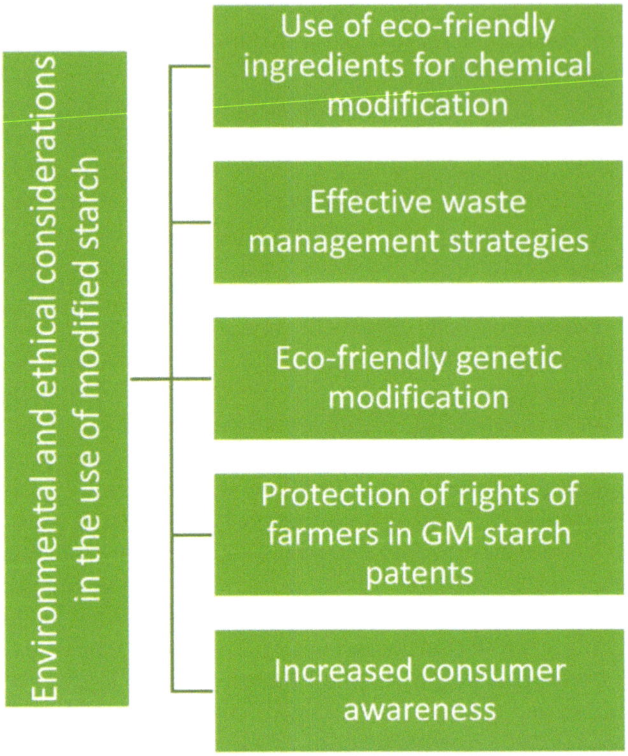

FIGURE 16.5 Environmental and ethical considerations in the use of modified starch.

chemicals. Lastly, environmentally and ethically sensitive production of and research on modified starch cannot be overemphasized. It is crucial to guarantee that approaches used in production cause no impact on the environment and support diversity. Likewise, the application of strategies to GM research, for example, access to information, conservation of biological diversity, and support for farmers, would enhance a responsible kind of modified starch. Solving these issues will involve cooperation among industry players, scientists, regulators, and consumers, focused on creativity without compromising the environment and ethics. Important environmental and ethical considerations are shown in Figure 16.5.

16.8 CASE STUDIES

Modified starch seems to have been widely adopted for application across possible industries, and this section looks at the use of the ingredient through various cases. The subsequent section provides successful examples that demonstrate how and why modified starches are used to improve functionality of food and pharmaceutical products, to manage environmental issues, and to tackle regulatory factors.

One example of the application of modified starchy substances is in gluten-free products, a new trend in the food market connected with trends in gluten intolerance and celiac disease. Bulkers used in conventional formulations seldom provide the extensibility, texture, and mouthfeel found in gluten-containing products. Due to this, food scientists have long relied on what is referred to as functional equivalencies to gluten, which include cross-linked and acetylated starches. These modifications help to enhance water absorption capacity and dough-handling properties, along with the shelf stability of products, and have led to the development of superior-quality gluten-free bread and other bakery products. Modified starches incorporated into gluten-free products also lead to improved feedback, as consumers consider such products to be close to the gluten-containing ones. However, legal regulations regarding labeling of these products differ from jurisdiction to jurisdiction. The European Union is especially specific about the extent of opacity on modified components. This case highlights the use of the modified starch in the development of products and accurate labeling to promote customer confidence in the products (Rai et al., 2018).

Another contribution of modified starches is to tablet formulation in the pharmaceutical sector. Pre-gelatinized and cross-linked starches are most commonly applied as binders and as disintegrating agents in tablet preparation. Such starches also contribute to the requirements of the tablet core in terms regulation of release rate and digestion of the pill in the gastrointestinal tract, which in turn improves bioavailability and other pharmaceutical uses. For instance, a large progressive drug manufacturer surveyed and incorporated pre-gelatinized starch in pain-eradicating tablets to enhance the tablets' copresence, particularly under conditions of high humidity. Officials in some countries have set strict rules governing use in drugs, requiring comprehensive testing that would ensure the modified starch meets certain standards of safety, stability, and efficacy, especially on how the starch affects the release and diffusing ability of the active ingredients. This success therefore underlines the critical role of modified starch in the handling and design of pharmaceutical products, apart from the importance of observing quality measures for these substances in order to promote safety for the end users: the patients (Apeji et al., 2022).

The use of modified starch in environmental applications has also attracted interest as evidenced in the case of biodegradable packaging. In response to increasing consciousness about pollution by plastics, firms have considered modified starches as a suitable feedstock for bio-based plastics. For example, a packaging firm created a nourishment-relevant bioplastic utilizing chemically altered starches to enhance the flexibility as well as the toughness of the final product compared to normal plastics. This bioplastic derived from modified starch was employed in disposable utensils for foods such as by Learnello Bioplastic Limited to reduce the use of petroleum products plastics. The flip side is a lack of homogenous quality or establishment that the virtues of biodegradable packaging are well mirrored under these circumstances. Thus, the company spent its efforts on research in order to develop the qualities of the material that would enhance water resistance and durability of the cloth; however, it has to be biodegradable. This case demonstrates that modified starches have the potential to be used in sustainable innovation, but at the same time, it has to be emphasized that further research must be conducted to improve bioplastics' characteristics for usage or make them more compatible with the environment (Gałkowska et al., 2023).

One case involving GM starches involves a major agricultural biotechnology firm that created a genetically modified potato with a starch modified to generate higher levels of amylopectin. This potato was directed to the food industry since high-amylopectin starch is useful in products such as sauces and soups as well as freeze–thaw products. This new starch was intended to have improved texture, reduced syneresis, and enhanced freeze–thaw stability compared to the native starches. However, challenges were faced related to the sporadic acceptance of the GMOs in the market due to limitations set by regulatory bodies and rejection of GMO products, especially in European markets. The company had to undertake many safety investigations and conduct consumer education in order to win the trust of the consumer. The product achieved regulatory approval in specific territories but failed to gain market acceptance, proving that much work remains to be done to resolve regulatory and consumer acceptance issues for GMO starch. This case serves as an example of how opportunities for application of new technologies, demands for improved regulation of GMOs, and consumer decision making should be managed in areas of the globe where GMO application is highly vulnerable to sanctions (Gałkowska et al., 2023).

The last real-world application is the use of modified starch in meat products, especially in processed ones such as sausages where starch functions as a binder with a role in enhancing texture, water holding capacity, and product yield. Modified potato starch was added in the sausage production line of a food manufacturer to aid moisture retention and produce sausages that consumers would prefer: juicier and firmer sausages. Further, the incorporation of modified starch enhanced the product's freeze–thaw stability, which is an important quality in frozen meat products. However, the firm encountered regulatory issues because of variations in the use of additives permitted in processed meats around the globe. For example, although modified starches are permitted by the FDA in processed meats, there may be usage limit laws implemented in European countries. Ready-to-use formulations also need to be adapted to local requirements, and this sometimes leads to a different product profile across different domestic markets. Modified starch in the food industry faces challenges dealing with varying standards for nutritional additives across the world (Owusu-Ansah et al., 2022).

These case studies indicate the many applications of modified starches across industries, including the improvement of gluten-free foods, pharmaceutical tablets, environmentally friendly packaging, and food processing. They also show that several types of regulatory, environmental, and consumer acceptance issues still exist in relation to modified starches, particularly as new technologies such as genetic modification enter the market. These cases also reveal that testing, labeling, and consumer education as well as environmentally safe and ethical modifications may greatly enhance the use of modified starches for enhanced functionality and improved nutrition.

16.9 FUTURE DIRECTIONS

The strategic direction of modified starch development is based on food science technology, sustainability, budgets, trends among health-conscious consumers, and overriding legislation. Modified starches are likely to experience innovation over the next few years as industries continue to turn to low-pollutant multifunctional products for use in foods and non-foods. Goals include development of the desired characteristics of functional and naturally occurring starches, novel enzyme application in starch production, starches that conform to clean-label and other food market trends and demands, and uses in the pharmaceutical and packaging industries. Another area with the potential for further advancement of modified starches is green chemistry and enzymatic methods of modifying. These reactions may use acids, alkalis, and other reagents that are dangerous and that may be environment- and health-threatening, producing byproducts. To this end, companies are funding other, less damaging modification techniques that minimize the use of or do not require aggressive compounds. For instance, enzymatic modification applies natural enzymes to change the properties of starches, and it is more sustainable and almost perfect as a clean-label technology. Also emerging are green chemistry strategies that can be used to accomplish functional properties with less detrimental influence on the environment. These changes indicate that manufacturers and businesses are gradually shifting toward sustainability. Manufactures that offer environmentally sustainable markets are preferred by environmentally conscious customers.

Another research interest in modified starch is to broaden the health-beneficial characteristics of the products. As consumers become more conscious about health problems such as blood sugar, digestion, and dietary fiber, nutritionally enhanced modified starches are on the table. The emphasis is this area is on resistant starches, which behave like dietary fibers. They remain undigested up to the small intestine then reach the colon, where they enhance the growth of beneficial bacteria, produce SCFAs, and have immunomodulatory effects through the production of butyrate. Future modified starches with such health benefits could meet the dietary requirements of people with diabetes, caloric restriction diets, and gastrointestinal health problems. These functional starches with prebiotic fibers enable food manufacturers to provide nutrition in processed foods without lessening taste or texture, to cater to the increasing need for healthy foods.

Sustainable packaging and biopolymers is another direction that will require mass use of modified starches. Replacing petroleum-based plastics through bioplastics derived through corn starch makes packaging and similar products renewable and responds to problems of plastic waste around the globe. To make starch-based bioplastics useful in applications, including food packaging, agricultural films, and disposable products, firms are innovating new ways of modifying starches to make them tougher, more flexible, and more waterproof. The improvement of cross-linking and esterification technologies is enabling the development of starch-based materials that can have comparable properties to commercial plastic products. With a growing emphasis on sustainable packaging, each of these modified starches will have a key role in contributing to the circular economy as current items are manufactured with the intent to be biodegraded.

The broad area of bioengineering and genetic modification of starches is a more contentious but exciting frontier of future starch development. By genetically modifying plants, one can produce plants with starches that have functional characteristics of interest, for instance, slow digestibility or stability at high temperatures and under extreme and suboptimal pH. Starches with high amylose or high amylopectin content, for instance, can be developed directly in the plant without passing through strenuous post-harvest methods, thus cutting costs of production. Nevertheless, the genetic modification of starches has not left the doubtful chapters in the book of legislation debates and consumer sentiment, especially in those countries that are strictly against GMO products. But organic techniques in gene editing, such as CRISPR, may hold potential to create such starches without the unpleasant connotations tied to GMOs, as editing is seen as less adverse. Consumer acceptance will, however, be determined by approval from regulatory agencies, consumer education, and commercialized bioengineered starch product safety information.

Last but not the least, the future prospects for modified starches will involve digital traceability and transparency, driven by consumer needs for information about the sourcing, processing, and sustainability of commodities used in foods. Modern digital tracking technologies such as blockchain are already being investigated for greater levels of traceability in food additives containing modified starches. Transparency of these systems enables consumers and regulators to review data on the sourcing, processing, and safety testing of these starches, increasing traceability and responsibility in the food chain. It has been found that increasing transparency helps to deal with ethical issues and give more confidence about environmental friendliness of a production or a service. By establishing better control over emerging digital tracking, modified starch producers will be able to better integrate with transparency and sustainability imperatives, thus better ensuring consumer trust.

16.10 SUMMARY

This chapter has discussed the following:

- The existing and potential issues related to modified starch, including its safety profile.
- Recent developments in the uses and development of modified starches.
- Existing legal guidelines and regulatory policies regarding the use of modified starch.
- Environmental and ethical considerations for the application of modified starch.
- Innovative approaches for the manufacture of environmentally friendly modified starch and their application.
- Proposed important steps that may be taken for the development of innovative strategies and effective regulatory framework to ensure the eco-friendly use of modified starch.

16.11 CONCLUSION

Modified starches are a highly promising material in the food, pharmaceutical, and packaging industries owing to multiple functionalities such as thickness and stability in food systems as well as biodegradable coatings for industrial applications. Recent advancements in green chemistry, bioengineering, and health-oriented modifications of starch satisfy customer demand for natural and eco-friendly products and meet the requirements for efficient industrial uses. As advancements emerge, packaging that is ethical, safe, and functional will be in high demand, particularly as the subject of GMOs and synthetic modifications causes controversy among regulators and consumers. In order to satisfy current requirements and gain consumer trust, manufacturers can implement sustainable solutions, enhance digital traceability, and focus on health aspects. Opportunities to expand the abilities of modified starches are based on their application in the technological development of primary global goals for sustainability and ethical standards as key assets for the development of environmentally friendly and healthy industries.

REFERENCES

Alcázar-Alay, S. C., & Meireles, M. A. A. (2015). Physicochemical properties, modifications and applications of starches from different botanical sources. *Food Science and Technology*, *35*(2), https://doi.org/10.1590/1678-457X.6749

Apeji, Y. E., Kaigama, R. T., Anyebe, S. N., Abdussalam, A. O., & Oyi, A. R.. (2022). Tableting performance of maize and potato starches used in combination as binder/disintegrant in metronidazole tablet formulation. *Turkish Journal of Pharmaceutical Science*,, *19*(5), 513–520. https://doi.org/10.4274/tjps.galenos.2021.47855

Bashir, K., & Aggarwal, M. (2019). Physicochemical, structural and functional properties of native and irradiated starch: A review. *Journal of Food Science and Technology*, *56*, 513–523. https://doi.org/10.1007/s13197-018-3530-2

Bawa, A. S., & Anilakumar, K. R. (2012). Genetically modified foods: Safety, risks and public concerns: A review. *Journal of Food Science and Technology*, *50*(6), 1035–1046. https://doi.org/10.1007/s13197-012-0899-1

Birt, D. F., Boylston, T., & Hendrich, S. et. al (2013). Resistant starch: Promise for improving human health. *Advances in Nutrition*, *4*(6), 587–601. https://doi.org/10.3945/an.113.004325

Biswas, R., Alam, M., Sarkar, A., Haque, M. I., Hasan, M. M., & Hoque, M. (2022). Application of nanotechnology in food:

Processing, preservation, packaging and safety assessment. *Heliyon*, *8*(11), e11795. https://doi.org/10.1016/j.heliyon.2022.e11795

Bojarczuk, A., Skąpska, S., Khaneghah, A. M., & Marszałek, K. (2015). Health benefits of resistant starch: A review of the literature. *Journal of Functional Foods*, *93*, 105094. https://doi.org/10.1016/j.jff.2022.105094

Buiatti, M., Christou, P., & Pastore, G. (2012). The application of GMOs in agriculture and in food production for a better nutrition: Two different scientific points of view. *Genes & Nutrition*, *8*, 255–270. https://doi.org/10.1007/s12263-012-0316-4

Compart, J., Singh, A., Fettke, J., & Apriyanto, A. (2023). Customizing starch properties: A review of starch modifications and their applications. *Polymers*, *15*(16), 3491. https://doi.org/10.3390/polym15163491

da Rosa Zavareze, E., & Dias, A. R. G. (2011). Impact of heat-moisture treatment and annealing in starches: A review. *Carbohydrate Polymers*, *83*(2), 317–328. https://doi.org/10.1016/j.carbpol.2010.08.064

Díaz, L. D., Fernández-Ruiz, V., & Cámara, M. (2020). An international regulatory review of food health-related claims in functional food products labelling. *Journal of Functional Foods*, *68*, 103896. https://doi.org/10.1016/j.jff.2020.103896

Dobranowski, P. A., & Stintzi, A. (2022). Resistant starch, microbiome, and precision modulation. *Gut Microbes*, *13*(1), 1926842. https://doi.org/10.1080/19490976.2021.1926842

Erhirhie, E. O., Ihekwereme, C. P., & Ilodigwe, E. E. (2018). Advances in acute toxicity testing: Strengths, weaknesses and regulatory acceptance. *Interdisciplinary Toxicology*, *11*(1), 5–12. https://doi.org/10.2478/intox-2018-0001

European Food and Safety Authority (2024, March 19). *Food additives*. https://www.efsa.europa.eu/en/topics/topic/food-additives

Fan, Y., & Picchioni, F. (2020). Modification of starch: A review on the application of "green" solvents and controlled functionalization. *Carbohydrate Polymers*, *241*, 116350. https://doi.org/10.1016/j.carbpol.2020.116350

Farooq, M. A., & Yu, J. (2024). Recent advances in physical processing techniques to enhance the resistant starch content in foods: A review. *Foods*, *13*(17), 2770. https://doi.org/10.3390/foods13172770

Gałkowska, D., Kapuśniak, K., & Juszczak, L. (2023). Chemically modified starches as food additives. *Molecules*, *28*(22), 7543. https://doi.org/10.3390/molecules28227543

González-Bermúdez, C. A., López-Nicolás, R., Peso-Echarri, P., Frontela-Saseta, C., & Martínez-Graciá, C. (2018). Effects of different thickening agents on infant gut microbiota. *Food & Nutrition*, *3*. http://dx.doi.org/10.1039/C7FO01992K

Goodman, R. E., & Tetteh, A. O. (2011). Suggested improvements for the allergenicity assessment of genetically modified plants used in foods. *Current Allergy and Asthma Reports*, *11*, 317–324. https://doi.org/10.1007/s11882-011-0195-6

Gupta, R. K., Guha, P., & Srivastav, P. P. (2022). Natural polymers in bio-degradable/edible film: A review on environmental concerns, cold plasma technology and nanotechnology application on food packaging-A recent trends. *Food Chemistry Advances*, *1*, 100135. https://doi.org/10.1016/j.focha.2022.100135

Gupta, R. K., Guha, P., & Srivastav, P. P. (2023). Effect of high voltage dielectric barrier discharge (DBD) atmospheric cold plasma treatment on physicochemical and functional properties of taro (*Colocasia esculenta*) starch. *International Journal of Biological Macromolecules*, *253*, 126772. https://doi.org/10.1016/j.ijbiomac.2023.126772

Gupta, R. K., Guha, P., & Srivastav, P. P. (2024a). Insights into the physicochemical, crystalline, multistructural, thermal and digestibility behavior of taro starch (*Colocasia esculenta*) modified by cold plasma treatment with high voltage dielectric barrier discharge (HVDBD): An alternative technique for chemical modification of starch and environmental sustainability. *Journal of Cleaner Production*, *461*, 142710. https://doi.org/10.1016/j.jclepro.2024.142710

Gupta, R. K., Guha, P., & Srivastav, P. P. (2024b). Dielectric barrier discharge plasma: A green and novel method to change the structure of taro peel starch and improve the physicochemical properties of taro peel starch films. *Plasma Processes and Polymers*, *21*(9), 2400047. https://doi.org/10.1002/ppap.202400047

Gupta, R. K., Guha, P., & Srivastav, P. P. (2024c). Exploring the potential of taro (*Colocasia esculenta*) starch: Recent developments in modification, health benefits, and food industry applications. *Food Bioengineering, 3*(3), 365–379. https://doi.org/10.1002/fbe2.12103

Gupta, R. K., Guha, P., & Srivastav, P. P. (2024d). Physical action of nonthermal cold plasma technology for starch modification. *Food Physics*, *1*, 100011. https://doi.org/10.1016/j.foodp.2024.100011

Gupta, R. K., Guha, P., & Srivastav, P. P. (2024e). Effect of cold plasma treatment and plasma-activated water on physicochemical and structural properties of starch: A green and novel approach for environmental sustainability. *Plasma Processes and Polymers*, *21*(4), 2300204. https://doi.org/10.1002/ppap.202300204

Gupta, R. K., Gnana Moorthy Eswaran, U., Kumar, S., Pipliya, S., Kaur, B., & Srivastav, P. P. (2025a). Synergistic modification of starch polysaccharide via sequential treatments with ultrasonic and cold plasma technology. *Journal of Food Processing and Preservation*, *2025*(1), 9063007. https://doi.org/10.1155/jfpp/9063007

Gupta, R. K., Pipliya, S., Patel, J., Srivastav, P. P., Castro-Muñoz, R., Wang, C. K., & Nemţanu, M. R. (2025b). Essential oil-embedded starch based films for active packaging: Insights into migration, stability, and preservation efficiency. *Trends in Food Science & Technology*, *163*, 105149. https://doi.org/10.1016/j.tifs.2025.105149

Jiang, T., Duan, Q., Zhu, J., Liu, H., & Yu, L. (2020). Starch-based biodegradable materials: Challenges and opportunities. *Advanced Industrial and Engineering Polymer Research*, *3* (1), 8–18. https://doi.org/10.1016/j.aiepr.2019.11.003

Karalis, D. T., Karalis, T., Karalis, S., & Kleisiari, A. S. (2020). Genetically modified products, perspectives and challenges. *Cureus*, *12*(3), e7306. https://doi.org/10.7759/cureus.7306

Magallanes-Cruz, P. A., Flores-Silva, P. C., & Bello-Perez, L. A. (2017). Starch structure influences its digestibility: A review. *Journal of Food Science*, *82*(9). https://doi.org/10.1111/1750-3841.13809

Makovicky, P., Makovicky, P., Caja, F., Rimarova, K., Samasca, G., & Vannucci, L. (2020). Celiac disease and gluten-free diet: Past, present, and future. *Gastroenterology and Hepatology from Bed to Bench*, *13*(1), 1–7

Manek, R. V., Builders, P. F., Kolling, W. M., Emeje, M., & Kunle, O. O. (2012). Physicochemical and binder properties of starch obtained from *Cyperus esculentus*. *AAPS PharmSciTech*, *13*, 379–388. https://doi.org/10.1208/s12249-012-9761-z

Mortensen, A., Aguilar, F., Crebelli, R., & Di Domenico, A. et. al (2017). Re-evaluation of oxidised starch (E 1404), monostarch phosphate (E 1410), distarch phosphate (E 1412), phosphated distarch phosphate (E 1413), acetylated distarch phosphate (E 1414), acetylated starch (E 1420), acetylated distarch adipate (E 1422), hydroxypropyl starch (E 1440), hydroxypropyl distarch phosphate (E 1442), starch sodium octenyl succinate (E 1450), acetylated oxidised starch (E 1451) and starch aluminium octenyl succinate (E 1452) as food additives. *EFSA Journal*, *15*(10), e04911. https://doi.org/10.2903/j.efsa.2017.4911

Nasrabadi, M. N., Doost, A. S., & Mezzenga, R. (2021). Modification approaches of plant-based proteins to improve their techno-functionality and use in food products. *Food Hydrocolloids*, *118*, 106789; https://doi.org/10.1016/j.foodhyd.2021.106789

Novasol Ingredients. (2021, July 26). *Modified starches: Application in food industry.* https://novasolingredients.com/application-of-modified-starches-in-food-industry/

Owusu-Ansah, P., Besiwah, E. K., Bonah, E., & Amagloh, F. K. (2022). Non-meat ingredients in meat products: A scoping review. *Applied Food Research*, *2*(1), 100044. https://doi.org/10.1016/j.afres.2022.100044

Priya, K. M., & Alur, S. (2023). Analyzing consumer behaviour towards food and nutrition labeling: A comprehensive review. *Heliyon*, *9*(9), e19401. https://doi.org/10.1016/j.heliyon.2023.e19401

Qin, Y., Pillidge, C., Harrison, B., & Adhikari, B. (2024). Pathways in formulating foods for the elderly. *Food Research International*, *186*, 114324. https://doi.org/10.1016/j.foodres.2024.114324

Rai, S., Kaur, A., & Chopra, C. S. (2018). Gluten-free products for celiac susceptible people. *Frontiers in Nutrition*, *5*, 116. https://doi.org/10.3389/fnut.2018.00116

Rashwan, A. K., Younis, H. A., Abdelshafy, A. M., Osman, A. I., Eletmany, M. R., Hafouda, M. A., & Chen, W. (2024). Plant starch extraction, modification, and green applications: A review. *Environmental Chemistry Letters*, *22*, 2483–2530.

Rong, L., Fei, W., Wang, Z., Chen, X., Wen, H., Xie, J., & Shen, M. (2024). Modulation of starch digestibility using non-thermal processing techniques: A review. *Grain & Oil Science and Technology*, *7*(3), 209–218. https://doi.org/10.1016/j.gaost.2024.05.002

Shrivastava, A., Gupta, R. K., & Srivastav, P. P. (2024). Exploring novel frontiers of advancements in purple yam (*Dioscorea alata* L.) starch extraction, modification, characterization, applications in food and other industries. *Measurement: Food*, *15*, 100196. https://doi.org/10.1016/j.meafoo.2024.100196

Singh, J., Dartois, A., & Kaur, L. (2018). Starch digestibility in food matrix: A review. *Trends in Food Science & Technology*, *21*(4), 168–180. https://doi.org/10.1016/j.tifs.2009.12.001

Ueda, M., Shimamura, N., Yoshida, M., & Kamimura, T. (2024). Consumer protection laws and regulations Japan 2024. In *Consumer protection 2024.* ICLG. https://iclg.com/practice-areas/consumer-protection-laws-and-regulations/japan

Yudhistira, B., Husnayain, N., Punthi, M., Chang, C. K., & Hsieh, C. W. (2024). Progress in the application of emerging technology for the improvement of starch-based active packaging properties: A review. *ACS Food Science & Technology*, *4*(9). https://doi.org/10.1021/acsfoodscitech.4c00260

Index